Probability and Measure Theory

SECOND EDITION

ROBERT B. ASH

with contributions from
Catherine Doléans-Dade

A Harcourt Science and Technology Company

San Diego   San Francisco   New York   Boston
London   Toronto   Sydney   Tokyo

ACADEMIC PRESS
A Harcourt Science and Technology Company
525 B Street, 1900, San Diego, California 92101-4495, USA
http://www.apnet.com

Academic Press
24–28 Oval Road, London NW1 7DX, UK
http://www.hbuk.co.uk/ap/

Harcourt/Academic Press
200 Wheeler Road, Burlington, Massachusetts 01803
http://www.harcourt.ap.com

Library of Congress Catalog Card Number: 99-65669
International Standard Book Number: 0-12-065202-1

Printed in the United States of America
99 00 01 02 03 IP 9 8 7 6 5 4 3 2 1

Contents

Preface

It is a pleasure to accept the invitation of Harcourt/Academic Press to publish a second edition. The first edition has been used mainly in graduate courses in measure and probability, offered by departments of mathematics and statistics and frequently taken by engineers. We have prepared the present text with this audience in mind, and the title has been changed from *Real Analysis and Probability* to *Probability and Measure Theory* to reflect the revisions we have made.

Chapters 1 and 2 develop the fundamentals of measure and integration theory. Included are several results that are crucial in constructing the foundations of probability: the Radon–Nikodym theorem, the product measure theorem, the Kolmogorov extension theorem and the theory of weak convergence of measures. We remain convinced that it is best to assemble a complete set of measure-theoretic tools before going into probability, rather than try to develop both areas simultaneously. The gain in efficiency far outweighs any temporary loss in motivation. Those who wish to reach probability as quickly as possible may omit Chapter 3, which gives a brief introduction to functional analysis, and Section 2.3, which gives some applications to real analysis. In addition, instructors may wish to summarize or sketch some of the intricate constructions in Sections 1.3, 1.4, and 2.7.

The study of probability begins with Chapter 4, which offers a summary of an undergraduate probability course from a measure–theoretic point of view. Chapter 5 is concerned with the general concept of conditional probability and expectation. The approach to problems that involve conditioning, given events of probability zero, is the gateway to many areas of probability theory. Chapter 6 deals with strong laws of large numbers, first from the classical viewpoint, and then via martingale theory. Basic properties and applications of martingale sequences are developed systematically. Chapter 7 considers the central limit problem, emphasizing the fundamental role of Prokhorov's weak compactness theorem. The last two sections of this chapter cover some material (not in the first edition) of special interest to statisticians: Slutsky's theorem, the Skorokhod construction, convergence of transformed sequences and a k-dimensional central limit theorem.

Chapters 8 and 9 have been added in the second edition, and should be of interest to the entire prospective audience: mathematicians, statisticians, and engineers. Chapter 8 covers ergodic theory, which is developed far enough so that connections with information theory are clearly visible. The Shannon–McMillan theorem is proved and the isomorphism problem for Bernoulli shifts is discussed. Chapter 9 treats the one-dimensional Brownian motion in detail, and then introduces stochastic integrals and the Itô differentiation formula.

To make room for the new material, the appendix on general topology and the old Chapter 4 on the interplay between measure theory and topology have been removed, along with the section on topological vector spaces in Chapter 3. We assume that the reader has had a course in basic analysis and is familiar with metric spaces, but not with general topology. All the necessary background appears in *Real Variables With Basic Metric Space Topology* by Robert B. Ash, IEEE Press, 1993. (The few exercises that require additional background are marked with an asterisk.)

It is theoretically possible to read the text without any prior exposure to probability, picking up the necessary equipment in Chapter 4. But we expect that in practice, almost all readers will have taken a standard undergraduate probability course. We believe that discrete time, discrete state Markov chains, and random walks are best covered in a second undergraduate probability course, without measure theory. But instructors and students usually find this area appealing, and we discuss the symmetric random walk on $\mathbb{R}^k$ in Appendix 1.

Problems are given at the end of each section. Fairly detailed solutions are given to many problems, and instructors may obtain solutions to those problems in Chapters 1–8 not worked out in the text by writing to the publisher.

Catherine Doleans–Dade wrote Chapter 9, and offered valuable advice and criticism for the other chapters. Mel Gardner kindly allowed some material from *Topics in Stochastic Processes* by Ash and Gardner to be used in Chapter 8. We appreciate the encouragement and support provided by the staff at Harcourt/Academic Press.

Robert B. Ash
Catherine Doleans–Dade
Urbana, Illinois, 1999

Summary of Notation

We indicate here the notational conventions to be used throughout the book. The numbering system is standard; for example, 2.7.4 means Chapter 2, Section 7, Part 4. In the appendices, the letter A is used; thus A2.3 means Part 3 of Appendix 2.

The symbol $\square$ is used to mark the end of a proof.

1   SETS

If A and B are subsets of a set Ω, $A \cup B$ will denote the *union* of A and B, and $A \cap B$ the *intersection* of A and B. The union and intersection of a family of sets A_i are denoted by $\bigcup_i A_i$ and $\bigcap_i A_i$. The *complement* of A (relative to Ω) is denoted by A^c.

The statement "B is a *subset* of A" is denoted by $B \subset A$; the inclusion need not be proper, that is, we have $A \subset A$ for any set A. We also write $B \subset A$ as $A \supset B$, to be read "A is an *overset* (or *superset*) of B."

The notation $A - B$ will always mean, unless otherwise specified, the set of points that belong to A but not to B. It is referred to as the *difference* between A and B; a *proper difference* is a set $A - B$, where $B \subset A$.

The *symmetric difference* between A and B is by definition the union of $A - B$ and $B - A$; it is denoted by $A \, \Delta \, B$.

If $A_1 \subset A_2 \subset \cdots$ and $\bigcup_{n=1}^{\infty} A_n = A$, we say that the A_n form an *increasing* sequence of sets (increasing to A) and write $A_n \uparrow A$. Similarly, if $A_1 \supset A_2 \supset \cdots$ and $\bigcap_{n=1}^{\infty} A_n = A$, we say that the A_n form a *decreasing* sequence of sets (decreasing to A) and write $A_n \downarrow A$.

The word "includes" will always imply a subset relation, and the word "contains" a membership relation. Thus if $\mathscr{C}$ and $\mathscr{D}$ are collections of sets, "$\mathscr{C}$ includes $\mathscr{D}$" means that $\mathscr{D} \subset \mathscr{C}$. Equivalently, we may say that $\mathscr{C}$ contains all sets in $\mathscr{D}$, in other words, each $A \in \mathscr{D}$ is also a member of $\mathscr{C}$.

A *countable* set is one that is either finite or countably infinite.

The *empty set* $\emptyset$ is the set with no members. The sets A_i, $i \in I$, are *disjoint* if $A_i \cap A_j = \emptyset$ for all $i \neq j$.

2 REAL NUMBERS

The set of real numbers will be denoted by $\mathbb{R}$, and $\mathbb{R}^n$ will denote n-dimensional Euclidean space. In $\mathbb{R}$, the interval $(a, b]$ is defined as $\{x \in \mathbb{R}: a < x \leq b\}$, and (a, ∞) as $\{x \in \mathbb{R}: x > a\}$; other types of intervals are defined similarly. If $a = (a_1, \ldots, a_n)$ and $b = (b_1, \ldots, b_n)$ are points in $\mathbb{R}^n$, $a \leq b$ will mean $a_i \leq b_i$ for all i. The interval $(a, b]$ is defined as $\{x \in R^n: a_i < x_i \leq b_i, i = 1, \ldots, n\}$, and other types of intervals are defined similarly.

The set of *extended real numbers* is the two-point compactification $\mathbb{R} \cup \{\infty\} \cup \{-\infty\}$, denoted by $\overline{\mathbb{R}}$; the set of n-tuples $(x_1, \ldots, x_n)$, with each $x_i \in \overline{\mathbb{R}}$, is denoted by $\overline{\mathbb{R}}^n$. We adopt the following rules of arithmetic in $\overline{\mathbb{R}}$:

$$a + \infty = \infty + a = \infty, \qquad a - \infty = -\infty + a = -\infty, \qquad a \in \mathbb{R},$$

$$\infty + \infty = \infty, \qquad -\infty - \infty = -\infty \qquad (\infty - \infty \text{ is not defined}),$$

$$b \cdot \infty = \infty \cdot b = \begin{cases} \infty & \text{if } b \in \overline{\mathbb{R}} \quad b > 0, \\ -\infty & \text{if } b \in \overline{\mathbb{R}}, \quad b < 0, \end{cases}$$

$$\frac{a}{\infty} = \frac{a}{-\infty} = 0, \quad a \in \mathbb{R} \quad \left(\frac{\infty}{\infty} \text{ is not defined}\right),$$

$$0 \cdot \infty = \infty \cdot 0 = 0.$$

The rules are convenient when developing the properties of the abstract Lebesgue integral, but it should be emphasized that $\overline{\mathbb{R}}$ is not a field under these operations.

Unless otherwise specified, *positive* means (strictly) greater than zero, and *nonnegative* means greater than or equal to zero.

The set of *complex numbers* is denoted by $\mathbb{C}$, and the set of n-tuples of complex numbers by $\mathbb{C}^n$.

3 FUNCTIONS

If f is a function from Ω to Ω' (written as $f: \Omega \to \Omega'$) and $B \subset \Omega'$, the *preimage* of B under f is given by $f^{-1}(B) = \{\omega \in \Omega: f(\omega) \in B\}$. It follows from the definition that $f^{-1}(\bigcup_i B_i) = \bigcup_i f^{-1}(B_i)$, $f^{-1}(\bigcap_i B_i) = \bigcap_i f^{-1}(B_i)$, $f^{-1}(A - B) = f^{-1}(A) - f^{-1}(B)$; hence $f^{-1}(A^c) = [f^{-1}(A)]^c$. If $\mathscr{C}$ is a class of sets, $f^{-1}(\mathscr{C})$ means the collection of sets $f^{-1}(B), B \in \mathscr{C}$.

If $f: \mathbb{R} \to \mathbb{R}$, f is *increasing* iff $x < y$ implies $f(x) \leq f(y)$; *decreasing* iff $x < y$ implies $f(x) \geq f(y)$. Thus, "increasing" and "decreasing" do not have the strict connotation. If $f_n: \Omega \to \overline{\mathbb{R}}$, $n = 1, 2, \ldots$, the f_n are said to form an *increasing sequence* iff $f_n(\omega) \leq f_{n+1}(\omega)$ for all n and ω; a *decreasing sequence* is defined similarly.

If f and g are functions from Ω to $\overline{\mathbb{R}}$, statements such as $f \leq g$ are always interpreted as holding pointwise, that is, $f(\omega) \leq g(\omega)$ for all $\omega \in \Omega$. Similarly, if $f_i \colon \Omega \to \overline{\mathbb{R}}$ for each $i \in I$, $\sup_i f_i$ is the function whose value at ω is $\sup\{f_i(\omega) \colon i \in I\}$.

If $f_1, f_2, \ldots$ form an increasing sequence of functions with limit f [that is, $\lim_{n \to \infty} f_n(\omega) = f(\omega)$ for all ω], we write $f_n \uparrow f$. (Similarly, $f_n \downarrow f$ is used for a decreasing sequence.)

Sometimes, a set such as $\{\omega \in \Omega \colon f(\omega) \leq g(\omega)\}$ is abbreviated as $\{f \leq g\}$; similarly, the preimage $\{\omega \in \Omega \colon f(\omega) \in B\}$ is written as $\{f \in B\}$.

If $A \subset \Omega$, the *indicator* of A is the function defined by $I_A(\omega) = 1$ if $\omega \in A$ and by $I_A(\omega) = 0$ if $\omega \notin A$. The phrase "characteristic function" is often used in the literature, but we shall not adopt this term here.

If f is a function of two variables x and y, the symbol $f(x, \cdot)$ is used for the mapping $y \to f(x, y)$ with x fixed.

The *composition* of two functions $X \colon \Omega \to \Omega'$ and $f \colon \Omega' \to \Omega''$ is denoted by $f \circ X$ or $f(X)$.

If $f \colon \Omega \to \overline{\mathbb{R}}$, the *positive* and *negative parts* of f are defined by $f^+ = \max(f, 0)$ and $f^- = \max(-f, 0)$, that is,

$$f^+(\omega) = \begin{cases} f(\omega) & \text{if} & f(\omega) \geq 0, \\ 0 & \text{if} & f(\omega) < 0, \end{cases}$$

$$f^-(\omega) = \begin{cases} -f(\omega) & \text{if} & f(\omega) \leq 0, \\ 0 & \text{if} & f(\omega) > 0. \end{cases}$$

4   TOPOLOGY

A *metric space* is a set Ω with a function d (called a *metric*) from $\Omega \times \Omega$ to the nonnegative reals, satisfying $d(x, y) \geq 0, d(x, y) = 0$ iff $x = y, d(x, y) = d(y, x)$, and $d(x, z) \leq d(x, y) + d(y, z)$. If $d(x, y)$ can be 0 for $x \neq y$, but d satisfies the remaining properties, d is called a *pseudometric* (the term *semimetric* is also used in the literature).

A *ball* (or *open ball*) in a metric or pseudometric space is a set of the form $B(x, r) = \{y \in \Omega \colon d(x, y) < r\}$ where x, the *center* of the ball, is a point of Ω, and r, the *radius*, is a positive real number. A *closed ball* is a set of the form $\overline{B}(x, r) = \{y \in \Omega \colon d(x, y) \leq r\}$.

Sequences in Ω are denoted by $\{x_n, n = 1, 2, \ldots\}$. The term "lower semicontinuous" is abbreviated LSC, and "upper semicontinuous" is abbreviated USC.

No knowledge of general topology (beyond metric spaces) is assumed, and the few comments that refer to general topological spaces can safely be ignored.

5   VECTOR SPACES

The terms "vector space" and "linear space" are synonymous. All vector spaces are over the real or complex field, and the complex field is assumed unless the term "real vector space" is used.

A *Hamel basis* for a vector space L is a maximal linearly independent subset B of L. (Linear independence means that if $x_1, \ldots, x_n \in B$, $n = 1, 2, \ldots$, and $c_1, \ldots, c_n$ are scalars, then $\sum_{i=1}^{n} c_i x_i = 0$ iff all $c_i = 0$.) Alternatively, a Hamel basis is a linearly independent subset B with the property that each $x \in L$ is a finite linear combination of elements in B. [An *orthonormal basis* for a Hilbert space (Chapter 3) is a different concept.]

The terms "subspace" and "linear manifold" are synonymous, each referring to a subset M of a vector space L that is itself a vector space under the operations of addition and scalar multiplication in L. If there is a metric on L and M is a closed subset of L, then M is called a *closed subspace*.

If B is an arbitrary subset of L, the *linear manifold generated by B*, denoted by $L(B)$, is the smallest linear manifold containing all elements of B, that is, the collection of finite linear combinations of elements of B. Assuming a metric on L, the *space spanned by B*, denoted by $S(B)$, is the smallest closed subspace containing all elements of B. Explicitly, $S(B)$ is the closure of $L(B)$.

6   ZORN'S LEMMA

A *partial ordering* on a set S is a relation "$\leq$" that is

(1)   *reflexive*: $a \leq a$;
(2)   *antisymmetric*: if $a \leq b$ and $b \leq a$, then $a = b$; and
(3)   *transitive*: if $a \leq b$ and $b \leq c$, then $a \leq c$.
(All elements a, b, c belong to S.)

If $C \subset S$, C is said to be *totally ordered* iff for all $a, b \in C$, either $a \leq b$ or $b \leq a$. A totally ordered subset of S is also called a *chain* in S.

The form of Zorn's lemma that will be used in the text is as follows.

Let S be a set with a partial ordering "$\leq$." Assume that every chain C in S has an upper bound; in other words, there is an element $x \in S$ such that $x \geq a$ for all $a \in C$. Then S has a maximal element, that is, an element m such that for each $a \in S$ it is not possible to have $m \leq a$ and $m \neq a$.

Zorn's lemma is actually an axiom of set theory, equivalent to the axiom of choice.

1

FUNDAMENTALS OF MEASURE AND INTEGRATION THEORY

In this chapter we give a self-contained presentation of the basic concepts of the theory of measure and integration. The principles discussed here and in Chapter 2 will serve as background for the study of probability as well as harmonic analysis, linear space theory, and other areas of mathematics.

1.1 INTRODUCTION

It will be convenient to start with a little practice in the algebra of sets. This will serve as a refresher and also as a way of collecting a few results that will often be useful.

Let $A_1, A_2, \ldots$ be subsets of a set Ω. If $A_1 \subset A_2 \subset \cdots$ and $\bigcup_{n=1}^{\infty} A_n = A$, we say that the A_n form an *increasing* sequence of sets with limit A, or that the A_n increase to A; we write $A_n \uparrow A$. If $A_1 \supset A_2 \supset \cdots$ and $\bigcap_{n=1}^{\infty} A_n = A$, we say that the A_n form a *decreasing* sequence of sets with limit A, or that the A_n decrease to A; we write $A_n \downarrow A$.

The *De Morgan laws*, namely, $\left(\bigcup_n A_n \right)^c = \bigcap_n A_n^c$, $\left(\bigcap_n A_n \right)^c = \bigcup_n A_n^c$, imply that

(1) if $A_n \uparrow A$, then $A_n^c \downarrow A^c$; if $A_n \downarrow A$, then $A_n^c \uparrow A^c$.

It is sometimes useful to write a union of sets as a disjoint union. This may be done as follows:

Let $A_1, A_2, \ldots$ be subsets of Ω. For each n we have

(2) $\bigcup_{i=1}^{n} A_i = A_1 \cup (A_1^c \cap A_2) \cup (A_1^c \cap A_2^c \cap A_3)$

$\cup \cdots \cup (A_1^c \cap \cdots A_{n-1}^c \cap A_n)$.

Furthermore,

(3) $\bigcup_{n=1}^{\infty} A_n = \bigcup_{n=1}^{\infty} (A_1^c \cap \cdots \cap A_{n-1}^c \cap A_n)$.

In (2) and (3), the sets on the right are disjoint. If the A_n form an increasing sequence, the formulas become

(4) $\bigcup_{i=1}^{n} A_i = A_1 \cup (A_2 - A_1) \cup \cdots \cup (A_n - A_{n-1})$

and

(5) $\bigcup_{n=1}^{\infty} A_n = \bigcup_{n=1}^{\infty} (A_n - A_{n-1})$

(take A_0 as the empty set).

The results (1)–(5) are proved using only the definitions of union, intersection, and complementation; see Problem 1.

The following set operation will be of particular interest. If $A_1, A_2, \ldots$ are subsets of Ω, we define

(6) $\limsup_n A_n = \bigcap_{n=1}^{\infty} \bigcup_{k=n}^{\infty} A_k$.

Thus $\omega \in \limsup_n A_n$ iff for every n, $\omega \in A_k$ for some $k \geq n$, in other words,

(7) $\omega \in \limsup_n A_n$ iff $\omega \in A_n$ for infinitely many n.

Also define

(8) $\liminf_n A_n = \bigcup_{n=1}^{\infty} \bigcap_{k=n}^{\infty} A_k$.

Thus $\omega \in \liminf_n A_n$ iff for some n, $\omega \in A_k$ for all $k \geq n$, in other words,

(9) $\omega \in \liminf_n A_n$ iff $\omega \in A_n$ eventually, that is, for all but finitely many n.

We shall call $\limsup_n A_n$ the *upper limit* of the sequence of sets A_n, and $\liminf_n A_n$ the *lower limit*. The terminology is, of course, suggested by the analogous concepts for sequences of real numbers

$$\limsup_n x_n = \inf_n \sup_{k \geq n} x_k,$$

$$\liminf_n x_n = \sup_n \inf_{k \geq n} x_k.$$

See Problem 4 for a further development of the analogy.

The following facts may be verified (Problem 5):

(10) $(\limsup_n A_n)^c = \liminf_n A_n^c$

(11) $(\liminf_n A_n)^c = \limsup_n A_n^c$

(12) $\liminf_n A_n \subset \limsup_n A_n$

(13) If $A_n \uparrow A$ or $A_n \downarrow A$, then $\liminf_n A_n = \limsup_n A_n = A$.

In general, if $\liminf_n A_n = \limsup_n A_n = A$, then A is said to be the *limit* of the sequence $A_1, A_2, \ldots$; we write $A = \lim_n A_n$.

Problems

1. Establish formulas (1)–(5).

2. Define sets of real numbers as follows. Let $A_n = (-1/n, 1]$ if n is odd, and $A_n = (-1, 1/n]$ if n is even. Find $\limsup_n A_n$ and $\liminf_n A_n$.

3. Let $\Omega = \mathbb{R}^2$, A_n the interior of the circle with center at $((-1)^n/n, 0)$ and radius 1. Find $\limsup_n A_n$ and $\liminf_n A_n$.

A   4.   Let $\{x_n\}$ be a sequence of real numbers, and let $A_n = (-\infty, x_n)$. What is the connection between $\limsup_{n \to \infty} x_n$ and $\limsup_n A_n$ (similarly for $\liminf$)?

5.   Establish formulas (10)–(13).

6.   Let $A = (a, b)$ and $B = (c, d)$ be disjoint open intervals of $\mathbb{R}$, and let $C_n = A$ if n is odd, $C_n = B$ if n is even. Find $\limsup_n C_n$ and $\liminf_n C_n$.

1.2   FIELDS, σ-FIELDS, AND MEASURES

Length, area, and volume, as well as probability, are instances of the measure concept that we are going to discuss. A measure is a *set function*, that is, an assignment of a number $\mu(A)$ to each set A in a certain class. Some structure must be imposed on the class of sets on which μ is defined, and probability considerations provide a good motivation for the type of structure required. If Ω is a set whose points correspond to the possible outcomes of a random experiment, certain subsets of Ω will be called "events" and assigned a probability. Intuitively, A is an event if the question "Does ω belong to A?" has a definite yes or no answer after the experiment is performed (and the outcome corresponds to the point $\omega \in \Omega$). Now if we can answer the question "Is $\omega \in A$?" we can certainly answer the question "Is $\omega \in A^c$?," and if, for each $i = 1, \ldots, n$, we can decide whether or not ω belongs to A_i, then we can determine whether or not ω belongs to $\bigcup_{i=1}^{n} A_i$ (and similarly for $\bigcap_{i=1}^{n} A_i$). Thus it is natural to require that the class of events be closed under complementation, finite union, and finite intersection; furthermore, as the answer to the question "Is $\omega \in \Omega$?" is always "yes," the entire space Ω should be an event. Closure under *countable* union and intersection is difficult to justify physically, and perhaps the most convincing reason for requiring it is that a richer mathematical theory is obtained. Specifically, we are able to assert that the limit of a sequence of events is an event; see 1.2.1.

1.2.1   Definitions. Let $\mathscr{F}$ be a collection of subsets of a set Ω. Then $\mathscr{F}$ is called a *field* (the term *algebra* is also used) iff $\Omega \in \mathscr{F}$ and $\mathscr{F}$ is closed under complementation and finite union, that is,

(a) $\Omega \in \mathscr{F}$.

(b)   If $A \in \mathscr{F}$, then $A^c \in \mathscr{F}$.

(c)   If $A_1, A_2, \ldots, A_n \in \mathscr{F}$, then $\bigcup_{i=1}^{n} A_i \in \mathscr{F}$.

It follows that $\mathscr{F}$ is closed under finite intersection. For if $A_1, \ldots, A_n \in \mathscr{F}$, then

$$\bigcap_{i=1}^{n} A_i = \left(\bigcup_{i=1}^{n} A_i^c \right)^c \in \mathscr{F}.$$

If (c) is replaced by closure under *countable* union, that is,

(d)   If $A_1, A_2, \ldots \in \mathscr{F}$, then $\bigcup_{i=1}^{\infty} A_i \in \mathscr{F}$,

$\mathscr{F}$ is called a *σ-field* (the term *σ-algebra* is also used). Just as above, $\mathscr{F}$ is also closed under countable intersection.

If $\mathscr{F}$ is a field, a countable union of sets in $\mathscr{F}$ can be expressed as the limit of an increasing sequence of sets in $\mathscr{F}$, and conversely. To see this, note that if $A = \bigcup_{n=1}^{\infty} A_n$, then $\bigcup_{i=1}^{n} A_i \uparrow A$; conversely, if $A_n \uparrow A$, then $A = \bigcup_{n=1}^{\infty} A_n$. This shows that a σ-field is a field that is closed under limits of increasing sequences.

1.2.2   Examples. The largest σ-field of subsets of a fixed set Ω is the collection of all subsets of Ω. The smallest σ-field consists of the two sets $\emptyset$ and Ω.

Let A be a nonempty proper subset of Ω, and let $\mathscr{F} = \{\emptyset, \Omega, A, A^c\}$. Then $\mathscr{F}$ is the smallest σ-field containing A. For if $\mathscr{G}$ is a σ-field and $A \in \mathscr{G}$, then by definition of a σ-field, Ω, $\emptyset$, and A^c belong to $\mathscr{G}$, hence $\mathscr{F} \subset \mathscr{G}$. But $\mathscr{F}$ is a σ-field, for if we form complements or unions of sets in $\mathscr{F}$, we invariably obtain sets in $\mathscr{F}$. Thus $\mathscr{F}$ is a σ-field that is included in any σ-field containing A, and the result follows.

If $A_1, \ldots, A_n$ are arbitrary subsets of Ω, the smallest σ-field containing $A_1, \ldots, A_n$ may be described explicitly; see Problem 8.

If $\mathscr{S}$ is a class of sets, the smallest σ-field containing the sets of $\mathscr{S}$ will be written as $\sigma(\mathscr{S})$, and sometimes called the *minimal σ-field over* $\mathscr{S}$. We also call $\sigma(\mathscr{S})$ *the σ-field generated by* $\mathscr{S}$, and currently this is probably the most common terminology.

Let Ω be the set $\mathbb{R}$ of real numbers. Let $\mathscr{F}$ consist of all finite disjoint unions of right-semiclosed intervals. (A right-semiclosed interval is a set of the form $(a, b] = \{x: a < x \le b\}, -\infty \le a < b < \infty$; by convention we also count (a, ∞) as right-semiclosed for $-\infty \le a < \infty$. The convention is necessary because $(-\infty, a]$ belongs to $\mathscr{F}$, and if $\mathscr{F}$ is to be a field, the complement (a, ∞) must also belong to $\mathscr{F}$.) It may be verified that conditions (a)–(c) of 1.2.1 hold; and thus $\mathscr{F}$ is a field. But $\mathscr{F}$ is not a σ-field; for example, $A_n = (0, 1 - (1/n)] \in \mathscr{F}, n = 1, 2, \ldots$, and $\bigcup_{n=1}^{\infty} A_n = (0, 1) \notin \mathscr{F}$.

If Ω is the set $\overline{\mathbb{R}} = [-\infty, \infty]$ of extended real numbers, then just as above, the collection of finite disjoint unions of right-semiclosed intervals forms a field but not a σ-field. Here, the right-semiclosed intervals are sets of the form $(a, b] = \{x: a < x \le b\}, -\infty \le a < b \le \infty$, and, by convention, the sets $[-\infty, b] = \{x: -\infty \le x \le b\}, -\infty \le b \le \infty$. (In this case the convention is necessary because $(b, \infty]$ must belong to $\mathscr{F}$, and therefore the complement $[-\infty, b]$ also belongs to $\mathscr{F}$.)

There is a type of reasoning that occurs so often in problems involving σ-fields that it deserves to be displayed explicitly, as in the following typical illustration.

If $\mathscr{C}$ is a class of subsets of Ω and $A \subset \Omega$, we denote by $\mathscr{C} \cap A$ the class $\{B \cap A \colon B \in \mathscr{C}\}$. If the minimal σ-field over $\mathscr{C}$ is $\sigma(\mathscr{C}) = \mathscr{F}$, let us show that

$$\sigma_A(\mathscr{C} \cap A) = \mathscr{F} \cap A,$$

where $\sigma_A(\mathscr{C} \cap A)$ is the minimal σ-field of *subsets of A* over $\mathscr{C} \cap A$. (In other words, A rather than Ω is regarded as the entire space.)

Now $\mathscr{C} \subset \mathscr{F}$, hence $\mathscr{C} \cap A \subset \mathscr{F} \cap A$, and it is not hard to verify that $\mathscr{F} \cap A$ is a σ-field of subsets of A. Therefore $\sigma_A(\mathscr{C} \cap A) \subset \mathscr{F} \cap A$.

To establish the reverse inclusion we must show that $B \cap A \in \sigma_A(\mathscr{C} \cap A)$ for all $B \in \mathscr{F}$. This is not obvious, so we resort to the following basic reasoning process, which might be called the *good sets principle*. Let $\mathscr{S}$ be the class of good sets, that is, let $\mathscr{S}$ consist of those sets $B \in \mathscr{F}$ such that

$$B \cap A \in \sigma_A(\mathscr{C} \cap A).$$

Since $\mathscr{F}$ and $\sigma_A(\mathscr{C} \cap A)$ are σ-fields, it follows quickly that $\mathscr{S}$ is a σ-field. But $\mathscr{C} \subset \mathscr{S}$, so that $\sigma(\mathscr{C}) \subset \mathscr{S}$, hence $\mathscr{F} = \mathscr{S}$ and the result follows. Briefly, every set in $\mathscr{C}$ is good and the class of good sets forms a σ-field; consequently, every set in $\sigma(\mathscr{C})$ is good.

One other comment: If $\mathscr{C}$ is closed under finite intersection and $A \in \mathscr{C}$, then $\mathscr{C} \cap A = \{C \in \mathscr{C} \colon C \subset A\}$. (Observe that if $C \subset A$, then $C = C \cap A$.)

1.2.3   Definitions and Comments. A *measure* on a σ-field $\mathscr{F}$ is a nonnegative, extended real-valued function μ on $\mathscr{F}$ such that whenever $A_1, A_2, \ldots$ form a finite or countably infinite collection of disjoint sets in $\mathscr{F}$, we have

$$\mu\left(\bigcup_n A_n\right) = \sum_n \mu(A_n).$$

If $\mu(\Omega) = 1$, μ is called a *probability measure*.

A *measure space* is a triple $(\Omega, \mathscr{F}, \mu)$ where Ω is a set, $\mathscr{F}$ is a σ-field of subsets of Ω, and μ is a measure on $\mathscr{F}$. If μ is a probability measure, $(\Omega, \mathscr{F}, \mu)$ is called a *probability space*.

It will be convenient to have a slight generalization of the notion of a measure on a σ-field. Let $\mathscr{F}$ be a *field*, μ a set function on $\mathscr{F}$ (a map from $\mathscr{F}$ to $\overline{\mathbb{R}}$). We say that μ is *countably additive* on $\mathscr{F}$ iff whenever $A_1, A_2, \ldots$ form a finite or countably infinite collection of disjoint sets in $\mathscr{F}$ whose union also belongs to $\mathscr{F}$ (this will always be the case if $\mathscr{F}$ is a σ-field) we have

$$\mu\left(\bigcup_n A_n\right) = \sum_n \mu(A_n).$$

If this requirement holds only for finite collections of disjoint sets in $\mathscr{F}$, μ is said to be *finitely additive* on $\mathscr{F}$. To avoid the appearance of terms of the form

$+\infty - \infty$ in the summation, we always assume that $+\infty$ and $-\infty$ cannot both belong to the range of μ.

If μ is countably additive and $\mu(A) \geq 0$ for all $A \in \mathscr{F}$, μ is called a *measure* on $\mathscr{F}$, a *probability measure* if $\mu(\Omega) = 1$.

Note that countable additivity actually implies finite additivity. For if $\mu(A) = +\infty$ for all $A \in \mathscr{F}$, or if $\mu(A) = -\infty$ for all $A \in \mathscr{F}$, the result is immediate; therefore assume $\mu(A)$ finite for some $A \in \mathscr{F}$. By considering the sequence $A, \emptyset, \emptyset, \ldots$, we find that $\mu(\emptyset) = 0$, and finite additivity is now established by considering the sequence $A_1, \ldots, A_n, \emptyset, \emptyset, \ldots$, where $A_1, \ldots, A_n$ are disjoint sets in $\mathscr{F}$.

Although the set function given by $\mu(A) = +\infty$ for all $A \in \mathscr{F}$ satisfies the definition of a measure, and similarly $\mu(A) = -\infty$ for all $A \in \mathscr{F}$ defines a countably additive set function, we shall from now on exclude these cases. Thus by the above discussion, we always have $\mu(\emptyset) = 0$.

If $A \in \mathscr{F}$ and $\mu(A^c) = 0$, we can frequently ignore A^c; we say that μ is *concentrated* on A.

1.2.4 Examples. Let Ω be any set, and let $\mathscr{F}$ consist of all subsets of Ω. Define $\mu(A)$ as the number of points of A. Thus if A has n members, $n = 0, 1, 2, \ldots$, then $\mu(A) = n$; if A is an infinite set, $\mu(A) = \infty$. The set function μ is a measure on $\mathscr{F}$, called *counting measure* on Ω.

A closely related measure is defined as follows. Let $\Omega = \{x_1, x_2, \ldots\}$ be a finite or countably infinite set, and let $p_1, p_2, \ldots$ be nonnegative numbers. Take $\mathscr{F}$ as all subsets of Ω, and define

$$\mu(A) = \sum_{x_i \in A} p_i.$$

Thus if $A = \{x_{i_1}, x_{i_2}, \ldots\}$, then $\mu(A) = p_{i_1} + p_{i_2} + \cdots$. The set function μ is a measure on $\mathscr{F}$ and $\mu\{x_i\} = p_i, i = 1, 2, \ldots$. A probability measure will be obtained iff $\sum_i p_i = 1$; if all $p_i = 1$, then μ is counting measure.

Now if A is a subset of $\mathbb{R}$, we try to arrive at a definition of the *length* of A. If A is an interval (open, closed, or semiclosed) with endpoints a and b, it is reasonable to take the length of A to be $\mu(A) = b - a$. If A is a complicated set, we may not have any intuition about its length, but we shall see in Section 1.4 that the requirements that $\mu(a, b] = b - a$ for all $a, b \in \mathbb{R}$, $a < b$, and that μ be a measure, determine μ on a large class of sets.

Specifically, μ is determined on the collection of *Borel sets* of $\mathbb{R}$, denoted by $\mathscr{B}(\mathbb{R})$ and defined as the smallest σ-field of subsets of $\mathbb{R}$ containing all intervals $(a, b], a, b \in \mathbb{R}$.

Note that $\mathscr{B}(\mathbb{R})$ is guaranteed to exist; it may be described (admittedly in a rather ethereal way) as the intersection of all σ-fields containing the intervals

$(a, b]$. Also, if a σ-field contains, say, all open intervals, it must contain all intervals $(a, b]$, and conversely. For

$$(a, b] = \bigcap_{n=1}^{\infty} \left(a, b + \frac{1}{n} \right) \qquad \text{and} \qquad (a, b) = \bigcup_{n=1}^{\infty} \left(a, b - \frac{1}{n} \right].$$

Thus $\mathscr{B}(\mathbb{R})$ is the smallest σ-field containing all open intervals. Similarly we may replace the intervals $(a, b]$ by other classes of intervals, for instance,

all closed intervals,
all intervals $[a, b), a, b \in \mathbb{R}$,
all intervals $(a, \infty), a \in \mathbb{R}$,
all intervals $[a, \infty), a \in \mathbb{R}$,
all intervals $(-\infty, b), b \in \mathbb{R}$,
all intervals $(-\infty, b], b \in \mathbb{R}$.

Since a σ-field that contains all intervals of a given type contains all intervals of any other type, $\mathscr{B}(\mathbb{R})$ may be described as the smallest σ-field that contains the class of all intervals of $\mathbb{R}$. Similarly, $\mathscr{B}(\mathbb{R})$ is the smallest σ-field containing all open sets of $\mathbb{R}$. (To see this, recall that an open set is a countable union of open intervals.) Since a set is open iff its complement is closed, $\mathscr{B}(\mathbb{R})$ is the smallest σ-field containing all closed sets of $\mathbb{R}$. Finally, if $\mathscr{F}_0$ is the field of finite disjoint unions of right-semiclosed intervals (see 1.2.2), then $\mathscr{B}(\mathbb{R})$ is the smallest σ-field containing the sets of $\mathscr{F}_0$.

Intuitively, we may think of generating the Borel sets by starting with the intervals and forming complements and countable unions and intersections in all possible ways. This idea is made precise in Problem 11.

The class of Borel sets of $\overline{\mathbb{R}}$, denoted by $\mathscr{B}(\overline{\mathbb{R}})$, is defined as the smallest σ-field of subsets of $\overline{\mathbb{R}}$ containing all intervals $(a, b], a, b \in \overline{\mathbb{R}}$. The above discussion concerning the replacement of the right-semiclosed intervals by other classes of sets applies equally well to $\overline{\mathbb{R}}$.

If $E \in \mathscr{B}(\mathbb{R})$, $\mathscr{B}(E)$ will denote $\{B \in \mathscr{B}(\mathbb{R}): B \subset E\}$; this coincides with $\{A \cap E: A \in \mathscr{B}(\mathbb{R})\}$ (see 1.2.2).

We now begin to develop some properties of set functions.

1.2.5 Theorem. Let μ be a finitely additive set function on the field $\mathscr{F}$.

(a) $\mu(\emptyset) = 0$.
(b) $\mu(A \cup B) + \mu(A \cap B) = \mu(A) + \mu(B)$ for all $A, B \in \mathscr{F}$.
(c) If $A, B \in \mathscr{F}$ and $B \subset A$, then $\mu(A) = \mu(B) + \mu(A - B)$

(hence $\mu(A - B) = \mu(A) - \mu(B)$ if $\mu(B)$ is finite, and $\mu(B) \leq \mu(A)$ if $\mu(A - B) \geq 0$).

(d)   If μ is nonnegative,

$$\mu\left(\bigcup_{i=1}^{n} A_i\right) \leq \sum_{i=1}^{n} \mu(A_i) \qquad \text{for all} \qquad A_1, \dots, A_n \in \mathscr{F}.$$

If μ is a measure,

$$\mu\left(\bigcup_{n=1}^{\infty} A_n\right) \leq \sum_{n=1}^{\infty} \mu(A_n)$$

for all $A_1, A_2, \dots \in \mathscr{F}$ such that $\bigcup_{n=1}^{\infty} A_n \in \mathscr{F}$.

PROOF.   (a)   Pick $A \in \mathscr{F}$ such that $\mu(A)$ is finite; then

$$\mu(A) = \mu(A \cup \emptyset) = \mu(A) + \mu(\emptyset).$$

(b)   By finite additivity,

$$\mu(A) = \mu(A \cap B) + \mu(A - B),$$
$$\mu(B) = \mu(A \cap B) + \mu(B - A).$$

Add the above equations to obtain

$$\mu(A) + \mu(B) = \mu(A \cap B) + [\mu(A - B) + \mu(B - A) + \mu(A \cap B)]$$
$$= \mu(A \cap B) + \mu(A \cup B).$$

(c)   We may write $A = B \cup (A - B)$, hence $\mu(A) = \mu(B) + \mu(A - B)$.
(d)   We have

$$\bigcup_{i=1}^{n} A_i = A_1 \cup (A_1^c \cap A_2) \cup (A_1^c \cap A_2^c \cap A_3) \cup \cdots \cup (A_1^c \cap \cdots \cap A_{n-1}^c \cap A_n)$$

[see Section 1.1, formula (2)]. The sets on the right are disjoint and

$$\mu(A_1^c \cap \cdots \cap A_{n-1}^c \cap A_n) \leq \mu(A_n) \qquad \text{by (c)}.$$

The case in which μ is a measure is handled using identity (3) of Section 1.1. $\square$

1.2.6   Definitions.   A set function μ defined on $\mathscr{F}$ is said to be *finite* iff $\mu(A)$ is finite, that is, not $\pm\infty$, for each $A \in \mathscr{F}$. If μ is finitely additive, it is sufficient to require that $\mu(\Omega)$ be finite; for $\Omega = A \cup A^c$, and if $\mu(A)$ is, say, $+\infty$, so is $\mu(\Omega)$.

A nonnegative, finitely additive set function μ on the field $\mathscr{F}$ is said to be *σ-finite* on $\mathscr{F}$ iff Ω can be written as $\bigcup_{n=1}^{\infty} A_n$ where the A_n belong to $\mathscr{F}$ and $\mu(A_n) < \infty$ for all n. [By formula (3) of Section 1.1, the A_n may be assumed disjoint.] We shall see that many properties of finite measures can be extended quickly to σ-finite measures.

It follows from 1.2.5(c) that a nonnegative, finitely additive set function μ on a field $\mathscr{F}$ is finite iff it is bounded; that is, $\sup\{|\mu(A)|: A \in \mathscr{F}\} < \infty$. This no longer holds if the nonnegativity assumption is dropped (see Problem 4). It is true, however, that a *countably* additive set function on a σ-field is finite iff it is bounded; this will be proved in 2.1.3.

Countably additive set functions have a basic continuity property, which we now describe.

1.2.7 **Theorem.** Let μ be a countably additive set function on the σ-field $\mathscr{F}$.

 (a)   If $A_1, A_2, \ldots \in \mathscr{F}$ and $A_n \uparrow A$, then $\mu(A_n) \to \mu(A)$ as $n \to \infty$.

 (b)   If $A_1, A_2, \ldots \in \mathscr{F}$, $A_n \downarrow A$, and $\mu(A_1)$ is finite [hence $\mu(A_n)$ is finite for all n since $\mu(A_1) = \mu(A_n) + \mu(A_1 - A_n)$], then $\mu(A_n) \to \mu(A)$ as $n \to \infty$.

The same results hold if $\mathscr{F}$ is only assumed to be a field, if we add the hypothesis that the limit sets A belong to $\mathscr{F}$. [If $A \notin \mathscr{F}$ and $\mu \geq 0$, 1.2.5(c) implies that $\mu(A_n)$ increases to a limit in part (a), and decreases to a limit in part (b), but we cannot identify the limit with $\mu(A)$.]

PROOF.   (a)   If $\mu(A_n) = \infty$ for some n, then $\mu(A) = \mu(A_n) + \mu(A - A_n)$ $= \infty + \mu(A - A_n) = \infty$. Replacing A by A_k we find that $\mu(A_k) = \infty$ for all $k \geq n$, and we are finished. In the same way we eliminate the case in which $\mu(A_n) = -\infty$ for some n. Thus we may assume that all $\mu(A_n)$ are finite.

Since the A_n form an increasing sequence, we may use identity (5) of Section 1.1:

$$A = A_1 \cup (A_2 - A_1) \cup \cdots \cup (A_n - A_{n-1}) \cup \cdots.$$

Therefore, by 1.2.5(c),

$$\mu(A) = \mu(A_1) + \mu(A_2) - \mu(A_1) + \cdots + \mu(A_n) - \mu(A_{n-1}) + \cdots$$

$$= \lim_{n \to \infty} \mu(A_n).$$

 (b)   If $A_n \downarrow A$, then $A_1 - A_n \uparrow A_1 - A$, hence $\mu(A_1 - A_n) \to \mu(A_1 - A)$ by (a). The result now follows from 1.2.5(c). ☐

We shall frequently encounter situations in which finite additivity of a particular set function is easily established, but countable additivity is more difficult. It is useful to have the result that finite additivity plus continuity implies countable additivity.

1.2.8 Theorem. Let μ be a finitely additive set function on the field $\mathscr{F}$.

(a) Assume that μ is *continuous from below* at each $A \in \mathscr{F}$, that is, if $A_1, A_2, \ldots \in \mathscr{F}$, $A = \bigcup_{n=1}^{\infty} A_n \in \mathscr{F}$, and $A_n \uparrow A$, then $\mu(A_n) \to \mu(A)$. It follows that μ is countably additive on $\mathscr{F}$.

(b) Assume that μ is *continuous from above* at the empty set, that is, if $A_1, A_2, \ldots, \in \mathscr{F}$ and $A_n \downarrow \emptyset$, then $\mu(A_n) \to 0$. It follows that μ is countably additive on $\mathscr{F}$.

PROOF. (a) Let $A_1, A_2, \ldots$ be disjoint sets in $\mathscr{F}$ whose union A belongs to $\mathscr{F}$. If $B_n = \bigcup_{i=1}^{n} A_i$ then $B_n \uparrow A$, hence $\mu(B_n) \to \mu(A)$ by hypothesis. But $\mu(B_n) = \sum_{i=1}^{n} \mu(A_i)$ by finite additivity, hence $\mu(A) = \lim_{n \to \infty} \sum_{i=1}^{n} \mu(A_i)$, the desired result.

(b) Let $A_1, A_2, \ldots$ be disjoint sets in $\mathscr{F}$ whose union A belongs to $\mathscr{F}$, and let $B_n = \bigcup_{i=1}^{n} A_i$. By 1.2.5(c), $\mu(A) = \mu(B_n) + \mu(A - B_n)$; but $A - B_n \downarrow \emptyset$, so by hypothesis, $\mu(A - B_n) \to 0$. Thus $\mu(B_n) \to \mu(A)$, and the result follows as in (a). $\square$

If μ_1 and μ_2 are measures on the σ-field $\mathscr{F}$, then $\mu = \mu_1 - \mu_2$ is countably additive on $\mathscr{F}$, assuming either μ_1 or μ_2 is finite-valued. We shall see later (in 2.1.3) that any countably additive set function on a σ-field can be expressed as the difference of two measures.

For examples of finitely additive set functions that are not countably additive, see Problems 1, 3, and 4.

Problems

1. Let Ω be a countably infinite set, and let $\mathscr{F}$ consist of all subsets of Ω. Define $\mu(A) = 0$ if A is finite, $\mu(A) = \infty$ if A is infinite.

   (a) Show that μ is finitely additive but not countably additive.

   (b) Show that Ω is the limit of an increasing sequence of sets A_n with $\mu(A_n) = 0$ for all n, but $\mu(\Omega) = \infty$.

2. Let μ be counting measure on Ω, where Ω is an infinite set. Show that there is a sequence of sets $A_n \downarrow \emptyset$ with $\lim_{n \to \infty} \mu(A_n) \neq 0$.

3. Let Ω be a countably infinite set, and let $\mathscr{F}$ be the field consisting of all finite subsets of Ω and their complements. If A is finite, set $\mu(A) = 0$, and if A^c is finite, set $\mu(A) = 1$.

   (a) Show that μ is finitely additive but not countably additive on $\mathscr{F}$.

    (b) Show that Ω is the limit of an increasing sequence of sets $A_n \in \mathscr{F}$ with $\mu(A_n) = 0$ for all n, but $\mu(\Omega) = 1$.

4.   Let $\mathscr{F}$ be the field of finite disjoint unions of right-semiclosed intervals of $\mathbb{R}$, and define the set function μ on $\mathscr{F}$ as follows.

$$\mu(-\infty, a] = a, \qquad\qquad a \in \mathbb{R},$$

$$\mu(a, b] = b - a, \qquad\quad a, b \in \mathbb{R}, \qquad a < b,$$

$$\mu(b, \infty) = -b, \qquad\qquad b \in \mathbb{R},$$

$$\mu(\mathbb{R}) = 0,$$

$$\mu\left(\bigcup_{i=1}^{n} I_i\right) = \sum_{i=1}^{n} \mu(I_i)$$

$\mu(a,o) = \mu(a,b] + \mu(b,o)$
$b - a \ -b = -a \ \checkmark$

                if $I_1, \ldots, I_n$ are disjoint right-semiclosed intervals.

    (a) Show that μ is finitely additive but not countably additive on $\mathscr{F}$.

    (b) Show that μ is finite but unbounded on $\mathscr{F}$.

5.   Let μ be a nonnegative, finitely additive set function on the field $\mathscr{F}$. If $A_1, A_2, \ldots$ are disjoint sets in $\mathscr{F}$ and $\bigcup_{n=1}^{\infty} A_n \in \mathscr{F}$, show that

$$\mu\left(\bigcup_{n=1}^{\infty} A_n\right) \geq \sum_{n=1}^{\infty} \mu(A_n).$$

6.   Let $f: \Omega \to \Omega'$, and let $\mathscr{C}$ be a class of subsets of Ω'. Show that

$$\sigma(f^{-1}(\mathscr{C})) = f^{-1}(\sigma(\mathscr{C})),$$

    where $f^{-1}(\mathscr{C}) = \{f^{-1}(A): A \in \mathscr{C}\}$. (Use the good sets principle.)

7.   If A is a Borel subset of $\mathbb{R}$, show that the smallest σ-field of subsets of A containing the sets open in A (in the relative topology inherited from $\mathbb{R}$) is $\{B \in \mathscr{B}(\mathbb{R}): B \subset A\}$.

8.   Let $A_1, \ldots, A_n$ be arbitrary subsets of a set Ω. Describe (explicitly) the smallest σ-field $\mathscr{F}$ containing $A_1, \ldots, A_n$. How many sets are there in $\mathscr{F}$? (Give an upper bound that is attainable under certain conditions.) List all the sets in $\mathscr{F}$ when n = 2.

9.   (a) Let $\mathscr{C}$ be an arbitrary class of subsets of Ω, and let $\mathscr{G}$ be the collection of all finite unions $\bigcup_{i=1}^{n} A_i$, $n = 1, 2, \ldots$, where each A_i is a finite intersection $\bigcap_{j=1}^{r} B_{ij}$, with B_{ij} or its complement a set in $\mathscr{C}$. Show that $\mathscr{G}$ is the minimal field (not σ-field) over $\mathscr{C}$.

    (b) Show that the minimal field can also be described as the collection $\mathscr{D}$ of all finite *disjoint* unions $\bigcup_{i=1}^{n} A_i$, where the A_i are as above.

(c) If $\mathscr{F}_1, \ldots, \mathscr{F}_n$ are fields of subsets of Ω, show that the smallest field including $\mathscr{F}_1, \ldots, \mathscr{F}_n$ consists of all finite (disjoint) unions of sets $A_1 \cap \cdots \cap A_n$ with $A_i \in \mathscr{F}_i$, $i = 1, \ldots, n$.

10. Let μ be a finite measure on the σ-field $\mathscr{F}$. If $A_n \in \mathscr{F}$, $n = 1, 2, \ldots$ and $A = \lim_n A_n$ (see Section 1.1), show that $\mu(A) = \lim_{n \to \infty} \mu(A_n)$.

11.* Let $\mathscr{C}$ be any class of subsets of Ω, with $\emptyset, \Omega \in \mathscr{C}$. Define $\mathscr{C}_0 = \mathscr{C}$, and for any ordinal $\alpha > 0$ write, inductively,

$$\mathscr{C}_\alpha = \left(\bigcup \{\mathscr{C}_\beta \colon \beta < \alpha\} \right)',$$

where $\mathscr{D}'$ denotes the class of all countable unions of differences of sets in $\mathscr{D}$.

Let $\mathscr{S} = \bigcup \{\mathscr{C}_\alpha \colon \alpha < \beta_1\}$, where β_1 is the first uncountable ordinal, and let $\mathscr{F}$ be the minimal σ-field over $\mathscr{C}$. Since each $\mathscr{C}_\alpha \subset \mathscr{F}$, we have $\mathscr{S} \subset \mathscr{F}$. Also, the $\mathscr{C}_\alpha$ increase with α, and $\mathscr{C} \subset \mathscr{C}_\alpha$ for all α.

(a) Show that $\mathscr{S}$ is a σ-field (hence $\mathscr{S} = \mathscr{F}$ by minimality of $\mathscr{F}$).

(b) If the cardinality of $\mathscr{C}$ is at most c, the cardinality of the reals, show that card $\mathscr{F} \leq c$ also.

12. Show that if μ is a finite measure, there cannot be uncountably many disjoint sets A such that $\mu(A) > 0$.

1.3 EXTENSION OF MEASURES

In 1.2.4, we discussed the concept of length of a subset of $\mathbb{R}$. The problem was to extend the set function given on intervals by $\mu(a, b] = b - a$ to a larger class of sets. If $\mathscr{F}_0$ is the field of finite disjoint unions of right-semiclosed intervals, there is no problem extending μ to $\mathscr{F}_0$: if $A_1, \ldots, A_n$ are disjoint right-semiclosed intervals, we set $\mu\left(\bigcup_{i=1}^n A_i\right) = \sum_{i=1}^n \mu(A_i)$. The resulting set function on $\mathscr{F}_0$ is finitely additive, but countable additivity is not clear at this point. Even if we can prove countable additivity on $\mathscr{F}_0$, we still have the problem of extending μ to the minimal σ-field over $\mathscr{F}_0$, namely, the Borel sets.

We are going to consider a generalization of the above problem. Instead of working only with length, we shall examine set functions given by $\mu(a, b] = F(b) - F(a)$ where F is an increasing right-continuous function from $\mathbb{R}$ to $\mathbb{R}$. The extension technique to be developed is not restricted to set functions defined on subsets of $\mathbb{R}$; we shall prove a general result concerning the extension of a measure from a field $\mathscr{F}_0$ to the minimal σ-field over $\mathscr{F}_0$.

It will be convenient to consider finite measures at first, and nothing is lost if we normalize and work with probability measures.

1.3.1 Lemma. Let $\mathscr{F}_0$ be a field of subsets of a set Ω, and let P be a probability measure on $\mathscr{F}_0$. Suppose that the sets $A_1, A_2, \ldots$ belong to $\mathscr{F}_0$ and increase to a limit A, and that the sets $A_1', A_2', \ldots$ belong to $\mathscr{F}_0$ and increase to A'. (A and A' need not belong to $\mathscr{F}_0$.) If $A \subset A'$, then

$$\lim_{m \to \infty} P(A_m) \le \lim_{n \to \infty} P(A_n').$$

Thus if A_n and A_n' both increase to the same limit A, then

$$\lim_{n \to \infty} P(A_n) = \lim_{n \to \infty} P(A_n').$$

PROOF. If m is fixed, $A_m \cap A_n' \uparrow A_m \cap A' = A_m$ as $n \to \infty$, hence

$$P(A_m \cap A_n') \to P(A_m)$$

by 1.2.7(a). But $P(A_m \cap A_n') \le P(A_n')$ by 1.2.5(c), hence

$$P(A_m) = \lim_{n \to \infty} P(A_m \cap A_n') \le \lim_{n \to \infty} P(A_n').$$

Let $m \to \infty$ to finish the proof. $\square$

We are now ready for the first extension of P to a larger class of sets.

1.3.2 *Lemma.* Let P be a probability measure on the field $\mathscr{F}_0$. Let $\mathscr{G}$ be the collection of all limits of increasing sequences of sets in $\mathscr{F}_0$, that is, $A \in \mathscr{G}$ iff there are sets $A_n \in \mathscr{F}_0$, $n = 1, 2, \ldots$, such that $A_n \uparrow A$. (Note that $\mathscr{G}$ can also be described as the collection of all countable unions of sets in $\mathscr{F}_0$; see 1.2.1.)

Define μ on $\mathscr{G}$ as follows. If $A_n \in \mathscr{F}_0$, $n = 1, 2, \ldots, A_n \uparrow A$ ($\in \mathscr{G}$), set $\mu(A) = \lim_{n \to \infty} P(A_n)$; μ is well defined by 1.3.1, and $\mu = P$ on $\mathscr{F}_0$. Then:

(a) $\emptyset \in \mathscr{G}$ and $\mu(\emptyset) = 0$; $\Omega \in \mathscr{G}$ and $\mu(\Omega) = 1$; $0 \le \mu(A) \le 1$ for all $A \in \mathscr{G}$.

(b) If $G_1, G_2 \in \mathscr{G}$, then $G_1 \cup G_2, G_1 \cap G_2 \in \mathscr{G}$ and

$$\mu(G_1 \cup G_2) + \mu(G_1 \cap G_2) = \mu(G_1) + \mu(G_2).$$

(c) If $G_1, G_2 \in \mathscr{G}$ and $G_1 \subset G_2$, then $\mu(G_1) \le \mu(G_2)$.

(d) If $G_n \in \mathscr{G}$, $n = 1, 2, \ldots$, and $G_n \uparrow G$,

then $G \in \mathscr{G}$ and $\mu(G_n) \to \mu(G)$.

PROOF. (a) This is clear since $\mu = P$ on $\mathscr{F}_0$ and P is a probability measure.

(b) Let $A_{n1} \in \mathscr{F}_0$, $A_{n1} \uparrow G_1$; $A_{n2} \in \mathscr{F}_0$, $A_{n2} \uparrow G_2$. We have $P(A_{n1} \cup A_{n2}) + P(A_{n1} \cap A_{n2}) = P(A_{n1}) + P(A_{n2})$ by 1.2.5(b); let $n \to \infty$ to complete the argument.

(c) This follows from 1.3.1.

(d)   Since G is a countable union of sets in $\mathscr{F}_0$, $G \in \mathscr{G}$. Now for each n we can find sets $A_{nm} \in \mathscr{F}_0$, $m = 1, 2, \ldots$, with $A_{nm} \uparrow G_n$ as $m \to \infty$. The situation may be represented schematically as follows:

$$
\begin{array}{cccccc}
A_{11} & A_{12} & \cdots & A_{1m} & \cdots & \uparrow G_1 \\
A_{21} & A_{22} & \cdots & A_{2m} & \cdots & \uparrow G_2 \\
\vdots & \vdots & \vdots & \vdots & \vdots & \vdots \\
A_{n1} & A_{n2} & \cdots & A_{nm} & \cdots & \uparrow G_n \\
\vdots & \vdots & \vdots & \vdots & \vdots & \vdots
\end{array}
$$

Let $D_m = A_{1m} \cup A_{2m} \cup \cdots \cup A_{mm}$ (the D_m form an increasing sequence). The key step in the proof is the observation that

$$A_{nm} \subset D_m \subset G_m \qquad \text{for} \qquad n \leq m \tag{1}$$

and, therefore,

$$P(A_{nm}) \leq P(D_m) \leq \mu(G_m) \qquad \text{for} \qquad n \leq m. \tag{2}$$

Let $m \to \infty$ in (1) to obtain $G_n \subset \bigcup_{m=1}^{\infty} D_m \subset G$; then let $n \to \infty$ to conclude that $D_m \uparrow G$, hence $P(D_m) \to \mu(G)$ by definition of μ. Now let $m \to \infty$ in (2) to obtain $\mu(G_n) \leq \lim_{m \to \infty} P(D_m) \leq \lim_{m \to \infty} \mu(G_m)$; then let $n \to \infty$ to conclude that $\lim_{n \to \infty} \mu(G_n) = \lim_{m \to \infty} P(D_m) = \mu(G)$. $\quad\square$

We now extend μ to the class of all subsets of Ω; however, the extension will not be countably additive on all subsets, but only on a smaller σ-field. The construction depends on properties (a)–(d) of 1.3.2, and not on the fact that μ was derived from a probability measure on a field. We express this explicitly as follows:

1.3.3   Lemma. Let $\mathscr{G}$ be a class of subsets of a set Ω, μ a nonnegative real-valued set function on $\mathscr{G}$ such that $\mathscr{G}$ and μ satisfy the four conditions (a)–(d) of 1.3.2. Define, for each $A \subset \Omega$,

$$\mu^*(A) = \inf\{\mu(G) \colon G \in \mathscr{G}, \quad G \supset A\}.$$

Then:

(a) $\mu^* = \mu$ on $\mathscr{G}$, $0 \leq \mu^*(A) \leq 1$ for all $A \subset \Omega$.

(b) $\mu^*(A \cup B) + \mu^*(A \cap B) \leq \mu^*(A) + \mu^*(B)$; in particular, $\mu^*(A) + \mu^*(A^c) \geq \mu^*(\Omega) + \mu^*(\emptyset) = \mu(\Omega) + \mu(\emptyset) = 1$ by 1.3.2(a).

(c)   If $A \subset B$, then $\mu^*(A) \leq \mu^*(B)$.

(d)   If $A_n \uparrow A$, then $\mu^*(A_n) \to \mu^*(A)$.

PROOF.   (a)   This is clear from the definition of μ^* and from 1.3.2(c).

(b)   If $\varepsilon > 0$, choose G_1, $G_2 \in \mathscr{G}$, $G_1 \supset A$, $G_2 \supset B$, such that $\mu(G_1) \leq \mu^*(A) + \varepsilon/2$, $\mu(G_2) \leq \mu^*(B) + \varepsilon/2$. By 1.3.2(b),

$$\mu^*(A) + \mu^*(B) + \varepsilon \geq \mu(G_1) + \mu(G_2) = \mu(G_1 \cup G_2) + \mu(G_1 \cap G_2)$$

$$\geq \mu^*(A \cup B) + \mu^*(A \cap B).$$

Since ε is arbitrary, the result follows.

(c)   This follows from the definition of μ^*.

(d)   By (c), $\mu^*(A) \geq \lim_{n \to \infty} \mu^*(A_n)$. If $\varepsilon > 0$, for each n we may choose $G_n \in \mathscr{G}$, $G_n \supset A_n$, such that

$$\mu(G_n) \leq \mu^*(A_n) + \varepsilon 2^{-n}.$$

Now $A = \bigcup_{n=1}^{\infty} A_n \subset \bigcup_{n=1}^{\infty} G_n \in \mathscr{G}$; hence

$$\mu^*(A) \leq \mu^* \left(\bigcup_{n=1}^{\infty} G_n \right) \qquad \text{by (c)}$$

$$= \mu \left(\bigcup_{n=1}^{\infty} G_n \right) \qquad \text{by (a)}$$

$$= \lim_{n \to \infty} \mu \left(\bigcup_{k=1}^{n} G_k \right) \qquad \text{by 1.3.2(d).}$$

The proof will be accomplished if we prove that

$$\mu \left(\bigcup_{i=1}^{n} G_i \right) \leq \mu^*(A_n) + \varepsilon \sum_{i=1}^{n} 2^{-i}, \qquad n = 1, 2, \ldots.$$

This is true for $n = 1$, by choice of G_1. If it holds for a given n, we apply 1.3.2(b) to the sets $\bigcup_{i=1}^{n} G_i$ and G_{n+1} to obtain

$$\mu \left(\bigcup_{i=1}^{n+1} G_i \right) = \mu \left(\bigcup_{i=1}^{n} G_i \right) + \mu(G_{n+1}) - \mu \left[\left(\bigcup_{i=1}^{n} G_i \right) \cap G_{n+1} \right].$$

Now $\left(\bigcup_{i=1}^{n} G_i\right) \cap G_{n+1} \supset G_n \cap G_{n+1} \supset A_n \cap A_{n+1} = A_n$, so that the induction hypothesis yields

$$\mu\left(\bigcup_{i=1}^{n+1} G_i\right) \leq \mu^*(A_n) + \varepsilon \sum_{i=1}^{n} 2^{-i} + \mu^*(A_{n+1}) + \varepsilon 2^{-(n+1)} - \mu^*(A_n)$$

$$\leq \mu^*(A_{n+1}) + \varepsilon \sum_{i=1}^{n+1} 2^{-i}. \quad \square$$

Our aim in this section is to prove that a σ-finite measure on a field $\mathscr{F}_0$ has a unique extension to the minimal σ-field over $\mathscr{F}_0$. In fact an arbitrary measure μ on $\mathscr{F}_0$ can be extended to $\sigma(\mathscr{F}_0)$, but the extension is not necessarily unique. In proving this more general result (see Problem 3), the following concept plays a key role.

1.3.4 Definition. An *outer measure* on Ω is a nonnegative, extended real-valued set function λ on the class of all subsets of Ω, satisfying

(a) $\lambda(\emptyset) = 0$,
(b) $A \subset B$ implies $\lambda(A) \leq \lambda(B)$ (monotonicity), and
(c) $\lambda\left(\bigcup_{n=1}^{\infty} A_n\right) \leq \sum_{n=1}^{\infty} \lambda(A_n)$ (countable subadditivity).

The set function μ^* of 1.3.3 is an outer measure on Ω. Parts 1.3.4(a) and (b) follow from 1.3.3(a), 1.3.2(a), and 1.3.3(c), and 1.3.4(c) is proved as follows:

$$\mu^*\left(\bigcup_{n=1}^{\infty} A_n\right) = \lim_{n \to \infty} \mu^*\left(\bigcup_{i=1}^{n} A_i\right) \qquad \text{by 1.3.3(d).}$$

$$\leq \lim_{n \to \infty} \sum_{i=1}^{n} \mu^*(A_i) \qquad \text{by 1.3.3(b),}$$

as desired.

We now identify a σ-field on which μ^* is countably additive:

1.3.5 Theorem. Under the hypothesis of 1.3.2, with μ^* defined as in 1.3.3, let $\mathscr{H} = \{H \subset \Omega: \mu^*(H) + \mu^*(H^c) = 1\}$
$[\mathscr{H} = \{H \subset \Omega: \mu^*(H) + \mu^*(H^c) \leq 1$ by 1.3.3(b).$]$
Then $\mathscr{H}$ is a σ-field and μ^* is a probability measure on $\mathscr{H}$.

PROOF.   First note that $\mathscr{G} \subset \mathscr{H}$. For if $A_n \in \mathscr{F}_0$ and $A_n \uparrow G \in \mathscr{G}$, then $G^c \subset A_n^c$, so $P(A_n) + \mu^*(G^c) \leq P(A_n) + P(A_n^c) = 1$. By 1.3.3(d), $\mu^*(G) + \mu^*(G^c) \leq 1$.

Clearly $\mathcal{H}$ is closed under complementation, and $\Omega \in \mathcal{H}$ by 1.3.3(a) and 1.3.2(a). If H_1, $H_2 \subset \Omega$, then by 1.3.3(b),

$$\mu^*(H_1 \cup H_2) + \mu^*(H_1 \cap H_2) \le \mu^*(H_1) + \mu^*(H_2) \tag{1}$$

and since

$$(H_1 \cup H_2)^c = H_1^c \cap H_2^c, \qquad (H_1 \cap H_2)^c = H_1^c \cup H_2^c,$$

we have

$$\mu^*(H_1 \cup H_2)^c + \mu^*(H_1 \cap H_2)^c \le \mu^*(H_1^c) + \mu^*(H_2^c). \tag{2}$$

If H_1, $H_2 \in \mathcal{H}$, add (1) and (2); the sum of the left sides is at least 2 by 1.3.3(b), and the sum of the right sides is 2. Thus the sum of the left sides is 2 as well. If $a = \mu^*(H_1 \cup H_2) + \mu^*(H_1 \cup H_2)^c$, $b = \mu^*(H_1 \cap H_2) + \mu^*(H_1 \cap H_2)^c$, then $a + b = 2$, hence $a \le 1$ or $b \le 1$. If $a \le 1$, then $a = 1$, so $b = 1$ also. Consequently $H_1 \cup H_2 \in \mathcal{H}$ and $H_1 \cap H_2 \in \mathcal{H}$. We have therefore shown that $\mathcal{H}$ is a field. Now equality holds in (1), for if not, the sum of the left sides of (1) and (2) would be less than the sum of the right sides, a contradiction. Thus μ^* is finitely additive on $\mathcal{H}$.

To show that $\mathcal{H}$ is a σ-field, let $H_n \in \mathcal{H}$, $n = 1, 2, \ldots, H_n \uparrow H$; $\mu^*(H)$ $+ \mu^*(H^c) \ge 1$ by 1.3.3(b). But $\mu^*(H) = \lim_{n \to \infty} \mu^*(H_n)$ by 1.3.3(d), hence for any $\varepsilon > 0$, $\mu^*(H) \le \mu^*(H_n) + \varepsilon$ for large n. Since $\mu^*(H^c) \le \mu^*(H_n^c)$ for all n by 1.3.3(c), and $H_n \in \mathcal{H}$, we have $\mu^*(H) + \mu^*(H^c) \le 1 + \varepsilon$. Since ε is arbitrary, $H \in \mathcal{H}$, making $\mathcal{H}$ a σ-field.

Since $\mu^*(H_n) \to \mu^*(H)$, μ^* is countably additive by 1.2.8(a). $\square$

We now have our first extension theorem.

1.3.6 Theorem. A finite measure on a field $\mathcal{F}_0$ can be extended to a measure on $\sigma(\mathcal{F}_0)$.

PROOF. Nothing is lost by considering a probability measure. (Replace μ by $\mu/\mu(\Omega)$ if necessary.) The result then follows from 1.3.1–1.3.5 if we observe that $\mathcal{F}_0 \subset \mathcal{G} \subset \mathcal{H}$, hence $\sigma(\mathcal{F}_0) \subset \mathcal{H}$. Thus μ^* restricted to $\sigma(\mathcal{F}_0)$ is the desired extension. $\square$

In fact there is very little difference between $\sigma(\mathcal{F}_0)$ and $\mathcal{H}$; if $B \in \mathcal{H}$, then B can be expressed as $A \cup N$, where $A \in \sigma(\mathcal{F}_0)$ and N is a subset of a set $M \in \sigma(\mathcal{F}_0)$ with $\mu^*(M) = 0$. To establish this, we introduce the idea of completion of a measure space.

1.3.7 Definitions. A measure μ on a σ-field $\mathcal{F}$ is said to be *complete* iff whenever $A \in \mathcal{F}$ and $\mu(A) = 0$ we have $B \in \mathcal{F}$ for all $B \subset A$.

In 1.3.5, μ^* on $\mathscr{H}$ is complete, for if $B \subset A \in \mathscr{H}$, $\mu^*(A) = 0$, then $\mu^*(B) + \mu^*(B^c) \leq \mu^*(A) + \mu^*(B^c) = \mu^*(B^c) \leq 1$; thus $B \in \mathscr{H}$.

The *completion* of a measure space $(\Omega, \mathscr{F}, \mu)$ is defined as follows. Let $\mathscr{F}_\mu$ be the class of sets $A \cup N$, where A ranges over $\mathscr{F}$ and N over all subsets of sets of measure 0 in $\mathscr{F}$.

Now $\mathscr{F}_\mu$ is a σ-field including $\mathscr{F}$, for it is clearly closed under countable union, and if $A \cup N \in \mathscr{F}$, $N \subset M \in \mathscr{F}$, $\mu(M) = 0$, then $(A \cup N)^c = A^c \cap N^c = (A^c \cap M^c) \cup (A^c \cap (N^c - M^c))$ and $A^c \cap (N^c - M^c) = A^c \cap (M - N) \subset M$, so $(A \cup N)^c \in \mathscr{F}_\mu$.

We extend μ to $\mathscr{F}_\mu$ by setting $\mu(A \cup N) = \mu(A)$. This is a valid definition, for if $A_1 \cup N_1 = A_2 \cup N_2 \in \mathscr{F}_\mu$, we have

$$\mu(A_1) = \mu(A_1 \cap A_2) + \mu(A_1 - A_2) = \mu(A_1 \cap A_2)$$

since $A_1 - A_2 \subset N_2$. Thus $\mu(A_1) \leq \mu(A_2)$, and by symmetry, $\mu(A_1) = \mu(A_2)$. The measure space $(\Omega, \mathscr{F}_\mu, \mu)$ is called the *completion* of $(\Omega, \mathscr{F}, \mu)$, and $\mathscr{F}_\mu$ the completion of $\mathscr{F}$ relative to μ.

Note that the completion is in fact complete, for if $M \subset A \cup N \in \mathscr{F}_\mu$ where $A \in \mathscr{F}$, $\mu(A) = 0$, $N \subset B \in \mathscr{F}$, $\mu(B) = 0$, then $M \subset A \cup B \in \mathscr{F}$, $\mu(A \cup B) = 0$; hence $M \in \mathscr{F}_\mu$.

1.3.8   Theorem. In 1.3.6, $(\Omega, \mathscr{H}, \mu^*)$ is the completion of $(\Omega, \sigma(\mathscr{F}_0), \mu^*)$.

PROOF.   We must show that $\mathscr{H} = \mathscr{F}_{\mu^*}$ where $\mathscr{F} = \sigma(\mathscr{F}_0)$. If $A \in \mathscr{H}$, by definition of $\mu^*(A)$ and $\mu^*(A^c)$ we can find sets G_n, $G_n' \in \sigma(\mathscr{F}_0)$, $n = 1, 2, \ldots$, with $G_n \subset A \subset G_n'$ and $\mu^*(G_n) \to \mu^*(A)$, $\mu^*(G_n') \to \mu^*(A)$. Let $G = \bigcup_{n=1}^\infty G_n$, $G' = \bigcap_{n=1}^\infty G_n'$. Then $A = G \cup (A - G)$, $G \in \sigma(\mathscr{F}_0)$, $A - G \subset G' - G \in \sigma(\mathscr{F}_0)$, $\mu^*(G' - G) \leq \mu^*(G_n' - G_n) \to 0$, so that $\mu^*(G' - G) = 0$. Thus $A \in \mathscr{F}_{\mu^*}$.

Conversely if $B \in \mathscr{F}_{\mu^*}$, then $B = A \cup N$, $A \in \mathscr{F}$, $N \subset M \in \mathscr{F}$, $\mu^*(M) = 0$. Since $\mathscr{F} \subset \mathscr{H}$ we have $A \in \mathscr{H}$, and since $(\Omega, \mathscr{H}, \mu^*)$ is complete we have $N \in \mathscr{H}$. Thus $B \in \mathscr{H}$. $\square$

To prove the uniqueness of the extension from $\mathscr{F}_0$ to $\mathscr{F}$, we need the following basic result.

1.3.9   Monotone Class Theorem. Let $\mathscr{F}_0$ be a field of subsets of Ω, and $\mathscr{C}$ a class of subsets of Ω that is monotone (if $A_n \in \mathscr{C}$ and $A_n \uparrow A$ or $A_n \downarrow A$, then $A \in \mathscr{C}$). If $\mathscr{C} \supset \mathscr{F}_0$, then $\mathscr{C} \supset \sigma(\mathscr{F}_0)$, the minimal σ-field over $\mathscr{F}_0$.

PROOF.   The technique of the proof might be called "boot strapping." Let $\mathscr{F} = \sigma(\mathscr{F}_0)$ and let $\mathscr{M}$ be the smallest monotone class containing all sets of

$\mathscr{F}_0$. We show that $\mathscr{M} = \mathscr{F}$, in other words, *the smallest monotone class and the smallest σ-field over a field coincide*. The proof is completed by observing that $\mathscr{M} \subset \mathscr{C}$.

Fix $A \in \mathscr{M}$ and let $\mathscr{M}_A = \{B \in \mathscr{M}: A \cap B, A \cap B^c \text{ and } A^c \cap B \in \mathscr{M}\}$; then $\mathscr{M}_A$ is a monotone class. In fact $\mathscr{M}_A = \mathscr{M}$; for if $A \in \mathscr{F}_0$, then $\mathscr{F}_0 \subset \mathscr{M}_A$ since $\mathscr{F}_0$ is a field, hence $\mathscr{M} \subset \mathscr{M}_A$ by minimality of $\mathscr{M}$; consequently $\mathscr{M}_A = \mathscr{M}$. But this shows that for any $B \in \mathscr{M}$ we have $A \cap B$, $A \cap B^c$, $A^c \cap B \in \mathscr{M}$ for any $A \in \mathscr{F}_0$, so that $\mathscr{M}_B \supset \mathscr{F}_0$. Again by minimality of $\mathscr{M}$, $\mathscr{M}_B = \mathscr{M}$.

Now $\mathscr{M}$ is a field (for if $A, B \in \mathscr{M} = \mathscr{M}_A$, then $A \cap B$, $A \cap B^c$, $A^c \cap B \in \mathscr{M}$) and *a monotone class that is also a field is a σ-field* (see 1.2.1), hence $\mathscr{M}$ is a σ-field. Thus $\mathscr{F} \subset \mathscr{M}$ by minimality of $\mathscr{F}$, and in fact $\mathscr{F} = \mathscr{M}$ because $\mathscr{F}$ is a monotone class including $\mathscr{F}_0$. □

We now prove the fundamental extension theorem.

1.3.10 Carathéodory Extension Theorem. Let μ be a measure on the field $\mathscr{F}_0$ of subsets of Ω, and assume that μ is σ-finite on $\mathscr{F}_0$, so that Ω can be decomposed as $\bigcup_{n=1}^{\infty} A_n$, where $A_n \in \mathscr{F}_0$ and $\mu(A_n) < \infty$ for all n. Then μ has a unique extension to a measure on the minimal σ-field $\mathscr{F}$ over $\mathscr{F}_0$.

PROOF. Since $\mathscr{F}_0$ is a field, the A_n may be taken as disjoint [replace A_n by $A_1^c \cap \cdots \cap A_{n-1}^c \cap A_n$, as in formula (3) of 1.1]. Let $\mu_n(A) = \mu(A \cap A_n)$, $A \in \mathscr{F}_0$; then μ_n is a finite measure on $\mathscr{F}_0$, hence by 1.3.6 it has an extension μ_n^* to $\mathscr{F}$. As $\mu = \sum_n \mu_n$, the set function $\mu^* = \sum_n \mu_n^*$ is an extension of μ, and it is a measure on $\mathscr{F}$ since the order of summation of any double series of nonnegative terms can be reversed.

Now suppose that λ is a measure on $\mathscr{F}$ and $\lambda = \mu$ on $\mathscr{F}_0$. Define $\lambda_n(A) = \lambda(A \cap A_n)$, $A \in \mathscr{F}$. Then λ_n is a finite measure on $\mathscr{F}$ and $\lambda_n = \mu_n = \mu_n^*$ on $\mathscr{F}_0$, and it follows that $\lambda_n = \mu_n^*$ on $\mathscr{F}$. For $\mathscr{C} = \{A \in \mathscr{F}: \lambda_n(A) = \mu_n^*(A)\}$ is a monotone class (by 1.2.7) that contains all sets of $\mathscr{F}_0$, hence $\mathscr{C} = \mathscr{F}$ by 1.3.9. But then $\lambda = \sum_n \lambda_n = \sum_n \mu_n^* = \mu^*$, proving uniqueness. □

The intuitive idea of constructing a minimal σ-field by forming complements and countable unions and intersections in all possible ways suggests that if $\mathscr{F}_0$ is a field and $\mathscr{F} = \sigma(\mathscr{F}_0)$, sets in $\mathscr{F}$ can be approximated in some sense by sets in $\mathscr{F}_0$. The following result formalizes this notion.

1.3.11 Approximation Theorem. Let $(\Omega, \mathscr{F}, \mu)$ be a measure space, and let $\mathscr{F}_0$ be a field of subsets of Ω such that $\sigma(\mathscr{F}_0) = \mathscr{F}$. Assume that μ is σ-finite on $\mathscr{F}_0$, and let $\varepsilon > 0$ be given. If $A \in \mathscr{F}$ and $\mu(A) < \infty$, there is a set $B \in \mathscr{F}_0$ such that $\mu(A \Delta B) < \varepsilon$.

PROOF.   Let $\mathscr{G}$ be the class of all countable unions of sets of $\mathscr{F}_0$. The conclusion of 1.3.11 holds for any $A \in \mathscr{G}$, by 1.2.7(a). By 1.3.3, if μ is finite and $A \in \mathscr{F}$, A can be approximated arbitrarily closely (in the sense of 1.3.11) by a set in $\mathscr{G}$, and therefore 1.3.11 is proved for finite μ. In general, let Ω be the disjoint union of sets $A_n \in \mathscr{F}_0$ with $\mu(A_n) < \infty$, and let $\mu_n(C) = \mu(C \cap A_n)$, $C \in \mathscr{F}$.

Then μ_n is a finite measure on $\mathscr{F}$, hence if $A \in \mathscr{F}$, there is a set $B_n \in \mathscr{F}_0$ such that $\mu_n(A \,\Delta\, B_n) < \varepsilon 2^{-n}$. Since

$$\mu_n(A \,\Delta\, B_n) = \mu((A \,\Delta\, B_n) \cap A_n)$$
$$= \mu[(A \,\Delta\, (B_n \cap A_n)) \cap A_n] = \mu_n(A \,\Delta\, (B_n \cap A_n)),$$

and $B_n \cap A_n \in \mathscr{F}_0$, we may assume that $B_n \subset A_n$. (The observation that $B_n \cap A_n \in \mathscr{F}_0$ is the point where we use the hypothesis that μ is σ-finite on $\mathscr{F}_0$, not merely on $\mathscr{F}$.) If $C = \bigcup_{n=1}^{\infty} B_n$, then $C \cap A_n = B_n$, so that

$$\mu_n(A \,\Delta\, C) = \mu((A \,\Delta\, C) \cap A_n) = \mu((A \,\Delta\, B_n) \cap A_n) = \mu_n(A \,\Delta\, B_n),$$

hence

$$\mu(A \,\Delta\, C) = \sum_{n=1}^{\infty} \mu_n(A \,\Delta\, C) < \varepsilon. \text{ But } \bigcup_{k=1}^{N} B_k - A \uparrow C - A \text{ as } N \to \infty,$$

and $A - \bigcup_{k=1}^{N} B_k \downarrow A - C$. If $A \in \mathscr{F}$ and $\mu(A) < \infty$, it follows from 1.2.7 that $\mu(A \,\Delta\, \bigcup_{k=1}^{N} B_k) \to \mu(A \,\Delta\, C)$ as $N \to \infty$, hence is less than ε for large enough N. Set $B = \bigcup_{k=1}^{N} B_k \in \mathscr{F}_0$. $\square$

1.3.12   Example. Let Ω be the rationals, $\mathscr{F}_0$ the field of finite disjoint unions of right-semiclosed intervals $(a, b] = \{\omega \in \Omega: a < \omega \leq b\}$, a, b rational [counting (a, ∞) and Ω itself as right-semiclosed; see 1.2.2]. Let $\mathscr{F} = \sigma(\mathscr{F}_0)$. Then:

(a) $\mathscr{F}$ consists of all subsets of Ω.
(b)   If $\mu(A)$ is the number of points in A (μ is counting measure), then μ is σ-finite on $\mathscr{F}$ but not on $\mathscr{F}_0$.
(c)   There are sets $A \in \mathscr{F}$ of finite measure that cannot be approximated by sets in $\mathscr{F}_0$, that is, there is no sequence $A_n \in \mathscr{F}_0$ with $\mu(A \,\Delta\, A_n) \to 0$.
(d)   If $\lambda = 2\mu$, then $\lambda = \mu$ on $\mathscr{F}_0$ but not on $\mathscr{F}$.

Thus both the approximation theorem and the Carathéodory extension theorem fail in this case.

PROOF. (a)   We have $\{x\} = \bigcap_{n=1}^{\infty} (x - (1/n), x]$, and therefore all singletons are in $\mathscr{F}$. But then all sets are in $\mathscr{F}$ since Ω is countable.

(b)   Since Ω is a countable union of singletons, μ is σ-finite on $\mathscr{F}$. But every nonempty set in $\mathscr{F}_0$ has infinite measure, so μ is not σ-finite on $\mathscr{F}_0$.

(c)   If A is any finite nonempty subset of Ω, then $\mu(A \, \Delta \, B) = \infty$ for all nonempty $B \in \mathscr{F}_0$, because any nonempty set in $\mathscr{F}_0$ must contain infinitely many points not in A.

(d)   Since $\lambda\{x\} = 2$ and $\mu\{x\} = 1, \lambda \neq \mu$ on $\mathscr{F}$. But $\lambda(A) = \mu(A) = \infty, A \in \mathscr{F}_0$ (except for $A = \emptyset$). $\square$

Problems

1.   Let $(\Omega, \mathscr{F}, \mu)$ be a measure space, and let $\mathscr{F}_\mu$ be the completion of $\mathscr{F}$ relative to μ. If $A \subset \Omega$, define:

$$\mu_0(A) = \sup\{\mu(B) \colon B \in \mathscr{F}, B \subset A\}, \quad \mu^0(A) = \inf\{\mu(B) \colon B \in \mathscr{F}, B \supset A\}.$$

If $A \in \mathscr{F}_\mu$, show that $\mu_0(A) = \mu^0(A) = \mu(A)$. Conversely, if $\mu_0(A) = \mu^0(A) < \infty$, show that $A \in \mathscr{F}_\mu$.

2.   Show that the monotone class theorem (1.3.9) fails if $\mathscr{F}_0$ is not assumed to be a field.

3.   This problem deals with the extension of an arbitrary (not necessarily σ-finite) measure on a field.

(a)   Let λ be an outer measure on the set Ω (see 1.3.4). We say that the set E is λ-*measurable* iff

$$\lambda(A) = \lambda(A \cap E) + \lambda(A \cap E^c) \qquad \text{for all} \qquad A \subset \Omega.$$

(The equals sign may be replaced by "$\geq$" by subadditivity of λ.) If $\mathscr{M}$ is the class of all λ-measurable sets, show that $\mathscr{M}$ is a σ-field, and that if $E_1, E_2, \ldots$ are disjoint sets in $\mathscr{M}$ whose union is E, and $A \subset \Omega$, we have

$$\lambda(A \cap E) = \sum_n \lambda(A \cap E_n). \tag{1}$$

In particular, λ is a measure on $\mathscr{M}$. [Use the definition of λ-measurability to show that $\mathscr{M}$ is a field and that (1) holds for finite sequences. If $E_1, E_2, \ldots$ are disjoint sets in $\mathscr{M}$ and $F_n = \bigcup_{i=1}^{n} E_i \uparrow E$, show that

$$\lambda(A) \geq \lambda(A \cap F_n) + \lambda(A \cap E^c) = \sum_{i=1}^{n} \lambda(A \cap E_i) + \lambda(A \cap E^c),$$

and then let $n \to \infty$.]

(b)   Let μ be a measure on a field $\mathscr{F}_0$ of subsets of Ω. If $A \subset \Omega$, define

$$\mu^*(A) = \inf\left\{\sum_n \mu(E_n)\colon A \subset \bigcup_n E_n, E_n \in \mathscr{F}_0\right\}.$$

Show that μ^* is an outer measure on Ω and that $\mu^* = \mu$ on $\mathscr{F}_0$.

(c)   In (b), if $\mathscr{M}$ is the class of μ^*-measurable sets, show that $\mathscr{F}_0 \subset \mathscr{M}$. Thus by (a) and (b), μ may be extended to the minimal σ-field over $\mathscr{F}_0$.

(d)   In (b), if μ is σ-finite on $\mathscr{F}_0$, show that $(\Omega, \mathscr{M}, \mu^*)$ is the completion of $[\Omega, \sigma(\mathscr{F}_0), \mu^*]$.

1.4   Lebesgue–Stieltjes Measures and Distribution Functions

We are now in a position to construct a large class of measures on the Borel sets of $\mathbb{R}$. If F is an increasing, right-continuous function from $\mathbb{R}$ to $\mathbb{R}$, we set $\mu(a, b] = F(b) - F(a)$; we then extend μ to a finitely additive set function on the field $\mathscr{F}_0(\mathbb{R})$ of finite disjoint unions of right-semiclosed intervals. If we can show that μ is countably additive on $\mathscr{F}_0(\mathbb{R})$, the Carathéodory extension theorem extends μ to $\mathscr{B}(\mathbb{R})$.

1.4.1   Definitions.   A *Lebesgue–Stieltjes measure* on $\mathbb{R}$ is a measure μ on $\mathscr{B}(\mathbb{R})$ such that $\mu(I) < \infty$ for each bounded interval I. A *distribution function* on $\mathbb{R}$ is a map $F\colon \mathbb{R} \to \mathbb{R}$ that is *increasing* $[a < b$ implies $F(a) \leq F(b)]$ and *right-continuous* $[\lim_{x \to x_0^+} F(x) = F(x_0)]$. We are going to show that the formula $\mu(a, b] = F(b) - F(a)$ sets up a one-to-one correspondence between Lebesgue–Stieltjes measures and distribution functions, where two distribution functions that differ by a constant are identified.

1.4.2   Theorem.   Let μ be a Lebesgue–Stieltjes measure on $\mathbb{R}$. Let $F\colon \mathbb{R} \to \mathbb{R}$ be defined, up to an additive constant, by $F(b) - F(a) = \mu(a, b]$. [For example, fix $F(0)$ arbitrarily and set $F(x) - F(0) = \mu(0, x], x > 0$; $F(0) - F(x) = \mu(x, 0], x < 0$.] Then F is a distribution function.

PROOF.   If $a < b$, then $F(b) - F(a) = \mu(a, b] \geq 0$. If $\{x_n\}$ is a sequence of points such that $x_1 > x_2 > \cdots \to x$, then $F(x_n) - F(x) = \mu(x, x_n] \to 0$ by 1.2.7(b). $\square$

Now let F be a distribution function on $\mathbb{R}$. It will be convenient to work in the compact space $\overline{\mathbb{R}}$, so we extend F to a map of $\overline{\mathbb{R}}$ into $\overline{\mathbb{R}}$ by defining $F(\infty) = \lim_{x \to \infty} F(x)$, $F(-\infty) = \lim_{x \to -\infty} F(x)$; the limits exist by monotonicity. Define $\mu(a, b] = F(b) - F(a), a, b \in \overline{\mathbb{R}}, a < b$, and let $\mu[-\infty, b] = F(b) - F(-\infty) = \mu(-\infty, b]$; then μ is defined on all right-semiclosed intervals of $\overline{\mathbb{R}}$ (counting $[-\infty, b]$ as right-semiclosed; see 1.2.2).

If $I_1, \ldots, I_k$ are disjoint right-semiclosed intervals of $\overline{\mathbb{R}}$, we define $\mu(\bigcup_{j=1}^{k} I_j) = \sum_{j=1}^{k} \mu(I_j)$. Thus μ is extended to the field $\mathscr{F}_0(\overline{\mathbb{R}})$ of finite disjoint unions of right-semiclosed intervals of $\overline{\mathbb{R}}$, and μ is finitely additive on $\mathscr{F}_0(\overline{\mathbb{R}})$. To show that μ is in fact countably additive on $\mathscr{F}_0(\overline{\mathbb{R}})$, we make use of 1.2.8(b), as follows.

_1.4.3_ **_Lemma._** The set function μ is countably additive on $\mathscr{F}_0(\overline{\mathbb{R}})$.

PROOF. First assume that $F(\infty) - F(-\infty) < \infty$, so that μ is finite. Let $A_1, A_2, \ldots$ be a sequence of sets in $\mathscr{F}_0(\overline{\mathbb{R}})$ decreasing to $\emptyset$. If $(a, b]$ is one of the intervals of A_n, then by right continuity of F, $\mu(a', b] = F(b) - F(a')$ $\rightarrow F(b) - F(a) = \mu(a, b]$ as $a' \rightarrow a$ from above.

Thus we can find sets $B_n \in \mathscr{F}_0(\overline{\mathbb{R}})$ whose closures $\overline{B}_n$ (in $\overline{\mathbb{R}}$) are included in A_n, with $\mu(B_n)$ approximating $\mu(A_n)$. If $\varepsilon > 0$ is given, the finiteness of μ allows us to choose the B_n so that $\mu(A_n) - \mu(B_n) < \varepsilon 2^{-n}$. Now $\bigcap_{n=1}^{\infty} \overline{B}_n = \emptyset$, and it follows that $\bigcap_{k=1}^{n} \overline{B}_k = \emptyset$ for sufficiently large n. (Perhaps the easiest way to see this is to note that the sets $\overline{\mathbb{R}} - \overline{B}_n$ form an open covering of the compact set $\overline{\mathbb{R}}$, hence there is a finite subcovering, so that $\bigcup_{k=1}^{n} (\overline{\mathbb{R}} - \overline{B}_k) = \overline{\mathbb{R}}$ for some n. Therefore $\bigcap_{k=1}^{n} \overline{B}_k = \emptyset$.) Now

$$\mu(A_n) = \mu\left(A_n - \bigcap_{k=1}^{n} B_k\right) + \mu\left(\bigcap_{k=1}^{n} B_k\right)$$

$$= \mu\left(A_n - \bigcap_{k=1}^{n} B_k\right)$$

$$\leq \mu\left(\bigcup_{k=1}^{n} (A_k - B_k)\right) \qquad \text{since} \qquad A_n \subset A_{n-1} \subset \cdots \subset A_1$$

$$\leq \sum_{k=1}^{n} \mu(A_k - B_k) \qquad \text{by 1.2.5(d)}$$

$$< \varepsilon.$$

Thus $\mu(A_n) \rightarrow 0$.

Now if $F(\infty) - F(-\infty) = \infty$, define $F_n(x) = F(x)$, $|x| \leq n$; $F_n(x) = F(n)$, $x \geq n$; $F_n(x) = F(-n)$, $x \leq -n$. If μ_n is the set function corresponding to F_n, then $\mu_n \leq \mu$ and $\mu_n \rightarrow \mu$ on $\mathscr{F}_0(\overline{\mathbb{R}})$). Let $A_1, A_2, \ldots$ be disjoint sets in $\mathscr{F}_0(\overline{\mathbb{R}})$ such that $A = \bigcup_{n=1}^{\infty} A_n \in \mathscr{F}_0(\overline{\mathbb{R}})$. Then $\mu(A) \geq \sum_{n=1}^{\infty} \mu(A_n)$

(Problem 5, Section 1.2) so if $\sum_{n=1}^{\infty} \mu(A_n) = \infty$, we are finished. If $\sum_{n=1}^{\infty} \mu(A_n) < \infty$, then

$$\mu(A) = \lim_{n \to \infty} \mu_n(A)$$

$$= \lim_{n \to \infty} \sum_{k=1}^{\infty} \mu_n(A_k)$$

since the μ_n are finite. Now as $\sum_{k=1}^{\infty} \mu(A_k) < \infty$, we may write

$$0 \leq \mu(A) - \sum_{k=1}^{\infty} \mu(A_k)$$

$$= \lim_{n \to \infty} \sum_{k=1}^{\infty} [\mu_n(A_k) - \mu(A_k)]$$

$$\leq 0 \qquad \text{since} \qquad \mu_n \leq \mu. \quad \square$$

We now complete the construction of Lebesgue–Stieltjes measures.

1.4.4   Theorem. Let F be a distribution function on $\mathbb{R}$, and let $\mu(a, b] = F(b) - F(a), a < b$. There is a unique extension of μ to a Lebesgue–Stieltjes measure on $\mathbb{R}$.

PROOF.   Extend μ to a countably additive set function on $\mathscr{F}_0(\overline{\mathbb{R}})$ as above. Let $\mathscr{F}_0(\mathbb{R})$ be the field of all finite disjoint unions of right-semiclosed intervals of $\mathbb{R}$ [counting (a, ∞) as right-semiclosed; see 1.2.2], and extend μ to $\mathscr{F}_0(\mathbb{R})$ as in the discussion that follows 1.4.2. [Take $\mu(a, \infty) = F(\infty) - F(a)$; $\mu(-\infty, b] = F(b) - F(-\infty), a, b \in \mathbb{R}$; $\mu(\mathbb{R}) = F(\infty) - F(-\infty)$; note that there is no other possible choice for μ on these sets, by 1.2.7(a).] Now the map

$$(a, b] \to (a, b]. \quad \text{if} \quad a, b \in \mathbb{R} \quad \text{or if} \quad b \in \mathbb{R}, \quad a = -\infty,$$

$$(a, \infty] \to (a, \infty) \quad \text{if} \quad a \in \mathbb{R} \quad \text{or if} \quad a = -\infty$$

sets up a one-to-one, μ-preserving correspondence between a subset of $\mathscr{F}_0(\overline{\mathbb{R}})$ (everything in $\mathscr{F}_0(\overline{\mathbb{R}})$ except sets including intervals of the form $[-\infty, b]$) and $\mathscr{F}_0(\mathbb{R})$. It follows that μ is countably additive on $\mathscr{F}_0(\mathbb{R})$. Furthermore, μ is σ-finite on $\mathscr{F}_0(\mathbb{R})$ since $\mu(-n, n] < \infty$; note that μ need not be σ-finite on $\mathscr{F}_0(\overline{\mathbb{R}})$ since the sets $(-n, n]$ do not cover $\overline{\mathbb{R}}$. By the Carathéodory extension theorem, μ has a unique extension to $\mathscr{B}(\mathbb{R})$. The extension is a Lebesgue–Stieltjes measure because $\mu(a, b] = F(b) - F(a) < \infty$ for $a, b \in \mathbb{R}, a < b$. $\quad \square$

1.4.5 Comments and Examples. If F is a distribution function and μ the corresponding Lebesgue–Stieltjes measure, we have seen that $\mu(a, b]$ $= F(b) - F(a)$, $a < b$. The measure of any interval, right-semiclosed or not, may be expressed in terms of F. For if $F(x^-)$ denotes $\lim_{y \to x^-} F(y)$, then

(1) $\mu(a, b] = F(b) - F(a)$,   (3) $\mu[a, b] = F(b) - F(a^-)$,

(2) $\mu(a, b) = F(b^-) - F(a)$,   (4) $\mu[a, b) = F(b^-) - F(a^-)$.

(Thus if F is continuous at a and b, all four expressions are equal.) For example, to prove (2), observe that

$$\mu(a, b) = \lim_{n \to \infty} \mu\left(a, b - \frac{1}{n}\right] = \lim_{n \to \infty}\left[F\left(b - \frac{1}{n}\right) - F(a)\right] = F(b^-) - F(a).$$

Statement (3) follows because

$$\mu[a, b] = \lim_{n \to \infty} \mu\left(a - \frac{1}{n}, b\right] = \lim_{n \to \infty}\left[F(b) - F\left(a - \frac{1}{n}\right)\right] = F(b) - F(a^-);$$

(4) is proved similarly. The proof of (3) works even if $a = b$, so that $\mu\{x\} = F(x) - F(x^-)$. Thus

(5) F is continuous at x iff $\mu\{x\} = 0$; the magnitude of a discontinuity of F at x coincides with the measure of $\{x\}$.

The following formulas are obtained from (1)–(3) by allowing a to approach $-\infty$ or b to approach $+\infty$.

(6) $\mu(-\infty, x] = F(x) - F(-\infty)$,   (9) $\mu[x, \infty) = F(\infty) - F(x^-)$,

(7) $\mu(-\infty, x) = F(x^-) - F(-\infty)$,   (10) $\mu(\mathbb{R}) = F(\infty) - F(-\infty)$.

(8) $\mu(x, \infty) = F(\infty) - F(x)$,

(The formulas (6), (8), and (10) have already been observed in the proof of 1.4.4.)

If μ is finite, then F is bounded; since F may always be adjusted by an additive constant, nothing is lost in this case if we set $F(-\infty) = 0$.

We may now generate a large number of measures on $\mathscr{B}(\mathbb{R})$. For example, if $f: \mathbb{R} \to \mathbb{R}$, $f \geq 0$, and f is integrable (Riemann for now) on any finite interval, then if we fix $F(0)$ arbitrarily and define

$$F(x) - F(0) = \int_0^x f(t)\,dt, \qquad x > 0;$$

$$F(0) - F(x) = \int_x^0 f(t)\,dt, \qquad x < 0,$$

then F is a (continuous) distribution function and thus gives rise to a Lebesgue–Stieltjes measure; specifically,

$$\mu(a, b] = \int_a^b f(x)\, dx.$$

In particular, we may take $f(x) = 1$ for all x, and $F(x) = x$; then $\mu(a, b] = b - a$. The set function μ is called the *Lebesgue measure* on $\mathscr{B}(\mathbb{R})$. The completion of $\mathscr{B}(\mathbb{R})$ relative to Lebesgue measure is called the class of *Lebesgue measurable sets*, written $\overline{\mathscr{B}}(\mathbb{R})$. Thus a Lebesgue measurable set is the union of a Borel set and a subset of a Borel set of Lebesgue measure 0. The extension of Lebesgue measure to $\overline{\mathscr{B}}(\mathbb{R})$ is also called "Lebesgue measure."

Now let μ be a Lebesgue–Stieltjes measure that is concentrated on a countable set $S = \{x_1, x_2, \ldots\}$, that is, $\mu(\mathbb{R} - S) = 0$. [In general if $(\Omega, \mathscr{F}, \mu)$ is a measure space and $B \in \mathscr{F}$, we say that μ is concentrated on B iff $\mu(\Omega - B) = 0$.] In the present case, such a measure is easily constructed: If $a_1, a_2, \ldots$ are nonnegative numbers and $A \subset \mathbb{R}$, set $\mu(A) = \sum\{a_i \colon x_i \in A\}$; μ is a measure on all subsets of $\mathbb{R}$, not merely on the Borel sets (see 1.2.4). If $\mu(I) < \infty$ for each bounded interval I, μ will be a Lebesgue–Stieltjes measure on $\mathscr{B}(\mathbb{R})$; if $\sum_i a_i < \infty$, μ will be a finite measure. The distribution function F corresponding to μ is continuous on $\mathbb{R} - S$; if $\mu\{x_n\} = a_n > 0$, F has a jump at x_n of magnitude a_n. If $x, y \in S$ and no point of S lies between x and y, then F is constant on $[x, y)$. For if $x \le b < y$, then $F(b) - F(x) = \mu(x, b] = 0$.

Now if we take S to be the rational numbers, the above discussion yields a monotone function F from $\mathbb{R}$ to $\mathbb{R}$ that is continuous at each irrational point and discontinuous at each rational point.

If F is an increasing, right-continuous, real-valued function defined on a closed bounded interval $[a, b]$, there is a corresponding finite measure μ on the Borel subsets of $[a, b]$; explicitly, μ is determined by the requirement that $\mu(a', b'] = F(b') - F(a')$, $a \le a' < b' \le b$. The easiest way to establish the correspondence is to extend F by defining $F(x) = F(b), x \ge b$; $F(x) = F(a), x \le a$; then take μ as the Lebesgue–Stieltjes measure corresponding to F, restricted to $\mathscr{B}[a, b]$.

We are going to consider Lebesgue–Stieltjes measures and distribution functions in Euclidean n-space. First, some terminology is required.

1.4.6 Definitions and Comments. If $a = (a_1, \ldots, a_n)$, $b = (b_1, \ldots, b_n) \in \mathbb{R}^n$, the interval $(a, b]$ is defined as $\{x = (x_1, \ldots, x_n) \in \mathbb{R}^n \colon a_i < x_i \le b_i$ for all $i = 1, \ldots, n\}$; (a, ∞) is defined as $\{x \in \mathbb{R}^n \colon x_i > a_i$ for all $i = 1, \ldots, n\}$, $(-\infty, b]$ as $\{x \in \mathbb{R}^n \colon x_i \le b_i$ for all $i = 1, \ldots, n\}$; other types of intervals are defined similarly. The smallest σ-field containing all intervals $(a, b]$, $a, b \in \mathbb{R}^n$, is called the class of *Borel sets* of $\mathbb{R}^n$, written $\mathscr{B}(\mathbb{R}^n)$. The Borel sets form the minimal σ-field over many other classes of sets, for example, the open sets, the intervals $[a, b)$, and so on, exactly as in the discussion of the one-dimensional case in 1.2.4. The class of Borel sets of $\overline{\mathbb{R}}^n$, written $\mathscr{B}(\overline{\mathbb{R}}^n)$, is defined similarly.

A *Lebesgue–Stieltjes measure* on $\mathbb{R}^n$ is a measure μ on $\mathscr{B}(\mathbb{R}^n)$ such that $\mu(I) < \infty$ for each bounded interval I.

The notion of a distribution function on $\mathbb{R}^n$, $n \geq 2$, is more complicated than in the one-dimensional case. To see why, assume for simplicity that $n = 3$, and let μ be a finite measure on $\mathscr{B}(\mathbb{R}^3)$. Define

$$F(x_1, x_2, x_3) = \mu\{\omega \in R^3 \colon \omega_1 \leq x_1, \quad \omega_2 \leq x_2, \quad \omega_3 \leq x_3\}, \quad (x_1, x_2, x_3) \in \mathbb{R}^3.$$

By analogy with the one-dimensional case, we expect that F is a distribution function corresponding to μ [see formula (6) of 1.4.5]. This will turn out to be correct, but the correspondence is no longer by means of the formula $\mu(a, b] = F(b) - F(a)$. To see this, we compute $\mu(a, b]$ in terms of F.

Introduce the *difference operator* $\triangle$ as follows:

If $G \colon \mathbb{R}^n \to \mathbb{R}$, $\triangle_{b_i a_i} G(x_1, \ldots, x_n)$ is defined as

$$G(x_1, \ldots, x_{i-1}, b_i, x_{i+1}, \ldots, x_n) - G(x_1, \ldots, x_{i-1}, a_i, x_{i+1}, \ldots, x_n).$$

1.4.7   Lemma. If $a \leq b$, that is, $a_i \leq b_i$, $i = 1, 2, 3$, then

(a) $\mu(a, b] = \triangle_{b_1 a_1} \triangle_{b_2 a_2} \triangle_{b_3 a_3} F(x_1, x_2, x_3)$, where

(b) $\triangle_{b_1 a_1} \triangle_{b_2 a_2} \triangle_{b_3 a_3} F(x_1, x_2, x_3)$

$$= F(b_1, b_2, b_3) - F(a_1, b_2, b_3) - F(b_1, a_2, b_3) - F(b_1, b_2, a_3)$$

$$+ F(a_1, a_2, b_3) + F(a_1, b_2, a_3) + F(b_1, a_2, a_3) - F(a_1, a_2, a_3)$$

Thus $\mu(a, b]$ is not simply $F(b) - F(a)$.

PROOF.

(a) $\triangle_{b_3 a_3} F(x_1, x_2, x_3) = F(x_1, x_2, b_3) - F(x_1, x_2, a_3)$

$$= \mu\{\omega \colon \omega_1 \leq x_1, \quad \omega_2 \leq x_2, \quad \omega_3 \leq b_3\}$$

$$- \mu\{\omega \colon \omega_1 \leq x_1, \quad \omega_2 \leq x_2, \quad \omega_3 \leq a_3\}$$

$$= \mu\{\omega \colon \omega_1 \leq x_1, \quad \omega_2 \leq x_2, \quad a_3 < \omega_3 \leq b_3\}$$

$$\text{since} \quad a_3 \leq b_3.$$

Similarly,

$$\Delta_{b_2 a_2} \Delta_{b_3 a_3} F(x_1, x_2, x_3) = \mu\{\omega: \omega_1 \leq x_1, \quad a_2 < \omega_2 \leq b_2, \quad a_3 < \omega_3 \leq b_3\}$$

and

$$\Delta_{b_1 a_1} \Delta_{b_2 a_2} \Delta_{b_3 a_3} F(x_1, x_2, x_3) = \mu\{\omega: a_1 < \omega_1 \leq b_1, \quad a_2 < \omega_2 \leq b_2,$$
$$a_3 < \omega_3 \leq b_3\}.$$

(b) $\Delta_{b_3 a_3} F(x_1, x_2, x_3) = F(x_1, x_2, b_3) - F(x_1, x_2, a_3),$

$$\Delta_{b_2 a_2} \Delta_{b_3 a_3} F(x_1, x_2, x_3) = F(x_1, b_2, b_3) - F(x_1, a_2, b_3)$$
$$- F(x_1, b_2, a_3) + F(x_1, a_2, a_3).$$

Thus $\Delta_{b_1 a_1} \Delta_{b_2 a_2} \Delta_{b_3 a_3} F(x_1, x_2, x_3)$ is the desired expression. □

The extension of 1.4.7 to n dimensions is clear.

1.4.8   Theorem. Let μ be a finite measure on $\mathscr{B}(\mathbb{R}^n)$ and define

$$F(x) = \mu(-\infty, x] = \mu\{\omega: \omega_i \leq x_i, i = 1, \ldots, n\}.$$

If $a \leq b$, then

(a) $\mu(a, b] = \Delta_{b_1 a_1} \cdots \Delta_{b_n a_n} F(x_1, \ldots, x_n)$, where

(b) $\Delta_{b_1 a_1} \cdots \Delta_{b_n a_n} F(x_1, \ldots, x_n) = F_0 - F_1 + F_2 - \cdots + (-1)^n F_n;$
F_i is the sum of all $\binom{n}{i}$ terms of the form $F(c_1, \ldots, c_n)$ with $c_k = a_k$ for exactly i integers in $\{1, 2, \ldots, n\}$, and $c_k = b_k$ for the remaining $n - i$ integers.

PROOF. Apply the computations of 1.4.7. □

We know that a distribution function of $\mathbb{R}$ determines a corresponding Lebesgue–Stieltjes measure. This is true in n dimensions if we change the definition of increasing.
   Let $F: \mathbb{R}^n \to \mathbb{R}$, and, for $a \leq b$, let $F(a, b]$ denote

$$\Delta_{b_1 a_1} \cdots \Delta_{b_n a_n} F(x_1, \ldots, x_n).$$

The function F is said to be *increasing* iff $F(a, b] \geq 0$ whenever $a \leq b$; F is *right-continuous* iff it is right-continuous in all variables together, in other words, for any sequence $x^1 \geq x^2 \geq \cdots \geq x^k \geq \cdots \to x$ we have $F(x^k) \to F(x)$.

An increasing right-continuous $F: \mathbb{R}^n \to \mathbb{R}$ is said to be a *distribution function* on $\mathbb{R}^n$. (Note that if F arises from a measure μ as in 1.4.8, F is a distribution function.)

If F is a distribution function on $\mathbb{R}^n$, we set $\mu(a, b] = F(a, b]$ [this reduces to $F(b) - F(a)$ if $n = 1$]. We are going to show that μ has a unique extension to a Lebesgue–Stieltjes measure on $\mathbb{R}^n$. The technique of the proof is the same in any dimension, but to avoid cumbersome notation and to capture the essential ideas, we sometimes specialize to the case $n = 2$. We break the argument into several steps:

(1) If $a \le a' \le b' \le b, I = (a, b]$ is the union of the nine disjoint intervals $I_1, \ldots, I_9$ formed by first constraining the first coordinate in one of the following three ways:

$$a_1 < x \le a_1', \qquad a_1' < x \le b_1', \qquad b_1' < x \le b_1,$$

and then constraining the second coordinate in one of the following three ways:

$$a_2 < y \le a_2', \qquad a_2' < y \le b_2', \qquad b_2' < y \le b_2.$$

For example, a typical set in the union is

$$\{(x, y): b_1' < x \le b_1, \qquad a_2 < y \le a_2'\};$$

in n dimensions we would obtain 3^n such sets.

Result (1) may be verified by looking at Fig. 1.4.1.

(2) In (1), $F(I) = \sum_{j=1}^9 F(I_j)$, hence $a \le a' \le b' \le b$ implies $F(a', b'] \le F(a, b]$.

This is verified by brute force, using 1.4.8.

Now a *right-semiclosed interval* $(a, b]$ in $\overline{\mathbb{R}}^n$ is, by convention, a set of the form $\{(x_1, \ldots, x_n): a_i < x_i \le b_i, i = 1, \ldots, n\}, a, b \in \overline{\mathbb{R}}^n$, with the proviso that $a_i < x_i \le b_i$ can be replaced by $a_i \le x_i \le b_i$ if $a_i = -\infty$. With this assumption, the set $\mathscr{F}_0(\overline{\mathbb{R}}^n)$ of finite disjoint unions of right-semiclosed intervals is a field. (The corresponding convention in $\mathbb{R}^n$ is that $a_i < x_i \le b_i$ can be replaced by $a_i < x_i < b_i$ if $b_i = +\infty$. Both conventions are dictated by considerations similar to those of the one-dimensional case; see 1.2.2.)

(3) If a and b belong to $\overline{\mathbb{R}}^n$ but not to $\mathbb{R}^n$, we define $F(a, b]$ as the limit of $F(a', b']$ where $a', b' \in \mathbb{R}^n, a'$ decreases to a, and b' increases to b. [The definition is sensible because of the monotonicity property in (2).] Similarly if $a \in \mathbb{R}^n, b \in \overline{\mathbb{R}}^n - \mathbb{R}^n$, we take $F(a, b] = \lim_{b' \uparrow b} F(a, b']$; if $a \in \overline{\mathbb{R}}^n - \mathbb{R}^n$, $b \in \mathbb{R}^n$, $F(a, b] = \lim_{a' \downarrow a} F(a', b]$.

Thus we define μ on right-semiclosed intervals of $\overline{\mathbb{R}}^n$; μ extends to a finitely additive set function on $\mathscr{F}_0(\overline{\mathbb{R}}^n)$, as in the discussion after 1.4.2. [There is a

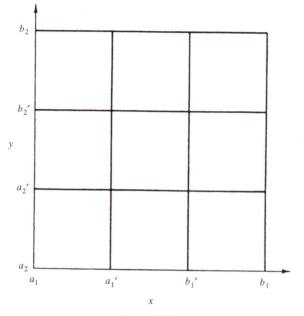

Figure 1.4.1.

slight problem here; a given interval I may be expressible as a finite disjoint union of intervals $I_1, \ldots, I_r$, so that for the extension to be well defined we must have $F(I) = \sum_{j=1}^{r} F(I_j)$; but this follows just as in (2).]

(4)   The set function μ is countably additive on $\mathscr{F}_0(\overline{\mathbb{R}}^n)$.

First assume that $\mu(\overline{\mathbb{R}}^n)$ is finite. If $a \in \mathbb{R}^n$, $F(a', b] \to F(a, b]$ as a' decreases to a by the right-continuity of F and 1.4.8(b); if $a \in \overline{\mathbb{R}}^n - \mathbb{R}^n$, the same result holds by (3). The argument then proceeds word for word as in 1.4.3.

Now assume $\mu(\overline{\mathbb{R}}^n) = \infty$. Then F, restricted to $C_k = \{x: -k < x_i \leq k, i = 1, \ldots, n\}$, induces a finite-valued set function μ_k on $\mathscr{F}_0(\overline{\mathbb{R}}^n)$ that is concentrated on C_k, so that $\mu_k(B) = \mu_k(B \cap C_k)$, $B \in \mathscr{F}_0(\overline{\mathbb{R}}^n)$. Since $\mu_k \leq \mu$ and $\mu_k \to \mu$ on $\mathscr{F}_0(\overline{\mathbb{R}}^n)$, the proof of 1.4.3 applies verbatim.

1.4.9   Theorem. Let F be a distribution function on $\mathbb{R}^n$, and let $\mu(a, b] = F(a, b]$, $a, b \in R^n$, $a \leq b$. There is a unique extension of μ to a Lebesgue–Stieltjes measure on $\mathbb{R}^n$.

PROOF.   Repeat the proof of 1.4.4, with appropriate notational changes. For example, in extending μ to $\mathscr{F}_0(\mathbb{R}^n)$, the field of finite disjoint unions of right-semiclosed intervals of $\mathbb{R}^n$, we take (say for $n = 3$)

$$\mu\{(x, y, z): a_1 < x \leq b_1, \quad a_2 < y < \infty, \quad a_3 < z < \infty\} = \lim_{b_2, b_3 \to \infty} F(a, b].$$

The one-to-one μ-preserving correspondence is given by

$$(a, b] \to (a, b] \qquad \text{if} \qquad a, b \in \mathbb{R}^n$$
$$\text{or if} \qquad b \in \mathbb{R}^n \text{ and at least one component of}$$
$$a \text{ is } -\infty;$$

also, if the interval $\{(x_1, \ldots, x_n): a_i < x_i \le b_i: i = 1, \ldots, n\}$ has some $b_i = \infty$, the corresponding interval in $\mathbb{R}^n$ has $a_i < x_i < \infty$. The remainder of the proof is as before. $\square$

1.4.10   Examples. (a)   Let $F_1, F_2, \ldots, F_n$ be distribution functions on $\mathbb{R}$, and define $F(x_1, \ldots, x_n) = F_1(x_1)F_2(x_2) \cdots F_n(x_n)$. Then F is a distribution function on $\mathbb{R}^n$ since

$$F(a, b] = \prod_{i=1}^{n} [F_i(b_i) - F_i(a_i)].$$

In particular, if $F_i(x_i) = x_i$, $i = 1, \ldots, n$, then each F_i corresponds to Lebesgue measure on $\mathscr{B}(\mathbb{R})$. In this case we have $F(x_1, \ldots, x_n) = x_1 x_2 \cdots x_n$ and $\mu(a, b] = F(a, b] = \prod_{i=1}^{n}(b_i - a_i)$. Thus the measure of any rectangular box is its volume; μ is called *Lebesgue measure* on $\mathscr{B}(\mathbb{R}^n)$. Just as in one dimension, the completion of $\mathscr{B}(\mathbb{R}^n)$ relative to Lebesgue measure is called the class of *Lebesgue measurable sets* in R^n, written $\overline{\mathscr{B}}(\mathbb{R}^n)$.

(b)   Let f be a nonnegative function from $\mathbb{R}^n$ to $\mathbb{R}$ such that

$$\int_{-\infty}^{\infty} \cdots \int_{-\infty}^{\infty} f(x_1, \ldots, x_n) \, dx_1 \cdots dx_n < \infty.$$

(For now, we assume the integration is in the Riemann sense.) Define

$$F(x) = \int_{(-\infty, x]} f(t) \, dt,$$

that is,

$$F(x_1, \ldots, x_n) = \int_{-\infty}^{x_1} \cdots \int_{-\infty}^{x_n} f(t_1, \ldots, t_n) \, dt_1 \cdots dt_n.$$

Then

$$\triangle_{b_n a_n} F(x_1, \ldots, x_n) = \int_{-\infty}^{x_1} \cdots \int_{-\infty}^{x_{n-1}} \int_{a_n}^{b_n} f(t_1, \ldots, t_n) \, dt_1 \cdots dt_n,$$

and we find by repeating this computation that

$$F(a, b] = \int_{a_1}^{b_1} \cdots \int_{a_n}^{b_n} f(t_1, \ldots, t_n) \, dt_1 \cdots dt_n.$$

Thus F is a distribution function. If μ is the Lebesgue–Stieltjes measure determined by F, we have

$$\mu(a, b] = \int_{(a,b]} f(x) \, dx.$$

We have seen that if F is a distribution function on $\mathbb{R}^n$, there is a unique Lebesgue–Stieltjes measure determined by $\mu(a, b] = F(a, b], a \le b$. Also, if μ is a finite measure on $\mathscr{B}(\mathbb{R}^n)$ and $F(x) = \mu(-\infty, x], x \in \mathbb{R}^n$, then F is a distribution function on $\mathbb{R}^n$ and $\mu(a, b] = F(a, b], a \le b$. It is possible to associate a distribution function with an arbitrary Lebesgue–Stieltjes measure on $\mathbb{R}^n$, and thus establish a one-to-one correspondence between Lebesgue–Stieltjes measures and distribution functions, provided distribution functions with the same increments $F(a, b], a, b \in \mathbb{R}^n, a \le b$, are identified. The result will not be needed, and the details are quite tedious and will be omitted.

The following result shows that under appropriate conditions, a Borel set can be approximated from below by a compact set, and from above by an open set.

1.4.11 Theorem. If μ is a σ-finite measure on $\mathscr{B}(\mathbb{R}^n)$, then for each $B \in \mathscr{B}(\mathbb{R}^n)$,

(a) $\mu(B) = \sup\{\mu(K): K \subset B, K \text{ compact}\}$.
If μ is in fact a Lebesgue–Stieltjes measure, then
(b) $\mu(B) = \inf\{\mu(V): V \supset B, B \text{ open}\}$.
(c)   There is an example of a σ-finite measure on $\mathscr{B}(\mathbb{R}^n)$ that is not a Lebesgue–Stieltjes measure and for which (b) fails.

PROOF.
(a)   First assume that μ is finite. Let $\mathscr{C}$ be the class of subsets of $\mathbb{R}^n$ having the desired property; we show that $\mathscr{C}$ is a monotone class. Indeed, let $B_n \in \mathscr{C}, B_n \uparrow B$. Let K_n be a compact subset of B_n with $\mu(B_n) \le \mu(K_n) + \varepsilon, \varepsilon > 0$ preassigned. By replacing K_n by $\bigcup_{i=1}^n K_i$, we may assume the K_n form an increasing sequence. Then $\mu(B) = \lim_{n\to\infty} \mu(B_n) \le \lim_{n\to\infty} \mu(K_n) + \varepsilon$, so that

$$\mu(B) = \sup\{\mu(K): K \text{ a compact subset of } B\},$$

and $B \in \mathscr{C}$. If $B_n \in \mathscr{C}$, $B_n \downarrow B$, let K_n be a compact subset of B_n such that $\mu(B_n) \leq \mu(K_n) + \varepsilon 2^{-n}$, and set $K = \bigcap_{n=1}^{\infty} K_n$. Then

$$\mu(B) - \mu(K) = \mu(B - K) \leq \mu\left(\bigcup_{n=1}^{\infty}(B_n - K_n)\right) \leq \sum_{n=1}^{\infty} \mu(B_n - K_n) \leq \varepsilon;$$

thus $B \in \mathscr{C}$. Therefore $\mathscr{C}$ is a monotone class containing all finite disjoint unions of right-semiclosed intervals (a right-semiclosed interval is the limit of an increasing sequence of compact intervals). Hence $\mathscr{C}$ contains all Borel sets.

If μ is σ-finite, each $B \in \mathscr{B}(\mathbb{R}^n)$ is the limit of an increasing sequence of sets B_i of finite measure. Each B_i can be approximated from within by compact sets [apply the previous argument to the measure given by $\mu_i(A) = \mu(A \cap B_i)$, $A \in \mathscr{B}(\mathbb{R}^n)$], and the preceding argument that $\mathscr{C}$ is closed under limits of increasing sequences shows that $B \in \mathscr{C}$.

(b) We have $\mu(B) \leq \inf\{\mu(V): V \supset B, \quad V \text{ open}\}$
$\leq \inf\{\mu(W): W \supset B, \quad W = K^c, \quad K \text{ compact}\}$.

If μ is finite, this equals $\mu(B)$ by (a) applied to B^c, and the result follows.

Now assume μ is an arbitrary Lebesgue–Stieltjes measure, and write $\mathbb{R}^n = \bigcup_{k=1}^{\infty} B_k$, where the B_k are disjoint bounded sets; then $B_k \subset C_k$ for some bounded open set C_k. The measure $\mu_k(A) = \mu(A \cap C_k)$, $A \in \mathscr{B}(\mathbb{R}^n)$, is finite; hence if B is a Borel subset of B_k and $\varepsilon > 0$, there is an open set $W_k \supset B$ such that $\mu_k(W_k) \leq \mu_k(B) + \varepsilon 2^{-k}$. Now $W_k \cap C_k$ is an open set V_k and $B \cap C_k = B$ since $B \subset B_k \subset C_k$; hence $\mu(V_k) \leq \mu(B) + \varepsilon 2^{-k}$. For any $A \in \mathscr{B}(\mathbb{R}^n)$, let V_k be an open set with $V_k \supset A \cap B_k$ and $\mu(V_k) \leq \mu(A \cap B_k) + \varepsilon 2^{-k}$. Then $V = \bigcup_{k=1}^{\infty} V_k$ is open, $V \supset A$, and $\mu(V) \leq \sum_{k=1}^{\infty} \mu(V_k) \leq \mu(A) + \varepsilon$.

(c) Construct a measure μ on $\mathscr{B}(\mathbb{R})$ as follows. Let μ be concentrated on $S = \{1/n: n = 1, 2, \ldots\}$ and take $\mu\{1/n\} = 1/n$ for all n. Since $\mathbb{R} = \bigcup_{n=1}^{\infty}\{1/n\} \cup S^c$ and $\mu(S^c) = 0$, μ is σ-finite. Since

$$\mu[0, 1] = \sum_{n=1}^{\infty} \frac{1}{n} = \infty,$$

μ is not a Lebesgue–Stieltjes measure. Now $\mu\{0\} = 0$, but if V is an open set containing 0, we have

$$\mu(V) \geq \mu(-\varepsilon, \varepsilon) \qquad \text{for some } \varepsilon$$
$$\geq \sum_{k=r}^{\infty} \frac{1}{k} \qquad \text{for some } r$$
$$= \infty.$$

Thus (b) fails. [Another example: Let $\mu(A)$ be the number of rational points in A.]

Problems

1. Let F be the distribution function on $\mathbb{R}$ given by $F(x) = 0$, $x < -1$; $F(x) = 1 + x$, $-1 \le x < 0$; $F(x) = 2 + x^2$, $0 \le x < 2$; $F(x) = 9$, $x \ge 2$. If μ is the Lebesgue–Stieltjes measure corresponding to F, compute the measure of each of the following sets:

   (a) $\{2\}$,
   (b) $\left[-\frac{1}{2}, 3\right)$,
   (c) $(-1, 0] \cup (1, 2)$,
   (d) $\left[0, \frac{1}{2}\right) \cup (1, 2]$,
   (e) $\{x: |x| + 2x^2 > 1\}$.

2. Let μ be a Lebesgue–Stieltjes measure on $\mathbb{R}$ corresponding to a continuous distribution function.

   (a) If A is a countable subset of $\mathbb{R}$, show that $\mu(A) = 0$.
   (b) If $\mu(A) > 0$, must A include an open interval?
   (c) If $\mu(A) > 0$ and $\mu(\mathbb{R} - A) = 0$, must A be dense in $\mathbb{R}$?
   (d) Do the answers to (b) or (c) change if μ is restricted to be Lebesgue measure?

3. If B is a Borel set in $\mathbb{R}^n$ and $a \in \mathbb{R}^n$, show that $a + B = \{a + x: x \in B\}$ is a Borel set, and $-B = \{-x: x \in B\}$ is a Borel set. (Use the good sets principle.)

4. Show that if $B \in \bar{\mathscr{B}}(\mathbb{R}^n)$, $a \in \mathbb{R}^n$, then $a + B \in \bar{\mathscr{B}}(\mathbb{R}^n)$ and $\mu(a + B) = \mu(B)$, where μ is Lebesgue measure. Thus Lebesgue measure is translation-invariant. (The good sets principle works here also, in conjunction with the monotone class theorem.)

5. Let μ be a Lebesgue–Stieltjes measure on $\mathscr{B}(\mathbb{R}^n)$ such that $\mu(a + I) = \mu(I)$ for all $a \in R^n$ and all (right-semiclosed) intervals I in R^n. In other words, μ is translation-invariant on intervals. Show that μ is a constant times Lebesgue measure.

6. (*A set that is not Lebesgue measurable*) Call two real numbers x and y equivalent iff $x - y$ is rational. Choose a member of each distinct equivalence class $B_x = \{y: y - x \text{ rational}\}$ to form a set A (this requires the axiom of choice). Assume that the representatives are chosen so that $A \subset [0, 1]$. Establish the following:

   (a) If r and s are distinct rational numbers, $(r + A) \cap (s + A) = \emptyset$; also $\mathbb{R} = \bigcup \{r + A: r \text{ rational}\}$.
   (b) If A is Lebesgue measurable (so that $r + A$ is Lebesgue measurable by Problem 4), then $\mu(r + A) = 0$ for all rational r (μ is Lebesgue measure). Conclude that A cannot be Lebesgue measurable.

The only properties of Lebesgue measure needed in this problem are translation-invariance and finiteness on bounded intervals. Therefore, the result implies that there is no translation-invariant measure λ (except $\lambda \equiv 0$) on the class of all subsets of $\mathbb{R}$ such that $\lambda(I) < \infty$ for each bounded interval I.

7. (*The Cantor ternary set*) Let E_1 be the middle third of the interval $[0, 1]$, that is, $E_1 = \left(\frac{1}{3}, \frac{2}{3}\right)$; thus $x \in [0, 1] - E_1$ iff x can be written in ternary form using 0 or 2 in the first digit. Let E_2 be the union of the middle thirds of the two intervals that remain after E_1 is removed, that is, $E_2 = \left(\frac{1}{9}, \frac{2}{9}\right) \cup \left(\frac{7}{9}, \frac{8}{9}\right)$; thus $x \in [0, 1] - (E_1 \cup E_2)$ iff x can be written in ternary form using 0 or 2 in the first two digits. Continue the construction; let E_n be the union of the middle thirds of the intervals that remain after $E_1, \ldots, E_{n-1}$ are removed. The Cantor ternary set C is defined as $[0, 1] - \bigcup_{n=1}^{\infty} E_n$; thus $x \in C$ iff x can be expressed in ternary form using only digits 0 and 2. Various topological properties of C follow from the definition: C is closed, perfect (every point of C is a limit point of C), and nowhere dense.

   Show that C is uncountable and has Lebesgue measure 0.

   Comment. In the above construction, we have $m(E_n) = \left(\frac{1}{3}\right)\left(\frac{2}{3}\right)^{n-1}$, $n = 1, 2, \ldots$, where m is Lebesgue measure. If $0 < \alpha < 1$, the procedure may be altered slightly so that $m(E_n) = \alpha\left(\frac{1}{2}\right)^n$. We then obtain a set $C(\alpha)$, homeomorphic to C, of measure $1 - \alpha$; such sets are called *Cantor sets of positive measure*.

8. Give an example of a function $F: \mathbb{R}^2 \to \mathbb{R}$ such that F is right-continuous and is increasing in each coordinate separately, but F is not a distribution function on $\mathbb{R}^2$.

9. A distribution function on $\mathbb{R}$ is monotone and thus has only countably many points of discontinuity. Is this also true for a distribution function on $\mathbb{R}^n$, $n > 1$?

10. (a) Let F and G be distribution functions on $\mathbb{R}^n$. If $F(a, b] = G(a, b]$ for all $a, b \in \mathbb{R}^n$, $a \le b$, does it follow that F and G differ by a constant?
    (b) Must a distribution function on $\mathbb{R}^n$ be increasing in each coordinate separately?

*11. If c is the cardinality of the reals, show that there are only c Borel subsets of $\mathbb{R}^n$, but 2^c Lebesgue measurable sets.

1.5   MEASURABLE FUNCTIONS AND INTEGRATION

If f is a real-valued function defined on a bounded interval $[a, b]$ of reals, we can talk about the Riemann integral of f, at least if f is piecewise continuous. We are going to develop a much more general integration process, one that applies to functions from an arbitrary set to the extended reals, provided that certain "measurability" conditions are satisfied.

Probability considerations may again be used to motivate the concept of measurability. Suppose that $(\Omega, \mathscr{F}, P)$ is a probability space, and that h is a function from Ω to $\overline{\mathbb{R}}$. Thus if the outcome of the experiment corresponds to the point $\omega \in \Omega$, we may compute the number $h(\omega)$. Suppose that we are interested in the probability that $a \leq h(\omega) \leq b$, in other words, we wish to compute $P\{\omega: h(\omega) \in B\}$ where $B = [a, b]$. For this to be possible, the set $\{\omega: h(\omega) \in B\} = h^{-1}(B)$ must belong to the σ-field $\mathscr{F}$. If $h^{-1}(B) \in \mathscr{F}$ for each interval B (and hence, as we shall see below, for each Borel set B), then h is a "measurable function," in other words, probabilities of events involving h can be computed. In the language of probability theory, h is a "random variable."

1.5.1   *Definitions and Comments.* If h: $\Omega_1 \to \Omega_2$, h is *measurable* relative to the σ-fields $\mathscr{F}_j$ of subsets of Ω_j, $j = 1, 2$, iff $h^{-1}(A) \in \mathscr{F}_1$ for each $A \in \mathscr{F}_2$.

It is sufficient that $h^{-1}(A) \in \mathscr{F}_1$ for each $A \in \mathscr{C}$, where $\mathscr{C}$ is a class of subsets of Ω_2 such that the minimal σ-field over $\mathscr{C}$ is $\mathscr{F}_2$. For $\{A \in \mathscr{F}_2: h^{-1}(A) \in \mathscr{F}_1\}$ is a σ-field that contains all sets of $\mathscr{C}$, hence coincides with $\mathscr{F}_2$. This is another application of the good sets principle.

The notation h: $(\Omega_1, \mathscr{F}_1) \to (\Omega_2, \mathscr{F}_2)$ will mean that h: $\Omega_1 \to \Omega_2$, measurable relative to $\mathscr{F}_1$ and $\mathscr{F}_2$.

If $\mathscr{F}$ is a σ-field of subsets of Ω, $(\Omega, \mathscr{F})$ is sometimes called a *measurable space*, and the sets in $\mathscr{F}$ are sometimes called *measurable sets*.

Notice that measurability of h does not imply that $h(A) \in \mathscr{F}_2$ for each $A \in \mathscr{F}_1$. For example, if $\mathscr{F}_2 = \{\emptyset, \Omega_2\}$, then any h: $\Omega_1 \to \Omega_2$ is measurable, regardless of $\mathscr{F}_1$, but if $A \in \mathscr{F}_1$ and $h(A)$ is a nonempty proper subset of Ω_2, then $h(A) \notin \mathscr{F}_2$. Actually, in measure theory, the inverse image is a much more desirable object than the direct image since the basic set operations are preserved by inverse images but not in general by direct images. Specifically, we have $h^{-1}\left(\bigcup_i B_i\right) = \bigcup_i h^{-1}(B_i)$, $h^{-1}\left(\bigcap_i B_i\right) = \bigcap_i h^{-1}(B_i)$, and $h^{-1}(B^c) = [h^{-1}(B)]^c$. We also have $h\left(\bigcup_i B_i\right) = \bigcup_i h(A_i)$, but $h\left(\bigcap_i A_i\right) \subset \bigcap_i h(A_i)$, and the inclusion may be proper. Furthermore, $h(A^c)$ need not equal $[h(A)]^c$, in fact when h is a constant function the two sets are disjoint.

If $(\Omega, \mathscr{F})$ is a measurable space and h: $\Omega \to \mathbb{R}^n$ (or $\overline{\mathbb{R}}^n$), h is said to be *Borel measurable* [on $(\Omega, \mathscr{F})$] iff h is measurable relative to the σ-fields $\mathscr{F}$ and $\mathscr{B}$, the class of Borel sets. If Ω is a Borel subset of $\mathbb{R}^k$ (or $\overline{\mathbb{R}}^k$) and we use the term "Borel measurable," we always assume that $\mathscr{F} = \mathscr{B}$.

A continuous map h from $\mathbb{R}^k$ to $\mathbb{R}^n$ is Borel measurable; for if $\mathscr{C}$ is the class of open subsets of $\mathbb{R}^n$, then $h^{-1}(A)$ is open, hence belongs to $\mathscr{B}(\mathbb{R}^k)$, for each $A \in \mathscr{C}$.

If A is a subset of $\mathbb{R}$ that is not a Borel set (Section 1.4, Problems 6 and 11) and I_A is the *indicator* of A, that is, $I_A(\omega) = 1$ for $\omega \in A$ and 0 for $\omega \notin A$, then I_A is not Borel measurable; for $\{\omega: I_A(\omega) = 1\} = A \notin \mathcal{B}(\mathbb{R})$.

To show that a function $h: \Omega \to \mathbb{R}$ (or $\overline{\mathbb{R}}$) is Borel measurable, it is sufficient to show that $\{\omega: h(\omega) > c\} \in \mathcal{F}$ for each real c. For if $\mathcal{C}$ is the class of sets $\{x: x > c\}, c \in \mathbb{R}$, then $\sigma(\mathcal{C}) = \mathcal{B}(\mathbb{R})$. Similarly, $\{\omega: h(\omega) > c\}$ can be replaced by $\{\omega: h(\omega) \geq c\}$, $\{\omega: h(\omega) < c\}$ or $\{\omega: h(\omega) \leq c\}$, or equally well by $\{\omega: a \leq h(\omega) \leq b\}$ for all real a and b, and so on.

If $(\Omega, \mathcal{F}, \mu)$ is a measure space the terminology "h is Borel measurable on $(\Omega, \mathcal{F}, \mu)$" will mean that h is Borel measurable on $(\Omega, \mathcal{F})$ and μ is a measure on $\mathcal{F}$.

1.5.2 Definition. Let $(\Omega, \mathcal{F})$ be a measurable space, fixed throughout the discussion. If $h: \Omega \to \overline{\mathbb{R}}$, h is said to be *simple* iff h is Borel measurable and takes on only finitely many distinct values. Equivalently, h is simple iff it can be written as a finite sum $\sum_{i=1}^{r} x_i I_{A_i}$ where the A_i are disjoint sets in $\mathcal{F}$ and I_{A_i} is the indicator of A_i; the x_i need not be distinct.

We assume the standard arithmetic of $\overline{\mathbb{R}}$; if $a \in \mathbb{R}$, $a + \infty = \infty, a - \infty = -\infty, a/\infty = a/-\infty = 0, a \cdot \infty = \infty$ if $a > 0, a \cdot \infty = -\infty$ if $a < 0$, $0 \cdot \infty = 0 \cdot (-\infty) = 0, \infty + \infty = \infty, -\infty - \infty = -\infty$, with commutativity of addition and multiplication. It is then easy to check that sums, differences, products, and quotients of simple functions are simple, as long as the operations are well-defined, in other words we do not try to add $+\infty$ and $-\infty$, divide by 0, or divide ∞ by ∞.

Let μ be a measure on $\mathcal{F}$, again fixed throughout the discussion. If $h: \Omega \to \overline{\mathbb{R}}$ is Borel measurable we are going to define the *abstract Lebesgue integral* of h with respect to μ, written as $\int_\Omega h \, d\mu, \int_\Omega h(\omega)\mu(d\omega)$, or $\int_\Omega h(\omega)d\mu(\omega)$.

1.5.3 Definition of the Integral. First let h be simple, say $h = \sum_{i=1}^{r} x_i I_{A_i}$ where the A_i are disjoint sets in $\mathcal{F}$. We define

$$\int_\Omega h \, d\mu = \sum_{i=1}^{r} x_i \mu(A_i)$$

as long as $+\infty$ and $-\infty$ do not both appear in the sum; if they do, we say that the integral does not exist. Strictly speaking, it must be verified that if h has a different representation, say $\sum_{j=1}^{s} y_j I_{B_j}$, then

$$\sum_{i=1}^{r} x_i \mu(A_i) = \sum_{j=1}^{s} y_j \mu(B_j).$$

(For example, if $A = B \cup C$, where $B \cap C = \emptyset$, then $xI_A = xI_B + xI_C$.) The proof is based on the observation that

$$h = \sum_{i=1}^{r} \sum_{j=1}^{s} z_{ij} I_{A_i \cap B_j},$$

where $z_{ij} = x_i = y_j$. Thus

$$\sum_{i,j} z_{ij} \mu(A_i \cap B_j) = \sum_i x_i \sum_j \mu(A_i \cap B_j)$$

$$= \sum_i x_i \mu(A_i)$$

$$= \sum_j y_j \mu(B_j) \qquad \text{by a symmetrical argument.}$$

If h is nonnegative Borel measurable, define

$$\int_\Omega h \, d\mu = \sup \left\{ \int_\Omega s \, d\mu \colon s \quad \text{simple}, \quad 0 \le s \le h \right\}.$$

This agrees with the previous definition if h is simple. Furthermore, we may if we like restrict s to be finite-valued.

Notice that according to the definition, the integral of a nonnegative Borel measurable function always exists; it may be $+\infty$.

Finally, if h is an arbitrary Borel measurable function, let $h^+ = \max(h, 0)$, $h^- = \max(-h, 0)$, that is,

$$h^+(\omega) = h(\omega) \quad \text{if } h(\omega) \ge 0; \qquad h^+(\omega) = 0 \quad \text{if } h(\omega) < 0;$$

$$h^-(\omega) = -h(\omega) \quad \text{if } h(\omega) \le 0; \qquad h^-(\omega) = 0 \quad \text{if } h(\omega) > 0.$$

The function h^+ is called the *positive part* of h, h^- the *negative part*. We have $|h| = h^+ + h^-$, $h = h^+ - h^-$, and h^+ and h^- are Borel measurable. For example, $\{\omega \colon h^+(\omega) \in B\} = \{\omega \colon h(\omega) \ge 0, h(\omega) \in B\} \cup \{\omega \colon h(\omega) < 0, 0 \in B\}$. The first set is $h^{-1}[0, \infty] \cap h^{-1}(B) \in \mathscr{F}$; the second is $h^{-1}[-\infty, 0)$ if $0 \in B$, and $\emptyset$ if $0 \notin B$. Thus $(h^+)^{-1}(B) \in \mathscr{F}$ for each $B \in \mathscr{B}(\mathbb{R})$, and similarly for h^-. Alternatively, if h_1 and h_2 are Borel measurable, then $\max(h_1, h_2)$ and $\min(h_1, h_2)$ are Borel measurable; to see this, note that

$$\{\omega \colon \max(h_1(\omega), h_2(\omega)) \le c\} = \{\omega \colon h_1(\omega) \le c\} \cap \{\omega \colon h_2(\omega) \le c\}$$

and $\{\omega \colon \min(h_1(\omega), h_2(\omega)) \le c\} = \{\omega \colon h_1(\omega) \le c\} \cup \{\omega \colon h_2(\omega) \le c\}$. It follows that if h is Borel measurable, so are h^+ and h^-.

We define

$$\int_\Omega h\, d\mu = \int_\Omega h^+\, d\mu - \int_\Omega h^-\, d\mu \qquad \text{if this is not of the form } +\infty -\infty;$$

if it is, we say that the integral does not exist. The function h is said to be μ-*integrable* (or simply *integrable* if μ is understood) iff $\int_\Omega h\, d\mu$ is finite, that is, iff $\int_\Omega h^+\, d\mu$ and $\int_\Omega h^-\, d\mu$ are both finite.

If $A \in \mathscr{F}$, we define

$$\int_A h\, d\mu = \int_\Omega hI_A\, d\mu.$$

(The proof that hI_A is Borel measurable is similar to the first proof above that h^+ is Borel measurable.)

If h is a step function from $\mathbb{R}$ to $\mathbb{R}$ and μ is Lebesgue measure, $\int_\mathbb{R} h\, d\mu$ agrees with the Riemann integral. However, the integral of h with respect to Lebesgue measure exists for many functions that are not Riemann integrable, as we shall see in 1.7.

Before examining the properties of the integral, we need to know more about Borel measurable functions. One of the basic reasons why such functions are useful in analysis is that a pointwise limit of Borel measurable functions is still Borel measurable.

1.5.4 Theorem. If $h_1, h_2, \ldots$ are Borel measurable functions from Ω to $\overline{\mathbb{R}}$ and $h_n(\omega) \to h(\omega)$ for all $\omega \in \Omega$, then h is Borel measurable.

PROOF. It is sufficient to show that $\{\omega \colon h(\omega) > c\} \in \mathscr{F}$ for each real c. We have

$$\{\omega \colon h(\omega) > c\} = \left\{\omega \colon \lim_{n\to\infty} h_n(\omega) > c\right\}$$

$$= \left\{\omega \colon h_n(\omega) \text{ is eventually } > c + \frac{1}{r} \text{ for some } r = 1, 2, \ldots\right\}$$

$$= \bigcup_{r=1}^{\infty} \left\{\omega \colon h_n(\omega) > c + \frac{1}{r} \text{ for all but finitely many } n\right\}$$

$$= \bigcup_{r=1}^{\infty} \liminf_n \left\{\omega \colon h_n(\omega) > c + \frac{1}{r}\right\}$$

$$= \bigcup_{r=1}^{\infty} \bigcup_{n=1}^{\infty} \bigcap_{k=n}^{\infty} \left\{\omega \colon h_k(\omega) > c + \frac{1}{r}\right\} \in \mathscr{F}. \quad \square$$

To show that the class of Borel measurable functions is closed under algebraic operations, we need the following basic approximation theorem.

1.5.5   Theorem. (a)   A nonnegative Borel measurable function h is the limit of an increasing sequence of nonnegative, finite-valued, simple functions h_n.

   (b)   An arbitrary Borel measurable function f is the limit of a sequence of finite-valued simple functions f_n, with $|f_n| \leq |f|$ for all n.

PROOF.   (a)   Define

$$h_n(\omega) = \frac{k-1}{2^n} \quad \text{if} \quad \frac{k-1}{2^n} \leq h(\omega) < \frac{k}{2^n}, \qquad k = 1, 2, \dots, n2^n,$$

and let $h_n(\omega) = n$ if $h(\omega) \geq n$. [Or equally well, $h_n(\omega) = (k-1)/2^n$ if $(k-1)/2^n < h(\omega) \leq k/2^n, k = 1, 2, \dots, n2^n$; $h_n(\omega) = n$ if $h(\omega) > n$; $h_n(\omega) = 0$ if $h(\omega) = 0$.] The h_n have the desired properties (Problem 1).

   (b)   Let g_n and h_n be nonnegative, finite-valued, simple functions with $g_n \uparrow f^+$ and $h_n \uparrow f^-$; take $f_n = g_n - h_n$.   □

1.5.6   Theorem. If h_1 and h_2 are Borel measurable functions from Ω to $\overline{\mathbb{R}}$, so are $h_1 + h_2, h_1 - h_2, h_1 h_2$, and h_1/h_2 [assuming these are well-defined, in other words, $h_1(\omega) + h_2(\omega)$ is never of the form $+\infty \, -\infty$ and $h_1(\omega)/h_2(\omega)$ is never of the form ∞/∞ or $a/0$].

PROOF.   As in 1.5.5, let s_{1n}, s_{2n} be finite-valued simple functions with $s_{1n} \to h_1, s_{2n} \to h_2$. Then $s_{1n} + s_{2n} \to h_1 + h_2$,

$$s_{1n} s_{2n} I_{\{h_1 \neq 0\}} I_{\{h_2 \neq 0\}} \to h_1 h_2,$$

and

$$\frac{s_{1n}}{s_{2n} + (1/n) I_{\{s_{2n}=0\}}} \to \frac{h_1}{h_2}.$$

Since

$$s_{1n} \pm s_{2n}, \qquad s_{1n} s_{2n} I_{\{h_1 \neq 0\}} I_{\{h_2 \neq 0\}}, \qquad s_{1n} \left(s_{2n} + \frac{1}{n} I_{\{s_{2n}=0\}} \right)^{-1}$$

are simple, the result follows from 1.5.4.   □

We are going to extend 1.5.4 and part of 1.5.6 to Borel measurable functions from Ω to $\overline{\mathbb{R}}^n$; to do this, we need the following useful result.

1.5.7   Lemma. A composition of measurable functions is measurable; specifically, if $g: (\Omega_1, \mathscr{F}_1) \to (\Omega_2, \mathscr{F}_2)$ and $h: (\Omega_2, \mathscr{F}_2) \to (\Omega_3, \mathscr{F}_3)$, then $h \circ g: (\Omega_1, \mathscr{F}_1) \to (\Omega_3, \mathscr{F}_3)$.

PROOF. If $B \in \mathscr{F}_3$, then $(h \circ g)^{-1}(B) = g^{-1}(h^{-1}(B)) \in \mathscr{F}_1$. $\square$

Since some books contain the statement "A composition of measurable functions need not be measurable," some explanation is called for. If h: $\mathbb{R} \to \mathbb{R}$, some authors call h "measurable" iff the preimage of a Borel set is a Lebesgue measurable set. We shall call such a function *Lebesgue measurable*. Note that every Borel measurable function is Lebesgue measurable, but not conversely. (Consider the indicator of a Lebesgue measurable set that is not a Borel set; see Section 1.4, Problem 11.) If g and h are Lebesgue measurable, the composition $h \circ g$ need not be Lebesgue measurable. Let $\mathscr{B}$ be the Borel sets, and $\overline{\mathscr{B}}$ the Lebesgue measurable sets. If $B \in \mathscr{B}$ then $h^{-1}(B) \in \overline{\mathscr{B}}$; but $g^{-1}(h^{-1}(B))$ is known to belong to $\overline{\mathscr{B}}$ only when $h^{-1}(B) \in \mathscr{B}$, so we cannot conclude that $(h \circ g)^{-1}(B) \in \overline{\mathscr{B}}$. For an explicit example, see Royden (1968, p. 70). If $g^{-1}(A) \in \overline{\mathscr{B}}$ for all $A \in \overline{\mathscr{B}}$, not just for all $A \in \mathscr{B}$, then we are in the situation described in Lemma 1.5.7, and $h \circ g$ is Lebesgue measurable; similarly, if h is Borel measurable (and g is Lebesgue measurable), then $h \circ g$ is Lebesgue measurable.

It is rarely necessary to replace Borel measurability of functions from $\mathbb{R}$ to $\mathbb{R}$ (or $\mathbb{R}^k$ to $\mathbb{R}^n$) by the slightly more general concept of Lebesgue measurability; in this book, the only instance is in 1.7. The integration theory that we are developing works for extended real-valued functions on an arbitrary measure space $(\Omega, \mathscr{F}, \mu)$. Thus there is no problem in integrating Lebesgue measurable functions; $\Omega = \mathbb{R}, \mathscr{F} = \overline{\mathscr{B}}$.

We may now assert that if $h_1, h_2, \ldots$ are Borel measurable functions from Ω to $\overline{\mathbb{R}}^n$ and h_n converges pointwise to h, then h is Borel measurable; furthermore, if h_1 and h_2 are Borel measurable functions from Ω to $\overline{\mathbb{R}}^n$, so are $h_1 + h_2$ and $h_1 - h_2$, assuming these are well-defined. The reason is that if $h(\omega) = (h_1(\omega), \ldots, h_n(\omega))$ describes a map from Ω to $\overline{\mathbb{R}}^n$, Borel measurability of h is equivalent to Borel measurability of all the component functions h_i.

1.5.8   Theorem. Let h: $\Omega \to \overline{\mathbb{R}}^n$; if p_i is the projection map of $\overline{\mathbb{R}}^n$ onto $\overline{\mathbb{R}}$, taking $(x_1, \ldots, x_n)$ to x_i, set $h_i = p_i \circ h, i = 1, \ldots, n$. Then h is Borel measurable iff h_i is Borel measurable for all $i = 1, \ldots, n$.

PROOF. Assume h Borel measurable. Since

$$p_i^{-1}\{x_i: a_i \leq x_i \leq b_i\} = \{x \in \overline{\mathbb{R}}^n: a_i \leq x_i \leq b_i, \quad -\infty \leq x_j \leq \infty, \quad j \neq i\},$$

which is an interval of $\overline{\mathbb{R}}^n$, p_i is Borel measurable. Thus

$$h: (\Omega, \mathscr{F}) \to (\overline{\mathbb{R}}^n, \mathscr{B}(\overline{\mathbb{R}}^n)), \qquad p_i: (\overline{\mathbb{R}}^n, \mathscr{B}(\overline{\mathbb{R}}^n)) \to (\overline{\mathbb{R}}, \mathscr{B}(\overline{\mathbb{R}})),$$

and therefore by 1.5.7, $h_i: (\Omega, \mathscr{F}) \to (\overline{\mathbb{R}}, \mathscr{B}(\overline{\mathbb{R}}))$.

Conversely, assume each h_i to be Borel measurable. Then

$$h^{-1}\{x \in \overline{\mathbb{R}}^n: a_i \le x_i \le b_i, \ i = 1, \ldots, n\}$$

$$= \bigcap_{i=1}^{n} \{\omega \in \Omega: a_i \le h_i(\omega) \le b_i\} \in \mathcal{F},$$

and the result follows. $\square$

We now proceed to some properties of the integral. In the following result, all functions are assumed Borel measurable from Ω to $\overline{\mathbb{R}}$.

1.5.9   Theorem. (a)   If $\int_\Omega h \, d\mu$ exists and $c \in \mathbb{R}$, then $\int_\Omega ch \, d\mu$ exists and equals $c \int_\Omega h \, d\mu$.

(b)   If $g(\omega) \ge h(\omega)$ for all ω, then $\int_\Omega g \, d\mu \ge \int_\Omega h \, d\mu$ in the sense that if $\int_\Omega h \, d\mu$ exists and is greater than $-\infty$, then $\int_\Omega g \, d\mu$ exists and $\int_\Omega g \, d\mu \ge \int_\Omega h \, d\mu$; if $\int_\Omega g \, d\mu$ exists and is less than $+\infty$, then $\int_\Omega h \, d\mu$ exists and $\int_\Omega h \, d\mu \le \int_\Omega g \, d\mu$. Thus if both integrals exist, $\int_\Omega g \, d\mu \ge \int_\Omega h \, d\mu$, whether or not the integrals are finite.

(c)   If $\int_\Omega h \, d\mu$ exists, then $\left| \int_\Omega h \, d\mu \right| \le \int_\Omega |h| \, d\mu$.

(d)   If $h \ge 0$ and $B \in \mathcal{F}$, then $\int_B h \, d\mu = \sup\{\int_B s \, d\mu: 0 \le s \le h, s \text{ simple}\}$.

(e)   If $\int_\Omega h \, d\mu$ exists, so does $\int_A h \, d\mu$ for each $A \in \mathcal{F}$; if $\int_\Omega h \, d\mu$ is finite, then $\int_A h \, d\mu$ is also finite for each $A \in \mathcal{F}$.

PROOF.   (a)   It is immediate that this holds when h is simple. If h is nonnegative and $c > 0$, then

$$\int_\Omega ch \, d\mu = \sup \left\{ \int_\Omega s \, d\mu; \quad 0 \le s \le ch, \quad s \quad \text{simple} \right\}$$

$$= c \sup \left\{ \int_\Omega \frac{s}{c} \, d\mu; \quad 0 \le \frac{s}{c} \le h, \quad \frac{s}{c} \text{ simple} \right\} = c \int_\Omega h \, d\mu.$$

In general, if $h = h^+ - h^-$ and $c > 0$, then $(ch)^+ = ch^+$, $(ch)^- = ch^-$; hence $\int_\Omega ch \, d\mu = c \int_\Omega h^+ \, d\mu - c \int_\Omega h^- \, d\mu$ by what we have just proved, so that $\int_\Omega ch \, d\mu = c \int_\Omega h \, d\mu$. If $c < 0$, then

$$(ch)^+ = -ch^-, \qquad (ch)^- = -ch^+,$$

so

$$\int_\Omega ch \, d\mu = -c \int_\Omega h^- \, d\mu + c \int_\Omega h^+ \, d\mu = c \int_\Omega h \, d\mu.$$

(b)   If g and h are nonnegative and $0 \le s \le h$, s simple, then $0 \le s \le g$; hence $\int_\Omega h \, d\mu \le \int_\Omega g \, d\mu$. In general, $h \le g$ implies $h^+ \le g^+$, $h^- \ge g^-$. If

$\int_\Omega h\,d\mu > -\infty$, we have $\int_\Omega g^-\,d\mu \le \int_\Omega h^-\,d\mu < \infty$; hence $\int_\Omega g\,d\mu$ exists and equals

$$\int_\Omega g^+\,d\mu - \int_\Omega g^-\,d\mu \ge \int_\Omega h^+\,d\mu - \int_\Omega h^-\,d\mu = \int_\Omega h\,d\mu.$$

The case in which $\int_\Omega g\,d\mu < \infty$ is handled similarly.

(c)   We have $-|h| \le h \le |h|$ so by (a) and (b), $-\int_\Omega |h|\,d\mu \le \int_\Omega h\,d\mu \le \int_\Omega |h|\,d\mu$ and the result follows. (Note that $|h|$ is Borel measurable by 1.5.6 since $|h| = h^+ + h^-$.)

(d)   If $0 \le s \le h$, then $\int_B s\,d\mu \le \int_B h\,d\mu$ by (b); hence

$$\int_B h\,d\mu \ge \sup\left\{ \int_B s\,d\mu\colon\ 0 \le s \le h\right\}.$$

If $0 \le t \le hI_B$, t simple, then $t = tI_B \le h$ so $\int_\Omega t\,d\mu \le \sup\{\int_\Omega sI_B\,d\mu\colon 0 \le s \le h,\ s\ \text{simple}\}$. Take the sup over t to obtain $\int_B h\,d\mu \le \sup\{\int_B s\,d\mu\colon 0 \le s \le h,\ s\ \text{simple}\}$.

(e)   This follows from (b) and the fact that $(hI_A)^+ = h^+I_A \le h^+$, $(hI_A)^- = h^-I_A \le h^-$. $\square$

Problems

1.   Show that the functions proposed in the proof of 1.5.5(a) have the desired properties. Show also that if h is bounded, the approximating sequence converges to h uniformly on Ω.

2.   Let f and g be extended real-valued Borel measurable functions on $(\Omega, \mathscr{F})$, and define

$$h(\omega) = f(\omega) \qquad \text{if} \qquad \omega \in A,$$
$$ = g(\omega) \qquad \text{if} \qquad \omega \in A^c,$$

where A is a set in $\mathscr{F}$. Show that h is Borel measurable.

3.   If $f_1, f_2, \ldots$ are extended real-valued Borel measurable functions on $(\Omega, \mathscr{F})$, $n = 1, 2, \ldots,$ show that $\sup_n f_n$ and $\inf_n f_n$ are Borel measurable (hence $\limsup_{n\to\infty} f_n$ and $\liminf_{n\to\infty} f_n$ are Borel measurable).

4.   Let $(\Omega, \mathscr{F}, \mu)$ be a complete measure space. If $f\colon (\Omega, \mathscr{F}) \to (\Omega', \mathscr{F}')$ and $g\colon \Omega \to \Omega'$, $g = f$ except on a subset of a set $A \in \mathscr{F}$ with $\mu(A) = 0$, show that g is measurable (relative to $\mathscr{F}$ and $\mathscr{F}'$).

*5.   (a)   Let f be a function from $\mathbb{R}^k$ to $\mathbb{R}^m$, not necessarily Borel measurable. Show that $\{x\colon f$ is discontinuous at $x\}$ is an F_σ (a countable union of closed subsets of $\mathbb{R}^k$), and hence is a Borel set. Does this result hold in spaces more general than the Euclidean space $\mathbb{R}^n$?

(b) Show that there is no function from $\mathbb{R}$ to $\mathbb{R}$ whose discontinuity set is the irrationals. (In 1.4.5 we constructed a distribution function whose discontinuity set was the rationals.)

*6. How many Borel measurable functions are there from $\mathbb{R}^n$ to $\mathbb{R}^k$?

7. We have seen that a pointwise limit of measurable functions is measurable. We may also show that under certain conditions, a pointwise limit of measures is a measure. The following result, known as Steinhaus' lemma, will be needed in the problem: If $\{a_{nk}\}$ is a double sequence of real numbers satisfying

(i) $\sum_{k=1}^{\infty} a_{nk} = 1$ for all n,
(ii) $\sum_{k=1}^{\infty} |a_{nk}| \le c < \infty$ for all n, and
(iii) $a_{nk} \to 0$ as $n \to \infty$ for all k,

there is a sequence $\{x_n\}$, with $x_n = 0$ or 1 for all n, such that $t_n = \sum_{k=1}^{\infty} a_{nk}x_k$ fails to converge to a finite or infinite limit.

To prove this, choose positive integers n_1 and k_1 arbitrarily; having chosen $n_1, \ldots, n_r, k_1, \ldots, k_r$, choose $n_{r+1} > n_r$ such that $\sum_{k \le k_r} |a_{n_{r+1}k}| < \frac{1}{8}$; this is possible by (iii). Then choose $k_{r+1} > k_r$ such that $\sum_{k > k_{r+1}} |a_{n_{r+1}k}| < \frac{1}{8}$; this is possible by (ii). Set $x_k = 0$, $k_{2s-1} < k \le k_{2s}$, $x_k = 1$, $k_{2s} < k \le k_{2s+1}$, $s = 1, 2, \ldots$. We may write $t_{n_{r+1}}$ as $h_1 + h_2 + h_3$, where h_1 is the sum of $a_{n_{r+1}k}x_k$ for $k \le k_r$, h_2 corresponds to $k_r < k \le k_{r+1}$, and h_3 to $k > k_{r+1}$. If r is odd, then $x_k = 0$, $k_r < k \le k_{r+1}$; hence $|t_{n_{r+1}}| < \frac{1}{4}$. If r is even, then $h_2 = \sum_{k_r < k \le k_{r+1}} a_{n_{r+1}k}$; hence by (i),

$$h_2 = 1 - \sum_{k \le k_r} a_{n_{r+1}k} - \sum_{k > k_{r+1}} a_{n_{r+1}k} > \frac{3}{4}.$$

Thus $t_{n_{r+1}} > \frac{3}{4} - |h_1| - |h_3| > \frac{1}{2}$, so $\{t_n\}$ cannot converge.

(a) *Vitali–Hahn–Saks Theorem.* Let $(\Omega, \mathscr{F})$ be a measurable space, and let $P_n, n = 1, 2, \ldots$, be probability measures on $\mathscr{F}$. If $P_n(A) \to P(A)$ for all $A \in \mathscr{F}$, then P is a probability measure on $\mathscr{F}$; furthermore, if $\{B_k\}$ is a sequence of sets in $\mathscr{F}$ decreasing to $\emptyset$, then $\sup_n P_n(B_k) \downarrow 0$ as $k \to \infty$. [Let A be the disjoint union of sets $A_k \in \mathscr{F}$; without loss of generality, assume $A = \Omega$ (otherwise add A^c to both sides). It is immediate that P is finitely additive, so by Problem 5, Section 1.2, $\alpha = \sum_k P(A_k) \le P(\Omega) = 1$. If $\alpha < 1$, set $a_{nk} = (1 - \alpha)^{-1}[P_n(A_k) - P(A_k)]$ and apply Steinhaus' lemma.]

(b) Extend the Vitali–Hahn–Saks theorem to the case where the P_n are not necessarily probability measures, but $P_n(\Omega) \leq c < \infty$ for all n. [For further extensions, see Dunford and Schwartz (1958).]

1.6 BASIC INTEGRATION THEOREMS

We are now ready to present the main properties of the integral. The results in this section will be used many times in the text. As above, $(\Omega, \mathscr{F}, \mu)$ is a fixed measure space, and all functions to be considered map Ω to $\overline{\mathbb{R}}$.

1.6.1 Theorem. Let h be a Borel measurable function such that $\int_\Omega h \, d\mu$ exists. Define $\lambda(B) = \int_B h \, d\mu, B \in \mathscr{F}$. Then λ is countably additive on $\mathscr{F}$; thus if $h \geq 0$, λ is a measure.

PROOF. Let h be a nonnegative simple function $\sum_{i=1}^n x_i I_{A_i}$. Then $\lambda(B) = \int_B h \, d\mu = \sum_{i=1}^n x_i \mu(B \cap A_i)$; since μ is countably additive, so is λ.

Now let h be nonnegative Borel measurable, and let $B = \bigcup_{n=1}^\infty B_n$, the B_n disjoint sets in $\mathscr{F}$. If s is simple and $0 \leq s \leq h$, then

$$\int_B s \, d\mu = \sum_{n=1}^\infty \int_{B_n} s \, d\mu$$

by what we have proved for nonnegative simple functions

$$\leq \sum_{n=1}^\infty \int_{B_n} h \, d\mu$$

by 1.5.9(b)   (or the definition of the integral).

Take the sup over s to obtain, by 1.5.9(d), $\lambda(B) \leq \sum_{n=1}^\infty \lambda(B_n)$.

Now $B_n \subset B$, hence $I_{B_n} \leq I_B$, so by 1.5.9(b), $\lambda(B_n) \leq \lambda(B)$. If $\lambda(B_n) = \infty$ for some n, we are finished, so assume all $\lambda(B_n)$ finite. Fix n and let $\varepsilon > 0$. It follows from 1.59(b), (d) and the fact that the maximum of a finite number of simple functions is simple that we can find a simple function s, $0 \leq s \leq h$, such that

$$\int_{B_i} s \, d\mu \geq \int_{B_i} h \, d\mu - \frac{\varepsilon}{n}, \qquad i = 1, 2, \ldots, n.$$

Now

$$\lambda(B_1 \cup \cdots \cup B_n) = \int_{\bigcup_{i=1}^n B_i} h \, d\mu \geq \int_{\bigcup_{i=1}^n B_i} s \, d\mu = \sum_{i=1}^n \int_{B_i} s \, d\mu$$

by what we have proved for nonnegative simple functions, hence

$$\lambda(B_1 \cup \cdots \cup B_n) \geq \sum_{i=1}^{n} \int_{B_i} h \, d\mu - \varepsilon = \sum_{i=1}^{n} \lambda(B_i) - \varepsilon.$$

Since $\lambda(B) \geq \lambda \left(\bigcup_{i=1}^{n} B_i \right)$ and ε is arbitrary, we have

$$\lambda(B) \geq \sum_{i=1}^{\infty} \lambda(B_i).$$

Finally let $h = h^+ - h^-$ be an arbitrary Borel measurable function. Then $\lambda(B) = \int_B h^+ \, d\mu - \int_B h^- \, d\mu$. Since $\int_\Omega h^+ \, d\mu < \infty$ or $\int_\Omega h^- \, d\mu < \infty$, the result follows. $\square$

The proof of 1.6.1 shows that λ is the difference of two measures λ^+ and λ^-, where $\lambda^+(B) = \int_B h^+ \, d\mu$, $\lambda^- = \int_B h^- \, d\mu$; at least one of the measures λ^+ and λ^- must be finite.

1.6.2 *Monotone Convergence Theorem.* Let $h_1, h_2, \ldots$ form an increasing sequence of nonnegative Borel measurable functions, and let $h(\omega) = \lim_{n \to \infty} h_n(\omega)$, $\omega \in \Omega$. Then $\int_\Omega h_n \, d\mu \to \int_\Omega h \, d\mu$. [Note that $\int_\Omega h_n \, d\mu$ increases with n by 1.5.9(b); for short, $0 \leq h_n \uparrow h$ implies $\int_\Omega h_n \, d\mu \uparrow \int_\Omega h \, d\mu$.]

PROOF. By 1.5.9(b), $\int_\Omega h_n \, d\mu \leq \int_\Omega h \, d\mu$ for all n, hence $k = \lim_{n \to \infty} \int_\Omega h_n \, d\mu \leq \int_\Omega h \, d\mu$. Let $0 < b < 1$, and let s be a nonnegative, finite-valued, simple function with $s \leq h$. Let $B_n = \{\omega: h_n(\omega) \geq bs(\omega)\}$. Then $B_n \uparrow \Omega$ since $h_n \uparrow h$ and s is finite-valued. Now $k \geq \int_\Omega h_n \, d\mu \geq \int_{B_n} h_n \, d\mu$ by 1.5.9(b), and $\int_{B_n} h_n \, d\mu \geq b \int_{B_n} s \, d\mu$ by 1.5.9(a) and (b). By 1.6.1 and 1.2.7, $\int_{B_n} s \, d\mu \to \int_\Omega s \, d\mu$, hence (let $b \to 1$) $k \geq \int_\Omega s \, d\mu$. Take the sup over s to obtain $k \geq \int_\Omega h \, d\mu$. $\square$

1.6.3 *Additivity Theorem.* Let f and g be Borel measurable, and assume that $f + g$ is well-defined. If $\int_\Omega f \, d\mu$ and $\int_\Omega g \, d\mu$ exist and $\int_\Omega f \, d\mu + \int_\Omega g \, d\mu$ is well-defined (not of the form $+\infty - \infty$ or $-\infty + \infty$), then

$$\int_\Omega (f + g) \, d\mu = \int_\Omega f \, d\mu + \int_\Omega g \, d\mu.$$

In particular, if f and g are integrable, so is $f + g$.

PROOF. If f and g are nonnegative simple functions, this is immediate from the definition of the integral. Assume f and g are nonnegative Borel measurable, and let t_n, u_n be nonnegative simple functions increasing to f and g, respectively. Then $0 \le s_n = t_n + u_n \uparrow f + g$. Now $\int_\Omega s_n \, d\mu = \int_\Omega t_n \, d\mu + \int_\Omega u_n \, d\mu$ by what we have proved for nonnegative simple functions; hence by 1.6.2, $\int_\Omega (f + g) \, d\mu = \int_\Omega f \, d\mu + \int_\Omega g \, d\mu$.

Now if $f \ge 0$, $g \le 0$, $h = f + g \ge 0$ (so g must be finite), we have $f = h + (-g)$; hence $\int_\Omega f \, d\mu = \int_\Omega h \, d\mu - \int_\Omega g \, d\mu$. If $\int_\Omega g \, d\mu$ is finite, then $\int_\Omega h \, d\mu = \int_\Omega f \, d\mu + \int_\Omega g \, d\mu$, and if $\int_\Omega g \, d\mu = -\infty$, then since $h \ge 0$,

$$\int_\Omega f \, d\mu \ge - \int_\Omega g \, d\mu = \infty,$$

contradicting the hypothesis that $\int_\Omega f \, d\mu + \int_\Omega g \, d\mu$ is well-defined. Similarly, if $f \ge 0$, $g \le 0$, $h \le 0$, we obtain $\int_\Omega h \, d\mu = \int_\Omega f \, d\mu + \int_\Omega g \, d\mu$ by replacing all functions by their negatives. (Explicitly, $-g \ge 0$, $-f \le 0$, $-h = -f - g \ge 0$, and the above argument applies.)

Let

$$E_1 = \{\omega: \ f(\omega) \ge 0, \qquad g(\omega) \ge 0\},$$

$$E_2 = \{\omega: \ f(\omega) \ge 0, \qquad g(\omega) < 0, \qquad h(\omega) \ge 0\},$$

$$E_3 = \{\omega: \ f(\omega) \ge 0, \qquad g(\omega) < 0, \qquad h(\omega) < 0\},$$

$$E_4 = \{\omega: \ f(\omega) < 0, \qquad g(\omega) \ge 0, \qquad h(\omega) \ge 0\},$$

$$E_5 = \{\omega: \ f(\omega) < 0, \qquad g(\omega) \ge 0, \qquad h(\omega) < 0\},$$

$$E_6 = \{\omega: \ f(\omega) < 0, \qquad g(\omega) < 0\}.$$

The above argument shows that $\int_{E_i} h \, d\mu = \int_{E_i} f \, d\mu + \int_{E_i} g \, d\mu$. Now $\int_\Omega f \, d\mu = \sum_{i=1}^6 \int_{E_i} f \, d\mu$, $\int_\Omega g \, d\mu = \sum_{i=1}^6 \int_{E_i} g \, d\mu$ by 1.6.1, so that $\int_\Omega f \, d\mu + \int_\Omega g \, d\mu = \sum_{i=1}^6 \int_{E_i} h \, d\mu$, and this equals $\int_\Omega h \, d\mu$ by 1.6.1, if we can show that $\int_\Omega h \, d\mu$ exists; that is, $\int_\Omega h^+ \, d\mu$ and $\int_\Omega h^- \, d\mu$ are not both infinite.

If this is the case, $\int_{E_i} h^+ \, d\mu = \int_{E_j} h^- \, d\mu = \infty$ for some i, j (1.6.1 again), so that $\int_{E_i} h \, d\mu = \infty$, $\int_{E_j} h \, d\mu = -\infty$. But then $\int_{E_i} f \, d\mu$ or $\int_{E_i} g \, d\mu = \infty$; hence $\int_\Omega f \, d\mu$ or $\int_\Omega g \, d\mu = \infty$. (Note that $\int_\Omega f^+ \, d\mu \ge \int_{E_i} f^+ \, d\mu$.) Similarly $\int_\Omega f \, d\mu$ or $\int_\Omega g \, d\mu = -\infty$, and this is a contradiction. $\square$

1.6.4   Corollaries. (a)   If $h_1, h_2, \ldots$ are nonnegative Borel measurable,

$$\int_\Omega \left(\sum_{n=1}^\infty h_n \right) d\mu = \sum_{n=1}^\infty \int_\Omega h_n \, d\mu.$$

Thus any series of nonnegative Borel measurable functions may be integrated term by term.

(b)   If h is Borel measurable, h is integrable iff $|h|$ is integrable.

(c)   If g and h are Borel measurable with $|g| \le h$, h integrable, then g is integrable.

PROOF.   (a) $\sum_{k=1}^n h_k \uparrow \sum_{k=1}^\infty h_k$, and the result follows from 1.6.2 and 1.6.3.

(b)   Since $|h| = h^+ + h^-$, this follows from the definition of the integral and 1.6.3.

(c)   By 1.5.9(b), $|g|$ is integrable, and the result follows from (b) above. □.

A condition is said to hold *almost everywhere* with respect to the measure μ (written a.e. $[\mu]$ or simply a.e. if μ is understood) iff there is a set $B \in \mathscr{F}$ of μ-measure 0 such that the condition holds outside of B. From the point of view of integration theory, functions that differ only on a set of measure 0 may be identified. This is established by the following result.

1.6.5   Theorem. Let f, g, and h be Borel measurable functions.

(a)   If $f = 0$ a.e. $[\mu]$, then $\int_\Omega f \, d\mu = 0$.

(b)   If $g = h$ a.e. $[\mu]$ and $\int_\Omega g \, d\mu$ exists, then so does $\int_\Omega h \, d\mu$, and $\int_\Omega g \, d\mu = \int_\Omega h \, d\mu$.

PROOF.

(a)   If $f = \sum_{i=1}^n x_i I_{A_i}$ is simple, then $x_i \ne 0$ implies $\mu(A_i) = 0$ by hypothesis, hence $\int_\Omega f \, d\mu = 0$. If $f \ge 0$ and $0 \le s \le f$, s simple, then $s = 0$ a.e. $[\mu]$, hence $\int_\Omega s \, d\mu = 0$; thus $\int_\Omega f \, d\mu = 0$. If $f = f^+ - f^-$, then f^+ and f^-, being less than or equal to $|f|$, are 0 a.e. $[\mu]$, and the result follows.

(b)   Let $A = \{\omega : g(\omega) = h(\omega)\}$, $B = A^c$. Then $g = gI_A + gI_B$, $h = hI_A + hI_B = gI_A + hI_B$. Since $gI_B = hI_B = 0$ except on B, a set of measure 0, the result follows from part (a) and 1.6.3. □

Thus in any integration theorem, we may freely use the phrase "almost everywhere." For example, if $\{h_n\}$ is an increasing sequence of nonnegative Borel measurable functions converging a.e. to the Borel measurable function h, then $\int_\Omega h_n \, d\mu \to \int_\Omega h \, d\mu$.

Another example: If g and h are Borel measurable and $g \ge h$ a.e., then $\int_\Omega g \, d\mu \ge \int_\Omega h \, d\mu$ [in the sense of 1.5.9(b)].

1.6.6   Theorem. Let h be Borel measurable.

(a)   If h is integrable, then h is finite a.e.
(b)   If $h \geq 0$ and $\int_\Omega h\,d\mu = 0$, then $h = 0$ a.e.

PROOF.   (a)   Let $A = \{\omega: |h(\omega)| = \infty\}$. If $\mu(A) > 0$, then $\int_\Omega |h|\,d\mu \geq \int_A |h|\,d\mu = \infty\mu(A) = \infty$, a contradiction.

(b)   Let $B = \{\omega: h(\omega) > 0\}$, $B_n = \{\omega: h(\omega) \geq 1/n\} \uparrow B$.   We have $0 \leq hI_{B_n} \leq hI_B = h$;   hence   by   1.5.9(b), $\int_{B_n} h\,d\mu = 0$.   But $\int_{B_n} h\,d\mu \geq (1/n)\mu(B_n)$, so that $\mu(B_n) = 0$ for all n, and thus $\mu(B) = 0$.   □

The monotone convergence theorem was proved under the hypothesis that all functions were nonnegative. This assumption can be relaxed considerably, as we now prove.

1.6.7   Extended Monotone Convergence Theorem. Let $g_1, g_2, \ldots, g, h$ be Borel measurable.

(a)   If $g_n \geq h$ for all n, where $\int_\Omega h\,d\mu > -\infty$, and $g_n \uparrow g$, then

$$\int_\Omega g_n\,d\mu \uparrow \int_\Omega g\,d\mu.$$

(b)   If $g_n \leq h$ for all n, where $\int_\Omega h\,d\mu < \infty$, and $g_n \downarrow g$, then

$$\int_\Omega g_n\,d\mu \downarrow \int_\Omega g\,d\mu.$$

PROOF.   (a)   If $\int_\Omega h\,d\mu = \infty$, then by 1.5.9(b), $\int_\Omega g_n\,d\mu = \infty$ for all n, and $\int_\Omega g\,d\mu = \infty$. Thus assume $\int_\Omega h\,d\mu < \infty$, so that by 1.6.6(a), h is a.e. finite; change h to 0 on the set where it is infinite. Then $0 \leq g_n - h \uparrow g - h$ a.e., hence by 1.6.2, $\int_\Omega (g_n - h)\,d\mu \uparrow \int_\Omega (g - h)\,d\mu$. The result follows from 1.6.3. (We must check that the additivity theorem actually applies. Since $\int_\Omega h\,d\mu > -\infty$, $\int_\Omega g_n\,d\mu$ and $\int_\Omega g\,d\mu$ exist and are greater than $-\infty$ by 1.5.9(b). Also, $\int_\Omega h\,d\mu$ is finite, so that $\int_\Omega g_n\,d\mu - \int_\Omega h\,d\mu$ and $\int_\Omega g\,d\mu - \int_\Omega h\,d\mu$ are well-defined.)

(b) $-g_n \geq -h$, $\int_\Omega -h\,d\mu > -\infty$, and $-g_n \uparrow -g$. By part (a), $-\int_\Omega g_n\,d\mu \uparrow -\int_\Omega g\,d\mu$, so $\int_\Omega g_n\,d\mu \downarrow \int_\Omega g\,d\mu$.   □

The extended monotone convergence theorem asserts that under appropriate conditions, the limit of the integrals of a sequence of functions is the integral of the limit function. More general theorems of this type can be obtained if we replace limits by upper or lower limits. If $f_1, f_2, \ldots$ are functions from Ω to $\overline{R}$, $\liminf_{n \to \infty} f_n$ and $\limsup_{n \to \infty} f_n$ are defined pointwise, that is,

$$\left(\liminf_{n \to \infty} f_n \right)(\omega) = \sup_n \inf_{k \geq n} f_k(\omega),$$

$$\left(\limsup_{n \to \infty} f_n \right)(\omega) = \inf_n \sup_{k \geq n} f_k(\omega).$$

1.6.8   Fatou's Lemma. Let $f_1, f_2, \ldots, f$ be Borel measurable.

(a)   If $f_n \geq f$ for all n, where $\int_\Omega f \, d\mu > -\infty$, then

$$\liminf_{n \to \infty} \int_\Omega f_n \, d\mu \geq \int_\Omega \left(\liminf_{n \to \infty} f_n \right) d\mu.$$

(b)   If $f_n \leq f$ for all n, where $\int_\Omega f \, d\mu < \infty$, then

$$\limsup_{n \to \infty} \int_\Omega f_n \, d\mu \leq \int_\Omega \left(\limsup_{n \to \infty} f_n \right) d\mu.$$

PROOF.   (a)   Let $g_n = \inf_{k \geq n} f_k$, $g = \liminf f_n$. Then $g_n \geq f$ for all n, $\int_\Omega f \, d\mu > -\infty$, and $g_n \uparrow g$. By 1.6.7, $\int_\Omega g_n \, d\mu \uparrow \int_\Omega (\liminf_{n \to \infty} f_n) \, d\mu$. But $g_n \leq f_n$, so

$$\lim_{n \to \infty} \int_\Omega g_n \, d\mu = \liminf_{n \to \infty} \int_\Omega g_n \, d\mu \leq \liminf_{n \to \infty} \int_\Omega f_n \, d\mu.$$

(b)   We may write

$$\int_\Omega \left(\limsup_{n \to \infty} f_n \right) d\mu = - \int_\Omega \liminf_{n \to \infty} (-f_n) \, d\mu$$

$$\geq - \liminf_{n \to \infty} \int_\Omega (-f_n) \, d\mu \qquad \text{by (a)}$$

$$= \limsup_{n \to \infty} \int_\Omega f_n \, d\mu. \quad \square$$

The following result is one of the "bread and butter" theorems of analysis; it will be used quite often in later chapters.

1.6.9   Dominated Convergence Theorem. If $f_1, f_2, \ldots, f, g$ are Borel measurable, $|f_n| \leq g$ for all n, where g is μ-integrable, and $f_n \to f$ a.e. $[\mu]$, then f is μ-integrable and $\int_\Omega f_n \, d\mu \to \int_\Omega f \, d\mu$.

PROOF.   We have $|f| \leq g$ a.e.; hence f is integrable by 1.6.4(c). By 1.6.8,

$$\int_\Omega \left(\liminf_{n\to\infty} f_n \right) d\mu \leq \liminf_{n\to\infty} \int_\Omega f_n \, d\mu \leq \limsup_{n\to\infty} \int_\Omega f_n \, d\mu$$

$$\leq \int_\Omega \left(\limsup_{n\to\infty} f_n \right) d\mu.$$

By hypothesis, $\liminf_{n\to\infty} f_n = \limsup_{n\to\infty} f_n = f$ a.e., so all terms of the above inequality are equal to $\int_\Omega f \, d\mu$. $\square$

1.6.10   Corollary. If $f_1, f_2, \ldots, f, g$ are Borel measurable, $|f_n| \leq g$ for all n, where $|g|^p$ is μ-integrable ($p > 0$, fixed), and $f_n \to f$ a.e. $[\mu]$, then $|f|^p$ is μ-integrable and $\int_\Omega |f_n - f|^p \, d\mu \to 0$ as $n \to \infty$.

PROOF.   We have $|f_n|^p \leq |g|^p$ for all n; so $|f|^p \leq |g|^p$, and therefore $|f|^p$ is integrable. Also $|f_n - f|^p \leq (|f_n| + |f|)^p \leq (2|g|)^p$, which is integrable, and the result follows from 1.6.9. $\square$

We have seen in 1.5.9(b) that $g \leq h$ implies $\int_\Omega g \, d\mu \leq \int_\Omega h \, d\mu$, and in fact $\int_A g \, d\mu \leq \int_A h \, d\mu$ for all $A \in \mathscr{F}$. There is a converse to this result.

1.6.11   Theorem. If μ is σ-finite on $\mathscr{F}$, g and h are Borel measurable, $\int_\Omega g \, d\mu$ and $\int_\Omega h \, d\mu$ exist, and $\int_A g \, d\mu \leq \int_A h \, d\mu$ for all $A \in \mathscr{F}$, then $g \leq h$ a.e. $[\mu]$.

PROOF.   It is sufficient to prove this when μ is finite. Let

$$A_n = \left\{ \omega \colon \, g(\omega) \geq h(\omega) + \frac{1}{n}, \qquad |h(\omega)| \leq n \right\}.$$

Then

$$\int_{A_n} h \, d\mu \geq \int_{A_n} g \, d\mu \geq \int_{A_n} h \, d\mu + \frac{1}{n} \mu(A_n).$$

But

$$\left| \int_{A_n} h \, d\mu \right| \leq \int_{A_n} |h| \, d\mu \leq n \mu(A_n) < \infty,$$

and thus we may subtract $\int_{A_n} h \, d\mu$ to obtain $(1/n)\mu(A_n) \leq 0$, hence $\mu(A_n) = 0$. Therefore $\mu(\bigcup_{n=1}^\infty A_n) = 0$; hence $\mu\{\omega \colon g(\omega) > h(\omega), h(\omega) \text{ finite}\} = 0$. Consequently $g \leq h$ a.e. on $\{\omega \colon h(\omega) \text{ finite}\}$. Clearly, $g \leq h$ everywhere on

$\{\omega: \ h(\omega) = \infty\}$, and by taking $C_n = \{\omega: \ h(\omega) = -\infty, g(\omega) \geq -n\}$ we obtain

$$-\infty\mu(C_n) = \int_{C_n} h \, d\mu \geq \int_{C_n} g \, d\mu \geq -n\mu(C_n);$$

hence $\mu(C_n) = 0$. Thus $\mu(\bigcup_{n=1}^{\infty} C_n) = 0$, so that

$$\mu\{\omega: \ g(\omega) > h(\omega), h(\omega) = -\infty\} = 0.$$

Therefore $g \leq h$ a.e. on $\{\omega: \ h(\omega) = -\infty\}$. $\square$

If g and h are integrable, the proof is simpler. Let $B = \{\omega: \ g(\omega) > h(\omega)\}$. Then $\int_B g \, d\mu \leq \int_B h \, d\mu \leq \int_B g \, d\mu$; hence all three integrals are equal. Thus by 1.6.3, $0 = \int_B (g-h) \, d\mu = \int_\Omega (g-h) I_B \, d\mu$, with $(g-h) I_B \geq 0$. By 1.6.6(b), $(g-h) I_B = 0$ a.e., so that $g = h$ a.e. on B. But $g \leq h$ on B^c, and the result follows. Note that in this case, μ need not be σ-finite.

The reader may have noticed that several integration theorems in this section were proved by starting with nonnegative simple functions and working up to nonnegative measurable functions and finally to arbitrary measurable functions. This technique is quite basic and will often be useful. A good illustration of the method is the following result, which introduces the notion of a measure-preserving transformation, a key concept in ergodic theory. In fact it is convenient here to start with indicators before proceeding to nonnegative simple functions.

1.6.12   Theorem. Let $T: (\Omega, \mathscr{F}) \to (\Omega_0, \mathscr{F}_0)$ be a measurable mapping, and let μ be a measure on $\mathscr{F}$. Define a measure $\mu_0 = \mu T^{-1}$ on $\mathscr{F}_0$ by

$$\mu_0(A) = \mu(T^{-1}(A)), \qquad A \in \mathscr{F}_0.$$

If $\Omega_0 = \Omega$, $\mathscr{F}_0 = \mathscr{F}$, and $\mu_0 = \mu$, T is said to *preserve* the measure μ.

If $f: (\Omega_0, \mathscr{F}_0) \to (\overline{R}, \mathscr{B}(\overline{R}))$ and $A \in \mathscr{F}_0$, then

$$\int_{T^{-1}A} f(T(\omega)) \, d\mu(\omega) = \int_A f(\omega) \, d\mu_0(\omega),$$

in the sense that if one of the integrals exists, so does the other, and the two integrals are equal.

PROOF.   If f is an indicator I_B, the desired formula states that

$$\mu(T^{-1}A \cap T^{-1}B) = \mu_0(A \cap B),$$

which is true by definition of μ_0. If f is a nonnegative simple function $\sum_{i=1}^{n} x_i I_{B_i}$, then

$$\int_{T^{-1}A} f(T(\omega)) \, d\mu(\omega) = \sum_{i=1}^{n} x_i \int_{T^{-1}A} I_{B_i}(T(\omega)) \, d\mu(\omega) \qquad \text{by 1.6.3}$$

$$= \sum_{i=1}^{n} x_i \int_{A} I_{B_i}(\omega) \, d\mu_0(\omega)$$

by what we have proved for indicators

$$= \int_{A} f(\omega) \, d\mu_0(\omega) \qquad \text{by 1.6.3.}$$

If f is a non-negative Borel measurable function, let $f_1, f_2, \ldots$ be nonnegative simple functions increasing to f. Then $\int_{T^{-1}A} f_n(T(\omega)) \, d\mu(\omega) = \int_A f_n(\omega) \, d\mu_0(\omega)$ by what we have proved for simple functions, and the monotone convergence theorem yields the desired result for f.

Finally, if $f = f^+ - f^-$ is an arbitrary Borel measurable function, we have proved that the result holds for f^+ and f^-. If, say, $\int_A f^+(\omega) \, d\mu_0(\omega) < \infty$, then $\int_{T^{-1}A} f^+(T(\omega)) \, d\mu(\omega) < \infty$, and it follows that if one of the integrals exists, so does the other, and the two integrals are equal. $\square$

If one is having difficulty proving a theorem about measurable functions or integration, it is often helpful to start with indicators and work upward. In fact it is possible to suspect that almost anything can be proved this way, but of course there are exceptions. For example, you will run into trouble trying to prove the proposition "All functions are indicators."

We shall adopt the following terminology: If μ is Lebesgue measure and A is an interval $[a, b]$, $\int_A f \, d\mu$, if it exists, will often be denoted by $\int_a^b f(x) \, dx$ (or $\int_{a_1}^{b_1} \cdots \int_{a_n}^{b_n} f(x_1, \cdots, x_n) dx_1 \cdots dx_n$ if we are integrating functions on $\mathbb{R}^n$). The endpoints may be deleted from the interval without changing the integral, since the Lebesgue measure of a single point is 0. If f is integrable with respect to μ, then we say that f is *Lebesgue integrable*. A different notation, such as $r_{ab}(f)$, will be used for the Riemann integral of f on $[a, b]$.

Problems

The first three problems give conditions under which some of the most commonly occurring operations in real analysis may be performed: taking a limit under the integral sign, integrating an infinite series term by term, and differentiating under the integral sign.

1. Let $f = f(x, y)$ be a real-valued function of two real variables, defined for $a < y < b, c < x < d$. Assume that for each x, $f(x, \cdot)$ is a Borel measurable function of y, and that there is a Borel measurable g: $(a, b) \rightarrow \mathbb{R}$ such that $|f(x, y)| \leq g(y)$ for all x, y, and $\int_a^b g(y)dy < \infty$. If $x_0 \in (c, d)$ and $\lim_{x \rightarrow x_0} f(x, y)$ exists for all $y \in (a, b)$, show that

$$\lim_{x \rightarrow x_0} \int_a^b f(x, y)dy = \int_a^b \left[\lim_{x \rightarrow x_0} f(x, y) \right] dy.$$

2. Let $f_1, f_2, \ldots$ be Borel measurable functions on $(\Omega, \mathscr{F}, \mu)$. If

$$\sum_{n=1}^{\infty} \int_{\Omega} |f_n| \, d\mu < \infty,$$

show that $\sum_{n=1}^{\infty} f_n$ converges a.e. $[\mu]$ to a finite-valued function, and $\int_{\Omega} \left(\sum_{n=1}^{\infty} f_n \right) d\mu = \sum_{n=1}^{\infty} \int_{\Omega} f_n \, d\mu$.

3. Let $f = f(x, y)$ be a real-valued function of two real variables, defined for $a < y < b, c < x < d$, such that f is a Borel measurable function of y for each fixed x. Assume that for each x, $f(x, \cdot)$ is integrable over (a, b) (with respect to Lebesgue measure). Suppose that the partial derivative $f_1(x, y)$ of f with respect to x exists for all (x, y), and suppose there is a Borel measurable h: $(a, b) \rightarrow \mathbb{R}$ such that $|f_1(x, y)| \leq h(y)$ for all x, y, where $\int_a^b h(y)dy < \infty$.

    Show that $d[\int_a^b f(x, y)dy]/dx$ exists for all $x \in (c, d)$, and equals $\int_a^b f_1(x, y)dy$. [It must be verified that $f_1(x, \cdot)$ is Borel measurable for each x.]

4. If μ is a measure on $(\Omega, \mathscr{F})$ and $A_1, A_2, \ldots$ is a sequence of sets in $\mathscr{F}$, use Fatou's lemma to show that

$$\mu \left(\liminf_n A_n \right) \leq \liminf_{n \rightarrow \infty} \mu(A_n).$$

If μ is finite, show that

$$\mu \left(\limsup_n A_n \right) \geq \limsup_{n \rightarrow \infty} \mu(A_n).$$

Thus if μ is finite and $A = \lim_n A_n$, then $\mu(A) = \lim_{n \rightarrow \infty} \mu(A_n)$. (For another proof of this, see Section 1.2, Problem 10.)

5. Give an example of a sequence of Lebesgue integrable functions f_n converging everywhere to a Lebesgue integrable function f, such that

$$\lim_{n \rightarrow \infty} \int_{-\infty}^{\infty} f_n(x) \, dx < \int_{-\infty}^{\infty} f(x) \, dx.$$

Thus the hypotheses of the dominated convergence theorem and Fatou's lemma cannot be dropped.

6.   (a)   Show that $\int_1^\infty e^{-t} \ln t \, dt = \lim_{n\to\infty} \int_1^n [1 - (t/n)]^n \ln t \, dt$.

      (b)   Show that $\int_0^1 e^{-t} \ln t \, dt = \lim_{n\to\infty} \int_0^1 [1 - (t/n)]^n \ln t \, dt$.

7.   If $(\Omega, \mathscr{F}, \mu)$ is the completion of $(\Omega, \mathscr{F}_0, \mu)$ and f is a Borel measurable function on $(\Omega, \mathscr{F})$, show that there is a Borel measurable function g on $(\Omega, \mathscr{F}_0)$ such that $f = g$, except on a subset of a set in $\mathscr{F}_0$ of measure 0. (Start with indicators.)

8.   If f is a Borel measurable function from $\mathbb{R}$ to $\mathbb{R}$ and $a \in \mathbb{R}$, show that

$$\int_{-\infty}^{\infty} f(x) \, dx = \int_{-\infty}^{\infty} f(x - a) \, dx$$

in the sense that if one integral exists, so does the other, and the two are equal. (Start with indicators.)

1.7   COMPARISON OF LEBESGUE AND RIEMANN INTEGRALS

In this section we show that integration with respect to Lebesgue measure is more general than Riemann integration, and we obtain a precise criterion for Riemann integrability.

Let $[a, b]$ be a bounded closed interval of reals, and let f be a bounded real-valued function on $[a, b]$, assumed fixed throughout the discussion. If $P: a = x_0 < x_1 < \cdots < x_n = b$ is a partition of $[a, b]$, we may construct the upper and lower sums of f relative to P as follows.

Let

$$M_i = \sup\{f(y): x_{i-1} < y \leq x_i\}, \qquad i = 1, \ldots, n,$$
$$m_i = \inf\{f(y): x_{i-1} < y \leq x_i\}, \qquad i = 1, \ldots, n,$$

and define step functions α and β, called the *upper* and *lower* functions corresponding to P, by

$$\alpha(x) = M_i \qquad \text{if} \qquad x_{i-1} < x \leq x_i, \qquad i = 1, \ldots, n,$$
$$\beta(x) = m_i \qquad \text{if} \qquad x_{i-1} < x \leq x_i, \qquad i = 1, \ldots, n$$

$[\alpha(a)$ and $\beta(a)$ may be chosen arbitrarily]. The upper and lower sums are given by

$$U(P) = \sum_{i=1}^{n} M_i(x_i - x_{i-1}),$$

$$L(P) = \sum_{i=1}^{n} m_i(x_i - x_{i-1}).$$

Now we take as a measure space $\Omega = [a, b]$, $\mathscr{F} = \overline{\mathscr{B}}[a, b]$, the *Lebesgue measurable* subsets of $[a, b]$, $\mu = $ Lebesgue measure. Since α and β are simple functions, we have

$$U(P) = \int_a^b \alpha \, d\mu, \qquad L(P) = \int_a^b \beta \, d\mu.$$

Now let $P_1, P_2, \ldots$ be a sequence of partitions of $[a, b]$ such that P_{k+1} is a refinement of P_k for each k, and such that $|P_k|$ (the length of the largest subinterval of P_k) approaches 0 as $k \to \infty$. If α_k and β_k are the upper and lower functions corresponding to P_k, then

$$\alpha_1 \geq \alpha_2 \geq \cdots \geq f \geq \cdots \geq \beta_2 \geq \beta_1.$$

Thus α_k and β_k approach limit functions α and β. If $|f|$ is bounded by M, then all $|\alpha_k|$ and $|\beta_k|$ are bounded by M as well, and the function that is constant at M is integrable on $[a, b]$ with respect to μ, since

$$\mu[a, b] = b - a < \infty.$$

By the dominated convergence theorem,

$$\lim_{k \to \infty} U(P_k) = \lim_{k \to \infty} \int_a^b \alpha_k \, d\mu = \int_a^b \alpha \, d\mu,$$

and

$$\lim_{k \to \infty} L(P_k) = \lim_{k \to \infty} \int_a^b \beta_k \, d\mu = \int_a^b \beta \, d\mu.$$

We shall need one other fact, namely that if x is not an endpoint of any of the subintervals of the P_k,

$$f \text{ is continuous at } x \qquad \text{iff} \qquad \alpha(x) = f(x) = \beta(x).$$

This follows by a standard ε–δ argument.

If $\lim_{k \to \infty} U(P_k) = \lim_{k \to \infty} L(P_k) = a$ finite number r, independent of the particular sequence of partitions, f is said to be *Riemann integrable* on $[a, b]$, and $r = r_{ab}(f)$ is said to be the (value of the) *Riemann integral* of f on $[a, b]$. The above argument shows that f is Riemann integrable iff

$$\int_a^b \alpha \, d\mu = \int_a^b \beta \, d\mu = r,$$

independent of the particular sequence of partitions. If f is Riemann integrable,

$$r_{ab}(f) = \int_a^b \alpha \, d\mu = \int_a^b \beta \, d\mu.$$

We are now ready for the main results.

1.7.1 Theorem. Let f be a bounded real-valued function on $[a, b]$.

(a) The function f is Riemann integrable on $[a, b]$ iff f is continuous almost everywhere on $[a, b]$ (with respect to Lebesgue measure).

(b) If f is Riemann integrable on $[a, b]$, then f is integrable with respect to Lebesgue measure on $[a, b]$, and the two integrals are equal.

PROOF. (a) If f is Riemann integrable,

$$r_{ab}(f) = \int_a^b \alpha \, d\mu = \int_a^b \beta \, d\mu.$$

As $\beta \le f \le \alpha$, 1.6.6(b) applied to $\alpha - \beta$ yields $\alpha = f = \beta$ a.e.; hence f is continuous a.e. Conversely, assume f is continuous a.e.; then $\alpha = f = \beta$ a.e. Now α and β are limits of simple functions, and hence are Borel measurable. Thus f differs from a measurable function on a subset of a set of measure 0, and therefore f is measurable because of the completeness of the measure space. (See Section 1.5, Problem 4.) Since f is bounded, it is integrable with respect to μ, and since $\alpha = f = \beta$ a.e., we have

$$\int_a^b \alpha \, d\mu = \int_a^b \beta \, d\mu = \int_a^b f \, d\mu, \tag{1}$$

independent of the particular sequence of partitions. Therefore f is Riemann integrable.

(b) If f is Riemann integrable, then f is continuous a.e. by part (a). But then Eq. (1) yields $r_{ab}(f) = \int_a^b f \, d\mu$, as desired. $\square$

Theorem 1.7.1 holds equally well in n dimensions, with $[a, b]$ replaced by a closed bounded interval of $\mathbb{R}^n$; the proof is essentially the same.

A somewhat more complicated situation arises with *improper integrals*; here the interval of integration is infinite or the function f is unbounded. Some results are given in Problem 3.

We have seen that convenient conditions exist that allow the interchange of limit operations on Lebesgue integrable functions. (For example, see Problems 1–3 of Section 1.6.) The corresponding results for Riemann integrable functions are more complicated, basically because the limit of a sequence of Riemann integrable functions need not be Riemann integrable, even if the entire sequence is uniformly bounded (see Problem 4). Thus Riemann integrability of the limit function must be added as a hypothesis, and this is a serious limitation on the scope of the results.

Problems

1. The function defined on [0, 1] by $f(x) = 1$ if x is irrational, and $f(x) = 0$ if x is rational, is the standard example of a function that is Lebesgue integrable (it is 1 a.e.) but not Riemann integrable. But what is wrong with the following reasoning?

    If we consider the behavior of f on the irrationals, f assumes the constant value 1 and is therefore continuous. Since the rationals have Lebesgue measure 0, f is therefore continuous almost everywhere and hence is Riemann integrable.

2. Let f be a bounded real-valued function on the bounded closed interval $[a, b]$. Let F be an increasing right-continuous function on $[a, b]$ with corresponding Lebesgue–Stieltjes measure μ (defined on the Borel subsets of $[a, b]$).

    Define M_i, m_i, α, and β as in 1.7, and take

    $$U(P) = \sum_{i=1}^{n} M_i(F(x_i) - F(x_{i-1})) = \int_a^b \alpha \, d\mu,$$

    $$L(P) = \sum_{i=1}^{n} m_i(F(x_i) - F(x_{i-1})) = \int_a^b \beta \, d\mu,$$

    where $\int_a^b$ indicates that the integration is over $(a, b]$. If $\{P_k\}$ is a sequence of partitions with $|P_k| \to 0$ and P_{k+1} refining P_k, with α_k and β_k the upper and lower functions corresponding to P_k,

    $$\lim_{k \to \infty} U(P_k) = \int_a^b \alpha \, d\mu,$$

    $$\lim_{k \to \infty} L(P_k) = \int_a^b \beta \, d\mu,$$

    where $\alpha = \lim_{k \to \infty} \alpha_k$, $\beta = \lim_{k \to \infty} \beta_k$. If $U(P_k)$ and $L(P_k)$ approach the same limit $r_{ab}(f; F)$ (independent of the particular sequence of partitions), this number is called the *Riemann–Stieltjes integral* of f with respect to F on $[a, b]$, and f is said to be *Riemann–Stieltjes integrable* with respect to F on $[a, b]$.

(a) Show that f is Riemann–Stieltjes integrable iff f is continuous a.e. $[\mu]$ on $[a, b]$.

(b) Show that if f is Riemann–Stieltjes integrable, then f is integrable with respect to the completion of the measure μ, and the two integrals are equal.

3. If $f \colon \mathbb{R} \to \mathbb{R}$, the improper Riemann integral of f may be defined as

$$r(f) = \lim_{\substack{a \to -\infty \\ b \to \infty}} r_{ab}(f)$$

if the limit exists and is finite.

(a) Show that if f has an improper Riemann integral, it is continuous a.e. [Lebesgue measure] on $\mathbb{R}$, but not conversely.

(b) If f is nonnegative and has an improper Riemann integral, show that f is integrable with respect to the completion of Lebesgue measure, and the two integrals are equal. Give a counterexample to this result if the nonnegativity hypothesis is dropped.

4. Give an example of a sequence of functions f_n on $[a, b]$ such that each f_n is Riemann integrable, $|f_n| \leq 1$ for all n, $f_n \to f$ everywhere, but f is not Riemann integrable.

Note: References on measure and integration will be given at the end of Chapter 2.

2

FURTHER RESULTS IN MEASURE AND INTEGRATION THEORY

2.1 INTRODUCTION

This chapter consists of a variety of applications of the basic integration theory developed in Chapter 1. Perhaps the most important result is the Radon–Nikodym theorem, which is fundamental in modern probability theory and other parts of analysis. It will be instructive to consider a special case of this result before proceeding to the general theory. Suppose that F is a distribution function on $\mathbb{R}$, and assume that F has a jump of magnitude a_k at the point x_k, $k = 1, 2, \ldots$. Let us subtract out the discontinuities of F; specifically let μ_1 be a measure concentrated on $\{x_1, x_2, \ldots\}$, with $\mu_1\{x_k\} = a_k$ for all k, and let F_1 be a distribution function corresponding to μ_1. Then $G = F - F_1$ is a continuous distribution function, so that the corresponding Lebesgue–Stieltjes measure λ satisfies $\lambda\{x\} = 0$ for all x. Now in any "practical" case, we can write $G(x) = \int_{-\infty}^{x} f(t)\,dt$, $x \in R$, for some nonnegative Borel measurable function f (the way to find f is to differentiate G). It follows that $\lambda(B) = \int_B f(x)\,dx$ for all $B \in \mathscr{B}(\mathbb{R})$. To see this, observe that if $\lambda'(B) = \int_B f(x)\,dx$, then λ' is a measure on $\mathscr{B}(\mathbb{R})$ and $\lambda'(a, b] = G(b) - G(a)$; thus λ' is the Lebesgue–Stieltjes measure determined by G; in other words, $\lambda' = \lambda$.

It is natural to conjecture that if λ is a measure on $\mathscr{B}(\mathbb{R})$ and $\lambda\{x\} = 0$ for all x, then we can write $\lambda(B) = \int_B f(x)\,dx$, $B \in \mathscr{B}(\mathbb{R})$, for some nonnegative Borel measurable f. However, as found by Lebesgue, the conjecture is false unless the hypothesis is strengthened. Not only must we assume that λ assigns measure 0 to singletons, but in fact we must assume that λ is *absolutely continuous* with respect to Lebesgue measure μ, that is, if $\mu(B) = 0$, then $\lambda(B) = 0$. In general, λ may be represented as the sum of two measures λ_1 and λ_2, where λ_1 is absolutely continuous with respect to μ and λ_2 is *singular* with respect to μ, which means that λ_2 is concentrated on a set of Lebesgue measure 0. A simple example of a measure singular with respect to μ is one that is concentrated on a countable set; however, as we shall see, more complicated examples exist.

The first step in the development of the general Radon–Nikodym theorem is the Jordan–Hahn decomposition, which represents a countably additive set function as the difference of two measures.

Let $(\Omega, \mathscr{F}, \mu)$ be a measure space, h a Borel measurable function such that $\int_\Omega h\,d\mu$ exists. If $\lambda(A) = \int_A h\,d\mu$, $A \in \mathscr{F}$, then by 1.6.1, λ is a countably additive set function on $\mathscr{F}$. We call λ the *indefinite integral* of h (with respect to μ). If μ is Lebesgue measure and $A = [a, x]$, then $\lambda(A) = \int_a^x h(y)\,dy$, the familiar indefinite integral of calculus. As we noted after the proof of 1.6.1, λ is the difference of two measures, at least one of which is finite. We are going to show that any countably additive set function can be represented in this way. First, a preliminary result.

2.1.1 Theorem. Let λ be a countably additive extended real-valued set function on the σ-field $\mathscr{F}$ of subsets of Ω. Then λ assumes a maximum and a minimum value, that is, there are sets, $C, D \in \mathscr{F}$ such that

$$\lambda(C) = \sup\{\lambda(A): A \in \mathscr{F}\} \quad \text{and} \quad \lambda(D) = \inf\{\lambda(A): A \in \mathscr{F}\}.$$

Before giving the proof, let us look at some special cases. If λ is a measure, the result is trivial: take $C = \Omega$, $D = \emptyset$. If λ is the indefinite integral of h with respect to μ, we may write

$$\lambda(A) = \int_A h\,d\mu = \int_{A\cap\{\omega: h(\omega)\geq 0\}} h\,d\mu + \int_{A\cap\{\omega: h(\omega)<0\}} h\,d\mu.$$

Thus

$$\int_{\{\omega: h(\omega)<0\}} h\,d\mu \leq \lambda(A) \leq \int_{\{\omega: h(\omega)\geq 0\}} h\,d\mu.$$

Therefore we may take $C = \{\omega: h(\omega) \geq 0\}$, $D = \{\omega: h(\omega) < 0\}$.

PROOF. First consider the sup. We may assume that $\lambda < \infty$, for if $\lambda(A_0) = \infty$ we take $C = A_0$. Let $A_n \in \mathscr{F}$ with $\lambda(A_n) \to \sup \lambda$, and let $A = \bigcup_{n=1}^\infty A_n \in \mathscr{F}$.

For each n, we may partition A into 2^n disjoint subsets A_{nm}, where each A_{nm} is of the form $A_1^* \cap A_2^* \cap \cdots \cap A_n^*$, with A_j^* either A_j or $A - A_j$. For example, if $n = 3$, we have (with intersections written as products)

$$A = A_1A_2A_3 \cup A_1A_2A_3' \cup A_1A_2'A_3 \cup A_1A_2'A_3' \cup A_1'A_2A_3$$
$$\cup A_1'A_2A_3' \cup A_1'A_2'A_3 \cup A_1'A_2'A_3', \quad \text{where} \quad A_j' = A - A_j.$$

Let $B_n = \bigcup_m \{A_{nm}: \lambda(A_{nm}) \geq 0\}$; set $B_n = \emptyset$ if $\lambda(A_{nm}) < 0$ for all m. Now each A_n is a finite union of some of the A_{nm}, hence $\lambda(A_n) \leq \lambda(B_n)$ by definition of B_n. Also, if $n' > n$, each $A_{n'm'}$ is either a subset of a given A_{nm} or disjoint from it (for example, $A_1A_2'A_3' \subset A_1A_2'$, and $A_1A_2'A_3'$ is disjoint from $A_1A_2, A_1'A_2$, and $A_1'A_2'$). Thus $\bigcup_{k=n}^{r} B_k$ can be expressed as a union of B_n and sets E disjoint from B_n such that $\lambda(E) \geq 0$. [Note that, for example, if $\lambda(A_1A_2') = \lambda(A_1A_2'A_3) + \lambda(A_1A_2'A_3') \geq 0$, it may happen that $\lambda(A_1A_2'A_3) < 0$, $\lambda(A_1A_2'A_3') > 0$, so the sequence $\{B_n\}$ need not be monotone.] Consequently,

$$\lambda(A_n) \leq \lambda(B_n) \leq \lambda\left(\bigcup_{k=n}^{r} B_k\right) \to \lambda\left(\bigcup_{k=n}^{\infty} B_k\right) \quad \text{as} \quad r \to \infty \quad \text{by 1.2.7(a).}$$

Let $C = \lim_n \sup B_n = \bigcap_{n=1}^{\infty}\bigcup_{k=n}^{\infty} B_k$. Now $\bigcup_{k=n}^{\infty} B_k \downarrow C$, and $0 \leq \lambda(\bigcup_{k=n}^{\infty} B_k) < \infty$ for all n. By 1.2.7(b), $\lambda(\bigcup_{k=n}^{\infty} B_k) \to \lambda(C)$ as $n \to \infty$. Thus

$$\sup \lambda = \lim_{n \to \infty} \lambda(A_n) \leq \lim_{n \to \infty} \lambda\left(\bigcup_{k=n}^{\infty} B_k\right) = \lambda(C) \leq \sup \lambda;$$

hence $\lambda(C) = \sup \lambda$. The above argument applied to $-\lambda$ yields $D \in \mathscr{F}$ with $\lambda(D) = \inf \lambda$. $\square$

We now prove the main theorem of this section.

2.1.2   Jordan–Hahn Decomposition Theorem. Let λ be a countably additive extended real-valued set function on the σ-field $\mathscr{F}$. Define

$$\lambda^+(A) = \sup\{\lambda(B) : B \in \mathscr{F}, B \subset A\},$$
$$\lambda^-(A) = -\inf\{\lambda(B) : B \in \mathscr{F}, B \subset A\}.$$

Then λ^+ and λ^- are measures on $\mathscr{F}$ and $\lambda = \lambda^+ - \lambda^-$.

PROOF.   We may assume λ never takes on the value $-\infty$. For if $-\infty$ belongs to the range of λ, $+\infty$ does not, by definition of a countably additive set function. Thus $-\lambda$ never takes on the value $-\infty$. But $(-\lambda)^+ = \lambda^-$ and $(-\lambda)^- = \lambda^+$, so that if the theorem is proved for $-\lambda$ it holds for λ as well.

Let D be a set on which λ attains its minimum, as in 2.1.1. Since $\lambda(\emptyset) = 0$, we have $-\infty < \lambda(D) \leq 0$. We claim that

$$\lambda(A \cap D) \leq 0, \qquad \lambda(A \cap D^c) \geq 0 \qquad \text{for all} \qquad A \in \mathscr{F}. \qquad (1)$$

For if $\lambda(A \cap D) > 0$, then $\lambda(D) = \lambda(A \cap D) + \lambda(A^c \cap D)$. Since $\lambda(D)$ is finite, so are $\lambda(A \cap D)$ and $\lambda(A^c \cap D)$; hence $\lambda(A^c \cap D) = \lambda(D) - \lambda(A \cap D)$

$< \lambda(D)$, contradicting the fact that $\lambda(D) = \inf \lambda$. If $\lambda(A \cap D^c) < 0$, then $\lambda(D \cup (A \cap D^c)) = \lambda(D) + \lambda(A \cap D^c) < \lambda(D)$, a contradiction.

We now show that

$$\lambda^+(A) = \lambda(A \cap D^c), \qquad \lambda^-(A) = -\lambda(A \cap D). \tag{2}$$

The theorem will follow from this. We have, for $B \in \mathscr{F}$, $B \subset A$,

$$\begin{aligned}
\lambda(B) &= \lambda(B \cap D) + \lambda(B \cap D^c) \\
&\leq \lambda(B \cap D^c) \qquad \text{by (1)} \\
&\leq \lambda(B \cap D^c) + \lambda((A - B) \cap D^c) \\
&= \lambda(A \cap D^c).
\end{aligned}$$

Thus $\lambda^+(A) \leq \lambda(A \cap D^c)$. But $\lambda(A \cap D^c) \leq \lambda^+(A)$ by definition of λ^+, proving the first assertion. Similarly,

$$\begin{aligned}
\lambda(B) &= \lambda(B \cap D) + \lambda(B \cap D^c) \\
&\geq \lambda(B \cap D) \\
&\geq \lambda(B \cap D) + \lambda((A - B) \cap D) \\
&= \lambda(A \cap D).
\end{aligned}$$

Hence $-\lambda^-(A) \geq \lambda(A \cap D)$. But $\lambda(A \cap D) \geq -\lambda^-(A)$ by definition of λ^-, completing the proof. $\square$

2.1.3 Corollaries. Let λ be a countably additive extended real-valued set function on the σ-field $\mathscr{F}$.

(a) The set function λ is the difference of two measures, at least one of which is finite.

(b) If λ is finite ($\lambda(A)$ is never $\pm\infty$ for any $A \in \mathscr{F}$), then λ is bounded.

(c) There is a set $D \in \mathscr{F}$ such that $\lambda(A \cap D) \leq 0$ and $\lambda(A \cap D^c) \geq 0$ for all $A \in \mathscr{F}$.

(d) If D is any set in $\mathscr{F}$ such that $\lambda(A \cap D) \leq 0$ and $\lambda(A \cap D^c) \geq 0$ for all $A \in \mathscr{F}$, then $\lambda^+(A) = \lambda(A \cap D^c)$ and $\lambda^-(A) = -\lambda(A \cap D)$ for all $A \in \mathscr{F}$.

(e) If E is another set in $\mathscr{F}$ such that $\lambda(A \cap E) \leq 0$ and $\lambda(A \cap E^c) \geq 0$ for all $A \in \mathscr{F}$, then $|\lambda|(D \triangle E) = 0$, where $|\lambda| = \lambda^+ + \lambda^-$.

PROOF. (a) If $\lambda > -\infty$, then in 2.1.2, λ^{-1} is finite; if $\lambda < +\infty$, λ^+ is finite [see Eq. (2)].

(b) In 2.1.2, λ^+ and λ^- are both finite; hence for any $A \in \mathscr{F}$, $|\lambda(A)| \leq \lambda^+(\Omega) + \lambda^-(\Omega) < \infty$.

(c) This follows from (1) of 2.1.2.

(d) Repeat the part of the proof of 2.1.2 after Eq. (2).

(e) By (d), $\lambda^+ (A) = \lambda (A \cap D^c)$, $A \in \mathscr{F}$; take $A = D \cap E^c$ to obtain $\lambda^+ (D \cap E^c) = 0$. Also by (d), $\lambda^+ (A) = \lambda(A \cap E^c)$, $A \in \mathscr{F}$; take $A = D^c \cap E$ to obtain $\lambda^+ (D^c \cap E) = 0$. Therefore $\lambda^+ (D \, \Delta \, E) = 0$. The same argument using $\lambda^- (A) = -\lambda(A \cap D) = -\lambda(A \cap E)$ shows that $\lambda^- (D \, \Delta \, E) = 0$. The result follows. □

Corollary 2.1.3(d) is often useful in finding the Jordan–Hahn decomposition of a particular set function (see Problems 1 and 2).

2.1.4   Terminology. We call λ^+ the *upper variation* or *positive part* of λ, λ^- the *lower variation* or *negative part*, $|\lambda| = \lambda^+ + \lambda^-$ the *total variation*. Since $\lambda = \lambda^+ - \lambda^-$, it follows that $|\lambda(A)| \leq |\lambda|(A)$, $A \in \mathscr{F}$. For a sharper result, see Problem 4.

Note that if $A \in \mathscr{F}$, then $|\lambda|(A) = 0$ iff $\lambda(B) = 0$ for all $B \in \mathscr{F}$, $B \subset A$.

The phrase *signed measure* is sometimes used for the difference of two measures. By 2.1.3(a), this is synonymous (on a σ-field) with *countably additive set function*.

Problems

1. Let P be an arbitrary probability measure on $\mathscr{B}(\mathbb{R})$, and let Q be point mass at 0, that is, $Q(B) = 1$ if $0 \in B$, $Q(B) = 0$ if $0 \notin B$. Find the Jordan–Hahn decomposition of the signed measure $\lambda = P - Q$.

2. Let $\lambda(A) = \int_A f \, d\mu$, A in the σ-field $\mathscr{F}$, where $\int_\Omega f \, d\mu$ exists; thus λ is a signed measure on $\mathscr{F}$. Show that

$$\lambda^+ (A) = \int_A f^+ \, d\mu, \qquad \lambda^- (A) = \int_A f^- \, d\mu, \qquad |\lambda|(A) = \int_A |f| \, d\mu.$$

3. If a signed measure λ on the σ-field $\mathscr{F}$ is the difference of two measures λ_1 and λ_2, show that $\lambda_1 \geq \lambda^+$, $\lambda_2 \geq \lambda^-$.

4. Let λ be a signed measure on the σ-field $\mathscr{F}$. Show that $|\lambda|(A) = \sup\{\sum_{i=1}^n |\lambda(E_i)|: E_1, E_2, \ldots, E_n$ disjoint measurable subsets of A, $n = 1, 2, \ldots\}$. Consequently, if λ_1 and λ_2 are signed measures on $\mathscr{F}$, then $|\lambda_1 + \lambda_2| \leq |\lambda_1| + |\lambda_2|$.

2.2   RADON–NIKODYM THEOREM AND RELATED RESULTS

If $(\Omega, \mathscr{F}, \mu)$ is a measure space, then $\lambda(A) = \int_A g \, d\mu$, $A \in \mathscr{F}$, defines a signed measure if $\int_\Omega g \, d\mu$ exists. Furthermore, if $A \in \mathscr{F}$ and $\mu(A) = 0$, then $\lambda(A) = 0$. For $g I_A = 0$ on A^c, so that $g I_A = 0$ a.e. $[\mu]$, and the result follows from 1.6.5(a).

If μ is a measure on the σ-field $\mathscr{F}$, and λ is a signed measure on $\mathscr{F}$, we say that λ is *absolutely continuous* with respect to μ (notation $\lambda \ll \mu$) iff $\mu(A) = 0$ implies $\lambda(A) = 0$ $(A \in \mathscr{F})$. Thus if λ is an indefinite integral with respect to μ, then $\lambda \ll \mu$. The Radon–Nikodym theorem is an assertion in the converse direction; if $\lambda \ll \mu$ (and μ is σ-finite on $\mathscr{F}$), then λ is an indefinite integral with respect to μ. As we shall see, large areas of analysis are based on this theorem.

2.2.1   Radon–Nikodym Theorem.   Let μ be a σ-finite measure and λ a signed measure on the σ-field $\mathscr{F}$ of subsets of Ω. Assume that λ is absolutely continuous with respect to μ. Then there is a Borel measurable function $g\colon \Omega \to \overline{\mathbb{R}}$ such that

$$\lambda(A) = \int_A g \, d\mu \qquad \text{for all} \qquad A \in \mathscr{F}.$$

If h is another such function, then $g = h$ a.e. $[\mu]$.

PROOF.   The uniqueness statement follows from 1.6.11. We break the existence proof into several parts.

(a)   Assume λ and μ are finite measures.

Let $\mathscr{S}$ be the set of all nonnegative μ-integrable functions f such that $\int_A f \, d\mu \leq \lambda(A)$ for all $A \in \mathscr{F}$. Partially order $\mathscr{S}$ by calling $f \leq g$ iff $f \leq g$ a.e. $[\mu]$. Let $s = \sup\{\int_\Omega f \, d\mu \colon f \in \mathscr{S}\} \leq \lambda(\Omega) < \infty$. ($\mathscr{S}$ is a nonempty collection since it contains the zero function.) We are going to produce a maximal element of $\mathscr{S}$, and we first note that if f and g belong to $\mathscr{S}$ and $h = \max(f, g)$ then $h \in \mathscr{S}$. For if B is the set on which $f \geq g$ and C the set on which $f < g$, then

$$\int_A h \, d\mu = \int_{A \cap B} f \, d\mu + \int_{A \cap C} g \, d\mu$$
$$\leq \lambda(A \cap B) + \lambda(A \cap C) = \lambda(A).$$

Now let $f_1, f_2, \ldots$ be a sequence in $\mathscr{S}$ such that $\int_\Omega f_n \, d\mu \to s$, and let $g_n = \max(f_1, \ldots, f_n)$. Then $g_n \in \mathscr{S}$, g_n increases to a limit g, and $\int_\Omega g \, d\mu = s$ (use the monotone convergence theorem and the fact that $g_n \geq f_n$). We claim that $g \in \mathscr{S}$; since $\int_\Omega g \, d\mu = s$, it will follow that g is a maximal element of $\mathscr{S}$. To show that g belongs to $\mathscr{S}$, let A be any set in $\mathscr{F}$. Then

$$0 \leq g_n I_A \uparrow g I_A, \qquad \text{so} \qquad \int_A g_n \, d\mu = \int_\Omega g_n I_A \, d\mu \uparrow \int_\Omega g I_A \, d\mu = \int_A g \, d\mu.$$

But $\int_A g_n \, d\mu \leq \lambda(A)$ for all n, and therefore $\int_A g \, d\mu \leq \lambda(A)$.

Now that we have our maximal element g, let $\lambda_1(A) = \lambda(A) - \int_A g \, d\mu$, $g \in \mathscr{F}$. Then λ_1 is a measure, $\lambda_1 \ll \mu$, and $\lambda_1(\Omega) < \infty$. If λ_1 is not identically 0, then $\lambda_1(\Omega) > 0$, hence

$$\mu(\Omega) - k\lambda_1(\Omega) < 0 \qquad \text{for some} \qquad k > 0. \tag{1}$$

Apply 2.1.3(c) to the signed measure $\mu - k\lambda_1$ to obtain $D \in \mathscr{F}$ such that for all $A \in \mathscr{F}$,

$$\mu(A \cap D) - k\lambda_1(A \cap D) \leq 0, \tag{2}$$

and

$$\mu(A \cap D^c) - k\lambda_1(A \cap D^c) \geq 0. \tag{3}$$

We claim that $\mu(D) > 0$. For if $\mu(D) = 0$, then $\lambda(D) = 0$ by absolute continuity, and therefore $\lambda_1(D) = 0$ by definition of λ_1. Take $A = \Omega$ in (3) to obtain

$$0 \leq \mu(D^c) - k\lambda_1(D^c)$$
$$= \mu(\Omega) - k\lambda_1(\Omega) \qquad \text{since} \qquad \mu(D) = \lambda_1(D) = 0$$
$$< 0 \qquad \text{by} \qquad (1),$$

a contradiction. Define $h(\omega) = 1/k$, $\omega \in D$; $h(\omega) = 0$, $\omega \notin D$. If $A \in \mathscr{F}$,

$$\int_A h \, d\mu = \frac{1}{k}\mu(A \cap D) \leq \lambda_1(A \cap D) \qquad \text{by} \qquad (2)$$

$$\leq \lambda_1(A) = \lambda(A) - \int_A g \, d\mu.$$

Thus $\int_A (h + g) \, d\mu \leq \lambda(A)$. But $h + g > g$ on the set D, with $\mu(D) > 0$, contradicting the maximality of g. Thus $\lambda_1 \equiv 0$, and the result follows.

(b)   Assume μ is a finite measure, λ a σ-finite measure.

Let Ω be the disjoint union of sets A_n with $\lambda(A_n) < \infty$, and let $\lambda_n(A) = \lambda(A \cap A_n)$, $A \in \mathscr{F}$, $n = 1, 2, \ldots$. By part (a) we find a nonnegative Borel measurable g_n with $\lambda_n(A) = \int_A g_n \, d\mu$, $A \in \mathscr{F}$. Thus $\lambda(A) = \int_A g \, d\mu$, where $g = \sum_n g_n$.

(c)   Assume μ is a finite measure, λ an arbitrary measure.

Let $\mathscr{C}$ be the class of sets $C \in \mathscr{F}$ such that λ on C (that is, λ restricted to $\mathscr{F}_c = \{A \cap C: A \in \mathscr{F}\}$) is σ-finite; note that $\emptyset \in \mathscr{C}$, so $\mathscr{C}$ is not empty. Let $s = \sup\{\mu(A): A \in \mathscr{C}\}$ and pick $C_n \in \mathscr{C}$ with $\mu(C_n) \to s$. If $C = \bigcup_{n=1}^{\infty} C_n$, then $C \in \mathscr{C}$ by definition of $\mathscr{C}$, and $s \geq \mu(C) \geq \mu(C_n) \to s$; hence $\mu(C) = s$.

By part (b), there is a nonnegative $g': C \to \overline{\mathbb{R}}$, measurable relative to $\mathscr{F}_c$ and $\mathscr{B}(\overline{\mathbb{R}})$, such that

$$\lambda(A \cap C) = \int_{A \cap C} g' \, d\mu \qquad \text{for all} \qquad A \in \mathscr{F}.$$

Now consider an arbitrary set $A \in \mathscr{F}$.

Case 1:   Let $\mu(A \cap C^c) > 0$. Then $\lambda(A \cap C^c) = \infty$, for if $\lambda(A \cap C^c) < \infty$, then $C \cup (A \cap C^c) \in \mathscr{C}$; hence

$$s \geq \mu(C \cup (A \cup C^c)) = \mu(C) + \mu(A \cap C^c) > \mu(C) = s,$$

a contradiction.

Case 2:   Let $\mu(A \cap C^c) = 0$. Then $\lambda(A \cap C^c) = 0$ by absolute continuity. Thus in either case, $\lambda(A \cap C^c) = \int_{A \cap C^c} \infty \, d\mu$. It follows that

$$\lambda(A) = \lambda(A \cap C) + \lambda(A \cap C^c) = \int_A g \, d\mu,$$

where $g = g'$ on C, $g = \infty$ on C^c.

(d)   Assume μ is a σ-finite measure, λ an arbitrary measure.

Let Ω be the union of disjoint sets A_n with $\mu(A_n) < \infty$. By part (c), there is a nonnegative function $g_n\colon A_n \to \overline{\mathbb{R}}$, measurable with respect to $\mathscr{F}_{A_n}$ and $\mathscr{B}(\overline{\mathbb{R}})$, such that $\lambda(A \cap A_n) = \int_{A \cap A_n} g_n \, d\mu$, $A \in \mathscr{F}$. We may write this as $\lambda(A \cap A_n) = \int_A g_n \, d\mu$ where $g_n(\omega)$ is taken as 0 for $\omega \notin A_n$. Thus $\lambda(A) = \sum_n \lambda(A \cap A_n) = \sum_n \int_A g_n \, d\mu = \int_A g \, d\mu$, where $g = \sum_n g_n$.

(e)   Assume μ is a σ-finite measure, λ an arbitrary signed measure.

Write $\lambda = \lambda^+ - \lambda^-$ where, say, λ^- is finite. By part (d), there are nonnegative Borel measurable functions g_1 and g_2 such that

$$\lambda^+(A) = \int_A g_1 \, d\mu, \qquad \lambda^-(A) = \int_A g_2 \, d\mu, \qquad A \in \mathscr{F}.$$

Since λ^- is finite, g_2 is integrable; hence by 1.6.3 and 1.6.6(a), $\lambda(A) = \int_A (g_1 - g_2) \, d\mu$. $\square$

2.2.2   Corollaries.   Under the hypothesis of 2.2.1,

(a)   If λ is finite, then g is μ-integrable, hence finite a.e. $[\mu]$.

(b)   If $|\lambda|$ is σ-finite, so that Ω can be expressed as a countable union of sets A_n such that $|\lambda|(A_n)$ is finite (equivalently $\lambda(A_n)$ is finite), then g is finite a.e. $[\mu]$.

(c)   If λ is a measure, then $g \geq 0$ a.e. $[\mu]$.

PROOF. All results may be obtained by examining the proof of 2.2.1. Alternatively, we may proceed as follows:

(a) Observe that $\lambda(\Omega) = \int_\Omega g \, d\mu$, finite by hypothesis.

(b) By (a), g is finite a.e. $[\mu]$ on each A_n, hence finite a.e. $[\mu]$ on Ω.

(c) Let $A = \{\omega: g(\omega) < 0\}$; then $0 \leq \lambda(A) = \int_A g \, d\mu \leq 0$. Thus $-gI_A$ is a nonnegative function whose integral is 0, so that $gI_A = 0$ a.e. $[\mu]$ by 1.6.6(b). Since $gI_A < 0$ on A, we must have $\mu(A) = 0$. $\square$

If $\lambda(A) = \int_A g \, d\mu$ for each $A \in \mathscr{F}$, g is called the *Radon–Nikodym derivative* or *density* of λ with respect to μ, written $d\lambda/d\mu$. If μ is Lebesgue measure, then g is often called simply the density of λ.

There are converse assertions to 2.2.2(a) and (c). Suppose that

$$\lambda(A) = \int_A g \, d\mu, \qquad A \in \mathscr{F},$$

where $\int_\Omega g \, d\mu$ is assumed to exist. If g is μ-integrable, then λ is finite; if $g \geq 0$ a.e. $[\mu]$, then $\lambda \geq 0$, so that λ is a measure. (Note that σ-finiteness of μ is not assumed.) However, the converse to 2.2.2(b) is false; if g is finite a.e. $[\mu]$, $|\lambda|$ need not be σ-finite (see Problem 1).

We now consider a property that is in a sense opposite to absolute continuity.

2.2.3 Definitions. Let μ_1 and μ_2 be measures on the σ-field $\mathscr{F}$. We say that μ_1 is *singular* with respect to μ_2 (written $\mu_1 \perp \mu_2$) iff there is a set $A \in \mathscr{F}$ such that $\mu_1(A) = 0$ and $\mu_2(A^c) = 0$; note μ_1 is singular with respect to μ_2 iff μ_2 is singular with respect to μ_1, so we may say that μ_1 and μ_2 are *mutually singular*. If λ_1 and λ_2 are signed measures on $\mathscr{F}$, we say that λ_1 and λ_2 are mutually singular iff $|\lambda_1| \perp |\lambda_2|$.

If $\mu_1 \perp \mu_2$, with $\mu_1(A) = \mu_2(A^c) = 0$, then μ_2 only assigns positive measure to subsets of A. Thus μ_2 concentrates its total effect on a set of μ_1-measure 0; on the other hand, if $\mu_2 \ll \mu_1$, μ_2 can have no effect on sets of μ_1-measure 0.

If λ is a signed measure with positive part λ^+ and negative part λ^-, we have $\lambda^+ \perp \lambda^-$ by 2.1.3(c) and (d).

Before establishing some facts about absolute continuity and singularity, we need the following lemma. Although the proof is quite simple, the result is applied very often in analysis, especially in probability theory.

2.2.4 Borel–Cantelli Lemma. If $A_1, A_2, \ldots \in \mathscr{F}$ and $\sum_{n=1}^\infty \mu(A_n) < \infty$, then $\mu(\limsup_n A_n) = 0$.

PROOF.   Recall that $\limsup_n A_n = \bigcap_{n=1}^{\infty} \bigcup_{k=n}^{\infty} A_k$; hence

$$\mu(\limsup_n A_n) \leq \mu\left(\bigcup_{k=n}^{\infty} A_k\right) \qquad \text{for all } n$$

$$\leq \sum_{k=n}^{\infty} \mu(A_k) \to 0 \qquad \text{as } n \to \infty. \quad \square$$

2.2.5   Lemma. Let μ be a measure, and λ_1 and λ_2 signed measures, on the σ-field $\mathscr{F}$.
   (a)   If $\lambda_1 \perp \mu$ and $\lambda_2 \perp \mu$, then $\lambda_1 + \lambda_2 \perp \mu$.
   (b)   If $\lambda_1 \ll \mu$, then $|\lambda_1| \ll \mu$, and conversely.
   (c)   If $\lambda_1 \ll \mu$ and $\lambda_2 \perp \mu$, then $\lambda_1 \perp \lambda_2$.
   (d)   If $\lambda_1 \ll \mu$ and $\lambda_1 \perp \mu$, then $\lambda_1 \equiv 0$.
   (e)   If λ_1 is finite, then $\lambda_1 \ll \mu$ iff $\lim_{\mu(A) \to 0} \lambda_1(A) = 0$.

PROOF.   (a)   Let $\mu(A) = \mu(B) = 0$, $|\lambda_1|(A^c) = |\lambda_2|(B^c) = 0$. Then $\mu(A \cup B) = 0$ and $\lambda_1(C) = \lambda_2(C) = 0$ for every $C \in \mathscr{F}$ with $C \subset A^c \cap B^c$; hence $|\lambda_1 + \lambda_2|[(A \cup B)^c] = 0$.
   (b)   Let $\mu(A) = 0$. If $\lambda_1^+(A) > 0$, then (see 2.1.2) $\lambda_1(B) > 0$ for some $B \subset A$; since $\mu(B) = 0$, this is a contradiction. It follows that λ_1^+, and similarly λ_1^-, are absolutely continuous with respect to μ; hence $|\lambda_1| \ll \mu$. (This may also be proved using Section 2.1, Problem 4.) The converse is clear.
   (c)   Let $\mu(A) = 0$, $|\lambda_2|(A^c) = 0$. By (b), $|\lambda_1|(A) = 0$, so $|\lambda_1| \perp |\lambda_2|$.
   (d)   By (c), $\lambda_1 \perp \lambda_1$; hence for some $A \in \mathscr{F}$, $|\lambda_1|(A) = |\lambda_1|(A^c) = 0$. Thus $|\lambda_1|(\Omega) = 0$.
   (e)   If $\mu(A_n) \to 0$ implies $\lambda_1(A_n) \to 0$, and $\mu(A) = 0$, set $A_n \equiv A$ to conclude that $\lambda_1(A) = 0$, so $\lambda_1 \ll \mu$.

Conversely, let $\lambda_1 \ll \mu$.
If $\lim_{\mu(A) \to 0} |\lambda_1|(A) \neq 0$ we can find, for some $\varepsilon > 0$, sets $A_n \in \mathscr{F}$ with $\mu(A_n) < 2^{-n}$ and $|\lambda_1|(A_n) \geq \varepsilon$ for all n. Let $A = \lim_n \sup A_n$; by 2.2.4, $\mu(A) = 0$. But $|\lambda_1|\left(\bigcup_{k=n}^{\infty} A_k\right) \geq |\lambda_1|(A_n) \geq \varepsilon$ for all n; hence by 1.2.7(b), $|\lambda_1|(A) \geq \varepsilon$, contradicting (b). Thus $\lim_{\mu(A) \to 0} |\lambda_1|(A) = 0$, and the result follows since $|\lambda_1(A)| \leq |\lambda_1|(A)$. $\quad \square$

If λ_1 is an indefinite integral with respect to μ (hence $\lambda_1 \ll \mu$), then 2.2.5(e) has an easier proof. If $\lambda_1(A) = \int_A f \, d\mu$, $A \in \mathscr{F}$, then

$$\int_A |f| \, d\mu = \int_{A \cap \{|f| \leq n\}} |f| \, d\mu + \int_{A \cap \{|f| > n\}} |f| \, d\mu \leq n\mu(A) + \int_{\{|f| > n\}} |f| \, d\mu.$$

By 1.6.1 and 1.2.7(b), $\int_{\{|f|>n\}} |f| \, d\mu$ may be made less than $\varepsilon/2$ for large n, say $n \geq N$. Fix $n = N$ and take $\mu(A) < \varepsilon/2N$, so that $\int_A |f| \, d\mu < \varepsilon$.

If μ is a measure and λ a signed measure on the σ-field $\mathscr{F}$, λ may be neither absolutely continuous nor singular with respect to μ. However, if $|\lambda|$ is σ-finite, the two concepts of absolute continuity and singularity are adequate to describe the relation between λ and μ, in the sense that λ can be written as the sum of two signed measures, one absolutely continuous and the other singular with respect to μ.

2.2.6 Lebesgue Decomposition Theorem. Let μ be a measure on the σ-field $\mathscr{F}$, λ a σ-finite signed measure (that is, $|\lambda|$ is σ-finite). Then λ has a unique decomposition as $\lambda_1 + \lambda_2$, where λ_1 and λ_2 are signed measures such that $\lambda_1 \ll \mu$, $\lambda_2 \perp \mu$.

PROOF.[1]   First assume that λ is a finite measure, and let $\mathscr{C} = \{A \in \mathscr{F}: \mu(A) = 0\}$ and $s = \sup\{\lambda(A): A \in \mathscr{C}\} \leq \lambda(\Omega) < \infty$. If $A_1, A_2, \ldots$ is a sequence of sets in $\mathscr{C}$ such that $\lambda(A_n) \to s$, then $A^* = \bigcup_{n=1}^{\infty} A_n \in \mathscr{C}$ and $\lambda(A^*) = s$. We claim that $\lambda(B - A^*) = 0$ for every $B \in \mathscr{C}$. For if $B \in \mathscr{C}$ and $\lambda(B - A^*) > 0$, then

$$\lambda(A^* \cup B) = \lambda(A^*) + \lambda(B - A^*) > s,$$

a contradiction. Now define

$$\lambda_1(A) = \lambda(A - A^*), \qquad \lambda_2(A) = \lambda(A \cap A^*), \qquad A \in \mathscr{F}.$$

If $\mu(B) = 0$ then $B \in \mathscr{C}$, so $\lambda_1(B) = \lambda(B - A^*) = 0$, hence $\lambda_1 \ll \mu$. Since $\mu(A^*) = 0$ and $\lambda_2(A^{*c}) = 0$, we have $\lambda_2 \perp \mu$.

If λ is a σ-finite measure, let Ω be the disjoint union of sets A_n such that $\lambda(A_n) < \infty$. If $\lambda_n(A) = \lambda(A \cap A_n)$, $A \in \mathscr{F}$, then by the above argument there are finite measures $\lambda_{n1} \ll \mu$ and $\lambda_{n2} \perp \mu$ such that $\lambda_n = \lambda_{n1} + \lambda_{n2}$. Sum on n to get $\lambda = \lambda_1 + \lambda_2$ with $\lambda_1 \ll \mu$, $\lambda_2 \perp \mu$.

Now if λ is a σ-finite signed measure, the above argument applied to λ^+ and λ^- proves the existence of the desired decomposition.

To prove uniqueness, first assume λ finite. If $\lambda = \lambda_1 + \lambda_2 = \lambda_1' + \lambda_2'$, where $\lambda_1, \lambda_1' \ll \mu$, $\lambda_2, \lambda_2' \perp \mu$, then $\lambda_1 - \lambda_1' = \lambda_2' - \lambda_2$ is both absolutely continuous and singular with respect to μ; hence it is identically 0 by 2.2.5(d). If λ is σ-finite and Ω is the disjoint union of sets A_n with $|\lambda|(A_n) < \infty$, apply the above argument to each A_n and put the results together to obtain uniqueness of λ_1 and λ_2. □

[1]   J. K. Brooks, American Mathematical Monthly, June–July 1971.

Note that as a corollary of the proof, if λ is a σ-finite measure (as opposed to a σ-finite signed measure), then λ_1 and λ_2 are measures. If λ is a probability measure, then $\lambda_1, \lambda_2 \leq 1$.

Problems

1. Give an example of a measure μ and a nonnegative finite-valued Borel measurable function g such that the measure λ defined by $\lambda(A) = \int_A g \, d\mu$ is not σ-finite.

2. If $\lambda(A) = \int_A g \, d\mu$, $A \in \mathscr{F}$, and g is μ-integrable, we know that λ is finite; in particular, $A = \{\omega: g(\omega) \neq 0\}$ has finite λ-measure. Show that A has σ-finite μ-measure, that is, it is a countable union of sets of finite μ-measure. Give an example to show that $\mu(A)$ need not be finite.

3. Give an example in which the conclusion of the Radon–Nikodym theorem fails; in other words, $\lambda \ll \mu$ but there is no Borel measurable g such that $\lambda(A) = \int_A g \, d\mu$ for all $A \in \mathscr{F}$. Of course μ cannot be σ-finite.

4. (*A chain rule*) Let $(\Omega, \mathscr{F}, \mu)$ be a measure space, and g a nonnegative Borel measurable function on Ω. Define a measure λ on $\mathscr{F}$ by

$$\lambda(A) = \int_A g \, d\mu, \qquad A \in \mathscr{F}.$$

Show that if f is a Borel measurable function on Ω,

$$\int_\Omega f \, d\lambda = \int_\Omega fg \, d\mu$$

in the sense that if one of the integrals exists, so does the other, and the two integrals are equal. (Intuitively, $d\lambda/d\mu = g$, so that $d\lambda = g \, d\mu$.)

5. Show that Theorem 2.2.5(e) fails if λ_1 is not finite.

6. (*Complex measures*) If $(\Omega, \mathscr{F})$ is a measurable space, a *complex measure* λ on $\mathscr{F}$ is a countably additive complex-valued set function; that is, $\lambda = \lambda_1 + i\lambda_2$, where λ_1 and λ_2 are finite signed measures.

    (a) Define the *total variation* of λ as

$$|\lambda|(A) = \sup \left\{ \sum_{i=1}^{n} |\lambda(E_i)|: E_1, \ldots, E_n \right.$$

$$\left. \text{disjoint measurable subsets of } A, n = 1, 2, \ldots \right\}.$$

Show that $|\lambda|$ is a measure on $\mathscr{F}$. (The definition is consistent with the earlier notion of total variation of a signed measure; see Section 2.1, Problem 4.)

In the discussion below, λ's, with various subscripts, denote arbitrary measures (real signed measures or complex measures), and μ denotes a nonnegative real measure. We define $\lambda \ll \mu$ in the usual way; if $A \in \mathscr{F}$ and $\mu(A) = 0$, then $\lambda(A) = 0$. Define $\lambda_1 \perp \lambda_2$ iff $|\lambda_1| \perp |\lambda_2|$. Establish the following results.

(b) $|\lambda_1 + \lambda_2| \le |\lambda_1| + |\lambda_2|; |a\lambda| = |a| \, |\lambda|$ for any complex number a. In particular if $\lambda = \lambda_1 + i\lambda_2$ is a complex measure, then

$$|\lambda| \le |\lambda_1| + |\lambda_2|; \qquad \text{hence} \qquad |\lambda|(\Omega) < \infty \qquad \text{by 2.1.3(b)}.$$

(c) If $\lambda_1 \perp \mu$ and $\lambda_2 \perp \mu$, then $\lambda_1 + \lambda_2 \perp \mu$.
(d) If $\lambda \ll \mu$, then $|\lambda| \ll \mu$, and conversely.
(e) If $\lambda_1 \ll \mu$ and $\lambda_2 \perp \mu$, then $\lambda_1 \perp \lambda_2$.
(f) If $\lambda \ll \mu$ and $\lambda \perp \mu$, then $\lambda = 0$.
(g) If λ is finite, then $\lambda \ll \mu$ iff $\lim_{\mu(A) \to 0} \lambda(A) = 0$.

2.3 APPLICATIONS TO REAL ANALYSIS

We are going to apply the concepts of the previous section to some problems involving functions of a real variable. If $[a, b]$ is a closed bounded interval of reals and $f: [a, b] \to \mathbb{R}$, f is said to be *absolutely continuous* iff for each $\varepsilon > 0$ there is a $\delta > 0$ such that for all positive integers n and all families $(a_1, b_1), \ldots, (a_n, b_n)$ of disjoint open subintervals of $[a, b]$ of total length at most δ, we have

$$\sum_{i=1}^{n} |f(b_i) - f(a_i)| \le \varepsilon.$$

It is immediate that this property holds also for countably infinite families of disjoint open intervals of total length at most δ. It also follows from the definition that f is continuous.

We can connect absolute continuity of functions with the earlier notion of absolute continuity of measures, as follows.

2.3.1 Theorem. Suppose that F and G are distribution functions on $[a, b]$, with corresponding (finite) Lebesgue–Stieltjes measures μ_1 and μ_2. Let $f = F - G$, $\mu = \mu_1 - \mu_2$, so that μ is a finite signed measure on $\mathscr{B}[a, b]$, with $\mu(x, y] = f(y) - f(x)$, $x < y$. If m is Lebesgue measure on $\mathscr{B}[a, b]$, then $\mu \ll m$ iff f is absolutely continuous.

PROOF. Assume $\mu \ll m$. If $\varepsilon > 0$, by 2.2.5(b) and (e), there is a $\delta > 0$ such that $m(A) \le \delta$ implies $|\mu|(A) \le \varepsilon$. Thus if $(a_1, b_1), \ldots, (a_n, b_n)$ are disjoint open intervals of total length at most δ,

$$\sum_{i=1}^{n} |f(b_i) - f(a_i)| = \sum_{i=1}^{n} |\mu(a_i, b_i]|$$

$$\leq \sum_{i=1}^{n} |\mu|(a_i, b_i] \leq |\mu|(A) \leq \varepsilon.$$

(Note that $\mu\{b_i\} = 0$ since $\mu \ll m$.) Therefore f is absolutely continuous.

Now assume f absolutely continuous; if $\varepsilon > 0$, choose $\delta > 0$ as in the definition of absolute continuity. If $m(A) = 0$, we must show that $\mu(A) = 0$. We use 1.4.11:

$$m(A) = \inf\{m(V) \colon V \supset A, \qquad V \text{ open}\},$$

$$\mu_i(A) = \inf\{\mu_i(V) \colon V \supset A, \qquad V \text{ open}\}, \qquad i = 1, 2.$$

(This problem assumes that the measures are defined on $\mathscr{B}(\mathbb{R})$ rather than $\mathscr{B}[a, b]$. The easiest way out is to extend all measures to $\mathscr{B}(\mathbb{R})$ by assigning measure 0 to $\mathbb{R} - [a, b]$.) Since a finite intersection of open sets is open, we can find a decreasing sequence $\{V_n\}$ of open sets such that $\mu(V_n) \to \mu(A)$ and $m(V_n) \to m(A) = 0$.

Choose n large enough so that $m(V_n) < \delta$; if V_n is the disjoint union of the open intervals (a_i, b_i), $i = 1, 2, \ldots$, then $|\mu(V_n)| \leq \sum_i |\mu(a_i, b_i)|$. But f is continuous, hence

$$\mu\{b_i\} = \lim_{n \to \infty} \mu(b_i - 1/n, b_i] = \lim_{n \to \infty} [f(b_i) - f(b_i - 1/n)] = 0.$$

Therefore

$$|\mu(V_n)| \leq \sum_i |\mu(a_i, b_i]| = \sum_i |f(b_i) - f(a_i)| \leq \varepsilon.$$

Since ε is arbitrary and $\mu(V_n) \to \mu(A)$, we have $\mu(A) = 0$. $\square$

If $f \colon \mathbb{R} \to \mathbb{R}$, absolute continuity of f is defined exactly as above. If F and G are bounded distribution functions on $\mathbb{R}$ with corresponding Lebesgue–Stieltjes measures μ_1 and μ_2, and $f = F - G$, $\mu = \mu_1 - \mu_2$ [a finite signed measure on $\mathscr{B}(\mathbb{R})$], then f is absolutely continuous iff μ is absolutely continuous with respect to Lebesgue measure; the proof is the same as in 2.3.1.

Any absolutely continuous function on $[a, b]$ can be represented as the difference of two absolutely continuous increasing functions. We prove this in a sequence of steps.

If $f\colon [a, b] \to \mathbb{R}$ and $P\colon a = x_0 < x_1 < \cdots < x_n = b$ is a partition of $[a, b]$, define

$$V(P) = \sum_{i=1}^{n} |f(x_i) - f(x_{i-1})|.$$

The sup of $V(P)$ over all partitions of $[a, b]$ is called the *variation* of f on $[a, b]$, written $V_f(a, b)$, or simply $V(a, b)$ if f is understood. We say that f is of *bounded variation* on $[a, b]$ iff $V(a, b) < \infty$. If $a < c < b$, a brief argument shows that $V(a, b) = V(a, c) + V(c, b)$.

2.3.2 Lemma. If $f\colon [a, b] \to \mathbb{R}$ and f is absolutely continuous on $[a, b]$, then f is of bounded variation on $[a, b]$.

PROOF.   Pick any $\varepsilon > 0$, and let $\delta > 0$ be chosen as in the definition of absolute continuity. If P is any partition of $[a, b]$, there is a refinement Q of P consisting of subintervals of length less than $\delta/2$. If $Q\colon a = x_0 < x_1 < \cdots < x_n = b$, let $i_0 = 0$, and let i_1 be the largest integer such that $x_{i_1} - x_{i_0} < \delta$; let i_2 be the largest integer greater than i_1 such that $x_{i_2} - x_{i_1} < \delta$, and continue in this fashion until the process terminates, say with $i_r = n$. Now $x_{i_k} - x_{i_{k-1}} \geq \delta/2$, $k = 1, 2, \ldots, r - 1$, by construction of Q; hence

$$(r - 1)\frac{\delta}{2} \leq b - a, \qquad \text{so} \qquad r \leq 1 + \frac{2(b - a)}{\delta} = M.$$

By absolute continuity, $V(Q) \leq M\varepsilon$. But $V(P) \leq V(Q)$ since the refining process can never decrease V; the result follows. $\square$

It is immediate that a monotone function F on $[a, b]$ is of bounded variation: $V_F(a, b) = |f(b) - f(a)|$. Thus if $f = F - G$, where F and G are increasing, then f is of bounded variation. The converse is also true.

2.3.3 Lemma. If $f\colon [a, b] \to \mathbb{R}$ and f is of bounded variation on $[a, b]$, then there are increasing functions F and G on $[a, b]$ such that $f = F - G$. If f is absolutely continuous, F and G may also be taken as absolutely continuous.

PROOF.   Let $f(x) = V_f(a, x)$, $a \leq x \leq b$; F is increasing, for if $h \geq 0$, $V(a, x + h) - V(a, x) = V(x, x + h) \geq 0$. If $G(x) = F(x) - f(x)$, then G is also increasing. For if $x_1 < x_2$, then

$$\begin{aligned}
G(x_2) - G(x_1) &= F(x_2) - F(x_1) - (f(x_2) - f(x_1)) \\
&= V(x_1, x_2) - (f(x_2) - f(x_1)) \\
&\geq V(x_1, x_2) - |f(x_2) - f(x_1)| \\
&\geq 0 \qquad \text{by definition of} \qquad V(x_1, x_2).
\end{aligned}$$

Now assume f absolutely continuous. If $\varepsilon > 0$, choose $\delta > 0$ as in the definition of absolute continuity. Let $(a_1, b_1), \ldots, (a_n, b_n)$ be disjoint open intervals with total length at most δ. If P_i is a partition of $[a_i, b_i]$, $i = 1, 2, \ldots, n$, then

$$\sum_{i=1}^{n} V(P_i) \leq \varepsilon \qquad \text{by absolute continuity of } f.$$

Take the sup successively over $P_1, \ldots, P_n$ to obtain

$$\sum_{i=1}^{n} V(a_i, b_i) \leq \varepsilon;$$

in other words,

$$\sum_{i=1}^{n} [F(b_i) - F(a_i)] \leq \varepsilon.$$

Therefore F is absolutely continuous. Since sums and differences of absolutely continuous functions are absolutely continuous, G is also absolutely continuous. $\square$

We have seen that there is a close connection between absolute continuity and indefinite integrals, via the Radon–Nikodym theorem. The connection carries over to real analysis, as follows.

2.3.4   Theorem. Let $f: [a, b] \to \mathbb{R}$. Then f is absolutely continuous on $[a, b]$ iff f is an indefinite integral, that is, iff

$$f(x) - f(a) = \int_a^x g(t)\, dt, \qquad a \leq x \leq b,$$

where $g: [a, b] \to \mathbb{R}$ is Borel measurable and integrable with respect to Lebesgue measure.

PROOF.   First assume f absolutely continuous. By 2.3.3, it is sufficient to assume f increasing. If μ is the Lebesgue–Stieltjes measure corresponding to f, and m is Lebesgue measure, then $\mu \ll m$ by 2.3.1. By the Radon–Nikodym theorem, there is an m-integrable function g such that $\mu(A) = \int_A g\, dm$ for all Borel subsets A of $[a, b]$. Take $A = [a, x]$ to obtain $f(x) - f(a) = \int_a^x g(t)\, dt$.

Conversely, assume $f(x) - f(a) = \int_a^x g(t)\,dt$. It is sufficient to assume $g \geq 0$ (if not, consider g^+ and g^- separately). Define $\mu(A) = \int_A g\,dm$, $A \in \mathscr{B}[a, b]$; then $\mu \ll m$, and if F is a distribution function corresponding to μ, F is absolutely continuous by 2.3.1. But

$$F(x) - F(a) = \mu(a, x] = \int_a^x g(t)\,dt = f(x) - f(a).$$

Therefore f is absolutely continuous. $\square$

If g is Lebesgue integrable on $\mathbb{R}$, the "if" part of the proof of 2.3.4 shows that the function defined by $\int_{-\infty}^x g(t)\,dt, x \in R$ is absolutely continuous, hence continuous, on $\mathbb{R}$. Another way of proving continuity is to observe that

$$\int_{-\infty}^{x+h} g(t)\,dt - \int_{-\infty}^x g(t)\,dt = \int_{-\infty}^\infty g(t) I_{(x,x+h)}(t)\,dt$$

if $h > 0$, and this approaches 0 as $h \to 0$, by the dominated convergence theorem.

If $f(x) - f(a) = \int_a^x g(t)\,dt$, $a \leq x \leq b$, and g is continuous at x, then f is differentiable at x and $f'(x) = g(x)$; the proof given in calculus carries over. If the continuity hypothesis is dropped, we can prove that $f'(x) = g(x)$ for *almost* every $x \in [a, b]$. One approach to this result is via the theory of differentiation of measures, which we now describe.

2.3.5   Definition.   For the remainder of this section, μ is a signed measure on the Borel sets of $\mathbb{R}^k$, assumed finite on bounded sets; thus if μ is nonnegative, it is a Lebesgue–Stieltjes measure. If m is Lebesgue measure, we define, for each $x \in \mathbb{R}^k$,

$$(\overline{D}\mu)(x) = \limsup_{r \to 0} {}_{C_r} \frac{\mu(C_r)}{m(C_r)}, \qquad (\underline{D}\mu)(x) = \liminf_{r \to 0} {}_{C_r} \frac{\mu(C_r)}{m(C_r)},$$

where the C_r range over all open cubes of diameter less than r that contain x. It will be convenient (although not essential) to assume that all cubes have edges parallel to the coordinate axes.

We say that μ is *differentiable* at x iff $\overline{D}\mu$ and $\underline{D}\mu$ are equal and finite at x; we write $(D\mu)(x)$ for the common value. Thus μ is differentiable at x iff for every sequence $\{C_n\}$ of open cubes containing x, with the diameter of C_n approaching 0, $\mu(C_n)/m(C_n)$ approaches a finite limit, independent of the particular sequence.

The following result will play an important role.

2.3.6 Lemma. If $\{C_1, \ldots, C_n\}$ is a family of open cubes in $\mathbb{R}^k$, there is a disjoint subfamily $\{C_{i_1}, \ldots, C_{i_s}\}$ such that $m(\bigcup_{j=1}^n C_j) \leq 3^k \sum_{p=1}^s m(C_{i_p})$.

PROOF. Assume that the diameter of C_i decreases with i. Set $i_1 = 1$, and take i_2 to be the smallest index greater than i_1 such that C_{i_2} is disjoint from C_{i_1}; let i_3 be the smallest index greater than i_2 such that C_{i_3} is disjoint from $C_{i_1} \cup C_{i_2}$. Continue in this fashion to obtain disjoint sets $C_{i_1}, \ldots, C_{i_s}$. Now for any $j = 1, \ldots, n$, we have $C_j \cap C_{i_p} \neq \emptyset$ for some $i_p \leq j$, for if not, j is not one of the i_p, hence $i_p < j < i_{p+1}$ for some p (or $i_s < j$). But $C_j \cap (C_{i_1} \cup \cdots \cup C_{i_p})$ is assumed empty, contradicting the definition of i_{p+1}.

If B_p is the open cube with the same center as C_{i_p} and diameter three times as large, then since $C_j \cap C_{i_p} \neq \emptyset$ and diameter $C_j \leq$ diameter C_{i_p}, we have $C_j \subset B_p$. Therefore,

$$m\left(\bigcup_{j=1}^n C_j\right) \leq m\left(\bigcup_{p=1}^s B_p\right) \leq \sum_{p=1}^s m(B_p) = 3^k \sum_{p=1}^s m(C_{i_p}). \quad \square$$

We now prove the first differentiation result.

2.3.7 Lemma. Let μ be a Lebesgue–Stieltjes measure on the Borel sets of $\mathbb{R}^k$. If $\mu(A) = 0$, then $D\mu = 0$ a.e. $[m]$ on A.

PROOF. If $a > 0$, let $B = \{x \in A: (\overline{D}\mu)(x) > a\}$. [Note that $\{x: \sup_{C_r} \mu(C_r)/m(C_r) > a\}$ is open, and it follows that B is a Borel set.] Fix $r > 0$, and let K be a compact subset of B. If $x \in K$, there is an open cube C_r of diameter less than r with $x \in C_r$ and $\mu(C_r) > am(C_r)$. By compactness, K is covered by finitely many of the cubes, say $C_1, \ldots, C_n$. If $\{C_{i_1}, \ldots, C_{i_s}\}$ is the subcollection of 2.3.6, we have

$$m(K) \leq m\left(\bigcup_{j=1}^n C_j\right) \leq 3^k \sum_{p=1}^s m(C_{i_p}) \leq \frac{3^k}{a} \sum_{p=1}^s \mu(C_{i_p})$$

$$= \frac{3^k}{a} \mu\left(\bigcup_{p=1}^s C_{i_p}\right) \leq \frac{3^k}{a} \mu(K_r),$$

where $K_r = \{x \in \mathbb{R}^k: \text{dist}(x, K) < r\}$. Since r is arbitrary, we have $m(K) \leq 3^k \mu(K)/a \leq 3^k \mu(A)/a = 0$. Take the sup over K to obtain, by 1.4.11, $m(B) = 0$, and since a is arbitrary, it follows that $\overline{D}\mu \leq 0$ a.e. $[m]$ on A. But $\mu \geq 0$; hence $0 \leq \underline{D}\mu \leq \overline{D}\mu$, so that $D\mu = 0$ a.e. $[m]$ on A. $\quad \square$

We are going to show that $D\mu$ exists a.e. $[m]$, and to do this the Lebesgue decomposition theorem is helpful. We write $\mu = \mu_1 + \mu_2$, where $\mu_1 \ll m$, $\mu_2 \perp m$. If $|\mu_2|(A) = 0$ and $m(A^c) = 0$, then by 2.3.7, $D\mu_2{}^+ = D\mu_2{}^- = 0$ a.e. $[m]$ on A; hence a.e. $[m]$ on $\mathbb{R}^k$. Thus $D\mu_2 = 0$ a.e. $[m]$ on $\mathbb{R}^k$.

By the Radon–Nikodym theorem, we have $\mu_1(E) = \int_E g \, dm$, $E \in \mathscr{B}(\mathbb{R}^k)$, for some Borel measurable function g. As might be expected intuitively, g is (a.e.) the derivative of μ_1; hence $D\mu = g$ a.e. $[m]$.

2.3.8   Theorem.   Let μ be a signed measure on $\mathscr{B}(\mathbb{R}^k)$ that is finite on bounded sets, and let $\mu = \mu_1 + \mu_2$, where $\mu_1 \ll m$ and $\mu_2 \perp m$. Then $D\mu$ exists a.e. $[m]$ and coincides a.e. $[m]$ with the Radon–Nikodym derivative $g = d\mu_1/dm$.

PROOF.   If $a \in R$ and C is an open cube of diameter less than r,

$$\mu_1(C) - am(C) = \int_C (g - a) \, dm \leq \int_{C \cap \{g \geq a\}} (g - a) \, dm.$$

If $\lambda(E) = \int_{E \cap \{g \geq a\}} (g - a) \, dm$, $E \in \mathscr{B}(\mathbb{R}^k)$, and $A = \{g < a\}$, then $\lambda(A) = 0$; so by 2.3.7, $D\lambda = 0$ a.e. $[m]$ on A. But

$$\frac{\mu_1(C)}{m(C)} \leq a + \frac{\lambda(C)}{m(C)};$$

hence $\overline{D}\mu_1 \leq a$ a.e. $[m]$ on A. Therefore, if $E_a = \{x \in \mathbb{R}^k: g(x) < a < (\overline{D}\mu_1) (x)\}$, then $m(E_a) = 0$. Since $\{\overline{D}\mu_1 > g\} \subset \bigcup \{E_a: a \text{ rational}\}$, we have $\overline{D}\mu_1 \leq g$ a.e. $[m]$. Replace μ_1 by $-\mu_1$ and g by $-g$ to obtain $\underline{D}\mu_1 \geq g$ a.e. $[m]$. By 2.2.2(b), g is finite a.e. $[m]$, and the result follows. $\square$

We now return to functions on the real line.

2.3.9   Theorem.   Let $f: [a, b] \to \mathbb{R}$ be an increasing function. Then the derivative of f exists at almost every point of $[a, b]$ (with respect to Lebesgue measure). Thus by 2.3.3, a function of bounded variation is differentiable almost everywhere.

PROOF.   Since f has only countably many discontinuities, we may assume without loss of generality that f takes the upper value at a discontinuity and is therefore a distribution function.

Let μ be the Lebesgue–Stieltjes measure corresponding to f; by 2.3.8, $D\mu$ exists a.e. $[m]$; we show that $D\mu = f'$ a.e. $[m]$. If $a \leq x \leq b$ and μ is differentiable at x, then f is continuous at x by definition of μ. If $\lim_{h \to 0} [f(x + h) - f(x)]/h \neq (D\mu)(x) = c$, there is an $\varepsilon > 0$ and a sequence $h_n \to 0$

with all h_n of the same sign and $|[f(x+h_n)-f(x)]/h_n - c| \geq \varepsilon$ for all n. Assuming all $h_n > 0$, we can find numbers $k_n > 0$ such that

$$\left| \frac{f(x+h_n)-f(x-k_n)}{h_n+k_n} - c \right| \geq \frac{\varepsilon}{2}$$

for all n, and since f has only countably many discontinuities, it may be assumed that f is continuous at $x+h_n$ and $x-k_n$. Thus we conclude that $\mu(x-k_n,x+h_n)/(h_n+k_n) \not\longrightarrow c$, a contradiction. $\square$

We now prove the main theorem on absolutely continuous functions.

2.3.10   Theorem. Let f be absolutely continuous on $[a,b]$, with $f(x) - f(a) = \int_a^x g(t)\,dt$, as in 2.3.4. Then $f' = g$ almost everywhere on $[a,b]$ (Lebesgue measure). Thus by 2.3.4, f is absolutely continuous iff f is the integral of its derivative, that is,

$$f(x) - f(a) = \int_a^x f'(t)\,dt, \qquad a \leq x \leq b.$$

PROOF. We may assume $g \geq 0$ (if not, consider g^+ and g^-). If $\mu_1(A) = \int_A g\,dm$, $A \in \mathscr{B}(\mathbb{R}^k)$, then $D\mu_1 = g$ a.e. $[m]$ by 2.3.8. But if $a \leq x \leq y \leq b$, then $\mu_1(x,y] = f(y) - f(x)$, so that μ_1 is the Lebesgue–Stieltjes measure corresponding to f. Thus by the proof of 2.3.9, $D\mu_1 = f'$ a.e. $[m]$. $\square$

Problems

1. Let F be a bounded distribution function on $\mathbb{R}$. Use the Lebesgue decomposition theorem to show that F may be represented uniquely (up to additive constants) as $F_1 + F_2 + F_3$, where the distribution functions F_j, $j = 1,2,3$ (and the corresponding Lebesgue–Stieltjes measures μ_j) have the following properties:

    (a) F_1 is discrete (that is, μ_1 is concentrated on a countable set of points).
    (b) F_2 is absolutely continuous (μ_2 is absolutely continuous with respect to Lebesgue measure; see 2.3.1).
    (c) F_3 is continuous and singular (that is, μ_3 is singular with respect to Lebesgue measure).

2. If f is an increasing function from $[a,b]$ to $\mathbb{R}$, show that $\int_a^b f'(x)\,dx \leq f(b) - f(a)$. The inequality may be strict, as Problem 3 shows. (Note that by 2.3.9, f' exists a.e.; for integration purposes, f' may be defined arbitrarily on the exceptional set of Lebesgue measure 0.)

3. (*The Cantor function*)   Let $E_1, E_2, \ldots$ be the sets removed from $[0,1]$ to form the Cantor ternary set (see Problem 7, Section 1.4). Define functions

F_n: $[0, 1] \to [0, 1]$ as follows: Let $A_1, A_2, \ldots, A_{2^n-1}$ be the subintervals of $\bigcup_{i=1}^{n} E_i$, arranged in increasing order. For example, if $n = 3$,

$$E_1 \cup E_2 \cup E_3 = \left(\tfrac{1}{27}, \tfrac{2}{27}\right) \cup \left(\tfrac{1}{9}, \tfrac{2}{9}\right) \cup \left(\tfrac{7}{27}, \tfrac{8}{27}\right) \cup \left(\tfrac{1}{3}, \tfrac{2}{3}\right)$$

$$\cup \left(\tfrac{19}{27}, \tfrac{20}{27}\right) \cup \left(\tfrac{7}{9}, \tfrac{8}{9}\right) \cup \left(\tfrac{25}{27}, \tfrac{26}{27}\right)$$

$$= A_1 \cup A_2 \cup \cdots \cup A_7.$$

Define

$$F_n(0) = 0,$$
$$F_n(x) = k/2^n \quad \text{if} \quad x \in A_k, \quad k = 1, 2, \ldots, 2^n - 1,$$
$$F_n(1) = 1.$$

Complete the specification of F_n by interpolating linearly. For $n = 2$, see Fig. 2.3.1, in this case,

$$E_1 \cup E_2 = \left(\tfrac{1}{9}, \tfrac{2}{9}\right) \cup \left(\tfrac{1}{3}, \tfrac{2}{3}\right) \cup \left(\tfrac{7}{9}, \tfrac{8}{9}\right)$$
$$= A_1 \cup A_2 \cup A_3.$$

Show that $F_n(x) \to F(x)$ for each x, where F, the Cantor function, has the following properties:

(a) F is continuous and increasing.
(b) $F' = 0$ almost everywhere (Lebesgue measure).
(c) F is not absolutely continuous.

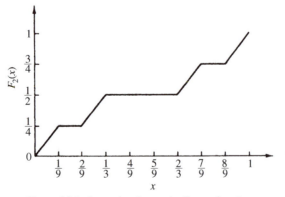

Figure 2.3.1. Approximation to the Cantor function.

In fact

(d) F is singular; that is, the corresponding Lebesgue–Stieltjes measure μ is singular with respect to Lebesgue measure.

4. Let f be a Lebesgue integrable real-valued function on $\mathbb{R}^k$ (or on an open subset of $\mathbb{R}^k$). If $\mu(E) = \int_E f(x)\,dx$, $E \in \mathscr{B}(\mathbb{R}^k)$, we know that $D\mu = f$ a.e. (Lebesgue measure). If $D\mu = f$ at x_0, then if C is an open cube containing x_0 and diam $C \to 0$, we have $\mu(C)/m(C) \to f(x_0)$; that is,

$$\frac{1}{m(C)} \int_C [f(x) - f(x_0)]\,dx \to 0 \qquad \text{as} \qquad \text{diam } C \to 0.$$

In fact, show that

$$\frac{1}{m(C)} \int_C |f(x) - f(x_0)|\,dx \to 0 \qquad \text{as} \qquad \text{diam } C \to 0$$

for almost every x_0. The set of favorable x_0 is called the *Lebesgue set* of f.

5. This problem relates various concepts discussed in Section 2.3. In all cases, f is a real-valued function defined on the closed bounded interval $[a, b]$. Establish the following:

(a) If f is continuous, f need not be of bounded variation.

(b) If f is continuous and increasing (hence of bounded variation), f need not be absolutely continuous.

(c) If f satisfies a Lipschitz condition, that is, $|f(x) - f(y)| \leq L|x - y|$ for some fixed positive number L and all x, $y \in [a, b]$, then f is absolutely continuous.

(d) If f' exists everywhere and is bounded, f is absolutely continuous. (It can also be shown that if f' exists everywhere and is Lebesgue integrable on $[a, b]$, then f is absolutely continuous; see Titchmarsh, 1939, p. 368.)

(e) If f is continuous and f' exists everywhere, f need not be absolutely continuous [consider $f(x) = x^2 \sin(1/x^2)$, $0 < x \leq 1$, $f(0) = 0$].

6. The following problem considers the change of variable formula in a multiple integral. Throughout the problem, T will be a map from V onto W, where V and W are open subsets of $\mathbb{R}^k$, T is assumed one-to-one, continuously differentiable, with a nonzero Jacobian. Thus T has a continuously differentiable inverse, by the inverse function theorem of advanced calculus [see, for example, Apostol (1957, p. 144)]. It also follows from standard advanced calculus results that for all $x \in V$,

$$\frac{1}{|h|}[|T(x+h) - T(x) - A(x)h|] \to 0 \qquad \text{as} \qquad h \to 0, \qquad (1)$$

where $A(x)$ is the linear transformation on $\mathbb{R}^k$ represented by the Jacobian matrix of T, evaluated at x. [See Apostol (1957, p. 118).]

(a) Let A be a nonsingular linear transformation on $\mathbb{R}^k$, and define a measure λ on $\mathscr{B}(\mathbb{R}^k)$ by $\lambda(E) = m(A(E))$ where m is Lebesgue measure. Show that $\lambda = c(A)m$ for some constant $c(A)$, and in fact $c(A)$ is the absolute value of the determinant of A. [Use translation-invariance of Lebesgue measure (Problem 5, Section 1.4) and the fact that any matrix can be represented as a product of matrices corresponding to elementary row operations.]

Now define a measure μ on $\mathscr{B}(V)$ by $\mu(E) = m(T(E))$. By continuity of T, if $\varepsilon > 0$, $x \in V$, and C is a sufficiently small open cube containing x, then $T(C)$ has diameter less than ε, in particular, $m(T(C)) < \infty$. It follows by a brief compactness argument that μ is a Lebesgue–Stieltjes measure on $\mathscr{B}(V)$.

Our objective is to show that μ is differentiable and $(D\mu)(x) = |J(x)|$ for every $x \in V$, where $J(x) = \det A(x)$, the Jacobian of the transformation T.

(b) Show that it suffices to prove that if $0 \in V$ and $T(0) = 0$, then $(D\mu)(0) = |\det A(0)|$.

(c) Show that it may be assumed without loss of generality that $A(0)$ is the identity transformation; hence $\det A(0) = 1$.

Now given $\varepsilon > 0$, choose $\alpha \in \left(0, \frac{1}{4}\right)$ such that

$$1 - \varepsilon < (1 - 2\alpha)^k < (1 + 2\alpha)^k < 1 + \varepsilon.$$

Under the assumptions of (b) and (c), by Eq. (1), there is a $\delta > 0$ such that if $|x| < \delta$, then $|T(x) - x| \le \alpha|x|/\sqrt{k}$.

(d) If C is an open cube containing 0 with edge length β and diameter $\sqrt{k}\beta < \delta$, take C_1, C_2 as open cubes concentric with C, with edge lengths $\beta_1 = (1 - 2\alpha)\beta$ and $\beta_2 = (1 + 2\alpha)\beta$. Establish the following:

   (i)   If $x \in \overline{C}$, then $T(x) \in C_2$.
   (ii)   If x belongs to the boundary of C, then $T(x) \notin C_1$.
   (iii) If x is the center of C, then $T(x) \in C_1$.
   (iv) $C_1 - T(C) = C_1 - T(\overline{C})$.

Use a connectedness argument to conclude that $C_1 \subset T(C) \subset C_2$, and complete the proof that $(D\mu)(0) = 1$.

(e) If λ is any measure on $\mathscr{B}(V)$ and $\overline{D}\lambda < \infty$ on V, show that λ is absolutely continuous with respect to Lebesgue measure. It therefore follows from Theorem 2.3.8 that

$$m(T(E)) = \int_E |J(x)|\,dx, \qquad E \in \mathscr{B}(V).$$

[If this is false, find a compact set K and positive integers n and j such that $m(K) = 0$, $\lambda(K) > 0$, and $\lambda(C) < nm(C)$ for all open cubes C containing a point of K and having diameter less than $1/j$. Essentially, the idea is to cover K by such cubes and conclude that $\lambda(K) = 0$, a contradiction.]

(f) If f is a real-valued Borel measurable function on W, show that

$$\int_W f(y)\, dy = \int_V f(T(x))|J(x)|\, dx$$

in the sense that if one of the two integrals exists, so does the other, and the two integrals are equal.

7. (*Fubini's differentiation theorem*) Let $f_1, f_2, \ldots$ be increasing functions from $\mathbb{R}$ to $\mathbb{R}$, and assume that for each x, $\sum_{n=1}^{\infty} f_n(x)$ converges to a finite number $f(x)$. Show that $\sum_{n=1}^{\infty} f_n'(x) = f'(x)$ almost everywhere (Lebesgue measure).

Outline:

(a) It suffices to restrict the domain of all functions to $[0, 1]$ and to assume all functions nonnegative. Use Fatou's lemma to show that $\sum_{n=1}^{\infty} f_n'(x) \le f'(x)$ a.e.; hence $f_n'(x) \to 0$ a.e.

(b) Choose $n_1, n_2, \ldots$ such that $\sum_{j>n_k} f_j(1) \le 2^{-k}$, $k = 1, 2, \ldots$ and apply part (a) to the functions $g_k(x) = f(x) - \sum_{j=1}^{n_k} f_j(x) = \sum_{j>n_k} f_j(x)$.

2.4 L^p SPACES

If $(\Omega, \mathscr{F}, \mu)$ is a measure space and p is a real number with $p \ge 1$, the set of all Borel measurable functions f such that $|f|^p$ is μ-integrable has many important properties. In order to fully develop these properties, it will be convenient to work with complex-valued functions.

2.4.1   Definitions. Let $(\Omega, \mathscr{F})$ be a measurable space, and let f be a complex-valued function on Ω, so that $f = \operatorname{Re} f + i \operatorname{Im} f$. We say that f is a *complex-valued Borel measurable function* on $(\Omega, \mathscr{F})$ if both $\operatorname{Re} f$ and $\operatorname{Im} f$ are real-valued Borel measurable functions. If μ is a measure on $\mathscr{F}$, we define

$$\int_\Omega f\, d\mu = \int_\Omega \operatorname{Re} f\, d\mu + i \int_\Omega \operatorname{Im} f\, d\mu,$$

provided $\int_\Omega \operatorname{Re} f\, d\mu$ and $\int_\Omega \operatorname{Im} f\, d\mu$ are *both* finite. In this case we say that f is μ-*integrable*. Thus in working with complex-valued functions, we do not consider any cases in which integrals exist but are not finite.

The following result was established earlier for real-valued f [see 1.5.9(c)]; it is still valid in the complex case, but the proof must be modified.

2.4.2 **Lemma.** If f is μ-integrable,

$$\left| \int_\Omega f \, d\mu \right| \leq \int_\Omega |f| \, d\mu.$$

PROOF. If $\int_\Omega f \, d\mu = re^{i\theta}$, $r \geq 0$, then $\int_\Omega e^{-i\theta} f \, d\mu = r = |\int_\Omega f \, d\mu|$. But if $f(\omega) = \rho(\omega)e^{i\varphi(\omega)}$ (taking $\rho \geq 0$), then

$$\int_\Omega e^{-i\theta} f \, d\mu = \int_\Omega \rho e^{i(\varphi - \theta)} \, d\mu$$

$$= \int_\Omega \rho \cos(\varphi - \theta) \, d\mu \qquad \text{since } r \text{ is real}$$

$$\leq \int_\Omega \rho \, d\mu = \int_\Omega |f| \, d\mu. \quad \square$$

Many other standard properties of the integral carry over to the complex case, in particular 1.5.5(b), 1.5.9(a) and (e), 1.6.1, 1.6.3, 1.6.4(b) and (c), 1.6.5, 1.6.9, 1.6.10, and 1.7.1. In almost all cases, the result is an immediate consequence of the fact that integrating a complex-valued function is equivalent to integrating the real and imaginary parts separately. Only two theorems require additional comment. To prove that h is integrable iff $|h|$ is integrable [1.6.4(b)], use the fact that $|\text{Re } h|, |\text{Im } h| \leq |h| \leq |\text{Re } h| + |\text{Im } h|$. Finally, to prove the dominated convergence theorem (1.6.9), apply the real version of the theorem to $|f_n - f|$, and note that $|f_n - f| \leq |f_n| + |f| \leq 2g$.

If $p > 0$, we define the space $L^p = L^p(\Omega, \mathscr{F}, \mu)$ as the collection of all complex-valued Borel measurable functions f such that $\int_\Omega |f|^p \, d\mu < \infty$. We set

$$\|f\|_p = \left(\int_\Omega |f|^p \, d\mu \right)^{1/p}, \qquad f \in L^p.$$

It follows that for any complex number a, $\|af\|_p = |a| \, \|f\|_p$, $f \in L^p$.

We are going to show that L^p forms a linear space over the complex field. The key steps in the proof are the Hölder and Minkowski inequalities, which we now develop.

2.4.3 **Lemma.** If $a, b, \alpha, \beta > 0$, $\alpha + \beta = 1$, then $a^\alpha b^\beta \leq \alpha a + \beta b$.

PROOF. The statement to be proved is equivalent to $-\log(\alpha a + \beta b) \leq \alpha(-\log a) + \beta(-\log b)$, which holds because $-\log$ is convex. [If g has a nonnegative second derivative on the interval $I \subset R$, then g is convex on I, that is $g(\alpha x + \beta y) \leq \alpha g(x) + \beta g(y)$, $x, y \in I$, $\alpha, \beta > 0$, $\alpha + \beta = 1$. To see this, assume $x < y$ and write

$$g(\alpha x + \beta y) - \alpha g(x) - \beta g(y) = \alpha[g(\alpha x + \beta y) - g(x)] + \beta[g(\alpha x + \beta y) - g(y)]$$
$$= \alpha\beta(y - x)[g'(u) - g'(v)] \qquad \text{for some} \qquad u, v$$

with $x \leq u \leq \alpha x + \beta y$, $\alpha x + \beta y \leq v \leq y$. But $g'(u) - g'(v) \leq 0$ since g' is increasing on I.] $\square$

2.4.4 Corollary. If $c, d > 0$, $p, q > 1$, $(1/p) + (1/q) = 1$, then $cd \leq (c^p/p) + (d^q/q)$.

PROOF. In 2.4.3, let $\alpha = 1/p$, $\beta = 1/q$, $a = c^p$, $b = d^q$. $\square$

2.4.5 Hölder Inequality. Let $1 < p < \infty$, $1 < q < \infty$, $(1/p) + (1/q) = 1$. If $f \in L^p$ and $g \in L^q$, then $fg \in L^1$ and $\|fg\|_1 \leq \|f\|_p \|g\|_q$.

PROOF. In 2.4.4, take $c = |f(\omega)|/\|f\|_p$, $d = |g(\omega)|/\|g\|_q$ (the inequality is immediate if $\|f\|_p$ or $\|g\|_q = 0$). Then

$$\frac{|f(\omega)g(\omega)|}{\|f\|_p\|g\|_q} \leq \frac{|f(\omega)|^p}{p\|f\|_p^p} + \frac{|g(\omega)|^q}{q\|g\|_q^q};$$

integrate to obtain

$$\frac{\int_\Omega |fg|\,d\mu}{\|f\|_p\|g\|_q} \leq \frac{1}{p} + \frac{1}{q} = 1. \quad \square$$

When $p = q = 2$, we obtain

$$\int_\Omega |fg|\,d\mu \leq \left[\int_\Omega |f|^2\,d\mu \int_\Omega |g|^2\,d\mu\right]^{1/2}$$

and thus, using 2.4.2, we have the *Cauchy–Schwarz inequality*: If f and $g \in L^2$, then $fg \in L^1$ and

$$\left|\int_\Omega f\bar{g}\,d\mu\right| \leq \left[\int_\Omega |f|^2\,d\mu \int_\Omega |g|^2\,d\mu\right]^{1/2},$$

where $\bar{g}$ is the complex conjugate of g. (The reason for replacing g by $\bar{g}$ is to make the inequality agree with the Hilbert space result to be discussed in Chapter 3.)

2.4.6   Lemma. If $a, b \geq 0$, $p \geq 1$, then $(a+b)^p \leq 2^{p-1}(a^p + b^p)$.

PROOF.   Let $h(x) = d[(a+x)^p - 2^{p-1}(a^p + x^p)]/dx = p(a+x)^{p-1} - 2^{p-1} px^{p-1}$; since $p \geq 1$,

$$h(x) > 0 \qquad \text{for} \qquad a + x > 2x, \qquad \text{that is,} \qquad x < a,$$

$$h(x) = 0 \qquad \text{at} \qquad x = a,$$

$$h(x) < 0 \qquad \text{for} \qquad x > a.$$

The maximum therefore occurs at $x = a$; hence

$$(a+b)^p - 2^{p-1}(a^p + b^p) \leq (a+a)^p - 2^{p-1}(a^p + a^p) = 0. \quad \square$$

2.4.7   Minkowski Inequality. If $f, g \in L^p (1 \leq p < \infty)$, then $f + g \in L^p$ and $\|f + g\|_p \leq \|f\|_p + \|g\|_p$.

PROOF.   By 2.4.6, $|f + g|^p \leq (|f| + |g|)^p \leq 2^{p-1}(|f|^p + |g|^p)$, hence f, $g \in L^p$ implies $f + g \in L^p$. Now the inequality is clear when $p = 1$, so assume $p > 1$ and choose q such that $(1/p) + (1/q) = 1$. Then

$$|f + g|^p = |f + g| \, |f + g|^{p-1} \leq |f| \, |f + g|^{p-1} + |g| \, |f + g|^{p-1}. \qquad (1)$$

Now $|f + g|^{p-1} \in L^q$; for

$$(p-1)q = \frac{p-1}{1/q} = \frac{p-1}{1 - 1/p} = p;$$

hence

$$\int_\Omega [|f + g|^{p-1}]^q \, d\mu = \int_\Omega |f + g|^p \, d\mu < \infty.$$

Since f and g belong to L^p and $|f + g|^{p-1} \in L^q$, Hölder's inequality implies that $|f| \, |f + g|^{p-1}$ and $|g| \, |f + g|^{p-1} \in L^1$, and

$$\int_\Omega |f| \, |f + g|^{p-1} \, d\mu \leq \|f\|_p \left[\int_\Omega (|f + g|^{p-1})^q \, d\mu \right]^{1/q}$$

$$= \|f\|_p \, \|f + g\|_p^{p/q}, \qquad (2)$$

$$\int_\Omega |g| \, |f + g|^{p-1} \, d\mu \leq \|g\|_p \, \|f + g\|_p^{p/q}. \qquad (3)$$

By Eq. (1), $\|f + g\|_p^p \leq (\|f\|_p + \|g\|_p)(\|f + g\|_p^{p/q})$. Since $p - (p/q) = 1$, the result follows. $\square$

By Minkowski's inequality and the fact that $\|af\|_p = |a| \, \|f\|_p$ for $f \in L^p$, $L^p(1 \le p < \infty)$ is a vector space over the complex field. Furthermore, there is a natural notion of distance in L^p, by virtue of the fact that $\| \ \|_p$ is a seminorm.

2.4.8   *Definitions and Comments.* A *seminorm* on a vector space L (over the real or complex field) is a real-valued function $\| \ \|$ on L, with the following properties:

$$\|f\| \ge 0,$$

$$\|af\| = |a| \, \|f\| \qquad \text{for each scalar } a;$$

$$\text{consequently, if } f = 0, \text{ then } \|f\| = 0.$$

$$\|f + g\| \le \|f\| + \|g\|$$

(f and g are arbitrary elements of L). If $\| \ \|$ is a seminorm with the additional property that $\|f\| = 0$ implies $f = 0$, $\| \ \|$ is said to be a *norm*.

Now $\| \ \|_p$ is a seminorm on L^p; the first two properties follow from the definition of $\| \ \|_p$, and the last property is a consequence of Minkowski's inequality.

We can, in effect, change $\| \ \|_p$ into a norm by passing to equivalence classes as follows.

If $f, g \in L^p(\Omega, \mathscr{F}, \mu)$, define $f \sim g$ iff $f = g$ a.e. $[\mu]$. Then $\|f\|_p$ is the same for all f in a given equivalence class, by 1.6.5(b). Thus if $\mathbf{L}^p$ is the collection of equivalence classes, $\mathbf{L}^p$ becomes a linear space, and $\| \ \|_p$ is a seminorm on $\mathbf{L}^p$. In fact $\| \ \|_p$ is a norm, since $\|f\|_p = 0$ implies $f = 0$ a.e. $[\mu]$, by 1.6.6(b).

If $\| \ \|$ is a seminorm on a vector space, we have a natural notion of distance: $d(f, g) = \|f - g\|$. By definition of seminorm we have

$$d(f, g) \ge 0,$$

$$d(f, g) = 0 \qquad \text{if} \qquad f = g,$$

$$d(f, g) = d(g, f),$$

$$d(f, h) \le d(f, g) + d(g, h).$$

Thus d has all the properties of a metric, except that $d(f, g) = 0$ does not necessarily imply $f = g$; we call d a *pseudometric*. If $\| \ \|$ is a norm, d is actually a metric. (There is an asymmetry of terminology between seminorm and pseudometric, but these terms seem to be most popular, although "pseudonorm" is sometimes used, as is "semimetric.")

One of the first questions that arises in any metric space is the problem of completeness; we ask whether or not Cauchy sequences converge. We are going to show that the L^p spaces are complete. The following result will be needed; students of probability are likely to recognize it immediately, but it appears in other parts of analysis as well.

2.4.9   Chebyshev's Inequality. Let f be a nonnegative, extended real-valued, Borel measurable function on $(\Omega, \mathscr{F}, \mu)$. If $0 < p < \infty$ and $0 < \varepsilon < \infty$,

$$\mu\{\omega:\ f(\omega) \geq \varepsilon\} \leq \frac{1}{\varepsilon^p} \int_\Omega f^p \, d\mu.$$

The following version is often applied in probability. If g is an extended real-valued Borel measurable function on $(\Omega, \mathscr{F})$ and P is a probability measure on $\mathscr{F}$, define

$$m = \int_\Omega g \, dP \qquad \text{(assumed finite, so that g is finite a.e. $[P]$),}$$

$$\sigma^2 = \int_\Omega (g - m)^2 \, dP.$$

If $0 < k < \infty$,

$$P\{\omega:\ |g(\omega) - m| \geq k\sigma\} \leq \frac{1}{k^2}.$$

This follows from the first version with $f = |g - m|$, $\varepsilon = k\sigma$, $p = 2$.

PROOF.

$$\int_\Omega f^p \, d\mu \geq \int_{\{\omega:\ f(\omega) \geq \varepsilon\}} f^p \, d\mu \geq \varepsilon^p \mu\{\omega:\ f(\omega) \geq \varepsilon\}. \quad \square$$

One more auxiliary result will be needed.

2.4.10   Lemma. If $g_1, g_2, \ldots \in L^p(p > 0)$ and $\|g_k - g_{k+1}\|_p < \left(\frac{1}{4}\right)^k$, $k = 1$, $2, \ldots$, then $\{g_k\}$ converges a.e.

PROOF.   Let $A_k = \{\omega:\ |g_k(\omega) - g_{k+1}(\omega)| \geq 2^{-k}\}$. Then by 2.4.9,

$$\mu(A_k) \leq 2^{kp}\|g_k - g_{k+1}\|_p^p < 2^{-kp}.$$

By 2.2.4, $\mu(\limsup_n A_n) = 0$. But if $\omega \notin \limsup_n A_n$, then $|g_k(\omega) - g_{k+1}(\omega)| < 2^{-k}$ for large k, so $\{g_k(\omega)\}$ is a Cauchy sequence of complex numbers, and therefore converges. $\square$

Now, the main result:

2.4.11 Completeness of L^p, $1 \leq p < \infty$. If $f_1, f_2, \ldots$ form a Cauchy sequence in L^p, that is, $\|f_n - f_m\|_p \to 0$ as $n, m \to \infty$, there is an $f \in L^p$ such that $\|f_n - f\|_p \to 0$.

PROOF. Let n_1 be such that $\|f_n - f_m\|_p < \frac{1}{4}$ for $n, m \geq n_1$, and let $g_1 = f_{n_1}$. In general, having chosen $g_1, \ldots, g_k$ and $n_1, \ldots, n_k$, let $n_{k+1} > n_k$ be such that $\|f_n - f_m\|_p < \left(\frac{1}{4}\right)^{k+1}$ for $n, m \geq n_{k+1}$, and let $g_{k+1} = f_{n_{k+1}}$. By 2.4.10, g_k converges a.e. to a limit function f.

Given $\varepsilon > 0$, choose N such that $\|f_n - f_m\|_p^p < \varepsilon$ for $n, m \geq N$. Fix $n \geq N$ and let $m \to \infty$ through values in the subsequence, that is, let $m = n_k$, $k \to \infty$. Then

$$\varepsilon \geq \liminf_{k \to \infty} \|f_n - f_{n_k}\|_p^p = \liminf_{k \to \infty} \int_\Omega |f_n - f_{n_k}|^p \, d\mu$$

$$\geq \int_\Omega \liminf_{k \to \infty} |f_n - g_k|^p \, d\mu \qquad \text{by Fatou's lemma}$$

$$= \|f_n - f\|_p^p.$$

Thus $\|f_n - f\|_p \to 0$. Since $f = f - f_n + f_n$, we have $f \in L^p$. $\square$

2.4.12 Examples and Comments. Let Ω be the positive integers; take $\mathscr{F}$ as all subsets of Ω, and let μ be counting measure. A real-valued function on Ω may be represented as a sequence of real numbers; we write $f = \{a_n, n = 1, 2, \ldots\}$. An integral on this space is really a sum [see Problem 1(a)]:

$$\int_\Omega f \, d\mu = \sum_{n=1}^\infty a_n,$$

where the series is interpreted as $\sum_{n=1}^\infty a_n{}^+ - \sum_{n=1}^\infty a_n{}^-$ if this is not of the form $+\infty - \infty$ (if it is, the integral does not exist). Thus the following cases occur:

(1) $\sum_{n=1}^\infty a_n{}^+ = \infty$, $\sum_{n=1}^\infty a_n{}^- < \infty$. The series diverges to ∞ and the integral is ∞.

(2) $\sum_{n=1}^\infty a_n{}^+ < \infty$, $\sum_{n=1}^\infty a_n{}^- = \infty$. The series diverges to $-\infty$ and the integral is $-\infty$.

(3) $\sum_{n=1}^\infty a_n{}^+ < \infty$, $\sum_{n=1}^\infty a_n{}^- < \infty$. The series is absolutely convergent and the integral equals the sum of the series.

(4) $\sum_{n=1}^\infty a_n{}^+ = \infty$, $\sum_{n=1}^\infty a_n{}^- = \infty$. The series is not absolutely convergent; it may or may not converge conditionally. Whether it does or not, the integral does not exist. Thus when summation is considered from the point

of view of Lebesgue integration theory, series that converge conditionally but not absolutely are ignored.

If μ is changed so that $\mu\{n\}$ is a nonnegative number p_n, not necessarily 1 as in the case of counting measure, the same analysis shows that

$$\int_\Omega f \, d\mu = \sum_{n=1}^\infty p_n a_n,$$

where the series is interpreted as $\sum_{n=1}^\infty p_n a_n^+ - \sum_{n=1}^\infty p_n a_n^-$.

If $f = \{a_n, n = 1, 2, \ldots\}$ is a sequence of complex numbers and μ is counting measure,

$$\int_\Omega f \, d\mu = \sum_{n=1}^\infty a_n = \sum_{n=1}^\infty \operatorname{Re} a_n + i \sum_{n=1}^\infty \operatorname{Im} a_n;$$

the integral is defined provided $\sum_{n=1}^\infty |a_n| < \infty$.

Now let Ω be an *arbitrary* set, and take $\mathscr{F}$ as all subsets of Ω and μ as counting measure. If $f = (f(\alpha), \alpha \in \Omega)$ is a nonnegative real-valued function on Ω, then [Problem 1(b)]

$$\int_\Omega f \, d\mu = \sum_\alpha f(\alpha), \tag{1}$$

where the series is defined as $\sup\{\sum_{\alpha \in F} f(\alpha): F \subset \Omega, F \text{ finite}\}$. If $f(\alpha) > 0$ for uncountably many α, then for some $\delta > 0$ we have $f(\alpha) \geq \delta$ for infinitely many α, so that $\sum_\alpha f(\alpha) = \infty$.

If the nonnegativity hypothesis is dropped, we apply the above results to f^+ and f^- to again obtain Eq. (1), where the series is interpreted as $\sum_\alpha f^+(\alpha) - \sum_\alpha f^-(\alpha)$. If f is complex-valued, Eq. (1) still applies, with the series interpreted as $\sum_\alpha \operatorname{Re} f(\alpha) + i \sum_\alpha \operatorname{Im} f(\alpha)$. The integral is defined provided $\sum_\alpha |f(\alpha)| < \infty$.

The space $L^p(\Omega, \mathscr{F}, \mu)$ will be denoted by $l^p(\Omega)$; it consists of all complex-valued functions $(f(\alpha), \alpha \in \Omega)$ such that $f(\alpha) = 0$ for all but countably many α, and

$$\|f\|_p^p = \sum_\alpha |f(\alpha)|^p < \infty.$$

If Ω is the set of positive integers, the space $l^p(\Omega)$ will be denoted simply by l^p; it consists of all sequences $f = \{a_n\}$ of complex numbers such that

$$\|f\|_p^p = \sum_{n=1}^\infty |a_n|^p < \infty.$$

It will be useful to state the Hölder and Minkowski inequalities for sums. If $f \in l^p(\Omega)$ and $g \in l^p(\Omega)$, where $1 < p < \infty$, $1 < q < \infty$, $(1/p) + (1/q) = 1$, then $fg \in l^1(\Omega)$ and

$$\sum_\alpha |f(\alpha)g(\alpha)| \le \left(\sum_\alpha |f(\alpha)|^p \right)^{1/p} \left(\sum_\alpha |g(\alpha)|^q \right)^{1/q}.$$

If $f, g \in l^p(\Omega)$, $1 \le p < \infty$, then $f + g \in l^p(\Omega)$ and

$$\left(\sum_\alpha |f(\alpha) + g(\alpha)|^p \right)^{1/p} \le \left(\sum_\alpha |f(\alpha)|^p \right)^{1/p} + \left(\sum_\alpha |g(\alpha)|^p \right)^{1/p}.$$

As in 2.4.5, we obtain the Cauchy–Schwarz inequality for sums from the Hölder inequality. If $f, g \in l^2(\Omega)$, then $fg \in l^1(\Omega)$ and

$$\left| \sum_\alpha f(\alpha)\overline{g(\alpha)} \right| \le \left(\sum_\alpha |f(\alpha)|^2 \right)^{1/2} \left(\sum_\alpha |g(\alpha)|^2 \right)^{1/2}.$$

If in the above discussion we replace Ω by $\{1, 2, \ldots, n\}$, all convergence difficulties are eliminated, and all the spaces $l^p(\Omega)$ coincide with $\mathbb{C}^n$.

If $0 < p < 1$, $\| \; \|_p$ is not a seminorm on $L^p(\Omega, \mathscr{F}, \mu)$. For let A and B be disjoint sets with $a = \mu(A)$ and $b = \mu(B)$ assumed finite and positive. If $f = I_A$, $g = I_B$, then

$$\|f + g\|_p = \left(\int_\Omega |f + g|^p \, d\mu \right)^{1/p} = \left(\int_\Omega (I_A + I_B) \, d\mu \right)^{1/p} = (a + b)^{1/p},$$

$$\|f\|_p = a^{1/p}, \qquad \|g\|_p = b^{1/p}.$$

But $(a + b)^{1/p} > a^{1/p} + b^{1/p}$ if $a, b > 0$, $0 < p < 1$, since $(a + x)^r - a^r - x^r$ is strictly increasing for $r > 1$, and has the value 0 when $x = 0$. Thus the triangle inequality fails. We can, however, describe convergence in L^p, $0 < p < 1$, in the following way. We use the inequality

$$(a + b)^p \le a^p + b^p, \qquad a, b \ge 0, \qquad 0 < p < 1,$$

which is proved by considering $(a + x)^p - a^p - x^p$. It follows that

$$\int_\Omega |f + g|^p \, d\mu \le \int_\Omega |f|^p \, d\mu + \int_\Omega |g|^p \, d\mu, \qquad f, g \in L^p, \qquad (2)$$

and therefore $d(f, g) = \int_\Omega |f - g|^p \, d\mu$ defines a pseudometric on L^p. In fact the pseudometric is complete (every Cauchy sequence converges); for Eq. (2)

implies that if $f, g \in L^p$, then $f + g \in L^p$, so that the proof of 2.4.11 goes through.

If Ω is an interval of reals, $\mathscr{F}$ is the class of Borel sets of Ω, and μ is Lebesgue measure, the space $L^p(\Omega, \mathscr{F}, \mu)$ will be denoted by $L^p(\Omega)$. Thus, for example, $L^p[a, b]$ is the set of all complex-valued Borel measurable functions f on $[a, b]$ such that

$$\| f \|_p^p = \int_a^b |f(x)|^p \, dx < \infty.$$

If f is a complex-valued Borel measurable function on $(\Omega, \mathscr{F}, \mu)$ and $f_1, f_2, \ldots \in L^p(\Omega, \mathscr{F}, \mu)$, we say that the sequence $\{f_n\}$ *converges to f in L^p* iff $\| f_n - f \|_p \to 0$, that is, iff $\int_\Omega |f_n - f|^p \, d\mu \to 0$ as $n \to \infty$. We use the notation $f_n \xrightarrow{L^p} f$. In Section 2.5, we shall compare various types of convergence of sequences of measurable functions. We show now that any $f \in L^p$ is an L^p-limit of simple functions.

2.4.13   Theorem. Let $f \in L^p$, $0 < p < \infty$. If $\varepsilon > 0$, there is a simple function $g \in L^p$ such that $\| f - g \|_p < \varepsilon$; g can be chosen to be finite-valued and to satisfy $|g| \leq |f|$. Thus the finite-valued simple functions are dense in L^p.

PROOF.   This follows from 1.5.5(b) and 1.6.10. $\square$

If we specialize to functions on $\mathbb{R}^n$ and Lebesgue–Stieltjes measures, we may obtain another basic approximation theorem.

2.4.14   Theorem. Let $f \in L^p(\Omega, \mathscr{F}, \mu)$, $0 < p < \infty$, where $\Omega = \mathbb{R}^n, \mathscr{F} = \mathscr{B}(\mathbb{R}^n)$, and μ is a Lebesgue–Stieltjes measure. If $\varepsilon > 0$, there is a continuous function $g \in L^p(\Omega, \mathscr{F}, \mu)$ such that $\| f - g \|_p < \varepsilon$; furthermore, g can be chosen so that $\sup |g| \leq \sup |f|$. Thus the continuous functions are dense in L^p.

PROOF.   By 2.4.13, it suffices to show that an indicator I_A in L^p can be approximated in the L^p sense by a continuous function with absolute value at most 1. Now $I_A \in L^p$ means that $\mu(A) < \infty$; hence by 1.4.11, there is a closed set $C \subset A$ and an open set $V \supset A$ such that $\mu(V - C) < \varepsilon^p 2^{-p}$. Let g be a continuous map of Ω into $[0, 1]$ with $g = 1$ on C and $g = 0$ on V^c (g exists by Urysohn's lemma). Then

$$\int_\Omega |I_A - g|^p \, d\mu = \int_{(I_A \neq g)} |I_A - g|^p \, d\mu.$$

But $\{I_A \neq g\} \subset V - C$ and $|I_A - g| \leq 2$; hence

$$\|I_A - g\|_p^p \leq 2^p \mu(V - C) < \varepsilon^p.$$

Since $g = g - I_A + I_A$, we have $g \in L^p$. $\square$

If $f_1, f_2, \ldots$ are continuous and f_n converges to f in L^p, it does not follow that f is continuous (see Problem 2).

2.4.15   The Space L^∞. If we wish to define L^p spaces for $p = \infty$, we must proceed differently. We define the *essential supremum* of the real-valued Borel measurable function g on $(\Omega, \mathscr{F}, \mu)$ as

$$\text{ess } \sup g = \inf\{c \in \overline{R}: \ \mu\{\omega: g(\omega) > c\} = 0\},$$

that is, the smallest number c such that $g \leq c$ a.e. $[\mu]$.

If f is a complex-valued Borel measurable function on $(\Omega, \mathscr{F}, \mu)$, we define

$$\|f\|_\infty = \text{ess } \sup |f|.$$

The space $L^\infty(\Omega, \mathscr{F}, \mu)$ is the collection of all f such that $\|f\|_\infty < \infty$. Thus $f \in L^\infty$ iff f is essentially bounded, that is, bounded outside a set of measure 0.

Now $|f + g| \leq |f| + |g| \leq \|f\|_\infty + \|g\|_\infty$ a.e.; hence

$$\|f + g\|_\infty \leq \|f\|_\infty + \|g\|_\infty.$$

In particular, $f, g \in L^\infty$ implies $f + g \in L^\infty$. The other properties of a seminorm are easily checked. Thus L^∞ is a vector space over the complex field, $\| \ \|_\infty$ is a seminorm on L^∞, and becomes a norm if we pass to equivalence classes as before.

If $f, f_1, f_2, \ldots \in L^\infty$ and $\|f_n - f\|_\infty \to 0$, we write $f_n \xrightarrow{L^\infty} f$; we claim that: $\|f_n - f\|_\infty \to 0$ iff there is a set $A \in \mathscr{F}$ with $\mu(A) = 0$ such that $f_n \to f$ uniformly on A^c.

Assume $\|f_n - f\|_\infty \to 0$. Given a positive integer m, $\|f_n - f\|_\infty \leq 1/m$ for sufficiently large n; hence $|f_n(\omega) - f(\omega)| \leq 1/m$ for almost every ω, say for $\omega \notin A_m$, where $\mu(A_m) = 0$. If $A = \bigcup_{m=1}^\infty A_m$, then $\mu(A) = 0$ and $f_n \to f$ uniformly on A^c. Conversely, assume $\mu(A) = 0$ and $f_n \to f$ uniformly on A^c. Given $\varepsilon > 0$, $|f_n - f| \leq \varepsilon$ on A^c for sufficiently large n, so that $|f_n - f| \leq \varepsilon$ a.e. Thus $\|f_n - f\|_\infty \leq \varepsilon$ for large enough n, and the result follows.

An identical argument shows that $\{f_n\}$ is a Cauchy sequence in L^∞ ($\|f_n - f_m\|_\infty \to 0$ as $n, m \to \infty$) iff there is a set $A \in \mathscr{F}$ with $\mu(A) = 0$ and $f_n - f_m \to 0$ uniformly on A^c.

It is immediate that the Hölder inequality still holds when $p = 1, q = \infty$, and we have shown above that the Minkowski inequality holds when $p = \infty$.

To show that L^∞ is complete, let $\{f_n\}$ be a Cauchy sequence in L^∞, and let A be a set of measure 0 such that $f_n(\omega) - f_m(\omega) \to 0$ uniformly for $\omega \in A^c$. But then $f_n(\omega)$ converges to a limit $f(\omega)$ for each $\omega \in A^c$, and the convergence is uniform on A^c. If we define $f(\omega) = 0$ for $\omega \in A$, we have $f \in L^\infty$ and $f_n \xrightarrow{\;L^\infty\;} f$.

Theorem 2.4.13 holds also when $p = \infty$. For if f is a function in L^∞, the standard approximating sequence $\{f_n\}$ of simple functions (see 1.5.5) converges to f uniformly, outside a set of measure 0. However, Theorem 2.4.14 fails when $p = \infty$ (see Problem 12).

If Ω is an arbitrary set, $\mathscr{F}$ consists of all subsets of Ω, and μ is counting measure, then $L^\infty(\Omega, \mathscr{F}, \mu)$ is the set of all bounded complex-valued functions $f = (f(\alpha), \alpha \in \Omega)$, denoted by $l^\infty(\Omega)$. The essential supremum is simply the supremum; in other words, $\|f\|_\infty = \sup\{|f(\alpha)|: \alpha \in \Omega\}$. If Ω is the set of positive integers, $l^\infty(\Omega)$ is the space of bounded sequences of complex numbers, denoted simply by l^∞.

Problems

1.　(a)　If $f = \{a_n, n = 1, 2, \ldots\}$, the a_n are real or complex numbers, and μ is counting measure on subsets of the positive integers, show that $\int_\Omega f \, d\mu = \sum_{n=1}^\infty a_n$, where the sum is interpreted as in 2.4.12.

　　(b)　If $f = (f(\alpha), \alpha \in \Omega)$ is a real- or complex-valued function on the arbitrary set Ω, and μ is counting measure on subsets of Ω, show that $\int_\Omega f \, d\mu = \sum_\alpha f(\alpha)$, where the sum is interpreted as in 2.4.12.

2.　Give an example of functions $f, f_1, f_2, \ldots$ from $\mathbb{R}$ to $[0, 1]$ such that

　　(a)　each f_n is continuous on $\mathbb{R}$,

　　(b)　$f_n(x)$ converges to $f(x)$ for all x, $\int_{-\infty}^\infty |f_n(x) - f(x)|^p \, dx \to 0$ for every $p \in (0, \infty)$, and

　　(c)　f is discontinuous at some point of $\mathbb{R}$.

3.　For each $n = 1, 2, \ldots$, let $f_n = \{a_1^{(n)}, a_2^{(n)}, \ldots\}$ be a sequence of complex numbers.

　　(a)　If the $a_k^{(n)}$ are real and $0 \le a_k^{(n)} \le a_k^{(n+1)}$ for all k and n, show that

$$\lim_{n \to \infty} \sum_{k=1}^\infty a_k^{(n)} = \sum_{k=1}^\infty \lim_{n \to \infty} a_k^{(n)}.$$

Show that the same conclusion holds if the $a_k^{(n)}$ are complex, $\lim_{n \to \infty} a_k^{(n)}$ exists for each k, and $|a_k^{(n)}| \le b_k$ for all k and n, where $\sum_{k=1}^\infty b_k < \infty$.

(b) If the $a_k^{(n)}$ are real and nonnegative, show that

$$\sum_{k=1}^{\infty}\sum_{n=1}^{\infty} a_k^{(n)} = \sum_{n=1}^{\infty}\sum_{k=1}^{\infty} a_k^{(n)}.$$

(c) If the $a_k^{(n)}$ are complex and $\sum_{n=1}^{\infty}\sum_{k=1}^{\infty} |a_k^{(n)}| < \infty$, show that $\sum_{n=1}^{\infty}\sum_{k=1}^{\infty} a_k^{(n)}$ and $\sum_{k=1}^{\infty}\sum_{n=1}^{\infty} a_k^{(n)}$ both converge to the same finite number.

4. Show that there is equality in the Hölder inequality iff $|f|^p$ and $|g|^q$ are linearly dependent, that is, iff $A|f|^p = B|g|^q$ a.e. for some constants A and B, not both 0.

5. If f is a complex-valued μ-integrable function, show that $|\int_{\Omega} f\, d\mu| = \int_{\Omega} |f|\, d\mu$ iff arg f is a.e. constant on $\{\omega\colon f(\omega) \neq 0\}$.

6. Show that equality holds in the Cauchy–Schwarz inequality iff f and g are linearly dependent.

7. (a) If $1 < p < \infty$, show that equality holds in the Minkowski inequality iff $Af = Bg$ a.e. for some nonnegative constants A and B, not both 0.
    (b) What are the conditions for equality if $p = 1$?

8. If $0 < r < s < \infty$, and $f \in L^s(\Omega, \mathcal{F}, \mu)$, μ finite, show that $\|f\|_r \leq k\|f\|_s$ for some finite positive constant k. Thus $L^s \subset L^r$ and L^s convergence implies L^r convergence. (We may take $k = 1$ if μ is a probability measure.) Note that finiteness of μ is essential here; if μ is Lebesgue measure on $\mathcal{B}(\mathbb{R})$ and $f(x) = 1/x$ for $x \geq 1$, $f(x) = 0$ for $x < 1$, then $f \in L^2$ but $f \notin L^1$.

9. If μ is finite, show that $\|f\|_p \to \|f\|_{\infty}$ as $p \to \infty$. Give an example to show that this fails if $\mu(\Omega) = \infty$.

10. (*Radon–Nikodym theorem, complex case*) If μ is a nonnegative, real measure, λ a complex measure on $(\Omega, \mathcal{F})$, and $\lambda \ll \mu$, show that there is a complex-valued μ-integrable function g such that $\lambda(A) = \int_A g\, d\mu$ for all $A \in \mathcal{F}$. If h is another such function, $g = h$ a.e.

Show also that the Lebesgue decomposition theorem holds if λ is a complex measure and μ is a σ-finite measure. (See Problem 6, Section 2.2, for properties of complex measures.)

11. (a) Let f be a complex-valued μ-integrable function, where μ is a nonnegative real measure. If S is a closed set of complex numbers and $[1/\mu(E)] \int_E f\, d\mu \in S$ for all measurable sets E such that $\mu(E) > 0$, show that $f(\omega) \in S$ for almost every ω. [If D is a closed disk with center at z and radius r, and $D \subset S^c$, take $E = f^{-1}(D)$. Show that $|\int_E (f - z)\, d\mu| \leq r\mu(E)$, and conclude that $\mu(E) = 0$.]

(b)   If λ is a complex measure, then $\lambda \ll |\lambda|$ by definition of $|\lambda|$; hence by the Radon–Nikodym theorem, there is a $|\lambda|$-integrable complex-valued function h such that $\lambda(E) = \int_E h \, d|\lambda|$ for all $E \in \mathscr{F}$. Show that $|h| = 1$ a.e. $[|\lambda|]$. [Let $A_r = \{\omega: |h(\omega)| < r\}$, $0 < r < 1$, and use the definition of $|\lambda|$ to show that $|h| \geq 1$ a.e. Use part (a) to show $|h| \leq 1$ a.e.]

(c)   Let μ be a nonnegative real measure, g a complex-valued μ-integrable function, and $\lambda(E) = \int_E g \, d\mu$, $E \in \mathscr{F}$. If $h = d\lambda/d|\lambda|$ as in part (b), show that $|\lambda|(E) = \int_E \bar{h}g \, d\mu$. (Intuitively, $\bar{h}g \, d\mu = \bar{h} \, d\lambda = \bar{h}h \, d|\lambda| = |h|^2 \, d|\lambda| = d|\lambda|$. Formally, show that $\int_\Omega f h \, d|\lambda| = \int_\Omega fg \, d\mu$ if f is a bounded, complex-valued, Borel measurable function, and set $f = \bar{h}I_E$.)

(d)   Under the hypothesis of (c), show that

$$|\lambda|(E) = \int_E |g| \, d\mu \qquad \text{for all} \qquad E \in \mathscr{F}.$$

12.   Give an example of a bounded real-valued function f on $\mathbb{R}$ such that there is no sequence of continuous functions f_n such that $\|f - f_n\|_\infty \to 0$. Thus the continuous functions are not dense in $L^\infty(R)$.

2.5   CONVERGENCE OF SEQUENCES OF MEASURABLE FUNCTIONS

In the previous section we introduced the notion of L^p convergence; we are also familiar with convergence almost everywhere. We now consider other types of convergence and make comparisons.

Let $f, f_1, f_2, \ldots$ be complex-valued Borel measurable functions on $(\Omega, \mathscr{F}, \mu)$. We say that $f_n \to f$ in *measure* (or in μ-measure if we wish to emphasize the dependence on μ) iff for every $\varepsilon > 0$, $\mu\{\omega: |f_n(\omega) - f(\omega)| \geq \varepsilon\} \to 0$ as $n \to \infty$. (Notation: $f_n \xrightarrow{\mu} f$.) When μ is a probability measure, the convergence is called *convergence in probability*.

The first result shows that L^p convergence is stronger than convergence in measure.

2.5.1   Theorem. If $f, f_1, f_2, \ldots \in L^p(0 < p < \infty)$, then, $f_n \xrightarrow{L^p} f$ implies $f_n \xrightarrow{\mu} f$.

PROOF.   Apply Chebyshev's inequality (2.4.9) to $|f_n - f|$. $\quad\square$

The same argument shows that if $\{f_n\}$ is a Cauchy sequence in L^p, then $\{f_n\}$ is *Cauchy in measure*, that is, given $\varepsilon > 0$, $\mu\{\omega: |f_n(\omega) - f_m(\omega)| \geq \varepsilon\} \to 0$ as $n, m \to \infty$.

If $f, f_1, f_2, \ldots$ are complex-valued Borel measurable functions on $(\Omega, \mathscr{F}, \mu)$, we say that $f_n \to f$ *almost uniformly* iff, given $\varepsilon > 0$, there is a set $A \in \mathscr{F}$ such that $\mu(A) < \varepsilon$ and $f_n \to f$ uniformly on A^c.

Almost uniform convergence is stronger than both a.e. convergence and convergence in measure, as we now prove.

2.5.2   Theorem. If $f_n \to f$ almost uniformly, then $f_n \to f$ in measure and almost everywhere.

PROOF.   If $\varepsilon > 0$, let $f_n \to f$ uniformly on A^c, with $\mu(A) < \varepsilon$. If $\delta > 0$, then eventually $|f_n - f| < \delta$ on A^c, so $\{|f_n - f| \geq \delta\} \subset A$. Therefore $\mu\{|f_n - f| \geq \delta\} \leq \mu(A) < \varepsilon$, proving convergence in measure.

To prove almost everywhere convergence, choose, for each positive integer k, a set A_k with $\mu(A_k) < 1/k$ and $f_n \to f$ uniformly on A_k^c. If $B = \bigcup_{k=1}^{\infty} A_k^c$, then $f_n \to f$ on B and $\mu(B^c) = \mu(\bigcap_{k=1}^{\infty} A_k) \leq \mu(A_k) \to 0$ as $k \to \infty$. Thus $\mu(B^c) = 0$ and the result follows.   □

The converse to 2.5.2 does not hold in general, as we shall see in 2.5.6(c), but we do have the following result.

2.5.3   Theorem. If $\{f_n\}$ is convergent in measure, there is a subsequence converging almost uniformly (in particular, a.e. and (of course) in measure) to the same limit function.

PROOF.   First note that $\{f_n\}$ is Cauchy in measure, because if $|f_n - f_m| \geq \varepsilon$, then either $|f_n - f| \geq \varepsilon/2$ or $|f - f_m| \geq \varepsilon/2$. Thus

$$\mu\{|f_n - f_m| \geq \varepsilon\} \leq \mu\left\{|f_n - f| \geq \frac{\varepsilon}{2}\right\}$$

$$+ \mu\left\{|f - f_m| \geq \frac{\varepsilon}{2}\right\} \to 0 \qquad \text{as} \qquad n, m \to \infty.$$

Now for each positive integer k, choose a positive integer N_k such that $N_{k+1} > N_k$ for all k and

$$\mu\{\omega \colon |f_n(\omega) - f_m(\omega)| \geq 2^{-k}\} \leq 2^{-k} \qquad \text{for} \qquad n, m \geq N_k.$$

Pick integers $n_k \geq N_k$ with $n_k < n_{k+1}$, $k = 1, 2, \ldots$; then if $g_k = f_{n_k}$,

$$\mu\{\omega \colon |g_k(\omega) - g_{k+1}(\omega)| \geq 2^{-k}\} \leq 2^{-k}.$$

Let $A_k = \{|g_k - g_{k+1}| \geq 2^{-k}\}$, $A = \limsup_k A_k$. Then $\mu(A) = 0$ by 2.2.4; but if $\omega \notin A$, then $\omega \in A_k$ for only finitely many k; hence $|g_k(\omega) - g_{k+1}(\omega)| < 2^{-k}$

for large k, and it follows that $g_k(\omega)$ converges to a limit $g(\omega)$. Since $\mu(A) = 0$ we have $g_k \to g$ a.e.

If $B_r = \bigcup_{k=r}^{\infty} A_k$, then $\mu(B_r) \le \sum_{k=r}^{\infty} \mu(A_k) < \varepsilon$ for large r. If $\omega \notin B_r$, then $|g_k(\omega) - g_{k+1}(\omega)| < 2^{-k}$, $k = r, r+1, r+2, \ldots$. By the Weierstrass M-test, $g_k \to g$ uniformly on B_r, which proves almost uniform convergence.

Now by hypothesis, we have $f_n \xrightarrow{\ \mu\ } f$ for some f, hence $f_{n_k} \xrightarrow{\ \mu\ } f$. But by 2.5.2, $f_{n_k} \xrightarrow{\ \mu\ } g$ as well, hence $f = g$ a.e. (see Problem 1). Thus f_{n_k} converges almost uniformly to f, completing the proof. $\square$

There is a partial converse to 2.5.2, but before discussing this it will be convenient to look at a condition equivalent to a.e. convergence.

2.5.4   Lemma. If μ is finite, then $f_n \to f$ a.e. iff for every $\delta > 0$,

$$\mu\left(\bigcup_{k=n}^{\infty} \{\omega \colon |f_k(\omega) - f(\omega)| \ge \delta\}\right) \to 0 \qquad \text{as} \qquad n \to \infty.$$

PROOF. Let $B_{n\delta} = \{\omega \colon |f_n(\omega) - f(\omega)| \ge \delta\}$, $B_\delta = \limsup_n B_{n\delta} = \bigcap_{n=1}^{\infty} \bigcup_{k=n}^{\infty} B_{k\delta}$. Now $\bigcup_{k=n}^{\infty} B_{k\delta} \downarrow B_\delta$; hence $\mu(\bigcup_{k=n}^{\infty} B_{k\delta}) \to \mu(B_\delta)$ as $n \to \infty$ by 1.2.7(b). Now

$$\{\omega \colon f_n(\omega) \nrightarrow f(\omega)\} = \bigcup_{\delta > 0} B_\delta$$

$$= \bigcup_{m=1}^{\infty} B_{1/m} \qquad \text{since} \qquad B_{\delta_1} \subset B_{\delta_2} \qquad \text{for} \qquad \delta_1 > \delta_2.$$

Therefore,

$$f_n \to f \quad \text{a.e. iff } \mu(B_\delta) = 0 \qquad \text{for all} \qquad \delta > 0$$

$$\text{iff } \mu\left(\bigcup_{k=n}^{\infty} B_{k\delta}\right) \to 0 \qquad \text{for all} \qquad \delta > 0. \ \square$$

2.5.5   Egoroff's Theorem. If μ is finite and $f_n \to f$ a.e., then $f_n \to f$ almost uniformly. Hence by 2.5.2, if μ is finite, then almost everywhere convergence implies convergence in measure.

PROOF.   It follows from 2.5.4 that given $\varepsilon > 0$ and a positive integer j, for sufficiently large $n = n(j)$, the set $A_j = \bigcup_{k=n(j)}^{\infty} \{|f_k - f| \ge 1/j\}$ has measure less than $\varepsilon/2^j$. If $A = \bigcup_{j=1}^{\infty} A_j$, then $\mu(A) \le \sum_{j=1}^{\infty} \mu(A_j) < \varepsilon$. Also, if $\delta > 0$

and j is chosen so that $1/j < \delta$, we have, for any $k \geq n(j)$ and $\omega \in A^c$ (hence $\omega \notin A_j$), $|f_k(\omega) - f(\omega)| < 1/j < \delta$. Thus $f_n \to f$ uniformly on A^c. $\square$

We now give some examples to illustrate the relations between the various types of convergence. In all cases, we assume that $\mathscr{F}$ is the class of Borel sets and μ is Lebesgue measure.

2.5.6 Examples. (a) Let $\Omega = [0, 1]$ and define

$$f_n(x) = \begin{cases} e^n & \text{if } 0 \leq x \leq \dfrac{1}{n}, \\ 0 & \text{elsewhere.} \end{cases}$$

Then $f_n \to 0$ a.e., hence in measure by 2.5.5. But for each $p \in (0, \infty]$, f_n fails to converge in L^p. For if $p < \infty$,

$$\|f_n\|_p^p = \int_0^1 |f_n(x)|^p \, dx = \frac{1}{n} e^{np} \to \infty,$$

and

$$\|f_n\|_\infty = e^n \to \infty.$$

(b) Let $\Omega = \mathbb{R}$, and define

$$f_n(x) = \begin{cases} \dfrac{1}{n} & \text{if } 0 \leq x \leq e^n, \\ 0 & \text{elsewhere.} \end{cases}$$

Then $f_n \to 0$ uniformly on $\mathbb{R}$, so that $f_n \xrightarrow{L^\infty} 0$. It follows quickly that $f_n \to 0$ a.e. and in measure. But for each $p \in (0, \infty)$, f_n fails to converge in L^p, since $\|f_n\|_p^p = n^{-p} e^n \to \infty$.

(c) Let $\Omega = [0, \infty)$ and define

$$f_n(x) = \begin{cases} 1 & \text{if } n \leq x \leq n + \dfrac{1}{n}, \\ 0 & \text{elsewhere.} \end{cases}$$

Then $f_n \to 0$ a.e. and in measure (as well as in L^p, $0 < p < \infty$), but does not converge almost uniformly. For, if $f_n \to 0$ uniformly on A and $\mu(A^c) < \varepsilon$, then eventually $f_n < 1$ on A; hence if $A_n = [n, n + (1/n)]$ we have $A \cap \bigcup_{k \geq n} A_k = \emptyset$ for sufficiently large n. Therefore, $A^c \supset \bigcup_{k \geq n} A_k$, and consequently $\mu(A^c) \geq \sum_{k=n}^\infty \mu(A_k) = \infty$, a contradiction. Note that if we change

$f_n(x)$ so that it is 1 for $n \le x \le n+1$ and 0 elsewhere, then f_n converges to 0 almost everywhere but not in measure, hence not almost uniformly.

(d)   Let $\Omega = [0, 1]$, and define

$$f_{nm}(x) = \begin{cases} 1 & \text{if } \dfrac{m-1}{n} < x \le \dfrac{m}{n}, & m = 1, \ldots, n, & n = 1, 2, \ldots, \\ 0 & \text{elsewhere.} \end{cases}$$

Then $\|f_{nm}\|_p^p = 1/n \to 0$, so for each $p \in (0, \infty)$, the sequence $f_{11}, f_{21}, f_{22}, f_{31}, f_{32}, f_{33}, \ldots$ converges to 0 in L^p (hence converges in measure by 2.5.1). But the sequence does not converge a.e., hence by 2.5.2, does not converge almost uniformly. To see this, observe that for any $x \ne 0$, the sequence $\{f_{nm}(x)\}$ has infinitely many zeros and infinitely many ones. Thus the set on which f_{nm} converges has measure 0. Also, f_{nm} does not converge in L^∞, for if $f_{nm} \xrightarrow{L^\infty} f$, then $f_{nm} \xrightarrow{\mu} f$, hence $f = 0$ a.e. (see Problem 1). But $\|f_{nm}\|_\infty \equiv 1$, a contradiction.

Problems

1.   If f_n converges to both f and g in measure, show that $f = g$ a.e.

2.   Show that a sequence is Cauchy in measure iff it is convergent in measure.

3.   (a)   If μ is finite, show that L^∞ convergence implies L^p convergence for all $p \in (0, \infty)$.

   (b)   Show that any real-valued function in $L^p[a, b]$, $-\infty < a < b < \infty$, $0 < p < \infty$, can be approximated in L^p by a polynomial, in fact by a polynomial with rational coefficients.

4.   If μ is finite, show that $\{f_n\}$ is Cauchy a.e. (for almost every ω, $\{f_n(\omega)\}$ is a Cauchy sequence) iff for every $\delta > 0$,

$$\mu\left(\bigcup_{j,k=n}^{\infty} \{\omega: |f_j(\omega) - f_k(\omega)| \ge \delta\} \right) \to 0 \qquad \text{as} \qquad n \to \infty.$$

5.   (*Extension of the dominated convergence theorem*) If $|f_n| \le g$ for all $n = 1, 2, \ldots$, where g is μ-integrable, and $f_n \xrightarrow{\mu} f$, show that f is μ-integrable and $\int_\Omega f_n \, d\mu \to \int_\Omega f \, d\mu$.

6.   A metric may be defined on the space of all measurable functions on $(\Omega, \mathscr{F}, \mu)$ by $d(f, g) = \int_\Omega \frac{|f-g|}{1+|f-g|} \, d\mu$. (Functions that agree almost everywhere are identified.)

(a)   If $d(f_n, f) \to 0$, show that $f_n \xrightarrow{\mu} f$.

(b)   If μ is finite, show that $f_n \xrightarrow{\mu} f$ implies $d(f_n, f) \to 0$.

(c)   Give an example in which $f_n \xrightarrow{\mu} f$ but $d(f_n, f)$ does not approach 0.

7.   If μ is a finite measure and for every $\varepsilon > 0$, $\sum_{n=1}^{\infty} P\{|f_n - f| \geq \varepsilon\} < \infty$, show that $f_n \to f$ almost everywhere. Thus by Chebyshev's inequality, $\sum_{n=1}^{\infty} \|f_n - f\|_p^p < \infty$ implies that $f_n \to f$ almost everywhere.

2.6   PRODUCT MEASURES AND FUBINI'S THEOREM

Lebesgue measure on $\mathbb{R}^n$ is in a sense the product of n copies of one-dimensional Lebesgue measure, because the volume of an n-dimensional rectangular box is the product of the lengths of the sides. In this section we develop this idea in a general setting. We shall be interested in two constructions. First, suppose that $(\Omega_j, \mathscr{F}_j, \mu_j)$ is a measure space for $j = 1, 2, \ldots, n$. We wish to construct a measure on subsets of $\Omega_1 \times \Omega_2 \times \cdots \times \Omega_n$ such that the measure of the "rectangle" $A_1 \times A_2 \times \cdots \times A_n$ [with each $A_j \in \mathscr{F}_j$] is $\mu_1(A_1)\mu_2(A_2) \cdots \mu_n(A_n)$. The second construction involves compound experiments in probability. Suppose that two observations are made, with the first observation resulting in a point $\omega_1 \in \Omega_1$, the second in a point $\omega_2 \in \Omega_2$. The probability that the first observation falls into the set A is, say, $\mu_1(A)$. Furthermore, if the first observation is ω_1, the probability that the second observation falls into B is, say, $\mu(\omega_1, B)$, where $\mu(\omega_1, \cdot)$ is a probability measure defined on $\mathscr{F}_2$ for each $\omega_1 \in \Omega_1$. The probability that the first observation will belong to A and the second will belong to B should be given by

$$\mu(A \times B) = \int_A \mu(\omega_1, B)\mu_1(d\omega_1),$$

and we would like to construct a probability measure on subsets of $\Omega_1 \times \Omega_2$ such that $\mu(A \times B)$ is given by this formula for each $A \in \mathscr{F}_1$ and $B \in \mathscr{F}_2$. [Intuitively, the probability that the first observation will fall near ω_1 is $\mu_1(d\omega_1)$; given that the first observation is ω_1, the second observation will fall in B with probability $\mu(\omega_1, B)$. Thus $\mu(\omega_1, B)\mu_1(d\omega_1)$ represents the probability of one possible favorable outcome of the experiment. The total probability is found by adding the probabilities of favorable outcomes, in other words, by integrating over A. Reasoning of this type may not appear natural at this point, since we have not yet talked in detail about probability theory. However, it may serve to indicate the motivation behind the theorems of this section.]

2.6.1   *Definition.*   Let $\mathscr{F}_j$ be a σ-field of subsets of Ω_j, $j = 1, 2, \ldots, n$, and let $\Omega = \Omega_1 \times \Omega_2 \times \cdots \times \Omega_n$. A *measurable rectangle* in Ω is a set $A = A_1 \times A_2 \times \cdots \times A_n$, where $A_j \in \mathscr{F}_j$ for each $j = 1, 2, \ldots, n$. The smallest σ-field containing the measurable rectangles is called the *product σ-field*, written $\mathscr{F}_1 \times \mathscr{F}_2 \times \cdots \times \mathscr{F}_n$. If all $\mathscr{F}_j$ coincide with a fixed σ-field $\mathscr{F}$, the product σ-field is denoted by $\mathscr{F}^n$. Note that in spite of the notation, $\mathscr{F}_1 \times \mathscr{F}_2 \times \cdots \times \mathscr{F}_n$ is not the Cartesian product of the $\mathscr{F}_j$; the Cartesian product is the set of measurable rectangles, while the product σ-field is the minimal σ-field over the measurable rectangles. Note also that the collection of finite disjoint unions of measurable rectangles forms a field (see Problem 1).

     The next theorem is stated in such a way that both constructions described above become special cases.

2.6.2   *Product Measure Theorem.*   Let $(\Omega_1, \mathscr{F}_1, \mu_1)$ be a measure space, with μ_1 σ-finite on $\mathscr{F}_1$, and let Ω_2 be a set with σ-field $\mathscr{F}_2$. Assume that for each $\omega_1 \in \Omega_1$ we are given a measure $\mu(\omega_1, \cdot)$ on $\mathscr{F}_2$. Assume that $\mu(\omega_1, B)$, besides being a measure in B for each fixed $\omega_1 \in \Omega_1$, is Borel measurable in ω_1 for each fixed $B \in \mathscr{F}_2$. Assume that the $\mu(\omega_1 \cdot)$ are uniformly σ-finite; that is, Ω_2 can be written as $\bigcup_{n=1}^{\infty} B_n$, where for some positive (finite) constants k_n we have $\mu(\omega_1, B_n) \leq k_n$ for all $\omega_1 \in \Omega_1$. [The case in which the $\mu(\omega_1, \cdot)$ are uniformly bounded, that is, $\mu(\omega_1, \Omega_2) \leq k < \infty$ for all ω_1, is of course included.]

     Then there is a unique measure μ on $\mathscr{F} = \mathscr{F}_1 \times \mathscr{F}_2$ such that

$$\mu(A \times B) = \int_A \mu(\omega_1, B)\mu_1(d\omega_1) \qquad \text{for all} \qquad A \in \mathscr{F}_1, \qquad B \in \mathscr{F}_2,$$

namely,

$$\mu(F) = \int_{\Omega_1} \mu(\omega_1, F(\omega_1))\mu_1(d\omega_1), \qquad F \in \mathscr{F},$$

where $F(\omega_1)$ denotes the *section* of F at ω_1:

$$F(\omega_1) = \{\omega_2 \in \Omega_2 \colon (\omega_1, \omega_2) \in F\}.$$

Furthermore, μ is σ-finite on $\mathscr{F}$; if μ_1 and all the $\mu(\omega_1, \cdot)$ are probability measures, so is μ.

PROOF.   First assume that the $\mu(\omega_1, \cdot)$ are finite.

     (1)   If $C \in \mathscr{F}$, then $C(\omega_1) \in \mathscr{F}_2$ for each $\omega_1 \in \Omega_1$.

     To prove this, let $\mathscr{C} = \{C \in \mathscr{F} \colon C(\omega_1) \in \mathscr{F}_2\}$. Then $\mathscr{C}$ is a σ-field since

$$\left(\bigcup_{n=1}^{\infty} C_n\right)(\omega_1) = \bigcup_{n=1}^{\infty} C_n(\omega_1), \qquad C^c(\omega_1) = (C(\omega_1))^c.$$

If $A \in \mathscr{F}_1$, $B \in \mathscr{F}_2$, then $(A \times B)(\omega_1) = B$ if $\omega_1 \in A$ and $\emptyset$ if $\omega_1 \notin A$. Thus $\mathscr{C}$ contains all measurable rectangles; hence $\mathscr{C} = \mathscr{F}$.

(2) If $C \in \mathscr{F}$, then $\mu(\omega_1, C(\omega_1))$ is Borel measurable in ω_1.

To prove this, let $\mathscr{C}$ be the class of sets in $\mathscr{F}$ for which the conclusion of (2) holds. If $C = A \times B$, $A \in \mathscr{F}_1$, $B \in \mathscr{F}_2$, then

$$\mu(\omega_1, C(\omega_1)) = \begin{cases} \mu(\omega_1, B) & \text{if} \quad \omega_1 \in A, \\ \mu(\omega_1, \emptyset) = 0 & \text{if} \quad \omega_1 \notin A. \end{cases}$$

Thus $\mu(\omega_1, C(\omega_1)) = \mu(\omega_1, B)I_A(\omega_1)$, and is Borel measurable by hypothesis. Therefore measurable rectangles belong to $\mathscr{C}$. If $C_1, \ldots, C_n$ are disjoint measurable rectangles,

$$\mu\left(\omega_1, \left(\bigcup_{i=1}^{n} C_i\right)(\omega_1)\right) = \mu\left(\omega_1, \bigcup_{i=1}^{n} C_i(\omega_1)\right) = \sum_{i=1}^{n} \mu(\omega_1, C_i(\omega_1))$$

is a finite sum of Borel measurable functions, and hence is Borel measurable in ω_1. Thus $\mathscr{C}$ contains the field of finite disjoint unions of measurable rectangles. But $\mathscr{C}$ is a monotone class, for if $C_n \in \mathscr{C}$, $n = 1, 2, \ldots$, and $C_n \uparrow C$, then $C_n(\omega_1) \uparrow C(\omega_1)$; hence $\mu(\omega_1, C_n(\omega_1)) \to \mu(\omega_1, C(\omega_1))$. Thus $\mu(\omega_1, C(\omega_1))$, a limit of measurable functions, is measurable in ω_1. If $C_n \downarrow C$, the same conclusion holds since the $\mu(\omega_1, \cdot)$ are finite. Thus $\mathscr{C} = \mathscr{F}$.

(3) Define

$$\mu(F) = \int_{\Omega_1} \mu(\omega_1, F(\omega_1))\mu_1(d\omega_1), \qquad F \in \mathscr{F}$$

[the integral exists by (2)]. Then μ is a measure on $\mathscr{F}$, and

$$\mu(A \times B) = \int_A \mu(\omega_1, B)\mu_1(d\omega_1) \qquad \text{for all} \qquad A \in \mathscr{F}_1, \qquad B \in \mathscr{F}_2.$$

To prove this, let $F_1, F_2, \ldots$ be disjoint sets in $\mathscr{F}$. Then

$$\mu\left(\bigcup_{n=1}^{\infty} F_n\right) = \int_{\Omega_1} \mu\left(\omega_1, \bigcup_{n=1}^{\infty} F_n(\omega_1)\right)\mu_1(d\omega_1)$$

$$= \int_{\Omega_1} \sum_{n=1}^{\infty} \mu(\omega_1, F_n(\omega_1))\mu_1(d\omega_1)$$

$$= \sum_{n=1}^{\infty} \int_{\Omega_1} \mu(\omega_1, F_n(\omega_1))\mu_1(d\omega_1) = \sum_{n=1}^{\infty} \mu(F_n),$$

proving that μ is a measure. Now

$$\mu(A \times B) = \int_{\Omega_1} \mu(\omega_1, (A \times B)(\omega_1))\mu_1(d\omega_1)$$

$$= \int_{\Omega_1} \mu(\omega_1, B)I_A(\omega_1)\mu_1(d\omega_1) \qquad \text{[see (2)]}$$

$$= \int_A \mu(\omega_1, B)\mu_1(d\omega_1)$$

as desired.

Now assume the $\mu(\omega_1, \cdot)$ uniformly σ-finite. Let $\Omega_2 = \bigcup_{n=1}^{\infty} B_n$, where the B_n are disjoint sets in $\mathcal{F}_2$ and $\mu(\omega_1, B_n) \leq k_n < \infty$ for all $\omega_1 \in \Omega_1$. If we set

$$\mu_n'(\omega_1, B) = \mu(\omega_1, B \cap B_n), \qquad B \in \mathcal{F}_2,$$

the $\mu_n'(\omega_1, \cdot)$ are finite, and the above construction gives a measure μ_n' on $\mathcal{F}$ such that

$$\mu_n'(A \times B) = \int_A \mu_n'(\omega_1, B)\mu_1(d\omega_1), \qquad A \in \mathcal{F}_1, \qquad B \in \mathcal{F}_2$$

$$= \int_A \mu(\omega_1, B \cap B_n)\mu_1(d\omega_1),$$

namely,

$$\mu_n'(F) = \int_{\Omega_1} \mu_n'(\omega_1, F(\omega_1))\mu_1(d\omega_1) = \int_{\Omega_1} \mu(\omega_1, F(\omega_1) \cap B_n)\mu_1(d\omega_1).$$

Let $\mu = \sum_{n=1}^{\infty} \mu_n'$; μ has the desired properties.

For the uniqueness proof, assume $\mu(\omega_1, \cdot)$ to be uniformly σ-finite. If λ is a measure on $\mathcal{F}$ such that $\lambda(A \times B) = \int_A \mu(\omega_1, B)\mu_1(d\omega_1)$ for all $A \in \mathcal{F}_1$, $B \in \mathcal{F}_2$, then $\lambda = \mu$ on the field $\mathcal{F}_0$ of finite disjoint unions of measurable rectangles. Now μ is σ-finite on $\mathcal{F}_0$, for if $\Omega_2 = \bigcup_{n=1}^{\infty} B_n$ with $B_n \in \mathcal{F}_2$ and $\mu(\omega_1, B_n) \leq k_n < \infty$ for all ω_1, and $\Omega_1 = \bigcup_{m=1}^{\infty} A_m$, where the A_m belong to $\mathcal{F}_1$ and $\mu_1(A_m) < \infty$, then $\Omega_1 \times \Omega_2 = \bigcup_{m,n=1}^{\infty}(A_m \times B_n)$ and

$$\mu(A_m \times B_n) = \int_{A_m} \mu(\omega_1, B_n)\mu_1(d\omega_1) \leq k_n \mu_1(A_m) < \infty.$$

Thus $\lambda = \mu$ on $\mathcal{F}$ by the Carathéodory extension theorem.

We have just seen that μ is σ-finite on $\mathscr{F}_0$, hence on $\mathscr{F}$. If μ_1 and all the $\mu(\omega_1, \cdot)$ are probability measures, it is immediate that μ is also. $\square$

2.6.3 *Corollary: Classical Product Measure Theorem.* Let $(\Omega_j, \mathscr{F}_j, \mu_j)$ be a measure space for $j = 1, 2$, with μ_j σ-finite on $\mathscr{F}_j$. If $\Omega = \Omega_1 \times \Omega_2$, $\mathscr{F} = \mathscr{F}_1 \times \mathscr{F}_2$, the set function given by

$$\mu(F) = \int_{\Omega_1} \mu_2(F(\omega_1)) \, d\mu_1(\omega_1) = \int_{\Omega_2} \mu_1(F(\omega_2)) \, d\mu_2(\omega_2)$$

is the unique measure on $\mathscr{F}$ such that $\mu(A \times B) = \mu_1(A)\mu_2(B)$ for all $A \in \mathscr{F}_1$, $B \in \mathscr{F}_2$. Furthermore, μ is σ-finite on $\mathscr{F}$, and is a probability measure if μ_1 and μ_2 are. The measure μ is called the product of μ_1 and μ_2, written $\mu = \mu_1 \times \mu_2$.

PROOF. In 2.6.2, take $\mu(\omega_1, \cdot) = \mu_2$ for all ω_1. The second formula for $\mu(F)$ is obtained by interchanging μ_1 and μ_2. $\square$

As a special case, Let $\Omega_1 = \Omega_2 = \mathbb{R}$, $\mathscr{F}_1 = \mathscr{F}_2 = \mathscr{B}(\mathbb{R})$, $\mu_1 = \mu_2 =$ Lebesgue measure. Then $\mathscr{F}_1 \times \mathscr{F}_2 = \mathscr{B}(\mathbb{R}^2)$ (Problem 2), and $\mu = \mu_1 \times \mu_2$ agrees with Lebesgue measure on intervals $(a, b] = (a_1, b_1] \times (a_2, b_2]$. By the Carathéodory extension theorem, μ is Lebesgue measure on $\mathscr{B}(\mathbb{R}^2)$, so we have another method of constructing two-dimensional Lebesgue measure. We shall generalize to n dimensions later in the section.

The integration theory we have developed thus far includes the notion of a multiple integral on $\mathbb{R}^n$; this is simply an integral with respect to n-dimensional Lebesgue measure. However, in calculus, integrals of this type are evaluated by computing iterated integrals. The general theorem which justifies this process is Fubini's theorem, which is a direct consequence of the product measure theorem.

2.6.4 *Fubini's Theorem.* Assume the hypothesis of the product measure theorem 2.6.2. Let $f: (\Omega, \mathscr{F}) \to (\overline{\mathbb{R}}, \mathscr{B}(\overline{\mathbb{R}}))$.

(a) If f is nonnegative, then $\int_{\Omega_2} f(\omega_1, \omega_2)\mu(\omega_1, d\omega_2)$ exists and defines a Borel measurable function of ω_1. Also

$$\int_{\Omega} f \, d\mu = \int_{\Omega_1} \left(\int_{\Omega_2} f(\omega_1, \omega_2)\mu(\omega_1, d\omega_2) \right) \mu_1(d\omega_1).$$

(b) If $\int_{\Omega} f \, d\mu$ exists (respectively, is finite), then $\int_{\Omega_2} f(\omega_1, \omega_2)\mu(\omega_1, d\omega_2)$ exists (respectively, is finite) for μ_1-almost every ω_1, and defines a Borel measurable function of ω_1 if it is taken as 0 (or as any Borel measurable function of ω_1) on the exceptional set. Also,

$$\int_\Omega f \, d\mu = \int_{\Omega_1} \left(\int_{\Omega_2} f(\omega_1, \omega_2) \mu(\omega_1, d\omega_2) \right) \mu_1(d\omega_1).$$

[The notation $\int_{\Omega_2} f(\omega_1, \omega_2) \mu(\omega_1, d\omega_2)$ indicates that for a fixed ω_1, the function given by $g(\omega_2) = f(\omega_1, \omega_2)$ is to be integrated with respect to the measure $\mu(\omega_1, \cdot)$.]

PROOF.   (a)   First note that:

(1)   For each fixed ω_1 we have $f(\omega_1, \cdot) : (\Omega_2, \mathscr{F}_2) \to (\overline{\mathbb{R}}, \mathscr{B}(\overline{\mathbb{R}}))$. In other words, if f is jointly measurable, that is, measurable relative to the product σ-field $\mathscr{F}_1 \times \mathscr{F}_2$, it is measurable in each variable separately. For if $B \in \mathscr{B}(\overline{\mathbb{R}})$, $\{\omega_2 \colon f(\omega_1, \omega_2) \in B\} = \{\omega_2 \colon (\omega_1, \omega_2) \in f^{-1}(B)\} = f^{-1}(B)(\omega_1)$ $\in \mathscr{F}_2$ by part (1) of the proof of 2.6.2. Thus $\int_{\Omega_2} f(\omega_1, \omega_2)\mu(\omega_1, d\omega_2)$ exists. Now let I_F, $F \in \mathscr{F}$, be an indicator. Then

$$\int_{\Omega_2} I_F(\omega_1, \omega_2)\mu(\omega_1, d\omega_2) = \int_{\Omega_2} I_{F(\omega_1)}(\omega_2)\mu(\omega_1, d\omega_2)$$

$$= \mu(\omega_1, F(\omega_1)),$$

and this is Borel measurable in ω_1 by part (2) of the proof of 2.6.2. Also

$$\int_\Omega I_F \, d\mu = \mu(F) = \int_{\Omega_1} \mu(\omega_1, F(\omega_1))\mu_1(d\omega_1) \qquad \text{by 2.6.2}$$

$$= \int_{\Omega_1} \int_{\Omega_2} I_F(\omega_1, \omega_2)\mu(\omega_1, d\omega_2)\mu_1(d\omega_1).$$

Now if $f = \sum_{j=1}^n x_j I_{F_j}$, the F_j disjoint sets in $\mathscr{F}$, is a nonnegative simple function, then

$$\int_{\Omega_2} f(\omega_1, \omega_2)\mu(\omega_1, d\omega_2) = \sum_{j=1}^n x_j \mu(\omega_1, F_j(\omega_1)),$$

Borel measurable in ω_1,

and

$$\int_\Omega f \, d\mu = \sum_{j=1}^n x_j \int_\Omega I_{F_j} \, d\mu = \sum_{j=1}^n x_j \int_{\Omega_1} \int_{\Omega_2} I_{F_j}(\omega_1, \omega_2) \mu(\omega_1, d\omega_2) \mu_1(d\omega_1)$$

by what we have proved for indicators

$$= \int_{\Omega_1} \int_{\Omega_2} f(\omega_1, \omega_2) \mu(\omega_1, d\omega_2) \mu_1(d\omega_1).$$

Finally, if $f : (\Omega, \mathscr{F}) \to (\overline{R}, \mathscr{B}(\overline{R}))$, $f \geq 0$, let $0 \leq f_n \uparrow f$, f_n simple. Then

$$\int_{\Omega_2} f(\omega_1, \omega_2) \mu(\omega_1, d\omega_2) = \lim_{n \to \infty} \int_{\Omega_2} f_n(\omega_1, \omega_2) \mu(\omega_1, d\omega_2),$$

which is Borel measurable in ω_1, and

$$\int_\Omega f \, d\mu = \lim_{n \to \infty} \int_\Omega f_n \, d\mu = \lim_{n \to \infty} \int_{\Omega_1} \int_{\Omega_2} f_n(\omega_1, \omega_2) \mu(\omega_1, d\omega_2) \mu_1(d\omega_1)$$

by what we have proved for simple functions

$$= \int_{\Omega_1} \int_{\Omega_2} f(\omega_1, \omega_2) \mu(\omega_1, d\omega_2) \mu_1(d\omega_1)$$

using the monotone convergence theorem twice.

This proves (a).

(b)   Suppose that $\int_\Omega f^- \, d\mu < \infty$. By (a),

$$\int_{\Omega_1} \int_{\Omega_2} f^-(\omega_1, \omega_2) \mu(\omega_1, d\omega_2) \mu_1(d\omega_1) = \int_\Omega f^- \, d\mu < \infty$$

so that $\int_{\Omega_2} f^-(\omega_1, \omega_2) \mu(\omega_1, d\omega_2)$ is μ_1-integrable, hence finite a.e. $[\mu_1]$; thus:

(2)   For μ_1-almost every ω_1 we may write:

$$\int_{\Omega_2} f(\omega_1, \omega_2) \mu(\omega_1, d\omega_2) = \int_{\Omega_2} f^+(\omega_1, \omega_2) \mu(\omega_1, d\omega_2)$$

$$- \int_{\Omega_2} f^-(\omega_1, \omega_2) \mu(\omega_1, d\omega_2).$$

If $\int_\Omega f \, d\mu$ is finite, both integrals on the right side of (2) are finite a.e. $[\mu_1]$. In any event, we may define all integrals in (2) to be 0 (or any other Borel measurable function of ω_1) on the exceptional set, and (2) will then be valid for all ω_1, and will define a Borel measurable function of ω_1. If we integrate (2) with respect to μ_1, we obtain, by (a) and the additivity theorem for integrals,

$$\int_{\Omega_1}\int_{\Omega_2} f(\omega_1,\omega_2)\mu(\omega_1,d\omega_2)\mu_1(d\omega_1) = \int_\Omega f^+\,d\mu - \int_\Omega f^-\,d\mu$$

$$= \int_\Omega f\,d\mu. \ \square$$

2.6.5   Corollary. If $f\colon (\Omega,\mathscr{F}) \to (\overline{R},\mathscr{B}\,(\overline{R}))$ and the iterated integral $\int_{\Omega_1}\int_{\Omega_2}|f(\omega_1,\omega_2)|\mu(\omega_1,d\omega_2)\mu_1(d\omega_1) < \infty$, then $\int_\Omega f\,d\mu$ is finite, and thus Fubini's theorem applies.

PROOF.   By 2.6.4(a), $\int_\Omega |f|\,d\mu < \infty$, and thus the hypothesis of 2.6.4(b) is satisfied. $\square$

As a special case, we obtain the following classical result.

2.6.6   Classical Fubini Theorem. Let $\Omega = \Omega_1 \times \Omega_2$, $\mathscr{F} = \mathscr{F}_1 \times \mathscr{F}_2$, $\mu = \mu_1 \times \mu_2$, where μ_j is a σ-finite measure on $\mathscr{F}_j$, $j = 1,2$. If f is a Borel measurable function on $(\Omega,\mathscr{F})$ such that $\int_\Omega f\,d\mu$ exists, then

$$\int_\Omega f\,d\mu = \int_{\Omega_1}\int_{\Omega_2} f\,d\mu_2\,d\mu_1$$

$$= \int_{\Omega_2}\int_{\Omega_1} f\,d\mu_1\,d\mu_2 \quad \text{by symmetry.}$$

PROOF.   Apply 2.6.4 with $\mu(\omega_1,\cdot) = \mu_2$ for all ω_1. $\square$

Note that by 2.6.5, if $\int_{\Omega_1}\int_{\Omega_2}|f|\,d\mu_2\,d\mu_1 < \infty$ (or $\int_{\Omega_2}\int_{\Omega_1}|f|\,d\mu_1\,d\mu_2 < \infty$), the iterated integration formula 2.6.6 holds.

In 2.6.4(b), if we wish to define $\int_{\Omega_2} f(\omega_1,\omega_2)\mu(\omega_1,d\omega_2)$ in a completely arbitrary fashion on the exceptional set where the integral does not exist, and still produce a Borel measurable function of ω_1, we should assume that $(\Omega_1,\mathscr{F}_1,\mu_1)$ is a complete measure space. The situation is as follows. We have $h\colon (\Omega_1,\mathscr{F}_1) \to (\overline{R},\mathscr{B}\,(\overline{R}))$, where h is the above integral, taken as 0 on the exceptional set A. We set $g(\omega_1) = h(\omega_1)$, $\omega_1 \notin A$; $g(\omega_1) = q(\omega_1)$ arbitrary, $\omega_1 \in A$ (q not necessarily Borel measurable). If B is a Borel subset of $\overline{R}$, then

$$\{\omega_1\colon g(\omega_1) \in B\} = [A^c \cap \{\omega_1\colon h(\omega_1) \in B\}] \cup [A \cap \{\omega_1\colon q(\omega_1) \in B\}].$$

The first set of the union belongs to $\mathscr{F}_1$, and the second is a subset of A, with $\mu_1(A) = 0$, and hence belongs to $\mathscr{F}_1$ by completeness. Thus g is Borel measurable.

In the classical Fubini theorem, if we want to define $\int_{\Omega_2} f(\omega_1, \omega_2) \, d\mu_2(\omega_2)$ and $\int_{\Omega_1} f(\omega_1, \omega_2) \, d\mu_1(\omega_1)$ in a completely arbitrary fashion on the exceptional sets, we should assume completeness of both spaces $(\Omega_1, \mathscr{F}_1, \mu_1)$ and $(\Omega_2, \mathscr{F}_2, \mu_2)$.

The product measure theorem and Fubini's theorem may be extended to n factors, as follows.

2.6.7   Theorem. Let $\mathscr{F}_j$ be a σ-field of subsets of Ω_j, $j = 1, \ldots, n$. Let μ_1 be a σ-finite measure on $\mathscr{F}_1$, and, for each $(\omega_1, \ldots, \omega_j) \in \Omega_1 \times \cdots \times \Omega_j$, let $\mu(\omega_1, \ldots, \omega_j, B)$, $B \in \mathscr{F}_{j+1}$, be a measure on $\mathscr{F}_{j+1}$ $(j = 1, 2, \ldots, n-1)$. Assume the $\mu(\omega_1, \ldots, \omega_j, \cdot)$ to be uniformly σ-finite, and assume that $\mu(\omega_1, \ldots, \omega_j, C)$ is measurable: $(\Omega_1 \times \cdots \times \Omega_j, \mathscr{F}_1 \times \cdots \times \mathscr{F}_j) \to (\bar{R}, \mathscr{B}(\bar{R}))$ for each fixed $C \in \mathscr{F}_{j+1}$.

Let $\Omega = \Omega_1 \times \cdots \times \Omega_n$, $\mathscr{F} = \mathscr{F}_1 \times \cdots \times \mathscr{F}_n$.

(a)   There is a unique measure μ on $\mathscr{F}$ such that for each measurable rectangle $A_1 \times \cdots \times A_n \in \mathscr{F}$,

$$\mu(A_1 \times \cdots \times A_n) = \int_{A_1} \mu_1(d\omega_1) \int_{A_2} \mu(\omega_1, d\omega_2)$$

$$\cdots \int_{A_{n-1}} \mu(\omega_1, \ldots, \omega_{n-2}, d\omega_{n-1}) \int_{A_n} \mu(\omega_1, \ldots, \omega_{n-1}, d\omega_n).$$

[Note that the last factor on the right is $\mu(\omega_1, \ldots, \omega_{n-1}, A_n)$.] The measure μ is σ-finite on $\mathscr{F}$, and is probability measure if μ_1 and all the $\mu(\omega_1, \ldots, \omega_j, \cdot)$ are probability measures.

(b)   Let $f \colon (\Omega, \mathscr{F}) \to (\bar{R}, \mathscr{B}(\bar{R}))$. If $f \geq 0$, then

$$\int_{\Omega} f \, d\mu = \int_{\Omega_1} \mu_1(d\omega_1) \int_{\Omega_2} \mu(\omega_1, d\omega_2) \cdots \int_{\Omega_{n-1}} \mu(\omega_1, \ldots, \omega_{n-2}, d\omega_{n-1})$$

$$\int_{\Omega_n} f(\omega_1, \ldots, \omega_n) \mu(\omega_1, \ldots, \omega_{n-1}, d\omega_n), \tag{1}$$

where, after the integration with respect to $\mu(\omega_1, \ldots, \omega_j, \cdot)$ is performed $(j = n-1, n-2, \ldots, 1)$, the result is a Borel measurable function of $(\omega_1, \ldots, \omega_j)$.

If $\int_{\Omega} f \, d\mu$ exists (respectively, is finite), then Eq. (1) holds in the sense that for each $j = n-1, n-2, \ldots, 1$, the integral with respect to $\mu(\omega_1, \ldots, \omega_j, \cdot)$ exists (respectively, is finite) except for $(\omega_1, \ldots, \omega_j)$ in a set of λ_j-measure

0, where λ_j is the measure determined [see (a)] by μ_1 and the measures $\mu(\omega_1, \cdot), \ldots, \mu(\omega_1, \ldots, \omega_{j-1}, \cdot)$. If the integral is defined on the exceptional set as 0 [or any Borel measurable function on the space $(\Omega_1 \times \cdots \times \Omega_j, \mathscr{F}_1 \times \cdots \times \mathscr{F}_j)$], it becomes Borel measurable in $(\omega_1, \ldots, \omega_j)$.

PROOF.   By 2.6.2 and 2.6.4, the result holds for $n = 2$. Assuming that (a) and (b) hold up to $n - 1$ factors, we consider the n-dimensional case. By the induction hypothesis, there is a unique measure λ_{n-1} on $\mathscr{F}_1 \times \mathscr{F}_2 \times \cdots \times \mathscr{F}_{n-1}$ such that for all $A_1 \in \mathscr{F}_1, \ldots, A_{n-1} \in \mathscr{F}_{n-1}$,

$$\lambda_{n-1}(A_1 \times \cdots \times A_{n-1}) = \int_{A_1} \mu_1(d\omega_1) \int_{A_2} \mu(\omega_1, d\omega_2)$$
$$\cdots \int_{A_{n-1}} \mu(\omega_1, \ldots, \omega_{n-2}, d\omega_{n-1})$$

and λ_{n-1} is σ-finite. By the $n = 2$ case, there is a unique measure μ on $(\mathscr{F}_1 \times \cdots \times \mathscr{F}_{n-1}) \times \mathscr{F}_n$ (which equals $\mathscr{F}_1 \times \cdots \times \mathscr{F}_n$; see Problem 3) such that for each $A \in \mathscr{F}_1 \times \cdots \times \mathscr{F}_{n-1}, A_n \in \mathscr{F}_n$,

$$\mu(A \times A_n) = \int_A \mu(\omega_1, \ldots, \omega_{n-1}, A_n) \, d\lambda_{n-1}(\omega_1, \ldots, \omega_{n-1})$$
$$= \int_{\Omega_1 \times \cdots \times \Omega_{n-1}} I_A(\omega_1, \ldots \omega_{n-1}) \mu(\omega_1, \ldots, \omega_{n-1}, A_n)$$
$$d\lambda_{n-1}(\omega_1, \ldots, \omega_{n-1}). \tag{2}$$

If A is a measurable rectangle $A_1 \times \cdots \times A_{n-1}$, then $I_A(\omega_1, \ldots, \omega_{n-1}) = I_{A_1}(\omega_1) \cdots I_{A_{n-1}}(\omega_{n-1})$; thus (2) becomes, with the aid of the induction hypothesis on (b),

$$\mu(A_1 \times \cdots \times A_n) = \int_{A_1} \mu_1(d\omega_1)$$
$$\cdots \int_{A_{n-1}} \mu(\omega_1, \ldots, \omega_{n-1}, A_n) \mu(\omega_1, \ldots, \omega_{n-2}, d\omega_{n-1})$$

which proves the existence of the desired measure μ on $\mathscr{F}$. To show that μ is σ-finite on $\mathscr{F}_0$, the field of finite disjoint unions of measurable rectangles, and, consequently, μ are unique, let $\Omega_j = \bigcup_{r=1}^{\infty} A_{jr}, j = 1, \ldots, n$, where $\mu(\omega_1, \ldots, \omega_{j-1}, A_{jr}) \leq k_{jr} < \infty$ for all $\omega_1, \ldots, \omega_{j-1}, j = 2, \ldots, n$, and $\mu_1(A_{1r}) = k_{1r} < \infty$. Then

$$\Omega = \bigcup_{i_1, \ldots, i_n = 1}^{\infty} (A_{1i_1} \times A_{2i_2} \times \cdots \times A_{ni_n}),$$

with

$$\mu(A_{1i_1} \times A_{2i_2} \times \cdots \times A_{ni_n}) \leq k_{1i_1} k_{2i_2} \cdots k_{ni_n} < \infty.$$

This proves (a).

To prove (b), note that the measure μ constructed in (a) is determined by λ_{n-1} and the measures $\mu(\omega_1, \ldots, \omega_{n-1}, \cdot)$. Thus by the $n = 2$ case,

$$\int_{\Omega} f \, d\mu = \int_{\Omega_1 \times \cdots \times \Omega_{n-1}} \int_{\Omega_n} f(\omega_1, \ldots, \omega_n) \mu(\omega_1, \ldots, \omega_{n-1}, d\omega_n)$$

$$d\lambda_{n-1}(\omega_1, \ldots, \omega_{n-1})$$

where the inner integral is Borel measurable in $(\omega_1, \ldots, \omega_{n-1})$, or becomes so after adjustment on a set of λ_{n-1}-measure 0. The desired result now follows by the induction hypothesis. $\square$

2.6.8   Comments. (a)   If we take $f = I_F$ in formula (1) of 2.6.7(b), we obtain an explicit formula for $\mu(F)$, $F \in \mathscr{F}$, namely,

$$\mu(F) = \int_{\Omega_1} (d\omega_1) \int_{\Omega_2} \mu(\omega_1, d\omega_2) \cdots \int_{\Omega_{n-1}} \mu(\omega_1, \ldots, \omega_{n-2}, d\omega_{n-1})$$

$$\int_{\Omega_n} I_F(\omega_1, \ldots, \omega_n) \mu(\omega_1, \ldots, \omega_{n-1}, d\omega_n).$$

(b)   We obtain the classical product measure and Fubini theorems by taking $\mu(\omega_1, \ldots, \omega_j, \cdot) \equiv \mu_{j+1}$, $j = 1, 2, \ldots, n - 1$ (with μ_{j+1} σ-finite). We obtain a unique measure μ on $\mathscr{F}$ such that on measurable rectangles,

$$\mu(A_1 \times \cdots \times A_n) = \mu_1(A_1) \mu_2(A_2) \cdots \mu_n(A_n).$$

If $f: (\Omega, \mathscr{F}) \to (\overline{R}, \mathscr{B}(\overline{R}))$ and $f \geq 0$ or $\int_{\Omega} f \, d\mu$ exists, then

$$\int_{\Omega} f \, d\mu = \int_{\Omega_1} d\mu_1 \int_{\Omega_2} d\mu_2 \cdots \int_{\Omega_n} f \, d\mu_n$$

and by symmetry, the integration may be performed in any order. The measure μ is called the *product* of $\mu_1, \ldots, \mu_n$, written $\mu = \mu_1 \times \cdots \times \mu_n$. In particular, if each μ_j is Lebesgue measure on $\mathscr{B}(\mathbb{R})$, then $\mu_1 \times \cdots \times \mu_n$ is Lebesgue measure on $\mathscr{B}(\mathbb{R}^n)$, just as in the discussion after 2.6.3.

Problems

1.   Show that the collection of finite disjoint unions of measurable rectangles in $\Omega_1 \times \cdots \times \Omega_n$ forms a field.

2. Show that $\mathscr{B}(\mathbb{R}^n) = \mathscr{B}(\mathbb{R}) \times \cdots \times \mathscr{B}(\mathbb{R})$ (n times).

3. If $\mathscr{F}_1, \ldots, \mathscr{F}_n$ are arbitrary σ-fields, show that

$$(\mathscr{F}_1 \times \cdots \times \mathscr{F}_{n-1}) \times \mathscr{F}_n = \mathscr{F}_1 \times \mathscr{F}_2 \times \cdots \times \mathscr{F}_n.$$

4. Let μ be the product of the σ-finite measures μ_1 and μ_2. If $C \in \mathscr{F}_1 \times \mathscr{F}_2$, show that the following are equivalent:

   (a) $\mu(C) = 0$,
   (b) $\mu_2(C(\omega_1)) = 0$ for μ_1-almost all $\omega_1 \in \Omega_1$,
   (c) $\mu_1(C(\omega_2)) = 0$ for μ_2-almost all $\omega_2 \in \Omega_2$.

5. In Problem 4, let $(\Omega', \mathscr{F}', \mu')$ be the completion of $(\Omega, \mathscr{F}, \mu)$, and assume μ_1, μ_2 complete. If $B \in \mathscr{F}'$, show that $B(\omega_1) \in \mathscr{F}_2$ for μ_1-*almost* all $\omega_1 \in \Omega_1$ [and $B(\omega_2) \in \mathscr{F}_1$ for μ_2-almost all $\omega_2 \in \Omega_2$]. Give an example in which $B(\omega_1) \notin \mathscr{F}_2$ for some $\omega_1 \in \Omega_1$.

6. (a) Let $\Omega_1 = \Omega_2 =$ the set of positive integers, $\mathscr{F}_1 = \mathscr{F}_2 =$ all subsets, $\mu_1 = \mu_2 =$ counting measure, $f(n, n) = n$, $f(n, n+1) = -n$, $n = 1, 2, \ldots, f(i, j) = 0$ if $j \neq i$ or $i + 1$. Show that $\int_{\Omega_1} \int_{\Omega_2} f \, d\mu_2 \, d\mu_1 = 0$, $\int_{\Omega_2} \int_{\Omega_1} f \, d\mu_1 \, d\mu_2 = \infty$. (Fubini's theorem fails since the integral of f with respect to $\mu_1 \times \mu_2$ does not exist.)

   (b) Let $\Omega_1 = \Omega_2 = \mathbb{R}$, $\mathscr{F}_1 = \mathscr{F}_2 = \mathscr{B}(\mathbb{R})$, $\mu_1 =$ Lebesgue measure, $\mu_2 =$ counting measure. Let $A = \{(\omega_1, \omega_2) \colon \omega_1 = \omega_2\} \in \mathscr{F}_1 \times \mathscr{F}_2$. Show that

$$\int_{\Omega_1} \int_{\Omega_2} I_A \, d\mu_2 \, d\mu_1 = \int_{\Omega_1} \mu_2(A(\omega_1)) \, d\mu_1(\omega_1) = \infty,$$

but

$$\int_{\Omega_2} \int_{\Omega_1} I_A \, d\mu_1 \, d\mu_2 = \int_{\Omega_2} \mu_1(A(\omega_2)) \, d\mu_2(\omega_2) = 0.$$

[Fubini's theorem fails since μ_2 is not σ-finite; the product measure theorem fails also since $\int_{\Omega_1} \mu_2(F(\omega_1)) \, d\mu_1(\omega_1)$ and

$$\int_{\Omega_2} \mu_1(F(\omega_2)) \, d\mu_2(\omega_2) \text{ do not agree on } \mathscr{F}_1 \times \mathscr{F}_2.]$$

*7.[2] Let $\Omega_1 = \Omega_2 =$ the first uncountable ordinal, $\mathscr{F}_1 = \mathscr{F}_2 =$ all subsets, $\Omega = \Omega_1 \times \Omega_2$, $\mathscr{F} = \mathscr{F}_1 \times \mathscr{F}_2$. Assume the continuum hypothesis, which identifies Ω_1 and Ω_2 with $[0, 1]$.

---

[2] Rao, B. V., *Bull. Amer. Math. Soc.* **75**, 614 (1969).

(a) If f is any function from Ω_1 (or from a subset of Ω_1) to $[0, 1]$ and $G = \{(x, y): x \in \Omega_1, y = f(x)\}$ is the graph of f, show that $G \in \mathscr{F}$.

(b) Let $C_1 = \{(x, y) \in \Omega: y \le x\}$, $C_2 = \{(x, y) \in \Omega: y > x\}$. If $B \subset C_1$ or $B \subset C_2$, show that $B \in \mathscr{F}$. (The relation $y \le x$ refers to the ordering of y and x as *ordinals*, not as real numbers.)

(c) Show that $\mathscr{F}$ consists of all subsets of Ω.

8. Show that a measurable function of one variable is jointly measurable. Specifically, if $g: (\Omega_1, \mathscr{F}_1) \to (\Omega', \mathscr{F}')$ and we define $f: \Omega_1 \times \Omega_2 \to \Omega'$ by $f(\omega_1, \omega_2) = g(\omega_1)$, then f is measurable relative to $\mathscr{F}_1 \times \mathscr{F}_2$ and $\mathscr{F}'$, regardless of the nature of $\mathscr{F}_2$.

*9. Give an example of a function $f: [0.1] \times [0, 1] \to [0, 1]$ such that

(a) $f(x, y)$ is Borel measurable in y for each fixed x and Borel measurable in x for each fixed y,

(b) f is not jointly measurable, that is, f is not measurable relative to the product σ-field $\mathscr{B}[0, 1] \times \mathscr{B}[0, 1]$, and

(c) $\int_0^1 (\int_0^1 f(x, y) \, dy) \, dx$ and $\int_0^1 (\int_0^1 f(x, y) \, dx) \, dy$ exist but are unequal. (One example is suggested by Problem 7.)

2.7 MEASURES ON INFINITE PRODUCT SPACES

The n-dimensional product measure theorem formalizes the notion of an n-stage random experiment, where the probability of an event associated with the nth stage depends on the result of the first $n - 1$ trials. It will be convenient later to have a single probability space which is adequate to handle n-stage experiments for n arbitrarily large (not fixed in advance). Such a space can be constructed if the product measure theorem can be extended to infinitely many dimensions. Our task is to construct the product of infinitely many σ-fields, and we first consider the countably infinite case.

2.7.1 Definitions. For each $j = 1, 2, \ldots$, let $(\Omega_j, \mathscr{F}_j)$ be a measurable space. Let $\Omega = \prod_{j=1}^{\infty} \Omega_j$, the set of all sequences $(\omega_1, \omega_2, \ldots)$ such that $\omega_j \in \Omega_j$, $j = 1, 2, \ldots$. If $B^n \subset \prod_{j=1}^{n} \Omega_j$, we define

$$B_n = \{\omega \in \Omega: (\omega_1, \ldots, \omega_n) \in B^n\}.$$

The set B_n is called the *cylinder* with *base* B^n; the cylinder is said to be *measurable* if $B^n \in \prod_{j=1}^{n} \mathscr{F}_j$. If $B^n = A_1 \times \cdots \times A_n$, where $A_i \subset \Omega_i$ for each i, B_n is called a *rectangle*, a *measurable rectangle* if $A_i \in \mathscr{F}_i$ for each i.

A cylinder with an n-dimensional base may always be regarded as having a higher dimensional base. For example, if

$$B = \{\omega \in \Omega: (\omega_1, \omega_2, \omega_3) \in B^3\},$$

then

$$B = \{\omega \in \Omega: \ (\omega_1, \omega_2, \omega_3) \in B^3, \omega_4 \in \Omega_4\}$$
$$= \{\omega \in \Omega: \ (\omega_1, \omega_2, \omega_3, \omega_4) \in B^3 \times \Omega_4\}.$$

It follows that the measurable cylinders form a field. It is also true that finite disjoint unions of measurable rectangles form a field; the argument is the same as in Problem 1 of Section 2.6.

The minimal σ-field over the measurable cylinders is called the *product* of the σ-fields $\mathscr{F}_j$, written $\prod_{j=1}^{\infty} \mathscr{F}_j$; $\prod_{j=1}^{\infty} \mathscr{F}_j$ is also the minimal σ-field over the measurable rectangles (see Problem 1). If all $\mathscr{F}_j$ coincide with a fixed σ-field $\mathscr{F}$, then $\prod_{j=1}^{\infty} \mathscr{F}_j$ is denoted by $\mathscr{F}^{\infty}$, and if all Ω_j coincide with a fixed set S, $\prod_{j=1}^{\infty} \Omega_j$ is denoted by S^{∞}.

The infinite-dimensional version of the product measure theorem will be used only for probability measures, and is therefore stated in that context. (In fact the construction to be described below runs into trouble for nonprobability measures.)

2.7.2   Theorem. Let $(\Omega_j, \mathscr{F}_j)$, $j = 1, 2, \ldots$, be arbitrary measurable spaces; let $\Omega = \prod_{j=1}^{\infty} \Omega_j$, $\mathscr{F} = \prod_{j=1}^{\infty} \mathscr{F}_j$. Suppose that we are given an arbitrary probability measure P_1 on $\mathscr{F}_1$, and for each $j = 1, 2, \ldots$ and each $(\omega_1, \ldots, \omega_j)$ $\in \Omega_1 \times \cdots \times \Omega_j$ we are given a probability measure $P(\omega_1, \ldots, \omega_j, \cdot)$ on $\mathscr{F}_{j+1}$. Assume that $P(\omega_1, \ldots, \omega_j, C)$ is measurable: $(\prod_{i=1}^{j} \Omega_i, \prod_{i=1}^{j} \mathscr{F}_i)$ $\to (R, \mathscr{B}(\mathbb{R}))$ for each fixed $C \in \mathscr{F}_{j+1}$.

If $B^n \in \prod_{j=1}^{n} \mathscr{F}_j$, define

$$P_n(B^n) = \int_{\Omega_1} P_1(d\omega_1) \int_{\Omega_2} P(\omega_1, d\omega_2) \ldots \int_{\Omega_{n-1}} P(\omega_1, \ldots, \omega_{n-2}, d\omega_{n-1})$$

$$\int_{\Omega_n} I_{B^n}(\omega_1, \ldots, \omega_n) P(\omega_1, \ldots, \omega_{n-1}, d\omega_n).$$

Note that P_n is a probability measure on $\prod_{j=1}^{n} \mathscr{F}_j$ by 2.6.7 and 2.6.8(a).

There is a unique probability measure P on $\mathscr{F}$ such that for all n, P agrees with P_n on n-dimensional cylinders, that is, $P\{\omega \in \Omega: \ (\omega_1, \ldots, \omega_n) \in B^n\}$ $= P_n(B^n)$ for all $n = 1, 2, \ldots$ and all $B^n \in \prod_{j=1}^{n} \mathscr{F}_j$.

PROOF. Any measurable cylinder can be represented in the form B_n $= \{\omega \in \Omega: \ (\omega_1, \ldots, \omega_n) \in B^n\}$ for some n and some $B^n \in \prod_{j=1}^{n} \mathscr{F}_j$; define $P(B_n) = P_n(B^n)$. We must show that P is well-defined on measurable cylinders. For suppose that B_n can also be expressed as $\{\omega \in \Omega: \ (\omega_1, \ldots, \omega_m)$ $\in C^m\}$ where $C^m \in \prod_{j=1}^{m} \mathscr{F}_j$; we must show that $P_n(B^n) = P_m(C^m)$. Say

$m < n$; then $(\omega_1, \ldots, \omega_m) \in C^m$ iff $(\omega_1, \ldots, \omega_n) \in B^n$, hence $B^n = C^m \times \Omega_{m+1} \times \cdots \times \Omega_n$. It follows from the definition of P_n that $P_n(B^n) = P_m(C^m)$. (The fact that the $P(\omega_1, \ldots, \omega_j, \cdot)$ are probability measures is used here.)

Since P_n is a measure on $\prod_{j=1}^n \mathscr{F}_j$, it is immediate that P is finitely additive on the field $\mathscr{F}_0$ of measurable cylinders. If we can show that P is continuous from above at the empty set, 1.2.8(b) implies that P is countably additive on $\mathscr{F}_0$, and the Carathéodory extension theorem extends P to a probability measure on $\prod_{j=1}^\infty \mathscr{F}_j$; by construction, P agrees with P_n on n-dimensional cylinders.

Let $\{B_n, n = n_1, n_2, \ldots\}$ be a sequence of measurable cylinders decreasing to $\emptyset$ (we may assume $n_1 < n_2 < \cdots$, and in fact nothing is lost if we take $n_i = i$ for all i). Assume $\lim_{n\to\infty} P(B_n) > 0$. Then for each $n > 1$,

$$P(B_n) = \int_{\Omega_1} g_n^{(1)}(\omega_1) P_1(d\omega_1),$$

where

$$g_n^{(1)}(\omega_1) = \int_{\Omega_2} P(\omega_1, d\omega_2) \cdots \int_{\Omega_n} I_{B^n}(\omega_1, \ldots, \omega_n) P(\omega_1, \ldots, \omega_{n-1}, d\omega_n).$$

Since $B_{n+1} \subset B_n$, it follows that $B^{n+1} \subset B^n \times \Omega_{n+1}$; hence

$$I_{B^{n+1}}(\omega_1, \ldots, \omega_{n+1}) \leq I_{B^n}(\omega_1, \ldots, \omega_n).$$

Therefore $g_n^{(1)}(\omega_1)$ decreases as n increases (ω_1 fixed); say $g_n^{(1)}(\omega_1) \to h_1(\omega_1)$. By the extended monotone convergence theorem (or the dominated convergence theorem), $P(B_n) \to \int_{\Omega_1} h_1(\omega_1) P_1(d\omega_1)$. If $\lim_{n\to\infty} P(B_n) > 0$, then $h_1(\omega_1') > 0$ for some $\omega_1' \in \Omega_1$. In fact $\omega_1' \in B^1$, for if not, $I_{B^n}(\omega_1', \omega_2, \ldots, \omega_n) = 0$ for all n; hence $g_n^{(1)}(\omega_1') = 0$ for all n, and $h_1(\omega_1') = 0$, a contradiction.

Now for each $n > 2$,

$$g_n^{(1)}(\omega_1') = \int_{\Omega_2} g_n^{(2)}(\omega_2) P(\omega_1', d\omega_2),$$

where

$$g_n^{(2)}(\omega_2) = \int_{\Omega_3} P(\omega_1', \omega_2, d\omega_3)$$

$$\cdots \int_{\Omega_n} I_{B^n}(\omega_1', \omega_2, \ldots, \omega_n) P(\omega_1', \ldots, \omega_{n-1}, d\omega_n).$$

As above, $g_n^{(2)}(\omega_2) \downarrow h_2(\omega_2)$; hence

$$g_n^{(1)}(\omega_1') \to \int_{\Omega_2} h_2(\omega_2) P(\omega_1', d\omega_2).$$

Since $g_n^{(1)}(\omega_1') \to h_1(\omega_1') > 0$, we have $h_2(\omega_2') > 0$ for some $\omega_2' \in \Omega_2$, and as above we have $(\omega_1', \omega_2') \in B^2$.

The process may be repeated inductively to obtain points $\omega_1', \omega_2', \ldots$ such that for each n, $(\omega_1', \ldots, \omega_n') \in B^n$. But then $(\omega_1', \omega_2', \ldots) \in \bigcap_{n=1}^\infty B_n = \emptyset$, a contradiction. This proves the existence of the desired probability measure P. If Q is another such probability measure, then $P = Q$ on measurable cylinders, hence $P = Q$ on $\mathscr{F}$ by the uniqueness part of the Carathéodory extension theorem. $\square$

The classical product measure theorem extends as follows:

2.7.3   Corollary. For each $j = 1, 2, \ldots$, let $(\Omega_j, \mathscr{F}_j, P_j)$ be an arbitrary probability space. Let $\Omega = \prod_{j=1}^\infty \Omega_j$, $\mathscr{F} = \prod_{j=1}^\infty \mathscr{F}_j$. There is a unique probability measure P on $\mathscr{F}$ such that

$$P\{\omega \in \Omega : \omega_1 \in A_1, \ldots, \omega_n \in A_n\} = \prod_{j=1}^n P_j(A_j)$$

for all $n = 1, 2, \ldots$ and all $A_j \in \mathscr{F}_j$, $j = 1, 2, \ldots$. We call P the *product* of the P_j, and write $P = \prod_{j=1}^\infty P_j$.

PROOF.   In 2.7.2, take $P(\omega_1, \ldots, \omega_j, B) = P_{j+1}(B)$, $B \in \mathscr{F}_{j+1}$. Then $P_n(A_1 \times \cdots \times A_n) = \prod_{j=1}^n P_j(A_j)$, and thus the probability measure P of 2.7.2 has the desired properties. If Q is another such probability measure, then $P = Q$ on the field of finite disjoint unions of measurable rectangles; hence $P = Q$ on $\mathscr{F}$ by the Carathéodory extension theorem. $\square$

Thus far we have considered probability measures on countably infinite product spaces. The results may be extended to uncountable products if certain assumptions are made about the individual factor spaces. We will be completely general at the beginning, but when we reach the main result, the Kolmogorov extension theorem, we will assume that all the factor spaces are the reals, with the σ-field of Borel sets.

2.7.4   Definitions and Comments. For t in the arbitrary index set T, let $(\Omega_t, \mathscr{F}_t)$ be a measurable space. Let $\prod_{t \in T} \Omega_t$ be the set of all functions $\omega = (\omega(t), t \in T)$ on T such that $\omega(t) \in \Omega_t$ for each $t \in T$. If $t_1, \ldots, t_n \in T$ and

$B^n \subset \prod_{i=1}^n \Omega_{t_i}$, we define the set $B^n(t_1, \ldots, t_n)$ as $\{\omega \in \prod_{t \in T} \Omega_t: (\omega(t_1), \ldots, \omega(t_n)) \in B^n\}$. We call $B^n(t_1, \ldots, t_n)$ the *cylinder* with *base* B^n at $(t_1, \ldots, t_n)$; the cylinder is said to be *measurable* iff $B^n \in \prod_{i=1}^n \mathscr{F}_{t_i}$. If $B^n = B_1 \times \cdots \times B_n$, the cylinder is called a *rectangle*, a *measurable rectangle* iff $B_i \in \mathscr{F}_{t_i}$, $i = 1, \ldots, n$. Note that if all $\Omega_t = \Omega$, then $\prod_{t \in T} \Omega_t = \Omega^T$, the set of all functions from T to Ω.

For example, let $T = [0, 1]$, $\Omega_t = \mathbb{R}$ for all $t \in T$, $B^2 = \{(u, v): u > 3, 1 < v < 2\}$. Then

$$B^2\left(\tfrac{1}{2}, \tfrac{3}{4}\right) = \{x \in \mathbb{R}^T: x\left(\tfrac{1}{2}\right) > 3, \qquad 1 < x\left(\tfrac{3}{4}\right) < 2\}$$

[see Fig. 2.7.1, where $x_1 \in B^2\left(\tfrac{1}{2}, \tfrac{3}{4}\right)$ and $x_2 \notin B^2\left(\tfrac{1}{2}, \tfrac{3}{4}\right)$].

Exactly as in 2.7.1, the measurable cylinders form a field, as do the finite disjoint unions of measurable rectangles. The minimal σ-field over the measurable cylinders is denoted by $\prod_{t \in T} \mathscr{F}_t$, and called the *product* of the σ-fields $\mathscr{F}_t$. If $\Omega_t = S$ and $\mathscr{F}_t = \mathscr{S}$ for all t, $\prod_{t \in T} \mathscr{F}_t$ is denoted by $\mathscr{S}^T$. Again as in 2.7.1, $\prod_{t \in T} \mathscr{F}_t$ is also the minimal σ-field over the measurable rectangles.

We now consider the problem of constructing probability measures on $\prod_{t \in T} \mathscr{F}_t$. The approach will be as follows: Let $v = \{t_1, \ldots, t_n\}$ be a finite subset of T, where $t_1 < t_2 < \cdots < t_n$. (If T is not a subset of $\mathbb{R}$, some fixed total ordering is put on T.) Assume that for each such v we are given a probability measure P_v on $\prod_{i=1}^n \mathscr{F}_{t_i}$; $P_v(B)$ is to represent $P\{\omega \in \prod_{t \in T} \Omega_t: (\omega(t_1), \ldots, \omega(t_n)) \in B\}$. We shall require that the P_v be "consistent"; to see what kind of consistency is needed, consider an example.

Suppose T is the set of positive integers and $\Omega_t = \mathbb{R}$, $\mathscr{F}_t = \mathscr{B}(\mathbb{R})$ for all t. Suppose we know $P_{12345}(B^5) = P\{\omega: (\omega_1, \omega_2, \omega_3, \omega_4, \omega_5) \in B^5\}$ for

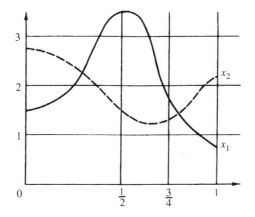

Figure 2.7.1.

all $B^5 \in \mathscr{B}(\mathbb{R}^5)$. Then $P\{\omega: (\omega_2, \omega_3) \in B^2\} = P\{\omega: (\omega_1, \omega_2, \omega_3, \omega_4, \omega_5) \in \mathbb{R} \times B^2 \times R^2\} = P_{12345}(\mathbb{R} \times B^2 \times \mathbb{R}^2)$, $B^2 \in \mathscr{B}(\mathbb{R}^2)$. Thus once probabilities of sets involving the first five coordinates are specified, probabilities of sets involving (ω_2, ω_3) [as well as $(\omega_1, \omega_3, \omega_4)$, and so on], are determined. Thus the original specification of P_{23} must agree with the measure induced from P_{12345}. We hope that a consistent family of probability measures P_v will determine a unique probability measure on $\prod_{t \in T} \mathscr{F}_t$.

Now to formalize: If $v = \{t_1, \ldots, t_n\}$, $t_1 < \cdots < t_n$, the space $(\prod_{i=1}^n \Omega_{t_i}, \prod_{i=1}^n \mathscr{F}_{t_i})$ is denoted by $(\Omega_v, \mathscr{F}_v)$. If $u = \{t_{i1}, \ldots, t_{ik}\}$ is a nonempty subset of v and $y = (y(t_1), \ldots, y(t_n)) \in \Omega_v$, the k-tuple $(y(t_{i1}), \ldots, y(t_{ik}))$ is denoted by y_u. Similarly if $\omega = (\omega(t), t \in T)$ belongs to $\prod_{t \in T} \Omega_t$, the notation ω_v will be used for $(\omega(t_1), \ldots, \omega(t_n))$. If $B \in \mathscr{F}_v$, the measurable cylinder with base B will be written as $B(v)$.

If P_v is a probability measure on $\mathscr{F}_v$, the *projection* of P_v on $\mathscr{F}_u$ is the probability measure $\pi_u(P_v)$ on $\mathscr{F}_u$ defined by

$$[\pi_u(P_v)](B) = P_v\{y \in \Omega_v: \ y_u \in B\}, \qquad B \in \mathscr{F}_u.$$

Similarly, if Q is a probability measure on $\prod_{t \in T} \mathscr{F}_t$, the projection of Q on $\mathscr{F}_v$ is defined by

$$[\pi_v(Q)](B) = Q\left\{\omega \in \prod_{t \in T} \Omega_t: \ \omega_v \in B\right\} = Q(B(v)), \qquad B \in \mathscr{F}_v.$$

Our main result, the Kolmogorov extension theorem, can be proved when each Ω_t is a complete, separable metric space, with $\mathscr{F}_t$ the class of Borel sets (the σ-field generated by the open sets). However, to avoid serious technical complications, we will take all Ω_t to be $\mathbb{R}$ and $\mathscr{F}_t = \mathscr{B}(\mathbb{R})$.

2.7.5   Kolmogorov Extension Theorem.   For each t in the arbitrary index set T, let $\Omega_t = \mathbb{R}$ and $\mathscr{F}_t$ the Borel sets of $\mathbb{R}$.

Assume that for each finite nonempty subset v of T, we are given a probability measure P_v on $\mathscr{F}_v$. Assume the P_v are consistent, that is, $\pi_u(P_v) = P_u$ for each nonempty $u \subset v$.

Then there is a unique probability measure P on $\mathscr{F} = \prod_{t \in T} \mathscr{F}_t$ such that $\pi_v(P) = P_v$ for all v.

PROOF.   We define the hoped-for measure on measurable cylinders by $P(B^n(v)) = P_v(B^n)$, $B^n \in \mathscr{F}_v$.

We must show that this definition makes sense since a given measurable cylinder can be represented in several ways. For example, suppose that

$B^2 = (-\infty, 3) \times (4, 5)$. Then

$$
\begin{aligned}
B^2(t_1, t_2) &= \{\omega: \omega(t_1) < 3, \quad 4 < \omega(t_2) < 5\} \\
&= \{\omega: \omega(t_1) < 3, \quad 4 < \omega(t_2) < 5, \quad \omega(t_3) \in \mathbb{R}\} \\
&= B^3(t_1, t_2, t_3) \quad \text{where} \quad B^3 = (-\infty, 3) \times (4, 5) \times \mathbb{R}.
\end{aligned}
$$

It is sufficient to consider dual representation of the same measurable cylinder in the form $B^n(v) = B^k(u)$ where $k < n$ and $u \subset v$. But then

$$
\begin{aligned}
P_u(B^k) &= [\pi_u(P_v)](B^k) \quad \text{by the consistency hypothesis} \\
&= P_v\{y \in \Omega_v: y_u \in B^k\} \quad \text{by definition of projection.}
\end{aligned}
$$

But the assumption $B^n(v) = B^k(u)$ implies that if $y \in \Omega_v$, then $y \in B^n$ iff $y_u \in B^k$, hence $P_u(B^k) = P_v(B^n)$, as desired.

Thus, P is well-defined on measurable cylinders; the class $\mathscr{F}_0$ of measurable cylinders forms a field, and $\sigma(\mathscr{F}_0) = \mathscr{F}$.

Now if $A_1, \ldots, A_m$ are disjoint sets in $\mathscr{F}_0$, we may write (by introducing extra factors as in the above example) $A_i = B_i^n(v)$, $i = 1, \ldots, m$, where $v = \{t_1, \ldots, t_n\}$ is fixed and the B_i^n, $i = 1, \ldots, m$, are disjoint sets in $\mathscr{F}_v$. Thus

$$
\begin{aligned}
P\left(\bigcup_{i=1}^{m} A_i\right) &= P\left(\bigcup_{i=1}^{m} B_i^n(v)\right) \\
&= P_v\left(\bigcup_{i=1}^{m} B_i^n\right) \quad \text{by definition of } P \\
&= \sum_{i=1}^{m} P_v(B_i^n) \quad \text{since } P_v \text{ is a measure, and} \\
&= \sum_{i=1}^{m} P(A_i) \quad \text{again by definition of } P.
\end{aligned}
$$

Therefore P is finitely additive on $\mathscr{F}_0$. To show that P is countably additive on $\mathscr{F}_0$, we must verify that P is continuous from above at $\emptyset$ and invoke 1.2.8(b). The Carathéodory extension theorem (1.3.10) then extends P to $\mathscr{F}$.

Let A_k, $k = 1, 2, \ldots$ be a sequence of measurable cylinders decreasing to $\emptyset$. If $P(A_k)$ does not approach 0, we have, for some $\varepsilon > 0$, $P(A_k) \geq \varepsilon > 0$ for all k. Suppose $A_k = B^{n_k}(v_k)$; by tacking on extra factors, we may assume that the numbers n_k and the sets v_k increase with k.

It follows from 1.4.11 that we can find a compact set $C^{n_k} \subset B^{n_k}$ such that $P_{v_k}(B^{n_k} - C^{n_k}) < \varepsilon/2^{k+1}$. Define $A_k' = C^{n_k}(v_k) \subset A_k$. Then $P(A_k - A_k')$

$= P_{v_k}(B^{n_k} - C^{n_k}) < \varepsilon/2^{k+1}$. In this way we approximate the given cylinders by cylinders with compact bases.

Now take

$$D_k = A_1{}' \cap A_2{}' \cap \cdots \cap A_k{}' \subset A_1 \cap A_2 \cap \cdots \cap A_k = A_k.$$

Then

$$P(A_k - D_k) = P\left(A_k \cap \bigcup_{i=1}^{k} A_i^{\prime c}\right) \leq \sum_{i=1}^{k} P(A_k \cap A_i^{\prime c})$$

$$\leq \sum_{i=1}^{k} P(A_i - A_i{}') < \sum_{i=1}^{k} 2\varepsilon^{i+1} < \frac{\varepsilon}{2}.$$

Since $D_k \subset A_k{}' \subset A_k$, $P(A_k - D_k) = P(A_k) - P(D_k)$, consequently $P(D_k) > P(A_k) - \varepsilon/2$. In particular, D_k is not empty.

Now pick $x^k \in D_k$, $k = 1, 2, \ldots$. Let $A_1{}' = C^{n_1}(t_1, \ldots, t_{n_1}) = C^{n_1}(v_1)$ [note all $D_k \subset A_1{}'$]. Consider the sequence

$$\left(x_{t_1}^1, \ldots, x_{t_{n_1}}^1\right), \quad \left(x_{t_1}^2, \ldots, x_{t_{n_1}}^2\right), \quad \left(x_{t_1}^3, \ldots, x_{t_{n_1}}^3\right), \ldots,$$

that is, $x_{v_1}^1, x_{v_1}^2, x_{v_1}^3, \ldots$.

Since the $x_{v_1}^n$ belong to C^{n_1}, a compact subset of Ω_{v_1}, we have a convergent subsequence $x_{v_1}^{r_{1n}}$ approaching some $x_{v_1} \in C^{n_1}$. If $A_2{}' = C^{n_2}(v_2)$ (so $D_k \subset A_2{}'$ for $k \geq 2$), consider the sequence $x_{v_2}^{r_{11}}, x_{v_2}^{r_{12}}, x_{v_2}^{r_{13}}, \ldots \in C^{n_2}$ (eventually), and extract a convergent subsequence $x_{v_2}^{r_{2n}} \to x_{v_2} \in C^{n_2}$.

Note that $(x_{v_2}^{r_{2n}})_{v_1} = x_{v_1}^{r_{2n}}$; as $n \to \infty$, the left side approaches $(x_{v_2})_{v_1}$, and as $\{r_{2n}\}$ is a subsequence of $\{r_{1n}\}$, the right side approaches x_{v_1}. Hence $(x_{v_2})_{v_1} = x_{v_1}$.

Continue in this fashion; at step i we have a subsequence

$$x_{v_i}^{r_{in}} \to x_{v_i} \in C^{n_i}, \quad \text{and} \quad (x_{v_i})_{v_j} = x_{v_j} \quad \text{for} \quad j < i.$$

Pick any $\omega \in \prod_{t \in T} \Omega_t$ such that $\omega_{v_j} = x_{v_j}$ for all $j = 1, 2, \ldots$ (such a choice is possible since $(x_{v_i})_{v_j} = x_{v_j}, j < i$). Then $\omega_{v_j} \in C^{n_j}$ for each j; hence

$$\omega \in \bigcap_{j=1}^{\infty} A_j{}' \subset \bigcap_{j=1}^{\infty} A_j = \emptyset,$$

a contradiction. Thus P extends to a measure on $\mathscr{F}$, and by construction, $\pi_v(P) = P_v$ for all v.

Finally, if P and Q are two probability measures on $\mathscr{F}$ such that $\pi_v(P)$
$= \pi_v(Q)$ for all finite $v \subset T$, then for any $B^n \in \mathscr{F}_v$,

$$P(B^n(v)) = [\pi_v(P)](B^n) = [\pi_v(Q)](B^n) = Q(B^n(v)).$$

Thus P and Q agree on measurable cylinders, and hence on $\mathscr{F}$ by the uniqueness part of the Carathéodory extension theorem.   □

Problems

1.   Show that $\prod_{j=1}^{\infty} \mathscr{F}_j$ is the minimal σ-field over the measurable rectangles.

2.   Let $\mathscr{F} = \mathscr{B}(\mathbb{R})$; show that the following sets belong to $\mathscr{F}^{\infty}$:
   (a) $\{x \in \mathbb{R}^{\infty}: \sup_n x_n < a\}$,
   (b) $\{x \in \mathbb{R}^{\infty}: \sum_{n=1}^{\infty} |x_n| < a\}$,
   (c) $\{x \in \mathbb{R}^{\infty}: \lim_{n \to \infty} x_n$ exists and is finite$\}$,
   (d) $\{x \in \mathbb{R}^{\infty}: \limsup_{n \to \infty} x_n \le a\}$, and
   (e) $\{x \in \mathbb{R}^{\infty}: \sum_{k=1}^{n} x_k = 0$ for at least one $n > 0\}$.

3.   Let $\mathscr{F}$ be a σ-field of subsets of a set S, and assume $\mathscr{F}$ is countably
   generated, that is, there is a sequence of sets $A_1, A_2, \ldots$ in $\mathscr{F}$ such that the
   smallest σ-field containing the A_j is $\mathscr{F}$. Show that $\mathscr{F}^{\infty}$ is also countably
   generated; in particular, $\mathscr{B}(\mathbb{R})^{\infty}$ is countably generated; take the A_j as
   intervals with rational endpoints.

*4.   How many sets are there in $\mathscr{B}(\mathbb{R})^{\infty}$?

5.   Define $f: \mathbb{R}^{\infty} \to \overline{\mathbb{R}}$ as follows:

$$f(x_1, x_2, \ldots) = \begin{cases} \text{the smallest positive integer } n \\ \text{such that } x_1 + \cdots + x_n \ge 1, \\ \text{if such an } n \text{ exists,} \\ \infty \text{ if } x_1 + \cdots + x_n < 1 \text{ for all } n. \end{cases}$$

Show that $f: (\mathbb{R}^{\infty}, \mathscr{B}(\mathbb{R})^{\infty}) \to (\overline{\mathbb{R}}, \mathscr{B}(\overline{\mathbb{R}}))$.

2.8   WEAK CONVERGENCE OF MEASURES

In 2.5 we studied convergence of sequence of measurable functions. We
now examine a somewhat different notion of convergence. The results form
the starting point for the study of the central limit theorem of probability.

We will need some properties of semicontinuous functions; proofs of all
necessary results are given in Appendix 2.

If Ω is a metric space, the class of *Borel sets* of Ω, denoted by $\mathscr{B}(\Omega)$, is
defined as the σ-field generated by the open subsets of Ω. In our applications
to probability, we will need only the case $\Omega = \mathbb{R}^k$.

2.8.1   Theorem. Let $\mu, \mu_1, \mu_2, \ldots$ be finite measures on the Borel sets of a metric space Ω. The following conditions are equivalent:

(a) $\int_\Omega f \, d\mu_n \to \int_\Omega f \, d\mu$ for all bounded continuous $f \colon \Omega \to \mathbb{R}$.

(b) $\liminf_{n \to \infty} \int_\Omega f \, d\mu_n \geq \int_\Omega f \, d\mu$ for all bounded lower semicontinuous $f \colon \Omega \to \mathbb{R}$.

(b′) $\limsup_{n \to \infty} \int_\Omega f \, d\mu_n \leq \int_\Omega f \, d\mu$ for all bounded upper semicontinuous $f \colon \Omega \to \mathbb{R}$.

(c) $\int_\Omega f \, d\mu_n \to \int_\Omega f \, d\mu$ for all bounded $f \colon (\Omega, \mathscr{B}(\Omega)) \to (\mathbb{R}, \mathscr{B}(\mathbb{R}))$ such that f is continuous a.e. $[\mu]$.

(d) $\liminf_{n \to \infty} \mu_n(A) \geq \mu(A)$ for every open set $A \subset \Omega$, and $\mu_n(\Omega) \to \mu(\Omega)$.

(d′) $\limsup_{n \to \infty} \mu_n(A) \leq \mu(A)$ for every closed set $A \subset \Omega$, and $\mu_n(\Omega) \to \mu(\Omega)$.

(e) $\mu_n(A) \to \mu(A)$ for every $A \in \mathscr{B}(\Omega)$ such that $\mu(\partial A) = 0$ (∂A denotes the boundary of A).

PROOF.   (a) *implies* (b): If $g \leq f$ and g is bounded continuous,

$$\liminf_{n \to \infty} \int_\Omega f \, d\mu_n \geq \liminf_{n \to \infty} \int_\Omega g \, d\mu_n = \int_\Omega g \, d\mu \qquad \text{by (a).}$$

But since f is lower semicontinuous (LSC), it is the limit of a sequence of continuous functions, and if $|f| \leq M$, all functions in the sequence can also be taken less than or equal to M in absolute value. Thus if we take the sup over g in the above equation, we obtain (b).

(b) *is equivalent to* (b′): Note that f is LSC iff $-f$ is upper semicontinuous (USC).

(b) *implies* (c): Let $\underline{f}$ be the lower envelope of f (the sup of all LSC functions g such that $g \leq f$) and $\overline{f}$ the upper envelope (the inf of all USC functions g such that $g \geq f$). Since $\underline{f}(x) = \liminf_{y \to x} f(y)$ and $\overline{f}(x) = \limsup_{y \to x} f(y)$, continuity of f at x implies $\underline{f}(x) = f(x) = \overline{f}(x)$. Furthermore, $\underline{f}$ is LSC and $\overline{f}$ is USC. Thus if f is bounded and continuous a.e. $[\mu]$,

$$\int_\Omega f \, d\mu = \int_\Omega \underline{f} \, d\mu \leq \liminf_{n \to \infty} \int_\Omega \underline{f} \, d\mu_n \qquad \text{by (b)}$$

$$\leq \liminf_{n \to \infty} \int_\Omega f \, d\mu_n \qquad \text{since} \qquad \underline{f} \leq f$$

$$\leq \limsup_{n \to \infty} \int_\Omega f \, d\mu_n \leq \limsup_{n \to \infty} \int_\Omega \overline{f} \, d\mu_n \quad \text{since } f \leq \overline{f}$$

$$\leq \int_\Omega \overline{f} \, d\mu \qquad \text{by (b′)}$$

$$= \int_\Omega f \, d\mu, \qquad \text{proving (c).}$$

(c) *implies* (d): Clearly (c) implies (a), which in turn implies (b). If A is open, then I_A is LSC, so by (b), $\liminf_{n\to\infty} \mu_n(A) \geq \mu(A)$. Now $I_\Omega \equiv 1$, so $\mu_n(\Omega) \to \mu(\Omega)$ by (c).

(d) *is equivalent to* (d'): Take complements.

(d) *implies* (e): Let A^0 be the interior of A, $\overline{A}$ the closure of A. Then

$$\limsup_{n\to\infty} \mu_n(A) \leq \limsup_{n\to\infty} \mu_n(\overline{A}) \leq \mu(\overline{A}) \qquad \text{by (d')}$$

$$= \mu(A) \qquad \text{by hypothesis.}$$

Also, using (d),

$$\liminf_{n\to\infty} \mu_n(A) \geq \liminf_{n\to\infty} \mu_n(A^0) \geq \mu(A^0) = \mu(A).$$

(e) *implies* (a): Let f be a bounded continuous function on Ω. If $|f| < M$, let $A = \{c \in R: \mu(f^{-1}\{c\}) \neq 0\}; A$ is countable since the $f^{-1}\{c\}$ are disjoint and μ is finite. Construct a partition of $[-M, M]$, say $-M = t_0 < t_1 < \cdots < t_j = M$, with $t_i \notin A, i = 0, 1, \ldots, j$ (M may be increased if necessary). If $B_i = \{x: t_i \leq f(x) < t_{i+1}\}, i = 0, 1, \ldots, j - 1$, it follows from (e) that

$$\sum_{i=0}^{j-1} t_i \mu_n(B_i) \to \sum_{i=0}^{j-1} t_i \mu(B_i).$$

As $f^{-1}(t_i, t_{i+1})$ is open, $\partial f^{-1}[t_i, t_{i+1}) \subset f^{-1}\{t_i, t_{i+1}\}$, and $\mu f^{-1}\{t_i, t_{i+1}\} = 0$ as $t_i, t_{i+1} \notin A$.] Now

$$\left| \int_\Omega f \, d\mu_n - \int_\Omega f \, d\mu \right| \leq \left| \int_\Omega f \, d\mu_n - \sum_{i=0}^{j-1} t_i \mu_n(B_i) \right|$$

$$+ \left| \sum_{i=0}^{j-1} t_i \mu_n(B_i) - \sum_{i=0}^{j-1} t_i \mu(B_i) \right|$$

$$+ \left| \sum_{i=0}^{j-1} t_i \mu(B_i) - \int_\Omega f \, d\mu \right|.$$

The first term on the right may be written as

$$\left| \sum_{i=0}^{j-1} \int_{B_i} (f(x) - t_i) \, d\mu_n(x) \right|$$

---

and this is bounded by $\max_i(t_{i+1} - t_i)\mu_n(\Omega)$, which can be made arbitrarily small by choice of the partition since $\mu_n(\Omega) \to \mu(\Omega) < \infty$ by (e). The third term on the right is bounded by $\max_i(t_{i+1} - t_i)\mu(\Omega)$, which can also be made arbitrarily small. The second term approaches 0 as $n \to \infty$, proving (a). $\square$

2.8.2 Comments. Another condition equivalent to those of 2.8.1 is that $\int_\Omega f\, d\mu_n \to \int_\Omega f\, d\mu$ for all bounded *uniformly* continuous $f \colon \Omega \to R$ (see Problem 1).

The convergence described in 2.8.1 is sometimes called *weak* or *vague* convergence of measures. We shall write $\mu_n \xrightarrow{w} \mu$.

If the measures μ_n and μ are defined on $\mathscr{B}(\mathbb{R})$, there are corresponding distribution functions F_n and F on $\mathbb{R}$. We may relate convergence of measures to convergence of distribution functions.

2.8.3 Definition. A *continuity point* of a distribution function F on $\mathbb{R}$ is a point $x \in \mathbb{R}$ such that F is continuous at x, or $\pm\infty$ (thus by convention, ∞ and $-\infty$ are continuity points).

2.8.4 Theorem. Let $\mu, \mu_1, \mu_2, \ldots$ be finite measures on $\mathscr{B}(\mathbb{R})$, with corresponding distribution functions $F, F_1, F_2, \ldots$. The following are equivalent:

(a) $\mu_n \xrightarrow{w} \mu$.

(b) $F_n(a, b] \to F(a, b]$ at all continuity points a, b of F, where $F(a, b] = F(b) - F(a)$, $F(\infty) = \lim_{x \to \infty} F(x)$, $F(-\infty) = \lim_{x \to -\infty} F(x)$.

If all distribution functions are 0 at $-\infty$, condition (b) is equivalent to the statement that $F_n(x) \to F(x)$ at all points $x \in R$ at which F is continuous, and $F_n(\infty) \to F(\infty)$.

PROOF. (a) *implies* (b): If a and b are continuity points of F in $\mathbb{R}$, then $(a, b]$ is a Borel set whose boundary has μ-measure 0. By 2.8.1 (e), $\mu_n(a, b] \to \mu(a, b]$, that is, $F_n(a, b] \to F(a, b]$. If $a = -\infty$, the argument is the same, and if $b = \infty$, then (a, ∞) is a Borel set whose boundary has μ-measure 0, and the proof proceeds as before.

(b) *implies* (a): Let A be an open subset of $\mathbb{R}$; write A as the disjoint union of open intervals $I_1, I_2, \ldots$. Then

$$\liminf_{n\to\infty} \mu_n(A) = \liminf_{n\to\infty} \sum_k \mu_n(I_k)$$

$$\geq \sum_k \liminf_{n\to\infty} \mu_n(I_k) \qquad \text{by Fatou's lemma.}$$

Let $\varepsilon > 0$ be given. For each k, let $I_k{}'$ be a right-semiclosed subinterval of I_k such that the endpoints of $I_k{}'$ are continuity points of F, and $\mu(I_k{}') \geq \mu(I_k)$ $- \varepsilon 2^{-k}$; the $I_k{}'$ can be chosen since F has only countably many discontinuities. Then

$$\liminf_{n\to\infty} \mu_n(I_k) \geq \liminf_{n\to\infty} \mu_n(I_k{}') = \mu(I_k{}') \qquad \text{by (b).}$$

Thus

$$\liminf_{n\to\infty} \mu_n(A) \geq \sum_k \mu(I_k{}') \geq \sum_k \mu(I_k) - \varepsilon = \mu(A) - \varepsilon.$$

Since ε is arbitrary, we have $\mu_n \xrightarrow{w} \mu$ by 2.8.1(d). $\square$

Condition (b) of 2.8.4 is sometimes called weak convergence of the sequence $\{F_n\}$ to F, written $F_n \xrightarrow{w} F$.

Problems

1. (a) If F is a closed subset of the metric space Ω, show that I_F is the limit of a decreasing sequence of uniformly continuous functions f_n, with $0 \leq f_n \leq 1$ for all n.
    (b) Show that in 2.8.1, $\mu_n \xrightarrow{w} \mu$ iff $\int_\Omega f \, d\mu_n \to \int_\Omega f \, d\mu$ for all bounded uniformly continuous $f \colon \Omega \to \mathbb{R}$.

2. Show that in 2.8.1, $\mu_n \xrightarrow{w} \mu$ iff $\mu_n(A) \to \mu(A)$ for all *open* sets A such that $\mu(\partial A) = 0$.

2.9 REFERENCES

The presentation in Chapters 1 and 2 has been strongly influenced by several sources. The first systematic presentation of measure theory appeared in Halmos (1950). Halmos achieves slightly greater generality at the expense of technical complications by replacing σ-fields by σ-rings. (A σ-ring is a class of sets closed under differences and countable unions.) However, σ-fields will be completely adequate for our purposes. The first account of measure theory specifically oriented toward probability was given by Loève (1955).

Several useful refinements were made by Royden (1963), Neveu (1965), and Rudin (1966). Neveu's book emphasizes probability while Rudin's book is particularly helpful as a preparation for work in harmonic analysis.

For further properties of finitely additive set functions, and a development of integration theory for functions with values in a Banach space, see Dunford and Schwartz (1958).

A more recent treatment of measure and integration theory is that of Folland (1984), who discusses applications to Fourier analysis and probability. A development of measure theory that emphasizes the connection with probability is given by Doob (1994).

3

INTRODUCTION TO FUNCTIONAL ANALYSIS

3.1 INTRODUCTION

An important part of analysis consists of the study of vector spaces endowed with an additional structure of some kind. In Chapter 2, for example, we studied the vector space $L^p(\Omega, \mathscr{F}, \mu)$. If $1 \le p \le \infty$, the seminorm $\| \ \|_p$ allowed us to talk about such notions as distance, convergence, and completeness.

In this chapter, we look at various structures that can be defined on vector spaces. The most general concept, which we will not study in detail, is that of a topological vector space, which is a vector space endowed with a topology compatible with the algebraic operations, that is, the topology makes vector addition and scalar multiplication continuous. We will concentrate on two special cases, Banach and Hilbert spaces. In a Banach space there is a notion of length of a vector, and in a Hilbert space, length is in turn determined by a "dot product" of vectors. Hilbert spaces are a natural generalization of finite-dimensional Euclidean spaces.

We now list the spaces we are going to study. The term "vector space" will always mean vector space over the complex field $\mathbb{C}$; "real vector space" indicates that the scalar field is $\mathbb{R}$; no other fields will be considered.

3.1.1 Definitions. Let L be a vector space. A *seminorm* on L is a function $\| \ \|$ from L to the nonnegative reals satisfying

$$\|ax\| = |a| \ \|x\| \qquad \text{for all} \qquad a \in \mathbb{C}, \quad x \in L,$$

$$\|x + y\| \le \|x\| + \|y\| \qquad \text{for all} \qquad x, y \in L.$$

The first property is called *absolute homogeneity*, the second *subadditivity*. Note that absolute homogeneity implies that $\|0\| = 0$. (We use the same symbol for the zero vector and the zero scalar.) If, in addition, $\|x\| = 0$ implies that $x = 0$, the seminorm is called a *norm* on L and L is said to be a *normed linear space*.

If $\| \ \|$ is a seminorm on L, and $d(x, y) = \|x - y\|, x, y \in L, d$ is a pseudometric on L; a metric if $\| \ \|$ is a norm. A *Banach space* is a complete normed linear space, that is, relative to the metric d induced by the norm, every Cauchy sequence converges.

An *inner product* on L is a function from $L \times L$ to $\mathbb{C}$, denoted by (x, y) $\rightarrow \langle x, y \rangle$, satisfying

$$\langle ax + by, z \rangle = a\langle x, z \rangle + b\langle y, z \rangle \qquad \text{for all} \qquad a, b \in \mathbb{C}, \quad x, y, z \in L,$$

$$\langle x, y \rangle = \overline{\langle y, x \rangle} \qquad \text{for all} \qquad x, y \in L$$

$$\langle x, x \rangle \geq 0 \qquad \text{for all} \qquad x \in L,$$

$$\langle x, x \rangle = 0 \qquad \text{if and only if} \qquad x = 0$$

(the over-bar indicates complex conjugation). A vector space endowed with an inner product is called an *inner product space* or *pre-Hilbert space*. If L is an inner product space, $\|x\| = (\langle x, x \rangle)^{1/2}$ defines a norm on L; this is a consequence of the Cauchy–Schwarz inequality, to be proved in Section 3.2. If, with this norm, L is complete, L is said to be a *Hilbert space*. Thus a Hilbert space is a Banach space whose norm is determined by an inner product.

Finally, a *topological vector space* is a vector space L with a topology such that addition and scalar multiplication are continuous, in other words, the mappings

$$(x, y) \rightarrow x + y \qquad \text{of} \quad L \times L \quad \text{into} \quad L$$

and

$$(a, x) \rightarrow ax \qquad \text{of} \quad \mathbb{C} \times L \quad \text{into} \quad L$$

are continuous, with the product topology on $L \times L$ and $\mathbb{C} \times L$.

A Banach space is a topological vector space with the topology induced by the metric $d(x, y) = \|x - y\|$. For if $x_n \rightarrow x$ and $y_n \rightarrow y$, then

$$\|x_n + y_n - (x + y)\| \leq \|x_n - x\| + \|y_n - y\| \rightarrow 0;$$

if $a_n \rightarrow a$ and $x_n \rightarrow x$, then

$$\|a_n x_n - ax\| \leq \|a_n x_n - a_n x\| + \|a_n x - ax\|$$

$$= |a_n| \, \|x_n - x\| + |a_n - a| \, \|x\| \rightarrow 0.$$

The above definitions remain unchanged if L is a real vector space, except of course that $\mathbb{C}$ is replaced by $\mathbb{R}$. Also, we may drop the complex conjugate in the symmetry requirement for inner product and simply write $\langle x, y \rangle = \langle y, x \rangle$ for all $x, y \in L$.

3.1.2 Examples. (a) If $(\Omega, \mathscr{F}, \mu)$ is a measure space and $1 \leq p \leq \infty$, $\| \ \|_p$ is a seminorm on the vector space $L^p(\Omega, \mathscr{F}, \mu)$. If we pass to equivalence classes by identifying functions that agree a.e. $[\mu]$, we obtain $\mathbf{L}^p(\Omega, \mathscr{F}, \mu)$, a Banach space (see 2.4). When $p = 2$, the norm $\| \ \|_p$ is determined by an inner product

$$\langle f, g \rangle = \int_\Omega f \bar{g} d\mu.$$

Hence $\mathbf{L}^2(\Omega, \mathscr{F}, \mu)$ is a Hilbert space.

If $\mathscr{F}$ consists of all subsets of Ω and μ is counting measure, then $f = g$ a.e. $[\mu]$ implies $f \equiv g$. Thus it is not necessary to pass to equivalence classes; $L^p(\Omega, \mathscr{F}, \mu)$ is a Banach space, denoted for simplicity by $l^p(\Omega)$.

By 2.4.12, if $1 \leq p < \infty$, then $l^p(\Omega)$ consists of all functions $f = (f(\alpha), \alpha \in \Omega)$ from Ω to $\mathbb{C}$ such that $f(\alpha) = 0$ for all but countably many α, and $\|f\|_p^p = \sum_\alpha |f(\alpha)|^p < \infty$. When $p = 2$, the norm on $l^2(\Omega)$ is induced by the inner product

$$\langle f, g \rangle = \sum_\alpha f(\alpha)\overline{g(\alpha)}.$$

When $p = \infty$, the situation is slightly different. The space $l^\infty(\Omega)$ is the collection of all bounded complex-valued functions on Ω, with the *sup norm*

$$\|f\| = \sup\{|f(x)|: x \in \Omega\}.$$

Similarly, if Ω is a metric space (or more generally, a topological space) and L is the class of all bounded continuous complex-valued functions on Ω, then L is a Banach space under the sup norm, for we may verify directly that the sup norm is actually a norm, or equally well we may use the fact that $L \subset l^\infty(\Omega)$. Thus we need only check completeness, and this follows because a uniform limit of continuous functions is continuous.

(b) Let c be the set of all convergent sequences of complex numbers, and put the sup norm on c; if $f = \{a_n, n \geq 1\} \in c$, then

$$\|f\| = \sup\{|a_n|: n = 1, 2, \ldots\}.$$

Again, to show that c is a Banach space we need only establish completeness. Let $\{f_n\}$ be a Cauchy sequence in c; if $f_n = \{a_{nk}, k \geq 1\}$, then $\lim_{k\to\infty} a_{nk}$ exists from each n since $f_n \in c$, and $b_k = \lim_{n\to\infty} a_{nk}$ exists, uniformly in k, since $|a_{nk} - a_{mk}| \leq \|f_n - f_m\| \to 0$ as $n, m \to \infty$. By the standard double limit theorem,

$$\lim_{n\to\infty} \lim_{k\to\infty} a_{nk} = \lim_{k\to\infty} \lim_{n\to\infty} a_{nk}.$$

In particular, $\lim_{k\to\infty} b_k$ exists, so if $f = \{b_k, k \geq 1\}$, then $f \in c$. But $\|f_n - f\| = \sup_k |a_{nk} - b_k| \to 0$ as $n \to \infty$ since $a_{nk} \to b_k$ uniformly in k. This proves completeness.

(c) (If you are unfamiliar with general topology, you may skip this example.)

Let L be the collection of all complex valued functions on S, where S is an arbitrary set. Put the topology of pointwise convergence on L, so that a sequence or net $\{f_n\}$ of functions in L converges to the function $f \in L$ if and only if $f_n(x) \to f(x)$ for each $x \in S$. With this topology, L is a topological vector space. To show that addition is continuous, observe that if $f_n \to f$ and $g_n \to g$ pointwise, then $f_n + g_n \to f + g$ pointwise. Similarly if $a_n \in \mathbb{C}$, $n = 1, 2, \ldots, a_n \to a$, and $f_n \to f$ pointwise, then $a_n f_n \to af$ pointwise.

3.2 BASIC PROPERTIES OF HILBERT SPACES

Hilbert spaces are a natural generalization of finite-dimensional Euclidean spaces in the sense that many of the familiar geometric results in $\mathbb{R}^n$ carry over. First recall the definition of the inner product (or "dot product") on $\mathbb{R}^n$: If $x = (x_1, \ldots, x_n)$ and $y = (y_1, \ldots, y_n)$, then $\langle x, y \rangle = \sum_{j=1}^{n} x_j y_j$. (This becomes $\sum_{j=1}^{n} x_j \bar{y}_j$ in the space $\mathbb{C}^n$ of all n-tuples of complex numbers.) The length of a vector in $\mathbb{R}^n$ is given by $\|x\| = (\langle x, x \rangle)^{1/2} = (\sum_{j=1}^{n} x_j^2)^{1/2}$, and the distance between two points of $\mathbb{R}^n$ is $d(x, y) = \|x - y\|$. In order to show that d is a metric, the triangle inequality must be established; this in turn follows from the Cauchy–Schwarz inequality $|\langle x, y \rangle| \leq \|x\| \, \|y\|$. In fact the Cauchy–Schwarz inequality holds in any inner product space, as we now prove:

3.2.1 Cauchy–Schwarz Inequality. If L is an inner product space, and $\|x\| = (\langle x, x \rangle)^{1/2}, x \in L$, then

$$|\langle x, y \rangle| \leq \|x\| \, \|y\| \qquad \text{for all} \qquad x, y \in L.$$

Equality holds iff x and y are linearly dependent.

PROOF. For any $a \in \mathbb{C}$,

$$0 \leq \langle x + ay, x + ay \rangle = \langle x + ay, x \rangle + \langle x + ay, ay \rangle$$

$$= \langle x, x \rangle + a\langle y, x \rangle + \bar{a}\langle x, y \rangle + |a|^2 \langle y, y \rangle.$$

Set $a = -\langle x, y \rangle / \langle y, y \rangle$ (if $\langle y, y \rangle = 0$, then $y = 0$ and the result is trivial). Since $\langle y, x \rangle = \overline{\langle x, y \rangle}$, we have

$$0 \leq \langle x, x \rangle - 2\frac{|\langle x, y \rangle|^2}{\langle y, y \rangle} + \frac{|\langle x, y \rangle|^2}{\langle y, y \rangle},$$

proving the inequality.

As $\langle x + ay, x + ay \rangle = 0$ iff $x + ay = 0$, equality holds iff x and y are linearly dependent. $\square$

3.2.2 Corollary. If L is an inner product space and $\|x\| = (\langle x, x \rangle)^{1/2}$, $x \in L$, then $\| \ \|$ is a norm on L.

PROOF.   It is immediate that $\|x\| \geq 0$, $\|ax\| = |a| \ \|x\|$, and $\|x\| = 0$ iff $x = 0$. Now

$$\|x + y\|^2 = \langle x + y, x + y \rangle = \|x\|^2 + \|y\|^2 + \langle x, y \rangle + \langle y, x \rangle$$
$$= \|x\|^2 + \|y\|^2 + 2 \operatorname{Re} \langle x, y \rangle$$
$$\leq \|x\|^2 + \|y\|^2 + 2\|x\| \ \|y\| \qquad \text{by 3.2.1.}$$

Therefore $\|x + y\|^2 \leq (\|x\| + \|y\|)^2$. $\square$

3.2.3 Corollary. An inner product is (jointly) continuous in both variables, that is, $x_n \to x$, $y_n \to y$ implies $\langle x_n, y_n \rangle \to \langle x, y \rangle$.

PROOF.

$$|\langle x_n, y_n \rangle - \langle x, y \rangle| = |\langle x_n, y_n - y \rangle + \langle x_n - x, y \rangle|$$
$$\leq \|x_n\| \ \|y_n - y\| + \|x_n - x\| \ \|y\| \qquad \text{by 3.2.1.}$$

However, by subadditivity of the norm, $\|x_n\| \leq \|x_n - x\| + \|x\|$ and $\|x\| \leq \|x - x_n\| + \|x_n\|$, and, therefore,

$$| \ \|x_n\| - \|x\| \ | \leq \|x_n - x\| \to 0;$$

hence $\|x_n\| \to \|x\|$. It follows that $\langle x_n, y_n \rangle \to \langle x, y \rangle$. $\square$

The computation of 3.2.2 establishes the following result, which says geometrically that the sum of the squares of the lengths of the diagonals of a parallelogram is twice the sum of the squares of the lengths of the sides.

3.2.4 Parallelogram Law. In an inner product space,

$$\|x + y\|^2 + \|x - y\|^2 = 2(\|x\|^2 + \|y\|^2).$$

PROOF.

$$\|x + y\|^2 = \|x\|^2 + \|y\|^2 + 2 \operatorname{Re}\langle x, y \rangle,$$

and

$$\|x - y\|^2 = \|x\|^2 + \|y\|^2 - 2 \operatorname{Re}\langle x, y \rangle. \quad \square$$

Now suppose that $x_1, \ldots, x_n$ are mutually perpendicular unit vectors in $\mathbb{R}^k$, $k \geq n$. If x is an arbitrary vector in $\mathbb{R}^k$, we try to approximate x by a linear combination $\sum_{j=1}^{n} a_j x_j$. The reader may recall that $\sum_{j=1}^{n} a_j x_j$ will be closest to x in the sense of Euclidean distance when $a_j = \langle x, x_j \rangle$. This result holds in an arbitrary inner product space.

3.2.5   Definition. Two elements x and y in an inner product space L are said to be *orthogonal* or *perpendicular* iff $\langle x, y \rangle = 0$. If $B \subset L$, B is said to be *orthogonal* iff $\langle x, y \rangle = 0$ for all $x, y \in B$ such that $x \neq y$; B is *orthonormal* iff it is orthogonal and $\|x\| = 1$ for all $x \in B$.

The computation of 3.2.2 shows that if $x_1, x_2, \ldots, x_n$ are orthogonal, the *Pythagorean relation* holds: $\|\sum_{i=1}^{n} x_i\|^2 = \sum_{i=1}^{n} \|x_i\|^2$.

3.2.6   Theorem. If $\{x_1, \ldots, x_n\}$ is an orthonormal set in the inner product space L, and $x \in L$,

$$\left\| x - \sum_{j=1}^{n} a_j x_j \right\| \quad \text{is minimized when} \quad a_j = \langle x, x_j \rangle, \quad j = 1, \ldots, n.$$

PROOF.

$$\left\| x - \sum_{j=1}^{n} a_j x_j \right\|^2 = \left\langle x - \sum_{j=1}^{n} a_j x_j, \; x - \sum_{k=1}^{n} a_k x_k \right\rangle$$

$$= \|x\|^2 - \sum_{k=1}^{n} \bar{a}_k \langle x, x_k \rangle - \sum_{j=1}^{n} a_j \langle x_j, x \rangle$$

$$+ \left\langle \sum_{j=1}^{n} a_j x_j, \; \sum_{k=1}^{n} a_k x_k \right\rangle.$$

The last term on the right is $\sum_{j=1}^{n} |a_j|^2$ since the x_j are orthonormal. Furthermore, $-\bar{a}_j \langle x, x_j \rangle - a_j \langle x_j, x \rangle + |a_j|^2 = -|\langle x, x_j \rangle|^2 + |a_j - \langle x, x_j \rangle|^2$. Thus

$$0 \leq \left\| x - \sum_{j=1}^{n} a_j x_j \right\|^2 = \|x\|^2 - \sum_{j=1}^{n} |\langle x, x_j \rangle|^2 + \sum_{j=1}^{n} |a_j - \langle x, x_j \rangle|^2, \qquad (1)$$

so that we can do no better than to take $a_j = \langle x, x_j \rangle$. $\quad \square$

The above computation establishes the following important inequality.

3.2.7 Bessel's Inequality. If B is an arbitrary orthonormal subset of the inner product space L and x is an element of L, then

$$\|x\|^2 \geq \sum_{y \in B} |\langle x, y \rangle|^2.$$

In other words, $\langle x, y \rangle = 0$ for all but countably many $y \in B$, say $y = x_1, x_2, \ldots$, and

$$\|x\|^2 \geq \sum_j |\langle x, x_j \rangle|^2.$$

Equality holds iff $\sum_{j=1}^n \langle x, x_j \rangle x_j \to x$ as $n \to \infty$.

PROOF. If $x_1, \ldots, x_n \in B$, set $a_j = \langle x, x_j \rangle$ in Eq. (1) of 3.2.6 to obtain $\|x - \sum_{j=1}^n \langle x, x_j \rangle x_j\|^2 = \|x\|^2 - \sum_{j=1}^n |\langle x, x_j \rangle|^2 \geq 0.$ ☐

We now consider another basic geometric idea, that of *projection*. If M is a subspace of $\mathbb{R}^n$ and x is any vector in $\mathbb{R}^n$, x can be resolved into a component in M and a component perpendicular to M. In other words, $x = y + z$ where $y \in M$ and z is orthogonal to every vector in M. Before generalizing to an arbitrary space, we indicate some terminology.

3.2.8 Definitions. A *subspace* or *linear manifold* of a vector space L is a subset M of L that is also a vector space; that is, M is closed under addition and scalar multiplication. The subset M is said to be a *closed subspace* of L if M is a subspace and is also a closed set in the metric of L.

A subset M of the vector space L is said to be *convex* iff for all $x, y \in M$, we have $ax + (1 - a)y \in M$ for all real $a \in [0, 1]$.

A key fact is required: If M is a closed convex subset of the Hilbert space H and x is an arbitrary point of H, there is a unique point of M closest to x.

3.2.9 Theorem. Let M be a nonempty closed convex subset of the Hilbert space H. If $x \in H$, there is a unique element $y_0 \in M$ such that

$$\|x - y_0\| = \inf\{\|x - y\|: \ y \in M\}.$$

PROOF. Let $d = \inf\{\|x - y\|: \ y \in M\}$, and pick points $y_1, y_2, \ldots \in M$ with $\|x - y_n\| \to d$ as $n \to \infty$; we show that $\{y_n\}$ is a Cauchy sequence.

Since $\|u + v\|^2 + \|u - v\|^2 = 2\|u\|^2 + 2\|v\|^2$ for all $u, v \in H$ by the parallelogram law 3.2.4, we may set $u = y_n - x$, $v = y_m - x$ to obtain

$$\|y_n + y_m - 2x\|^2 + \|y_n - y_m\|^2 = 2\|y_n - x\|^2 + 2\|y_m - x\|^2$$

or

$$\|y_n - y_m\|^2 = 2\|y_n - x\|^2 + 2\|y_m - x\|^2 - 4\|\tfrac{1}{2}(y_n + y_m) - x\|^2.$$

Since $\tfrac{1}{2}(y_n + y_m) \in M$ by convexity, $\|\tfrac{1}{2}(y_n + y_m) - x\|^2 \geq d^2$, and it follows that $\|y_n - y_m\| \to 0$ as $n, m \to \infty$.

By completeness of H, y_n approaches a limit y_0, hence $\|x - y_n\| \to \|x - y_0\|$. But then $\|x - y_0\| = d$, and $y_0 \in M$ since M is closed; this finishes the existence part of the proof.

To prove uniqueness, let $y_0, z_0 \in M$, with $\|x - y_0\| = \|x - z_0\| = d$. In the parallelogram law, take $u = y_0 - x$, $v = z_0 - x$, to obtain

$$\|y_0 + z_0 - 2x\|^2 + \|y_0 - z_0\|^2 = 2\|y_0 - x\|^2 + 2\|z_0 - x\|^2 = 4d^2.$$

But $\|y_0 + z_0 - 2x\|^2 = 4\|\tfrac{1}{2}(y_0 + z_0) - x\|^2 \geq 4d^2$; hence $\|y_0 - z_0\| = 0$, so $y_0 = z_0$. $\square$

If M is a closed subspace of H, the element y_0 found in 3.2.9 is called the *projection* of x on M. The following result helps to justify this terminology.

3.2.10   Theorem. Let M be a closed subspace of the Hilbert space H, and y_0 an element of M. Then

$$\|x - y_0\| = \inf\{\|x - y\|: \ y \in M\} \qquad \text{iff} \qquad x - y_0 \perp M,$$

that is, $\langle x - y_0, y \rangle = 0$ for all $y \in M$.

PROOF.   Assume $x - y_0 \perp M$. If $y \in M$, then

$$
\begin{aligned}
\|x - y\|^2 &= \|x - y_0 - (y - y_0)\|^2 \\
&= \|x - y_0\|^2 + \|y - y_0\|^2 - 2\,\mathrm{Re}\langle x - y_0, y - y_0 \rangle \\
&= \|x - y_0\|^2 + \|y - y_0\|^2 \qquad \text{since} \qquad y - y_0 \in M \\
&\geq \|x - y_0\|^2.
\end{aligned}
$$

Therefore, $\|x - y_0\| = \inf\{\|x - y\|: \ y \in M\}$.

Conversely, assume $\|x - y_0\| = \inf\{\|x - y\|: \ y \in M\}$. Let $y \in M$ and let c be an arbitrary complex number. Then $y_0 + cy \in M$ since M is a subspace, hence $\|x - y_0 - cy\| \geq \|x - y_0\|$. But

$$\|x - y_0 - cy\|^2 = \|x - y_0\|^2 + |c|^2\|y\|^2 - 2\,\mathrm{Re}\langle x - y_0, cy \rangle;$$

hence
$$|c|^2 \|y\|^2 - 2 \operatorname{Re}\langle x - y_0, cy \rangle \geq 0.$$

Take $c = b\langle x - y_0, y \rangle$, b real. Then $\langle x - y_0, cy \rangle = b|\langle x - y_0, y \rangle|^2$. Thus $|\langle x - y_0, y \rangle|^2 [b^2 \|y\|^2 - 2b] \geq 0$. But the expression in square brackets is negative if b is positive and sufficiently close to 0; hence $\langle x - y_0, y \rangle = 0$. $\quad\square$

We may give still another way of characterizing the projection of x on M.

3.2.11 Projection Theorem. Let M be a closed subspace of the Hilbert space H. If $x \in H$, then x has a unique representation $x = y + z$ where $y \in M$ and $z \perp M$. Furthermore, y is the projection of x on M.

PROOF. Let y_0 be the projection of x on M, and take $y = y_0$, $z = x - y_0$. By 3.2.10, $z \perp M$, proving the existence of the desired representation. To prove uniqueness, let $x = y + z = y' + z'$ where $y, y' \in M, z, z' \perp M$. Then $y - y' \in M$ since M is a subspace, and $y - y' \perp M$ since $y - y' = z' - z$. Thus $y - y'$ is orthogonal to itself, hence $y = y'$. But then $z = z'$, proving uniqueness. $\quad\square$

If M is any subset of H, the set $M^\perp = \{x \in H: x \perp M\}$ is a closed subspace by definition of the inner product and 3.2.3. If M is a closed subspace, $M^\perp$ is called the *orthogonal complement* of H, and the projection theorem is expressed by saying that H is the *orthogonal direct sum* of M and $M^\perp$, written $H = M \oplus M^\perp$.

In $\mathbb{R}^n$, it is possible to construct an orthonormal basis, that is, a set $\{x_1, \ldots, x_n\}$ of n mutually perpendicular unit vectors. Any vector x in $\mathbb{R}^n$ may then be represented as $x = \sum_{i=1}^{n} \langle x, x_i \rangle x_i$, so that $\langle x, x_i \rangle$ is the component of x in the direction of x_i. We are now able to generalize this idea to an arbitrary Hilbert space. The following terminology will be used.

3.2.12 Definitions. If B is a subset of the normed linear space (or more generally, the topological vector space) L, the *space spanned by B*, denoted by $S(B)$, is the smallest closed subspace of L containing all elements of B. If $L(B)$ is the linear manifold generated by B, that is, $L(B)$ consists of all elements $\sum_{i=1}^{n} a_i x_i, a_i \in \mathbb{C}, x_i \in B, i = 1, \ldots, n, n = 1, 2, \ldots$, then $S(B) = \overline{L(B)}$.

If B is a subset of the Hilbert space H, B is said to be an *orthonormal basis* for H iff B is a maximal orthonormal subset of H, in other words, B is not a proper subset of any other orthonormal subset of H. An orthonormal set $B \subset H$ is maximal iff $S(B) = H$, and there are several other conditions equivalent to this, as we now prove.

3.2.13 ***Theorem.*** Let $B = \{x_\alpha, \alpha \in I\}$ be an orthonormal subset of the Hilbert space H. The following conditions are equivalent:

(a) B is an orthonormal basis.

(b) B is a "complete orthonormal set," that is, the only $x \in H$ such that $x \perp B$ is $x = 0$.

(c) B spans H, that is, $S(B) = H$.

(d) For all $x \in H$, $x = \sum_\alpha \langle x, x_\alpha \rangle x_\alpha$. (Let us explain this notation. By 3.2.7, $\langle x, x_\alpha \rangle = 0$ for all but countably many x_α, say for $x_1, x_2, \ldots$; the assertion is that $\sum_{j=1}^n \langle x, x_j \rangle x_j \to x$, and this holds regardless of the order in which the x_j are listed.)

(e) For all $x, y \in H$, $\langle x, y \rangle = \sum_\alpha \langle x, x_\alpha \rangle \langle x_\alpha, y \rangle$.

(f) For all $x \in H$, $\|x\|^2 = \sum_\alpha |\langle x, x_\alpha \rangle|^2$.

Condition (f) [and sometimes (e) as well] is referred to as the Parseval relation.

PROOF. (a) *implies* (b): If $x \perp B$, $x \neq 0$, let $y = x/\|x\|$. Then $B \cup \{y\}$ is an orthonormal set, contradicting the maximality of B.

(b) *implies* (c): If $x \in H$, write $x = y + z$ where $y \in S(B)$ and $z \perp S(B)$ (see 3.2.11). By (b), $z = 0$; hence $x \in S(B)$.

(c) *implies* (d): Since $S(B) = \overline{L(B)}$, given $x \in H$ and $\varepsilon > 0$ there is a finite set $F \subset I$ and complex numbers $a_\alpha, \alpha \in F$, such that

$$\left\| x - \sum_{\alpha \in F} a_\alpha x_\alpha \right\| \leq \varepsilon.$$

By 3.2.6, if G is any finite subset of I such that $F \subset G$,

$$\left\| x - \sum_{\alpha \in G} \langle x, x_\alpha \rangle x_\alpha \right\| \leq \left\| x - \sum_{\alpha \in G} a_\alpha x_\alpha \right\| \qquad \text{where} \quad a_\alpha = 0 \quad \text{for} \quad \alpha \notin F$$

$$= \left\| x - \sum_{\alpha \in F} a_\alpha x_\alpha \right\| \leq \varepsilon.$$

We may assume that $\langle x, x_\alpha \rangle \neq 0$ for every $\alpha \in F$.

Thus if $x_1, x_2, \ldots$ is any ordering of the points $x_\alpha \in B$ for which $\langle x, x_\alpha \rangle \neq 0$, $\|x - \sum_{j=1}^n \langle x, x_j \rangle x_j\| \leq \varepsilon$ for sufficiently large n, as desired.

(d) *implies* (e): This is immediate from 3.2.3.

(e) *implies* (f): Set $x = y$ in (e).

(f) *implies* (a): Let C be an orthonormal set with $B \subset C, B \neq C$. If $x \in C, x \notin B$, we have $\|x\|^2 = \sum_\alpha |\langle x, x_\alpha \rangle|^2 = 0$ since by orthonormality of

C, x is orthogonal to everything in B. This is a contradiction because $\|x\| = 1$ for all $x \in C$. $\quad\square$

3.2.14 Corollary. Let $B = \{x_\alpha, \alpha \in I\}$ be an orthonormal subset of H, not necessarily a basis.

(a) B is an orthonormal basis for $S(B)$. [Note that $S(B)$ is a closed subspace of H, hence is itself a Hilbert space with the same inner product.]
(b) If $x \in H$ and y is the projection of x on $S(B)$, then

$$y = \sum_\alpha \langle x, x_\alpha\rangle x_\alpha$$

[see 3.2.13(d) for the interpretation of the series].

PROOF.   (a)   Note that the space spanned by B in $S(B)$ is $S(B)$ itself.
(b)   By part (a) and 3.2.13(d), $y = \sum_\alpha \langle y, x_\alpha\rangle x_\alpha$. But $x - y \perp S(B)$ by 3.2.11, hence $\langle x, x_\alpha\rangle = \langle y, x_\alpha\rangle$ for all α. $\quad\square$

A standard application of Zorn's lemma shows that every Hilbert space has an orthonormal basis; an additional argument shows that any two orthonormal bases have the same cardinality (see Problem 5). This fact may be used to classify all possible Hilbert spaces, as follows.

3.2.15 Theorem. Let S be an arbitrary set, and let H be a Hilbert space with an orthonormal basis B having the same cardinality as S. Then there is an isometric isomorphism (a one-to-one-onto, linear, norm-preserving map) between H and $l^2(S)$.

PROOF.   We may write $B = \{x_\alpha, \alpha \in S\}$. If $x \in H$, 3.2.13(d) then gives $x = \sum_\alpha \langle x, x_\alpha\rangle x_\alpha$, where $\sum_\alpha |\langle x, x_\alpha\rangle|^2 = \|x\|^2 < \infty$ by 3.2.13(f). The map $x \to (\langle x, x_\alpha\rangle, \alpha \in S)$ of H into $l^2(S)$ is therefore norm-preserving; since it is also linear, it must be one-to-one. To show that the map is onto, consider any collection of complex numbers $a_\alpha, \alpha \in S$, with $\sum_\alpha |a_\alpha|^2 < \infty$. Say $a_\alpha = 0$ except for $\alpha = \alpha_1, \alpha_2, \ldots$, and let $x = \sum_\alpha a_{\alpha_j} x_{\alpha_j}$. [The series converges to an element of H because of the following fact, which occurs often enough to be stated separately: If $\{y_1, y_2, \ldots\}$ is an orthonormal subset of H, the series $\sum_j c_j y_j$ converges to some element of H iff $\sum_j |c_j|^2 < \infty$. To see this, observe that $\| \sum_{j=n}^m c_j y_j \|^2 = \sum_{j=n}^m |c_j|^2 \|y_j\|^2 = \sum_{j=n}^m |c_j|^2$; thus the partial sums form a Cauchy sequence iff $\sum_j |c_j|^2 < \infty$.]
Since the x_α are orthonormal, it follows that $\langle x, x_\alpha\rangle = a_\alpha$ for all α, so that x maps onto $(a_\alpha, \alpha \in S)$. $\quad\square$

Theorem 3.2.15 is not as useful as it looks. For example, when working in $L^2[0, 1]$, we usually take advantage of what we know about [0,1].

We may also characterize Hilbert spaces that are separable, that is, have a countable dense set.

3.2.16   Theorem. A Hilbert space H is separable iff it has a countable orthonormal basis. If the orthonormal basis has n elements, H is isometrically isomorphic to $\mathbb{C}^n$; if the orthonormal basis is infinite, H is isometrically isomorphic to l^2, that is, $l^2(S)$ with $S = \{1, 2, \ldots\}$.

PROOF.   Let B be an orthonormal basis for H. Now $\|x - y\|^2 = \|x\|^2 + \|y\|^2 = 2$ for all $x, y \in B$, $x \neq y$, hence the balls $A_x = \{y: \|y - x\| < \frac{1}{2}\}, x \in B$, are disjoint. If D is dense in H, D must contain a point in each A_x, so that if B is uncountable, D must be also, and therefore H cannot be separable.

Now assume B is a countable set $\{x_1, x_2, \ldots\}$. If U is a nonempty open subset of $H[= S(B) = \overline{L(B)}]$, U contains an element of the form $\sum_{j=1}^n a_j x_j$ with the $a_j \in \mathbb{C}$; in fact the a_j may be assumed to be rational, in other words, to have rational numbers as real and imaginary parts. Thus

$$D = \left\{ \sum_{j=1}^n a_j x_j: n = 1, 2, \ldots, \quad \text{the } a_j \text{ rational} \right\}$$

is a countable dense set, so that H is separable. The remaining statements of the theorem follow from 3.2.15.   □

A linear norm-preserving map from one Hilbert space to another automatically preserves inner products; this is a consequence of the following proposition.

3.2.17   Polarization Identity. In any inner product space,

$$4\langle x, y \rangle = \|x + y\|^2 - \|x - y\|^2 + i\|x + iy\|^2 - i\|x - iy\|^2.$$

PROOF.

$$\|x + y\|^2 = \|x\|^2 + \|y\|^2 + 2\,\text{Re}\langle x, y \rangle$$
$$\|x - y\|^2 = \|x\|^2 + \|y\|^2 - 2\,\text{Re}\langle x, y \rangle$$
$$\|x + iy\|^2 = \|x\|^2 + \|y\|^2 + 2\,\text{Re}\langle x, iy \rangle$$
$$\|x - iy\|^2 = \|x\|^2 + \|y\|^2 - 2\,\text{Re}\langle x, iy \rangle$$

But $\text{Re}\langle x, iy \rangle = \text{Re}[-i\langle x, y \rangle] = \text{Im}\langle x, y \rangle$, and the result follows.   □

Problems

1. In the Hilbert space $l^2(S)$, show that the elements e_α, $\alpha \in S$, form an orthonormal basis, where

$$e_\alpha(s) = \begin{cases} 0, & s \neq \alpha, \\ 1, & s = \alpha. \end{cases}$$

2. (a) If A is an arbitrary subset of the Hilbert space H, show that $A^{\perp\perp} = S(A)$.
   (b) If M is a linear manifold of H, show that M is dense in H iff $M^\perp = \{0\}$.

3. Let $x_1, \ldots, x_n$ be elements of a Hilbert space. Show that the x_i are linearly dependent iff the Gramian (the determinant of the inner products $\langle x_i, x_j \rangle$, $i, j = 1, \ldots, n$) is 0.

4. (*Gram–Schmidt process*) Let $B = \{x_1, x_2, \ldots\}$ be a countable linearly independent subset of the Hilbert space H. Define $e_1 = x_1/\|x_1\|$; having chosen orthonormal elements $e_1, \ldots, e_n$, let y_{n+1} be the projection of x_{n+1} on the space spanned by $e_1, \ldots, e_n$:

$$y_{n+1} = \sum_{i=1}^{n} \langle x_{n+1}, e_i \rangle e_i.$$

Define

$$e_{n+1} = \frac{x_{n+1} - y_{n+1}}{\|x_{n+1} - y_{n+1}\|}.$$

   (a) Show that $L\{e_1, \ldots, e_n\} = L\{x_1, \ldots, x_n\}$ for all n, hence $x_{n+1} \neq y_{n+1}$ and the process is well defined.
   (b) Show that the e_n form an orthonormal basis for $S(B)$.

   Comments. Consider the space $H = \mathbf{L}^2(-1, 1)$; if we take $x_n(t) = t^n$, $n = 0, 1, \ldots$, the Gram–Schmidt process yields the *Legendre polynomials* $e_n(t) = a_n d^n[(t^2 - 1)^n]/dt^n$, where a_n is chosen so that $\|e_n\| = 1$. Similarly, if in $\mathbf{L}^2(-\infty, \infty)$ we take $x_n(t) = t^n e^{-t^2/2}$, $n = 0, 1, \ldots$, we obtain the *Hermite polynomials* $e_n(t) = a_n(-1)^n e^{t^2} d^n(e^{-t^2})/dt^n$.

5. (a) If you are familiar with Zorn's Lemma, show that every Hilbert space has an orthonormal basis.
   (b) If you are familiar with cardinal arithmetic, show that any two orthonormal bases have the same cardinality.

6. Let U be an open subset of the complex plane, and let $H(U)$ be the collection of all functions f analytic on U such that

$$\|f\|^2 = \iint_U |f(x + iy)|^2 \, dx \, dy < \infty.$$

If we define

$$\langle f, g \rangle = \iint_U f(x + iy)\overline{g}(x + iy) \, dx \, dy, \qquad f, g \in H(U),$$

$H(U)$ becomes an inner product space.

(a) If K is a compact subset of U and $f \in H(U)$, show that

$$\sup\{|f(z)|\colon z \in K\} \leq \|f\|/\sqrt{\pi} \, d_0$$

where d_0 is the Euclidean distance from K to the complement of U. Therefore convergence in $H(U)$ implies uniform convergence on compact subsets of U. (If $z \in K$, the Cauchy integral formula yields

$$f(z) = (2\pi)^{-1} \int_0^{2\pi} f(z + \mathrm{re}^{i\theta}) d\theta, \qquad r < d_0.$$

Integrate this equation with respect to r, $0 \leq r \leq d < d_0$. Note also that if U is the entire plane, we may take $d_0 = \infty$, and it follows that $H(C) = \{0\}$.)

(b) Show that $H(U)$ is complete, and hence is a Hilbert space.

7. (a) If f is analytic on the unit disk $D = \{z\colon |z| < 1\}$ with Taylor expansion $f(z) = \sum_{n=0}^{\infty} a_n z^n$, show that

$$\sup_{0 \leq r < 1} \frac{1}{2\pi} \int_0^{2\pi} |f(re^{i\theta})|^2 d\theta = \sum_{n=0}^{\infty} |a_n|^2.$$

It follows that if H^2 is the collection of all functions f analytic on D such that

$$N^2(f) = \sup_{0 \leq r < 1} \frac{1}{2\pi} \int_0^{2\pi} |f(re^{i\theta})|^2 \, d\theta < \infty,$$

then H^2, with norm N, is a pre-Hilbert space.

(b) If $f \in H^2$, show that

$$\iint_D |f(x + iy)|^2 \, dx \, dy \leq \pi N^2(f);$$

hence $H^2 \subset H(D)$ and convergence in H^2 implies convergence in $H(D)$.

(c) If $f_n(z) = z^n$, $n = 0, 1, \ldots$, show that $f_n \to 0$ in $H(D)$ but not in H^2.

(d) Show that H^2 is complete, and hence is a Hilbert space. [By (a), H^2 is isometrically isomorphic to a subspace of l^2.]

(e) If $e_n(z) = z^n$, $n = 0, 1, \ldots$, show that the e_n form an orthonormal basis for H^2.

(f) If $e_n(z) = [(2n + 2)/2\pi]^{1/2} z^n$, $n = 0, 1, \ldots$, show that the e_n form an orthonormal basis for $H(D)$.

8. Let M be a closed convex subset of the Hilbert space H, and y_0 an element of M. If $x \in H$, show that

$$\|x - y_0\| = \inf\{\|x - y\|\colon \ y \in M\}$$

iff

$$\operatorname{Re}\langle x - y_0, y - y_0 \rangle \leq 0 \qquad \text{for all} \qquad y \in M.$$

*9. (a) If g is a continuous complex-valued function on $[0, 2\pi]$ with $g(0) = g(2\pi)$, use the Stone–Weierstrass theorem to show that g can be uniformly approximated by trigonometric polynomials $\sum_{k=-n}^{n} c_k e^{ikt}$. Conclude that the trigonometric polynomials are dense in $L^2[0, 2\pi]$.

(b) If $f \in L^2[0, 2\pi]$, show that the Fourier series $\sum_{n=-\infty}^{\infty} a_n e^{int}$, $a_n = (1/2\pi) \int_0^{2\pi} f(t) e^{-int}\, dt$, converges to f in L^2, that is,

$$\int_0^{2\pi} \left| f(t) - \sum_{k=-n}^{n} a_k e^{ikt} \right|^2 dt \to 0 \qquad \text{as} \qquad n \to \infty.$$

(c) Show that $\{e^{int}/\sqrt{2\pi}, n = 0, \pm 1, \pm 2, \ldots\}$ yields an orthonormal basis for $L^2[0, 2\pi]$.

10. (a) Give an example to show that if M is a nonempty, closed, but not convex subset of a Hilbert space H, there need not be an element of minimum norm in M. Thus the convexity hypothesis cannot be dropped from Theorem 3.2.9, even if we restrict ourselves to existence and forget about uniqueness.

(b) Show that convexity is not necessary in the existence part of 3.2.9 if H is finite-dimensional.

3.3 LINEAR OPERATORS ON NORMED LINEAR SPACES

The idea of a linear transformation from one Euclidean space to another is familiar. If A is a linear map from $\mathbb{R}^n$ to $\mathbb{R}^m$, then A is completely specified by giving its values on a basis $e_1, \ldots, e_n$: $A(\sum_{i=1}^{n} c_i e_i) = \sum_{i=1}^{n} c_i A(e_i)$; furthermore, A is always continuous. If elements of $\mathbb{R}^n$ and $\mathbb{R}^m$ are represented by column vectors, A is represented by an $m \times n$ matrix. If $n = m$, so that A is a linear transformation on $\mathbb{R}^n$, A is one-to-one iff it is onto, and if A^{-1} exists, it is always continuous (as well as linear).

Linear transformations on infinite-dimensional spaces have many features not found on the finite-dimensional case, as we shall see.

In this section, we study mappings A from one normed linear space L to another such space M. The mapping A will be a *linear operator*, that is, $A(ax + by) = aA(x) + bA(y)$ for all $x, y \in L$, $a, b \in \mathbb{C}$. We use the symbol $\| \ \|$ for the norm on both spaces; no confusion should result. Linear operators can of course be defined on arbitrary vector spaces, but in this section, it is always understood that the domain and range are normed.

Linearity does not imply continuity; to study this idea, we introduce a new concept.

3.3.1   Definitions and Comments. If A is a linear operator, the *norm* of A is defined by:

(a) $\|A\| = \sup\{\|Ax\|: x \in L, \|x\| \leq 1\}$. We may express $\|A\|$ in two other ways.

(b) $\|A\| = \sup\{\|Ax\|: x \in L, \|x\| = 1\}$.

(c) $\|A\| = \sup\{\|Ax\|/\|x\|: x \in L, x \neq 0\}$.

To see this, note that (b) $\leq$ (a) is clear; if $x \neq 0$, then $\|Ax\|/\|x\| = \|A(x/\|x\|)\|$, and $x/\|x\|$ has norm 1; hence (c) $\leq$ (b). Finally if $\|x\| \leq 1$, $x \neq 0$, then $\|Ax\| \leq \|Ax\|/\|x\|$, so (a) $\leq$ (c).

It follows from (c) that $\|Ax\| \leq \|A\| \ \|x\|$, and in fact $\|A\|$ is the smallest number k such that $\|Ax\| \leq k\|x\|$ for all $x \in L$.

The linear operator A is said to be *bounded* iff $\|A\| < \infty$. Boundedness is often easy to check, a very fortunate circumstance because we can show that boundedness is equivalent to continuity.

3.3.2   Theorem. A linear operator A is continuous iff it is bounded.

PROOF.   If A is bounded and $\{x_n\}$ is a sequence in L converging to 0, then $\|Ax_n\| \leq \|A\| \ \|x_n\| \to 0$. (We use here the fact that the mapping $x \to \|x\|$ of L into the nonnegative reals is continuous; this follows because $| \ \|x\| - \|y\| \ | \leq \|x - y\|$.) Thus A is continuous at 0, and therefore, by linearity, is continuous everywhere. On the other hand, if A is unbounded, we can find elements $x_n \in L$ with $\|x_n\| \leq 1$ and $\|Ax_n\| \to \infty$. Let $y_n = x_n/\|Ax_n\|$; then $y_n \to 0$, but $\|Ay_n\| = 1$ for all n, hence Ay_n does not converge to 0. Consequently, A is discontinuous. $\square$

We are going to show that the set of all bounded linear operators from L to M is itself a normed linear space, but first we consider some examples.

3.3.3 Examples. (a) Let $L = C[a, b]$, the set of all continuous complex-valued functions on the closed bounded interval $[a, b]$ of reals. Put the *sup norm* on L:

$$\|x\| = \sup\{|x(t)|: a \le t \le b\}, \qquad x \in C[a, b].$$

With this norm, L is a Banach space [see 3.1.2(a)]. Let $K = K(s, t)$ be a continuous complex-valued function on $[a, b] \times [a, b]$, and define a linear operator on L by

$$(Ax)(s) = \int_a^b K(s, t)x(t)\, dt, \qquad a \le s \le b.$$

(By the dominated convergence theorem, Ax actually belongs to L.) We show that A is bounded:

$$\|Ax\| = \sup_{a \le s \le b} \left| \int_a^b K(s, t)x(t)\, dt \right|$$

$$\le \sup_{a \le t \le b} |x(t)| \sup_{a \le s \le b} \int_a^b |K(s, t)|\, dt$$

$$= \|x\| \max_{a \le s \le b} \int_a^b |K(s, t)|\, dt$$

(note that $\int_a^b |K(s, t)|\, dt$ is a continuous function of s, by the dominated convergence theorem). Thus

$$\|A\| \le \max_{a \le s \le b} \int_a^b |K(s, t)|\, dt < \infty.$$

In fact $\|A\| = \max_{a \le s \le b} \int_a^b |K(s, t)|\, dt$ (see Problem 3).

(b)   Let $L = l^p(S)$, where S is the set of all integers and $1 \le p < \infty$. Define a linear operator T on L by $(Tf)_n = f_{n+1}, n \in S$; T is called the *two-sided shift* or the *bilateral shift*. It follows from the definition that T is one-to-one onto, and $(T^{-1}f)_n = f_{n-1}, n \in S$. Also, $\|Tf\| = \|f\|$ for all $f \in L$; hence $\|T^{-1}f\| = \|f\|$ for all $f \in L$, so that T is an isometric isomorphism of L with itself. (In particular, $\|T\| = \|T^{-1}\| = 1$.)

If we replace S by the positive integers and define T as above, the resulting operator is called the *one-sided* or *unilateral* shift. The one-sided shift is onto with norm 1, but is not one-to-one; $T(f_1, f_2, \ldots) = (f_2, f_3, \ldots)$.

The shift operators we have defined are shifts to the left. We may also define shifts to the right; in the two-sided case we take $(Af)_n = f_{n-1}, n \in S$ (so that $A = T^{-1}$). In the one-sided case, we set $(Af)_n = f_{n-1}, n \geq 2; (Af)_1 = 0$; thus $A(f_1, f_2, \ldots) = (0, f_1, f_2, \ldots)$. The operator A is one-to-one but not onto; $A(L)$ is the closed subspace of L consisting of those sequences whose first coordinate is 0.

(c) Assume that the Banach space L is the direct sum of the two closed subspaces M and N, in other words each $x \in L$ can be represented in a unique way as $y + z$ for some $y \in M$, $z \in N$. (We have already encountered this situation with L a Hilbert space, M a closed subspace, $N = M^\perp$.) We define a linear operator P on L by

$$Px = y.$$

P is called a *projection*; specifically, P is the projection of L on M and it has the following properties:

(1) P is *idempotent*; that is, $P^2 = P$, where P^2 is the composition of P with itself.

(2) P is continuous.

Property (1) follows from the definition of P; property (2) will be proved later, as a consequence of the closed graph theorem (see after 3.4.16).

Conversely, let P be a continuous idempotent linear operator on L. Define

$$M = \{x \in L: Px = x\}, \qquad N = \{x \in L: Px = 0\}.$$

Then we can prove that M and N are closed subspaces, L is the direct sum of M and N, and P is the projection of L on M.

By continuity of P, M and N are closed subspaces. If $x \in L$, then $x = Px + (I - P)x = y + z$ where $y \in M$, $z \in N$. Since $M \cap N = \{0\}$, L is the direct sum of M and N. Furthermore, $Px = y$ by definition of y, so that P is the projection of L on M.

If f is a linear operator from a vector space L to the scalar field, f is called a *linear functional*. (The norm of a scalar b is taken as $|b|$.) Considerable insight is gained about normed linear spaces by studying ways of representing continuous linear functionals on such spaces. We give some examples.

3.3.4 *Representations of Continuous Linear Functionals.* (a) Let f be a continuous linear functional on the Hilbert space H. We show that there is a unique element $y \in H$ such that

$$f(x) = \langle x, y \rangle \qquad \text{for all} \qquad x \in H.$$

This is one of several results called the *Riesz representation theorem*.

If the desired y exists, it must be unique, for if $\langle x, y \rangle = \langle x, z \rangle$ for all x, then $y - z$ is orthogonal to everything in H (including itself), so $y = z$.

To prove existence, let N be the *null space* of f, that is, $N = \{x \in H: f(x) = 0\}$. If $N^\perp = \{0\}$, then $N = H$ by the projection theorem; hence $f \equiv 0$, and we may take $y = 0$. Thus assume we have an element $u \in N^\perp$ with $u \neq 0$. Then $u \notin N$, and if we define $z = u/f(u)$, we have $z \in N^\perp$ and $f(z) = 1$.

If $x \in H$ and $f(x) = a$, then

$$x = (x - az) + az, \qquad \text{with} \qquad x - az \in N, \quad az \perp N.$$

If $y = z/\|z\|^2$, then

$$
\begin{aligned}
\langle x, y \rangle &= \langle x - az, y \rangle + a\langle z, y \rangle \\
&= a\langle z, y \rangle \qquad \text{since} \qquad y \perp N \\
&= a = f(x)
\end{aligned}
$$

as desired.

The above argument shows that if f is not identically 0, then $N^\perp$ is one-dimensional. For if $x \in N^\perp$ and $f(x) = a$, then $x - az \in N \cap N^\perp$, hence $x = az$. Therefore $N^\perp = \{az: a \in C\}$.

Notice also that if $\|f\|$ is the norm of f, considered as a linear operator, then

$$\|f\| = \|y\|.$$

For $|f(x)| = |\langle x, y \rangle| \leq \|x\| \, \|y\|$ for all x, hence $\|f\| \leq \|y\|$; but $|f(y)| = |\langle y, y \rangle| = \|y\| \, \|y\|$, so $\|f\| \geq \|y\|$.

Now consider the space H^* of all continuous linear functionals on H; H^* is a vector space under the usual operations of addition and scalar multiplication. If to each $f \in H^*$ we associate the element of H given by the Riesz representation theorem, we obtain a map $\psi: H^* \to H$ that is one-to-one onto, norm-preserving, and *conjugate linear*; that is,

$$\psi(af + bg) = \bar{a}\psi(f) + \bar{b}\psi(g), \qquad f, g \in H^*, \quad a, b \in C.$$

[Note that if $f(x) = \langle x, y \rangle$ for all x, then $af(x) = \langle x, \bar{a}y \rangle$.] Such a map is called a *conjugate isometry*.

(b)   Let f be a continuous linear functional on l^p ($= l^p(S)$), where S is the set of positive integers), $1 < p < \infty$. We show that if q is defined by $(1/p) + (1/q) = 1$, there is a unique element $y = (y_1, y_2, \ldots) \in l^q$ such that

$$f(x) = \sum_{k=1}^{\infty} x_k y_k \qquad \text{for all} \qquad x \in l^p.$$

Furthermore,

$$\|f\| = \|y\| = \left(\sum_{k=1}^{\infty} |y_k|^q\right)^{1/q}.$$

To prove this, let e_n be the sequence in l^p defined by $e_n(j) = 0$, $j \neq n$; $e_n(n) = 1$. If $x \in l^p$, then $\|x - \sum_{k=1}^{n} x_k e_k\|^p = \sum_{k=n+1}^{\infty} |x_k|^p \to 0$; hence $x = \sum_{k=1}^{\infty} x_k e_k$, where the series converges in l^p. By continuity of f,

$$f(x) = \sum_{k=1}^{\infty} x_k y_k, \qquad \text{where} \qquad y_k = f(e_k). \tag{1}$$

Now write y_k in polar form, that is, $y_k = r_k e^{i\theta_k}$, $r_k \geq 0$. Let

$$z_n = (r_1^{q-1} e^{-i\theta_1}, \ldots, r_n^{q-1} e^{-i\theta_n}, 0, 0, \ldots);$$

by (1),

$$f(z_n) = \sum_{k=1}^{n} r_k^{q-1} e^{-i\theta_k} r_k e^{i\theta_k} = \sum_{k=1}^{n} |y_k|^q. \tag{2}$$

But

$$|f(z_n)| \leq \|f\| \, \|z_n\|$$

$$= \|f\| \left(\sum_{k=1}^{n} |y_k|^{[q-1]p}\right)^{1/p}$$

$$= \|f\| \left(\sum_{k=1}^{n} |y_k|^q\right)^{1/p}.$$

By Eq. (2),

$$\left(\sum_{k=1}^{n} |y_k|^q\right)^{1/q} \leq \|f\|;$$

hence $y \in l^q$ and $\|y\| \leq \|f\|$. To prove that $\|f\| \leq \|y\|$, observe that Eq. (1) is of the form $f(x) = \int_\Omega xy \, d\mu$ where Ω is the positive integers and μ is counting measure. By Hölder's inequality, $|f(x)| \leq \|x\| \, \|y\|$, so that $\|f\| \leq \|y\|$.

Finally, we prove uniqueness. If $y, z \in l^q$ and $f(x) = \sum_{k=1}^{\infty} x_k y_k = \sum_{k=1}^{\infty} x_k z_k$ for all $x \in l^p$, then $g(x) = \sum_{k=1}^{\infty} x_k (y_k - z_k) = 0$ for all $x \in l^p$.

The above argument with f replaced by g shows that $\|g\| = \|y - z\|$; hence $\|y - z\| = 0$, and thus $y = z$.

(c) Let f be a continuous linear functional on l^1. We show that there is a unique element $y \in l^\infty$ such that

$$f(x) = \sum_{k=1}^{\infty} x_k y_k \qquad \text{for all} \qquad x \in l^1.$$

Furthermore, $\|f\| = \|y\| = \sup_k |y_k|$.

The argument of (b) may be repeated up to Eq. (1). In this case, however, if $y_k = r_k e^{i\theta_k}$, $k = 1, 2, \ldots$, we take

$$z_n = (0, \ldots, 0, e^{-i\theta_n}, 0, \ldots), \qquad \text{with} \qquad e^{-i\theta_n} \text{ in position } n.$$

Thus by (1),

$$f(z_n) = e^{-i\theta_n} r_n e^{i\theta_n} = |y_n|.$$

But

$$|f(z_n)| \le \|f\| \, \|z_n\| = \|f\|.$$

Therefore $|y_n| \le \|f\|$ for all n, so that $y \in l^\infty$ and $\|y\| \le \|f\|$. But by (1),

$$|f(x)| \le \left(\sup_k |y_k| \right) \sum_{k=1}^{\infty} |x_k| = \|y\| \, \|x\|;$$

hence $\|f\| \le \|y\|$. Uniqueness is proved as in (b).

In 3.3.4(b) and (c), the map $f \to y$ of $(l^p)^*$ to $l^q [q = \infty$ in 3.3.4(c)] is linear and norm-preserving, hence one-to-one. To show that it is onto, observe that any linear functional of the form $f(x) = \sum_{k=1}^{\infty} x_k y_k$, $x \in l^p$, $y \in l^q$, satisfies $\|f\| \le \|y\|$ by the analysis of 3.3.4(b) and (c). Therefore f is continuous, so that every $y \in l^q$ is the image of some $f \in (l^p)^*$. Since $\|f\| = \|y\|$, we have an isometric isomorphism of $(l^p)^*$ and l^q.

If we replace y_k by $\bar{y}_k$, we obtain the result that there is a unique $y \in l^q$ such that $f(x) = \sum_{k=1}^{\infty} x_k \bar{y}_k$ for all $x \in l^p$. This makes the map $f \to y$ a conjugate isometry rather than an isometric isomorphism. If $p = 2$, then $q = 2$ also, and thus we have another proof of the Riesz representation theorem for separable Hilbert spaces (see 3.2.16). In fact, essentially the same argument may be used in an arbitrary Hilbert space if the e_n are replaced by an arbitrary orthonormal basis. (Other examples of representation of continuous linear functionals are given in Problems 10 and 11.)

We now show that the set of bounded linear operators from one normed linear space to another can be made into a normed linear space.

3.3.5   ***Theorem.*** Let L and M be normed linear spaces, and let $[L, M]$ be the collection of all bounded linear operators from L to M. The operator norm defined by 3.3.1 is a norm on $[L, M]$, and if M is complete, then $[L, M]$ is complete. In particular, the set L^* of all continuous linear functionals on L is a Banach space (whether or not L is complete).

PROOF. It follows from 3.3.1 that $\|A\| \geq 0$ and $\|aA\| = |a| \, \|A\|$ for all $A \in [L, M]$ and $a \in \mathbb{C}$. Also by 3.3.1, if $\|A\| = 0$, then $Ax = 0$ for all $x \in L$, hence $A = 0$. If $A, B \in [L, M]$, then again by 3.3.1,

$$\|A + B\| = \sup\{\|(A + B)x\|: \; x \in L, \|x\| \leq 1\}.$$

Since $\|(A + B)x\| \leq \|Ax\| + \|Bx\|$ for all x,

$$\|A + B\| \leq \|A\| + \|B\|$$

and it follows that $[L, M]$ is a vector space and the operator norm is in fact a norm on $[L, M]$.

Now let $A_1, A_2, \ldots$ be a Cauchy sequence in $[L, M]$. Then

$$\|(A_n - A_m)x\| \leq \|A_n - A_m\| \, \|x\| \to 0 \qquad \text{as} \qquad n, m \to \infty. \qquad (1)$$

Therefore $\{A_n x\}$ is a Cauchy sequence in M for each $x \in L$, hence A_n converges pointwise on L to an operator A. Since the A_n are linear, so is A (observe that $A_n(ax + by) = aA_n x + bA_n y$, and let $n \to \infty$). Now given $\varepsilon > 0$, choose N such that $\|A_n - A_m\| \leq \varepsilon$ for $n, m \geq N$. Fix $n \geq N$ and let $m \to \infty$ in Eq. (1) to conclude that $\|(A_n - A)x\| \leq \varepsilon \|x\|$ for $n \geq N$; therefore $\|A_n - A\| \to 0$ as $n \to \infty$. Since $\|A\| \leq \|A - A_n\| + \|A_n\|$, we have $A \in [L, M]$ and $A_n \to A$ in the operator norm. $\square$

In the above proof we have talked about two different types of convergence of sequences of operators.

3.3.6   ***Definitions and Comments.*** Let $A, A_1, A_2, \ldots \in [L, M]$. We say that A_n converges *uniformly* to A iff $\|A_n - A\| \to 0$ (notation: $A_n \xrightarrow{u} A$). Since $\|(A_n - A)x\| \leq \|A_n - A\| \, \|x\|$, uniform operator convergence means that $A_n x \to Ax$, uniformly for $\|x\| \leq 1$ (or equally well for $\|x\| \leq k$, k any positive real number).

We say that A_n converges *pointwise* to A iff $A_n x \to Ax$ for each $x \in L$. Thus, pointwise operator convergence is pointwise convergence on all of L.

Uniform convergence implies strong convergence, but not conversely. For example, let $\{e_1, e_2, \ldots\}$ be an orthonormal set in a Hilbert space, and let

$A_n x = \langle x, e_n \rangle$, $n = 1, 2, \ldots$. Then A_n converges pointwise to 0 by Bessel's inequality, but A_n does not converge uniformly to 0. In fact $\|A_n e_n\| = 1$ for all n, hence $\|A_n\| \equiv 1$.

There is an important property of finite-dimensional spaces that we are now in a position to discuss. In the previous section, we regarded $\mathbb{C}^n$ as a Hilbert space, so that if $x \in \mathbb{C}^n$, the norm of x was taken as the Euclidean norm $\|x\| = \left(\sum_{i=1}^{n} |x_i|^2 \right)^{1/2}$. The metric associated with this norm yields the standard topology on $\mathbb{C}^n$. However, we may put various other norms on $\mathbb{C}^n$, for example, the L^p norm $\|x\|_p = \left(\sum_{i=1}^{n} |x_i|^p \right)^{1/p}$, $1 \leq p < \infty$, or the sup norm $\|x\|_\infty = \max(|x_1|, \ldots, |x_n|)$. Since the space is finite-dimensional, there are no convergence difficulties and all elements have finite norm. Fortunately, the proliferation of norms causes no confusion because *all norms on a given finite-dimensional space induce the same topology*. In other words, the open sets in $\mathbb{C}^n$ will be the same, regardless of which norm we use. The proof of this result is outlined in Problems 6 and 7.

Problems

1. Let f be a linear functional on the normed linear space L. If f is not identically 0, show that the following are equivalent:

   (a) f is continuous.
   (b) The null space $N = f^{-1}\{0\}$ is closed.
   (c) N is not dense in L.
   (d) f is bounded on some neighborhood of 0.

   [To prove that (c) implies (d), show that if $B(x, \varepsilon) \cap N = \phi$, and f is unbounded on $B(0, \varepsilon)$, then $f(B(0, \varepsilon)) = \mathbb{C}$. In particular, there is a point $z \in B(0, \varepsilon)$ such that $f(z) = -f(x)$, and this leads to a contradiction.]

2. Show that any infinite-dimensional normed linear space had a discontinuous linear functional. (Let $e_1, e_2, \ldots$ be an infinite sequence of linearly independent elements such that $\|e_n\| = 1$ for all n. Define f appropriately on the e_n and extend f to the whole space using linearity.)

3. In Example 3.3.3(a), show that $\|A\| = \max_{a \leq s \leq b} \int_b^a |K(s, t)| \, dt$. [If $\int_a^b K(s, t) \, dt$ assumes a maximum at the point u, and $K(u, t) = r(t)e^{i\theta(t)}$, let $z(t) = r(t)e^{-i\theta(t)}$. Let $x_1, x_2, \ldots$ be continuous functions such that $|x_n(t)| \leq 1$ for all n and t, and $\int_q^b |x_n(t) - z(t)| \, dt \to 0$ as $n \to \infty$. Since $\|Ax_n\| \leq \|A\|$ and $(Ax_n)(s) \to \int_a^b K(s, t) z(t) \, dt$ as $n \to \infty$, it follows that $\int_a^b |K(u, t)| \, dt \leq \|A\|$.]

   The same argument, with integrals replaced by sums, shows that if A is a matrix operator on $\mathbb{C}^n$, with sup norm, $\|A\| = \max_{1 \leq i \leq n} \sum_{j=1}^{n} |a_{ij}|$.

4. Let M be a linear manifold in the normed linear space L. Denote by $[x]$ the coset of x modulo M, that is, $\{x + y\colon y \in M\}$. Define

$$\|[x]\| = \inf\{\|y\|\colon \ y \in [x]\};$$

note that $\|[x]\| = \inf\{\|x - z\|\colon z \in M\} = \text{dist}(x, M)$. Show that the above formula defines a seminorm on the quotient space L/M, a norm if M is closed.

5. If L is a Banach space and M is a closed subspace, show that L/M is a Banach space.

6. (a) Let A be a linear operator from L to M. Show that A is one-to-one with A^{-1} continuous on its domain $A(L)$, iff there is a finite number $m > 0$ such that $\|Ax\| \geq m\|x\|$ for all $x \in L$.

   (b) Let $\| \ \|_1$ and $\| \ \|_2$ be norms on the linear space L. Show that the norms induce the same topology (in other words, the open sets are the same for each norm) iff there are finite numbers $m, M > 0$ such that

   $$m\|x\|_1 \leq \|x\|_2 \leq M\|x\|_1 \qquad \text{for all} \qquad x \in L.$$

   [This may be done using part (a), or it may be shown directly that, for example, if $\|x\|_1 \leq (1/m)\|x\|_2$ for all x, then the topology induced by $\| \ \|_1$ is weaker than (that is, included in) the topology induced by $\| \ \|_2$.]

7. Let L be a finite-dimensional normed linear space, with basis $e_1, \ldots, e_n$. Let $\| \ \|_1$ be any norm on L, and define

$$\|x\|_2 = \left(\sum_{i=1}^{n} |x_i|^2 \right)^{1/2}, \qquad \text{where} \qquad x = \sum_{i=1}^{n} x_i e_i.$$

   (a) Show that for some positive real number k, $\|x\|_1 \leq k\|x\|_2$ for all $x \in L$.

   (b) Show that for some positive real number m, $\|x\|_1 \geq m$ on $\{x\colon \|x\|_2 = 1\}$. [By (a), the map $x \to \|x\|_1$ is continuous in the topology induced by $\| \ \|_2$.]

   (c) Show that $\|x\|_1 \geq m\|x_2\|$ for all $x \in L$, and conclude that all norms on a finite-dimensional space induce the same topology.

   (d) Let M be an *arbitrary* normed linear space, and let L be a finite-dimensional subspace of M. Show that L is closed in M.

8. (*Riesz lemma*) Let M be a closed proper subspace of the normed linear space L. Show that if $0 < \delta < 1$, there is an element $x_\delta \in L$

such that $\|x_\delta\| = 1$ and $\|x - x_\delta\| \geq \delta$ for all $x \in M$. [Choose $x_1 \notin M$, and let $d = \text{dist } (x_1, M) > 0$ since M is closed. Choose $x_0 \in M$ such that $\|x_1 - x_0\| \leq d/\delta$, and set $x_\delta = (x_1 - x_0)/\|x_1 - x_0\|$.]

9. Let L be a normed linear space. Show that the following are equivalent:

(a) L is finite-dimensional.

(b) L is topologically isomorphic to a Euclidean space $\mathbb{C}^n$ (or $\mathbb{R}^n$ if L is a real vector space), that is, there is a one-to-one, onto, linear, bicontinuous map between the two spaces.

(c) L is locally compact (every point of L has a neighborhood whose closure is compact).

(d)   Every closed bounded subset of L is compact.

(e)   The set $\{x \in L: \|x\| = 1\}$ is compact.

(f)   The set $\{x \in L: \|x\| = 1\}$ is totally bounded, that is, can be covered by a finite number of open balls of any preassigned radius.

[Problem 7 shows that (a) implies (b), (b) implies (c), (d) implies (e), and (e) implies (f) are obvious, and (c) implies (d) is easy. To prove that (f) implies (a), use Problem 8.]

10. Let c be the space of convergent sequences of complex numbers [see 3.1.2(b)]. If f is a continuous linear functional on c, show that there is a unique element $y = (y_0, y_1, \ldots) \in l^1$ such that for all $x \in c$,

$$f(x) = \left(\lim_{n \to \infty} x_n \right) \left(y_0 - \sum_{k=1}^{\infty} y_k \right) + \sum_{k=1}^{\infty} x_k y_k.$$

Furthermore, $\|f\| = |y_0 - \sum_{k=1}^{\infty} y_k| + \sum_{k=1}^{\infty} |y_k|$. If c_0 is the closed subspace of c consisting of those sequences converging to 0, the representation of a continuous linear functional on c_0 is simpler: $f(x) = \sum_{k=1}^{\infty} x_k y_k$, $\|f\| = \sum_{k=1}^{\infty} |y_k|$. Thus c_0^* is isometrically isomorphic to l^1.

11. Let $(\Omega, \mathscr{F}, \mu)$ be a measure space.

(a)   Assume μ finite, $1 < p < \infty$, $(1/p) + (1/q) = 1$. If f is a continuous linear functional on $L^p = L^p(\Omega, \mathscr{F}, \mu)$, show that there is an element $y \in L^q$ such that

$$f(x) = \int_\Omega xy \, d\mu \qquad \text{for all} \qquad x \in L^p.$$

Furthermore, $\|f\| = \|y\|_q$. If y_1 is another such function, show that $y = y_1$ a.e. $[\mu]$. [Define $\lambda(A) = f(I_A)$, $A \in \mathscr{F}$, and apply the Radon–Nikodym theorem.]

(b) Drop the finiteness assumption on μ. If $A \in \mathscr{F}$ and $\mu(A) < \infty$, part (a) applied to $(A, \mathscr{F} \cap A, \mu)$ provides an essentially unique $y_A \in L^q$ such that $y_A = 0$ on A^c and

$$ f(xI_A) = \int_\Omega xy_A \, d\mu \qquad \text{for all} \qquad x \in L^p; $$

also, $\|y_A\|_q$ is the L^p norm of the restriction of f to A, so

$$ \|y_A\|_q \leq \|f\|. $$

(i) If $\mu(A) < \infty$ and $\mu(B) < \infty$, show that $y_A = y_B$ a.e. $[\mu]$ on $A \cap B$. Thus $y_{A \cup B}$ may be obtained by piecing together y_A and y_B.

(ii) Let A_n, $n = 1, 2, \ldots$, be sets with $\|y_{A_n}\|_q \to k$ $= \sup\{\|y_A\|_q : A \in \mathscr{F}, \mu(A) < \infty\} \leq \|f\|$. Show that y_{A_n} converges in L_q to a limit function y, and y is essentially independent of the particular sequence $\{A_n\}$. Furthermore, $\|y\|_q \leq \|f\|$.

(iii) Show that $f(x) = \int_\Omega xy \, d\mu$ for all $x \in L^p$. Since $\|y\|_q$ $\leq \|f\|$ by (ii) and $\|f\| \leq \|y\|_q$ by Hölder's inequality, we have $\|f\| = \|y\|_q$. Thus the result of (a) holds for arbitrary μ.

(c) Prove (a) with μ finite, $p = 1$, $q = \infty$.
(d) Prove (a) with μ σ-finite, $p = 1$, $q = \infty$. [For an extension of this result, see Kelley and Namioka (1963, Problem 14M).]

It follows that there is an isometric isomorphism of $(L^p)^*$ and L^q if $1 < p < \infty$, $1 < q < \infty$, $(1/p) + (1/q) = 1$; if μ is σ-finite, this is true also for $p = 1$, $q = \infty$.

3.4 BASIC THEOREMS OF FUNCTIONAL ANALYSIS

Almost every area of functional analysis leans heavily on at least one of the three basic results of this section: the Hahn–Banach theorem, the uniform boundedness principle, and the open mapping theorem. We are going to establish these results and discuss applications.

We first consider an extension problem. If f is a linear functional defined on a subspace M of a vector space L, there is no difficulty in extending f to a linear functional on all of L; simply extend a Hamel basis of M to a Hamel basis for L, define f arbitrarily on the basis vectors not belonging to

M, and extend by linearity. However, if L is normed and we require that the extension of f have the same norm as the original functional, the problem becomes more difficult. We first prove a preliminary result.

3.4.1   Lemma. Let L be a real vector space, not necessarily normed, and let p be a map from L to $\mathbb{R}$ satisfying

$$p(x + y) \leq p(x) + p(y) \qquad \text{for all} \quad x, y \in L$$

$$p(ax) = a p(x) \qquad \text{for all} \quad x \in L \quad \text{and all} \quad a > 0.$$

[The first property is called *subadditivity*, the second *positive-homogeneity*. Note that positive-homogeneity implies that $p(0) = 0$ [set $x = 0$ to obtain $p(0) = a p(0)$ for all $a > 0$]. A subadditive, positive-homogeneous map is sometimes called a *sublinear functional*.]

Let M be a subspace of L and g a linear functional defined on M such that $g(x) \leq p(x)$ for all $x \in M$. Let x_0 be a fixed element of L. For any real number c, the following are equivalent:

(1) $g(x) + \lambda c \leq p(x + \lambda x_0)$ for all $x \in M$ and all $\lambda \in \mathbb{R}$.
(2) $-p(-x - x_0) - g(x) \leq c \leq p(x + x_0) - g(x)$ for all $x \in M$.

Furthermore, there is a real number c satisfying (2), and hence (1).

PROOF.   To prove that (1) implies (2), first set $\lambda = 1$, and then set $\lambda = -1$ and replace x by $-x$ [note $g(-x) = -g(x)$ by linearity]. Conversely, if (2) holds and $\lambda > 0$, replace x by x/λ in the right-hand inequality of (2); if $\lambda < 0$, replace x by x/λ in the left-hand inequality. In either case, the positive homogeneity of p yields (1). If $\lambda = 0$, (1) is true by hypothesis.

To produce the desired c, let x and y be arbitrary elements of M. Then

$$g(x) - g(y) = g(x - y) \leq p(x - y)$$

$$\leq p(x + x_0) + p(-y - x_0) \qquad \text{by subadditivity.}$$

It follows that

$$\sup_{y \in M}[-p(-y - x_0) - g(y)] \leq \inf_{x \in M}[p(x + x_0) - g(x)].$$

Any c between the sup and the inf will work. $\square$

We may now prove the main extension theorem.

3.4.2   Hahn–Banach Theorem. Let p be a subadditive, positive-homogeneous functional on the real linear space L, and g a linear functional on the subspace M, with $g \le p$ on M. There is a linear functional f on L such that $f = g$ on M and $f \le p$ on all of L.

PROOF.   If $x_0 \notin M$, consider the subspace $M_1 = L(M \cup \{x_0\})$, consisting of all elements $x + \lambda x_0$, $x \in M$, $\lambda \in \mathbb{R}$. We may extend g to a linear functional on M_1 by defining $g_1(x + \lambda x_0) = g(x) + \lambda c$, where c is any real number. If we choose c to satisfy (1) of 3.4.1, then $g_1 \le p$ on M_1.

Now let $\mathscr{E}$ be the collection of all pairs (h, H) where h is an extension of g to the subspace $H \supset M$, and $h \le p$ on H. Partially order $\mathscr{E}$ by $(h_1, H_1) \le (h_2, H_2)$ iff $H_1 \subset H_2$ and $h_1 = h_2$ on H_1; then every chain in $\mathscr{E}$ has an upper bound (consider the union of all subspaces in the chain). By Zorn's lemma, $\mathscr{E}$ has a maximal element (f, F). If $F \ne L$, the first part of the proof yields an extension of f to a larger subspace, contradicting maximality. $\square$

There is a version of the Hahn–Banach theorem for complex spaces. First, we observe that if L is a vector space over $\mathbb{C}$, L is automatically a vector space over $\mathbb{R}$, since we may restrict scalar multiplication to real scalars. For example, $\mathbb{C}^n$ is an n-dimensional space over $\mathbb{C}$, with basis vectors $(1, 0, \ldots, 0), \ldots, (0, \ldots, 0, 1)$. If $\mathbb{C}^n$ is regarded as a vector space over $\mathbb{R}$, it becomes $2n$-dimensional, with basis vectors

$$(1, 0, \ldots, 0), \ldots, (0, \ldots, 0, 1), \qquad (i, 0, \ldots, 0), \ldots, (0, \ldots, 0, i).$$

Now if f is a linear functional on L, with $f_1 = \operatorname{Re} f$, $f_2 = \operatorname{Im} f$, then f_1 and f_2 are linear functionals on L', where L' is L regarded as a vector space over $\mathbb{R}$. Also, for all $x \in L$,

$$f(ix) = f_1(ix) + if_2(ix).$$

But

$$f(ix) = if(x) = -f_2(x) + if_1(x).$$

Thus $f_1(ix) = -f_2(x)$, $f_2(ix) = f_1(x)$; consequently

$$f(x) = f_1(x) - if_1(ix) = f_2(ix) + if_2(x).$$

Therefore f is determined by f_1 (or by f_2). Conversely, let f_1 be a linear functional on L'. Then f_1 is a map from L to $\mathbb{R}$ such that $f_1(ax + by) = af_1(x) + bf_1(y)$ for all $x, y \in L$ and all $a, b \in \mathbb{R}$. Define $f(x) = f_1(x) - if_1(ix)$, $x \in L$. It follows that f is a linear functional on L (and $f_1 = \operatorname{Re} f$). For f_1 is additive, hence so is f, and if $a, b \in \mathbb{R}$, we have

$$f((a + ib)x) = f_1(ax + ibx) - if_1(-bx + iax)$$
$$= af_1(x) + bf_1(ix) + ibf_1(x) - iaf_1(ix)$$

$$= (a + ib)[f_1(x) - if_1(ix)]$$
$$= (a + ib)f(x).$$

Note that homogeneity of f_1 does not immediately imply homogeneity of f. For $f_1(\lambda x) = \lambda f_1(x)$ for real λ but not in general for complex λ. [For example, let $L = \mathbb{C}$, $f(x) = x$, $f_1(x) = \text{Re } x$.]

We now prove the complex version of the Hahn–Banach theorem.

3.4.3 Theorem. Let L be a vector space over $\mathbb{C}$, and p a seminorm on L. If g is a linear functional on the subspace M, and $|g| \leq p$ on M, there is a linear functional f on L such that $f = g$ on M and $|f| \leq p$ on L.

PROOF.   Since p is a seminorm, it is subadditive and absolutely homogeneous, and hence positive-homogeneous. If $g_1 = \text{Re } g$, then $g_1 \leq |g| \leq p$; so by 3.4.2, there is an extension of g_1 to a linear map f_1 of L to R such that $f_1 = g_1$ on M and $f_1 \leq p$ on L. Define $f(x) = f_1(x) - if_1(ix)$, $x \in L$. Then f is a linear functional on L and $f = g$ on M. Fix $x \in L$, and let $f(x) = re^{i\theta}$, $r \geq 0$. Then

$$\begin{aligned}
|f(x)| = r &= f(e^{-i\theta}x) \\
&= f_1(e^{-i\theta}x) \quad &&\text{since } r \text{ is real} \\
&\leq p(e^{-i\theta}x) \quad &&\text{since } f_1 \leq p \quad \text{on} \quad L \\
&= p(x) \quad &&\text{by absolute homogeneity.} \quad \square
\end{aligned}$$

3.4.4 Corollary. Let g be a continuous linear functional on the subspace M of the normed linear space L. There is an extension of g to a continuous linear functional f on L such that $\|f\| = \|g\|$.

PROOF.   Let $p(x) = \|g\| \, \|x\|$; then p is a seminorm on L and $|g| \leq p$ on M by definition of $\|g\|$. The result follows from 3.4.3. $\square$

A direct application of the Hahn–Banach theorem is the result that in a normed linear space, there are enough continuous linear functionals to distinguish points; in other words, if $x \neq y$, there is a continuous linear functional f such that $f(x) \neq f(y)$. We now prove this, along with other related results.

3.4.5 Theorem. Let M be a subspace of the normed linear space L, and let L^* be the collection of all continuous linear functionals on L.

(a) If $x_0 \notin \overline{M}$, there is an $f \in L^*$ such that $f = 0$ on M, $f(x_0) = 1$, and $\|f\| = 1/d$, where d is the distance from x_0 to M

(b) $x_0 \in \overline{M}$ iff every $f \in L^*$ that vanishes on M also vanishes at x_0.

(c)   If $x_0 \neq 0$, there is an $f \in L^*$ such that $\|f\| = 1$ and $f(x_0) = \|x_0\|$; thus the maximum value of $|f(x)|/\|x\|$, $x \neq 0$, is achieved at x_0. In particular, if $x \neq y$, there is an $f \in L^*$ such that $f(x) \neq f(y)$.

PROOF.   (a)   First note that $L(M \cup \{x_0\})$ is the set of all elements $y = x + ax_0$, $x \in M$, $a \in \mathbb{C}$, and since $x_0 \notin M$, a is uniquely determined by y. Define f on $N = L(M \cup \{x_0\})$ by $f(x + ax_0) = a$; f is linear, and furthermore, $\|f\| = 1/d$, as we now prove. By 3.3.1 we have

$$\|f\| = \sup \left\{ \frac{|f(y)|}{\|y\|} : y \in N, \quad y \neq 0 \right\}$$

$$= \sup \left\{ \frac{|a|}{\|x + ax_0\|} : x \in M, \quad a \in \mathbb{C}, \quad x \neq 0 \quad \text{or} \quad a \neq 0 \right\}$$

$$= \sup \left\{ \frac{|a|}{\|x + ax_0\|} : x \in M, \quad a \in \mathbb{C}, \quad a \neq 0 \right\}$$

since $f(y) = 0$ when $a = 0$. Now

$$\frac{|a|}{\|x + ax_0\|} = \frac{1}{\left\| x_0 + \dfrac{x}{a} \right\|} = \frac{1}{\|x_0 - z\|} \quad \text{for some} \quad z \in M;$$

hence $\|f\| = (\inf\{\|x_0 - z\| : z \in M\})^{-1} = 1/d < \infty$. The result now follows from 3.4.4.

(b)   This is immediate from (a).

(c)   Apply (a) with $M = \{0\}$, to obtain $g \in L^*$ with $g(x_0) = 1$ and $\|g\| = 1/\|x_0\|$; set $f = \|x_0\|g$. $\square$

The Hahn–Banach theorem is basic in the study of the concept of reflexivity, which we now discuss. Let L be a normed linear space, and L^* the set of continuous linear functionals on L; L^* is sometimes called the *conjugate space* of L. By 3.3.5, L^* is a Banach space, so that we may talk about L^{**}, the conjugate space of L^*, or the *second conjugate space* of L. We may identify L with a subspace of L^{**} as follows: If $x \in L$, we define $x^{**} \in L^{**}$ by

$$x^{**}(f) = f(x), \quad f \in L^*.$$

If $\|f_n - f\| \to 0$, then $f_n(x) \to f(x)$; hence x^{**} is in fact a continuous linear functional on L^*. Let us examine the map $x \to x^{**}$ of L into L^{**}.

3.4.6   Theorem. Define $h: L \to L^{**}$ by $h(x) = x^{**}$. Then h is an isometric isomorphism of L and the subspace $h(L)$ of L^{**}; therefore, if $x \in L$, we have, by 3.3.1, $\|x\| = \|x^{**}\| = \sup\{|f(x)| : f \in L^*, \|f\| \leq 1\}$.

PROOF. To show that h is linear, we write

$$[h(ax + by)](f) = f(ax + by) = af(x) + bf(y) = [ah(x)](f) + [bh(y)](f).$$

We now prove that h is norm-preserving ($\|h(x)\| = \|x\|$ for all $x \in L$); consequently, h is one-to-one. If $x \in L$, $|[h(x)](f)| = |f(x)| \le \|x\| \, \|f\|$, and hence $\|h(x)\| \le \|x\|$. On the other hand, by 3.4.5(c), there is an $f \in L^*$ such that $\|f\| = 1$ and $|f(x)| = \|x\|$. Thus

$$\sup\{|[h(x)](f)|\colon \; f \in L^*, \quad \|f\| = 1\} \ge \|x\|,$$

so that $\|h(x)\| \ge \|x\|$, and consequently $\|h(x)\| = \|x\|$. □

If $h(L) = L^{**}$, L is said to be *reflexive*. Note that L^{**} is complete by 3.3.5 and so, by 3.4.6, a reflexive normed linear space is necessarily complete. We shall now consider some examples.

3.4.7 Examples. (a) Every Hilbert space is reflexive. For if ψ is the conjugate isometry of 3.3.4(a), H^* becomes a Hilbert space if we take $\langle f, g \rangle = \langle \psi(g), \psi(f) \rangle$. Thus if $q \in H^{**}$, we have, for some $g \in H^*$, $q(f) = \langle f, g \rangle = \langle \psi(g), \psi(f) \rangle = f(x)$, where $x = \psi(g)$. Therefore $q = h(x)$.

(b) If $1 < p < \infty$, l^p is reflexive. For by 3.3.4(b), $(l^p)^*$ is isometrically isomorphic to l^q, where $(1/p) + (1/q) = 1$. Thus if $t \in (l^p)^{**}$ we have $t(y) = \sum_{k=1}^{\infty} y_k z_k$, $y \in l^q$, where z is an element of $(l^q)^* = l^p$. But then $t = h(z)$.

Essentially the same argument, with the aid of Problem 11 of Section 3.3, shows that if $(\Omega, \mathscr{F}, \mu)$ is an arbitrary measure space, $L^p(\Omega, \mathscr{F}, \mu)$ is reflexive for $1 < p < \infty$.

(c) The space l^1 is not reflexive. This depends on the following result.

3.4.8 Theorem. If L is a normed linear space and L^* is separable, so is L. Thus if L is reflexive and separable (so that L^{**} is separable), then so is L^*.

PROOF. Let $f_1, f_2, \ldots$ form a countable dense subset of $\{f \in L^*\colon \|f\| = 1\}$ (note that any subset of a separable metric space is separable). Since $\|f_n\| = 1$, we can find points $x_n \in L$ with $\|x_n\| = 1$ and $|f_n(x_n)| \ge \frac{1}{2}$ for all n. Let M be the space spanned by the x_n; we claim that $M = L$. If not, 3.4.5(a) yields an $f \in L^*$ with $f = 0$ on M and $\|f\| = 1$. But then $\frac{1}{2} \le |f_n(x_n)| = |f_n(x_n) - f(x_n)| \le \|f_n - f\|$ for all n, contradicting the assumption that $\{f_1, f_2, \ldots\}$ is dense. □

To return to 3.4.7(c), we note that l^1 is separable since $\{x \in l^1\colon x_k = 0$ for all but finitely many $k\}$ is dense. [If $x \in l^1$ and $x^{(n)} = (x_1, \ldots, x_n, 0, 0, \ldots)$ then

$x^{(n)} \to x$ in l^1.] But $(l^1)^* = l^\infty$ by 3.3.4(c), and this space is not separable. For if $S = \{x \in l^\infty: x_k = 0 \text{ or } 1 \text{ for all } k\}$, then S is uncountable and $\|x - y\| = 1$ for all $x, y \in S, x \neq y$. Thus the sets $B_x = \{y \in l^\infty: \|y - x\| < \frac{1}{2}\}, x \in S$, form an uncountable family of disjoint open sets. If there were a countable dense set D, there would be at least one point of D in each B_x, a contradiction. Thus l^1 is not reflexive.

We now consider the second basic result of this section. Suppose that the A_i, i belonging to the arbitrary index set I, are bounded linear operators from L to M, where L and M are normed linear spaces. The uniform boundedness principle asserts that if L is complete and the A_i are pointwise bounded, that is, $\sup\{\|A_i x\|: i \in I\} < \infty$ for each $x \in L$, then the A_i are uniformly bounded, that is, $\sup\{\|A_i\|: i \in I\} < \infty$. Completeness of L is essential; to see this, let L be the set of all sequences $x = (x_1, x_2, \ldots)$ of complex numbers such that $x_k = 0$ for all but finitely many k, with the l^p norm, $1 \leq p \leq \infty$. Take $M = \mathbb{C}$, and $A_n x = n x_n$, $n = 1, 2, \ldots$. For any $x, A_n x = 0$ for sufficiently large n, so the A_n are pointwise bounded, although $\|A_n\| = n \to \infty$.

The proof that we shall give uses the Baire category theorem, which states that if a complete metric space is a countable union of closed sets, one of the sets must have a nonempty interior.

3.4.9   *Principle of Uniform Boundedness.* Let A_i, $i \in I$, be bounded linear operators from the Banach space L to the normed linear space M. If the A_i are pointwise bounded, they are uniformly bounded.

PROOF.   Let $C_n = \{x \in L: \sup_i \|A_i x\| \leq n\}$, $n = 1, 2, \ldots$. Since the A_i are pointwise bounded, $\bigcup_{n=1}^\infty C_n = L$, and since the A_i are continuous, each C_n is closed. By the Baire category theorem, for some n there is a closed ball $\bar{B} = \{x \in L: \|x - x_0\| \leq r\} \subset C_n$. Now if $\|y\| \leq 1$ and $i \in I$, we have

$$\|A_i y\| = \frac{1}{r}\|A_i z\| \qquad \text{where} \quad z = ry$$

$$\leq \frac{1}{r}\|A_i(x_0 + z)\| + \frac{1}{r}\|A_i x_0\|$$

$$\leq \frac{2n}{r} \qquad \text{since} \quad x_0 + z \quad \text{and} \quad x_0 \quad \text{belong to} \quad \bar{B}.$$

Thus $\|A_i\| \leq 2n/r$. $\quad\square$

The uniform boundedness principle is used in an important way in the study of weak convergence, a concept that we now describe.

3.4.10   Definitions and Comments. The sequence $\{x_n\}$ in the normed linear space L is said to converge *weakly* to $x \in L$ iff $f(x_n) \to f(x)$ for every $f \in L^*$ (notation: $x_n \xrightarrow{w} x$). Convergence in the metric of $L(\|x_n - x\| \to 0)$ will be called *strong convergence* and will be written simply as $x_n \to x$. [In this terminology, pointwise convergence of the sequence of linear operators A_n to the linear operator A means strong convergence of $A_n x$ to Ax for each x (see 3.3.6).]

It follows from the definitions that strong convergence implies weak convergence. To see that the converse does not hold, let $\{e_1, e_2, \ldots\}$ be an infinite orthonormal sequence in a Hilbert space H. If $x \in H$, then $\langle e_n, x \rangle \to 0$ as $n \to \infty$ by Bessel's inequality, hence $e_n \xrightarrow{w} 0$. But $\|e_n\| \equiv 1$, so e_n does not converge strongly to 0.

In a finite-dimensional space, strong and weak convergence coincide. For if $x_n = a_{1n}e_1 + \cdots + a_{kn}e_k$ converges weakly to $x = a_1 e_1 + \cdots + a_k e_k$ (where the e_i are basis vectors for $\mathbb{C}^k$), let f be a continuous linear functional that is 1 at e_i and 0 at e_j, $j \neq i$. Then $f(x_n) \to f(x)$, so that $a_{in} \to a_i$ as $n \to \infty$ $(i = 1, \ldots, k)$. Thus $x_n \to x$ strongly.

We give a few properties of weak convergence.

3.4.11   Theorem. (a)   A weakly convergent sequence $\{x_n\}$ is bounded, that is, $\sup_n \|x_n\| < \infty$. In fact if $x_n \xrightarrow{w} x_0$, then $\|x_0\| \leq \liminf_{n \to \infty} \|x_n\|$.

(b)   If A is a bounded linear operator from L to M and the sequence $\{x_n\}$ converges weakly to x_0 in L, then Ax_n converges weakly to Ax_0 in M.

(c)   If the sequence $\{x_n\}$ converges weakly to x_0, then x_0 belongs to the subspace spanned by the x_n.

(d)   Let M be a linear manifold of L. If x is the weak limit of some sequence in M, then x is the strong limit of some sequence in M.

PROOF.   (a)   The x_n may be regarded as continuous linear functionals on L^* with $x_n(f) = f(x_n)$ (see 3.4.6). The x_n are pointwise bounded on L^* since $f(x_n) \to f(x_0)$; hence by the uniform boundedness principle, $\sup_n \|x_n\| < \infty$. Also, if $f \in L^*$,

$$|f(x_0)| = \lim_{n \to \infty} |f(x_n)| = \liminf_{n \to \infty} |f(x_n)| \leq \|f\| \liminf_{n \to \infty} \|x_n\|.$$

Since $\|x_0\| = \sup\{|f(x_0)|\colon f \in L^*, \|f\| \leq 1\}$, we may conclude that $\|x_0\| \leq \liminf_{n \to \infty} \|x_n\|$.

(b)   If $g \in M^*$, define $f = g \circ A$; as A is continuous, we have $f \in L^*$, so that $f(x_n) \to f(x_0)$, that is, $g(Ax_n) \to g(Ax_0)$. But g is arbitrary; hence $Ax_n \xrightarrow{w} Ax_0$.

(c)   If this is false, 3.4.5(a) yields an $f \in L^*$ with $f(x_0) = 1$ and $f(x_i) \equiv 0$, contradicting $x_i \xrightarrow{w} x_0$.

(d)   Let $\{x_n\}$ be a sequence in M converging weakly to x. By (c), x is a strong limit of a sequence of finite linear combinations of the x_n. However, these finite linear combinations belong to M since M is a subspace. Thus x is the strong limit of a sequence in M. $\square$

In order to characterize weak convergence in specific spaces, the following result is useful.

3.4.12   Theorem. Let E be a subset of L^* such that $S(E) = L^*$, and let $\{x_n\}$ be a sequence in L. If $x_0 \in L$, then $x_n \xrightarrow{w} x_0$ iff $\sup_n \|x_n\| < \infty$ and $f(x_n) \to f(x_0)$ for all $f \in E$.

PROOF.   The "only if" part follows from 3.4.11(a) and the definition of weak convergence. For the "if" part, let $f \in L^*$, and choose elements $f_k \in L(E)$ with $\|f - f_k\| \to 0$. Then

$$|f(x_n) - f(x_0)| \leq |f(x_n) - f_k(x_n)| + |f_k(x_n) - f_k(x_0)| + |f_k(x_0) - f(x_0)|$$

$$\leq \|f - f_k\| \, \|x_n\| + |f_k(x_n) - f_k(x_0)| + \|f_k - f\| \, \|x_0\|.$$

Since the x_n are bounded in norm, given $\varepsilon > 0$, we may choose k such that the right-hand side is at most $|f_k(x_n) - f_k(x_0)| + \varepsilon$; but $f_k(x_n) \to f_k(x_0)$ as $n \to \infty$ since $f_k \in L(E)$, and since ε is arbitrary, the result follows. $\square$

We now describe weak convergence in l^p and $\mathbf{L}^p$.

3.4.13   Theorem. Assume $1 < p < \infty$.

(a)   Let $x_n = (x_{n1}, x_{n2}, \ldots) \in l^p$, $n = 1, 2, \ldots$. If $z \in l^p$, then $x_n \xrightarrow{w} z$ iff $\sup_n \|x_n\| < \infty$ and $x_{nk} \xrightarrow{w} z_k$ as $n \to \infty$ for each k.

(b)   Let $x_n \in \mathbf{L}^p(\Omega, \mathscr{F}, \mu)$, $n = 1, 2, \ldots$, where μ is assumed finite. If $z \in \mathbf{L}^p(\Omega, \mathscr{F}, \mu)$, then $x_n \xrightarrow{w} z$ iff $\sup_n \|x_n\| < \infty$ and $\int_A x_n \, d\mu \to \int_A z \, d\mu$ for each $A \in \mathscr{F}$. (It will often be convenient to blur the distinction between L^p and $\mathbf{L}^p$, and treat the elements of $\mathbf{L}^p$ as functions rather than equivalence classes.)

PROOF.   (a)   Define $f_k \in (l^p)^*$ by $f_k(x) = x_k$; then f_k corresponds to the sequence in l^q with a one in position k and zeros elsewhere [see 3.3.4(b)]. Take $E = \{f_1, f_2, \ldots\}$ and apply 3.4.12.

(b)   For each $A \in \mathscr{F}$, define $f_A \in (\mathbf{L}^p)^*$ by $f_A(x) = \int_A x \, d\mu$; f_A corresponds to the indicator function $I_A \in \mathbf{L}^q$ (see Problem 11, Section 3.3). Take

E as the set of all f_A, $A \in \mathscr{F}$; E spans $(\mathbf{L}^p)^*$ because the simple functions are dense in $\mathbf{L}^q$, and the result follows from 3.4.12. □

Note that 3.4.13 (b) holds also when $p = 1$ since simple functions are dense in $\mathbf{L}^\infty$.

We now consider the third basic result, the open mapping theorem. This will allow us to conclude that under certain conditions, the inverse of a one-to-one continuous linear operator is continuous. We cannot make this assertion in general, as the following example shows: Let L be the set of all continuous complex-valued functions x on [0, 1] such that $x(0) = 0$, $M = \{x \in L: x$ has a continuous derivative on [0, 1]$\}$; put the sup norm on L and M. If A is defined by $(Ax)(t) = \int_0^t x(s)\,ds$, $0 \le t \le 1$, then A is a one-to-one, bounded, linear operator from L onto M. [The condition $x(0) = 0$ is used in showing that A is onto.]

But A^{-1} is discontinuous; for example, if $x_n(t) = \sin nt$, then $y_n(t) = (Ax_n)(t) = (1 - \cos nt)/n$, so that $y_n \to 0$ in M, but x_n has no limit in L. A hypothesis under which continuity of the inverse holds is the completeness of both spaces L and M. In the above example, M is not complete; for example, a sequence of polynomials may converge uniformly to a continuous function without a continuous derivative (in fact to a continuous nowhere differentiable function).

We now state the third basic theorem.

3.4.14 Open Mapping Theorem. Let A be a bounded linear operator from the Banach space L onto the Banach space M. Then A is an open map, that is, if D is an open subset of L, then $A(D)$ is an open subset of M. Consequently if A is also one-to-one, then A^{-1} is bounded.

PROOF. Let $B_r = B(0, r)$ be the open ball with center at 0 and radius r in L. If we can show that $A(B_r)$ contains a ball with center at 0 in M, we are finished, since neighborhoods of an arbitrary point are translations of neighborhoods of 0. Now $L = \bigcup_{n=1}^\infty B_n$ and A maps onto M, so $M = \bigcup_{n=1}^\infty A(B_n)$. Since M is complete, we can conclude from the Baire category theorem that some $A(B_n)$ is not nowhere dense. Since $A(B_1)$ and $A(B_n)$ differ only by a scale factor, $A(B_1)$ is not nowhere dense. Thus for some $y_0 \in M$ and $r > 0$, the ball $B(y_0, 4r)$ is contained in the closure $\overline{A(B_1)}$. It follows that we may select $y_1 = Ax_1$ in $A(B_1)$ such that $\|y_1 - y_0\| < 2r$. [If y_1 unluckily ends up on the boundary of $A(B_1)$, approximate y_1 by a very close element in $A(B_1)$; this element can be chosen so that its distance to y_0 is still $< 2r$.] By the triangle inequality,

$$B(y_1, 2r) \subset B(y_0, 4r) \subset \overline{A(B_1)}. \tag{1}$$

We claim that

$$\text{if } \|y\| < 2r \qquad \text{then} \qquad y \in \overline{A(B_2)}. \tag{2}$$

To see this, note that $y = Ax_1 + (y - Ax_1)$, and since $y_1 = Ax_1$, the second term belongs to $\overline{A(B_1)}$ by (1). Therefore y is Ax_1 plus the limit of a sequence in $A(B_1)$, so that $y \in \overline{A(x_1 + B_1)}$. But $x_1 \in B_1$, so $\|x_1\| < 1$, and consequently $x_1 + B_1 \subset B_2$, proving (2).

Again because the B_i differ only by a scale factor, we can repeat the above argument with B_2 replaced by B_k for any $k > 0$; $2r$ becomes kr, and (2) becomes:

$$\text{if } \|y\| < kr \qquad \text{then} \qquad y \in \overline{A(B_k)}. \tag{3}$$

We are going to show that if $\|y\| < r/2$ then $y \in A(B_1)$. Thus $A(B_1)$, and hence by scaling the image of any ball with center at 0 in L, contains a ball with center at 0 in M, completing the proof. We use (3) to generate an inductive procedure:

Set $k = 1/2$; choose $x_1 \in B_{1/2}$ such that $\|y - Ax_1\| < r/4$.

Now apply (3) with $k = 1/4$; choose $x_2 \in B_{1/4}$ such that $\|y - Ax_1 - Ax_2\| < r/8$.

Now apply (3) with $k = 1/8$; choose $x_3 \in B_{1/8}$ such that $\|y - Ax_1 - Ax_2 - Ax_3\| < r/16$, and continue in this fashion. In general, we select $x_n \in B_{1/2^n}$ such that

$$\left\| y - \sum_{i=1}^{n} Ax_i \right\| < \frac{r}{2^{n+1}}. \tag{4}$$

If $s_n = \sum_{i=1}^{n} x_i$ then for $n > m$ we have $\|s_n - s_m\| \leq \sum_{i=m+1}^{n} \|x_i\| \to 0$ as $n, m \to \infty$, since $\|x_i\| < 2^{-i}$. The completeness of L implies that $\sum_{n=1}^{\infty} x_n$ converges. If x is the sum of the series then $\|x\| \leq \sum_{n=1}^{\infty} \|x_n\| < \sum_{n=1}^{\infty} 2^{-n} = 1$, and by (4), $Ax = y$. Therefore $y \in A(B_1)$. $\square$

The open mapping theorem allows us to prove the closed graph theorem, which is often useful in proving that a particular linear operator is bounded.

First we observe that if L and M are normed linear spaces, we may define a norm on the product space $L \times M$ by $\|(x, y)\| = (\|x\|^p + \|y\|^p)^{1/p}$, $x \in L$, $y \in M$, where p is any fixed real number in $[1, \infty)$. For if $x, x' \in L$, $y, y' \in M$,

$$(\|x + x'\|^p + \|y + y'\|^p)^{1/p} \leq [(\|x\| + \|x'\|)^p + (\|y\| + \|y'\|)^p]^{1/p}$$

$$\leq (\|x\|^p + \|y\|^p)^{1/p} + (\|x'\|^p + \|y'\|^p)^{1/p}$$

by Minkowski's inequality applied to $\mathbb{R}^2$.

Therefore the triangle inequality is satisfied and we have defined a norm on $L \times M$ (the other requirements for a norm are immediate). Furthermore, $(x_n, y_n) \to (x, y)$ iff $x_n \to x$ and $y_n \to y$; thus regardless of the value of p, the norm on $L \times M$ induces the product topology; also, if L and M are complete, so is $L \times M$. [The same result is obtained using the analog of the L^∞ norm, that is, $\|(x, y)\| = \max(\|x\|, \|y\|)$.]

3.4.15 Definition. Let A be a linear operator from L to M, where L and M are normed linear spaces. We say that A is *closed* iff the graph $G(A)$ $= \{(x, Ax): x \in L\}$ is a closed subset of $L \times M$. Equivalently, A is closed iff the following condition holds:

If $x_n \in L$, $x_n \to x$, and $Ax_n \to y$, then $(x, y) \in G(A)$; in other words, $y = Ax$. This formulation shows that every bounded linear operator is closed. The converse holds if L and M are Banach spaces.

3.4.16 Closed Graph Theorem. If A is a closed linear operator from the Banach space L to the Banach space M, then A is bounded.

PROOF. Since $G(A)$ is a closed subspace of $L \times M$, it is a Banach space. Define $P\colon G(A) \to L$ by $P(x, Ax) = x$. Then P is linear and maps onto L, and $\|P(x, Ax)\| = \|x\| \le \|(x, Ax)\|$; hence $\|P\| \le 1$ so that P is bounded. [Alternatively, if $(x_n, Ax_n) \to (x, y)$, then $x_n \to x$, proving continuity of P.] Similarly, the linear operator $Q\colon G(A) \to M$ given by $Q(x, Ax) = Ax$ is bounded. If $P(x, Ax) = 0$, then $x = Ax = 0$, so P is one-to-one. By 3.4.14, P^{-1} is bounded, and since $A = Q \circ P^{-1}$, A is bounded. $\square$

As an application of the closed graph theorem, we show that if P is the projection of a Banach space L on a closed subspace M, then P is continuous [see 3.3.3(c)]. Let $\{x_n\}$ be a sequence of points in L with $x_n \to x$, and assume Px_n converges to the element $y \in M$. Recall that in defining a projection operator it is assumed that L is the direct sum of closed subspaces M and N; thus

$$x_n = y_n + z_n \qquad \text{where} \qquad y_n = Px_n \in M, \quad z_n \in N.$$

Since $x_n \to x$ and $y_n \to y$, it follows that $z_n \to z = x - y$, necessarily in N. Therefore $x = y + z$, $y \in M$, $z \in N$, so that $y = Px$, proving P closed. By 3.4.16, P is continuous.

Problems

1. Show that a subadditive, absolutely homogeneous functional on a vector space must be nonnegative, and hence a seminorm. Give an example of a subadditive, positive-homogeneous functional that fails to be nonnegative.

2. Let $(\Omega, \mathscr{F}, \mu)$ be a measure space, and assume $\mathscr{F}$ is countably generated, that is, there is a countable set $\mathscr{C} \subset \mathscr{F}$ such that $\sigma(\mathscr{C}) = \mathscr{F}$. (Note that

the minimal field $\mathscr{F}_0$ over $\mathscr{C}$ is also countable; see Problem 9, 1.2.) If μ is σ-finite on $\mathscr{F}_0$ and $1 \le p < \infty$, show that $\mathbf{L}^p(\Omega, \mathscr{F}, \mu)$ is separable. If in addition there is an infinite collection of disjoint sets $A \in \mathscr{F}$ with $\mu(A) > 0$, show that $\mathbf{L}^1(\Omega, \mathscr{F}, \mu)$ is not reflexive.

3. If L and M are normed linear spaces and $[L, M]$ is complete, show that M must be complete.

4. Let $A \in [L, M]$, where L and M are normed linear spaces. The *adjoint* of A is an operator $A^*: M^* \to L^*$, defined as follows: If $f \in M^*$ we take $(A^* f)(x) = f(Ax)$, $x \in L$. Establish the following results:

   (a) $\|A^*\| = \|A\|$.
   (b) $(aA + bB)^* = aA^* + bB^*$ for all $a, b \in C, A, B \in [L, M]$.
   (c) If $A \in [L, M]$, $B \in [M, N]$, then $(BA)^* = A^*B^*$, where BA is the composition of A and B.
   (d) If $A \in [L, M]$, A maps onto M, and A^{-1} exists and belongs to $[M, L]$, then $(A^{-1})^* = (A^*)^{-1}$.

5. Define the *annihilator* of the subset K of the normed linear space L as $K^\perp = \{f \in L^*: f(x) = 0 \text{ for all } x \in K\}$. Similarly, if $J \subset L^*$, define $J^\perp = \{x \in L: f(x) = 0 \text{ for all } f \in J\}$. If $A \in [L, M]$, we denote by $N(A)$ the null space $\{x \in L: Ax = 0\}$, and by $R(A)$ the closure of the range of A, that is, the closure of $\{Ax: x \in L\}$. Establish the following:

   (a) For any $K \subset L$, $K^{\perp\perp} = S(K)$, the space spanned by K.
   (b) $R(A)^\perp = N(A^*)$ and $R(A) = N(A^*)^\perp$.
   (c) $R(A) = M$ iff A^* is one-to-one.
   (d) $R(A^*)^\perp = N(A)$.
   (e) For any $J \subset L^*$, $S(J) \subset J^{\perp\perp}$; $S(J) = J^{\perp\perp}$ if L is reflexive.
   (f) $R(A^*) \subset N(A)^\perp$; $R(A^*) = N(A)^\perp$ if L is reflexive.
   (g) If $R(A^*) = L^*$, then A is one-to-one; the converse holds if L is reflexive.

6. Consider the Hahn–Banach theorem 3.4.2, with the additional assumption that L is a normed linear space (or more generally, a topological vector space) and p is continuous at 0; hence continuous on all of L since $|p(x) - p(y)| \le p(x - y)$. Show that if L is separable, the theorem may be proved without Zorn's lemma. It follows that 3.4.3 and 3.4.4 do not require Zorn's lemma under the above hypothesis.

7. If μ_0 is a finitely additive, nonnegative real-valued set function on a field $\mathscr{F}_0$ of subsets of a set Ω, use the Hahn–Banach theorem to show that μ_0 has an extension to a finitely additive, nonnegative real-valued set function on the class of *all* subsets of Ω. Thus in one respect, at least, finite additivity is superior to countable additivity.

8. If the sequence of bounded linear operators A_n on a Banach space converges pointwise to the (necessarily linear) operator A, show that A is bounded; in fact

$$\|A\| \leq \liminf_{n \to \infty} \|A_n\| \leq \sup_n \|A_n\| < \infty.$$

9. Let $\{A_i, i \in I\}$ be a family of continuous linear operators from the Banach space L to the normed linear space M. Assume the A_i are weakly bounded, that is, $\sup_i |f(A_i x)| < \infty$ for all $x \in L$ and all $f \in M^*$. Show that the A_i are uniformly bounded, that is, $\sup_i \|A_i\| < \infty$.

10. (a) If the elements x_i, $i \in I$, belong to the normed linear space L, and $\sup_i |f(x_i)| < \infty$ for each $f \in L^*$, show that $\sup_i \|x_i\| < \infty$.

    (b) If A is a linear operator from the normed linear space L to the normed linear space M, and $f \circ A$ is continuous for each $f \in M^*$, show that A is continuous.

11. Let L, M, and N be normed linear spaces, with L or M complete, and let $B: L \times M \to N$ be a bilinear form, that is, $B(x, y)$ is linear in x for each fixed y, and linear in y for each fixed x. If for each $f \in N^*$, $f(B(x, y))$ is continuous in x for each fixed y, and continuous in y for each fixed x, show that B is bounded, that is,

$$\sup\{|B(x, y)|: \ \|x\|, \|y\| \leq 1\} < \infty.$$

   Equivalently, for some positive constant k we have $|B(x, y)| \leq k\|x\|\|y\|$ for all x, y.

12. Give an example of a closed unbounded operator from one normed linear space to another.

13. Let A be a bounded linear operator from the Banach space L onto the Banach space M. Show that there is a positive number k such that for each $y \in M$ there is an $x \in L$ with $y = Ax$ and $\|x\| \leq k\|y\|$. This result is sometimes called the *solvability theorem*.

3.5 REFERENCES

There is a vast literature on functional analysis, and we give only a few representative titles. Readable introductory treatments are given in Liusternik and Sobolev (1961), Taylor (1958), Bachman and Narici (1966), and Halmos (1951); the last deals exclusively with Hilbert spaces. Among the more advanced treatments, Dunford and Schwartz (1958, 1963, 1970) emphasize normed spaces, Kelley and Namioka (1963) and Schaefer (1966) emphasize topological vector spaces. Yosida (1968) gives a broad survey of applications to differential equations, semigroup theory, and other areas of analysis.

More recent works are by Conway (1990) and Wojtaszczyk (1991).

4

BASIC CONCEPTS OF PROBABILITY

4.1 INTRODUCTION

The starting point for probability theory is a set Ω called the *sample space* whose points are in one-to-one correspondence with the possible outcomes of a given performance of a random experiment. For example, if two dice are tossed, we may take Ω to have 36 points, one for each ordered pair (i, j), $i, j = 1, \ldots, 6$. The sample space for a given experiment is not unique. For example, if two dice are tossed and N is the sum of the faces, we may take Ω to consist of 11 points, corresponding to the outcomes $N = 2, 3, \ldots, 12$. The particular sample space to be used will be determined by the problem at hand. For example, in the dice-tossing example above, if we are interested in the result of the first toss, the sample space corresponding to $N = 2, 3, \ldots, 12$ will not be of value.

An "event" in a random experiment corresponds to a question that can be answered "yes" or "no." For example, in the dice-tossing example, let Ω be the set of ordered pairs (i, j), $i, j = 1, \ldots, 6$. We may ask the question "Is the maximum of the two coordinates i and j less than or equal to 2?" The subset of Ω associated with a "yes" answer is $A = \{(1, 1), (1, 2), (2, 1), (2, 2)\}$; the subset associated with a "no" answer is A^c.

Thus it is reasonable to define an event as a subset of Ω. However, in some situations not all subsets may be regarded as events. As an example, admittedly somewhat artificial, suppose that a coin is tossed four times, and Ω consists of the 16 sequences of length 4 with components H and T. Assume that only the results of the first two tosses are written down. If A is the set of points of Ω corresponding to at least three heads, then A is not "measurable," that is, the given information concerning ω is not sufficient to determine whether or not $\omega \in A$. More serious problems arise when Ω is $\mathbb{R}^n$; in this case we are almost always forced by mathematical consistency requirements to take the event class to be the Borel sets rather than the collection of all subsets of Ω.

The development of the mathematical theory will be facilitated if we require that the event class form a σ-field. Thus we may form countable unions, countable intersections, and complements of events and be assured that the resulting sets are also events.

Finally, we must talk about the probability $P(A)$ assigned to an event A. The basic physical requirement is that $P(A)$ correspond to the relative frequency of A in a very large number of independent repetitions of the random experiment. It follows that P should be a nonnegative, finitely additive set function, with $P(\Omega) = 1$. In order to be able to calculate the probability of a limit of events, we must require P to be countably additive.

The above discussion may be summarized by saying that P is a probability measure on the σ-field $\mathscr{F}$. Thus the basic mathematical object we are to study is a probability space $(\Omega, \mathscr{F}, P)$.

4.2   DISCRETE PROBABILITY SPACES

If the sample space Ω is a finite or countably infinite set, measure-theoretic difficulties do not arise. We take $\mathscr{F}$ to consist of all subsets of Ω, and assign probabilities in the following canonical way. Let $\Omega = \{\omega_1, \omega_2, \ldots\}$, and let $p_1, p_2, \ldots$ be nonnegative numbers whose sum is 1. If A is any subset of Ω, we define

$$P(A) = \sum_{\omega_i \in A} p_i.$$

In particular,

$$P\{\omega_i\} = p_i.$$

Then P is a probability measure, and the probability of the event A is computed simply by adding the probabilities of the individual points of A.

If Ω is countable, $\mathscr{F}$ consists of all subsets of Ω, and P is defined as above, $(\Omega, \mathscr{F}, P)$ is called a *discrete probability space*.

Classical probability theory was concerned with the special case $\Omega = \{\omega_1, \ldots, \omega_n\}$, $p_i = 1/n, i = 1, \ldots, n$. In this case,

$$P(A) = \frac{\text{number of points in } A}{\text{number of points in } \Omega}.$$

Thus to find $P(A)$ we count the number of favorable outcomes and divide by the total number of outcomes.

Unless otherwise specified, if Ω is countable, we always take $\mathscr{F}$ to contain all subsets of Ω. Thus all subsets of Ω are events, all functions on Ω are measurable, and measure-theoretic machinery is not needed.

4.3   INDEPENDENCE

Intuitively, two events A and B are independent if a statement concerning the occurrence or nonoccurrence of one of the events does not change the odds about the other event. Let us translate the physical requirement into mathematical terms. Suppose, for example, that $P(A) = 0.4$ and $P(B) = 0.3$. In a long sequence of repetitions of the random experiment, A will occur approximately 40% of the time. If B is to be independent of A, the occurrence of A will not influence the odds about B, so that if we examine only those trials on which A occurs, B will occur roughly 30% of the time; hence $P(A \cap B) = (0.4)(0.3) = P(A)P(B)$. Similarly, the nonoccurrence of A will not influence the odds about B; hence $P(A^c \cap B) = (0.6)(0.3) = P(A^c)P(B)$.

Conversely, if $P(A \cap B) = P(A)P(B)$ and $P(A^c \cap B) = P(A^c)P(B)$, B will be independent of A. If we examine only the trials on which A occurs, B must occur roughly 30% of the time in order to have $P(A \cap B) = P(A)P(B)$. Thus the occurrence of A does not influence the odds about B, and similarly, the occurrence of A^c does not change the odds about B.

The above discussion suggests that we call the event B independent of A if and only if $P(A \cap B) = P(A)P(B)$ and $P(A^c \cap B) = P(A^c)P(B)$. However, the first condition implies the second. Suppose $P(A \cap B) = P(A)P(B)$. Then

$$P(A^c \cap B) = P(B - A) = P(B - (A \cap B))$$

$$= P(B) - P(A \cap B) \qquad \text{since} \quad A \cap B \subset B$$

$$= P(B) - P(A)P(B) \qquad \text{by hypothesis}$$

$$= (1 - P(A))P(B)$$

$$= P(A^c)P(B).$$

Therefore B is independent of A if and only if $P(A \cap B) = P(A)P(B)$. But this condition is not altered if A and B are interchanged; hence B is independent of A if and only if A is independent of B. We may therefore formulate the definition of independence as follows.

4.3.1 Definition. Two events A and B are *independent* iff $P(A \cap B) = P(A)P(B)$.

We now consider independence of more than two events. If the events $A_i, i \in I$, are to be independent, and $i_1, \ldots, i_k$ are distinct indices, a statement about one or more of the events $A_{i_1}, \ldots, A_{i_k}$ should not change the odds about any of the remaining events. The physical discussion at the beginning of the section leads to the following requirement.

4.3.2 Definition. Let I be an arbitrary index set, and let $A_i, i \in I$, be events in a given probability space. The A_i are *independent* iff for all finite collections $\{i_1, \ldots, i_k\}$ of distinct indices in I, we have

$$P(A_{i_1} \cap A_{i_2} \cap \cdots \cap A_{i_k}) = P(A_{i_1})P(A_{i_2}) \cdots P(A_{i_k}).$$

4.3.3 ***Comments.*** (a)   If the events A_i, $i \in I$, are independent and any event is replaced by its complement, independence is maintained, that is,

$$P(B_{i_1} \cap \cdots \cap B_{i_k}) = P(B_{i_1}) \cdots P(B_{i_k}),$$

where $i_1, \ldots, i_k$ are distinct indices and B_{i_j} is either A_{i_j} or $A_{i_j}^c$, $j = 1, \ldots, k$.

We have essentially proved this in the discussion preceding the definition of independence by showing that $P(A \cap B) = P(A)P(B)$ implies $P(A^c \cap B) = P(A^c)P(B)$.

(b)   If $A_1, \ldots, A_n$ are events such that $A_{i_1}, \ldots, A_{i_k}$ are independent for all distinct indices $i_1, \ldots, i_k$, $k = 2, \ldots, n - 1$, it does not follow that $A_1, \ldots, A_n$ are independent. For example, let a coin be tossed twice, and assign probability $\frac{1}{4}$ to each of the outcomes HH, HT, TH, TT. Let $A = \{$first toss is a head$\}$, $B = \{$second toss is a head$\}$, $C = \{$first toss = second toss$\} = \{$HH, TT$\}$. Then

$$P(A \cap B) = \tfrac{1}{4} = P(A)P(B),$$

$$P(A \cap C) = \tfrac{1}{4} = P(A)P(C),$$

$$P(B \cap C) = \tfrac{1}{4} = P(B)P(C).$$

But

$$P(A \cap B \cap C) = \tfrac{1}{4} \neq P(A)P(B)P(C).$$

Thus A and B are independent, as are A and C, and also B and C, but A, B, and C are not independent.

Conversely, if $P(A_1 \cap \cdots \cap A_n) = P(A_1) \cdots P(A_n)$, it does not follow that $P(A_{i_1} \cap \cdots \cap A_{i_k}) = P(A_{i_1}) \cdots P(A_{i_k})$ when $k < n$. For example, suppose that two dice are tossed, and let Ω be all ordered pairs (i, j), $i, j = 1, 2, \ldots, 6$, with probability $\frac{1}{36}$ assigned to each point. Let

$$A = \{\text{second die is } 1, 2 \text{ or } 5\},$$

$$B = \{\text{second die is } 4, 5 \text{ or } 6\},$$

$$C = \{\text{the sum of the faces is } 9\}.$$

Then

$$P(A \cap B) = \tfrac{1}{6} \neq P(A)P(B) = \tfrac{1}{4},$$

$$P(A \cap C) = \tfrac{1}{36} \neq P(A)P(C) = \tfrac{1}{18},$$

$$P(B \cap C) = \tfrac{1}{12} \neq P(B)P(C) = \tfrac{1}{18},$$

but

$$P(A \cap B \cap C) = \tfrac{1}{36} = P(A)P(B)P(C).$$

4.4   BERNOULLI TRIALS

A sequence of n *Bernoulli trials* consists of n independent observations, with the property that each observation has only two possible results, called "success" and "failure." The probability of success on a given trial is p and the probability of failure is $q = 1 - p$.

To construct a probability space that meets the given physical requirements, we take Ω to be all 2^n ordered sequences of length n with components 1 and 0, with 1 indicating success and 0 failure. Consider a typical sample point ω with ones in positions $i_1, \ldots, i_k$ and zeros in positions $i_{k+1}, \ldots, i_n$. If A_i is the event of obtaining a success on trial i, so that $A_i = \{\omega: \text{the } i\text{th coordinate of } \omega \text{ is } 1\}$, we have

$$\{\omega\} = A_{i_1} \cap \cdots \cap A_{i_k} \cap A^c_{i_{k+1}} \cap \cdots \cap A^c_{i_n}.$$

Since the trials are independent and $P(A_i) = p$ for all i, the probability assigned to ω is determined; it must be

$$P\{\omega\} = P(A_{i_1}) \cdots P(A_{i_k})P(A^c_{i_{k+1}}) \cdots P(A^c_{i_n}) = p^k q^{n-k}.$$

Now there are $\binom{n}{k} = n!/k!(n-k)!$ sequences in Ω having exactly k ones, because such a sequence is determined by selecting k positions out of n for the ones to occur. Thus the probability of obtaining exactly k successes is

$$p(k) = \binom{n}{k} p^k q^{n-k}, \qquad k = 0, 1, \ldots, n. \tag{1}$$

By the binomial theorem, $\sum_{k=0}^{n} p(k) = (p + q)^n = 1$; therefore the sum of the probabilities assigned to all points is 1, and we have a legitimate probability measure.

A sequence of n *generalized Bernoulli trials* consists of n independent observations, such that each observation has k possible outcomes $(k > 2)$. If the k outcomes are labeled $b_1, \ldots, b_k$, the probability that b_i will occur on a given trial is p_i, where the p_i are nonnegative and $\sum_{i=1}^{k} p_i = 1$.

To construct an appropriate probability space, we take Ω to be all k^n ordered sequences of length n with components $b_1, \ldots, b_k$. If ω is a sample point having n_i occurrences of $b_i, i = 1, \ldots, k$, the independence of the trials and the assumption that the probability of obtaining b_i on a given trial is p_i leads us to assign to ω the probability $p_1^{n_1} p_2^{n_2} \cdots p_k^{n_k}$.

Now to find the number of sequences in Ω in which b_i occurs exactly n_i times, $i = 1, \ldots, k$, we reason as follows. Such a sequence is determined by selecting n_1 positions out of n to be occupied by b_1's, then n_2 positions from the remaining $n - n_1$ for the b_2's, and so on. Thus the total number of sequences is

$$\binom{n}{n_1}\binom{n-n_1}{n_2}\binom{n-n_1-n_2}{n_3}\cdots\binom{n-n_1-\cdots-n_{k-2}}{n_{k-1}}\binom{n_k}{n_k}$$

$$= \frac{n!}{n_1!n_2!\cdots n_k!}.$$

The total probability assigned to all points is

$$\sum \frac{n!}{n_1!n_2!\cdots n_k!} p_1^{n_1} p_2^{n_2} \cdots p_k^{n_k},$$

where the sum is taken over all nonnegative integers $n_1, \ldots, n_k$ whose sum is n. But this is $(p_1 + \cdots + p_k)^n = 1$, using the multinomial theorem.

The probability that b_1 will occur n_1 times, b_2 will occur n_2 times, $\ldots$, and b_k will occur n_k times, is

$$p(n_1, \ldots, n_k) = \frac{n!}{n_1!\cdots n_k!} p_1^{n_1} \cdots p_k^{n_k}, \tag{2}$$

where $n_1, \ldots, n_k = 0, 1, \cdots, n_1 + \cdots + n_k = n$.

4.5   CONDITIONAL PROBABILITY

If two events A and B are independent, a statement about the occurrence or nonoccurrence of one of the events does not change the odds about the other. In the absence of independence, the odds are altered, and the concept of conditional probability gives a quantitative measure of the change.

For example, suppose that the probability of A is 0.4 and the probability of $A \cap B$ is 0.1. If we repeat the experiment independently a large number of times and examine only the trials on which A has occurred, B will occur roughly 25% of the time. In general, the ratio $P(A \cap B)/P(A)$ is a measure of the probability of B under the condition that A is known to have occurred.

We therefore define the *conditional probability of B given A*, as

$$P(B \mid A) = \frac{P(A \cap B)}{P(A)} \tag{1}$$

provided $P(A) > 0$.

In the next chapter, we shall discuss in detail the concept of conditional probability $P(B|A)$ when the event A has probability 0. This is not a degenerate case; there are many natural and intuitive examples. Of course, the definition (1) no longer makes sense, and the approach will be somewhat indirect. At this point we shall only derive a few consequences of (1).

4.5.1   Theorem. (a)   If $P(A) > 0$, A and B are independent iff $P(B|A)$ $= P(B)$. [Similarly, independence is equivalent to $P(A|B) = P(A)$ if $P(B) > 0$.]
   (b)   If $P(A_1 \cap \cdots \cap A_{n-1}) > 0$, then

$$P(A_1 \cap \cdots \cap A_n) = P(A_1)P(A_2|A_1)P(A_3|A_1 \cap A_2) \cdots P(A_n|A_1 \cap \cdots \cap A_{n-1}).$$

PROOF.   Part (a) follows from the definitions of independence and conditional probability. To prove (b), observe that $P(A_1 \cap \cdots \cap A_{n-1}) > 0$ implies that $P(A_1)$, $P(A_1 \cap A_2)$, $\ldots$, $P(A_1 \cap \cdots \cap A_{n-2}) > 0$, so all conditional probabilities are well defined. Now by the definition of conditional probability,

$$P(A_1 \cap \cdots \cap A_n) = P(A_1 \cap \cdots \cap A_{n-1})P(A_n|A_1 \cap \cdots \cap A_{n-1}).$$

An induction argument completes the proof.   □

The following result will be quite useful.

4.5.2   Theorem of Total Probability. Let B_1, B_2, $\ldots$ form a finite or countably infinite family of mutually exclusive and exhaustive events, that is, the B_i are disjoint and their union is Ω.

   (a)   If A is any event, then $P(A) = \sum_i P(A \cap B_i)$. Thus $P(A)$ is calculated by making a list of mutually exclusive exhaustive ways in which A can happen, and adding the individual probabilities.
   (b) $P(A) = \sum_i P(B_i)P(A|B_i)$, where the sum is taken over those i for which $P(B_i) > 0$. Thus $P(A)$ is a weighted average of the conditional probabilities $P(A|B_i)$.

PROOF.
   (a) $P(A) = P(A \cap \Omega) = P(A \cap \bigcup_i B_i) = P(\bigcup_i (A \cap B_i)) = \sum_i P(A \cap B_i)$.
   (b)   This follows from (a) and the fact that $P(A \cap B_i) = 0$ if $P(B_i) = 0$, and equals $P(B_i)P(A|B_i)$ if $P(B_i) > 0$.   □

4.5.3   Example. A positive integer I is selected, with $P\{I = n\} = \left(\frac{1}{2}\right)^n$, $n = 1, 2, \ldots$. If I takes the value n, a coin with probability e^{-n} of heads is tossed once. Find the probability that the resulting toss is a head.
   Here we have specified $P(B_n) = \left(\frac{1}{2}\right)^n$, where $B_n = \{I = n\}$, $n = 1, 2, \ldots$. If A is the event that the coin comes up heads, we have specified $P(A|B_n)$ $= e^{-n}$. By the theorem of total probability, this is enough to determine $P(A)$. Formally, we may take Ω to consist of all ordered pairs (n, m), $n = 1, 2, \ldots$, $m = 0$ (tail) or 1 (head). We assign to the point $(n, 1)$ the probability $\left(\frac{1}{2}\right)^n e^{-n}$,

and to $(n, 0)$ the probability $\left(\frac{1}{2}\right)^n (1 - e^{-n})$. We may then verify that $P(B_n)$ $= \left(\frac{1}{2}\right)^n$, $P(A \mid B_n) = e^{-n}$. Hence

$$P(A) = \sum_{n=1}^{\infty} \left(\frac{1}{2}\right)^n e^{-n} = \frac{e^{-1}/2}{1 - (e^{-1}/2)}.$$

4.6 RANDOM VARIABLES

Intuitively, a random variable is a quantity that is measured in connection with a random experiment. If $(\Omega, \mathscr{F}, P)$ is a probability space and the outcome of the experiment corresponds to the point $\omega \in \Omega$, a measuring process is carried out to obtain a number $X(\omega)$. Thus X is a function from the sample space Ω to the reals (or the extended reals). For example, if $(\Omega, \mathscr{F}, P)$ corresponds to a sequence of four Bernoulli trials (4.4) and X is the number of successes, then $X(1\ 0\ 1\ 1) = 3$, $X(0\ 1\ 0\ 0) = 1$, and so on.

If we are interested in a random variable X defined on a given probability space, we generally want to know the probability of events involving X; for example, the probability that in a given performance of the experiment the value of X will belong to B, where B is a set of real numbers. In particular, we will be interested in $P\{\omega: a < X(\omega) \le b\}$ for all real a, b. Thus $\mathscr{F}$ must contain all sets of the form $X^{-1}(a, b]$, and therefore all sets $X^{-1}(B)$, B a Borel set in $\mathbb{R}$.

4.6.1 Definitions. A *random variable* X on a probability space $(\Omega, \mathscr{F}, P)$ is a Borel measurable function from Ω to $\mathbb{R}$. In the terminology of 1.5, we have $X: (\Omega, \mathscr{F}) \to (\mathbb{R}, \mathscr{B}(\mathbb{R}))$. In many situations it is convenient to allow X to take on the values $\pm\infty$; X is said to be an *extended random variable* iff X is a Borel measurable function from Ω to $\overline{\mathbb{R}}$, that is, $X: (\Omega, \mathscr{F}) \to (\overline{\mathbb{R}}, \mathscr{B}(\overline{\mathbb{R}}))$.

If X is a random variable on $(\Omega, \mathscr{F}, P)$ the *probability measure induced* by X is the probability measure P_X on $\mathscr{B}(\mathbb{R})$ given by

$$P_X(B) = P\{\omega: X(\omega) \in B\}, \qquad B \in \mathscr{B}(\mathbb{R}).$$

The numbers $P_X(B)$, $B \in \mathscr{B}(\mathbb{R})$, completely characterize the random variable X in the sense that they provide the probabilities of all events involving X. It is useful to know that this information may be captured by a single function from $\mathbb{R}$ to $\mathbb{R}$.

4.6.2 Definition. The *distribution function* of a random variable X is the function $F = F_X$ from $\mathbb{R}$ to $[0, 1]$ given by

$$F(x) = P\{\omega: X(\omega) \le x\}, \qquad x \text{ real.}$$

Since, for $a < b$, $F(b) - F(a) = P\{\omega : a < X(\omega) \le b\} = P_X(a, b]$, F is a distribution function corresponding to the Lebesgue–Stieltjes measure P_X (1.4). In particular, F is increasing and right-continuous. By 1.2.7, $F(x) \to 1$ as $x \to \infty$ and $F(x) \to 0$ as $x \to -\infty$. Thus among all distribution functions corresponding to P_X, we choose the one with $F(\infty) = 1$, $F(-\infty) = 0$.

Very often, the following statement is made: "Let X be a random variable with distribution function F," where F is a given function from $\mathbb{R}$ to $[0, 1]$ that is increasing and right-continuous, with $F(\infty) = 1$, $F(-\infty) = 0$. There is no reference to the underlying probability space $(\Omega, \mathscr{F}, P)$, and actually the nature of the underlying space is not important. The distribution function F determines the probability measure P_X, which in turn determines the probability of all events involving X. The only thing we have to check is that there be at least one $(\Omega, \mathscr{F}, P)$ on which a random variable X with distribution function F can be defined. In fact we can always supply the probability space in a canonical way; take $\Omega = \mathbb{R}$, $\mathscr{F} = \mathscr{B}(\mathbb{R})$, with P the Lebesgue–Stieltjes measure corresponding to F, and define $X(\omega) = \omega$, $\omega \in \Omega$, that is, X is the identity map. Since $P_X(B) = P\{\omega : X(\omega) \in B\} = P(B)$, X has induced probability measure P and therefore distribution function F.

Thus if $F \colon \mathbb{R} \to [0, 1]$ is increasing and right-continuous, with $F(\infty) = 1$, $F(-\infty) = 0$, then F is the distribution function of some random variable.

We isolate some particularly important classes of random variables.

4.6.3   Definitions and Comments. Let X be a random variable on $(\Omega, \mathscr{F}, P)$. We say that X is *simple* iff X can take on only finitely many possible values, *discrete* iff the set of values of X is finite or countably infinite. (Any random variable on a discrete probability space is discrete, since Ω is countable.)

If X is discrete, and the values $\{x_n\}$ of X can be arranged so that $x_n < x_{n+1}$ for all n, then the distribution function F is a step function with a discontinuity at each x_n, of magnitude $p_n = P\{X = x_n\}$; F is constant between the x_n and takes the upper value at each discontinuity. To see this, observe that if $x_{n-1} < a < x_n \le b < x_{n+1}$, then $F(b) - F(a) = P\{a < X \le b\} = p_n$; if $x_n \le c < d < x_{n+1}$, then $F(d) - F(c) = 0$.

If X is an arbitrary discrete random variable, the properties of X are completely determined by the *probability function* p_x, defined by

$$p_X(x) = P\{X = x\}, \qquad x \in \mathbb{R}.$$

Explicitly,

$$P_X(B) = \sum_{x \in B} p_X(x).$$

This is a countable sum since p_X is 0 except at the x_n.

Thus a discrete random variable may be specified by giving a countable set $\{x_n, n = 1, 2, \ldots\}$ of real numbers and a set of probabilities $\{p_n, n = 1, 2, \ldots\}$ ($p_n \geq 0$, $\sum_n p_n = 1$), where p_n is to serve as $P\{X = x_n\}$. The probability that X belongs to B is found by summing the p_n for those x_n which belong to B.

The random variable X is said to be *absolutely continuous* iff there is a nonnegative real-valued Borel measurable function f on $\mathbb{R}$ such that

$$F(x) = \int_{-\infty}^{x} f(t)dt, \qquad x \in \mathbb{R}.$$

We call f the *density* or *density function* of X; because $F(x) \to 1$ as $x \to \infty$, we have $\int_{-\infty}^{\infty} f(x)\,dx = 1$.

If X is absolutely continuous with density f, it follows that

$$P_X(B) = \int_B f(x)\,dx \qquad \text{for each} \qquad B \in \mathscr{B}(\mathbb{R}).$$

For the measure μ defined by $\mu(B) = \int_B f(x)\,dx$, $B \in \mathscr{B}(\mathbb{R})$, satisfies $\mu(a, b]$ $= F(b) - F(a)$, $a < b$. Thus μ is the Lebesgue–Stieltjes measure corresponding to F; hence $\mu = P_X$.

Thus absolute continuity of X means that $P_X \ll$ Lebesgue measure, or equivalently, by 2.3.1, F_X is an absolutely continuous function.

Any nonnegative Borel measurable function f on $\mathbb{R}$ with $\int_{-\infty}^{\infty} f(x)\,dx = 1$ is the density of some absolutely continuous random variable X. Let $F(x)$ $= \int_{-\infty}^{x} f(t)\,dt$; F is clearly increasing, and by 2.3.4, F is absolutely continuous, hence continuous, on $\mathbb{R}$. Since $F(\infty) = 1$, $F(-\infty) = 0$, F is the distribution function of some random variable X, and X must have density f.

The random variable X is said to be *continuous* iff its distribution function F is continuous on all of $\mathbb{R}$. Equivalently, X is continuous iff $P\{X = x\} = 0$ for all x [see (5) of 1.4.5].

We have seen that absolute continuity implies continuity. If F is the Cantor function (Problem 3, Section 2.3), extended so that $F(x) = 1$ for $x \geq 1$, and $F(x) = 0$ for $x < 0$, a random variable with distribution function F is continuous but not absolutely continuous.

Some typical examples of density functions are:

(1)   Uniform density on $[a, b]$:

$$f(x) = \begin{cases} (b - a)^{-1}, & a \leq x \leq b, \\ 0, & \text{elsewhere.} \end{cases}$$

(2)   Exponential density:

$$f(x) = \begin{cases} \lambda e^{-\lambda x}, & x \geq 0, \\ 0, & x < 0, \end{cases}$$

where $\lambda > 0$.

(3)   Two-sided exponential density:

$$f(x) = \tfrac{1}{2}\lambda e^{-\lambda |x|}, \qquad \lambda > 0.$$

(4)   Normal density:

$$f(x) = \frac{1}{\sqrt{2\pi}\sigma} \exp\left[-\frac{(x-m)^2}{2\sigma^2}\right], \sigma > 0, \qquad m \text{ real.}$$

(5)   Cauchy density:

$$f(x) = \frac{\theta}{\pi(x^2 + \theta^2)}, \qquad \theta > 0.$$

Finally, some remarks on terminology. We often abbreviate $\{\omega: X(\omega) \in B\}$ by $\{X \in B\}$; note that this set is also $X^{-1}(B)$, the preimage of B under the mapping X. Similarly, $\{\omega: a < X(\omega) \le b\}$ will be abbreviated by $\{a < X \le b\}$. The letter $\mathscr{B}$ will always stand for the Borel sets of an appropriate space. Thus $f: (\mathbb{R}^n, \mathscr{B}) \to (\mathbb{R}, \mathscr{B})$ means that $f^{-1}(B) \in \mathscr{B}(\mathbb{R}^n)$ for each $B \in \mathscr{B}(\mathbb{R})$. The phrase "almost surely," abbreviated a.s., is often used in the literature. It means "almost everywhere" with respect to a specified probability measure.

4.7   RANDOM VECTORS

We now consider situations involving more than one random variable associated with the same experiment.

An n-dimensional *random vector* on a probability space $(\Omega, \mathscr{F}, P)$ is a Borel measurable map from Ω to $\mathbb{R}^n$.

If $X: \Omega \to \mathbb{R}^n$ and $X_i = p_i \circ X$, where p_i is the projection of $\mathbb{R}^n$ onto the ith coordinate space, then X is Borel measurable iff each X_i is Borel measurable (see 1.5.8). Thus a random vector may be regarded as an n-tuple $(X_1, \ldots, X_n)$ of random variables.

Much of the development of the previous section carries over. As before, the *probability measure induced by the random vector X* is defined by

$$P_X(B) = P\{\omega: X(\omega) \in B\}, \qquad B \in \mathscr{B}(\mathbb{R}^n).$$

The *distribution function* of X is the function $F = F_X$ from $\mathbb{R}^n$ to $[0, 1]$ defined by

$$F(x) = P_X(-\infty, x] = P\{\omega: X_i(\omega) \le x_i, \qquad i = 1, \ldots, n\};$$

F is also called the *joint distribution function* of $X_1, \ldots, X_n$; F is increasing and right-continuous on $\mathbb{R}^n$, and P_X is the Lebesgue–Stieltjes measure determined by F (see 1.4).

By 1.2.7, we have

$$F(x_1, \ldots, x_n) \rightarrow \begin{cases} 1 & \text{as} & x_i \uparrow \infty & \text{for } all \ i; \\ 0 & \text{as} & x_i \downarrow -\infty & \text{for } any \ particular \ i \\ & & & \text{(with all other coordinates fixed).} \end{cases} \qquad (1)$$

If F is a distribution function on $\mathbb{R}^n$ that satisfies (1), then F is the distribution function of some random vector X, and the underlying probability space can be constructed in a canonical way. Take $\Omega = \mathbb{R}^n$, $\mathscr{F} = \mathscr{B}(\mathbb{R}^n)$, with P the Lebesgue–Stieltjes measure determined by F, and X the identity function on Ω. Since $P_X(B) = P\{\omega : X(\omega) \in B\} = P(B)$, X has induced probability measure P. Furthermore, the distribution function of X is

$$F_X(x) = P_X(-\infty, x] = P(-\infty, x]$$

$$= \lim_{a \downarrow -\infty} P(a, x] \qquad \text{by 1.2.7}$$

$$= \lim_{a \downarrow -\infty} F(a, x] \qquad \begin{array}{l} \text{since } P \text{ is the Lebesgue–Stieltjes} \\ \text{measure determined by } F \end{array}$$

$$= F(x) \qquad \begin{array}{l} \text{by 1.4.8(b) and the hypothesis} \\ \text{that } F(x) \rightarrow 0 \text{ if any } x_i \downarrow -\infty. \end{array}$$

[The hypothesis that $F(x) \rightarrow 1$ as $x \uparrow \infty$ implies that P is actually a probability measure.]

The random vector X is said to be *discrete* iff the set of values of X is finite or countably infinite, or equivalently, iff the component random variables $X_1, \ldots, X_n$ are all discrete. In this case, the properties of X are determined by the *probability function* p, given by $p(x) = p\{X = x\}$; explicitly, $P\{X \in B\} = \sum_{x \in B} p(x)$.

The random vector X is said to be *absolutely continuous* iff there is a nonnegative Borel measurable function f on $\mathbb{R}^n$, called the *density* or *density function* of X, such that

$$F(x) = \int_{(-\infty, x]} f(t) \, dt, \qquad x \in \mathbb{R}^n.$$

It follows that

$$P_X(B) = \int_B f(x) \, dx \cdot \qquad \text{for all} \qquad B \in \mathscr{B}(\mathbb{R}^n).$$

For the measure μ defined by $\mu(B) = \int_B f(x) \, dx$, $B \in \mathscr{B}(\mathbb{R}^n)$, satisfies $\mu(a, b] = \int_{(a,b]} f(x) \, dx = F(a, b]$ [see 1.4.10(b)]. Thus μ is the Lebesgue–Stieltjes measure determined by F; hence $\mu = P_x$.

Just as in the previous section, any nonnegative Borel measurable function f on $\mathbb{R}^n$ with $\int_{R^n} f(x)\,dx = 1$ is the density of some absolutely continuous random vector X.

4.8   INDEPENDENT RANDOM VARIABLES

We have talked previously about independence of events; we now consider independent random variables. Intuitively, independence of $X_1, \ldots, X_n$ means that a statement about one or more of the X_i does not affect the odds concerning the remaining X_i. Now a statement about X_i corresponds to an event of the form $A_i = \{X_i \in B_i\}$; thus the events $A_1, \ldots, A_n$ will be independent. The formal definition is as follows.

4.8.1   Definition.   Let $X_1, \ldots, X_n$ be random variables on $(\Omega, \mathscr{F}, P)$. Then $X_1, \ldots, X_n$ are said to be *independent* iff for all sets $B_1, \ldots, B_n \in \mathscr{B}(\mathbb{R})$, we have

$$P\{X_1 \in B_1, \ldots, X_n \in B_n\} = P\{X_1 \in B_1\} \cdots P\{X_n \in B_n\}.$$

By 2.6.8(b), independence of $X_1, \ldots, X_n$ may be expressed by saying that if $X = (X_1, \ldots, X_n)$, then P_X is the product of the P_{X_i}, $i = 1, \ldots, n$.

4.8.2   Comments.   (a)   If $X_1, \ldots, X_n$ are independent, so are $X_1, \ldots, X_k$ for $k < n$. To see this, let $B_1, \ldots, B_k \in \mathscr{B}(\mathbb{R})$. Then

$$P\{X_1 \in B_1, \ldots, X_k \in B_k\} = P\{X_1 \in B_1, \ldots, X_k \in B_k, X_{k+1} \in \mathbb{R}, \ldots, X_n \in \mathbb{R}\}$$
$$= P\{X_1 \in B_1\} \cdots P\{X_k \in B_k\}.$$

Thus it is not necessary to check all subfamilies of the collection of events $\{X_i \in B_i\}$, $i = 1, \ldots, n$, as in the definition of independent events in 4.3.

(b)   Independence of extended random variables is defined exactly as above, with $\mathscr{B}(\overline{\mathbb{R}})$ replacing $\mathscr{B}(\mathbb{R})$. In fact, suppose that each X_i is a *random object*; that is, a map $X_i \colon (\Omega, \mathscr{F}) \to (\Omega_i, \mathscr{F}_i)$, where Ω_i is an arbitrary set and $\mathscr{F}_i$ is a σ-field of subsets of Ω_i. Then the $X_1, \ldots, X_n$ are said to be independent iff for all sets $B_1 \in \mathscr{F}_1, \ldots, B_n \in \mathscr{F}_n$ we have

$$P\{X_1 \in B_1, \ldots, X_n \in B_n\} = P\{X_1 \in B_1\} \cdots P\{X_n \in B_n\}.$$

(c)   Let X_i, $i \in I$ (an arbitrary index set) be an arbitrary family of random objects. The X_i are said to be independent iff $X_{i_1}, \ldots, X_{i_n}$ are independent for each finite set $\{i_1, \ldots, i_n\}$ of distinct indices in I.

(d)   If the $X_i \colon (\Omega, \mathscr{F}) \to (\Omega_i, \mathscr{F}_i)$ are independent random objects, and $g_i \colon (\Omega_i, \mathscr{F}_i) \to (\Omega_i{}', \mathscr{F}_i{}')$, then the random objects $g_i \circ X_i$, $i \in I$, are independent.

("Functions of independent random objects are independent.") This follows from the definition of independence and the fact that

$$\{g_i \circ X_i \in B_i\} = \{X_i \in g_i^{-1}(B_i)\}.$$

Independence of random variables may be characterized in terms of distribution functions as follows.

4.8.3   Theorem. Let $X_1, \ldots, X_n$ be random variables on $(\Omega, \mathscr{F}, P)$. Let F_i be the distribution function of X_i, $i = 1, \ldots, n$, and F the distribution function of $X = (X_1, \ldots, X_n)$. Then $X_1, \ldots, X_n$ are independent iff

$$F(x_1, x_2, \ldots, x_n) = F_1(x_1)F_2(x_2) \cdots F_n(x_n) \qquad \text{for all real} \quad x_1, \ldots, x_n.$$

PROOF.   If $X_1, \ldots, X_n$ are independent, then

$$F(x_1, \ldots, x_n) = P\{X_1 \leq x_1, \ldots, X_n \leq x_n\} = \prod_{i=1}^{n} P\{X_i \leq x_i\} = \prod_{i=1}^{n} F_i(x_i).$$

Conversely, assume $F(x_1, \ldots, x_n) = \prod_{i=1}^{n} F_i(x_i)$ for all $x_1, \ldots, x_n$. Then

$$P_X(a, b] = F(a, b] = \prod_{i=1}^{n} [F_i(b_i) - F_i(a_i)] = \prod_{i=1}^{n} P_{X_i}(a_i, b_i]$$

[see 1.4.10(a)]. Thus

$$P\{X_1 \in B_1, \ldots, X_n \in B_n\} = P\{X_1 \in B_1\} \cdots P\{X_n \in B_n\} \qquad (1)$$

when the B_i are right-semiclosed intervals of reals. Now fix the intervals $B_2, \ldots, B_n$. The collection $\mathscr{C}$ of sets $B_1 \in \mathscr{B}(\mathbb{R})$ for which (1) holds is a monotone class including the field of finite disjoint unions of right-semiclosed intervals, and therefore $\mathscr{C} = \mathscr{B}(\mathbb{R})$ by the monotone class theorem. Applying the same reasoning to each coordinate in turn, we obtain the independence of the X_i. (Explicitly, we prove by induction that if $B_1, \ldots, B_i$ are arbitrary Borel sets and $B_{i+1}, \ldots, B_n$ are right-semiclosed intervals, then (1) holds for $B_1, \ldots, B_n$.) $\square$

We may also characterize independence in terms of densities.

4.8.4   Theorem. If $X = (X_1, \ldots, X_n)$ has a density f, then each X_i has a density f_i. Furthermore, in this case $X_1, \ldots, X_n$ are independent iff $f(x_1, \ldots, x_n)$

$= f_1(x_1) \cdots f_n(x_n)$ for all $(x_1, \ldots, x_n)$ except possibly for a Borel subset of $\mathbb{R}^n$ with Lebesgue measure zero.

PROOF.

$$F_1(x_1) = P\{X_1 \le x_1\} = P\{X_1 \le x_1, X_2 \in \mathbb{R}, \ldots, X_n \in \mathbb{R}\}$$

$$= \int_{-\infty}^{x_1} \int_{-\infty}^{\infty} \cdots \int_{-\infty}^{\infty} f(t_1, \ldots, t_n) \, dt_1 \cdots dt_n.$$

By definition of absolute continuity, X_1 has a density given by

$$f_1(x_1) = \int_{-\infty}^{\infty} \cdots \int_{-\infty}^{\infty} f(x_1, \ldots, x_n) \, dx_2 \cdots dx_n$$

(in other words, we integrate out the unwanted variables). Borel measurability of f_1 follows from Fubini's theorem. Similarly, each X_i has a density f_i, obtained by integrating out all variables except x_i.

Now if $f(x_1, \ldots, x_n) = f_1(x_1) \cdots f_n(x_n)$ a.e., then

$$F(x_1, \ldots, x_n) = \int_{-\infty}^{x_1} \cdots \int_{-\infty}^{x_n} f(t_1, \ldots, t_n) \, dt_1 \cdots dt_n = F_1(x_1) \cdots F_n(x_n),$$

so by 4.8.3, the X_i are independent. Conversely, if the X_i are independent, then

$$F(x_1, \ldots, x_n) = F_1(x_1) \cdots F_n(x_n)$$

$$= \int_{-\infty}^{x_1} \cdots \int_{-\infty}^{x_n} f_1(t_1) \cdots f_n(t_n) \, dt_1 \cdots dt_n. \qquad (1)$$

Thus (see the end of 4.7) if $g(x_1, \ldots, x_n) = f_1(x_1) \cdots f_n(x_n)$, then

$$P_X(B) = \int_B g(x) \, dx, \qquad B \in \mathscr{B}(\mathbb{R}^n).$$

But $P_X(B) = \int_B f(x) dx$, and it follows that $f = g$ a.e. (Lebesgue measure) by 1.6.11. $\square$

4.8.5   Corollary. If $X_1, \ldots, X_n$ are independent and X_i has density f_i, $i = 1, \ldots, n$, then X has a density f given by $f(x_1, \ldots, x_n) = f_1(x_1) \cdots f_n(x_n)$.

PROOF.   Equation (1) of 4.8.4 applies. $\square$

If $X_1, \ldots, X_n$ each have a density, it does not follow that $(X_1, \ldots, X_n)$ has a density; thus 4.8.5 is false without the independence hypothesis (see Problem 1).

Problems

1. Give an example of random variables X and Y (on the same probability space) such that X and Y each have densities, but (X, Y) does not.

2. Give an example to show that even if (X, Y) has a density, it is not determined by the individual densities of X and Y.

3. Let $X_1, \ldots, X_n$ be discrete random variables. Show that the X_i are independent iff

$$P\{X_1 = x_1, \ldots, X_n = x_n\} = \prod_{i=1}^{n} P\{X_i = x_i\} \qquad \text{for all real} \qquad x_1, \ldots, x_n.$$

4. Let $(\Omega, \mathscr{F}, P)$ be a probability space. The classes $\mathscr{C}_i, i \in I$, of sets in $\mathscr{F}$, are said to be *independent* iff given any choice of $C_i \in \mathscr{C}_i, i \in I$, the events C_i are independent. (Thus the random objects $X_i \colon (\Omega, \mathscr{F}) \to (\Omega_i', \mathscr{F}_i')$ are independent iff the classes $X_i^{-1}(\mathscr{F}_i')$ are independent.)

    If the $\mathscr{C}_i, i \in I$, are independent, show that if the following sets are added to each $\mathscr{C}_i$, the enlarged classes still remain independent.

    (a) Proper differences $A - B, A, B \in \mathscr{C}_i, B \subset A$.
    (b) The sets $\emptyset, \Omega$.
    (c) Countable disjoint unions of sets in $\mathscr{C}_i$.
    (d) Limits of monotone sequences in $\mathscr{C}_i$.

    Give an example to show that finite intersections *cannot* be added.
    If you are familiar with Zorn's lemma, show that if the $\mathscr{C}_i$ are independent classes, each closed under finite intersection, the minimal σ-fields over the $\mathscr{C}_i$ are also independent.

4.9 SOME EXAMPLES FROM BASIC PROBABILITY

In this section we give a few illustrations to show how some of the computations done in elementary probability courses fit in with the present measure-theoretic framework.

4.9.1 Example. Two numbers X and Y are picked at random between 0 and 1. Assume that X and Y are independent and that each is uniformly distributed (that is, X and Y have densities f_1 and f_2 given by $f_1(x) = f_2(x) = 1$, $0 \le x \le 1$, and 0 elsewhere). Let Z be the product XY, and let us find the distribution function of Z.

We take $\Omega = \mathbb{R}^2$, $\mathscr{F} = \mathscr{B}(\mathbb{R}^2)$, $X(x, y) = x$, $Y(x, y) = y$. By 4.8.5, (X, Y) must have density $f(x, y) = f_1(x)f_2(y)$; hence

$$P\{(X, Y) \in B\} = \iint_B f(x, y)\, dx\, dy.$$

Thus we take our probability measure to be

$$P(B) = \iint_B f_1(x)f_2(y)\, dx\, dy, \qquad B \in \mathscr{B}(\mathbb{R}^2).$$

Now

$$F(z) = P\{Z \le z\} = P\{(x, y): xy \le z\} = \iint_{xy \le z} f_1(x)f_2(y)\, dx\, dy.$$

Since X and Y are between 0 and 1 (with probability 1), $F(z) = 1$ for $z \ge 1$ and $F(z) = 0$ for $z \le 0$. Since $f_1(x)f_2(y)$ is 1 for $0 \le x \le 1$, $0 \le y \le 1$, and 0 elsewhere,

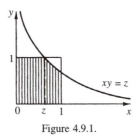

Figure 4.9.1.

$F(z)(0 < z < 1)$ is the shaded area in Fig. 4.9.1; that is, $z + \int_z^1 (z/x)\, dx$ $= z - z \ln z$. Thus Z has a density

$$f(z) = F'(z) = \begin{cases} -\ln z, & 0 < z < 1, \\ 0 & \text{elsewhere.} \end{cases}$$

Note that although f is unbounded, its *integral*, namely, F, is always finite.

4.9.2 Example. Let X, Y, and Z be independent random variables, each normally distributed with $m = 0$, $\sigma = 1$; that is, X, Y, and Z each have the normal density

$$g(x) = \frac{1}{\sqrt{2\pi}\sigma} \exp\left[\frac{-(x - m)^2}{2\sigma^2}\right] \qquad \text{with} \quad m = 0, \quad \sigma = 1.$$

Let $W = (X^2 + Y^2 + Z^2)^{1/2}$ (take the positive square root so that $W \geq 0$). Find the distribution function of W.

We take $\Omega = \mathbb{R}^3$, $\mathscr{F} = \mathscr{B}(\mathbb{R}^3)$, $X(x, y, z) = x$, $Y(x, y, z) = y$, $Z(x, y, z) = z$, and

$$P(B) = \iiint_B f(x, y, z)\, dx\, dy\, dz, \qquad B \in \mathscr{B}(\mathbb{R}^3),$$

where

$$f(x, y, z) = f_1(x)f_2(y)f_3(z) = g(x)g(y)g(z)$$
$$= (2\pi)^{-3/2} \exp\left[-\tfrac{1}{2}(x^2 + y^2 + z^2)\right].$$

Thus

$$F(w) = P\{W \leq w\} = P\{X^2 + Y^2 + Z^2 \leq w^2\}$$

If $w \geq 0$,

$$F(w) = \iiint_{x^2+y^2+z^2 \leq w^2} (2\pi)^{-3/2} \exp\left[-\frac{1}{2}(x^2 + y^2 + z^2)\right]\, dx\, dy\, dz$$

or in spherical coordinates,

$$F(w) = \int_0^{2\pi} d\theta \int_0^{\pi} d\phi \int_0^w (2\pi)^{-3/2} \exp\left[-\frac{1}{2}r^2\right] r^2 \sin\phi\, dr$$
$$= (2\pi)^{-3/2}(2\pi)(2) \int_0^w r^2 \exp\left[-\frac{1}{2}r^2\right] dr.$$

Thus W is absolutely continuous, with density

$$f(w) = \begin{cases} \dfrac{2}{\sqrt{2\pi}} w^2 \exp\left[-\dfrac{1}{2}w^2\right], & w \geq 0, \\ 0, & w < 0. \end{cases}$$

4.9.3   *Example.* Let $X_1, \ldots, X_n$ be independent random variables, each with density f and distribution function F; that is, $\Omega = \mathbb{R}^n$, $\mathscr{F} = \mathscr{B}(\mathbb{R}^n)$, $X_i(x_1, \ldots, x_n) = x_i$, $1 \leq i \leq n$,

$$P(B) = \int_B f(x_1) \cdots f(x_n)\, dx_1 \cdots dx_n, \qquad B \in \mathscr{B}(\mathbb{R}^n).$$

Let T_k be the kth smallest of the X_i; for example, if $n = 4$, $X_1(\omega) = 2$, $X_2(\omega) = 1.4$, $X_3(\omega) = -7$, $X_4(\omega) = 8$, then $T_1(\omega) = \min_i X_i(\omega) = X_3(\omega)$

$= -7, \quad T_2(\omega) = X_2(\omega) = 1.4, \quad T_3(\omega) = X_1(\omega) = 2, \quad T_4(\omega) = \max_i X_i(\omega)$
$= X_4(\omega) = 8.$ [Note that

$$P\{X_i = X_j\} = \iint\limits_{x_i = x_j} f(x_i) f(x_j) \, dx_i \, dx_j = 0 \qquad \text{for} \qquad i \neq j,$$

and therefore

$$P\{X_i = X_j \text{ for at least one } i \neq j\} \leq \sum_{i \neq j} P\{X_i = X_j\} = 0.$$

Thus ties occur with probability 0 and can be ignored.]

Find the individual distribution functions of the T_k, and the joint distribution function of $(T_1, \ldots, T_n)$.

Now

$$P\{T_k \leq x\} = \sum_{i=1}^{n} P\{T_k \leq x, T_k = X_i\} \qquad \text{by 4.5.2} \tag{1}$$

and, for example,

$P\{T_k \leq x, T_k = X_1\} = P\{X_1 \leq x, \text{ exactly } k-1 \text{ of the random}$
$\qquad\qquad\qquad\qquad\qquad\text{variables } X_2, \ldots, X_n \text{ are less than } X_1$
$\qquad\qquad\qquad\qquad\qquad\text{and the remaining } n-k \text{ random variables}$
$\qquad\qquad\qquad\qquad\qquad\text{are greater than } X_1\}. \tag{2}$

But, using Fubini's theorem,

$$P\{X_1 \leq x, X_2 < X_1, \ldots, X_k < X_1, X_{k+1} > X_1, \ldots, X_n > X_1\}$$

$$= \int_{x_1 = -\infty}^{x} \int_{x_2 = -\infty}^{x_1} \cdots \int_{x_k = -\infty}^{x_1} \int_{x_{k+1} = x_1}^{\infty}$$

$$\cdots \int_{x_n = x_1}^{\infty} f(x_1) \cdots f(x_n) \, dx_1 \cdots dx_n$$

$$= \int_{-\infty}^{x} f(x_1)(F(x_1))^{k-1}(1 - F(x_1))^{n-k} \, dx_1. \tag{3}$$

Now by symmetry, (2) is the sum of $\binom{n-1}{k-1}$ terms, each of which has the same value as (3), since we may select the $k-1$ random variables to be less than X_1 in $\binom{n-1}{k-1}$ ways. Also, each term in the summation (1) has the same value as (2). Thus

$$P\{T_k \leq x\} = \int_{-\infty}^{x} n\binom{n-1}{k-1} f(x_1)(F(x_1))^{k-1}(1 - F(x_1))^{n-k}\,dx_1$$

so that T_k is absolutely continuous, with density

$$f_{T_k}(x) = n\binom{n-1}{k-1} f(x)(F(x))^{k-1}(1 - F(x))^{n-k}.$$

We now find the joint distribution function of $T_1, \ldots, T_n$. Let $b_1 < b_2 < \cdots < b_n$. Then

$$
\begin{aligned}
P\{T_1 &\leq b_1, \ldots, T_n \leq b_n\} \\
&= n!\,P\{T_1 \leq b_1, \ldots, T_n \leq b_n, X_1 < X_2 < \cdots < X_n\} \qquad \text{by symmetry} \\
&= n!\,P\{X_1 \leq b_1, X_1 < X_2 \leq b_2, X_2 < X_3 \leq b_3, \ldots, X_{n-1} < X_n \leq b_n\} \\
&= n! \int_{-\infty}^{b_1} f(x_1)\,dx_1 \int_{x_1}^{b_2} f(x_2)\,dx_2 \cdots \int_{x_{n-1}}^{b_n} f(x_n)\,dx_n \\
&= \int_{-\infty}^{b_1} \cdots \int_{-\infty}^{b_n} g(x_1, \ldots, x_n)\,dx_1, \cdots dx_n,
\end{aligned}
$$

where

$$g(x_1, \ldots, x_n) = \begin{cases} n!\,f(x_1) \cdots f(x_n), & x_1 < x_2 < \cdots < x_n, \\ 0 & \text{elsewhere.} \end{cases}$$

Thus $(T_1, \ldots, T_n)$ is absolutely continuous with density g. (Note that f_{T_k} can be found from g (see 4.8.4), but the calculation is not any simpler than the direct method we have used above.)

4.9.4 *Example.*

Let X be an absolutely continuous random variable with density f, assumed to be piecewise-continuous. Let D be a Borel subset of $\mathbb{R}$ such that D includes the range of X, and let g be a Borel measurable function from D to $\mathbb{R}$.

If $Y = g \circ X$, we wish to find the distribution of Y. [*Distribution* is a generic term; to say that we know the distribution of Y means that we know how to calculate $P\{Y \in B\}$ for all Borel sets B. Thus the distribution may be specified by giving the induced probability measure P_Y or the distribution function F_Y. If Y is absolutely continuous, its density is adequate, and if Y is discrete, the probability function suffices. If $Y: (\Omega, \mathscr{F}) \to (\Omega', \mathscr{F}')$ is an arbitrary random object, the distribution of Y means the probability measure P_Y, defined by $P_Y(B) = P\{Y \in B\}, B \in \mathscr{F}'.$]

Assume that D is an open interval I, and g is either strictly increasing or strictly decreasing, with inverse h. Assume also that g has a continuous

nonzero derivative (hence so does h). We show that Y is absolutely continuous
with density

$$f_2(y) = \begin{cases} f_1(h(y)) \, |h'(y)|, & y \in g(I), \\ 0 & \text{elsewhere.} \end{cases}$$

We compute, for $y \in g(I)$,

$$F_2(y) = P\{Y \le y\} = P\{\omega:\ g(X(\omega)) \le y\}$$
$$= P\{X \le h(y)\} \quad \text{if} \quad g \text{ is increasing}$$
$$= P\{X \ge h(y)\} \quad \text{if} \quad g \text{ is decreasing}$$

(see Fig. 4.9.2).

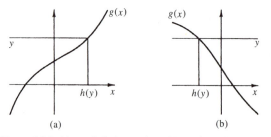

Figure 4.9.2. (a) g strictly increasing; (b) g strictly decreasing.

Thus

$$F_2(y) = \begin{cases} \displaystyle\int_{-\infty}^{h(y)} f_1(x) \, dx & \text{if } g \text{ is increasing,} \\ \displaystyle\int_{h(y)}^{\infty} f_1(x) \, dx & \text{if } g \text{ is decreasing.} \end{cases}$$

Therefore

$$\frac{dF_2(y)}{dy} = \begin{cases} f_1(h(y))h'(y) & \text{if } g \text{ is increasing,} \\ f_1(h(y))(-h'(y)) & \text{if } g \text{ is decreasing,} \\ f_1(h(y))|h'(y)| & \text{in either case.} \end{cases}$$

Now F_2 is continuous everywhere, and has a piecewise-continuous derivative;
it follows that F_2 is the integral of its derivative:

$$F_2(y) = \int_{-\infty}^{y} \frac{dF_2(z)}{dz} dz$$

(apply the fundamental theorem of calculus). Thus Y is absolutely continuous
with density $f_1(h(y))|h'(y)|$.

Examples of this type in which g is more complicated will be considered
in the problems.

Problems

1. Consider Example 4.9.4, but weaken the hypothesis so that the domain of g is the union of closed intervals $I_1, \ldots, I_n$, such that on the interior of each I_j, g has a continuous nonzero derivative and is either strictly increasing or strictly decreasing, with inverse h_j. Show that Y is absolutely continuous with density

$$f_2(y) = \sum_{j=1}^{n} f_1(h_j(y))|h_j'(y)|,$$

where $f_1(h_j(y))|h_j'(y)|$ is interpreted as 0 if y does not belong to the domain of h_j.

2. Let X be an absolutely continuous random variable with density $f_1(x) = x^3/64, 0 \leq x \leq 4; f_1(x) = 0$ elsewhere. Define a random variable Y by $Y = \min(\sqrt{X}, 2 - \sqrt{X})$. Find the density of Y.

3. Let X, Y, and Z be independent random variables, each uniformly distributed between 0 and 1. Find the probability that $Z^2 \leq XY$.

4. Let X be a random n-vector with density f_1, and let $Y = g \circ X$, where $g: \mathbb{R}^n \to \mathbb{R}^n$ (or $g: D \to \mathbb{R}^n$, where D is open in $\mathbb{R}^n$ and $P\{X \in D\} = 1$). Assume g to be one-to-one and continuously differentiable with a nonzero Jacobian J_g (hence g has a continuously differentiable inverse h). Show that Y is absolutely continuous with density

$$f_2(y) = f_1(h(y))|J_h(y)| = \frac{f_1(h(y))}{|J_g(x)|_{x=h(y)}}.$$

5. Let X and Y be independent random variables, each normally distributed with $m = 0$ and the same σ. Define random variables R and Θ by $X = R \cos \Theta, Y = R \sin \Theta$. Show that R and Θ are independent, and find their density functions (use Problem 4).

6. Let $X_1, \ldots, X_n$ be independent random variables, each with density f. Let X_0 be the number of random variables among $X_1, \ldots, X_n$ that exceed the smallest of the X_i by more than 2. Find $\int_\Omega X_0 \, dP$. (Leave the answer in the form of an integral on the real line.) Hint: Express X_0 as a sum of indicators.

7. Let $\Omega = [0, 1], \mathscr{F} =$ Borel sets, $P =$ Lebesgue measure. Show that $(\Omega, \mathscr{F}, P)$ is a universal probability space in the sense that if F is any proper distribution function on $\mathbb{R}$ ["proper" means that $F(\infty) = 1, F(-\infty) = 0$], there is a random variable X on $(\Omega, \mathscr{F}, P)$ with distribution function F. (Define $F^{-1}(y) = \sup\{x: F(x) < y\}, 0 < y < 1$, and take $X(\omega) = F^{-1}(\omega)$, with $X(0)$ and $X(1)$ arbitrary.)

4.10   EXPECTATION

Let X be a simple random variable on $(\Omega, \mathscr{F}, P)$, taking the values $x_1, \ldots,$ x_n with probabilities $p_1, \ldots, p_n$. If the random experiment is repeated independently N times, N very large, X will take the value x_i roughly Np_i times, so the arithmetic average of the values of X in the N observations is roughly

$$\frac{1}{N}[Np_1x_1 + Np_2x_2 + \cdots + Np_nx_n] = \sum_{i=1}^{n} p_i x_i.$$

This is a reasonable figure for the average value of X. If X is represented as $\sum_{i=1}^{n} x_i I_{B_i}$, where the B_i are disjoint sets in $\mathscr{F}$, the average value may be expressed as $\sum_{i=1}^{n} x_i P(B_i)$, which is $\int_{\Omega} X \, dP$. Since arbitrary random variables are ultimately built up from simple ones, it is reasonable to take $\int_{\Omega} X \, dP$ as the definition of the average value (henceforth to be called the "expectation") of X.

4.10.1   Definition.   If X is a random variable on $(\Omega, \mathscr{F}, P)$, the *expectation* of X is defined by

$$E(X) = \int_{\Omega} X \, dP$$

provided the integral exists. Thus $E(X)$ is the integral of the Borel measurable function X with respect to the probability measure P, so that all the results of integration theory are applicable. The same definition is used if X is an extended random variable.

In many situations it is inconvenient to compute $E(X)$ by integrating over Ω; the following result expresses $E(X)$ as an integral with respect to the induced probability measure P_X, which in turn is determined by the distribution function F.

First, a word about notation. If F is a distribution function on $\mathbb{R}^n$ with corresponding Lebesgue–Stieltjes measure μ, and $g: (\mathbb{R}^n, \mathscr{B}) \to (\mathbb{R}, \mathscr{B})$, then $\int_{\mathbb{R}^n} g(x) \, dF(x)$ means $\int_{\mathbb{R}^n} g \, d\mu$; it is *not* a Riemann–Stieltjes integral.

4.10.2   Theorem.   Let X be a random variable on $(\Omega, \mathscr{F}, P)$, with distribution function F. Let g be a Borel measurable function from $\mathbb{R}$ to $\mathbb{R}$. If $Y = g \circ X$, then

$$E(Y) = \int_{\mathbb{R}} g(x) \, dF(x) \left(= \int_{\mathbb{R}} g \, dP_X \right)$$

in the sense that if either of the two sides exists, so does the other, and the two sides are equal.

PROOF.   We use the basic technique of starting with indicators and proceeding to more complicated functions.

Let g be an indicator I_B, $B \in \mathscr{B}(\mathbb{R})$. Then

$$E(Y) = E(I_B \circ X) = E(I_{\{X \in B\}}) = P_X(B) = \int_{\mathbb{R}} g \, dP_X$$

so that $E(Y)$ and $\int_{\mathbb{R}} g \, dP_X$ exist and are equal.

Now let g be a nonnegative simple function, say, $g(x) = \sum_{j=1}^{n} x_j I_{B_j}(x)$, the B_j disjoint sets in $\mathscr{B}(\mathbb{R})$. Then

$$E(Y) = \sum_{j=1}^{n} x_j E(I_{B_j} \circ X) = \sum_{j=1}^{n} x_j \int_{\mathbb{R}} I_{B_j} \, dP_X \qquad \text{by what we have just proved}$$

$$= \int_{\mathbb{R}} \left(\sum_{j=1}^{n} x_j I_{B_j} \right) dP_X \qquad \text{since } g \geq 0$$

$$= \int_{\mathbb{R}} g \, dP_X.$$

Again, both integrals exist and are equal.

If g is a nonnegative Borel measurable function, let $g_1, g_2, \ldots$ be nonnegative simple functions with $g_n \uparrow g$. We have just proved that

$$E(g_n \circ X) = \int_{\mathbb{R}} g_n \, dP_X;$$

hence by the monotone convergence theorem,

$$E(g \circ X) = \int_{\mathbb{R}} g \, dP_X,$$

and again both integrals exist and are equal.

Finally, if $g = g^+ - g^-$ is an arbitrary Borel measurable function and $Y = g \circ X$, we have

$$E(Y) = E(Y^+) - E(Y^-) = E(g^+ \circ X) - E(g^- \circ X)$$

$$= \int_{\mathbb{R}} g^+ \, dP_X - \int_{\mathbb{R}} g^- \, dP_X \qquad \text{by what we have already proved}$$

$$= \int_{\mathbb{R}} g \, dP_X.$$

If $E(Y)$ exists and, say, $E(Y^-)$ is finite, then $\int_{\mathbb{R}} g^- \, dP_X$ is finite, and hence $\int_{\mathbb{R}} g \, dP_X$ exists; by the same reasoning, the existence of $\int_{\mathbb{R}} g \, dP_X$ implies that of $E(Y)$. □

190   4 BASIC CONCEPTS OF PROBABILITY

4.10.3 Corollaries and Extensions. (a) Let X be a random vector on $(\Omega, \mathscr{F}, P)$, and let g be a Borel measurable function from $\mathbb{R}^n$ to $\mathbb{R}$. Then $E(g \circ X) = \int_{\mathbb{R}^n} g(x)\, dF(x)$ in the sense that if either integral exists, so does the other, and the two are equal.

The proof is exactly as in 4.10.2, with $\mathbb{R}$ replaced by $\mathbb{R}^n$.

(b) More generally, let X be a random object on $(\Omega, \mathscr{F}, P)$, that is, $X \colon (\Omega, \mathscr{F}) \to (\Omega', \mathscr{F}')$, where $(\Omega', \mathscr{F}')$ is an arbitrary measurable space. Let g be a Borel measurable real (or extended real) valued function on $(\Omega', \mathscr{F}')$. Let P_X be the probability measure induced by X:

$$P_X(B) = P\{\omega \colon X(\omega) \in B\}, \qquad B \in \mathscr{F}'.$$

Then

$$E(g \circ X) = \int_{\Omega'} g\, dP_X$$

in the sense that if either integral exists, so does the other, and the two are equal.

Again, the proof is just as in 4.10.2, with $\mathbb{R}$ replaced by Ω' and $\mathscr{B}(\mathbb{R})$ by $\mathscr{F}'$.

(c) If X is a random variable (or random vector) with density f, then

$$\int g(x)\, dF(x) = \int g(x) f(x)\, dx$$

(integration over $\mathbb{R}$ in the case of a random variable, and over $\mathbb{R}^n$ in the case of a random vector) in the sense that if either integral exists, then so does the other, and the two are equal.

When g is an indicator I_B, this says that $P_X(B) = \int_B f(x)\, dx$, which holds for any Borel set B. The proof is completed by passing in turn to nonnegative simple functions, nonnegative Borel measurable functions, and arbitrary Borel measurable functions.

(d) If X is a discrete random variable with probability function p, then

$$\int g(x)\, dF(x) = \sum_x g(x) p(x),$$

where the series is interpreted as $\sum_x g^+(x) p(x) - \sum_x g^-(x) p(x)$, and again the interpretation is that the integral exists iff the sum exists (that is,

$$\sum_x g^+(x) p(x) < \infty \qquad \text{or} \qquad \sum_x g^-(x) p(x) < \infty),$$

and in this case the two are equal.

This is proved by starting with indicators as before.

Expectations of certain functions of X are of special interest.

4.10.4 Definition. Let X be a random variable on $(\Omega, \mathscr{F}, P)$. If $k > 0$, the number $E(X^k)$ is called the *kth moment* of X; $E[|X|^k]$ is called the *kth absolute moment* of X. $E[(X - E(X))^k]$ is called the *kth central moment*; $E[|X - E(X)|^k]$ the *kth absolute central moment*; central moments are defined only when $E(X)$ is finite.

The first moment $(k = 1)$ is $E(X)$, sometimes called the *mean* of X, and the first central moment (if it exists) is always 0. The second central moment $\sigma^2 = E[(X - E(X))^2]$ is called the *variance* of X, sometimes written Var X, and the positive square root σ the *standard deviation*.

Note that $E(X^k)$ is finite iff $E[|X|^k]$ is finite, by 1.6.4(b). Also, finiteness of the kth moments implies finiteness of lower moments, as we now prove.

4.10.5 Lemma. If $k > 0$ and $E(X^k)$ is finite, then $E(X^j)$ is finite for $0 < j < k$.

FIRST PROOF.

$$E[|X|^j] = \int_\Omega |X|^j \, dP = \int_{\{|X|^j < 1\}} |X|^j \, dP + \int_{\{|X|^j \geq 1\}} |X|^j \, dP$$

$$\leq P\{|X|^j < 1\} + \int_\Omega |X|^k \, dP < \infty. \quad \square$$

SECOND PROOF. We have $\|X\|_j \leq \|X\|_k$ for $0 < j < k$ (8.2.4). $\square$

Central moments of integral order can be obtained from moments, as follows.

4.10.6 Lemma. If n is a positive integer greater than 1, $E(X^{n-1})$ is finite, and $E(X^n)$ exists, then $E[(X - E(X))^n] = \sum_{k=0}^n \binom{n}{k} [-E(X)]^{n-k} E(X^k)$. In particular, if $E(X)$ is finite [$E(X^2)$ always exists since $X^2 \geq 0$], then

$$\text{Var } X = E(X^2) - [E(X)]^2.$$

PROOF. Use the binomial theorem and the additivity theorem for integrals (1.6.3). $\square$

A similar formula expresses moments in terms of central moments. (Write $X^n = (X - E(X) + E(X))^n$ and use the binomial theorem.)

We now restate a result proved earlier in a measure-theoretic context.

4.10.7   Chebyshev's Inequality. (a)   If X is a nonnegative random variable, $0 < p < \infty$ and $0 < \varepsilon < \infty$,

$$P\{X \geq \varepsilon\} \leq \frac{E(X^p)}{\varepsilon^p}.$$

(b)   If X is a random variable with finite mean m and variance σ^2, and $0 < k < \infty$,

$$P\{|X - m| \geq k\sigma\} \leq \frac{1}{k^2}.$$

This is a quantitative result to the effect that a random variable with small variance is likely to be close to its mean.

PROOF.   See 2.4.9. $\square$

A normally distributed random variable has the useful property that the distribution is completely determined by the mean and variance. Specifically, if X has the normal density, that is,

$$f(x) = \frac{1}{\sqrt{2\pi}\sigma} \exp\left[\frac{-(x - m)^2}{2\sigma^2}\right],$$

then $m = E(X)$ and $\sigma^2 = \operatorname{Var} X$; the computation is straightforward, using the standard integrals

$$\int_{-\infty}^{\infty} \exp(-x^2)\, dx = \sqrt{\pi} \quad \text{and} \quad \int_{-\infty}^{\infty} x^2 \exp(-x^2)\, dx = \frac{1}{2}\sqrt{\pi}.$$

The phrase "normal (m, σ^2)" is used for a random variable that is normally distributed with mean m and variance σ^2.

The following result on the expectation of a product of independent random variables is a direct consequence of Fubini's theorem.

4.10.8   Theorem. Let $X_1, \ldots, X_n$ be independent random variables on $(\Omega, \mathcal{F}, P)$. If all X_i are nonnegative or if $E(X_i)$ is finite for all i, then $E(X_1 \cdots X_n)$ exists and equals $E(X_1)E(X_2) \cdots E(X_n)$.

PROOF.   If all $X_i \geq 0$, then by 4.10.3(a),

$$E(X_1 \cdots X_n) = \int_{\mathbb{R}} x_1 \cdots x_n\, dP_X(x_1, \cdots, x_n) \qquad \text{where} \quad X = (X_1, \cdots, X_n).$$

Since P_X is the product of the P_{X_i} (see 4.8.1), Fubini's theorem yields

$$E(X_1 \cdots X_n) = \int_{\mathbb{R}} x_1 dP_{X_1}(x_1) \cdots \int_{\mathbb{R}} x_n dP_{X_n}(x_n) = E(X_1) \cdots E(X_n).$$

(This can also be proved without Fubini's theorem by starting with indicators and proceeding to nonnegative simple functions and then nonnegative measurable functions, but the present proof is faster. Note also that the result holds for extended random variables, with the same proof.)

If all $E(X_i)$ are finite, the above argument shows that

$$E(|X_1 \cdots X_n|) = \prod_{i=1}^{n} E(|X_i|) < \infty,$$

and thus Fubini's theorem may be applied just as in the first part of the proof. $\square$

Theorem 4.10.8 can be extended to complex-valued random variables. Recall from 2.4 that a complex-valued random variable X on $(\Omega, \mathscr{F}, P)$ is given by $X = X_1 + iX_2$ where $X_1 = \operatorname{Re} X$ and $X_2 = \operatorname{Im} X$ are (real-valued) random variables. In view of the discussion at the beginning of 4.7, we may regard X as simply a two-dimensional random vector. We define $E(X) = E(X_1) + iE(X_2)$ provided $E(X_1)$ and $E(X_2)$ are both finite.

4.10.9   Theorem. If $X_1, \ldots, X_n$ are independent complex-valued random variables and $E(X_i)$ is finite for all i, then $E(X_1 \cdots X_n)$ is finite and equals $E(X_1) \cdots E(X_n)$.

PROOF.   First, let $n = 2$, $X_1 = Y_1 + iZ_1$, $X_2 = Y_2 + iZ_2$. By 4.8.2(d), Y_1 and Y_2 are independent, as are Y_1 and Z_2, Z_1 and Y_2, and Z_1 and Z_2. By 4.10.8,

$$E(X_1 X_2) = E(Y_1)E(Y_2) - E(Z_1)E(Z_2) + iE(Y_1)E(Z_2) + iE(Z_1)E(Y_2)$$

$$= (E(Y_1) + iE(Z_1))(E(Y_2) + iE(Z_2)) = E(X_1)E(X_2).$$

Now let $n > 2$, $X_j = Y_j + iZ_j$, $j = 1, \ldots, n$, and assume the result has been established for $n - 1$ random variables. If $V = (Y_1, Z_1, Y_2, Z_2, \ldots, Y_{n-1}, Z_{n-1})$ and $W = (Y_n, Z_n)$, we claim that V and W are independent. By independence of $X_1, \ldots, X_n$, $P\{V \in A, W \in B\} = P\{V \in A\}P\{W \in B\}$ when A and B are measurable rectangles. Pass from measurable rectangles to finite disjoint unions of measurable rectangles, and then, by means of the monotone class theorem, to arbitrary Borel sets.

The independence of V and W implies, by 4.8.2(d), that $X_1 \cdots X_{n-1}$ and X_n are independent, so that $E(X_1 \cdots X_n) = E(X_1 \cdots X_{n-1})E(X_n) = E(X_1) \cdots E(X_n)$ by the induction hypothesis. $\square$

Theorem 4.10.8 implies that the variance of a sum of independent random variables is the sum of the variances. Actually, a somewhat more general result may be derived. We first introduce some new terminology.

4.10.10   Definitions and Comments. Let X and Y be random variables with finite expectation, and assume $E(XY)$ is also finite. (In particular, if X and Y have finite second moments, $E(XY)$ is finite by the Cauchy–Schwarz inequality.) The *covariance* of X and Y is defined by

$$\text{Cov}(X, Y) = E[(X - E(X))(Y - E(Y))] = E(XY) - E(X)E(Y).$$

If X and Y are independent, then $\text{Cov}(X, Y) = 0$ by 4.10.8; however, the converse is not true (consider $X = \cos\theta$, $Y = \sin\theta$, where θ is uniformly distributed between 0 and 2π).

If the variances $\sigma_X{}^2$ and $\sigma_Y{}^2$ are finite and greater than 0, the *correlation coefficient* between X and Y is defined by

$$\rho(X, Y) = (\text{Cov}(X, Y)/\sigma_X\sigma_Y).$$

By the Cauchy–Schwarz inequality applied to $X - E(X)$ and $Y - E(Y)$, $-1 \leq \rho(X, Y) \leq 1$. Furthermore (see Problem 6, Section 2.4), $|\rho(X, Y)| = 1$ iff $X' = X - E(X)$ and $Y' = Y - E(Y)$ are linearly dependent, that is, $P\{aX' + bY' = 0\} = 1$ for some real numbers a and b, not both 0.

We now look at the variance of a sum.

4.10.11   Theorem.

If $X_1, \ldots, X_n$ are random variables with finite expectation, and $E(X_iX_j)$ is finite for all i, j with $i \neq j$,

$$\text{Var}(X_1 + \cdots + X_n) = \sum_{i=1}^{n} \text{Var}\, X_i + 2 \sum_{\substack{i,j=1 \\ i<j}}^{n} \text{Cov}(X_i, X_j).$$

Thus if $X_1, \ldots, X_n$ are mutually uncorrelated, that is, $\text{Cov}(X_i, X_j) = 0$ for $i \neq j$; in particular, if $X_1, \ldots, X_n$ are independent, then

$$\text{Var}(X_1 + \cdots + X_n) = \sum_{i=1}^{n} \text{Var}\, X_i.$$

PROOF.

$$\text{Var}(X_1 + \cdots + X_n)$$
$$= E[(X_1 + \cdots + X_n - E(X_1) - \cdots - E(X_n))^2]$$
$$= \sum_{i=1}^{n} E[(X_i - E(X_i))^2] + 2 \sum_{\substack{i,j=1 \\ i<j}}^{n} E[(X_i - E(X_i))(X_j - E(X_j))]. \quad \square$$

4.10.12 Corollary. Under the hypothesis of 4.10.11, if $a_1, \ldots, a_n, b$ are arbitrary real numbers,

$$\operatorname{Var}(a_1 X_1 + \cdots + a_n X_n + b) = \sum_{i=1}^{n} a_i^2 \operatorname{Var} X_i + 2 \sum_{\substack{i,j=1 \\ i<j}}^{n} a_i a_j \operatorname{Cov}(X_i, X_j).$$

PROOF. This follows from 4.10.11, along with the observations, which may be verified from the definitions, that $\operatorname{Var}(aX + b) = a^2 \operatorname{Var} X$ and $\operatorname{Cov}(a_i X_i, a_j X_j) = a_i a_j \operatorname{Cov}(X_i, X_j)$. $\square$

Problems

1. Let X be a discrete random variable, with $P\{X = n\} = (\frac{1}{2})^n$, $n = 1, 2, \ldots$. Let $Y = g \circ X$, where $g(n) = (-1)^{n+1} 2^n / n$. Show that $E(Y)$ does not exist, although the series $\sum_{n=1}^{\infty} g(n) P\{X = n\}$ is conditionally convergent.

2. Let X be a random variable with the distribution function shown in Fig. 4.10.1. Compute $E(X^2)$.

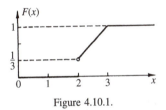

Figure 4.10.1.

3. Suppose X is a random variable with distribution function F_X, and $Y = g \circ X$, $g \colon \mathbb{R} \to \mathbb{R}$, Borel measurable. It is desired to find the expectation of Y. One student evaluates $\int_{-\infty}^{\infty} g(x) \, dF_X(x)$; another first finds the distribution function of Y, that is,

$$F_Y(y) = P\{Y \le y\} = P\{X \in g^{-1}(-\infty, y]\}$$

and then evaluates $\int_{-\infty}^{\infty} y \, dF_Y(y)$. Will the answers be the same?

4. Show that the random variables $X_1, \ldots, X_n$ are independent iff for all Borel measurable $g_i \colon \mathbb{R} \to \mathbb{R}$ such that $g_i \ge 0$, we have

$$E[g_1(X_1) \cdots g_n(X_n)] = \prod_{i=1}^{n} E[g_i(X_i)], \qquad \text{where} \qquad g_i(X_i) = g_i \circ X_i. \tag{1}$$

Also show that $X_1, \ldots, X_n$ are independent iff (1) holds for all Borel measurable $g_i \colon \mathbb{R} \to \mathbb{R}$ such that $E[g_i(X_i)]$ is finite for each i.

5. Let X be an extended random variable on $(\Omega, \mathscr{F}, P)$ and let g be a Borel mea-
    surable function from $\overline{\mathbb{R}}$ to $\mathbb{R}$. Define $F(x) = P\{-\infty < X \le x\}$,
    $x \in \mathbb{R}$; thus $F(\infty) - F(-\infty) = P\{X \text{ finite}\} \le 1$. Show that if $E[g \circ X]$ exists,
    $E[g \circ X] = \int_{-\infty}^{\infty} g(x) \, dF(x) + g(\infty)P\{X = \infty\} + g(-\infty)P\{X = -\infty\}$.

4.11 INFINITE SEQUENCES OF RANDOM VARIABLES

Very often we shall be interested in a sequence of random variables X_1,
$X_2, \ldots, X_n$, where n is arbitrarily large and not fixed in advance. Thus it is
convenient to have a single probability space on which we can define an infinite
sequence of random variables. The discussion of measures on infinite product
spaces (2.7) provides the necessary machinery. In particular, we can require
that $X_1, X_2, \ldots$ be independent, with X_i having a specified distribution. In fact
the X_i can be random objects with values in an arbitrary measurable space.

4.11.1   Theorem. Let $(\Omega_j, \mathscr{F}_j, P_j)$, $j = 1, 2, \ldots$ be an arbitrary sequence of
probability spaces. There exists a probability space $(\Omega, \mathscr{F}, P)$ and a sequence
of independent random objects $X_j \colon (\Omega, \mathscr{F}) \to (\Omega_j, \mathscr{F}_j)$ such that

$$P\{X_j \in B\} = P_j(B) \qquad \text{for all} \qquad B \in \mathscr{F}_j, j = 1, 2, \ldots.$$

To obtain a sequence of independent random variables with specified dis-
tribution functions $F_1, F_2, \ldots$, we take $\Omega_j = \mathbb{R}$, $\mathscr{F}_j = \mathscr{F}(\mathbb{R})$, with P_j the
Lebesgue–Stieltjes measure corresponding to F_j.

PROOF.   Let $\Omega = \prod_{j=1}^{\infty} \Omega_j$, $\mathscr{F} = \prod_{j=1}^{\infty} \mathscr{F}_j$, $P = \prod_{j=1}^{\infty} P_j$ (see 2.7.3). If
$\omega = (\omega_1, \omega_2, \ldots) \in \Omega$, let $X_j(\omega) = \omega_j$, $j = 1, 2, \ldots$. If $B \in \mathscr{F}_j$, then
$\{\omega \colon X_j(\omega) \in B\} = \{\omega \colon \omega_j \in B\}$, a measurable rectangle. Thus $\{\omega \colon X_j(\omega) \in B\}$
$\in \mathscr{F}$, so that the X_j are random objects. By 2.7.3,

$$P\{X_1 \in A_1, \ldots, X_n \in A_n\} = \prod_{j=1}^{n} P_j(A_j) \qquad \text{if} \qquad A_j \in \mathscr{F}_j, \quad 1 \le j \le n.$$

Take $A_i = \Omega_i$, $i \ne j$, to conclude that $P\{X_j \in A_j\} = P_j(A_j)$. Therefore
$P\{X_1 \in A_1, \ldots, X_n \in A_n\} = \prod_{j=1}^{n} P\{X_j \in A_j\}$, proving independence. $\square$

The results of 2.7 may also be used to provide the underlying probability
space for a *Markov chain*. Let S be a finite or countably infinite set, called the
state space; for convenience we may take S to be a subset of the integers. Let
$\Pi = [p_{ij}]$, $i, j \in S$, be a *stochastic matrix*, that is, $p_{ij} \ge 0$ for all $i, j \in S$, and
$\sum_j p_{ij} = 1$ for all $i \in S$. Let p_i, $i \in S$ be a set of nonnegative numbers adding
to 1 (the *initial distribution*). We envision a process that starts at time $t = 0$
in an initial state X_0, where $P\{X_0 = i\} = p_i$, $i \in S$, and makes transitions at

times $t = 1, 2, \ldots$ in accordance with the following rule. If X_n denotes the state at time n, and we know that $X_n = i$, then regardless of the past history, in other words, regardless of how the process happened to arrive at state i, the probability of moving to state j at time $n + 1$ is p_{ij}. We expect from 4.5.1(b) that for all $i_0, i_1, \ldots, i_n \in S$, $n = 0, 1, 2, \ldots,$

$$P\{X_0 = i_0, X_1 = i_1, \ldots, X_n = i_n\} = p_{i_0} p_{i_0 i_1} \cdots p_{i_{n-1} i_n}.$$

We now show that it is possible to construct a sequence of random variables satisfying this requirement.

4.11.2   Theorem. For a given state space S, stochastic matrix $\Pi = [p_{ij}]$, $i, j \in S$, and initial distribution p_i, $i \in S$, there is a sequence of random variables $X_0, X_1, \ldots$, all defined on the same probability space and taking values in S, such that

$$P\{X_0 = i_0, X_1 = i_1, \ldots, X_n = i_n\} = p_{i_0} p_{i_0 i_1} \cdots p_{i_{n-1} i_n} \qquad (1)$$

for all $i_0, \ldots, i_n \in S$ and all $n = 0, 1, \ldots$.

PROOF.   Let $\mathscr{S}$ consist of all subsets of S, and take $\Omega = S^\infty$, $\mathscr{F} = \mathscr{S}^\infty$. Define

$$P_0(B) = \sum_{i \in B} p_i, \qquad B \in \mathscr{S},$$

$$P(i_0, \ldots, i_{n-1}, B) = \sum_{j \in B} p_{i_{n-1} j}, \qquad B \in \mathscr{S}, \quad i_0, \ldots, i_{n-1} \in S.$$

Since $\mathscr{S}^n$ consists of all subsets of S^n, the measurability requirements of Theorem 2.7.2 are automatically satisfied. If we define $X_n(\omega_0, \omega_1, \ldots) = \omega_n$, $n = 0, 1, \ldots$, we obtain

$$P\{X_0 = i_0, \ldots, X_n = i_n\} = \int_S P_0(d\omega_0) \int_S P(\omega_0, d\omega_1)$$

$$\cdots \int_S I_B(\omega_0, \ldots, \omega_n) P(\omega_0, \ldots, \omega_{n-1}, d\omega_n),$$

where $B = \{(\omega_0, \ldots, \omega_n) \colon \omega_0 = i_0, \ldots, \omega_n = i_n\}$. Now if μ is a measure on $\mathscr{S}$ with $\mu\{j\} = q_j$, $j \in S$, and $f \colon S \to R$, then $\int_S f \, d\mu = \sum_j q_j f(j)$ (see 2.4.12). Thus

$$\int_S I_B(\omega_0, \ldots, \omega_n) P(\omega_0, \ldots, \omega_{n-1}, d\omega_n)$$

$$= I_B(\omega_0, \ldots, \omega_{n-1}, i_n) P(\omega_0, \ldots, \omega_{n-1}, \{i_n\})$$

$$= I_B(\omega_0, \ldots, \omega_{n-1}, i_n) p_{\omega_{n-1} i_n}.$$

We may continue this process to obtain

$$P\{X_0 = i_0, \ldots, X_n = i_n\} = p_{i_0} p_{i_0 i_1} \cdots p_{i_{n-1} i_n}. \quad \square$$

A sequence of random variables $\{X_n\}$ satisfying (1) of 4.11.2 is called a *Markov chain* corresponding to the matrix Π and the initial distribution $\{p_i\}$; Π is called the *transition matrix* of the chain and the numbers p_{ij} are called *transition probabilities*.

The basic properties of Markov chains are discussed by Ash (1970, Chapters 6 and 7). The symmetric random walk in $\mathbb{R}^k$, an important special case, is considered in Appendix 1.

If $X_1, \ldots, X_n$ are random variables on $(\Omega, \mathscr{F}, P)$, the random vector $X = (X_1, \ldots, X_n)$ is a Borel measurable map from Ω to $\mathbb{R}^n$. The same interpretation is possible for an infinite sequence of random variables, as follows.

4.11.3   Theorem. Let $(\Omega_j, \mathscr{F}_j)$, $j = 1, 2, \ldots$, be arbitrary measurable spaces, and let p_j be the projection of $\prod_{j=1}^{\infty} \Omega_j$ onto the jth coordinate space. If $(\Omega, \mathscr{F})$ is a measurable space and $X \colon \Omega \to \prod_{j=1}^{\infty} \Omega_j$, let $X_j = p_j \circ X$, $j = 1, 2, \ldots$. Then X is measurable (relative to $\mathscr{F}$ and $\prod_{j=1}^{\infty} \mathscr{F}_j$) iff each X_j is measurable (relative to $\mathscr{F}$ and $\mathscr{F}_j$). In particular, if $X_1, X_2, \ldots$ are random variables, then $X = (X_1, X_2, \ldots)$ is measurable: $(\Omega, \mathscr{F}) \to (\mathbb{R}^{\infty}, \mathscr{B}^{\infty})$, $\mathscr{B} = \mathscr{B}(\mathbb{R})$. For this reason X is sometimes called a *random sequence*. The same result holds when Ω is an arbitrary (possibly uncountable) product space.

PROOF.   If X is measurable, each X_j is a composition of measurable maps and is therefore measurable. Conversely, assume each X_j to be measurable. Let $B = \{(\omega_1, \omega_2, \ldots) \colon \omega_j \in A_j, j = 1, \ldots, n\}$, the $A_j \in \mathscr{F}_j$, be a measurable rectangle in $\prod_{j=1}^{\infty} \mathscr{F}_j$. Then

$$\{\omega \in \Omega \colon X(\omega) \in B\} = \bigcap_{j=1}^{n} \{\omega \in \Omega \colon X_j(\omega) \in A_j\} \in \mathscr{F}.$$

The proof for uncountable product spaces is essentially the same, with B replaced by a measurable rectangle in $\Pi_j \mathscr{F}_j$. $\square$

The proof of 4.11.3 shows also that if $X \colon \Omega \to \prod_{j=1}^{n} \Omega_j$, then X is measurable iff each X_j, $1 \leq j \leq n$, is measurable.

We close the chapter with an introduction to one of the basic limit theorems of probability.

4.11.4   Weak Law of Large Numbers. Let $X_1, X_2, \ldots$ be independent random variables (not necessarily with the same distribution), each with finite mean and variance. Assume the variances to be uniformly bounded by $M < \infty$. Let $S_n = X_1 + \cdots + X_n$. Then $[S_n - E(S_n)]/n$ converges in probability to 0, that is, given $\varepsilon > 0$,

$$P\left\{ \left| \frac{S_n - E(S_n)}{n} \right| \geq \varepsilon \right\} \to 0 \qquad \text{as} \qquad n \to \infty.$$

PROOF.   By Chebyshev's inequality,

$$P\{|(S_n - ES_n)/n| \geq \varepsilon\} \leq \frac{1}{\varepsilon^2} E\left[\left(\frac{S_n - ES_n}{n} \right)^2 \right]$$

$$= \frac{1}{\varepsilon^2 n^2} \operatorname{Var} S_n$$

$$= \frac{1}{\varepsilon^2 n^2} \sum_{k=1}^{n} \operatorname{Var} X_k \qquad \text{by 4.10.11}$$

$$\leq \frac{M}{\varepsilon^2 n} \to 0. \quad \square$$

There are two special cases of particular interest.

1.   If $E(X_i) = m$ for all i, then $[S_n - E(S_n)]/n = (S_n/n) - m$; hence $S_n/n \to m$ in probability. Thus for large n, the arithmetic average of n independent random variables, each with finite expectation m (and with the variances uniformly bounded) is quite likely to be very close to m.

2.   If $X_1, X_2, \ldots$ are independent, and for each i, $P\{X_i = 1\} = p$, $P\{X_i = 0\} = q = 1 - p$ (thus we have an infinite sequence of Bernoulli trials), then $X_1 + \cdots + X_n$ is the number of successes in n trials, hence S_n/n is the relative frequency of successes. Since $EX_i = p$, we have $S_n/n \to p$ in probability. Thus for large n, the relative frequency of successes is quite likely to be very close to p.

Intuitively, the weak law of large numbers says the following. If we regard observation of $X_1, \ldots, X_n$ as one performance of an experiment, where n is very large but *fixed*, then if we repeat the experiment independently, $(S_n - ES_n)/n$ will be close to 0 a very high percentage of the time.

But physically we expect something more than this. If a coin with probability p of heads is tossed over and over again, we expect the relative frequency to approach p in the ordinary sense of convergence of a sequence of real numbers; in other words, given $\varepsilon > 0$, eventually the relative frequency S_n/n gets and *remains* within ε of p. Here we are considering observation of the infinite sequence $X_1, X_2, \ldots$ as one performance of the experiment, and what we must show is that $\lim_{n \to \infty} S_n(\omega)/n = p$ for almost every ω. A statement of this type is called a *strong law of large numbers*. This subject will be considered in detail later.

Problems

1. (a) If $Y_1, Y_2, \ldots$ are independent random objects, show that for each n, $(Y_1, \ldots, Y_n)$ and $(Y_{n+1}, Y_{n+2}, \ldots)$ are independent.

   (b) In part (a), if $Y_i : (\Omega, \mathscr{F}) \to (S, \mathscr{S})$ for all i, and all Y_i have the same distribution, show that $(Y_1, Y_2, \ldots)$ and $(Y_n, Y_{n+1}, \ldots)$ have the same distribution for all n.

2. Consider the gambler's ruin problem, that is, the simple random walk with absorbing barriers at 0 and b. In this problem, $Y_1, Y_2, \ldots$ are independent random variables, with $P\{Y_i = 1\} = p, P\{Y_i = -1\} = q = 1 - p$. Let $X_n = \sum_{k=1}^{n} Y_k$, and let x be an arbitrary integer between 1 and $b - 1$. We wish to find the probability $h(x)$ of eventual ruin starting from x, in other words, the probability that $x + X_n$ will reach 0 before it reaches b. Intuitive reasoning based on the theorem of total probability leads to the result that

$$h(x) = ph(x + 1) + qh(x - 1).$$

   Give a formal proof of this result. (For further details, see Ash, 1970, Chapter 6.)

4.12 REFERENCES

The general outline of this chapter is based on Ash (1970), which is a text for an undergraduate course in probability. Measure theory is not used in the book, although some of the underlying measure-theoretic ideas are sketched. Many additional examples and problems can be found in Feller (1950) and Parzen (1960).

To develop intuitive skills, we recommend *The Probability Tutoring Book* by C. Ash (1993). Another useful reference is Ross (1993).

CHAPTER

5

CONDITIONAL PROBABILITY AND EXPECTATION

5.1 INTRODUCTION

In Chapter 4, we defined the conditional probability $P(B|A)$ only when $P(A) > 0$. However, conditional probabilities given events of probability zero are in no sense degenerate cases; they occur naturally in many problems. For example, consider the following two-stage random experiment. A random variable X is observed, where X has distribution function F. If X takes the value x, a random variable Y is observed, where the distribution of Y depends on x. (For example, if $0 \le x \le 1$, a coin with probability of heads x might be tossed independently n times, with Y the resulting number of heads.) Thus $P(x, B) = P\{Y \in B | X = x\}$ is prescribed in the statement of the problem, although the event $\{X = x\}$ may have probability zero for all values of x.

Let us try to construct a model for the above situation. Let $\Omega = \mathbb{R}^2$, $\mathscr{F} = \mathscr{B}(\mathbb{R}^2)$, $X(x, y) = x$, $Y(x, y) = y$. Instead of specifying the joint distribution function of X and Y, we specify the distribution function of X, and thus the corresponding probability measure P_X; also, for each x we are given a probability measure $P(x, \cdot)$ defined on $B(\mathbb{R})$; $P(x, B)$ is interpreted (informally for now) as $P\{Y \in B | X = x\}$.

We claim that the probability of any event of the form $\{(X, Y) \in C\}$ is determined. Reasoning intuitively, the probability that X falls into $(x, x + dx]$ is $dF(x)$. Given that this occurs, in other words (roughly), given $X = x$, (X, Y) will lie in C iff Y belongs to the section $C(x) = \{y: (x, y) \in C\}$. The probability of this event is $P(x, C(x))$. The total probability that (X, Y) will belong to C is

$$P(C) = \int_{-\infty}^{\infty} P(x, C(x)) \, dF(x). \tag{1}$$

201

In the special case $C = \{(x, y): x \in A, y \in B\} = A \times B, C(x) = B$ if $x \in A$ and $C(x) = \emptyset$ if $x \notin A$; therefore

$$P(C) = P(A \times B) = \int_A P(x, B)\, dF(x). \tag{2}$$

Now if $P(x, B)$ is Borel measurable in x for each fixed $B \in \mathscr{B}(\mathbb{R})$, then by the product measure theorem, there is a unique (probability) measure on $\mathscr{B}(\mathbb{R}^2)$ satisfying (2) for all $A, B \in \mathscr{B}(\mathbb{R})$, namely, the measure given by (1). Thus in the mathematical formulation of the problem, we take the probability measure P on $\mathscr{F} = \mathscr{B}(\mathbb{R}^2)$ to be the unique measure determined by P_X and the measures $P(x, \cdot), x \in \mathbb{R}$.

5.2 APPLICATIONS

We apply the results of 5.1 to some typical situations in probability.

5.2.1 Example. Let X be uniformly distributed between 0 and 1. If $X = x$, a coin with probability x of heads is tossed independently n times. If Y is the resulting number of heads, find $P\{Y = k\}, k = 0, 1, \ldots, n$.

Let us translate this into mathematical terms. Let $\Omega_1 = [0, 1], \mathscr{F}_1 = \mathscr{B}[0, 1]$. We have specified $P_X(A) = \int_A dx = $ Lebesgue measure of $A, A \in \mathscr{F}_1$.

For each x, we are given $P(x, B)$, to be interpreted as the conditional probability that $Y \in B$, given $X = x$. We may take $\Omega_2 = \{0, 1, \ldots, n\}, \mathscr{F}_2$ the class of all subsets of Ω_2; then $P(x, \{k\}) = \binom{n}{k} x^k (1 - x)^{n-k}, k = 0, 1, \ldots, n$ (this is Borel measurable in x). We take $\Omega = \Omega_1 \times \Omega_2, \mathscr{F} = \mathscr{F}_1 \times \mathscr{F}_2, P$ the unique probability measure determined by P_X and the $P(x, \cdot)$, namely,

$$P(C) = \int_0^1 P(x, C(x)) dP_X(x) = \int_0^1 P(x, C(x)) dx.$$

Now let $X(x, y) = x, Y(x, y) = y$. Then

$$P\{Y = k\} = P(\Omega_1 \times \{k\}) = \int_0^1 P(x, \{k\})\, dx$$

$$= \int_0^1 \binom{n}{k} x^k (1 - x)^{n-k}\, dx = \binom{n}{k} \beta(k + 1, n - k + 1),$$

where $\beta(r, s) = \int_0^1 x^{r-1}(1 - x)^{s-1} dx, r, s > 0$, is the *beta function*. We can express $\beta(r, s)$ as $\Gamma(r)\Gamma(s)/\Gamma(r + s)$, where $\Gamma(r) = \int_0^\infty x^{r-1} e^{-x} dx, r > 0$, is the *gamma function*. Since $\Gamma(n + 1) = n!, n = 0, 1, \ldots$, we have

$$P\{Y = k\} = \frac{\binom{n}{k} k! (n - k)!}{(n + 1)!} = \frac{1}{n + 1}, \qquad k = 0, 1, \ldots, n.$$

In solving a problem of this type, intuitive reasoning serves as a useful check on the formal development. Thus, the probability that X falls near

x is dx; given that $X = x$, the probability that k heads will be obtained is $\binom{n}{k} x^k (1-x)^{n-k}$. Integrate this from 0 to 1 to obtain the total probability.

The next example involves an n-stage random experiment.

5.2.2 Example. Let X_1 be uniformly distributed between 0 and 1. If $X_1 = x_1$, let X_2 be uniformly distributed between 0 and x_1. In general, if $X_1 = x_1, \ldots, X_k = x_k$, let X_{k+1} be uniformly distributed between 0 and x_k $(k = 1, \ldots, n-1)$. Find the expectation of X_n.

Here we have $\Omega_j = \mathbb{R}$, $\mathscr{F}_j = \mathscr{B}(\mathbb{R})$, $\Omega = \Pi_{j=1}^n \Omega_j$, $\mathscr{F} = \Pi_{j=1}^n \mathscr{F}_j$, $X_j(x_1, \ldots, x_n) = x_j$, $j = 1, \ldots, n$.

Set $P_1 = $ Lebesgue measure on $(0, 1)$, and for each $x_1 \in (0, 1)$,

$$P(x_1, \cdot) = \frac{1}{x_1}[\text{Lebesgue measure on } (0, x_1)],$$

that is,

$$P(x_1, B) = \frac{1}{x_1} \int_{B \cap (0, x_1)} dx_2.$$

In general, for each $x_1, \ldots, x_k \in (0, 1)$, $k = 1, \ldots, n-1$, take

$$P(x_1, \ldots, x_k, \cdot) = \frac{1}{x_k}[\text{Lebesgue measure on } (0, x_k)].$$

(We use open intervals to avoid division by zero.)

Let P be the unique measure on $\mathscr{F}$ determined by P_1 and the $P(x_1, \ldots, x_k, \cdot)$. We may find the expectation of a Borel measurable function g from $\mathbb{R}^n$ to $\mathbb{R}$ by Fubini's theorem:

$$\int_\Omega g \, dP = \int_{\Omega_1} P_1(dx_1) \int_{\Omega_2} P(x_1, dx_2) \cdots \int_{\Omega_n} g(x_1, \ldots, x_n)$$
$$\times P(x_1, \ldots, x_{n-1}, dx_n).$$

In the present case we have $g(x_1, \ldots, x_n) = X_n(x_1, \ldots, x_n) = x_n$. Thus

$$E(X_n) = \int_0^1 dx_1 \int_0^{x_1} \frac{1}{x_1} dx_2 \cdots \int_0^{x_{n-2}} \frac{1}{x_{n-2}} dx_{n-1} \int_0^{x_{n-1}} \frac{x_n}{x_{n-1}} dx_n$$

$$= \int_0^1 \frac{x_1}{2^{n-1}} dx_1 = 2^{-n}.$$

This example has an alternative interpretation. Let $Y_1, \ldots, Y_n$ be independent random variables, each uniformly distributed between 0 and 1. Let Z_k be the product $Y_1 \cdots Y_k$, $1 \le k \le n$. It turns out (see Problem 2, Section 5.6) that $(Z_1, \ldots, Z_n)$ has the same distribution as $(X_1, \ldots, X_n)$; hence $E(X_n)$

$= E(Z_n)$. Since the expectation of a product of independent nonnegative random variables is the product of the expectations, $E(Z_n) = \prod_{k=1}^{n} E(Y_k) = 2^{-n}$ as before.

5.2.3 Example. Let X be a discrete random variable, taking on the positive integer values $1, 2, \ldots$ with probabilities $p_1, p_2, \ldots$ ($p_i \geq 0, \sum_{i=1}^{\infty} p_i = 1$). If $X = n$, a nonnegative number Y is selected according to the density f_n. Find the probability that $1 \leq X + Y \leq 3$.

Here we have

$$\Omega_1 = \{1, 2, \ldots\}, \qquad \mathscr{F}_1 = \text{the class of all subsets of } \Omega,$$

$$P_1\{k\} = p_k, \qquad k = 1, 2, \ldots,$$

$$\Omega_2 = R, \qquad \mathscr{F}_2 = \mathscr{B}(\mathbb{R}), \qquad P(n, B) = \int_B f_n(x)\, dx,$$

$$X(\omega_1, \omega_2) = \omega_1, \qquad Y(\omega_1, \omega_2) = \omega_2.$$

The measure determined by P_1 and the $P(n, \cdot)$ is given by

$$P(F) = \int_{\Omega_1} P(\omega_1, F(\omega_1)) P_1(d\omega_1) \qquad F \in \mathscr{F}_1 \times \mathscr{F}_2$$

$$= \sum_{n=1}^{\infty} P(n, F(n)) p_n \qquad [\text{see } 4.10.3(\text{d})].$$

If $F = \{(\omega_1, \omega_2): 1 \leq \omega_1 + \omega_2 \leq 3\}$, then $F(n) = \{\omega_2: 1 - n \leq \omega_2 \leq 3 - n\}$; hence

$$P(F) = P\{1 \leq X + Y \leq 3\} = \sum_{n=1}^{\infty} p_n \int_{1-n}^{3-n} f_n(x)\, dx$$

$$= p_1 \int_0^2 f_1(x)\, dx + p_2 \int_0^1 f_2(x)\, dx \quad \text{since} \quad f_n(x) = 0 \quad \text{for} \quad x < 0.$$

Note that if $p_n > 0$, then $P\{Y \in B | X = n\}$ is defined and equals $P\{X = n, Y \in B\}/P\{X = n\}$. But $P\{X = n, Y \in B\} = P\{(\omega_1, \omega_2): \omega_1 = n, \omega_2 \in B\}$ $= P(n, B) p_n$. Thus $P\{Y \in B | X = n\} = P(n, B)$, as we would expect intuitively.

For additional examples, see Ash (1970, Chapter 4).

5.3 THE GENERAL CONCEPT OF CONDITIONAL PROBABILITY AND EXPECTATION

We have seen that specification of the distribution of a random variable X, together with $P(x, B)$, x real, $B \in \mathscr{B}(\mathbb{R})$, interpreted intuitively as the conditional probability that $Y \in B$ given $X = x$, determines a unique probability measure on $\mathcal{B}(\mathbb{R}^2)$, so that there is only one reasonable joint distribution of X and Y consistent with the given data. However, this somewhat oblique approach has not resolved the difficulty of defining conditional probabilities given events with probability zero. For example, if X is a random object on $(\Omega, \mathcal{F}, P)$, that is, $X\colon (\Omega, \mathcal{F}) \to (\Omega', \mathcal{F}')$, and $B \in \mathcal{F}$, we may ask whether it is possible to define in a meaningful way the conditional probability $P(x, B) = P(B|X = x)$, even though the event $\{X = x\}$ may have probability zero for some, in fact perhaps for all, x.

By the discussion in 5.1, if we have a reasonable conditional probability $P(x, B)$, it should satisfy

$$P(\{X \in A\} \cap B) = \int_A P(x, B)\, dP_X(x),$$

where P_X is the probability measure induced by X, namely,

$$P_X(A) = P\{\omega\colon X(\omega) \in A\}, A \in \mathcal{F}'.$$

In fact, this requirement determines $P(x, B)$ in the following sense.

5.3.1 Theorem. Let $X\colon (\Omega, \mathcal{F}) \to (\Omega', \mathcal{F}')$ be a random object on $(\Omega, \mathcal{F}, P)$, and let B be a fixed set in $\mathcal{F}$. Then there is a real-valued Borel measurable function g on $(\Omega', \mathcal{F}')$ such that for each $A \in \mathcal{F}'$,

$$P(\{X \in A\} \cap B) = \int_A g(x)\, dP_X(x).$$

Furthermore, if h is another such function then $g = h$ a.e. $[P_X]$. [We define $P(B|X = x)$ as $g(x)$; it is essentially unique *for a given B*.]

PROOF. Let $\lambda(A) = P(\{X \in A\} \cap B), A \in \mathcal{F}'$. Then λ is a finite measure on $\mathcal{F}'$, absolutely continuous with respect to $P_X[P_X(A) = 0$ implies $\lambda(A) = 0]$. The result follows from the Radon–Nikodym theorem. ☐

Let us verify that the conditional probability we have just introduced coincides with our intuition in simple cases.

5.3.2 Examples. (a) Let X take on only countably many values $x_1, x_2, \ldots$, with

$$p_i = P\{X = x_i\} > 0, \qquad \sum_{i=1}^{\infty} p_i = 1.$$

We claim that

$$g(x_i) = P(B|X = x_i) = \frac{P(B \cap \{X = x_i\})}{P\{X = x_i\}}, \qquad i = 1, 2, \ldots.$$

(Since P_X is concentrated on the x_i, we need not bother to specify $g(x)$ for x unequal to any of the x_i.) Thus the general definition reduces to the elementary definition in the discrete case. To prove this, let $\Omega' = \{x_1, x_2, \ldots\}$, with $\mathscr{F}'$ the collection of all subsets of Ω'. If $A \in \mathscr{F}'$ and g is defined as above, then

$$\int_A g(x)\, dP_X(x) = \int_{\Omega'} g(x) I_A(x)\, dP_X(x) = \sum_{i=1}^{\infty} g(x_i) I_A(x_i) P_X\{x_i\}$$

[see 4.10.3(d)]

$$= \sum_{x_i \in A} g(x_i) P\{X = x_i\} = \sum_{x_i \in A} P(B \cap \{X = x_i\})$$

$$= P(\{X \in A\} \cap B).$$

Since there is essentially only one g satisfying

$$\int_A g\, dP_X = P(\{X \in A\} \cap B), \quad A \in \mathscr{F}',$$

the g we proposed must be correct.

(b) Let X and Y be random variables with joint density f

$$[\Omega = R^2, \qquad \mathscr{F} = \mathscr{B}(\mathbb{R}^2), \qquad X(x, y) = x, \qquad Y(x, y) = y,$$

$$P(A) = \iint_A f(x, y)\, dx\, dy, \qquad A \in \mathscr{F}].$$

Now $\{X = x\}$ has probability zero for each x, but there is a reasonable approach to the conditional probability $P\{Y \in C | X = x\}$, as follows:

$$P\{Y \in C | x - h < X < x + h\} = \frac{P\{x - h < X < x + h, Y \in C\}}{P\{x - h < X < x + h\}}$$

$$= \frac{\int_{u=x-h}^{x+h} \int_{y \in C} f(u, y)\, du\, dy}{\int_{x-h}^{x+h} f_1(u)\, du},$$

where $f_1(x) = \int_{-\infty}^{\infty} f(x, y)\, dy$ is the density of X.
For small h, this is (hopefully) approximately

$$\frac{2h \int_C f(x, y)\, dy}{2h f_1(x)} = \int_C \frac{f(x, y)}{f_1(x)}\, dy.$$

We are led to define

$$h(y|x) = \frac{f(x, y)}{f_1(x)}$$

as the *conditional density* of Y, given $X = x$ (or for short, the *conditional density of Y given X*). Note that h is defined only when $f_1(x) \neq 0$; however, if $S = \{(x, y): f_1(x) = 0\}$, then $P\{(X, Y) \in S\} = 0$, since

$$P\{(X, Y) \in S\} = \iint_S f(x, y) \, dx \, dy = \int_{\{x: f_1(x)=0\}} \left[\int_{y=-\infty}^{\infty} f(x, y) \, dy \right] dx$$

$$= \int_{\{x: f_1(x)=0\}} f_1(x) \, dx = 0.$$

Thus we may essentially ignore those (x, y) for which the conditional density is not defined.

We expect that $P\{Y \in C | X = x\} = \int_C h(y|x) \, dy$. More generally, if $B \in \mathscr{F}$ and $X = x$, then B will occur iff $Y \in B(x)$. To find $P\{Y \in B(x) | X = x\}$, we integrate $h(y|x)$ over $y \in B(x)$. Thus we propose

$$g(x) = \int_{B(x)} h(y|x) \, dy, \qquad B \in \mathscr{F}, \qquad x \in \mathbb{R},$$

as the conditional probability of B given $X = x$. To prove this, first note that

$$g(x) = \int_{-\infty}^{\infty} I_B(x, y) h(y|x) \, dy;$$

hence g is Borel measurable by Fubini's theorem. Also, if $A \in \mathscr{B}(\mathbb{R})$,

$$P(\{X \in A\} \cap B) = \iint_{\substack{x \in A \\ (x, y) \in B}} f(x, y) \, dx \, dy$$

$$= \int_{-\infty}^{\infty} \left[\int_{-\infty}^{\infty} I_B(x, y) h(y|x) \, dy \right] I_A(x) f_1(x) \, dx$$

$$= \int_{x \in A} f_1(x) \int_{y \in B(x)} h(y|x) \, dy \, dx$$

$$= \int_A g(x) f_1(x) \, dx$$

$$= \int_A g(x) \, dP_X(x) \qquad \text{[see 4.10.3(c)]}.$$

Therefore $g(x) = P(B|X = x)$.

In this example we may look at the formula $f(x, y) = f_1(x)h(y|x)$ in two ways. If (X, Y) has density f, we have a notion of conditional probability: $P\{Y \in C|X = x\} = \int_C h(y|x)\,dy$. On the other hand, suppose that we specify that X has density f_1, and whenever $X = x$, we select Y according to the density $h(\cdot|x)$; in other words, we specify $P(x, B) = \int_B h(y|x)\,dy$, $B \in \mathcal{B}(\mathbb{R})$. A unique measure P on $\mathcal{B}(\mathbb{R}^2)$ is determined, satisfying, for $A, B \in \mathcal{B}(\mathbb{R})$,

$$P\{X \in A, Y \in B\} = \int_A P(x, B)f_1(x)\,dx$$

$$= \iint_{\substack{x \in A \\ y \in B}} f_1(x)h(y|x)\,dx\,dy.$$

Therefore (X, Y) has density $f(x, y) = f_1(x)h(y|x)$.

Thus we have two points of view. We may regard the conditional density of Y given $X = x$ as ultimately derived from the joint density of X and Y. On the other hand, we may regard the observation of X and Y as a two-stage random experiment, where the distribution of Y at stage 2 depends on the value of X at stage 1. The above discussion shows that the assignment of probabilities to events involving (X, Y) is the same in either case.

We may also define conditional densities in higher dimensions. For example, if X, Y, Z, W have joint density f, we define (say) the conditional density of (Z, W), given (X, Y), as

$$h(z, w|x, y) = \frac{f(x, y, z, w)}{f_{XY}(x, y)},$$

where $f_{XY}(x, y) = \int_{-\infty}^{\infty} \int_{-\infty}^{\infty} f(x, y, z, w)\,dz\,dw$. If $B \in \mathcal{B}(\mathbb{R}^4)$, then, exactly as before,

$$P(B|X = x, Y = y) = \iint_{B(x, y)} h(z, w|x, y)\,dz\,dw.$$

This is verified by proving that, for $A \in \mathcal{B}(\mathbb{R}^2)$,

$$P(\{(X, Y) \in A\} \cap B) = \int_A P(B|X = x, Y = y)f_{XY}(x, y)\,dx\,dy.$$

(c)   Let $(\Omega_1, \mathcal{F}_1)$ and $(\Omega_2, \mathcal{F}_2)$ be given, with no probability defined as yet. Take $\Omega = \Omega_1 \times \Omega_2$, $\mathcal{F} = \mathcal{F}_1 \times \mathcal{F}_2$, $X(\omega_1, \omega_2) = \omega_1$, $Y(\omega_1, \omega_2) = \omega_2$. Assume that we are given P_X, a probability measure on $(\Omega_1, \mathcal{F}_1)$, and also that we are given $P(x, B), x \in \Omega_1, B \in \mathcal{F}_2$, a probability measure in B for each fixed x, and a Borel measurable function of x for each fixed B. (We are specifying the distribution of X and the conditional distribution of Y, given $X = x$.) By the product measure theorem, there is a unique measure P on $\mathscr{F}$ such that

$$P\{X \in A, Y \in B\} = P(A \times B) = \int_A P(x, B) \, dP_X(x).$$

It follows that $P(x, B)$ is in fact the conditional probability $P\{Y \in B | X = x\}$.

We now consider conditional expectation. Let X and Y be random variables on $(\Omega, \mathscr{F}, P)$; we ask for a reasonable definition of the expectation of Y given that $X = x$, written $E(Y | X = x)$. Intuitively, $E(Y | X = x)$ should reflect the long-run average value of Y in a sequence of independent trials when we look only at those observations on which $\{X = x\}$ has occurred.

If X and Y are discrete and we are given that $X = x$, the conditional probability of an event involving Y is governed by the set of conditional probabilities $p(y|x) = P\{Y = y | X = x\}$. Thus a reasonable figure for $E(Y | X = x)$ is $\sum_y y p(y|x)$. Similarly, if (X, Y) is absolutely continuous, and $h = h(y|x)$ is the conditional density of Y given $X = x$, we expect that $E(Y | X = x)$ should be $\int_{-\infty}^{\infty} y h(y|x) \, dy$. What we need is a general framework that includes these special cases.

Let Y be a random variable (or an extended random variable) on $(\Omega, \mathscr{F}, P)$, and let $X \colon (\Omega, \mathscr{F}) \to (\Omega', \mathscr{F}')$ be a random object. Our general definition of conditional probability hinges on a version of the theorem of total probability:

$$P(\{X \in A\} \cap B) = \int_A P(B | X = x) \, dP_X(x), \qquad A \in \mathscr{F}', \qquad B \in \mathscr{F}.$$

There is a closely related "theorem of total expectation," which may be developed intuitively as follows. The probability that X falls near x is $dP_X(x)$; given that $X = x$, the average value of Y is what we are looking for, namely, $E(Y | X = x)$. It is reasonable to hope that the total expectation may be found by adding all the contributions:

$$E(Y) = \int_{\Omega'} E(Y | X = x) \, dP_X(x).$$

To develop this further, we replace Y by $YI_{\{X \in A\}}$, where $A \in \mathscr{F}'$. If $x \in A$, we expect that $E(YI_{\{X \in A\}} | X = x) = E(Y | X = x)$ since $X(\omega) = x \in A$ implies $I_{\{X \in A\}}(\omega) = 1$. If $x \notin A$, we expect that $E(YI_{\{X \in A\}} | X = x) = 0$. Replacing Y by $YI_{\{X \in A\}}$ in the above version of the theorem of total expectation, we obtain

$$E(YI_{\{X \in A\}}) = \int_{\Omega'} E(YI_{\{X \in A\}} | X = x) \, dP_X(x)$$

or

$$\int_{\{X \in A\}} Y \, dP = \int_A E(Y | X = x) \, dP_X(x).$$

In fact, this requirement essentially determines $E(Y | X = x)$.

5.3.3   Theorem. Let Y be an extended random variable on $(\Omega, \mathscr{F}, P)$, and $X: (\Omega, \mathscr{F}) \rightarrow (\Omega', \mathscr{F}')$, a random object. If $E(Y)$ exists, there is a function $g: (\Omega', \mathscr{F}') \rightarrow (\overline{\mathbb{R}}, \mathscr{B})$ such that for each $A \in \mathscr{F}'$,

$$\int_{\{X \in A\}} Y \, dP = \int_A g(x) \, dP_X(x).$$

(As usual, $\mathscr{B}$ denotes the class of Borel sets.) Furthermore, if h is another such function, then $g = h$ a.e. $[P_X]$. [We define $E(Y|X = x)$ as $g(x)$; it is essentially unique *for a given* Y.]

PROOF.   Let

$$\lambda(A) = \int_{\{X \in A\}} Y \, dP = \int_{X^{-1}(A)} Y \, dP, \qquad A \in \mathscr{F}'.$$

Then λ is a countably additive set function on $\mathscr{F}'$ by 1.6.1, and is absolutely continuous with respect to P_X since $P_X(A) = P\{X \in A\}$. The result follows from the Radon–Nikodym theorem.   □

Conditional expectation includes conditional probability as a special case, as we now prove.

5.3.4   Corollary. If X is a random object on $(\Omega, \mathscr{F}, P)$ and $B \in \mathscr{F}$, then

$$E(I_B|X = x) = P(B|X = x) \qquad \text{a.e. } [P_X].$$

PROOF.   In 5.3.3, set $Y = I_B$; the defining equation for conditional expectation becomes

$$P(\{X \in A\} \cap B) = \int_A E(I_B|X = x) \, dP_X(x).$$

The result now follows from 5.3.1.   □

Let us compare the general definition with the intuitive concept in special cases.

5.3.5   Examples. (a)   Let X take on only countably many values $x_1, x_2, \ldots$ (assume all $P\{X = x_i\} > 0$). We have seen that

$$P(B|X = x_i) = \frac{P(B \cap \{X = x_i\})}{P\{X = x_i\}}, \qquad B \in \mathscr{F}.$$

Thus we should expect that

$$E(I_B|X = x_i) = \frac{1}{P\{X = x_i\}} \int_{\{X=x_i\}} I_B \, dP.$$

Proceeding from indicators to nonnegative simple functions to nonnegative measurable functions to arbitrary measurable functions, we should like to believe that if $E(Y)$ exists,

$$E(Y|X = x_i) = \frac{1}{P\{X = x_i\}} \int_{\{X=x_i\}} Y \, dP, \qquad i = 1, 2, \dots . \tag{1}$$

$\Big[$We are not proving anything here since we do not yet know, for example, that

$$E\left(\sum_{i=1}^{n} Y_i \Big| X = x\right) = \sum_{i=1}^{n} E(Y_i|X = x).\Big]$$

To establish (1), let

$$g(x_i) = \frac{1}{P\{X = x_i\}} \int_{\{X=x_i\}} Y \, dP, \qquad i = 1, 2, \dots .$$

(We may assume $\Omega' = \{x_1, x_2, \dots\}$, with $\mathscr{F}'$ the class of all subsets of Ω'.) Then

$$\int_{\{X \in A\}} Y \, dP = \sum_{x_i \in A} P\{X = x_i\} \frac{1}{P\{X = x_i\}} \int_{\{X=x_i\}} Y \, dP$$

$$= \sum_{x_i \in A} P\{X = x_i\} g(x_i) = \int_A g(x) \, dP_x(x), \qquad A \in \mathscr{F}',$$

as desired.

In the special case when Y is discrete, (1) assumes a simpler form. If Y takes on the values $y_1, y_2, \dots$, we obtain (using countable additivity of the integral)

$$E(Y|X = x_i) = \sum_{j} y_j \frac{P\{X = x_i, Y = y_j\}}{P\{X = x_i\}}$$

$$= \sum_{j} y_j P\{Y = y_j|X = x_i\}. \tag{2}$$

(b) Let $B \in \mathscr{F}$, and assume $P(B) > 0$. If $E(Y)$ exists, we define the *conditional expectation of Y given B*, as follows. Let $X = I_B$, and set $E(Y|B) = E(Y|X = 1)$. This is a special case of (a); we obtain [see (1)]

$$E(Y|B) = \frac{1}{P(B)} \int_B Y \, dP,$$

in other words,

$$E(Y|B) = \frac{E(YI_B)}{P(B)}. \tag{3}$$

(c) Let X and Y be random variables having a joint density f, and let $h = h(y|x)$ be the conditional density of Y given X. We claim that if $E(Y)$ exists,

$$E(Y|X = x) = \int_{-\infty}^{\infty} yh(y|x) \, dy. \tag{4}$$

To prove this, note that

$$\int_{\{X \in A\}} Y \, dP = \iint_{\{(x,y):\, x \in A\}} yf(x, y) \, dx \, dy$$

$$= \int_{x \in A} f_1(x) \left[\int_{-\infty}^{\infty} yh(y|x) \, dy \right] dx \qquad \text{by Fubini's theorem}$$

$$= \int_A \left[\int_{-\infty}^{\infty} yh(y|x) \, dy \right] dP_X(x),$$

proving (4).

Notice also that if q is a Borel measurable function from $\mathbb{R}$ to $\mathbb{R}$ and $E[q(Y)]$ exists, then

$$E(q(Y)|X = x) = \int_{-\infty}^{\infty} q(y)h(y|x) \, dy \tag{5}$$

by the same argument as above. Similarly, if X and Y are discrete [see (a), (2)] and $E[q(Y)]$ exists, then

$$E(q(Y)|X = x_i) = \sum_j q(y_j)P\{Y = y_j|X = x_i\}. \tag{6}$$

(d)   Let $(\Omega_1, \mathscr{F}_1)$ and $(\Omega_2, \mathscr{F}_2)$ be given, with no probability defined as yet. Let $\Omega = \Omega_1 \times \Omega_2, \mathscr{F} = \mathscr{F}_1 \times \mathscr{F}_2, X(x, y) = x, Y(x, y) = y$. Assume that a probability measure P_X on $\mathscr{F}_1$ is given, and also that we are given $P(x, B)$, $x \in \Omega_1, B \in \mathscr{F}_2$, a probability measure in B for each fixed x, and a Borel measurable function of x for each fixed B. Let P be the unique measure on $\mathscr{F}$ determined by P_X and the $P(x, \cdot)$.
   If $f \colon (\Omega_2, \mathscr{F}_2) \to (\overline{\mathbb{R}}, \mathscr{B}\,(\overline{\mathbb{R}}))$ and $E[f(Y)]$ exists, we claim that

$$E(f(Y)|X = x) = \int_{\Omega_2} f(y) P(x, dy). \qquad (7)$$

To see this, we note, with the aid of Fubini's theorem, that

$$\int_{\{X \in A\}} f(Y)\, dP = \int_{\Omega} f(Y) I_{\{X \in A\}}\, dP$$

$$= \int_{\Omega_1} \int_{\Omega_2} f(Y(x, y)) I_A(x) P(x, dy)\, dP_X(x)$$

$$= \int_A \left[\int_{\Omega_2} f(y) P(x, dy) \right] dP_X(x).$$

Problems

1. Let X and Y be random variables with joint density $f(x, y)$. Indicate how to compute the following quantities.
   (a) $E(g(X)|Y = y)$, where g is a Borel measurable function from R to R such that $E[g(X)]$ exists;
   (b) $E(Y|A)$, where $A = \{X \in B\}, \mathscr{B} \in (\mathbb{R})$;
   (c) $E(X|A)$, where $A = \{X + Y \in B\}, \mathscr{B} \in (\mathbb{R}^2)$.
2. Let X be a random variable with density $f_0(\lambda)$. If $X = \lambda, n$ independent observations $X_1, \ldots, X_n$ are taken, where each X_i has density $f_\lambda(x)$. Indicate how to compute the conditional expectation of $g(X)$, given $X_1 = x_1, \ldots, X_n = x_n$.
3. Let X be a discrete random variable; if $X = x$, let Y have a conditional density $h(y|x)$. Show that

$$P\{X = x|Y = y\} = \frac{P\{X = x\} h(y|x)}{\sum_{x'} P\{X = x'\} h(y|x')}.$$

4. Let X be an absolutely continuous random variable. If $X = x$, let Y be discrete, with $P\{Y = y | X = x\} = p(y|x)$ specified. Show that there is a conditional density of X given Y, namely,

$$h(x|y) = \frac{f_X(x)p(y|x)}{p_Y(y)},$$

where

$$p_y(y) = P\{Y = y\} = \int_{-\infty}^{\infty} f_X(x)p(y|x)\,dx.$$

5. (a) Let X be a discrete random variable: If $X = \lambda$, n independent observations $X_1, \ldots, X_n$ are taken, where each X_i has density $f_\lambda(x)$. Indicate how to compute $E(g(X)|X_1 = x_1, \ldots, X_n = x_n)$.
    (b) Let X be an absolutely continuous random variable. If $X = \lambda$, n independent observations $X_1, \ldots, X_n$ are taken, where each X_i is discrete with probability function $p_\lambda(x)$. Indicate how to compute $E(g(X)|X_1 = x_1, \ldots, X_n = x_n)$.

6. If X is a random vector with density f, and $A = \{X \in B_0\}$, $B_0 \in \mathcal{B}(\mathbb{R}^n)$, show that there is a conditional density for X given A, namely,

$$f(x|A) = \begin{cases} \dfrac{f(x)}{P(A)} & \text{if} \quad x \in B_0, \\ 0 & \text{if} \quad x \notin B_0. \end{cases}$$

The interpretation of the conditional density is that

$$P\{X \in B|A\} = \int_B f(x|A)\,dx, \qquad B \in \mathcal{B}(\mathbb{R}^n).$$

7. Let $B_1, B_2, \ldots$ be mutually exclusive and exhaustive events with strictly positive probability. Establish the following version of the theorem of total expectation: If $E(X)$ exists, then

$$E(X) = \sum_{n=1}^{\infty} P(B_n)E(X|B_n).$$

8. Let X and Y be nonnegative random variables, such that (X, Y) has induced probability measure

$$P_{XY} = \tfrac{1}{4}P_1 + \tfrac{3}{4}P_2,$$

where, for $B \in \mathscr{B}([0, \infty) \times [0, \infty))$, $P_1 = $ point mass at $(1, 2)$, that is,

$$P_1(B) = \begin{cases} 1 & \text{if} & (1, 2) \in B, \\ 0 & \text{if} & (1, 2) \notin B, \end{cases}$$

and

$$P_2(B) = \iint_B e^{-x} e^{-y} \, dx \, dy.$$

Thus with probability $\frac{1}{4}$, $(X, Y) = (1, 2)$; with probability $\frac{3}{4}$, X and Y are chosen independently, each with density $e^{-x}, x \geq 0$. Calculate $P\{Y \in B | X = x\}, x \geq 0, B \in \mathscr{B}[0, \infty)$. (Hint: think like a statistician; if you observe that $X = 1$, it is a moral certainty that you are operating under P_1; if $X \neq 1$, it is an absolute certainty that you are operating under P_2.)

5.4 CONDITIONAL EXPECTATION GIVEN A σ-FIELD

It will be very convenient to regard conditional expectations as functions defined on the sample space Ω. Let us first recall the main result of the previous section.

If Y is an extended random variable on $(\Omega, \mathscr{F}, P)$ whose expectation exists, and X: $(\Omega, \mathscr{F}) \to (\Omega', \mathscr{F}')$ is a random object, then $g(x) = E(Y|X = x)$ is characterized as the a.e. $[P_X]$ unique function: $(\Omega', \mathscr{F}') \to (\overline{R}, \mathscr{B})$ satisfying

$$\int_{\{X \in A\}} Y \, dP = \int_A E(Y|X = x) \, dP_x(x), \qquad A \in \mathscr{F}'. \tag{1}$$

Now let $h(\omega) = g(X(\omega))$; then h: $(\Omega, \mathscr{F}) \to (\overline{R}, \mathscr{B})$ [see Fig. 5.4.1].

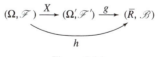

Figure 5.4.1

Thus $h(\omega)$ is the conditional expectation of Y, given that X takes the value $x = X(\omega)$; consequently, h measures the average value of Y given X, but h is defined on Ω rather than Ω'.

It will be useful to have an analog on (1) for h. We claim that

$$\int_{\{X \in A\}} h \, dP = \int_{\{X \in A\}} Y \, dP \qquad \text{for each } A \in \mathscr{F}'. \tag{2}$$

To prove this, note that

$$\int_{\{X \in A\}} h \, dP = \int_\Omega g(X(\omega)) I_A(X(\omega)) \, dP(\omega)$$

$$= \int_{\Omega'} g(x) I_A(x) \, dP_X(x) \qquad \text{[by 4.10.3(b)]}$$

$$= \int_A g(x) \, dP_X(x)$$

$$= \int_{\{X \in A\}} Y \, dP \qquad \text{[by (1)].}$$

Since $\{X \in A\} = X^{-1}(A) = \{\omega \in \Omega \colon X(\omega) \in A\}$, we may express (2) as follows:

$$\int_C h \, dP = \int_C Y \, dP \qquad \text{for each } C \in X^{-1}(\mathscr{F}'), \tag{3}$$

where $X^{-1}(\mathscr{F}') = \{X^{-1}(A) \colon A \in \mathscr{F}'\}$.

The σ-field $X^{-1}(\mathscr{F}')$ will be very important for us, and we shall look at some of its properties before proceeding.

5.4.1 Definition. Let $X \colon (\Omega, \mathscr{F}) \to (\Omega', \mathscr{F}')$ be a random object. The σ-field induced by X is given by

$$\sigma(X) = X^{-1}(\mathscr{F}').$$

Thus a set in $\sigma(X)$ is of the form $\{X \in A\}$ for some $A \in \mathscr{F}'$. In particular, if $X = (X_1, \ldots, X_n)$, a random vector, $\sigma(X)$ consists of all sets $\{X \in B\}$, $B \in \mathscr{B}(R^n)$.

The induced σ-field has the following properties.

5.4.2 Theorem. Let $X \colon (\Omega, \mathscr{F}) \to (\Omega', \mathscr{F}')$.
  (a) The induced σ-field $\sigma(X)$ is the smallest σ-field $\mathscr{G}$ of subsets of Ω making X measurable relative to $\mathscr{G}$ and $\mathscr{F}'$.
  (b) If $\Omega' = \Pi_j \Omega_j$, $\mathscr{F}' = \Pi_j \mathscr{F}_j$ so that $X = (X_1, X_2, \ldots)$, where $X_j \colon (\Omega, \mathscr{F}) \to (\Omega_j, \mathscr{F}_j)$ is the jth coordinate of X, then $\sigma(X)$ is the smallest σ-field $\mathscr{G}$ of subsets of Ω making each X_j measurable relative to $\mathscr{G}$ and $\mathscr{F}_j$.
  (c) If $Z \colon (\Omega, \sigma(X)) \to (\overline{\mathbb{R}}, \mathscr{B})$ (or $(\mathbb{R}, \mathscr{B})$), then $Z = f \circ X$ for some $f \colon (\Omega', \mathscr{F}') \to (\overline{\mathbb{R}}, \mathscr{B})$. Conversely, if $Z = f \circ X$ and $f \colon (\Omega', \mathscr{F}') \to (\overline{\mathbb{R}}, \mathscr{B})$, then $Z \colon (\Omega, \sigma(X)) \to (\overline{\mathbb{R}}, \mathscr{B})$.

PROOF. (a) If $A \in \mathscr{F}'$, then $X^{-1}(A) \in \sigma(X)$ by definition of $\sigma(X)$; hence $\sigma(X)$ makes X measurable. If $\mathscr{G}$ is any σ-field making X measurable, $X^{-1}(A) \in \mathscr{G}$ for all $A \in \mathscr{F}'$; hence $\sigma(X) \subset \mathscr{G}$.

(b) By 4.11.3, X is measurable relative to $\mathscr{G}$ and $\mathscr{F}'$ iff each X_j is measurable relative to $\mathscr{G}$ and $\mathscr{F}_j$. The result follows from (a).

(c) Assume Z: $(\Omega, \sigma(X)) \rightarrow (\overline{\mathbb{R}}, \mathscr{B})$. If Z is an indicator $I_C, C \in \sigma(X)$, then $C = X^{-1}(A)$ for some $A \in \mathscr{F}'$. If $f = I_A$, then $f \circ X = I_{\{X \in A\}} = I_C = Z$. If $Z = \sum_{k=1}^{n} z_k I_{C_k}$ is a finite-valued simple function and $I_{C_k} = f_k \circ X$ as above, then $Z = f \circ X$, where $f = \sum_{k=1}^{n} z_k f_k$.

In general, let $Z_1, Z_2, \dots$ be finite-valued simple functions such that $Z_n \rightarrow Z$. We can express $Z_n = f_n \circ X$ as above; define $f = \lim_{n \to \infty} f_n$ where the limit exists, and 0 elsewhere. Then

$$Z(\omega) = \lim_{n \to \infty} Z_n(\omega) = \lim_{n \to \infty} f_n(X(\omega)) = f(X(\omega)).$$

The converse holds because a composition of measurable functions is measurable. □

Note that 5.4.2(b) holds equally well for uncountable product spaces, because, as we observed at the time, 4.11.3 extends to arbitrary products. For an extension of 5.4.2(c), see Problem 1.

Now let us return to Eq. (3) at the beginning of this section:

$$\int_C h \, dP = \int_C Y \, dP, \qquad C \in \sigma(X),$$

where $h = g \circ X$, $g(x) = E(Y|X = x)$. Since g: $(\Omega', \mathscr{F}') \rightarrow (\overline{\mathbb{R}}, \mathscr{B})$, we have h: $(\Omega, \sigma(X)) \rightarrow (\overline{\mathbb{R}}, \mathscr{B})$ by 5.4.2(c). This fact, along with (3), characterizes h, and gives us the concept of conditional expectation given a σ-field.

5.4.3 Theorem. Let Y be an extended random variable on $(\Omega, \mathscr{F}, P)$, $\mathscr{G}$ a sub σ-field of $\mathscr{F}$. Assume that $E(Y)$ exists. Then there is a function $E(Y|\mathscr{G})$: $(\Omega, \mathscr{G}) \rightarrow (\overline{\mathbb{R}}, \mathscr{B})$, called *the conditional expectation of Y given $\mathscr{G}$*, such that

$$\int_C Y \, dP = \int_C E(Y|\mathscr{G}) \, dP \qquad \text{for each } C \in \mathscr{G}.$$

Any two such functions must coincide a.e. [P]. [Note that we cannot simply set $E(Y|\mathscr{G}) = Y$, as $E(Y|\mathscr{G})$ is required to be measurable relative to $\mathscr{G}$.]

PROOF. Let $\lambda(C) = \int_C Y \, dP$, $C \in \mathscr{G}$. Then λ is a countably additive set function on $\mathscr{G}$, absolutely continuous with respect to P; the result follows from the Radon-Nikodym theorem. $\square$

5.4.4 Comment. If $g(x) = E(Y|X = x)$ and $h(\omega) = g(X(\omega))$, then by 5.4.3, $h = E(Y|\mathscr{G})$, where $\mathscr{G} = \sigma(X)$; for convenience we shall usually write

$$h = E(Y|X).$$

[For example, if $E(Y|X = x) = x^2$, then $E(Y|X) = X^2$.]

We have seen that the conditional expectation $g(x) = E(Y|X = x)$, $x \in \Omega'$, can be transferred to Ω by forming $h(\omega) = g(X(\omega))$. Conversely, any condition expectational $E(Y|\mathscr{G})$, $\mathscr{G}$ an arbitrary sub σ-field of $\mathscr{F}$, arises from a random object X in this way. Simply take $X: (\Omega, \mathscr{F}) \to (\Omega, \mathscr{G})$ to be the identity map: $X(\omega) = \omega$, $\omega \in \Omega$. Then $X^{-1}(\mathscr{G}) = \mathscr{G}$, so if $g(x) = E(Y|X = x)$, then $h = E(Y|\sigma(X)) = E(Y|\mathscr{G})$.

Now intuitively, $E(Y|\mathscr{G}) = E(Y|X)$ is the average value of Y, given that X is known. But what does it mean to "know" $X: (\Omega, \mathscr{F}) \to (\Omega, \mathscr{G})$? The events involving X are sets of the form $\{X \in G\}$, $G \in \mathscr{G}$, and since X is the identity map, $\{X \in G\} = G$. Since an event corresponds to a question that has a yes or no answer, $E(Y|\mathscr{G})$ may be interpreted as the average value of $Y(\omega)$, given that we know, for each $G \in \mathscr{G}$, whether or not $\omega \in G$. Some examples may help to make this clear.

5.4.5 Example. (a) Let X be discrete, with values $x_1, x_2, \ldots$; take $\Omega' = \{x_1, x_2, \ldots\}$, with $\mathscr{F}'$ the class of all subsets of Ω', and assume $P\{X = x_i\} > 0$ for all i. We have seen in 5.3.5(a) that

$$g(x_i) = E(Y|X = x_i) = \frac{1}{P\{X = x_i\}} \int_{\{X=x_i\}} Y \, dP.$$

Let $h = E(Y|X)$, that is, $h(\omega) = g(X(\omega))$. Then h has the constant value $g(x_i)$ on the set $\{X = x_i\}$, and $\mathscr{G} = X^{-1}(\mathscr{F}')$ consists of all unions of the sets $\{X = x_i\}$. Knowledge of the value of $X(\omega)$ is equivalent to knowledge, for each $G \in \mathscr{G}$, of the membership or nonmembership of $\omega \in G$.

(b) Let X and Y be random variables with a joint density f. Let $\Omega = R^2$, $\mathscr{F} = \mathscr{B}(R^2)$, $P(B) = \int \int_B f(x, y) \, dx \, dy$, $B \in \mathscr{F}$; $X(x, y) = x$, $Y(x, y) = y$. Take $\Omega' = \mathbb{R}$, $\mathscr{F}' = \mathscr{B}(\mathbb{R})$. We have seen in 5.3.5(c) that $g(x) = E(Y|X = x) = \int_{-\infty}^{\infty} y h_0(y|x) \, dy$, where h_0 is the conditional density of Y given X. Let $h = E(Y|X)$, that is, $h(\omega) = g(X(\omega))$ or $h(x, y) = g(x)$. Thus $E(Y|X)$ is constant on vertical strips; also, $X^{-1}(\mathscr{F}')$ consists of all sets $B \times \mathbb{R}$, $B \in \mathscr{B}(\mathbb{R})$. Since $x \in B$ iff $(x, y) \in B \times \mathbb{R}$, information about $X(\omega)$

[obtained intuitively by asking questions of the form "does $X(\omega)$ belong to B ?"] is equivalent to information about membership of ω in sets of $\mathscr{G}$.

We also have a general concept of conditional probability given a σ-field.

5.4.6 Theorem. Let $(\Omega, \mathscr{F}, P)$ be a probability space, $\mathscr{G}$ a sub σ-field of $\mathscr{F}$; fix $B \in \mathscr{F}$. There is a function $P(B|\mathscr{G})$: $(\Omega, \mathscr{G}) \to (R, \mathscr{B})$, called *the conditional probability of B given $\mathscr{G}$*, such that

$$P(C \cap B) = \int_C P(B|\mathscr{G}) \, dP \qquad \text{for each } C \in \mathscr{G}.$$

Any two such functions must coincide a.e. [P], and in fact

$$P(B|\mathscr{G}) = E(I_B|\mathscr{G}) \qquad \text{a.e. } [P].$$

PROOF. Let $\lambda(C) = P(C \cap B)$, $C \in \mathscr{G}$. Then λ is a countably additive finite-valued set function on $\mathscr{G}$, absolutely continuous with respect to P; the existence and a.e. [P] uniqueness of $P(B|\mathscr{G})$ follow from the Radon–Nikodym theorem. (Since λ is finite, the range of $P(B|\mathscr{G})$ may be taken as $\mathbb{R}$ rather than $\overline{\mathbb{R}}$.) The connection between conditional probability and conditional expectation follows from 5.4.3 with $Y = I_B$. □

5.4.7 Comment. If $g(x) = P(B|X = x)$, X a random object, then $g(x) = E(I_B|X = x)$ a.e. [P_X] by 5.3.4. If $h(\omega) = g(X(\omega))$, then $h = E(I_B|\sigma(X))$ by 5.4.4, hence $h = P(B|\sigma(X))$ by 5.4.6. To summarize: If $g(x) = P(B|X = x)$ and $h(\omega) = g(X(\omega))$, then

$$h = P(B|\mathscr{G}), \qquad \mathscr{G} = \sigma(X).$$

For convenience we shall usually write

$$h = P(B|X).$$

Problems

1. Let X: $(\Omega, \mathscr{F}) \to (\Omega', \mathscr{F}')$ and Z: $(\Omega, \mathscr{F}(X)) \to (\Omega'', \mathscr{F}'')$. We investigate conditions under which $Z = f \circ X$ for some f: $(\Omega', \mathscr{F}') \to (\Omega'', \mathscr{F}'')$. By 5.4.2(c), such an f can be found if $\Omega'' = \overline{R}, \mathscr{F}'' = \mathscr{B}(\overline{R})$.
    (a) Assume that $\mathscr{F}''$ separates points; in other words, if $a, b \in \Omega''$, $a \neq b$, there are disjoint sets $A, B \in \mathscr{F}''$ such that $a \in A, b \in B$ (this will always hold if $\mathscr{F}''$ contains all singletons). Show that there is a function f: $\Omega' \to \Omega''$ (not necessarily measurable) such that $Z = f \circ X$.
    (b) Assume that $Z = f \circ X$, where f: $\Omega' \to \Omega''$. If $X(\Omega) \in \mathscr{F}'$ and $f(\Omega' - X(\Omega))$ consists of a single point, show that f is measurable relative to $\mathscr{F}'$ and $\mathscr{F}''$.

Thus by (a) and (b), a measurable f such that $Z = f \circ X$ can be found if $\mathscr{F}''$ separates points and $X(\Omega) \in \mathscr{F}'$.

5.5 PROPERTIES OF CONDITIONAL EXPECTATION

The conditional expectation $E(Y|X = x)$ is a more intuitive object than the conditional expectation $E(Y|\mathscr{G})$; however, the intuition cannot easily be pushed beyond the case in which X is a finite-dimensional random vector. Thus in formal arguments in which $\mathscr{G}$ is an arbitrary σ-field, we are forced to use $E(Y|\mathscr{G})$.

For convenience, we develop the basic properties of conditional expectation in pairs, one argument for $E(Y|\mathscr{G})$ and another (usually very similar) for $E(Y|X = x)$. Theorems about conditional probabilities are obtained by replacing Y by I_B, and results concerning $E(Y|X)$ are obtained by setting $\mathscr{G} = \mathscr{F}(X)$.

In the discussion to follow, $Y, Y_1, Y_2, \ldots$ are extended random variables on $(\Omega, \mathscr{F}, P)$, with all expectations assumed to exist; $X: (\Omega, \mathscr{F}) \to (\Omega', \mathscr{F}')$ is a random object, and $\mathscr{G}$ is a sub σ-field of $\mathscr{F}$. The phrase "a.e." with no measure specified will always mean a.e. $[P]$. If $Z: (\Omega, \mathscr{G}) \to (\overline{\mathbb{R}}, \mathscr{B})$, we say that Z is $\mathscr{G}$-measurable, and if $g: (\Omega', \mathscr{F}') \to (\overline{\mathbb{R}}, \mathscr{B})$, we say that g is $\mathscr{F}'$-measurable.

5.5.1   Theorem. If Y is a constant k a.e., then

(a) $E(Y|\mathscr{G}) = k$ a.e.

(a') $E(Y|X = x) = k$ a.e. $[P_X]$.

If $Y_1 \leq Y_2$ a.e., then

(b) $E(Y_1|\mathscr{G}) \leq E(Y_2|\mathscr{G})$ a.e.

(b') $E(Y_1|X = x) \leq E(Y_2|X = x)$ a.e. $[P_X]$.

[A statement such as $E(Y_1|\mathscr{G}) \leq E(Y_2|\mathscr{G})$ a.e. means that if Z_j is a version of $E(Y_j|\mathscr{G})$, in other words, Z_j satisfies the defining requirement 5.4.3, then $Z_1 \leq Z_2$ a.e.]

(c) $|E(Y|\mathscr{G})| \leq E(|Y| \, |\mathscr{G})$ a.e.

(c') $|E(Y|X = x)| \leq E(|Y| \, |X = x)$ a.e. $[P_X]$.

PROOF.   (a)   The function constant at k is $\mathscr{G}$-measurable and

$$\int_C Y \, dP = \int_C k \, dP, \qquad C \in \mathscr{G}.$$

(a')   If $g(x) \equiv k$, $x \in \Omega'$, then g is $\mathscr{F}'$-measurable and

$$\int_{\{X \in A\}} Y \, dP = \int_A k \, dP_X.$$

(b) $\int_C Y_1 dP \leq \int_C Y_2 \, dP$; hence

$$\int_C E(Y_1 | \mathscr{G}) \, dP \leq \int_C E(Y_2 | \mathscr{G}) \, dP \qquad \text{for each } C \in \mathscr{G}.$$

The result follows from 1.6.11.

(b') $\int_{\{X \in A\}} Y_1 \, dP \leq \int_{\{X \in A\}} Y_2 \, dP$; hence

$$\int_A E(Y_1 | X = x) \, dP_X \leq \int_A E(Y_2 | X = x) \, dP_X \qquad \text{for each } A \in \mathscr{F}'.$$

The result follows from 1.6.11.

Parts (c) and (c') follow from (b) and (b'), along with the observation that $-|Y| \leq Y \leq |Y|$. □

We now prove an additivity theorem for conditional expectations.

5.5.2   Theorem.   (a)   If $a, b \in \mathbb{R}$, and $aE(Y_1) + bE(Y_2)$ is well defined (not of the form $\infty - \infty$), then $E(aY_1 + bY_2 | \mathscr{G}) = aE(Y_1 | \mathscr{G}) + bE(Y_2 | \mathscr{G})$ a.e.

(a')   If $a, b \in \mathbb{R}$ and $aE(Y_1) + bE(Y_2)$ is well defined, then

$$E(aY_1 + bY_2 | X = x) = aE(Y_1 | X = x) + bE(Y_2 | X = x) \text{ a.e. } [P_X].$$

PROOF.   (a) If $C \in \mathscr{G}$,

$$\int_C (aY_1 + bY_2) \, dP = \int_C aY_1 \, dP + \int_C bY_2 \, dP$$

by the additivity theorem for integrals

$$= \int_C aE(Y_1 | \mathscr{G}) \, dP + \int_C bE(Y_2 | \mathscr{G}) \, dP$$

by definition of conditional expectation.

Thus $\int_C aE(Y_1 | \mathscr{G}) \, dP + \int_C bE(Y_2 | \mathscr{G}) \, dP$ is well defined, so again by the additivity theorem for integrals,

$$\int_C (aY_1 + bY_2) \, dP = \int_C [aE(Y_1 | \mathscr{G}) + bE(Y_2 | \mathscr{G})] \, dP,$$

as desired.

(a′)   This is done as in (a), with C replaced by $\{X \in A\}$ and

$$\int_C E(Y_j | \mathcal{G}) \, dP \qquad \text{by} \qquad \int_A E(Y_j | X = x) \, dP_X.$$

In the future, we shall dispose of proofs of this type with a phrase such as "same as (a)." $\square$

The monotone convergence theorem and the fact that a nonnegative series can be integrated term by term have exact analogs for conditional expectations.

5.5.3   **Theorem.** If $Y_n \geq 0$ for all n and $Y_n \uparrow Y$ a.e., then

(a) $E(Y_n | \mathcal{G}) \uparrow E(Y | \mathcal{G})$ a.e.

(a′) $E(Y_n | X = x) \uparrow E(Y | X = x)$ a.e. $[P_X]$.

If all $Y_n \geq 0$, then

(b) $E\left(\displaystyle\sum_{n=1}^{\infty} Y_n \Big| \mathcal{G}\right) = \displaystyle\sum_{n=1}^{\infty} E(Y_n | \mathcal{G})$ a.e.

(b′) $E\left(\displaystyle\sum_{n=1}^{\infty} Y_n \Big| X = x\right) = \displaystyle\sum_{n=1}^{\infty} E(Y_n | X = x)$ a.e. $[P_X]$.

In particular, if $B_1, B_2, \ldots$ are disjoint sets in $\mathcal{F}$,

(c) $P\left(\displaystyle\bigcup_{n=1}^{\infty} B_n \Big| \mathcal{G}\right) = \displaystyle\sum_{n=1}^{\infty} P(B_n | \mathcal{G})$ a.e.

(c′) $P\left(\displaystyle\bigcup_{n=1}^{\infty} B_n \Big| X = x\right) = \displaystyle\sum_{n=1}^{\infty} P(B_n | X = x)$ a.e. $[P_X]$.

PROOF.   (a) $\int_C Y_n \, dP = \int_C E(Y_n | \mathcal{G}) \, dP$, $C \in \mathcal{G}$; by 5.5.1(b), the $E(Y_n | \mathcal{G})$ increase to a $\mathcal{G}$-measurable function h. By the monotone convergence theorem, $\int_C Y \, dP = \int_C h \, dP$; hence $h = E(Y | \mathcal{G})$ a.e.

(a′)   Same as (a).

(b)   By 5.5.2(a), $E(\sum_{k=1}^{n} Y_k | \mathcal{G}) = \sum_{k=1}^{n} E(Y_k | \mathcal{G})$ a.e. Let $n \to \infty$ and apply part (a) to obtain the desired result.

(b′)   Same as (b).

Finally, (c) is a special case of (b), and (c′) of (b′). $\square$

If we take the expectation of a conditional expectation, the result is the same as if we were to take the expectation directly. This is actually a special case of the defining equations of Theorems 5.4.3 and 5.3.3.

5.5.4 Theorem. (a) $E[E(Y|\mathscr{G})] = E(Y)$; hence if Y is integrable, so is $E(Y|\mathscr{G})$.

(a') $\displaystyle\int_{\Omega'} E(Y|X = x)\,dP_X(x) = E(Y).$

PROOF. (a) $\int_\Omega Y\,dP = \int_\Omega E(Y|\mathscr{G})\,dP.$

(a') $\displaystyle\int_{\Omega'} E(Y|X = x)\,dP_X(x) = \int_{\{X\in\Omega'\}} Y\,dP = \int_\Omega Y\,dP.$ □

We now prove the dominated convergence theorem for conditional expectations.

5.5.5 Theorem. If $|Y_n| \le Z$ for all n, with $E(Z)$ finite, and $Y_n \to Y$ a.e., then

(a) $E(Y_n|\mathscr{G}) \to E(Y|\mathscr{G})$ a.e.
(a') $E(Y_n|X = x) \to E(Y|X = x)$ a.e. $[P_X]$.

PROOF. (a) Let $Z_n = \sup_{k\ge n}|Y_k - Y|$; then $Z_n \downarrow 0$ a.e. Now $E(Y_n|\mathscr{G})$ and $E(Y|\mathscr{G})$ are a.e. finite by 5.5.4(a), and $|E(Y_n|\mathscr{G}) - E(Y|\mathscr{G})| \le E[|Y_n - Y| \,|\,\mathscr{G}]$ by 5.5.1(c) and 5.5.2(a); this is less than or equal to $E(Z_n|\mathscr{G})$ by 5.5.1(b). Thus it suffices to show that $E(Z_n|\mathscr{G}) \to 0$ a.e. By 5.5.1(b), $E(Z_n|\mathscr{G}) \downarrow h$ a.e. for some $\mathscr{G}$-measurable function h. Since $0 \le Z_n \le 2Z$, which is integrable, we have

$$0 \le \int_\Omega h\,dP \le \int_\Omega E(Z_n|\mathscr{G})\,dP = \int_\Omega Z_n\,dP \to 0$$

by the dominated convergence theorem. Thus $h = 0$ a.e., as desired.
(a') Same as (a). □

The extended monotone convergence theorem and Fatou's lemma may be proved for conditional expectations, as follows.

5.5.6 Theorem. Assume $Y_n \ge Z$ for all n, where $E(Z) > -\infty$.

(a) If $Y_n \uparrow Y$ a.e., then $E(Y_n|\mathscr{G}) \uparrow E(Y|\mathscr{G})$ a.e.
(a') If $Y_n \uparrow Y$ a.e., then $E(Y_n|X = x) \uparrow E(Y|X = x)$ a.e. $[P_X]$.
(b) $\liminf_{n\to\infty} E(Y_n|\mathscr{G}) \ge E(\liminf_{n\to\infty} Y_n|\mathscr{G})$ a.e.
(b') $\liminf_{n\to\infty} E(Y_n|X = x) \ge E(\liminf_{n\to\infty} Y_n|X = x)$ a.e. $[P_X]$.
Now assume $Y_n \le Z$ for all n, where $E(Z) < +\infty$.

(c)   If $Y_n \downarrow Y$ a.e., then $E(Y_n|\mathscr{G}) \downarrow E(Y|\mathscr{G})$ a.e.

(c') If $Y_n \downarrow Y$ a.e., then $E(Y_n|X = x) \downarrow E(Y|X = x)$ a.e. $[P_X]$.

(d) $\limsup_{n \to \infty} E(Y_n|\mathscr{G}) \le E(\limsup_{n \to \infty} Y_n|\mathscr{G})$ a.e.

(d') $\limsup_{n \to \infty} \sup E(Y_n|X = x) \le E(\limsup_{n \to \infty} Y_n|X = x)$ a.e. $[P_X]$.

PROOF.   (a)   If $C \in \mathscr{G}$ then $\int_C Y_n \, dP = \int_C E(Y_n|\mathscr{G}) \, dP$, and $E(Y_n|\mathscr{G})$ in-
creases to a limit h a.e. By the extended monotone convergence theorem,
$\int_C Y \, dP = \int_C h \, dP$, and therefore $h = E(Y|\mathscr{G})$ a.e.

(b)   Let $Y_n{}' = \inf_{k \ge n} Y_k$; we then have $Y_n{}' \uparrow Y' = \liminf_{n \to \infty} Y_n$. By (a),
$E(Y_n{}'|\mathscr{G}) \uparrow E(Y'|\mathscr{G})$ a.e. But $Y_n{}' \le Y_n$, so

$$E(Y'|\mathscr{G}) = \lim_{n \to \infty} E(Y_n{}'|\mathscr{G}) = \liminf_{n \to \infty} E(Y_n{}'|\mathscr{G})$$

$$\le \liminf_{n \to \infty} E(Y_n|\mathscr{G}) \qquad \text{by 5.5.1(b).}$$

(c)   This follows from (a) upon replacing all extended random variables
by their negatives.

(d) $E(\limsup Y_n|\mathscr{G}) = -E(\liminf_{n \to \infty}(-Y_n)|\mathscr{G})$

$$\ge -\liminf_{n \to \infty} E(-Y_n|\mathscr{G}) \qquad \text{by (b)}$$

$$= \limsup_{n \to \infty} E(Y_n|\mathscr{G}).$$

The proofs of (a') to (d') are the same as the proofs of (a) to (d).   ☐

Thus far we have examined $E(Y|\mathscr{G})$ and $E(Y|X = x)$ under various hy-
potheses on Y; now we impose conditions on $\mathscr{G}$ and X.

5.5.7   Theorem.   (a) $E(Y|\{\emptyset, \Omega\}) = E(Y)$ a.e.

(a') If X is a constant b a.e., then $E(Y|X = x) = E(Y)$ a.e. $[P_X]$.

(b) $E(Y|\mathscr{F}) = Y$ a.e.

(b') If $X: (\Omega, \mathscr{F}) \to (\Omega, \mathscr{F})$ is the identity map, then $E(Y|X = \omega)$
$= Y(\omega)$ a.e. $[P]$.

PROOF.   (a) $\int_C Y \, dP = \int_C E(Y) \, dP$ if $C = \emptyset$ or Ω.

(a') If $b \in A$,

$$\int_{\{X \in A\}} X \, dP = \int_\Omega Y \, dP = E(Y) = \int_A E(Y) \, dP_X.$$

If $b \notin A$,

$$\int_{\{X \in A\}} Y \, dP = \int_\emptyset Y \, dP = 0 = \int_A E(Y) \, dP_X.$$

(b) $\int_C Y \, dP = \int_C Y \, dP$, $C \in \mathcal{F}$, and Y is $\mathcal{F}$-measurable.
(b') $\int_{\{X \in A\}} Y \, dP = \int_A Y \, dP$, and Y is $\mathcal{F}'(= \mathcal{F})$-measurable. $\quad\Box$

The following result is preparatory to the next theorem.

5.5.8   *Lemma.*   If $f \colon (\Omega, \mathcal{G}) \to (\overline{\mathbb{R}}, \mathcal{B})$, μ is a measure on $\mathcal{G}$, and B is an *atom* of $\mathcal{G}$ relative to μ; that is, $B \in \mathcal{G}$, $\mu(B) > 0$, and if $A \in \mathcal{G}, A \subset B$, then $\mu(A) = 0$ or $\mu(B - A) = 0$, then f is a.e. constant on B.

PROOF.   If $x \in \overline{\mathbb{R}}$ and $\mu\{\omega \in B \colon f(\omega) < x\} = 0$, then $\mu\{\omega \in B \colon f(\omega) < y\} = 0$ for all $y \leq x$. Let $k = \sup\{x \in \overline{\mathbb{R}} \colon \mu\{\omega \in B \colon f(\omega) < x\} = 0\}$. Then

$$\mu\{\omega \in B \colon f(\omega) < k\} = \mu \left[\bigcup_{\substack{r \text{ rational} \\ r < k}} \{\omega \in B \colon f(\omega) < r\} \right] = 0.$$

If $x > k$, then $\mu\{\omega \in B \colon f(\omega) < x\} > 0$; hence $\mu\{\omega \in B \colon f(\omega) \geq x\} = 0$ since B is an atom. Thus

$$\mu\{\omega \in B \colon f(\omega) > k\} = \mu \left[\bigcup_{\substack{r \text{ rational} \\ r > k}} \{\omega \in B \colon f(\omega) \geq r\} \right] = 0.$$

It follows that $f = k$ a.e. on B. $\quad\Box$

We now show that conditional expectation is an "averaging" or "smoothing" operation; if B is an atom of $\mathcal{G}, E(Y|\mathcal{G}) = k$ a.e. on B, where k is the average value of Y on B.

5.5.9   *Theorem.*   (a)   Let B be an atom of $\mathcal{G}$ relative to P. Then

$$E(Y|\mathcal{G}) = \frac{1}{P(B)} \int_B Y \, dP = \frac{E(YI_B)}{P(B)} \qquad \text{a.e. on } B.$$

As a special case, let $B_1, B_2, \ldots$ be disjoint sets in $\mathcal{F}$ whose union is Ω, with $P(B_n) > 0$ for all n. Let $\mathcal{G}$ be the minimal σ-field over the B_n, so that $\mathcal{G}$ is the collection of all unions formed from the B_n. Then

$$E(Y|\mathcal{G}) = \frac{1}{P(B_n)} \int_{B_n} Y \, dP \qquad \text{a.e. on } B_n, \qquad n = 1, 2, \ldots .$$

(a')   If $B = \{X = x_0\}$ and $P(B) > 0$, then

$$E(Y|X = x_0) = \frac{1}{P(B)} \int_B Y \, dP.$$

PROOF. (a) By 5.5.8, $E(Y|\mathscr{G})$ is a constant k a.e. on B. Since $\int_A Y\,dP = \int_A E(Y|\mathscr{G})\,dP$ for all $A \in \mathscr{G}$, in particular for $A = B$, we have $\int_B Y\,dP = kP(B)$, as asserted.

(a') We first show that B is an atom of $\mathscr{G} = \mathscr{F}(X)$. For assume $A \in \mathscr{F}'$ and $X^{-1}(A) \subset B = X^{-1}\{x_0\}$. If $x_0 \in A$, then $X^{-1}\{x_0\} \subset X^{-1}(A)$; hence $X^{-1}(A) = B$. If $x_0 \notin A$, then $X^{-1}(A) \cap X^{-1}\{x_0\} = \emptyset$; but $X^{-1}(A) \subset X^{-1}\{x_0\}$, so $X^{-1}(A) = \emptyset$.

Now let $g(x) = E(Y|X = x)$, $h(\omega) = g(X(\omega)) = E(Y|\mathscr{G})(\omega)$. By (a), $h(\omega) = k = E(YI_B)/P(B)$ a.e. on B. If ω is any point of B such that $h(\omega) = k$, then $g(X(\omega)) = g(x_0) = k$, the desired result. $\square$

We now consider successive conditioning relative to two σ-fields, one of which is coarser than (that is, a subset of) the other. The result is that no matter which conditioning operation is applied first, the result is the same as the conditioning with respect to the coarser σ-field alone. This is intuitively reasonable; for example, to find the average value of a real-valued function f defined on $[0, 3]$, we may compute a_1, the average of f on $[0, 1]$, and a_2, the average of f on $[1, 3]$; the average of f on $[0, 3]$, namely, $\frac{1}{3}\int_0^3 f(x)\,dx$, is then $\frac{1}{3}a_1 + \frac{2}{3}a_2$.

5.5.10   Theorem. (a) If $\mathscr{G}_1 \subset \mathscr{G}_2$, then $E[E(Y|\mathscr{G}_2)|\mathscr{G}_1] = E(Y|\mathscr{G}_1)$ a.e.
(a') If $f: (\Omega', \mathscr{F}') \to (\Omega'', \mathscr{F}'')$, then $E[E(Y|X)|f \circ X] = E(Y|f \circ X)$ a.e.
(b) If $\mathscr{G}_1 \subset \mathscr{G}_2$, then $E[E(Y|\mathscr{G}_1)|\mathscr{G}_2] = E(Y|\mathscr{G}_1)$ a.e.
(b') If $f: (\Omega', \mathscr{F}') \to (\Omega'', \mathscr{F}'')$, then $E[E(Y|f \circ X)|X] = E(Y|f \circ X)$ a.e.

PROOF. (a) If $C \in \mathscr{G}_1$, then $\int_C E(Y|\mathscr{G}_1)\,dP = \int_C Y\,dP = \int_C E(Y|\mathscr{G}_2)\,dP$ since $C \in \mathscr{G}_2$. Thus $E[E(Y|\mathscr{G}_2)|\mathscr{G}_1] = E(Y|\mathscr{G}_1)$ a.e. Alternatively, if $C \in \mathscr{G}_1$, then $\int_C E[E(Y|\mathscr{G}_2)|\mathscr{G}_1]\,dP = \int_C E(Y|\mathscr{G}_2)\,dP = \int_C Y\,dP$ since $C \in \mathscr{G}_2$; thus $E(Y|\mathscr{G}_1) = E[E(Y|\mathscr{G}_2)|\mathscr{G}_1]$ a.e.

(a') Let $\mathscr{G}_2 = X^{-1}(\mathscr{F}')$, $\mathscr{G}_1 = [f \circ X]^{-1}(\mathscr{F}'') = X^{-1}(f^{-1}(\mathscr{F}''))$, and apply (a).

(b) $E(Y|\mathscr{G}_1)$ is $\mathscr{G}_1$-measurable, hence $\mathscr{G}_2$-measurable, and

$$\int_C E(Y|\mathscr{G}_1)\,dP = \int_C E(Y|\mathscr{G}_1)\,dP, \qquad C \in \mathscr{G}_2.$$

(b') Take $\mathscr{G}_1$ and $\mathscr{G}_2$ as in (a'), and apply (b). $\square$

If we take the conditional expectation of a product of two random variables, under certain conditions one of the terms can be factored out, as follows.

5.5.11   Theorem. (a) If Z is $\mathscr{G}$-measurable and both Y and YZ are integrable, then $E(YZ|\mathscr{G}) = ZE(Y|\mathscr{G})$ a.e. In particular, $E(Z|\mathscr{G}) = Z$ a.e.

(a') If $f: (\Omega', \mathscr{F}') \to (\overline{R}, \mathscr{B})$ and both Y and $Y(f \circ X)$ are integrable, then $E(Y(f \circ X)|X = x) = f(x)E(Y|X = x)$ a.e. $[P_X]$. In particular,

$$E(f \circ X|X = x) = f(x) \qquad \text{a.e. } [P_X].$$

PROOF. (a) If Z is an indicator I_B, $B \in \mathscr{G}$, and $C \in \mathscr{G}$, we have

$$\int_C YZ \, dP = \int_{C \cap B} Y \, dP = \int_{C \cap B} E(Y|\mathscr{G}) \, dP = \int_C I_B E(Y|\mathscr{G}) \, dP$$
$$= \int_C ZE(Y|\mathscr{G}) \, dP,$$

and $ZE(Y|\mathscr{G})$ is $\mathscr{G}$-measurable. Thus the result holds for indicators.

Now let Z be simple, say

$$Z = \sum_{j=1}^{n} z_j I_{B_j} \qquad \text{with} \qquad B_j \in \mathscr{G}.$$

By 5.5.2(a),

$$E(YZ|\mathscr{G}) = \sum_{j=1}^{n} z_j E(Y I_{B_j}|\mathscr{G}) = \sum_{j=1}^{n} z_j I_{B_j} E(Y|\mathscr{G}) = ZE(Y|\mathscr{G}).$$

If Z is an arbitrary $\mathscr{G}$-measurable function, let $Z_1, Z_2, \ldots$ be simple (and $\mathscr{G}$-measurable) with $|Z_n| \leq |Z|$ and $Z_n \to Z$. Now $E(YZ_n|\mathscr{G}) = Z_n E(Y|\mathscr{G})$ by what we have just proved, and $E(YZ_n|\mathscr{G}) \to E(YZ|\mathscr{G})$ by 5.5.5(a). (The integrability of YZ is used here.) Since Y is integrable, so is $E(Y|\mathscr{G})$; hence $E(Y|\mathscr{G})$ is finite a.e., and consequently $Z_n E(Y|\mathscr{G}) \to ZE(Y|\mathscr{G})$ a.e. [Note that, for example, $1/n \to 0$ but $(1/n)(\infty) \not\longrightarrow 0(\infty) = 0$; thus finiteness of $E(Y|\mathscr{G})$ is important.] Therefore, $E(YZ|\mathscr{G}) = ZE(Y|\mathscr{G})$.

(a') Let $f = I_B$, $B \in \mathscr{F}'$. Then

$$\int_{\{X \in A\}} Y(f \circ X) \, dP = \int_{\{X \in A\}} Y I_{\{X \in B\}} \, dP = \int_{\{X \in A \cap B\}} Y \, dP$$
$$= \int_{A \cap B} E(Y|X = x) \, dP_X(x) = \int_A f(x)E(Y|X = x) \, dP_X(x).$$

Thus the result holds when f is an indicator. Passage to simple functions and then to arbitrary measurable functions is carried out just as in (a). $\square$

5.5.12 Comments. (a) Theorem 5.5.11(a) holds under the weaker assumption that $E(Y)$ and $E(YZ)$ exist [and 5.5.11(b) under the assumption that $E(Y)$ and $E[Y(f \circ X)]$ exist]; see Problem 4.

(b) Intuitively, if it is known that $X = x$, then $f \circ X$ may be treated as a constant $f(x)$, so that $E(Y(f \circ X)|X = x)$ should be $f(x)E(Y|X = x)$ as in 5.5.11(a′). A similar interpretation may be given to 5.5.11(a) as follows. We can express $E(YZ|\mathscr{G})$ as $E(YZ|\mathscr{F}(X)) = E(YZ|X)$ for an appropriate random object X (see 5.4.4). Since Z is $\mathscr{F}(X)$-measurable, $Z = f \circ X$ for some $f: (\Omega', \mathscr{F}') \to (\overline{\mathbb{R}}, \mathscr{B})$ by 5.4.2(c). Thus if X is given, Z may be treated as a constant and factored out.

Problems

1. (a) If X is a random object, Y a random variable such that $E(Y)$ exists, and X and Y are independent, show that $E(Y|X) = E(Y)$ a.e. $[P]$, and $E(Y|X = x) = E(Y)$ a.e. $[P_X]$.
    (b) Give an example of integrable random variables X and Y such that $E(Y|X) = E(Y)$, but X and Y are not independent.

2. If Y is an integrable random variable and X and Z are random objects, show that if (X, Y) and Z are independent, then $E(Y|X, Z) = E(Y|X)$.

3. This problem illustrates how to obtain a theorem about $E(Y|X = x)$ from a corresponding theorem about $E(Y|X)$ without writing a separate proof or saying "similarly." As a typical example, suppose we have proved that if $Y_1 \le Y_2$ a.e., then $E(Y_1|X) \le E(Y_2|X)$ a.e. Show that

$$P\{E(Y_1|X) > E(Y_2|X)\} = P_X\{E(Y_1|X = x) > E(Y_2|X = x)\}.$$

Conclude that if $Y_1 \le Y_2$ a.e., then $E(Y_1|X = x) \le E(Y_2|X = x)$ a.e. $[P_X]$.

4. Extend Theorem 5.5.11 to the case where $E(Y)$ and $E(YZ)$ exist but are not necessarily finite.

5. Let $(\Omega, \mathscr{F}, P)$ be a probability space, and $\mathscr{G}$ a sub σ-field of $\mathscr{F}$. If $Y \in L^1(\Omega, \mathscr{F}, P)$, then $E(Y|\mathscr{G})$ is also in $L^1(\Omega, \mathscr{F}, P)$, by 5.5.4(a). Thus $A(Y) = E(Y|\mathscr{G})$ defines a linear operator on L^1, and A may be transferred to the Banach space $\mathbf{L}^1(\Omega, \mathscr{F}, P)$.
    (a) Show that $\|A\| = 1$.
    (b) Define $\langle X, Y \rangle = \int_\Omega XY \, dP$, $X \in L^1$, $Y \in L^\infty$. Show that A has the "self-adjointness" property $\langle AX, Y \rangle = \langle X, AY \rangle$. [Note that $L^\infty \subset L^1$, and if $Y \in L^\infty$, then $AY \in L^\infty$, so $\langle X, AY \rangle$ is well defined.]

5.6 REGULAR CONDITIONAL PROBABILITIES

We have seen that if $B_1, B_2, \ldots$ are disjoint sets in $\mathscr{F}$, $n = 1, 2, \ldots$, then

$$P\left(\bigcup_{n=1}^{\infty} B_n \middle| \mathscr{G}\right) = \sum_{n=1}^{\infty} P(B_n|\mathscr{G}) \qquad \text{a.e.}$$

This does not imply that we will be able to choose $P(B|\mathscr{G})(\omega)$, $B \in \mathscr{F}$, $\omega \in \Omega$, so that it is a measure in B for all (or almost all) $\omega \in \Omega$. To clarify the problem, suppose that for each $B \in \mathscr{F}$, we choose a particular version of the conditional probability $P(B|\mathscr{G})(\omega)$, $\omega \in \Omega$; we now have a number $P(B|\mathscr{G})(\omega)$ for each $B \in \mathscr{F}$ and $\omega \in \Omega$. The difficulty is that for a fixed ω, $P(\cdot|\mathscr{G})(\omega)$ need not be countably additive on $\mathscr{F}$. Suppose that $B_1, B_2, \ldots$ are disjoint sets in $\mathscr{F}$. Then

$$P\left[\bigcup_{n=1}^{\infty} B_n | \mathscr{G}\right](\omega) = \sum_{n=1}^{\infty} P(B_n | \mathscr{G})(\omega)$$

except for ω belonging to a set $N(B_1, B_2, \ldots)$ of probability 0. Thus the set of ω's for which countable additivity fails is

$$M = \bigcup\{N(B_1, B_2, \ldots): B_1, B_2, \ldots \text{ disjoint sets in } \mathscr{F}\}.$$

In general, M is an uncountable union of sets of probability 0, and therefore M need not have probability 0 (or even be in $\mathscr{F}$). Thus there is no guarantee that we can specify the $P(B|\mathscr{G})$ to be countably additive in B. [Similarly, there is no guarantee that $P(B|X = x)$ will be countably additive in B for P_X-almost all $x \in \Omega'$.]

We are going to establish conditions under which the countable additivity requirement can be met.

5.6.1 Definition. Let Y be a random variable on $(\Omega, \mathscr{F}, P)$, $\mathscr{G}$ a sub σ-field of $\mathscr{F}$. The function $F = F(\omega, y)$, $\omega \in \Omega$, $y \in \mathbb{R}$, is called a *regular conditional distribution function* for Y given $\mathscr{G}$ iff the following two conditions are satisfied.

1. For each ω, $F(\omega, \cdot)$ is a proper distribution function on $\mathbb{R}$, that is, increasing and right-continuous, with $F(\omega, \infty) = 1$, $F(\omega, -\infty) = 0$.
2. For each y, $F(\omega, y) = P\{Y \le y|\mathscr{G}\}(\omega)$ for almost every ω.

The key step in the development is the result that a regular conditional distribution function for a given random variable always exists.

5.6.2 Theorem. If Y is a random variable on $(\Omega, \mathscr{F}, P)$, $\mathscr{G}$ a sub σ-field of $\mathscr{F}$, there is always a regular conditional distribution function for Y given $\mathscr{G}$.

PROOF. Select a version of $F_r(\omega) = P\{Y \le r|\mathscr{G}\}(\omega)$ for each fixed rational r. If $r_1, r_2, \ldots$ is an enumeration of the rationals, let

$$A_{ij} = \{\omega: F_{r_j}(\omega) < F_{r_i}(\omega)\}, \qquad A = \bigcup\{A_{ij}: r_i < r_j\}.$$

Then $P(A) = 0$ since $r_i < r_j$ implies $P\{Y \leq r_i|\mathscr{G}\} \leq P\{Y \leq r_j|\mathscr{G}\}$ a.e. by 5.5.1 (b).

Now let

$$B_i = \left\{\omega: \lim_{n\to\infty} F_{r_i+(1/n)}(\omega) \neq F_{r_i}(\omega)\right\}, \qquad B = \bigcup_{i=1}^{\infty} B_i.$$

Since

$$I_{\{Y \leq r_i+(1/n)\}} \downarrow I_{\{Y \leq r_i\}} \qquad \text{as} \qquad n \to \infty,$$

5.5.5(a) yields

$$F_{r_i+(1/n)}(\omega) \to F_{r_i}(\omega) \qquad \text{a.e.};$$

hence $P(B) = 0$.

If $C = \{\omega: \lim_{n\to\infty} F_n(\omega) \neq 1\}$, then $P(C) = 0$ since $\{Y \leq n\} \uparrow \Omega$, so that $P\{Y \leq n|\mathscr{G}\} \to 1$ a.e. Similarly, $D = \{\omega: \lim_{n\to-\infty} F_n(\omega) \neq 0\}$ has probability 0.

Define

$$F(\omega, y) = \begin{cases} \lim_{r\to y^+} F_r(\omega) & \text{if} \quad \omega \notin A \cup B \cup C \cup D \\ \text{any proper distribution function } G(y) \\ & \text{if} \quad \omega \in A \cup B \cup C \cup D. \end{cases}$$

Then F is well defined, for if $\omega \notin A$, then $F_r(\omega)$ is monotone in r, so that $\lim_{r\to y^+} F_r(\omega)$ exists. Note also that if $\omega \notin A \cup B$, then $\lim_{r\to y^+} F_r(\omega) = F_y(\omega)$ if y is rational, so that in this case, $F(\omega, y) = F_y(\omega)$. Similarly, if $\omega \notin A \cup C \cup D$, then $\lim_{r\to\infty} F_r(\omega) = 1$, $\lim_{r\to-\infty} F_r(\omega) = 0$.

We show that F is a regular conditional distribution function for Y given $\mathscr{G}$. Fix $\omega \notin A \cup B \cup C \cup D$; then $F(\omega, \cdot)$ is clearly increasing. If $y < y' \leq r$, then $F(\omega, y) \leq F(\omega, y') \leq F(\omega, r) = F_r(\omega) \to F(\omega, y)$ as $r \to y$. Thus $F(\omega, \cdot)$ is right-continuous. If $r \leq y$, then $F(\omega, y) \geq F(\omega, r) = F_r(\omega) \to 1$ as $r \to \infty$; hence $F(\omega, y) \to 1$ as $y \to \infty$; similarly, $F(\omega, y) \to 0$ as $y \to -\infty$. Thus the first requirement is satisfied.

Now $P\{Y \leq r|\mathscr{G}\}(\omega) = F_r(\omega) = F(\omega, r)$ by construction of F. As $r \downarrow y$, $F(\omega, r) \to F(\omega, y)$ for all ω by right-continuity, and

$$P\{Y \leq r|\mathscr{G}\} \to P\{Y \leq y|\mathscr{G}\} \quad \text{a.e.}$$

by 5.5.5(a). Thus $P\{Y \leq y|\mathscr{G}\}(\omega) = F(\omega, y)$ for almost every ω (y fixed), establishing the second requirement. $\square$

We are going to show that if Y is a random variable, $P\{Y \in B|\mathscr{G}\}$ can be chosen so as to be countably additive in B. This will follow from 5.6.2 and the fact that a distribution function determines a unique Lebesgue–Stieltjes measure.

5.6.3 Definition. Let $Y: (\Omega, \mathscr{F}) \to (\Omega', \mathscr{F}')$ be a random object, and $\mathscr{G}$ a sub σ-field of $\mathscr{F}$. The function $Q: \Omega \times \mathscr{F}' \to [0, 1]$ is called a *regular conditional probability* for Y given $\mathscr{G}$ iff

(1) $Q(\omega, B)$ is a probability measure in B for each fixed $\omega \in \Omega$, and
(2)   for each fixed $B \in \mathscr{F}'$, $Q(\omega, B) = P\{Y \in B | \mathscr{G}\}(\omega)$ a.e.

If Y is a random variable, so that $\Omega' = \mathbb{R}$, $\mathscr{F}' = \mathscr{B}(\mathbb{R})$, a regular conditional probability for Y given $\mathscr{G}$ always exists.

5.6.4 Theorem. Let Y be a random variable on $(\Omega, \mathscr{F}, P)$, $\mathscr{G}$ a sub σ-field of $\mathscr{F}$. There exists a regular conditional probability for Y given $\mathscr{G}$.

PROOF.   Let F be a regular conditional distribution function for Y given $\mathscr{G}$. Define

$$Q(\omega, B) = \int_{y \in B} dF(\omega, y).$$

Thus for each ω, $Q(\omega, \cdot)$ is the Lebesgue–Stieltjes measure corresponding to $F(\omega, \cdot)$; hence Q is a probability measure in B if ω is fixed.

Now let $\mathscr{C} = \{B \in \mathscr{B}(\mathbb{R}): Q(\omega, B) = P\{Y \in B | \mathscr{G}\}(\omega) \text{ a.e.}\}$. Then $\mathscr{C}$ contains all intervals $(-\infty, y]$ since $F(\omega, y) = P\{Y \le y | \mathscr{G}\}(\omega)$ a.e. If A, $B \in \mathscr{C}$, $A \subset B$, then $B - A \in \mathscr{C}$, and it follows that $\mathscr{C}$ contains all intervals $(a, b]$, hence all finite disjoint unions of right-semiclosed intervals. By the monotone class theorem, $\mathscr{C} = \mathscr{B}(\mathbb{R})$. Thus Q is a regular conditional probability for Y given $\mathscr{G}$. $\square$

We now extend this result to objects Y more general than random variables.

5.6.5 Theorem. Let $Y: (\Omega, \mathscr{F} \to (\Omega', \mathscr{F}')$ be a random object, and $\mathscr{G}$ a sub σ-field of $\mathscr{F}$. Suppose there is a map $\Psi: (\Omega', \mathscr{F}') \to (\mathbb{R}, \mathscr{B}(\mathbb{R}))$ such that Ψ is one-to-one, $E = \Psi(\Omega')$ is a Borel subset of $\mathbb{R}$, and Ψ^{-1} is measurable, that is, $\Psi^{-1}: (E, \mathscr{B}(E)) \to (\Omega', \mathscr{F}')$. Then there is a regular conditional probability for Y given $\mathscr{G}$.

PROOF.   Let $Q_0 = Q_0(\omega, B)$, $B \in \mathscr{B}(\mathbb{R})$, $\omega \in \Omega$, be a regular conditional probability for the random variable $\Psi(Y)$ given $\mathscr{G}$. Define $Q(\omega, A) = Q_0(\omega, \Psi(A))$, $A \in \mathscr{F}'$; since Ψ^{-1} is measurable, $\Psi(A) \in \mathscr{B}(E) \subset \mathscr{B}(\mathbb{R})$, and Q is well defined. Now Q is a probability measure in A for ω fixed, and if A is fixed, then

$$Q(\omega, A) = P\{\Psi(Y) \in \Psi(A) | \mathscr{G}\}(\omega) = P\{Y \in A | \mathscr{G}\}(\omega) \qquad \text{a.e.} \quad \square$$

A map Ψ of the type described in 5.6.5 is called a *Borel equivalence* of Ω' and E.

Problems

1. Let $X: (\Omega, \mathscr{F}) \to (\Omega', \mathscr{F}')$ and $Y: (\Omega, \mathscr{F}) \to (\Omega'', \mathscr{F}'')$ be random objects on $(\Omega, \mathscr{F}, P)$, and assume that $P_x(B) = P\{Y \in B | X = x\}$ can be chosen so as to be a probability measure in B for each fixed x. If $g: (\Omega' \times \Omega'', \mathscr{F}' \times \mathscr{F}'') \to (\overline{\mathbb{R}}, \mathscr{B})$ and $E[g(X, Y)]$ exists, show that

   (a) $E[g(X, Y) | X = x] = \int_{\Omega''} g(x, y) \, dP_x(y)$ a.e. $[P_X]$.
   In particular, if $C \in \mathscr{F}' \times \mathscr{F}''$, then

   (b) $P\{(X, Y) \in C | X = x\} = P_x(C(x))$ a.e. $[P_X]$.
   Conclude that

   (c) $P\{(X, Y) \in C\} = \int_{\Omega'} P_x(C(x)) \, dP_X(x)$.

2. Let $Y_1, \ldots, Y_n$ be independent random variables, each uniformly distributed between 0 and 1. Let Z_k be the product $Y_1 \cdots Y_k$, $1 \le k \le n$. Show that, given $Z_1 = z_1, \ldots, Z_k = z_k$, Z_{k+1} is uniformly distributed between 0 and z_k, that is,

$$P\{Z_{k+1} \in B | Z_1 = z_1, \ldots, Z_k = z_k\} = \int_B g_k(z) dz,$$

   where $g_k(z) = 1/z_k, 0 \le z \le z_k; g_k(z) = 0$ elsewhere. (This is another way of looking at Example 5.2.2.)

3. The following result is preparatory to the next problem; it is adapted from Halmos (1950).

   (a) Let E be a Lebesgue measurable subset of $\mathbb{R}$ with $0 < \mu(E) < \infty$, where μ is Lebesgue measure. If $0 < \delta < 1$, show that there is an open interval I such that $\mu(E \cap I) \ge \delta \mu(I)$.

   (b) With E as in part (a), let $D(E) = \{x - y: x, y \in E\}$. Show that $D(E)$ includes a neighborhood of 0. [This holds also if $\mu(E) = \infty$ since $0 < \mu(E \cap [-n, n]) < \infty$ for some n.]

   (c) Let ζ be an irrational number, and let $A = \{m + n\zeta: m, n \text{ integers}\}$. Show that A is dense in $\mathbb{R}$. Equivalently, the set of numbers $n\zeta$, n an integer, reduced modulo 1, is dense in $[0, 1)$. If $[0, 1)$ is identified

with the unit circle under the correspondence $\theta \to e^{i2\pi\theta}$, the problem
is as follows. If $\alpha/2\pi$ is irrational, the set $\{e^{in\alpha}: n \text{ an integer}\}$ is
dense in the circle. In fact more can be proved. Let z_0 be any point
on the circle and let $z_n = e^{in\alpha}z_0 = z_0$ rotated by $n\alpha$, $n = 1, 2, \ldots$
If z is an arbitrary point on the circle and $\varepsilon > 0$, $|z_n - z| < \varepsilon$ for
infinitely many values of n.

(d) If ζ is irrational, let $B = \{m + n\zeta: n \text{ an integer}, m \text{ an even integer}\}$,
$C = \{m + n\zeta: n \text{ an integer}, m \text{ an odd integer}\}$. Show that B and C
are dense in $\mathbb{R}$.

(e) Define an equivalence relation on $\mathbb{R}$ by $x \sim y$ iff $x - y \in A$, where
A is as defined in (c). Form a set E_0 by selecting one point from
each distinct equivalence class. Show that E_0 is not a Lebesgue
measurable set. [If F is a Borel set with $F \subset E_0$, show that $\mu(F) =
0$. Then show that $\mathbb{R}$ is a disjoint union of the sets $E_0 + a$, $a \in A$.
Use the translation-invariance of Lebesgue measure to show that E_0
is not Lebesgue measurable.]

(f) Let $M = \{x + y: x \in E_0, y \in B\}$, $M' = \{x + y: x \in E_0, y \in C\}$.
Show that $\mathbb{R} = M \cup M'$, and any Borel subset of M or of M' has
Lebesgue measure 0.

(g) Let E be an arbitrary Lebesgue measurable subset of $\mathbb{R}$. Show that
if F is a Borel subset of $E \cap M$, then $\mu(F) = 0$, and if G is a Borel
set such that $E \cap M \subset G \subset E$, then $\mu(E - G) = 0$.

4. Let H be a subset of $[0, 1]$ with inner Lebesgue measure 0 [sup$\{\mu(B)$:
B a Borel subset of $H\}=0$, $\mu=$ Lebesgue measure] and outer Lebesgue
measure 1 (inf$\{\mu(B)$: B a Borel overset of $H\} = 1$). To construct such a
set, take $E = [0, 1]$, $H = E \cap M$ in Problem 3(g).
Let $\Omega = [0, 1]$, and let $\mathscr{F}$ consist of all sets $(B_1 \cap H) \cup (B_2 \cap H^c)$,
where B_1, B_2 are Borel subsets of $[0, 1]$. Define

$$P[(B_1 \cap H) \cup (B_2 \cap H^c)] = \tfrac{1}{2}(\mu(B_1) + \mu(B_2)).$$

Thus $P = \mu$ on $\mathscr{B}[0, 1]$ [if $B \in \mathscr{B}[0, 1]$, $B = (B \cap H) \cup (B \cap H^c)$]. Take
$Y(\omega) = \omega$, $\omega \in \Omega$.

(a) Show that P is well defined, that is, if $(B_1 \cap H) \cup (B_2 \cap H^c)$
$= (B_1' \cap H) \cup (B_2' \cap H^c)$, then $\mu(B_1) = \mu(B_1')$, $\mu(B_2) = \mu(B_2')$.

(b) Suppose that Q is a regular conditional probability for Y given
$\mathscr{G} = \mathscr{B}[0, 1]$. Show that $Q(\omega, H) = Q(\omega, H^c) = \tfrac{1}{2}$ a.e.

(c) If $B \in \mathscr{G}$, show that $Q(\omega, B) = I_B(\omega)$ a.e.

(d)   Show that $Q(\omega, \{\omega\}) = 1$ for almost every ω, and thus arrive at a contradiction.

Conclude that there is no regular conditional probability for Y given $\mathscr{G}$.

Note: Books that discuss Martingale theory must also treat conditional probability and expectation. Thus for a selected bibliography on the material in this chapter, see Section 6.10.

6

STRONG LAWS OF LARGE NUMBERS AND MARTINGALE THEORY

6.1 INTRODUCTION

At the end of Chapter 4, we indicated that the physical fact of convergence of the relative frequency of heads in coin tossing is best expressed as a statement about almost everywhere convergence of S_n/n, where S_n is a sum of independent random variables $X_1, X_2, \ldots, X_n$; we attached the name "strong law of large numbers" to such a result. Now a "strong law of large numbers" in the most general sense is any statement about the almost everywhere convergence of a sequence of random variables, and this is the main subject matter of this chapter. A large class of convergence theorems will be developed with the aid of martingale theory, but before going into this it will be useful to consider the classical approach.

First we prove some results from real analysis that will be needed.

6.1.1 Lemma. Let $A = [a_{ij}]$ be an infinite matrix of real numbers; assume that $a_{nj} \to 0$ as $n \to \infty$ for each fixed j, and that for some nonnegative real number c, $\sum_{j=1}^{\infty} |a_{nj}| \le c$ for all n. If $\{x_n\}$ is a bounded sequence of real numbers, define

$$y_n = \sum_{j=1}^{\infty} a_{nj} x_j, \qquad n = 1, 2, \ldots.$$

Then:

   (a)   If $x_n \to 0$, then $y_n \to 0$.
   (b)   If $\sum_{j=1}^{\infty} a_{nj} \to 1$ and $x_n \to x$ (x real), then $y_n \to x$.

PROOF. (a) We may write

$$|y_n| \le \sum_{j=1}^{N} |a_{nj}|\, |x_j| + \sum_{j=N+1}^{\infty} |a_{nj}|\, |x_j|. \qquad (1)$$

Given $\varepsilon > 0$, choose N so that $|x_j| \leq \varepsilon/c$ for $j > N$; the second term on the right-hand side of (1) is at most $c(\varepsilon/c) = \varepsilon$. Since the first term approaches 0 as $n \to \infty$ for any fixed N, it follows that $y_n \to 0$.

(b) By (a), $\sum_{j=1}^{\infty} a_{nj}(x_j - x) \to 0$, and the result follows. $\square$

6.1.2 Toeplitz Lemma. Let $\{a_n\}$ be a sequence of nonnegative real numbers, and let $b_n = \sum_{j=1}^{n} a_j$; assume $b_n > 0$ for all n, and $b_n \to \infty$ as $n \to \infty$. If $\{x_n\}$ is a sequence of real numbers converging to the real number x, then

$$\frac{1}{b_n} \sum_{j=1}^{n} a_j x_j \to x.$$

PROOF. Form an infinite matrix A whose nth row is

$$\left(\frac{a_1}{b_n} \quad \frac{a_2}{b_n} \quad \cdots \quad \frac{a_n}{b_n} \quad 0 \quad 0 \quad \cdots \right)$$

and apply 6.1.1(b). $\square$

6.1.3 Kronecker Lemma. Let $\{b_n\}$ be an increasing sequence of positive real numbers with $b_n \to \infty$, and let $\{x_n\}$ be a sequence of real numbers with $\sum_{n=1}^{\infty} x_n = x$ (finite). Then

$$\frac{1}{b_n} \sum_{j=1}^{n} b_j x_j \to 0 \qquad \text{as} \qquad n \to \infty.$$

PROOF. If $s_n = \sum_{j=1}^{n} x_j$ (with $s_0 = 0$),

$$\sum_{j=1}^{n} b_j x_j = \sum_{j=1}^{n} b_j (s_j - s_{j-1})$$

$$= b_n s_n - b_0 s_0 - \sum_{j=1}^{n} s_{j-1}(b_j - b_{j-1}),$$

taking $b_0 = 0$. (This is the "summation by parts" formula; it is proved by brute force.) Thus

$$\frac{1}{b_n} \sum_{j=1}^{n} b_j x_j = s_n - \frac{1}{b_n} \sum_{j=1}^{n} a_j s_{j-1},$$

where $a_j = b_j - b_{j-1} \geq 0$. Since $s_n \to x$ as $n \to \infty$, and $s_{j-1} \to x$ as $j \to \infty$,

$$\frac{1}{b_n} \sum_{j=1}^{n} b_j x_j \to 0 \quad \text{by 6.1.2.} \square$$

Now if S_n is a random variable with finite expectation, Chebyshev's inequality implies that

$$P\{|S_n - E(S_n)| \geq \varepsilon\} \leq \frac{1}{\varepsilon^2}E[(S_n - E(S_n))^2] = \frac{\text{Var } S_n}{\varepsilon^2}.$$

If in fact S_n is a sum of independent random variables, this result can be strengthened considerably, as follows.

6.1.4   Kolmogorov's Inequality. Let $X_1, \ldots, X_n$ be independent random variables with finite expectation, and let $S_j = X_1 + \cdots + X_j$, $j = 1, \ldots, n$. Then for any $\varepsilon > 0$,

$$P\left\{\max_{1 \leq j \leq n} |S_j - E(S_j)| \geq \varepsilon\right\} \leq \frac{\text{Var } S_n}{\varepsilon^2}.$$

PROOF.   We may assume without loss of generality that $E(X_j) \equiv 0$, hence $E(S_j) \equiv 0$. Let

$$A_k = \{|S_j| < \varepsilon, j = 1, \ldots, k-1, |S_k| \geq \varepsilon\},$$

$$A = \left\{\max_{1 \leq j \leq n} |S_j| \geq \varepsilon\right\};$$

A is the disjoint union of the A_k, $k = 1, \ldots, n$. Now

$$\text{Var } S_n = \int_\Omega S_n^2 \, dP \geq \int_A S_n^2 \, dP = \sum_{k=1}^n \int_{A_k} S_n^2 \, dP. \tag{1}$$

But $S_n = S_k + Y_k$, where $Y_k = X_{k+1} + \cdots + X_n$; hence

$$\int_{A_k} S_n^2 \, dP = \int_{A_k} S_k^2 \, dP + 2\int_{A_k} S_k Y_k \, dP + \int_{A_k} Y_k^2 \, dP. \tag{2}$$

The second term on the right-hand side of (2) is $2E[I_{A_k} S_k Y_k]$ which is 0 since $I_{A_k} S_k$ and Y_k are functions of independent random variables, and are therefore independent [see 4.8.2(d) and 4.10.8]. Since the third term of the right-hand side of (2) nonnegative, we have

$$\int_{A_k} S_n^2 \, dP \geq \int_{A_k} S_k^2 \, dP \geq \varepsilon^2 P(A_k) \qquad \text{by definition of } A_k.$$

By (1),

$$\text{Var } S_n \geq \varepsilon^2 \sum_{k=1}^n P(A_k) = \varepsilon^2 P(A). \quad \square$$

We shall use the Borel–Cantelli lemma (2.2.4) quite often: If $A_1, A_2, \ldots$ are events such that $\sum_n P(A_n) < \infty$, then $\limsup_n A_n$ has probability 0. There is a partial converse which will also be needed [see (6.8.9)].

6.1.5   Second Borel–Cantelli Lemma. Let $(\Omega, \mathscr{F}, P)$ be a probability space, and let $A_1, A_2, \ldots$ be independent events in $\mathscr{F}$. If $\sum_{n=1}^{\infty} P(A_n) = \infty$, then $P(\limsup_n A_n) = 1$.

PROOF.

$$P\left(\limsup_n A_n\right) = P\left(\bigcap_n \bigcup_{k \geq n} A_k\right) = \lim_{n \to \infty} P\left(\bigcup_{k=n}^{\infty} A_k\right)$$

$$= \lim_{n \to \infty} \lim_{m \to \infty} P\left(\bigcup_{k=n}^{m} A_k\right).$$

Now

$$P\left(\bigcup_{k=n}^{m} A_k\right)^c = P\left(\bigcap_{k=n}^{m} A_k^c\right) = \prod_{k=n}^{m} P(A_k^c) \qquad \text{by independence}$$

$$\leq \prod_{k=n}^{m} \exp[-P(A_k)] \qquad \text{since} \qquad P(A_k^c) = 1 - P(A_k)$$

$$\leq \exp[-P(A_k)]$$

$$\to 0 \qquad \text{as } m \to \infty \qquad \text{since} \qquad \sum P(A_k) = \infty. \quad \square$$

Problems

1. (*Extension of the second Borel–Cantelli lemma*) Let $A_1, A_2, \ldots$ be events in a given probability space such that $\sum_{n=1}^{\infty} P(A_n) = \infty$ and

$$\liminf_{n \to \infty} \frac{\sum_{j,k=1}^{n} P(A_j \cap A_k)}{\left(\sum_{j=1}^{n} P(A_j)\right)^2} = 1.$$

(a) Show that the lim inf condition above is satisfied if the A_n are *pairwise independent* (A_j and A_k are independent whenever $j \neq k$) and $\sum_{n=1}^{\infty} P(A_n) = \infty$.

(b)   Use Chebyshev's inequality to show that if $I_n = I_{A_n}$, then

$$\liminf_{n\to\infty} P\left\{ \left| \sum_{k=1}^{n} I_k - \sum_{k=1}^{n} P(A_k) \right| > \frac{1}{2} \sum_{k=1}^{n} P(A_k) \right\} = 0.$$

(c)   Conclude from (b) that there is a sequence of integers $n_1 < n_2 < \cdots$
such that with probability 1,

$$\sum_{k=1}^{n_j} I_k \geq \frac{1}{2} \sum_{k=1}^{n_j} P(A_k) \qquad \text{for sufficiently large } j.$$

(d)   Show that $P(\limsup_n A_n) = 1$.

2.   Let θ be uniformly distributed on $[0, 2\pi]$ and define $X_k = \sin k\theta$
$(k = 1, 2, \ldots)$. Show that

$$\frac{X_1 + \cdots + X_n}{n} \to 0 \qquad \text{a.e.}$$

6.2   CONVERGENCE THEOREMS

We are now in a position to establish several basic results on convergence of sequences of random variables. We start with an example that motivated some of the early work in this subject, the problem of *random signs*. Let $a_1, a_2, \ldots$ be a fixed sequence of real numbers, and let an unbiased coin be tossed independently over and over again. If the nth toss results in heads, we write down the number $+a_n$, if tails, the number $-a_n$. In this way we generate a series such as $a_1 - a_2 - a_3 + a_4 + \cdots$, where the signs are chosen at random. Will the series converge?

The general question suggested by the random signs problem involves the convergence of a series of independent random variables. The following result gives considerable information.

6.2.1   Theorem. Let $X_1, X_2, \ldots$ be independent random variables with finite expectation. If $\sum_{n=1}^{\infty} \operatorname{Var} X_n < \infty$, then $\sum_{n=1}^{\infty} [X_n - E(X_n)]$ converges a.e. [All random variables are assumed to be defined on a fixed probability space $(\Omega, \mathscr{F}, P)$, and "almost everywhere" refers to the probability measure P. Also, throughout this chapter, convergence will always mean to a *finite* limit.]

PROOF.   We may assume that $E(X_n) \equiv 0$. Let $S_n = X_1 + \cdots + X_n$. Then S_n converges iff $S_j - S_k \to 0$ as $j, k \to \infty$, and this happens a.e. iff for each $\varepsilon > 0$,

$$P\left[\bigcup_{j,k\geq n} \{|S_j - S_k| \geq \varepsilon\} \right] \to 0 \qquad \text{as} \qquad n \to \infty$$

(see Section 2.5, Problem 4). Equivalently, we must prove that for each $\varepsilon > 0$,

$$P\left[\bigcup_{k=1}^{\infty}\{|S_{m+k} - S_m| \geq \varepsilon\}\right] \to 0 \quad \text{as} \quad m \to \infty.$$

We have

$$P\left[\bigcup_{k=1}^{\infty}\{|S_{m+k} - S_m| \geq \varepsilon\}\right] = \lim_{n\to\infty} P\left[\bigcup_{k=1}^{n}\{|S_{m+k} - S_m| \geq \varepsilon\}\right]$$

$$= \lim_{n\to\infty} P\left\{\max_{1\leq k\leq n} |S_{m+k} - S_m| \geq \varepsilon\right\}$$

$$\leq \frac{1}{\varepsilon^2} \limsup_{n\to\infty} \text{Var}(S_{m+n} - S_m) \qquad \text{by 6.1.4}$$

$$= \frac{1}{\varepsilon^2} \limsup_{n\to\infty} \sum_{j=1}^{n} \text{Var}(X_{m+j}) \qquad \text{by 4.10.11}$$

$$\to 0 \quad \text{as } m \to \infty \quad \text{since} \quad \sum_{n=1}^{\infty} \text{Var } X_n < \infty. \quad \square$$

In the random signs problem we have $X_n = a_n Y_n$, where the Y_n are independent, taking values $+1$ and -1 with equal probability. It follows that if $\sum_{n=1}^{\infty} a_n^2 < \infty$, the series $\sum_{n=1}^{\infty} X_n$ converges a.e. After we prove Theorem 6.8.7, we shall see that the condition $\sum_{n=1}^{\infty} a_n^2 < \infty$ is necessary as well as sufficient for a.e. convergence of the series.

If $X_1, X_2, \ldots$ are independent random variables, we proved in Chapter 4 that under appropriate conditions, $(S_n - E(S_n))/n$ converges to 0 in probability (the weak law of large numbers). We now consider almost everywhere convergence.

6.2.2   Kolmogorov Strong Law of Large Numbers. Let $X_1, X_2, \ldots$ be independent random variables, each with finite mean and variance, and let $\{b_n\}$ be an increasing sequence of positive real numbers with $b_n \to \infty$. If

$$\sum_{n=1}^{\infty} \frac{\text{Var } X_n}{b_n^2} < \infty,$$

then (with $S_n = X_1 + \cdots + X_n$)

$$\frac{S_n - E(S_n)}{b_n} \to 0 \qquad \text{a.e.}$$

PROOF.

$$\sum_{n=1}^{\infty} \mathrm{Var}\left(\frac{X_n - E(X_n)}{b_n}\right) = \sum_{n=1}^{\infty} \frac{\mathrm{Var}\, X_n}{b_n^2} < \infty \qquad \text{by hypothesis.}$$

By 6.2.1, $\sum_{n=1}^{\infty}(X_n - E(X_n))/b_n$ converges a.e. But

$$\frac{S_n - E(S_n)}{b_n} = \frac{1}{b_n}\sum_{k=1}^{n} b_k\left(\frac{X_k - E(X_k)}{b_k}\right),$$

and this approaches zero a.e. by 6.1.3. $\quad\square$

In particular, if the X_n are independent random variables, each with finite mean m and finite variance σ^2, then $S_n/n \to m$ a.e. (take $b_n = n$ in 6.2.2).

Another special case: If the X_n are independent and the fourth central moments are uniformly bounded, that is, for some finite M we have $E[(X_n - E(X_n))^4] \le M$ for all n, then $(S_n - E(S_n))/n \to 0$ a.e. For by the Cauchy–Schwarz inequality,

$$\mathrm{Var}\, X_n = E[(X_n - E(X_n))^2 \cdot 1] \le (E[(X_n - E(X_n))^4])^{1/2} \le M^{1/2},$$

and therefore 6.2.2 applies with $b_n = n$. This result, due to Cantelli, may in fact be proved without much machinery from measure theory; see Ash (1970, p. 206).

If the X_n are independent and all have the same distribution, in other words, for each Borel set $B \subset \mathbb{R}$, $P\{X_n \in B\}$ is the same for all n, a version of the strong law of large numbers may be proved under a hypothesis on the mean of the X_n but no assumptions about higher moments. We first indicate some terminology that will be used in the remainder of the book.

6.2.3 Definition. If the random variables X_n all have the same distribution, they will be called *identically distributed*. The phrase "independent and identically distributed" will be abbreviated *iid*.

We need one preliminary result.

6.2.4 Lemma. If Y is a nonnegative random variable,

$$\sum_{n=1}^{\infty} P\{Y \ge n\} \le E(Y) \le 1 + \sum_{n=1}^{\infty} P\{Y \ge n\}.$$

PROOF.

$$\sum_{n=1}^{\infty} P\{Y \geq n\}$$

$$= \sum_{n=1}^{\infty} \sum_{k=n}^{\infty} P\{k \leq Y < k+1\} = \sum_{k=1}^{\infty} \sum_{n=1}^{k} P\{k \leq Y < k+1\}$$

$$= \sum_{k=1}^{\infty} k P\{k \leq Y < k+1\} = \sum_{k=0}^{\infty} \int_{\{k \leq Y < k+1\}} k \, dP$$

$$\leq \sum_{k=0}^{\infty} \int_{\{k \leq Y < k+1\}} Y \, dP = E(Y) \leq \sum_{k=0}^{\infty} (k+1) P\{k \leq Y < k+1\}$$

$$= \sum_{n=1}^{\infty} P\{Y \geq n\} + \sum_{k=0}^{\infty} P\{k \leq Y < k+1\} = \sum_{n=1}^{\infty} P\{Y \geq n\} + 1. \quad \square$$

6.2.5 Strong Law of Large Numbers, iid Case. If $X_1, X_2, \ldots$ are iid random variables with finite expectation m, and $S_n = X_1 + \cdots + X_n$, then $S_n/n \to m$ a.e.

PROOF.   Since all X_n have the same distribution,

$$\sum_{n=1}^{\infty} P\{|X_n| \geq n\} = \sum_{n=1}^{\infty} P\{|X_1| \geq n\};$$

thus

$$\sum_{n=1}^{\infty} P\{|X_n| \geq n\} \leq E(|X_1|) < \infty \qquad \text{by 6.2.4.}$$

By the Borel–Cantelli lemma, $P\{|X_n| \geq n$ for infinitely many $n\} = 0$. Thus if we define $Y_n = X_n$ if $|X_n| < n$; $Y_n = 0$ if $|X_n| \geq n$, then except on a set of probability 0, $Y_n = X_n$ for sufficiently large n. Thus assuming (without loss of generality) that $m = 0$, it suffices to show that

$$\frac{1}{n} \sum_{j=1}^{n} Y_j \to 0 \qquad \text{a.e.}$$

Now

$$E(Y_n) = E(X_n I_{\{|X_n| < n\}})$$

$$= E(X_1 I_{\{|X_1| < n\}})$$

by the iid hypothesis

$$\to E(X_1) = 0 \quad \text{as} \quad n \to \infty$$

by the dominated convergence theorem.

Consequently,

$$\frac{1}{n}\sum_{j=1}^{n} E(Y_j) \to 0,$$

and therefore it is sufficient to show that

$$\frac{1}{n}\sum_{j=1}^{n}[Y_j - E(Y_j)] \to 0 \qquad \text{a.e.}$$

If we can show that

$$\sum_{n=1}^{\infty}\left[\frac{Y_n - E(Y_n)}{n}\right]$$

converges a.e., then the Kronecker lemma 6.1.3 with $b_n = n$ and

$$x_n = \frac{Y_n - E(Y_n)}{n},$$

yields

$$\frac{1}{n}\sum_{j=1}^{n}[Y_j - E(Y_j)] \to 0 \qquad \text{a.e.,}$$

as desired. Now the Y_n are functions of the independent random variables X_n, and hence are independent, so by 6.2.1, it suffices to show that

$$V = \sum_{n=1}^{\infty} \text{Var}\left(\frac{Y_n}{n}\right) < \infty.$$

(Note that $|Y_n| < n$, so Var Y_n is finite, although nothing is known about Var X_n.) But

$$V = \sum_{n=1}^{\infty}\frac{1}{n^2}\text{Var } Y_n$$

$$\leq \sum_{n=1}^{\infty}\frac{1}{n^2}E(Y_n{}^2)$$

$$\qquad \text{since Var } Y_n = E(Y_n{}^2) - [E(Y_n)]^2$$

$$= \sum_{n=1}^{\infty}\frac{1}{n^2}E(X_1^2 I_{\{|X_1|<n\}}) \qquad \text{by the iid hypothesis}$$

$$= \sum_{n=1}^{\infty}\frac{1}{n^2}\sum_{m=1}^{n}E(X_1^2 I_{\{m-1\leq|X_1|<m\}})$$

$$= \sum_{m=1}^{\infty}E(X_1^2 I_{\{m-1\leq|X_1|<m\}})\sum_{n=m}^{\infty}\frac{1}{n^2}.$$

By comparing $\sum(1/n^2)$ with $\int(1/x^2)dx$, we find that $\sum_{n=m}^{\infty}(1/n^2) \le K/m$ for some fixed positive constant K. Thus

$$V \le K \sum_{m=1}^{\infty} \frac{1}{m} E(X_1^2 I_{\{m-1 \le |X_1| < m\}}).$$

If $m - 1 \le |X_1| < m$, then $X_1^2 = |X_1| \, |X_1| \le m|X_1|$; hence

$$V \le K \sum_{m=1}^{\infty} E(|X_1| I_{\{m-1 \le |X_1| < m\}}) = KE(|X_1|) < \infty. \quad \square$$

If $E(X_i)$ exists but is not necessarily finite in Theorem 6.2.5, the result still holds. To see this, first assume that the X_i are nonnegative, with infinite expectation. If $M > 0$ and $S_n' = \sum_{i=1}^{n} X_i I_{\{X_i \le M\}}$, then, almost everywhere,

$$\liminf_{n \to \infty} \frac{S_n}{n} \ge \liminf_{n \to \infty} \frac{S_n'}{n} = E(X_1 I_{\{X_1 \le M\}}) \to E(X_1) = \infty \qquad \text{as} \qquad M \to \infty.$$

Therefore $n^{-1}S_n \to \infty$ a.e. The general case is handled by splitting the random variables X_i into positive and negative parts.

We conclude this section with a remarkable result about independent random variables. If $X_1, X_2, \ldots$ are independent, this question might arise: What is the probability that $\sum_{n=1}^{\infty} X_n$ converges? It might be expected that examples exist with any number between 0 and 1 as the probability of convergence, but in fact, the probability must be 0 or 1. Many other events defined in terms of independent random variables have this "zero–one" property, as we shall see.

6.2.6   Definitions and Comments. Let $X_1, X_2, \ldots$ be a sequence of random variables, and let $\mathscr{F}_n = \sigma(X_n, X_{n+1}, \ldots)$, $n = 1, 2, \ldots$; $\mathscr{F}_n$ may be thought of as the σ-field of events involving $X_n, X_{n+1}, \ldots$. The σ-field $\mathscr{F}_{\infty} = \bigcap_{n=1}^{\infty} \mathscr{F}_n$ is called the *tail σ-field* of the X_n, sets in $\mathscr{F}_{\infty}$ are called *tail events* and $\mathscr{F}_{\infty}$-measurable functions, that is, functions $f: (\Omega, \mathscr{F}_{\infty}) \to (\mathbb{R}, \mathscr{B}(\mathbb{R}))$ are called *tail functions* (relative to the X_n). Intuitively, a tail event is one whose occurrence or nonoccurrence is not affected by changing the values of finitely many of the X_i, and a tail function is one whose value is not affected by such a change. Thus $\{\lim_{n \to \infty} X_n \text{ exists}\}$, $\{\sum_{n=1}^{\infty} X_n \text{ converges}\}$, and $\{X_n < 1 \text{ for infinitely many } n\}$ are tail events, and $\limsup_{n \to \infty} X_n$ and $\liminf_{n \to \infty} X_n$ are tail functions. [Example of a formal proof: $\{\sum_{n=1}^{\infty} X_n \text{ converges}\} = \{\sum_{k=n}^{\infty} X_k \text{ converges}\} \in \mathscr{F}_n$ for each n; hence the event belongs to $\mathscr{F}_{\infty}$. Similarly,

$$\left\{\limsup_{n \to \infty} X_n < c\right\} = \left\{\limsup_{\substack{k \to \infty \\ k \ge n}} X_k < c\right\} \in \mathscr{F}_n \qquad \text{for all} \qquad n;$$

hence $\limsup_{n \to \infty} X_n$ is $\mathscr{F}_{\infty}$-measurable.]

6.2.7 Kolmogorov Zero–One Law. All tail events relative to a sequence of independent random variables have probability 0 or 1, and all tail functions are constant almost everywhere.

PROOF. Fix $A \in \mathscr{F}_\infty$; the idea is to show that A is independent of itself, so that $P(A \cap A) = P(A)P(A)$, and consequently $P(A) = 0$ or 1. Since $\mathscr{F}_\infty \subset \mathscr{F}_1$, A is of the form $\{(X_1, X_2, \ldots) \in A'\}$ for some $A' \in \mathscr{B}(\mathbb{R})^\infty$. Let $\mathscr{C}$ be the class of sets $C' \in \mathscr{B}(\mathbb{R})^\infty$ such that A and C are independent, where $C = \{(X_1, X_2, \ldots) \in C'\}$. If C' is a measurable cylinder, then C is of the form $\{(X_1, \ldots, X_n) \in B_n\}$; since $A \in \mathscr{F}_{n+1}$, A can be written in the form $\{(X_{n+1}, X_{n+2}, \ldots) \in A_{n+1}\}$, and it follows that A and C are independent. Thus $\mathscr{C}$ contains all measurable cylinders. But if $C_n' \in \mathscr{C}$, $C_n' \uparrow C'$ (or $C_n' \downarrow C'$), and $P(A \cap C_n) = P(A)P(C_n)$, $n = 1, 2, \ldots$, then $C_n \uparrow C$ (or $C_n \downarrow C$); hence $P(A \cap C) = P(A)P(C)$. Therefore $\mathscr{C}$ is a monotone class containing the measurable cylinders; hence $\mathscr{C}$ contains all sets in $\mathscr{B}(\mathbb{R})^\infty$, in particular, A'. But then A is independent of itself.

Finally, if f is a tail function, then for each $c \in \overline{\mathbb{R}}$, $\{\omega: f(\omega) < c\}$ is a tail event, and hence has probability 0 or 1. If $k = \sup\{c \in \overline{R}: P\{f < c\} = 0\}$, then $f = k$ a.e. □

If the X_n are identically distributed as well as independent, a wider class of events will be shown to have probability 0 or 1.

6.2.8 Definitions and Comments. Let $X_1, X_2, \ldots$ be a sequence of random variables, and define the σ-fields $\mathscr{F}_n$ as in 6.2.6. Let $A \in \mathscr{F}_1$, so that $A = \{(X_1, X_2, \ldots) \in A'\}$, $A' \in \mathscr{B}(\mathbb{R})^\infty$. The event A is said to be *symmetric* iff the occurrence or nonoccurrence of A is not affected by a permutation of finitely many of the X_i. Formally, if $T: \{1, 2, \ldots\} \to \{1, 2, \ldots\}$ is a permutation of finitely many coordinates, we require that $A = \{X_{T(1)}, X_{T(2)}, \ldots) \in A'\}$.

Any tail event A is symmetric, for if T permutes the first n coordinates, we may write A in the form $\{(X_{n+1}, X_{n+2}, \ldots) \in A_{n+1}\}$ since $A \in \mathscr{F}_{n+1}$. Thus

$$A = \{(X_1, X_2, \ldots) \in \mathbb{R}^n \times A_{n+1}\} = \{(X_{T(1)}, X_{T(2)}, \ldots) \in \mathbb{R}^n \times A_{n+1}\}$$

since $T(k) = k$, $k > n$.

There are, however, symmetric events that are not tail events, for example, $\{X_n = 0$ for all $n\}$ and $\{\lim_{n \to \infty}(X_1 + \cdots + X_n)$ exists and is less than $c\}$.

If $B = \{(X_1, X_2, \ldots) \in B'\} \in \mathscr{F}_1$, not necessarily symmetric, and T permutes finitely many coordinates, we denote by $X(T)$ the sequence $(X_{T(1)}, X_{T(2)}, \ldots)$ and we denote by $B(T)$ the event $\{X(T) \in B'\}$. Thus B is symmetric iff $B = B(T)$ for all T.

6.2.9   Hewitt–Savage Zero–One Law. Let $X_1, X_2, \ldots$ be iid random variables. If A is a symmetric set in $\sigma(X_1, X_2, \ldots)$, then $P(A) = 0$ or 1.

PROOF.   Let $A = \{(X_1, X_2, \ldots) \in A'\}$ so that $P(A) = P_X(A')$, $X = (X_1, X_2, \ldots)$. Find measurable cylinders $C_k{}'$ such that $P_X(A' \Delta C_k{}') \to 0$ as $k \to \infty$ (see 1.3.11), and let $C_k = \{X \in C_k{}'\}$; say $C_k = \{(X_1, \ldots, X_{n_k}) \in B_k\}$. Let T_k interchange $(1, \ldots, n_k)$ and $(n_k + 1, \ldots, 2n_k)$.

Since the X_n are iid, X and $X(T_k)$ have the same distribution; therefore

$$
\begin{aligned}
P(A \Delta C_k) &= P_X(A' \Delta C_k{}') = P_{X(T_k)}(A' \Delta C_k{}') \\
&= P[\{X(T_k) \in A'\} \Delta \{X(T_k) \in C_k{}'\}] \\
&= P[\{X \in A'\} \Delta \{X(T_k) \in C_k{}'\}] \qquad \text{since } A \text{ is symmetric} \\
&= P(A \Delta C_k(T_k)).
\end{aligned}
$$

Thus $P(A \Delta C_k)$ and $P(A \Delta C_k(T_k))$ approach 0, and therefore so does $P(A \Delta [C_k \cap C_k(T_k)])$. It follows that $P(C_k)$, $P(C_k(T_k))$ and $P[C_k \cap C_k(T_k)]$ all approach $P(A)$. But

$$
\begin{aligned}
P[C_k \cap C_k(T_k)] &= P[\{(X_1, \ldots, X_{n_k}) \in B_k, \quad (X_{n_k+1}, \ldots, X_{2n_k}) \in B_k\}] \\
&= P(C_k)P(C_k(T_k)).
\end{aligned}
$$

Let $k \to \infty$ to obtain $P(A) = P(A)P(A)$.   □

Problems

1.   Let $X_1, X_2, \ldots$ be iid random variables. If

$$
\frac{1}{n} \sum_{k=1}^{n} X_k
$$

converges a.e. to a finite limit, show that $E(X_1)$ is finite and the limit equals $E(X_1)$ a.e.

2.   The following result generalizes Lemma 6.2.4, which was used in the proof of the strong law of large numbers.

   Let F be an increasing, right-continuous function from the set of non-negative real numbers to itself, with $F(0) = 0$. If Y is an arbitrary non-negative random variable, show that

$$
E[F(Y)] = \int_0^\infty P\{Y \geq \lambda\} \, dF(\lambda).
$$

In particular, if $F(\lambda) = \lambda$ we obtain

$$E(Y) = \int_0^\infty P\{Y \geq \lambda\} \, d\lambda.$$

We obtain 6.2.4 by expressing the integral from 0 to ∞ as a sum of integrals from n to $n + 1$. [Let $h(\omega, \lambda) = 1$ if $Y(\omega) \geq \lambda$, and 0 if $Y(\omega) < \lambda$, and use Fubini's theorem. Note also that $\{Y \geq \lambda\}$ can be replaced by $\{Y > \lambda\}$ in the above formulas; the same proof applies.]

3. Give an example to show that the Hewitt–Savage zero–one law may fail when the X_n are independent but not identically distributed.

4. Let $X_1, X_2, \ldots$ be iid random variables, and let g: $(\mathbb{R}^\infty, \mathcal{B}^\infty) \to (\mathbb{R}^\infty, \mathcal{B}^\infty)$, where $\mathcal{B} = \mathcal{B}(\mathbb{R})$. If g is symmetric, in other words, $g(x_{T(1)}, x_{T(2)}, \ldots) = g(x_1, x_2, \ldots)$ whenever T is a permutation of finitely many coordinates, show that $g(X_1, X_2, \ldots)$ is constant a.e.

5. Let $X_1, X_2, \ldots$ be independent random variables, with $P\{X_n = 1\} = p_n$, $P\{X_n = 0\} = 1 - p_n$. Show that

$$X_n \xrightarrow{P} 0 \qquad \text{iff} \qquad \lim_{n \to \infty} p_n = 0$$

and

$$X_n \xrightarrow{a.e.} 0 \qquad \text{iff} \qquad \sum_{n=1}^\infty p_n < \infty.$$

6. Let $X_1, X_2, \ldots$ be nonnegative random variables, with X_n having density $\lambda_n \exp(-\lambda_n x)$, $x \geq 0$ $(\lambda_n > 0)$.
   (a) If $\sum_{n=1}^\infty \lambda_n^{-1} < \infty$, show that $\sum_{n=1}^\infty X_n < \infty$ a.e.
   (b) If the X_n are independent and $\sum_{n=1}^\infty \lambda_n^{-1} = \infty$, show that

$$\sum_{n=1}^\infty X_n = \infty \qquad \text{a.e.}$$

   [In (b), consider $\exp(-\sum_{j=1}^n X_j)$.]

7. Let $X_1, X_2, \ldots$ be iid random variables, with $P\{X_n = i\} = 1/r$, $i = 0, 1, \ldots, r - 1$, and define $X = \sum_{n=1}^\infty r^{-n} X_n$. Thus X is the number in $[0, 1]$ with r-adic expansion $.X_1 X_2 \cdots$.

   (a) Show that X is uniformly distributed, in other words, P_X is Lebesgue measure.
   (b) Show that for almost every $x \in [0, 1]$ (Lebesgue measure), the following condition holds:

       For every $r = 2, 3, \ldots$ and all $i = 0, 1, \ldots, r - 1$, the relative frequency of i in the first n digits of the r-adic expansion of x converges to $1/r$ as $n \to \infty$.

(c) If $x \in [0, 1]$, let $R_n(x) = 2x_n - 1$, where x_n is the nth digit of the binary expansion of x (to avoid ambiguity, eliminate expansions with only a finite number of zeros). The R_n are called the *Rademacher functions*. Use part (a) to show that $\int_0^1 R_n(x)\,dx = 0$, and

$$\int_0^1 R_n(x)R_m(x)\,dx = \begin{cases} 0, & n \neq m \\ 1, & n = m. \end{cases}$$

6.3 MARTINGALES

Probability theory has its roots in games of chance, and it is often profitable to interpret results in terms of a gambling situation. For example, if $X_1, X_2, \ldots$ is a sequence of random variables, we may think of X_n as our total winnings after n trials in a succession of games. Having survived the first n trials, our expected fortune after trial $n + 1$ is $E(X_{n+1}|X_1, \ldots, X_n)$. If this equals X_n, the game is "fair" since the expected gain on trial $n + 1$ is $E(X_{n+1} - X_n|X_1, \ldots, X_n) = X_n - X_n = 0$. If $E(X_{n+1}|X_1, \ldots, X_n) \geq X_n$, the game is "favorable," and if $E(X_{n+1}|X_1, \ldots, X_n) \leq X_n$, the game is "unfavorable."

We are going to study sequences of this type; the results to be obtained will have significance outside the casino as well as inside.

6.3.1 Definitions. Let $(\Omega, \mathscr{F}, P)$ be a probability space, $\{X_1, X_2, \ldots\}$ a sequence of integrable random variables on $(\Omega, \mathscr{F}, P)$, and $\mathscr{F}_1 \subset \mathscr{F}_2 \subset \cdots$ an increasing sequence of sub σ-fields of $\mathscr{F}$; X_n is assumed $\mathscr{F}_n$-measurable, that is, $X_n: (\Omega, \mathscr{F}_n) \to (\mathbb{R}, \mathscr{B}(\mathbb{R}))$. The sequence $\{X_n\}$ is said to be a *martingale* relative to the $\mathscr{F}_n$ (alternatively, we say that $\{X_n, \mathscr{F}_n\}$ is a martingale) iff for all $n = 1, 2, \ldots, E(X_{n+1}|\mathscr{F}_n) = X_n$ a.e., a *submartingale* iff $E(X_{n+1}|\mathscr{F}_n) \geq X_n$ a.e., a *supermartingale* iff $E(X_{n+1}|\mathscr{F}_n) \leq X_n$ a.e. (In statements involving conditional expectations, the "a.e." is always understood and will usually be omitted.)

Let $\{\mathscr{F}_n\}$ be a *decreasing* sequence of sub σ-fields of $\mathscr{F}$, with X_n assumed $\mathscr{F}_n$-measurable. If $E(X_n|\mathscr{F}_{n+1}) = X_{n+1}$, we say that $\{X_n, \mathscr{F}_n\}$ is a *reverse martingale*. Similarly, $E(X_n|\mathscr{F}_{n+1}) \geq X_{n+1}$ defines a *reverse submartingale*, and $E(X_n|\mathscr{F}_{n+1}) \leq X_{n+1}$ defines a *reverse supermartingale*.

6.3.2 Comments. (a) If $\{X_n, \mathscr{F}_n\}$ is a martingale, then

$$E(X_{n+k}|\mathscr{F}_n) = X_n, \qquad n, k = 1, 2, \ldots$$

(with corresponding statements for sub- and supermartingales). For

$$E(X_{n+2}|\mathscr{F}_n) = E[E(X_{n+2}|\mathscr{F}_{n+1})|\mathscr{F}_n] \qquad \text{by 5.5.10(a)}$$

$$= E(X_{n+1}|\mathscr{F}_n)$$

$$= X_n.$$

The general statement follows by induction.

(b) If $\{X_n, \mathscr{F}_n\}$ is a martingale, then

$$E(X_{n+1}|X_1, \ldots, X_n) = X_n, \qquad n = 1, 2, \ldots$$

Thus $\{X_n\}$ is automatically a martingale relative to the standard σ-fields $\sigma(X_1, \ldots, X_n)$ (with corresponding statements for sub- and supermartingales).

For $\mathscr{F}_1 \subset \cdots \subset \mathscr{F}_n$, and thus $X_1, \ldots, X_n$ are all $\mathscr{F}_n$-measurable. Since $\sigma(X_1, \ldots, X_n)$ is the smallest σ-field making $X_1, \ldots, X_n$ measurable [see 5.4.2(b)], we have $\sigma(X_1, \ldots, X_n) \subset \mathscr{F}_n$. If, in the defining relation

$$E(X_{n+1}|\mathscr{F}_n) = X_n,$$

we take conditional expectations with respect to $\sigma(X_1, \ldots, X_n)$, we obtain the desired result by 5.5.10(a) and 5.5.11(a).

If we say that $\{X_n\}$ is a martingale (or sub-, supermartingale) without mentioning the σ-fields $\mathscr{F}_n$, we shall always mean $\mathscr{F}_n = \sigma(X_1, \ldots, X_n)$, so that $E(X_{n+1}|X_1, \ldots, X_n) = X_n$.

(c) $\{X_n, \mathscr{F}_n\}$ is a martingale iff

$$\int_A X_n \, dP = \int_A X_{n+1} \, dP \qquad \text{for all} \qquad A \in \mathscr{F}_n, \quad n = 1, 2, \ldots.$$

This follows since the condition $E(X_{n+1}|\mathscr{F}_n) = X_n$ a.e. $[P]$ is equivalent to

$$\int_A E(X_{n+1}|\mathscr{F}_n) \, dP = \int_A X_n \, dP \qquad \text{for all} \qquad A \in \mathscr{F}_n$$

(see 1.6.11); also

$$\int_A E(X_{n+1}|\mathscr{F}_n) \, dP = \int_A X_{n+1} \, dP$$

by definition of conditional expectation.

Similarly, $\{X_n, \mathscr{F}_n\}$ is a submartingale iff

$$\int_A X_n \, dP \leq \int_A X_{n+1} \, dP \qquad \text{for all} \qquad A \in \mathscr{F}_n,$$

and a supermartingale iff

$$\int_A X_n \, dP \geq \int_A X_{n+1} \, dP \qquad \text{for all} \qquad A \in \mathscr{F}_n.$$

In particular, $E(X_n)$ is constant in a martingale, increases in a submartingale, and decreases in a supermartingale.

(d)   The defining condition for a martingale relative to the σ-fields $\sigma(X_1, \ldots, X_n)$ is equivalent to

$$E(X_{n+1}|X_1 = x_1, \ldots, X_n = x_n) = x_n \qquad \text{a.e. } [P_{(X_1, \ldots, X_n)}]$$

with similar statements for sub- and supermartingales.

For if $A \in \sigma(X_1, \ldots, X_n)$, then A is of the form $\{(X_1, \ldots, X_n) \in B\}$, $B \in \mathcal{B}(\mathbb{R}^n)$ (see 5.4.1). If $X = (X_1, \ldots, X_n)$, then

$$\int_A X_{n+1} \, dP = \int_{\{X \in B\}} X_{n+1} \, dP$$

$$= \int_B E(X_{n+1}|X_1 = x_1, \ldots, X_n = x_n) \, dP_X \qquad \text{by 5.3.3}$$

and

$$\int_A X_n \, dP = \int_B E(X_n|X_1 = x_1, \ldots, X_n = x_n) \, dP_X$$

$$= \int_B x_n \, dP_X \qquad \text{by 5.5.11(a').}$$

The result now follows from (c).

(e)   A finite sequence $\{X_k, \mathcal{F}_k, k = 1, \ldots, n\}$ is called a martingale iff $E(X_{k+1}|\mathcal{F}_k) = X_k$, $k = 1, 2, \ldots, n-1$; finite sub- and supermartingale sequences are defined similarly.

(f)   If $\{X_n, \mathcal{F}_n\}$ and $\{Y_n, \mathcal{F}_n\}$ are submartingales, so is $\{\max(X_n, Y_n), \mathcal{F}_n\}$.

For $E(\max(X_{n+1}, Y_{n+1})|\mathcal{F}_n) \geq E(X_{n+1}|\mathcal{F}_n) \geq X_n$, and similarly

$$E(\max(X_{n+1}, Y_{n+1})|\mathcal{F}_n) \geq Y_n.$$

The same approach shows that if $\{X_n, \mathcal{F}_n\}$ and $\{Y_n, \mathcal{F}_n\}$ are supermartingales, so is $\{\min(X_n, Y_n), \mathcal{F}_n\}$.

6.3.3   Examples. If $X_n \equiv X$, then $\{X_n\}$ is a martingale; if $X_1 \leq X_2 \leq \cdots$, then $\{X_n\}$ is a submartingale; if $X_1 \geq X_2 \geq \cdots$, then $\{X_n\}$ is a supermartingale (assuming all random variables integrable).

We give some more substantial examples.

(a)   Let $Y_1, Y_2, \ldots$ be independent random variables with zero mean, and set $X_n = \sum_{k=1}^n Y_k$, $\mathcal{F}_n = \sigma(Y_1, \ldots, Y_n)$. Then $\{X_n, \mathcal{F}_n\}$ is a martingale. For

$$E(X_{n+1}|\mathscr{F}_n) = E(X_n + Y_{n+1}|Y_1, \ldots, Y_n)$$

$$= X_n + E(Y_{n+1}|Y_1, \ldots, Y_n)$$
$$\text{since} \quad X_n \text{ is } \mathscr{F}_n\text{-measurable}$$

$$= X_n + E(Y_{n+1})$$
by independence (Problem 1, Section 5.5)

$$= X_n$$
$$\text{since} \quad E(Y_j) \equiv 0.$$

(b)   Let $Y_1, Y_2, \ldots$ be independent random variables with $E(Y_j) = a_j \neq 0$, and set $X_n = \prod_{j=1}^{n}(Y_j/a_j)$, $\mathscr{F}_n = \sigma(Y_1, \ldots, Y_n)$. Then $\{X_n, \mathscr{F}_n\}$ is a martingale. For

$$E(X_{n+1}|\mathscr{F}_n) = E\left[\left(\frac{X_n Y_{n+1}}{a_{n+1}}\right)\middle| Y_1, \ldots, Y_n\right]$$

$$= X_n E\left(\frac{Y_{n+1}}{a_{n+1}}\right)$$
by 5.5.11(a) and Problem 1, Section 5.5

$$= X_n.$$

(c)   Let Y be an integrable random variable on $(\Omega, \mathscr{F}, P)$.

If $\{\mathscr{F}_n\}$ is an *increasing* sequence of sub σ-fields of $\mathscr{F}$, and $X_n = E(Y|\mathscr{F}_n)$, then $\{X_n, \mathscr{F}_n\}$ is a martingale. For

$$E(X_{n+1}|\mathscr{F}_n) = E[E(Y|\mathscr{F}_{n+1})|\mathscr{F}_n]$$

$$= E(Y|\mathscr{F}_n) \quad \text{since} \quad \mathscr{F}_n \subset \mathscr{F}_{n+1}$$

$$= X_n.$$

If $\{\mathscr{F}_n\}$ is a *decreasing* sequence of sub σ-fields of $\mathscr{F}$, and $X_n = E(Y|\mathscr{F}_n)$, then $\{X_n, \mathscr{F}_n\}$ is a *reverse martingale*. For

$$E(X_n|\mathscr{F}_{n+1}) = E[E(Y|\mathscr{F}_n)|\mathscr{F}_{n+1}]$$

$$= E(Y|\mathscr{F}_{n+1}) \quad \text{since} \quad \mathscr{F}_{n+1} \subset \mathscr{F}_n$$

$$= X_{n+1}.$$

Note that as in 6.3.2(a), $E(X_n|\mathscr{F}_{n+k}) = X_{n+k}$, $n, k = 1, 2, \ldots$.

(d)   (Branching Processes) We define a Markov chain (see 4.11) with state space $S = \{0, 1, 2, \ldots\}$. The state at time n, denoted by X_n, is to represent the number of offspring after n generations. We take $X_0 = 1$, and, if

$X_n = k$, X_{n+1} is the sum of k independent, identically distributed, nonnegative integer valued random variables, say $Y_1, \ldots, Y_k$, where $P\{Y_i = l\} = p_l$, $l = 0, 1, 2, \ldots$. Thus p_l is the probability that a given being will produce exactly l offspring. (Formally, we take $p_{kj} = P\{Y_1 + \cdots + Y_k = j\}$, $k = 1$, $2, \ldots, j = 0, 1, \ldots; p_{00} = 1$.)

Let $m = E(Y_i) = \sum_{l=0}^{\infty} l\, p_l$. If m is finite and greater than 0, then $\{X_n/m^n\}$ is a martingale [relative to

$$\mathscr{F}_n = \sigma(X_0, \ldots, X_n) = \sigma\left(\frac{X_1}{m}, \frac{X_2}{m^2}, \ldots, \frac{X_n}{m^n}\right)\right].$$

For

$$E\left(\frac{X_{n+1}}{m^{n+1}}\,\bigg|\, X_0 = i_0, \ldots, X_n = i_n\right) = \sum_{j=0}^{\infty} p_{i_n j} \frac{j}{m^{n+1}}$$

$$\text{(see 5.3.5(a), Eq. (2) and the definition of a Markov chain)}$$

$$= \frac{1}{m^{n+1}} \sum_{j=0}^{\infty} j P\{Y_1 + \cdots + Y_{i_n} = j\}$$

$$= \frac{1}{m^{n+1}} E(Y_1 + \cdots + Y_{i_n})$$

$$= \frac{i_n m}{m^{n+1}} = \frac{i_n}{m^n}.$$

The result now follows from 6.3.2(d).

(e) Consider the branching process of part (d). Let $g(s) = \sum_j p_j s^j$, $s \geq 0$. If for some r we have $g(r) = r$, then $\{r^{X_n}\}$ is a martingale relative to the σ-fields $\mathscr{F}(X_0, \ldots, X_n)$. For as in (d),

$$E(r^{X_{n+1}} | X_0 = i_0, \ldots, X_n = i_n) = \sum_{j=0}^{\infty} p_{i_n j} r^j$$

$$= \sum_{j=0}^{\infty} r^j P\{Y_1 + \cdots + Y_{i_n} = j\}$$

$$= E[\exp_r(Y_1 + \cdots + Y_{i_n})]$$

$$= [E(r^{Y_1})]^{i_n} = [g(r)]^{i_n} = r^{i_n}.$$

If $\{X_n, \mathscr{F}_n\}$ is a martingale, what can we say about X_n^+ or $|X_n|$? In order to answer this question, we need some basic convexity theorems.

6.3.4 *Line of Support Theorem.* Let $g: I \to \mathbb{R}$, where I is an open interval of reals, bounded or unbounded. Assume g is convex, that is,

$$g(ax + (1 - a)y) \le ag(x) + (1 - a)g(y)$$

for all $x, y \in I$ and all $a \in [0, 1]$. Then there are sequences $\{a_n\}$ and $\{b_n\}$ of real numbers such that for all $y \in I$ we have $g(y) = \sup_n (a_n y + b_n)$.

PROOF.   In the course of the proof, we develop many of the basic properties of convex functions of one variable.

   Let g be a convex function from I to $\mathbb{R}, I$ an open interval of reals. If $0 < h_1 < h_2$, then by convexity,

$$g(x \pm h_1) \le \frac{(h_2 - h_1)}{h_2} g(x) + \frac{h_1}{h_2} g(x \pm h_2);$$

hence

$$\frac{1}{h_1}[g(x + h_1) - g(x)] \le \frac{1}{h_2}[g(x + h_2) - g(x)] \tag{1}$$

and

$$-\frac{1}{h_2}[g(x - h_2) - g(x)] \le -\frac{1}{h_1}[g(x - h_1) - g(x)]. \tag{2}$$

Also, if $h, h' > 0$,

$$g(x) \le \frac{h'}{h + h'} g(x - h) + \frac{h}{h + h'} g(x + h'),$$

so

$$\frac{g(x - h) - g(x)}{-h} \le \frac{g(x + h') - g(x)}{h'}. \tag{3}$$

By (1) and (2), *the right and left derivatives g_+' and g_-' exist on I*; by (3) they are *finite*. [Note that if $x \in I$, we have $x - h, x + h' \in I$ for small $h, h' > 0$ since I is open; thus, in fact, the difference quotients $[g(x + h') - g(x)]/h'$, which decrease as $h' \downarrow 0$ by (1), are bounded below by a finite constant.] Furthermore, by (1) and (2),

$$g_+'(x) = \inf_{y > x} \frac{g(y) - g(x)}{y - x}, \qquad g_-'(x) = \sup_{y < x} \frac{g(x) - g(y)}{x - y} \tag{4}$$

and by (3),

$$g_-'(x) \le g_+'(x). \tag{5}$$

If $x_1 < x_2$, then

$$g_+'(x_1) \leq \frac{g(x_2) - g(x_1)}{x_2 - x_1} = \frac{g(x_1) - g(x_2)}{x_1 - x_2} \leq g_-'(x_2); \qquad (6)$$

hence g_+' and g_-' are *increasing* on I. The existence of right and left derivatives implies that g has right and left limits at each point; thus g *is continuous*; for if not, g_+' or g_-' would be infinite at some point. If $y \geq x$, then $g(y) \geq g(x) + (y - x)g_+'(x)$, and if $y < x$, then $g(y) \geq g(x) + (y - x)g_-'(x)$ [by (4)]. We conclude that if $g_-'(x) \leq a_x \leq g_+'(x)$, then

$$g(y) \geq g(x) + (y - x)a_x \qquad \text{for all} \qquad y \in I. \qquad (7)$$

The function L_x given by $L_x(y) = g(x) + (y - x)a_x$, $y \in I$, is called a *line of support* for g at the point x. It is immediate that $g(x) = \sup\{L_s(x): s \in I\}$ [since $L_x(x) = g(x)$], but we are trying to prove that g is the sup of *countably* many lines of support. If $x \in I$, let r approach x through a sequence of rational numbers. Then

$$L_r(x) = g(r) + (x - r)a_r.$$

But $g_-'(r) \leq a_r \leq g_+'(r)$, and the g', being increasing functions, are bounded on any finite closed subinterval of I, so that the a_r form a bounded sequence. Consequently, $L_r(x) \to g(x)$. By (7), $L_r(y) \leq g(y)$ for all $y \in I$ and all r; hence $g(x) = \sup\{L_s(x): s \in I, s \text{ rational}\}$, that is, $g(x) = \sup\{g(s) + (x - s)a_s: s \in I, s \text{ rational}\}$. The proof is complete. $\square$

If I is not open, the theorem is false: consider $g(x) = 0$, $0 \leq x < 1$, $g(1) = 1$.

Now if X is a random variable with finite mean, we have seen that $E(X^2) \geq [E(X)]^2$ (4.10.6). This is a special case of the following general convexity theorem.

6.3.5 Jensen's Inequality. *Let g be a convex function from I to $\mathbb{R}$, where I is an open interval of reals, bounded or unbounded. Let X be a random variable on $(\Omega, \mathscr{F}, P)$, with $X(\omega) \in I$ for all ω. Assume $E(X)$ to be finite. If $\mathscr{H}$ is a sub σ-field of $\mathscr{F}$, then $E[g(X)|\mathscr{H}] \geq g[E(X|\mathscr{H})]$ a.e. In particular, $E[g(X)] \geq g[E(X)]$.*

PROOF.   First note that $E(X|\mathscr{H}) \in I$ a.e. For if, say, a is real and $X > a$, then $E(X|\mathscr{H}) > a$ a.e. because

$$0 \geq \int_{\{E(X|\mathscr{H}) \leq a\}} E(X - a|\mathscr{H}) \, dP = \int_{\{E(X|\mathscr{H}) \leq a\}} (X - a) \, dP \geq 0;$$

hence $X = a$ a.e. on $\{E(X|\mathscr{H}) \le a\}$, which implies that $P\{E(X|\mathscr{H}) \le a\} = 0$. Thus $g[E(X|\mathscr{H})]$ is well-defined.

By 6.3.4 we may write $g(y) = \sup_n (a_n y + b_n)$, $y \in I$, so $g(X) \ge a_n X + b_n$ for all n. Therefore $E[g(X)|\mathscr{H}] \ge a_n E(X|\mathscr{H}) + b_n$ a.e. Take the sup over n to finish the proof. $\square$

The proof of 6.3.5 shows that the hypothesis that $E(X)$ is finite may be dropped if it is known that $E(X)$ and $E[g(X)]$ exist, and $E(X|\mathscr{H}) \in I$ a.e.

We are now able to answer the question raised earlier about $X_n{}^+$ and $|X_n|$ when $\{X_n, \mathscr{F}_n\}$ is a martingale.

6.3.6 Theorem. (a)   Let $\{X_n, \mathscr{F}_n\}$ be a submartingale, g a convex, increasing function from $\mathbb{R}$ to $\mathbb{R}$. If $g(X_n)$ is integrable for all n, then $\{g(X_n), \mathscr{F}_n\}$ is a submartingale. Thus, for example, if $\{X_n\}$ is a submartingale, so is $\{X_n{}^+\}$.

(b)   Let $\{X_n, \mathscr{F}_n\}$ be a martingale, g a convex function from $\mathbb{R}$ to $\mathbb{R}$. If $g(X_n)$ is integrable for all n, then $\{g(X_n), \mathscr{F}_n\}$ is a submartingale. Thus if $r \ge 1$, $\{X_n\}$ is a martingale and $|X_n|^r$ is integrable for all n, then $\{|X_n|^r\}$ is a submartingale.

PROOF.   We have $E[g(X_{n+1})|\mathscr{F}_n] \ge g[E(X_{n+1}|\mathscr{F}_n)]$ by Jensen's inequality. In (a), $E(X_{n+1}|\mathscr{F}_n) \ge X_n$ by the submartingale property; hence $g[E(X_{n+1}|\mathscr{F}_n)] \ge g(X_n)$ since g is increasing. In (b), $E(X_{n+1}|\mathscr{F}_n) = X_n$ by the martingale property, so $g[E(X_{n+1}|\mathscr{F}_n)] = g(X_n)$. The result follows. $\square$

Problems

1.   Let $X_n = \sum_{k=1}^n Y_k$, where the Y_k are independent, with $P\{Y_k = 1\} = p$, $P\{Y_k = -1\} = q$ $(p, q > 0, p + q = 1)$. Show that $\{(q/p)^{X_n}\}$ is a martingale relative to the σ-fields $\sigma(X_1, \ldots, X_n)[= \sigma(Y_1, \ldots, Y_n)]$.

2.   Consider of Markov chain whose state space is the integers, and assume that p_{ij} depends only on the difference between j and i: $p_{ij} = q_{j-i}$, where $q_k \ge 0$, $\sum_k q_k = 1$.

(a)   Show that if X_n is the state at time n, X_n may be written as $X_0 + Y_1 + \cdots + Y_n$, where $X_0, Y_1, \ldots, Y_n$ are independent and the Y_i all have the same distribution, namely, $P\{Y_i = k\} = q_k$, k an integer.

(b)   If $\sum_j q_j r^j = 1$ (where the series is assumed to converge absolutely), show that $\{r^{X_n}\}$ is a martingale relative to the σ-fields $\sigma(X_0, \ldots, X_n)$.

3.   Let λ be a countably additive set function on the σ-field $\mathscr{F}$, and let $\mathscr{F}_n$ be generated by the sets $A_{n1}, A_{n2}, \ldots$, assumed to form a partition of Ω,

with $P(A_{nj}) > 0$ for all j. Assume that the $(n+1)$st partition refines the nth, so that $\mathscr{F}_n \subset \mathscr{F}_{n+1}$.

Define

$$X_n(\omega) = \frac{\lambda(A_{nj})}{P(A_{nj})} \quad \text{if} \quad \omega \in A_{nj}, \quad n, j = 1, 2, \ldots .$$

Show that $\{X_n, \mathscr{F}_n\}$ is a martingale.

4. Define a sequence of random variables as follows. Let X_1 be uniformly distributed between 0 and 1. Given that $X_1 = x_1$, $X_2 = x_2, \ldots, X_{n-1} = x_{n-1}$, let X_n be uniformly distributed between 0 and x_{n-1}. Show that $\{X_n\}$ is a supermartingale and $E(X_n) = 2^{-n}$. Conclude that $X_n \to 0$ a.e.

5. Let $X_1, X_2, \ldots$ be real-valued Borel measurable functions on $(\Omega, \mathscr{F})$. Assume that under the probability measure P on $\mathscr{F}$, $(X_1, \ldots, X_n)$ has density p_n, and under the probability measure Q on $\mathscr{F}$, $(X_1, \ldots, X_n)$ has density q_n. Define

$$Y_n = \begin{cases} \dfrac{q_n(X_1, \ldots, X_n)}{p_n(X_1, \ldots, X_n)} & \text{if the denominator is greater than 0,} \\ 0 & \text{otherwise.} \end{cases}$$

Show that if $\mathscr{F}_n = \mathscr{F}(X_1, \ldots, X_n)$, $\{Y_n, \mathscr{F}_n\}$ is a supermartingale on $(\Omega, \mathscr{F}, P)$ and $0 \le E(Y_n) \le 1$ for all n.

6. Let $\{X_n, \mathscr{F}_n, n \ge 0\}$ be a supermartingale, and define

$$Y_0 = X_0,$$
$$Y_1 = Y_0 + (X_1 - E(X_1 | \mathscr{F}_0)),$$
$$\vdots$$
$$Y_n = Y_{n-1} + (X_n - E(X_n | \mathscr{F}_{n-1})).$$

Also define

$$A_0 = 0,$$
$$A_1 = X_0 - E(X_1 | \mathscr{F}_0),$$
$$A_2 = A_1 + (X_1 - E(X_2 | \mathscr{F}_1)),$$
$$\vdots$$
$$A_n = A_{n-1} + (X_{n-1} - E(X_n | \mathscr{F}_{n-1})).$$

(a) Show that $X_n = Y_n - A_n$.
(b) Show that $\{Y_n, \mathscr{F}_n\}$ is a martingale.
(c) Show that for a.e. ω, $A_n(\omega)$ increases with n.

Thus a supermartingale can be expressed as the difference between a martingale and an increasing sequence. (Similarly, a submartingale can be expressed as the sum of a martingale and an increasing sequence.)

6.4 MARTINGALE CONVERGENCE THEOREMS

Under rather mild conditions, sub- and supermartingales converge almost everywhere. This result has very many ramifications in probability theory.

We first prove a theorem which has an interesting gambling interpretation.

6.4.1 Optional Skipping Theorem (Halmos). Let $\{X_n, \mathscr{F}_n\}$ be a submartingale. Let $\varepsilon_1, \varepsilon_2, \ldots$ be random variables defined by

$$\varepsilon_k = \begin{cases} 1 & \text{if} & (X_1, \ldots, X_k) \in B_k, \\ 0 & \text{if} & (X_1, \ldots, X_k) \notin B_k, \end{cases}$$

where the B_k are arbitrary sets in $\mathscr{B}(\mathbb{R}^n)$. Set

$$Y_1 = X_1,$$
$$Y_2 = X_1 + \varepsilon_1(X_2 - X_1)$$
$$\vdots$$
$$Y_n = X_1 + \varepsilon_1(X_2 - X_1) + \cdots + \varepsilon_{n-1}(X_n - X_{n-1}).$$

Then $\{Y_n \mathscr{F}_n\}$ is also a submartingale and $E(Y_n) \leq E(X_n)$ for all n. If $\{X_n, \mathscr{F}_n\}$ is a martingale, so is $\{Y_n, \mathscr{F}_n\}$ and $E(Y_n) = E(X_n)$ for all n.

Interpretation. Let X_n be the gambler's fortune after n trials; then Y_n is our fortune if we follow an optional skipping strategy. After observing $X_1, \ldots, X_k$, we may choose to bet with the gambler at trial $k + 1$ [in this case $\varepsilon_k = \varepsilon_k(X_1 \ldots, X_k) = 1$] or we may pass ($\varepsilon_k = 0$). Our gain on trial $k + 1$ is $\varepsilon_k(X_{k+1} - X_k)$. The theorem states that whatever strategy we employ, if the game is initially "fair" (a martingale) or "favorable" (a submartingale), it remains fair (or favorable), and no strategy of this type can increase the expected winning.

PROOF.

$$E(Y_{n+1}|\mathscr{F}_n) = E(Y_n + \varepsilon_n(X_{n+1} - X_n)|\mathscr{F}_n)$$
$$= Y_n + \varepsilon_n E[(X_{n+1} - X_n)|\mathscr{F}_n]$$

since ε_n is a Borel measurable function of $X_1, \ldots, X_n$, and hence is

$$\sigma(X_1, \ldots, X_n) \subset \mathscr{F}_n\text{-measurable}.$$

Therefore

$$E(Y_{n+1}|\mathscr{F}_n) = Y_n + \varepsilon_n(X_n - X_n) = Y_n \qquad \text{in the martingale case}$$
$$\geq Y_n + \varepsilon_n(X_n - X_n) = Y_n \qquad \text{in the submartingale case.}$$

Since $Y_1 = X_1$, we have $E(X_1) = E(Y_1)$. Having shown $E(X_k - Y_k) \geq 0$ ($= 0$ in the martingale case),

$$X_{k+1} - Y_{k+1} = X_{k+1} - Y_k - \varepsilon_k(X_{k+1} - X_k)$$
$$= (1 - \varepsilon_k)(X_{k+1} - X_k) + X_k - Y_k.$$

Thus

$$E(X_{k+1} - Y_{k+1}|\mathscr{F}_k) = (1 - \varepsilon_k)E(X_{k+1} - X_k|\mathscr{F}_k) + E(X_k - Y_k|\mathscr{F}_k)$$
$$\geq E(X_k - Y_k|\mathscr{F}_k) = X_k - Y_k,$$

with equality in the martingale case. Take expectations and use $E[E(X|\mathscr{G})] = E(X)$ to obtain

$$E(X_{k+1} - Y_{k+1}) \geq E(X_k - Y_k) \geq 0,$$

with equality in the martingale case. $\square$

The key step in the development is the following result, due in its original form to Doob (1940).

6.4.2   Upcrossing Theorem. Let $\{X_k, \mathscr{F}_k, k = 1, 2, \ldots, n\}$ be a submartingale. If a and b are real numbers, with $a < b$, let U_{ab} be the number of upcrossings of (a, b) by $X_1, \ldots, X_n$, defined as follows.

Let $T_1 = T_1(\omega)$ be the first integer in $\{1, 2, \ldots, n\}$ such that $X_{T_1} \leq a, T_2$ be the first integer greater than T_1 such that $X_{T_2} \geq b, T_3$ be the first integer greater than T_2 such that $X_{T_3} \leq a, T_4$ be the first integer greater than T_3 such that $X_{T_4} \geq b$, and so on. (Set $T_i = \infty$ if the condition cannot be satisfied.) If N is the number of finite, T_i, define $U_{ab} = N/2$ if N is even, and $(N - 1)/2$ if N is odd. Then

$$E(U_{ab}) \leq \frac{1}{b - a} E[(X_n - a)^+].$$

PROOF.   First assume $a = 0$, and all $X_j \geq 0$. Define the T_i as above ($X_{T_i} \leq a$ is now equivalent to $X_{T_i} = 0$. Let $\varepsilon_j = 0$ for $j < T_1$; $\varepsilon_j = 1$ for $T_1 \leq j < T_2$; $\varepsilon_j = 0$ for $T_2 \leq j < T_3$; $\varepsilon_j = 1$ for $T_3 \leq j < T_4$,   and   so   on   (see Fig. 6.4.1).

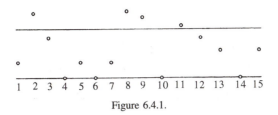

Figure 6.4.1.

In Figure 6.4.1, we have (with $n = 15$) $T_1 = 4, T_2 = 8, T_3 = 10, T_4 = 11,$
$T_5 = 14, T'_n = \infty, n > 5, U_{ab} = 2;$

$$\varepsilon_1 = \varepsilon_2 = \varepsilon_3 = 0, \qquad \varepsilon_4 = \varepsilon_5 = \varepsilon_6 = \varepsilon_7 = 1, \qquad \varepsilon_8 = \varepsilon_9 = 0,$$

$$\varepsilon_{10} = 1, \qquad \varepsilon_{11} = \varepsilon_{12} = \varepsilon_{13} = 0, \qquad \varepsilon_{14} = 1;$$

$$X_1 + \varepsilon_1(X_2 - X_1) + \cdots + \varepsilon_{14}(X_{15} - X_{14})$$
$$= X_1 + X_8 - X_4 + X_{11} - X_{10} + X_{15} - X_{14}.$$

Note that Y_n, as defined in 6.4.1, is the total increase during upcrossings, plus possibly a "partial upcrossing" at the end, plus a contribution due to X_1 (necessarily nonnegative). Thus $Y_n \geq bU$. But the ε_j can be expressed in terms of $X_1, \ldots, X_j$, so the optional skipping theorem applies; hence $\{Y_k, \mathscr{F}_k, k = 1, 2, \ldots, n\}$ is a submartingale, and $E(Y_n) \leq E(X_n)$. Thus

$$E(U_{ab}) \leq \frac{1}{b}E(Y_n) \leq \frac{1}{b}E(X_n),$$

as asserted.

In general, $\{(X_k - a)^+, \mathscr{F}_k, k = 1, 2, \ldots, n\}$ is a submartingale by 6.3.6(a), and the number of upcrossings of (a, b) by $\{X_j\}$ is the same as the number of upcrossings of $(0, b - a)$ by $\{(X_j - a)^+\}$ (note that $X_j \leq a, X_j - a \leq 0$, and $(X_j - a)^+ \leq 0$ are equivalent, as are $X_j \geq b, X_j - a \geq b - a$, and $(X_j - a)^+ \geq b - a$. The result follows from the above argument. $\square$

We now prove the main convergence theorem.

6.4.3 Submartingale Convergence Theorem. Let $\{X_n, \mathscr{F}_n, n = 1, 2, \ldots\}$ be a submartingale. If $\sup_n E(X_n^+) < \infty$, there is an integrable random variable X_∞ such that $X_n \to X_\infty$ almost everywhere.

PROOF. $P\{\omega: X_n(\omega)$ does not converge to a finite or infinite limit$\}$

$$P\left[\bigcup_{\substack{a < b \\ a, b \text{ rational}}} \left\{\omega: \liminf_{n \to \infty} X_n(\omega) < a < b < \limsup_{n \to \infty} X_n(\omega)\right\}\right].$$

If for some $a < b$, $P\{\liminf_{n\to\infty} X_n < a < b < \limsup_{n\to\infty} X_n\} > 0$, then $\{X_n\}$ has an infinite number of upcrossings of (a, b) on a set of positive probability; hence $E(U_{ab}) = \infty$. But U_{ab} is the limit of the monotone sequence $U_{ab;n}$ = the number of upcrossings of (a, b) by $X_i, \ldots, X_n$, so that $E(U_{ab;n}) \to E(U_{ab})$. But by 6.4.2,

$$E(U_{ab;n}) \leq (b - a)^{-1}E[(X_n - a)^+] \leq (b - a)^{-1}[\sup_n E(X_n^+) + a^-] < \infty,$$

a contradiction. Thus X_n converges to a limit X_∞ a.e. Now $|X_n| = X_n^+ + X_n^- = 2X_n^+ - X_n$, and $E(X_n) \geq E(X_1)$ by the submartingale property. Therefore

$$E(|X_n|) \leq 2\sup_n E(X_n^+) - E(X_1) < \infty.$$

By Fatou's lemma,

$$E(|X_\infty|) \leq \liminf_{n\to\infty} E(|X_n|) < \infty;$$

hence X_∞ is integrable, and therefore finite a.e. By changing X_∞ on a set of measure 0, if necessary, we may take X_∞ as a random variable (rather than an extended random variable). The theorem is proved. $\square$

6.4.4 Corollary. Let $\{X_n, \mathscr{F}_n, n = 1, 2, \ldots\}$ be a reverse submartingale [the $\mathscr{F}_n$ decrease as n increases, and $E(X_n|\mathscr{F}_{n+1}) \geq X_{n+1}$ a.e.]. If $\inf_n E(X_n) > -\infty$, there is an integrable random variable X_∞ such that $X_n \to X_\infty$ a.e. [Note that the hypothesis is satisfied for any reverse martingale since $E(X_n)$ is constant.]

PROOF. Proceed as in 6.4.3, but instead let $U_{ab;n}$ be the number of upcrossings of (a, b) by $\{X_n, X_{n-1}, \ldots, X_1\}$, which is a submartingale because

$$E(X_k|X_{k+1}, \ldots, X_n) = E[E(X_k|\mathscr{F}_{k+1})|X_{k+1}, \ldots, X_n] \geq X_{k+1}.$$

We obtain $E(U_{ab;n}) \leq (b - a)^{-1}E[(X_1 - a)^+] < \infty$, and thus $X_n \to X_\infty$ a.e. as before.

Now $|X_n| = 2X_n^+ - X_n$ and $E(X_n) \geq \inf_n E(X_n) > -\infty$. Also $\{X_n^+, \ldots, X_1^+\}$ is a submartingale by 6.3.6(a), so $E(X_n^+) \leq E(X_1^+)$. Thus $E(|X_n|) \leq 2E(X_1^+) - \inf_n E(X_n) < \infty$, so X_∞ is integrable by Fatou's lemma as before. $\square$

6.4.5 Comments. (a) In 6.4.3 and 6.4.4, the proofs show that $\{X_n\}$ must be L^1 *bounded*, that is, $\sup_n E(|X_n|) < \infty$. Thus for a submartingale, $\sup_n E(X_n^+) < \infty$ is equivalent to L^1 boundedness, and implies convergence. However, a submartingale may converge without being L^1 bounded (see Problems 1 and 2).

(b)   Results analogous to 6.4.3 and 6.4.4 hold for supermartingales: If $\{X_n, \mathscr{F}_n, n = 1, 2, \ldots\}$ is a supermartingale and $\sup_n E(X_n^-) < \infty$, then there is an integrable random variable X_∞ such that $X_n \to X_\infty$ a.e. In particular, a nonnegative supermartingale converges a.e. If $\{X_n, \mathscr{F}_n, n = 1, 2, \ldots\}$ is a reverse supermartingale and $\sup_n E(X_n) < \infty$, there is an integrable random variable X_∞ such that $X_n \to X_\infty$ a.e. The first statement follows from 6.4.3 since $\{-X_n, \mathscr{F}_n\}$ is a submartingale and $\sup_n E[(-X_n)^+] = \sup_n E(X_n^-)$. The second follows from 6.4.4 since $\{-X_n, \mathscr{F}_n\}$ is a reverse submartingale and $\inf_n E(-X_n) = -\sup_n E(X_n)$.

Problems

1.   Consider the following Markov chain. Take $X_1 = 0$. If $X_n = 0$ (regardless of $X_k, k < n$), then:

$$X_{n+1} = a_{n+1} \qquad \text{with probability} \qquad p_{n+1}$$
$$= -a_{n+1} \qquad \text{with probability} \qquad p_{n+1}$$
$$= 0 \qquad \text{with probability} \qquad 1 - 2p_{n+1},$$

where $0 < p_{n+1} < \frac{1}{2}$ and the a_n are distinct and greater than 0. If $X_n \neq 0$, take $X_{n+1} = X_n$ (thus if $X_n \neq 0$, we have $X_j = X_n$ for all $j \geq n$).

(a)   Show that $\{X_n\}$ is a martingale, and X_n converges everywhere.

(b)   If $\sum_{k=1}^\infty p_k < \infty$ and $\sum_{k=1}^\infty a_k p_k = \infty$, show that $\sup_n E(|X_n|) = \infty$.

2.   (Problem by W. F. Stout, personal communication.) Consider the following Markov chain. Take $X_0 = 0$, and let

$$P\{X_{n+1} = n + 1 | X_n = n\} = p_{n+1},$$
$$P\{X_{n+1} = -(n + 1) | X_n = n\} = 1 - p_{n+1},$$
$$P\{X_{n+1} = -k | X_n = -k\} = 1, \qquad (n = 0, 1, \ldots, \quad k = 1, 2, \ldots).$$

(a)   Show that if $p_{n+1} = (2n + 1)/(2n + 2)$ for all n, then $\{X_n\}$ is a martingale.

(b)   If the p_n are chosen as in (a), show that the martingale converges a.e. to a finite limit, although $E(|X_n|) \to \infty$.

(*Note:* In Problems 1 and 2, the Markov chain has nonstationary transition probabilities, in other words, the probability of moving from state i at time n to state j at time $n + 1$ depends on n. However, the basic construction of 4.11.2 carries over.)

3.   (Kemeny, Snell, and Knapp, 1966) Let $\{X_n\}$ be a Markov chain with state space $S =$ the set of rationals in $(0, 1)$, and the following transition probabilities:

Let $0 < b \le a < 1$, a, b rational. If $x \in S$ and $X_n = x$, then $X_{n+1} = bx$ with probability $1 - x$, and $X_{n+1} = bx + 1 - a$ with probability x.

   (a)   If $a = b$, show that $\{X_n\}$ is a martingale, and $X_n \to X_\infty$ a.e., where $X_\infty = 0$ or 1; also $P\{X_\infty = 1\} = E(X_0)$.

   (b)   If $b < a$, show that $\{X_n\}$ is a supermartingale and $X_n \to 0$ a.e.

4   (*Polya urn scheme*) An urn contains white and black balls; one ball is drawn, and then replaced by two of the same color, and the process is repeated. Thus if the urn contains c white and $r - c$ black balls, and a white ball is drawn, the fraction of white balls in the urn before the next drawing is $(c + 1)/(r + 1)$.

If X_n is the fraction of white balls in the urn just before the nth drawing, show that $\{X_n\}$ is a martingale, and X_n converges a.e. to a limit X_∞, where $E(X_\infty) = E(X_1)$.

5.   Martingales may be defined on a measure space $(\Omega, \mathscr{F}, \mu)$ if μ is finite; simply replace μ by the probability measure $P = \mu/[\mu(\Omega)]$. If $\mu(\Omega) = \infty$ we can use 6.3.2(c) in the definition: $\int_A X_n \, d\mu = \int_A X_{n+1} \, d\mu$ for all $A \in \mathscr{F}_n$ (of course, the X_n are still required to be $\mathscr{F}_n$-measurable and integrable). Sub- and supermartingales may be defined similarly.

Show that an L^1 bounded submartingale converges a.e. $[\mu]$ to an integrable limit.

6.5   UNIFORM INTEGRABILITY

We now introduce a concept that has important application to martingale theory, and in fact to integration theory in general.

6.5.1   *Definitions and Comments.*   Let $f_1, f_2, \ldots$ be real- or complex-valued Borel measurable functions on the measure space $(\Omega, \mathscr{F}, \mu)$, μ finite. The f_n are said to be *uniformly integrable* iff

$$\int_{\{|f_n| \ge c\}} |f_n| \, d\mu \to 0 \qquad \text{as} \qquad c \to \infty$$

uniformly in n. (The definition is the same for an uncountable family $\{f_i\}$.)

It is immediate that if the f_n are uniformly integrable, each f_n is integrable. Also, if $|f_n| \le g$ for all n, where g is integrable, in particular, if the f_n are uniformly bounded, then the f_n are uniformly integrable.

Furthermore, if the f_n are uniformly integrable, then $\sup_n \int_\Omega |f_n|\,d\mu < \infty$, because if $\varepsilon > 0$,

$$\int_\Omega |f_n|\,d\mu = \int_{\{|f_n|\geq c\}} |f_n|\,d\mu + \int_{\{|f_n|<c\}} |f_n|\,d\mu \leq \varepsilon + c\mu(\Omega)$$

for large n.

One basic application of uniform integrability is the following extension of Fatou's lemma and the dominated convergence theorem.

6.5.2 Theorem. Let $f_1, f_2, \ldots$ be real-valued and uniformly integrable.

(a) $$\int_\Omega (\liminf_n f_n)\,d\mu \leq \liminf_n \int_\Omega f_n\,d\mu$$

$$\leq \limsup_n \int_\Omega f_n\,d\mu \leq \int_\Omega (\limsup_n f_n)\,d\mu.$$

(b)   If $f_n \to f$ a.e. or in measure, then f is integrable and

$$\int_\Omega f_n\,d\mu \to \int_\Omega f\,d\mu.$$

PROOF.   (a)   We have

$$\int_\Omega f_n\,d\mu = \int_{\{f_n<-c\}} f_n\,d\mu + \int_{\{f_n\geq -c\}} f_n\,d\mu, \qquad c > 0.$$

By uniform integrability, c may be chosen so large that $|\int_{\{f_n\leq -c\}} f_n\,d\mu|$ $< \varepsilon$ for all n, where $\varepsilon > 0$ is preassigned. Since $f_n I_{\{f_n\geq -c\}} \geq -c$, which is integrable since μ is finite, Fatou's lemma yields

$$\liminf_n \int_{\{f_n\leq -c\}} f_n\,d\mu \geq \int_\Omega \liminf_n (f_n I_{\{f_n\geq -c\}})\,d\mu.$$

Since $f_n I_{\{f_n\geq -c\}} \geq f_n$, this integral is in turn greater than or equal to $\int_\Omega (\liminf_n f_n)\,d\mu$. Thus

$$\liminf_n \int_\Omega f_n\,d\mu \geq \int_\Omega \left(\liminf_n f_n\right)\,d\mu - \varepsilon,$$

proving the lim inf part. The lim sup part is done by a symmetrical argument.

(b) This is immediate from (a) if $f_n \to f$ a.e., so assume $f_n \to f$ in measure. By 2.5.3, there is a subsequence $f_{n_k} \to f$ a.e.; hence by (a) applied to the f_{n_k}, f is integrable and $\int_\Omega f_{n_k} \, d\mu \to \int_\Omega f \, d\mu$. If $\int_\Omega f_n \, d\mu$ does not converge to $\int_\Omega f \, d\mu$, then for some $\varepsilon > 0$ we have $\left| \int_\Omega f_n \, d\mu - \int_\Omega f \, d\mu \right| \geq \varepsilon$ for infinitely many n, and for convenience in notation we may assume this holds for all n. But then we find a subsequence $f_{m_j} \to f$ a.e., and as above, $\int_\Omega f_{m_j} \, d\mu \to \int_\Omega f \, d\mu$, a contradiction. □

We now establish a useful criterion for uniform integrability.

6.5.3   Theorem. The complex-valued Borel measurable functions f_n are uniformly integrable iff the integrals $\int_\Omega |f_n| \, d\mu$ are uniformly bounded and also uniformly continuous, that is, $\int_A |f_n| \, d\mu \to 0$ as $\mu(A) \to 0$, uniformly in n.

PROOF.   Assume the integrals are uniformly bounded and uniformly continuous. Then

$$\mu\{|f_n| \geq c\} \leq \frac{1}{c} \int_\Omega |f_n| \, d\mu$$

by Chebyshev's inequality, and this approaches 0 as $c \to \infty$, uniformly in n, by the uniform boundedness. Thus $\int_{\{|f_n| \geq c\}} |f_n| \, d\mu \to 0$ as $c \to \infty$, uniformly in n, by the uniform continuity.

Conversely, assume uniform integrability. We have

$$\int_A |f_n| \, d\mu = \int_{A \cap \{|f_n| \geq c\}} |f_n| \, d\mu + \int_{A \cap \{|f_n| < c\}} |f_n| \, d\mu$$

$$\leq \int_{\{|f_n| \geq c\}} |f_n| \, d\mu + c\mu(A). \tag{1}$$

Choose c so that $\int_{\{|f_n| \geq c\}} |f_n| \, d\mu < \varepsilon/2$ for all n; if $\mu(A) < \varepsilon/2c$, then by (1), $\int_A |f_n| \, d\mu < (\varepsilon/2) + (\varepsilon/2) = \varepsilon$ for all n, proving uniform continuity. Uniform boundedness was verified in 6.5.1. □

We have seen in 2.5.1 that L^p convergence implies convergence in measure. The converse holds under an additional hypothesis of uniform integrability.

6.5.4   Theorem. Let μ be a finite measure on $(\Omega, \mathscr{F})$, and let $0 < p < \infty$. If $f_n \xrightarrow{\mu} f$ and the $|f_n|^p$ are uniformly integrable, then $f_{n_i} \xrightarrow{L^p} f$.

PROOF.   First assume that the $|f_n - f|^p$ are uniformly integrable. By 2.5.3, there is a subsequence $\{f_{n_k}\}$ converging to f a.e. and in measure. By 6.5.2(b),

$\int_{\Omega} |f_{n_k} - f|^p \, d\mu \to 0$ as $k \to \infty$. The same argument shows that *any* subsequence of $\{f_n\}$ has a subsequence converging to f in L^p. Hence $f_n \xrightarrow{L^p} f$, for if not, there would be an $\varepsilon > 0$ and a subsequence $\{f_{n_i}\}$ such that $\int_{\Omega} |f_{n_i} - f|^p \, d\mu \geq \varepsilon$ for all i.

Now assume the $|f_n|^p$ to be uniformly integrable. Then $|f_n - f|^p \leq |f_n|^p + |f|^p$ if $p \leq 1$, and $|f_n - f|^p \leq 2^{p-1}(|f_n|^p + |f|^p)$ if $p \geq 1$. (See 2.4.6 and the end of 2.4.12.) As above, we have a subsequence $f_{n_k} \to f$ a.e. By 6.5.2(b), $|f|^p$ is integrable, and it follows that the integrals $\int_{\Omega} |f_n - f|^p \, d\mu$ are uniformly bounded and uniformly continuous. [Note that $\int_A |f|^p \, d\mu \to 0$ as $\mu(A) \to 0$ by 2.2.5(e).] By 6.5.3, the $|f_n - f|^p$ are uniformly integrable, and the previous argument applies. $\square$

6.5.5   Corollary. In 6.5.2(b), $f_n \xrightarrow{L^1} f$, that is, $\int_{\Omega} |f_n - f| \, d\mu \to 0$.

PROOF.   The $|f_n|$ are uniformly integrable by hypothesis, and $f_n \xrightarrow{\mu} f$ either by hypothesis or by 2.5.5 $\square$

The following result will be needed in the next section.

6.5.6   Lemma. Let f_i, $i \in I$, be integrable functions on $(\Omega, \mathscr{F}, \mu)$. If $h \colon [0, \infty) \to [0, \infty)$ is Borel measurable, $h(t)/t \to \infty$ as $t \to \infty$, and $\sup_{i \in I} \int_{\Omega} h(|f_i|) \, d\mu < \infty$, then the f_i are uniformly integrable. For example, take $h|t| = t^p$, $p > 1$; thus if $\int_{\Omega} |f_n|^p \, d\mu \leq M < \infty$ for all n, then the f_n are uniformly integrable.

PROOF.   Given $\varepsilon > 0$, let $M = \sup_{i \in I} \int_{\Omega} h(|f_i|) \, d\mu$ and set $a = M/\varepsilon$. There is a positive number c such that $h(t)/t \geq a$ for $t \geq c$. Thus

$$\int_{\{|f_i| \geq c\}} |f_i| \, d\mu \leq \int_{\{|f_i| \geq c\}} \frac{h(|f_i|)}{a} \, d\mu$$

$$\leq \int_{\Omega} \frac{h(|f_i|)}{a} \, d\mu \leq \frac{M}{a} = \varepsilon. \quad \square$$

Problems

1.   If $0 < p < \infty$, $f, f_1, f_2, \ldots \in L^p$, and $f_n \xrightarrow{L^p} f$, show that the $|f_n|^p$ are uniformly integrable.

2.   Give an example of a uniformly integrable sequence of functions f_n on a measure space $(\Omega, \mathscr{F}, \mu)$, μ finite, such that the $|f_n|$ cannot be bounded above by an integrable function.

6.6   Uniform Integrability and Martingale Theory

The application of the uniform integrability concept yields many additional facts about martingales. In particular, we shall find that a martingale is uniformly integrable iff it can be represented as $X_n = E(Y|\mathscr{F}_n)$ for some integrable random variable Y. Thus Y can be considered a "last element" of the martingale $\{X_n, \mathscr{F}_n\}$.

6.6.1   Lemma. Let Y be an integrable random variable on $(\Omega, \mathscr{F}, P)$ and let $\mathscr{G}_i$, $i \in I$, arbitrary sub σ-fields of $\mathscr{F}$. Then the random variables $X_i = E(Y|\mathscr{G}_i)$, $i \in I$, are uniformly integrable, that is,

$$\int_{\{|X_i| \geq c\}} |X_i| \, dP \to 0 \qquad \text{as} \qquad c \to \infty$$

uniformly in i.

PROOF.   As $|E(Y|\mathscr{G}_i)| \leq E(|Y| \, | \mathscr{G}_i)$,

$$\int_{\{|X_i| \geq c\}} |X_i| \, dP \leq \int_{\{|X_i| \geq c\}} E(|Y| \, | \mathscr{G}_i) \, dP = \int_{\{|X_i| \geq c\}} |Y| \, dP$$

since $\{|X_i| \geq c\} \in \mathscr{G}_i$ (remember X_i is $\mathscr{G}_i$-measurable). But by Chebyshev's inequality,

$$P\{|X_i| \geq c\} \leq c^{-1} E(|X_i|) \leq c^{-1} E[E(|Y| \, | \mathscr{G}_i)] = c^{-1} E(|Y|) \to 0$$

$$\text{as} \qquad c \to \infty$$

uniformly in i. $\square$

The following result, due to Lévy (1937), was historically the first of the martingale convergence theorems.

6.6.2   Theorem. Let $\{\mathscr{F}_n\}$ be an increasing sequence of sub σ-fields of $\mathscr{F}$, and let $\mathscr{F}_\infty$ be the σ-field generated by $\bigcup_{n=1}^\infty \mathscr{F}_n$. If Y is integrable and $X_n = E(Y|\mathscr{F}_n)$, $n = 1, 2, \ldots$, then $X_n \to E(Y|\mathscr{F}_\infty)$ a.e. and in L^1.

PROOF. $\{X_n, \mathscr{F}_n\}$ is a martingale by 6.3.3(c), and is uniformly integrable by 6.6.1. Since $E(|X_n|) \leq E(|Y|) < \infty$, X_n converges a.e. to an integrable random variable X_∞, by 6.4.3; L^1 convergence follows from 6.5.5. It remains to show that $X_\infty = E(Y|\mathscr{F}_\infty)$ a.e.

If $A \in \mathscr{F}_n$, then

$$\int_A Y \, dP = \int_A E(Y|\mathscr{F}_n) \, dP = \int_A X_n \, dP \to \int_A X_\infty \, dP$$

by L^1 convergence. Thus $\int_A Y\,dP = \int_A X_\infty\,dP$ for all A in the field $\bigcup_{n=1}^\infty \mathscr{F}_n$, and hence for all $A \in \mathscr{F}_\infty$ (monotone class theorem). Since X_n is $\mathscr{F}_n \subset \mathscr{F}_\infty$-measurable, X_∞ is $\mathscr{F}_\infty$-measurable, and therefore $X_\infty = E(Y|\mathscr{F}_\infty)$ a.e. by definition of conditional expectation. $\square$

A result similar to 6.6.2 holds if the σ-fields form a decreasing sequence.

6.6.3 Theorem. Let $\{\mathscr{F}_n\}$ be a decreasing sequence of sub σ-fields of $\mathscr{F}$, and let $\mathscr{F}_\infty = \bigcap_{n=1}^\infty \mathscr{F}_n$. If Y is integrable and $X_n = E(Y|\mathscr{F}_n)$, $n = 1, 2, \ldots$, then $X_n \to E(Y|\mathscr{F}_\infty)$ a.e. and in L^1.

PROOF. Just as in 6.6.2 (using 6.4.4 instead of 6.4.3), $X_n \to X_\infty$ a.e. and in L^1, so we must show that $X_\infty = E(Y|\mathscr{F}_\infty)$ a.e. If $A \in \mathscr{F}_\infty \subset \mathscr{F}_n$, then

$$\int_A Y\,dP = \int_A E(Y|\mathscr{F}_n)\,dP = \int_A X_n\,dP \to \int_A X_\infty\,dP.$$

Since X_n is $\mathscr{F}_n \subset \mathscr{F}_k$-measurable for $n \geq k$, X_∞ is $\mathscr{F}_k$-measurable for all k; hence X_∞ is $\mathscr{F}_\infty$- measurable. $\square$

6.6.4 Comments. Let $Z_i\colon (\Omega, \mathscr{F}) \to (\Omega_i', \mathscr{F}_i')$ be a random object ($i = 1, 2, \ldots$). In 6.6.2, if $\mathscr{F}_n = \sigma(Z_1, \ldots, Z_n)$, then $\mathscr{F}_\infty = \sigma(Z_1, Z_2, \ldots)$ (see Problem 1). Thus $E(Y|Z_1, \ldots, Z_n) \to E(Y|Z_1, Z_2, \ldots)$ a.e. and in L^1. In 6.6.3, if $\mathscr{F}_n = \sigma(Z_n, Z_{n+1}, \ldots)$, then $\mathscr{F}_\infty$ is the tail σ-field of the Z_n (see 6.2.6).

We now show that uniform integrability of a submartingale implies a.e. and L^1 convergence, and also implies that a last element can be attached.

6.6.5 Theorem. Let $\{X_n, \mathscr{F}_n, n = 1, 2, \ldots\}$ be a uniformly integrable submartingale. Then $\sup_n E(X_n{}^+) < \infty$, and X_n converges to a limit X_∞ a.e. and in L^1. Furthermore, if $\mathscr{F}_\infty$ is the σ-field generated by $\bigcup_{n=1}^\infty \mathscr{F}_n$, then $\{X_n, \mathscr{F}_n, n = 1, 2, \ldots, \infty\}$ is a submartingale. If $\{X_n, \mathscr{F}_n, n = 1, 2, \ldots\}$ is a uniformly integrable martingale, so is $\{X_n, \mathscr{F}_n, n = 1, 2, \ldots, \infty\}$. [If $\{X_n, \mathscr{F}_n, n = 1, 2, \ldots, \infty\}$ is a (sub- or super-) martingale, where $\mathscr{F}_\infty$ is the σ-field generated by the $\mathscr{F}_n$, X_∞ is said to be a *last element*.]

PROOF. By 6.5.3, $\sup_n E(|X_n|) < \infty$, so by 6.4.3, $X_n \to X_\infty$ a.e. By 6.5.5, $X_n \xrightarrow{L^1} X_\infty$.

Now if $A \in \mathscr{F}_n$ and $k \geq n$, then by 6.3.2(c), $\int_A X_n\,dP \leq \int_A X_k\,dP$. Let $k \to \infty$; the L^1-convergence yields $\int_A X_n\,dP \leq \int_A X_\infty\,dP$. Thus [6.3.2(c) again] $X_n \leq E(X_\infty|\mathscr{F}_n)$; hence $\{X_n, \mathscr{F}_n, n = 1, 2, \ldots, \infty\}$ is a submartingale. The last statement follows from the above argument with "$\leq$" replaced by "$=$." $\square$

Theorem 6.6.5 indicates that uniform integrability of a submartingale implies some rather strong conclusions. In fact, a uniformly integrable martingale must have a last element.

6.6.6 Theorem. $\{X_n, \mathscr{F}_n, n = 1, 2, \ldots\}$ is a uniformly integrable martingale iff there is an integrable random variable Y such that $X_n = E(Y|\mathscr{F}_n)$, $n = 1, 2, \ldots$; in this case, $X_n \to E(Y|\mathscr{F}_\infty)$ a.e. and in L^1, where $\mathscr{F}_\infty$ is the σ-field generated by $\bigcup_{n=1}^\infty \mathscr{F}_n$.

PROOF. The "if" part follows from 6.6.1 and 6.6.2. The "only if" part follows from 6.6.5 with $Y = X_\infty$. □

In 6.6.6, if we require that Y be $\mathscr{F}_\infty$-measurable, then Y is unique (up to sets of measure 0). For if $X_n = E(Y|\mathscr{F}_n)$, then $E(Y|\mathscr{F}_n) = E(X_\infty|\mathscr{F}_n)$, $n = 1, 2, \ldots$, so $\int_A Y\,dP = \int_A X_\infty\,dP$ for all $A \in \bigcup_{n=1}^\infty \mathscr{F}_n$, and hence for all $A \in \mathscr{F}_\infty$ (monotone class theorem). Thus $Y = X_\infty$ a.e. by 1.6.11.

A sub- or supermartingale with a last element need not be uniformly integrable, but there are partial results in the direction.

6.6.7 Theorem. Let $\{X_n, \mathscr{F}_n, n = 1, 2, \ldots, \infty\}$ be a nonnegative submartingale with a last element. Then the X_n are uniformly integrable.

PROOF.

$$\int_{\{X_n \geq c\}} X_n\,dP \leq \int_{\{X_n \geq c\}} X_\infty\,dP,$$

and

$$P\{X_n \geq c\} \leq \frac{E(X_n)}{c} \leq \frac{E(X_\infty)}{c} \to 0 \qquad \text{as} \qquad c \to \infty$$

uniformly in n. □

We now give an example of a supermartingale $\{X_n, \mathscr{F}_n\}$ with a last element that is not uniformly integrable. (Also, $\{-X_n, \mathscr{F}_n\}$ will be a submartingale with a last element that is not uniformly integrable.) The key feature of this example is that $\lim_{n\to\infty} X_n$ will be a last element when the sequence is regarded as a supermartingale, but not when it is regarded as a martingale.

6.6.8 Example. Let $Y_1, Y_2, \ldots$ be independent, with $P\{Y_j = 1\} = p$, $P\{Y_j = 0\} = 1 - p, 0 < p < 1$. Let $X_n = p^{-n}\prod_{j=1}^n Y_j$, $\mathscr{F}_n = \mathscr{F}(Y_1, \ldots, Y_n)$. Then $\{X_n, \mathscr{F}_n, n = 1, 2, \ldots\}$ is a martingale, and hence a supermartingale [see 6.3.3(b)]. Since all $X_n \geq 0$, we have $E(0|\mathscr{F}_n) = 0 \leq X_n$, so 0 is a last element when the sequence is regarded as a supermartingale, but not when it is regarded as a martingale. But the X_n are not uniformly integrable;

for $P\{Y_j = 1$ for all $j\} = \lim_{n\to\infty} p^n = 0$; hence (a.e.) $X_n = 0$ eventually, so $X_n \to 0$ a.e. If the X_n were uniformly integrable, then $X_n \xrightarrow{L^1} 0$ by 6.5.5; hence $E(X_n) \to 0$. But $E(X_n) = 1$ for all n, a contradiction. If we regard the sequence as a martingale, there can be no last element. For if X_∞ is a last element, then $X_n = E(X_\infty | \mathscr{F}_n)$ for all n; hence by 6.6.1, the X_n are uniformly integrable.

Since (a.e.) $X_n = 0$ eventually, we have an example of a "fair" game in which the gambler is almost certain to be wiped out. Thus the term "locally fair" is perhaps more appropriate than "fair."

Note that a sub- or supermartingale with a last element converges a.e. [In the submartingale case, for example, $\sup_n E(X_n^+) \le E(X_\infty^+) < \infty$.] But the limit need not coincide with the last element.

We now look at the problem of L^p convergence.

6.6.9 Theorem. Let $\{X_n, \mathscr{F}_n, n = 1, 2, \ldots\}$ be a martingale or a nonnegative submartingale with $E[|X_n|^p] \le M < \infty$ for all n, where $p > 1$. Then X_n converges to a limit X_∞ a.e. and in L^p.

PROOF. By 6.5.6, the X_n are uniformly integrable, so by 6.6.5, X_n converges a.e. to a limit X_∞, and X_∞ is a last element.

Now $\{|X_n|^p, \mathscr{F}_n, n = 1, 2, \ldots, \infty\}$ is a nonnegative submartingale. [In the case in which $\{X_n, \mathscr{F}_n\}$ is a martingale, use 6.3.6(b) with $g(x) = |x|^p$; in the nonnegative submartingale case, use 6.3.6(a) with $g(x) = x^p, x \ge 0$; $g(x) = 0, x < 0$.] By 6.6.7, the $|X_n|^p$ are uniformly integrable, and by 6.5.4, $X_n \to X_\infty$ in L^p. □

Problems

1. Let $Z_n\colon (\Omega, \mathscr{F}) \to (\Omega_n', \mathscr{F}_n'), n = 1, 2, \ldots,$ be random objects. If $\mathscr{F}_n = \sigma(Z_1, \ldots, Z_n), n = 1, 2, \ldots,$ show that the σ-field generated by $\bigcup_{n=1}^{\infty} \mathscr{F}_n$ is $\sigma(Z_1, Z_2, \ldots)$.

2. Let $\{X_n, \mathscr{F}_n\}$ be a nonnegative supermartingale, so that X_n converges a.e. to an integrable random variable X_∞.

   (a) Show that $E(X_n) \to E(X_\infty)$ iff the X_n are uniformly integrable. (This holds for any sequence of nonnegative integrable random variables converging a.e. to an integrable limit; the supermartingale property is not involved. Note also that we have $E[|X_n - X_\infty|] \to 0$ since we have a.e. convergence and uniform integrability.)

   (b) Show that X_∞ is a last element.

   (c) If $E(X_n) \to 0$, show that $X_n \to 0$ a.e.

3. Let $\{X_n, \mathcal{F}\}$, $n = 1, 2, \ldots$ be a submartingale (respectively, a martingale), and let $\mathcal{F}_\infty$ be the σ-field generated by the $\mathcal{F}_n$. If there is an $\mathcal{F}$-measurable, integrable random variable Y such that $E(Y|\mathcal{F}_n) \geq X_n$ a.e., $[E(Y|\mathcal{F}_n) = X_n$ a.e. in the martingale case], then there is an $\mathcal{F}_\infty$-measurable, integrable random variable X_∞ such that $E(X_\infty|\mathcal{F}_n) \geq X_n$ a.e. $[E(X_\infty|\mathcal{F}_n) = X_n$ a.e. in the martingale case.]

6.7 OPTIONAL SAMPLING THEOREMS

Let $\{X_n, \mathcal{F}_n, n = 1, 2, \ldots\}$ be a martingale, with X_n interpreted as a gambler's total capital after n plays of a game of chance. Suppose that after each trial, the gambler decides either to quit or to keep playing. If T is the time of quitting, what can be said about the final capital X_T?

First of all, the random variable T must have the property that if we observe $X_1, \ldots, X_n$, we can come to a definite decision as to whether or not $T = n$. A nonnegative random variable of this type is called a *stopping time*.

6.7.1 Definition. Let $\{\mathcal{F}_n, n = 0, 1, \ldots\}$ be an increasing sequence of sub σ-fields of $\mathcal{F}$. A *stopping time* for the $\mathcal{F}_n$ is a map $T: \Omega \to \{0, 1, \ldots, \infty\}$ such that $\{T \leq n\} \in \mathcal{F}_n$ for each nonnegative integer n. Since $\{T = n\} = \{T \leq n\} - \{T \leq n - 1\}$ and $\{T \leq n\} = \bigcup_{k=0}^{n}\{T = k\}$, the definition is equivalent to the requirement that $\{T = n\} \in \mathcal{F}_n$ for all $n = 0, 1, \ldots$. If $\{X_n, n = 0, 1, \ldots\}$ is a sequence of random variables, a stopping time for $\{X_n\}$ is, by definition, a stopping time relative to the σ-fields $\mathcal{F}_n = \sigma(X_0, \ldots, X_n)$. (The above definitions are modified in the obvious way if the index n starts from 1 rather than 0).

If S and T are stopping times, so are $S \vee T = \max(S, T)$ and $S \wedge T = \min(S, T)$. ($\{S \vee T \leq n\} = \{S \leq n\} \cap \{T \leq n\}, \{S \wedge T \leq n\} = \{S \leq n\} \cup \{T \leq n\}$. Also, if $T \equiv n$ then T is a stopping time.)

By far the most important example of a stopping time is the *hitting time* of a set. If $\{X_n\}$ is a sequence of random variables and $B \in \mathcal{B}(\mathbb{R})$, let $T(\omega) = \min\{n: X_n(\omega) \in B\}$ if $X_n(\omega) \in B$ for some n; $T(\omega) = \infty$ if $X_n(\omega)$ is never in B. T is a stopping time since $\{T \leq n\} = \bigcup_{k \leq n}\{X_k \in B\} \in \mathcal{F}(X_k, k \leq n)$.

If T is a stopping time for $\{X_n\}$, an event A is said to be "prior to T" iff, whenever $T = n$, we can tell by examination of the $X_k, k \leq n$, whether or not A has occurred. The formal definition is as follows.

6.7.2 Definition. Let T be a stopping time for the σ-fields $\mathcal{F}_n, n = 0, 1, \ldots$, and let A belong to $\mathcal{F}$. The set A is said to be *prior to T* iff $A \cap \{T \leq n\} \in \mathcal{F}_n$ for all $n = 0, 1, \ldots$ [Equivalently, as in 6.7.1, $A \cap \{T = n\} \in \mathcal{F}_n$ for all $n = 0, 1, \ldots$.] The collection of all sets prior to T will be denoted by $\mathcal{F}_T$; it follows quickly that $\mathcal{F}_T$ is a σ-field. Also, if $T \equiv n$ then $\mathcal{F}_T$ is simply $\mathcal{F}_n$.

If S and T are stopping times and $S \leq T$, then $\mathscr{F}_S \subset \mathscr{F}_T$. For if $A \in \mathscr{F}_S$ then

$$A \cap \{T \leq k\} = \bigcup_{i=1}^{k} [A \cap \{S = i\}] \cap \{T \leq k\}.$$

But $A \cap \{S = i\} \in \mathscr{F}_i \subset \mathscr{F}_k$, and $\{T \leq k\} \in \mathscr{F}_k$; hence $A \in \mathscr{F}_T$.

If the stopping time T is constant at n, then X_T is $\mathscr{F}_T$-measurable. We would like this idea to carry over to a general stopping time. Formally, let T be a finite stopping time for the σ-fields $\mathscr{F}_n$, and define X_T in the natural way; if $T(\omega) = n$, let $X_T(\omega) = X_n(\omega)$. If $B \in \mathscr{B}(\mathbb{R})$, then $\{X_T \in B\} \in \mathscr{F}_T$, in other words, X_T is $\mathscr{F}_T$-measurable. (Since $\mathscr{F}_T \subset \mathscr{F}$ by definition, if follows in particular that X_T is a random variable.) To see this, write

$$\{X_T \in B\} \cap \{T \leq n\} = \bigcup_{k=0}^{n} [\{X_k \in B\} \cap \{T = k\}].$$

Since $\{X_k \in B\} \cap \{T = k\} \in \mathscr{F}_k$ for $k \leq n$, we have

$$\{X_T \in B\} \cap \{T \leq n\} \in \mathscr{F}_n.$$

Also, as T is finite, we have $\bigcup_{n=0}^{\infty} \{T \leq n\} = \Omega$, so that $\{X_T \in B\} \in \mathscr{F}$, as desired.

If T is not necessarily finite, the same argument shows that $I_{\{T < \infty\}} X_T$ is $\mathscr{F}_T$-measurable.

Now in the gambling situation described at the beginning of the section, a basic quantity of interest is $E(X_T)$, the average accumulation at the quitting time. For example, if $E(X_T)$ turns out to be the same as $E(X_1)[= E(X_n)$ for all n by the martingale property], the gambler's strategy does not offer any improvement over the procedure of stopping at a fixed time. Now in comparing X_1 and X_T we are considering two stopping times S and T $(S \equiv 1)$ with $S \leq T$, and looking at X_S versus X_T. More generally, if $T_1 \leq T_2 \leq \cdots$ form an increasing sequence of finite stopping times, we may examine the sequence $X_{T_1}, X_{T_2}, \ldots$. If the sequence forms a martingale, then $E(X_{T_n}) = E(X_{T_1})$ for all n, and if $T_1 \equiv 1$, then $E(X_{T_1}) = E(X_1)$.

Thus if we sample the gambler's fortune at random times $T_1, T_2, \ldots$, the basic question is whether the martingale (or submartingale) property is preserved. This will always be the case when the sequence $\{X_n\}$ is finite.

6.7.3 Theorem. Let $\{X_n, \mathscr{F}_n, n = 1, \ldots, m\}$ be a submartingale, and let T_1, $T_2, \ldots$ be an increasing sequence of stopping times for the $\mathscr{F}_n$. [In other

words, the T_n take values in $\{1, \ldots, m\}$, and $\{T_n \leq k\} \in \mathscr{F}_k$, $k = 1, \ldots, m$.]
The σ-fields $\mathscr{F}_{T_i}$ are defined as before:

$$\mathscr{F}_{T_i} = \{A \in \mathscr{F} \colon A \cap \{T_i \leq k\} \in \mathscr{F}_k, k = 1, \ldots, m\}.$$

Then the X_{T_n} form a submartingale relative to the σ-fields $\mathscr{F}_{T_i}$, a martingale
if $\{X_n\}$ is a martingale.

PROOF.   We follow Breiman (1968). Define $Y_n = X_{T_n}$, and note that each Y_n
is integrable:

$$\int_\Omega |X_{T_n}| \, dP = \sum_{i=1}^m \int_{\{T_n=i\}} |X_i| \, dP \leq \sum_{i=1}^m E(|X_i|) < \infty.$$

As the T_n increase with n, so do the $\mathscr{F}_{T_i}$ (see the discussion after 6.7.2).

Now if $A \in \mathscr{F}_{T_n}$, we must show that $\int_A Y_{n+1} \, dP \geq \int_A Y_n \, dP$ (with equality
in the martingale case). Since $A = \bigcup_j [A \cap \{T_n = j\}]$, it suffices to replace A
by $D_j = A \cap \{T_n = j\}$, which belongs to $\mathscr{F}$; now if $k > j$, we note that $T_n = j$
implies $T_{n+1} \geq j$, so that

$$\int_{D_j} Y_{n+1} \, dP = \sum_{i=j}^k \int_{D_j \cap \{T_{n+1}=i)\}} Y_{n+1} \, dP + \int_{D_j \cap \{T_{n+1}>k\}} Y_{n+1} \, dP.$$

Thus

$$\int_{D_j} Y_{n+1} \, dP = \sum_{i=j}^k \int_{D_j \cap \{T_{n+1}=i\}} X_i \, dP + \int_{D_j \cap \{T_{n+1}>k\}} X_k \, dP$$

$$- \int_{D_j \cap \{T_{n+1}>k\}} (X_k - Y_{n+1}) \, dP. \tag{1}$$

Now combine the $i = k$ term in (1) with the $\int X_k \, dP$ term to obtain

$$\int_{D_j \cap \{T_{n+1}=k\}} X_k \, dP + \int_{D_j \cap \{T_{n+1}>k\}} X_k \, dP = \int_{D_j \cap \{T_{n+1} \geq k\}} X_k \, dP$$

$$\geq \int_{D_j \cap \{T_{n+1} \geq k\}} X_{k-1} \, dP$$

since $\{T_{n+1} \geq k\} = \{T_{n+1} \leq k - 1\}^c \in \mathscr{F}_{k-1}$ and $D_j \in \mathscr{F}_j \subset \mathscr{F}_{k-1}$. But

$$\int_{D_j \cap \{T_{n+1} \geq k\}} X_{k-1} \, dP = \int_{D_j \cap \{T_{n+1}>k-1\}} X_{k-1} \, dP,$$

so this term may be combined with the $i = k - 1$ term of (1) to obtain

$$\int_{D_j \cap \{T_{n+1}>k-2\}} X_{k-2} \, dP.$$

Proceeding inductively, we find

$$\int_{D_j} Y_{n+1}\, dP \geq \int_{D_j \cap \{T_{n+1} \geq j\}} X_j\, dP - \int_{D_j \cap \{T_{n+1} > k\}} (X_k - Y_{n+1})\, dP. \qquad (2)$$

Now $\{T_{n+1} > k\}$ is empty for $k \geq m$. Finally, $D_j \cap \{T_{n+1} \geq j\} = D_j$ since $D_j \subset \{T_n = j\}$, and $X_j = Y_n$ on D_j. Thus

$$\int_{D_j} Y_{n+1}\, dP \geq \int_{D_j} Y_n\, dP$$

as desired. In the martingale case, all inequalities in the proof become equalities. □

Theorem 6.7.3 extends immediately to the case of an infinite sequence if each T_n is bounded, that is, for each n there is a positive constant K_n such that $T_n \leq K_n$ a.e. The same proof may be used; the key point is that $\{T_{n+1} > k\}$ is still empty for sufficiently large k.

When $\{X_n\}$ is an infinite sequence, the martingale or submartingale property is not preserved in general, but the following result gives useful sufficient conditions.

6.7.4 Optional Sampling Theorem. Let $\{X_1, X_2 \cdots\}$ be a submartingale, and let $T_1, T_2, \ldots$ be an increasing sequence of finite stopping times for $\{X_n\}$, with $Y_n = X_{T_n}, n = 1, 2, \ldots$. If

(A) $E(|Y_n|) < \infty$ for all n, and
(B) $\liminf_{k \to \infty} \int_{\{T_n > k\}} |X_k|\, dP = 0$ for all n,

then $\{Y_n\}$ is a submartingale relative to the σ-fields $\mathscr{F}_{T_n}$. If $\{X_n\}$ is a martingale, so is $\{Y_n\}$.

PROOF.   Since integrability of the Y_n is now hypothesis (A), we can follow the proof of 6.7.3 to (2). The first integral on the right-hand side is $\int_{D_j} Y_n\, dP$ as before, but in the second integral, we no longer have $\{T_{n+1} > k\}$ empty for large k. But by hypothesis (B), $\int_{D_j \cap \{T_{n+1} > k\}} X_k\, dP \to 0$ as $k \to \infty$ through an appropriate subsequence, and $\int_{D_j \cap \{T_{n+1} > k\}} Y_{n+1}\, dP \to 0$ as $k \to \infty$ since $\{T_{n+1} > k\}$ decreases to the empty set. Thus

$$\int_{D_j} Y_{n+1}\, dP \geq \int_{D_j} Y_n\, dP$$

as desired. As before, all inequalities become equalities in the martingale case. □

If $\{X_n\}$ is a submartingale with a last element X_∞, we can define the random variable X_T for *any* stopping time T. On the set $\{T = \infty\}$, $X_T = X_\infty$. In this case, the optional sampling theorem holds.

6.7.5   Theorem.   If $\{T_n\}$ is an increasing sequence of stopping times (not necessarily finite) for a submartingale $\{X_n\}$ having a last element X_∞, then $\{X_{T_n}\}$ is a submartingale relative to the σ-fields $\mathscr{F}_{T_n}$; if $\{X_n\}$ is a martingale, so is $\{X_{T_n}\}$. In particular, this holds if the X_n are uniformly integrable (see 6.6.5).

PROOF.   *Case 1:* $X_n \le 0$ for all n, and $X_\infty = 0$. For any fixed n, let $S_k = T_n \wedge k = \min(T_n, k)$, $k = 1, 2, \ldots$; it is easily checked that S_k is a stopping time for the X_n. Now, if T_n is finite, $X_{S_k} \to Y_n = X_{T_n}$ as $k \to \infty$; hence by Fatou's lemma,

$$\limsup_{k \to \infty} \int_\Omega X_{S_k}\, dP \le \int_\Omega Y_n\, dP \le 0.$$

The same conclusion holds for arbitrary T_n, because on the set $\{T_n = \infty\}$ we have $X_{S_k} = X_k \le 0 = X_{T_n}$. But by 6.7.3, $\{X_{S_k}\}$ is a submartingale; hence

$$\int_\Omega X_{S_k}\, dP \ge \int_\Omega X_{S_1}\, dP = \int_\Omega X_1\, dP,$$

which is finite. Therefore Y_n is integrable.

Again by 6.7.3, $\{X_{T_n \wedge k}, \mathscr{F}_{T_n \wedge k}, n = 1, 2, \ldots\}$ is a submartingale (k fixed). Thus

$$\int_A X_{T_n \wedge k}\, dP \le \int_A X_{T_{n+1} \wedge k}\, dP$$

if $A \in \mathscr{F}_{T_n \wedge k}$. But if $A \in \mathscr{F}_{T_n}$, then $A \cap \{T_n \le k\} \in \mathscr{F}_{T_n \wedge k}$, for

$$A \cap \{T_n \le k\} \cap \{T_n \wedge k \le i\} = \begin{cases} A \cap \{T_n \le i\} & \text{for} \quad i \le k, \\ A \cap \{T_n \le k\} & \text{for} \quad i > k. \end{cases}$$

Thus

$$\int_{A \cap \{T_n \le k\}} X_{T_n \wedge k}\, dP \le \int_{A \cap \{T_n \le k\}} X_{T_{n+1} \wedge k}\, dP, \qquad A \in \mathscr{F}_{T_n}.$$

But on $\{T_n \le k\}$, $T_n \wedge k = T_n$; also, $\{T_{n+1} \le k\} \subset \{T_n \le k\}$ and $X_{T_{n+1} \wedge k} \le 0$; hence

$$\int_{A \cap \{T_n \le k\}} X_{T_n}\, dP \le \int_{A \cap \{T_{n+1} \le k\}} X_{T_{n+1}}\, dP.$$

Let $k \to \infty$ to obtain

$$\int_{A \cap \{T_n < \infty\}} X_{T_n} \, dP \le \int_{A \cap \{T_{n+1} < \infty\}} X_{T_{n+1}} \, dP$$

As $X_{T_n} = 0$ on $\{T_n = \infty\}$ and $X_{T_{n+1}} = 0$ on $\{T_{n+1} = \infty\}$, we get

$$\int_A X_{T_n} \, dP \le \int_A X_{T_{n+1}} \, dP,$$

which is the desired submartingale property.

Case 2: $X_n = E(X_\infty | \mathscr{F}_n)$, $n = 1, 2, \ldots$. In this case, $X_{T_n \wedge k}$ $= E(X_k | \mathscr{F}_{T_n \wedge k}) = E(X_\infty | \mathscr{F}_{T_n \wedge k})$ by 6.7.3 and 5.5.10(a). Therefore, the $X_{T_n \wedge k}$, $k = 1, 2, \ldots$ are uniformly integrable by 6.6.1.

Now if $B \in \mathscr{F}_{T_n}$ then $B \cap \{T_n \le k\} \in \mathscr{F}_{T_n \wedge k}$ (see case 1). Thus

$$\int_{B \cap \{T_n \le k\}} X_{T_n \wedge k} \, dP = \int_{B \cap \{T_n \le k\}} X_\infty \, dP.$$

Let $k \to \infty$ and use the uniform integrability of the $X_{T_n \wedge k}$ to obtain

$$\int_{B \cap \{T_n < \infty\}} X_{T_n} \, dP = \int_{B \cap \{T_n < \infty\}} X_\infty \, dP.$$

But on $\{T_n = \infty\}$ we have $X_{T_n} = X_\infty$, so

$$\int_B X_{T_n} \, dP = \int_B X_\infty \, dP \qquad \text{for every} \qquad B \in \mathscr{F}_{T_n}.$$

Therefore $X_{T_n} = E(X_\infty | \mathscr{F}_{T_n})$, so that $X_{T_1}, X_{T_2}, \ldots$ is a martingale.

General Case: Write $X_n = X_n' + X_n''$, where $X_n' = X_n - E(X_\infty | \mathscr{F}_n)$, $X_n'' = E(X_\infty | \mathscr{F}_n)$. The X_n' fall into case 1 and the X_n'' into case 2, and the result follows. Note that if $\{X_n\}$ is a martingale with last element X_∞, we must have $X_n = E(X_\infty | \mathscr{F}_n)$ so that $\{Y_n\}$ is a martingale by the analysis of case 2. □

To conclude this section we give an example of a situation in which the optional sampling theorem does not apply. Consider the problem of fair coin tossing, that is, let $Y_1, Y_2, \ldots$ be independent random variables, each taking on values ± 1 with equal probability. If $X_n = Y_1 + \cdots + Y_n$, the X_n form a martingale by 6.3.3(a). Now with probability 1, $X_n = 1$ for some n. (This is a standard random walk result; for a proof, see Ash, 1970, p. 185.) If T is the time that 1 is reached (the hitting time for $\{1\}$), and $S \equiv 1$, then S and T are (a.e.) finite stopping times, but $\{X_S, X_T\}$ is not a martingale. For if this were the case, we would have $E(X_T) = E(X_S) = E(X_1) = 0$. But $X_T \equiv 1$; hence $E(X_T) = 1$, a contradiction. In addition, we obtain from 6.7.5 the result that the X_n are not uniformly integrable.

Problems

1. Let $\{X_1, \ldots, X_n\}$ be a submartingale, and let T be a stopping time for $\{X_i, 1 \le i \le n\}$. Show that

    $$E(|X_T|) \le 2E(X_n^+) - E(X_1).$$

    The corresponding result for supermartingales, which may be obtained by replacing X_i by $-X_i$, is

    $$E(|X_T|) \le 2E(X_n^-) + E(X_1).$$

2. Let $\{X_1, X_2, \ldots\}$ be a submartingale, and T a finite stopping time for $\{X_n\}$. Show that

    $$E(|X_T|) \le 2 \sup_n E(X_n^+) - E(X_1).$$

    As in Problem 1, the analogous result for supermartingales is

    $$E(|X_T|) \le 2 \sup_n E(X_n^-) + E(X_1).$$

3. (*Sub- and supermartingale inequalities*) (a)   Let $\{X_1, \ldots, X_n\}$ be a submartingale. If $\lambda \ge 0$, show that

    $$\lambda P\left\{ \max_{1 \le i \le n} X_i \ge \lambda \right\} \le \int_{\left\{ \max_{1 \le i \le n} X_i \ge \lambda \right\}} X_n \, dP \le E(X_n^+).$$

    (b)   Let $\{X_1, \ldots, X_n\}$ be a submartingale. If $\lambda \ge 0$, show that

    $$\lambda P\left\{ \max_{1 \le i \le n} X_i \ge \lambda \right\} \le E(X_1) - \int_{\left\{ \max_{1 \le i \le n} X_i < \lambda \right\}} X_n \, dP$$

    $$\le E(X_1) + E(X_n^-).$$

    (Apply 6.7.3, with $T = \min\{i \colon X_i \ge \lambda\}$; $T = n$ if all $X_i < \lambda$.)
    (c)   If $\{X_1, X_2, \ldots\}$ is a submartingale and $\lambda \ge 0$, show that

    $$\lambda P\{\sup_n X_n > \lambda\} \le \sup_n E(X_n^+);$$

    if $\{X_1, X_2, \ldots\}$ is a supermartingale and $\lambda \ge 0$, show that

    $$\lambda P\{\sup_n X_n > \lambda\} \le E(X_1) + \sup_n E(X_n^-).$$

4. Use Problem 3 to give an alternative proof of Kolmogorov's inequality 6.1.4.

5. (*Wald's theorem on the sum of a random number of random variables*) Let $Y_1, Y_2, \ldots$ be independent, identically distributed random variables with finite mean m, and let $X_n = \sum_{k=1}^n Y_k$. If T is a finite stopping time for $\{X_n\}$, establish the following:

   (a) If all $Y_j \geq 0$, then $E(X_T) = mE(T)$.
   (b) If $E(T) < \infty$, then $E(|X_T|) < \infty$ and $E(X_T) = mE(T)$.

   [Let $T_n = T \wedge n$ and apply 6.7.3 to $\{X_n - nm\}$ to prove (a); use (a) to prove (b).]

   (c) If T is a positive integer-valued random variables that is independent of $(Y_1, Y_2, \ldots)$, but not necessarily a stopping time, show that the results (a) and (b) still hold.

6. [*Alternative proof of the upcrossing theorem* (Meyer, 1966)] Let

$$\{X_k, \mathscr{F}_k, k = 1, \ldots, n\}$$

be a nonnegative submartingale, and U the number of upcrossings of $(0, b)$ by $X_1, \ldots, X_n$. Define the stopping times T_i as in 6.4.2, and for convenience set $X_\infty = X_n$.

   (a) Show that $X_n = \sum_{k=1}^n (X_{T_k} - X_{T_{k-1}})$ (take $X_{T_0} = 0$).
   (b) Show that $E(X_n) \geq bE(U)$; the general upcrossing theorem is then

   obtained just as in 6.4.2. $\left[\vphantom{\sum}\right.$ Since $E(X_{T_k} - X_{T_{k-1}}) \geq 0$ for all k, and

   $X_{T_k} - X_{T_{k-1}} \geq b$ if k is even and $T_k < \infty$,

$$E(X_n) \geq \sum_{\substack{k=1 \\ k \text{ even}}}^n E(X_{T_k} - X_{T_{k-1}}) \geq bE(U). \left.\vphantom{\sum}\right]$$

6.8 APPLICATIONS OF MARTINGALE THEORY

Martingale ideas provide fresh insights and simplifications for many problems in probability; in this section we consider some important examples. First, we use the martingale convergence theorem to provide a short proof of the strong law of large numbers for iid random variables (see 6.2.5). We need two preliminary facts.

6.8.1 Lemma. If $X_1, \ldots, X_n$ are independent, identically distributed random variables with finite expectation, and $S_n = \sum_{k=1}^n X_k$, then

$$E(X_k|S_n) = \frac{S_n}{n} \qquad \text{a.e.}, \qquad k = 1, \ldots, n.$$

Intuitively, given $S_n = X_1 + \cdots + X_n$, the average contribution of each X_k is the same, and hence must be S_n/n.

PROOF.   If $B \in \mathscr{B}(\mathbb{R})$, then

$$\int_{\{S_n \in B\}} X_k \, dP = E[X_k I_{\{S_n \in B\}}]$$

$$= \int_{-\infty}^{\infty} \cdots \int_{-\infty}^{\infty} x_k I_B(x_1 + \cdots + x_u) \, dF(x_1) \cdots dF(x_n),$$

where F is the distribution function of the X_i. By Fubini's theorem, this is independent of k; hence

$$\int_{\{S_n \in B\}} X_k \, dP = \frac{1}{n} \int_{\{S_n \in B\}} \sum_{k=1}^{n} X_k \, dP = \int_{\{S_n \in B\}} \frac{S_n}{n} \, dP. \quad \square$$

6.8.2   Lemma. If $X_1, X_2, \ldots$ are random variables and $S_n = \sum_{k=1}^{n} X_k$, then $\sigma(S_n, S_{n+1}, S_{n+2}, \ldots) = \sigma(S_n, X_{n+1}, X_{n+2}, \ldots)$.

PROOF.   Since $X_{n+k} = S_{n+k} - S_{n+k-1}, S_n, X_{n+1}, X_{n+2}, \ldots$ are each $\sigma(S_n, S_{n+1}, S_{n+2}, \ldots)$-measurable; therefore $\sigma(S_n, X_{n+1}, X_{n+2}, \ldots) \subset \sigma(S_n, S_{n+1}, S_{n+2}, \ldots)$. Similarly, $S_n, S_{n+1}, S_{n+2}, \ldots$ are each $\sigma(S_n, X_{n+1}, X_{n+2}, \ldots)$-measurable; hence $\sigma(S_n, S_{n+1}, S_{n+2}, \ldots) \subset \sigma(S_n, X_{n+1}, X_{n+2}, \ldots)$. $\square$

6.8.3   Strong Law of Large Numbers, iid Case. If $X_1, X_2, \ldots$ are iid random variables with finite expectation m, and $S_n = X_1 + \cdots + X_n$, then $S_n/n \to m$ a.e. and in L^1.

PROOF. $(X_1, \ldots, X_n)$ and $(X_{n+1}, X_{n+2}, \ldots)$ are independent [Problem 1(a), Section 4.11]; hence (X_1, S_n) and $(X_{n+1}, X_{n+2}, \ldots)$, as functions of independent random objects, are independent by 4.8.2(d). Therefore

$$E(X_1|S_n) = E(X_1|S_n, X_{n+1}, X_{n+2}, \ldots) \qquad \text{by Problem 2, Section 5.5}$$

$$= E(X_1|S_n, S_{n+1}, S_{n+2}, \ldots) \qquad \text{by 6.8.2.}$$

Thus by 6.8.1,

$$E(X_1|S_n, S_{n+1}, \ldots) = \frac{S_n}{n} \qquad \text{a.e.}$$

But by 6.6.3 and 6.6.4, $E(X_1|S_n, S_{n+1}, \ldots) \to E(X_1|\mathscr{G}_\infty)$ a.e. and in L^1, where $\mathscr{G}_\infty$ is the tail σ-field of the S_n. Thus S_n/n converges a.e. and in L^1 to a finite limit.

Now to show that the limit is in fact m, we may proceed in two ways. One approach is to note that $\lim_{n\to\infty}(S_n/n)$ is a tail function of the X_n, and hence is a.e. constant by the Kolmogorov zero–one law 6.2.7. Since S_n/n is L^1-convergent and $E(S_n/n) \equiv m$, the constant must be m.

Alternatively, we may use the Hewitt–Savage zero–one law 6.2.9 to show that each set in $\mathscr{G}_\infty$ has probability 0 or 1. For if $A \in \mathscr{G}_\infty$ and T permutes n coordinates, then $A \in \sigma(S_n, S_{n+1}, \ldots)$; hence A is of the form $\{(S_n, S_{n+1}, \ldots) \in A'\}$ for some $A' \in [\mathscr{B}(\mathbb{R})]^\infty$. Since $S_k = X_1 + \cdots + X_k = X_{T(1)} + \cdots + X_{T(k)} = S_{T(k)}, k \geq n, A$ is symmetric, and, therefore, the Hewitt–Savage zero–one law is applicable. Thus $\mathscr{G}_\infty$ is trivial, and it follows that $E(X_1|\mathscr{G}_\infty) = E(X_1)$ a.e. by 5.5.7(a). $\quad\square$

Notice that we have obtained L^1 convergence in the strong law of large numbers; this would be more cumbersome to derive using the classical approach of Section 6.2.

We now consider the general problem of convergence of series of independent random variables. The following variation of the martingale convergence theorem will be proved first. The technique of the proof rather than the result itself will be used in the development, but the theorem does have applications to series of *dependent* random variables. (For example, see Problem 5.)

6.8.4 Theorem. Let $\{X_n, \mathscr{F}_n, n = 1, 2, \ldots\}$ be a submartingale, and let $Z = \sup_n(X_n - X_{n-1})$ (define $X_0 = 0$). If $E(Z) < \infty$ (for example, if $X_n - X_{n-1}$ is less than a constant for all $n \geq 2$), then X_n converges to a finite limit a.e. on the set $\{\sup_n X_n < \infty\}$. If $\{X_n, \mathscr{F}_n\}$ is a supermartingale and $E[\inf_n(X_n - X_{n-1})] > -\infty$, then X_n converges to a finite limit a.e. on $\{\inf_n X_n > -\infty\}$.

PROOF. Fix $M > 0$, and let $T = \inf\{n: X_n > M\}$; set $T = \infty$ if there is no such n. Define $T_n = T \wedge n$; if $Y_n = X_{T_n}, n = 1, 2, \ldots$, then $\{Y_n\}$ is a submartingale by 6.7.4(a). (This is sometimes called the *optional stopping theorem* since $Y_n = X_n$ if $n < T, Y_n = X_T$ if $n \geq T$; thus $\{Y_n\}$ is the original process stopped at time T.)

Now if $n < T$, then $Y_n = X_n = X_{n-1} + (X_n - X_{n-1}) \leq M + Z$, and if $n \geq T$, then $Y_n = X_{T-1} + (X_T - X_{T-1}) \leq M + Z$. Thus $Y_n \leq M + Z$ in any case, so $\sup_n E(Y_n^+) \leq M + E(Z^+) < \infty$ by hypothesis. By 6.4.3, Y_n converges a.e. to a finite limit. But if $T = \infty$, then $Y_n \equiv X_n$; hence X_n converges a.e. on $\{\sup_n X_n \leq M\}$. Since M is arbitrary, X_n converges a.e. on $\{\sup_n X_n < \infty\}$. The last statement of the theorem is proved by applying the above argument to $-X_n$. $\quad\square$

We have seen in 6.2.1 that if $Y_1, Y_2, \ldots$ are independent random variables with 0 mean, and $\sum_{k=1}^\infty E(Y_k^2) < \infty$, then $\sum_{k=1}^\infty Y_k$ converges a.e. There is a partial converse to this result, which we prove after one preliminary.

6.8.5   *Theorem.* If $\{X_1, X_2, \ldots\}$ is a martingale and $E(X_n{}^2) < \infty$ for all n, then the martingale differences $X_1, X_2 - X_1, \ldots, X_n - X_{n-1}, \ldots$ are orthogonal.

PROOF.   If $j < k$, and $\mathscr{F}_j = \sigma(X_1, \ldots, X_j)$,

$$E[(X_j - X_{j-1})(X_k - X_{k-1})] = E[E((X_j - X_{j-1})(X_k - X_{k-1})|\mathscr{F}_j)]$$
$$= E[(X_j - X_{j-1})E(X_k - X_{k-1}|\mathscr{F}_j)]$$

since $X_j - X_{j-1}$ is $\mathscr{F}_j$-measurable. But $E(X_k - X_{k-1}|\mathscr{F}_j) = X_j - X_j = 0$ by the martingale property, and the result follows. $\square$

6.8.6   *Theorem.* Let $Y_1, Y_2, \ldots$ be independent random variables with 0 mean, and assume $E[\sup_k Y_k{}^2] < \infty$. (For example, this holds if the Y_k are uniformly bounded.) If $\sum_{k=1}^{\infty} Y_k$ converges a.e., then $\sum_{k=1}^{\infty} E(Y_k{}^2) < \infty$. (As in Section 6.2, "convergence" of a series means covergence to a finite limit.)

PROOF.   The $X_n = \sum_{k=1}^{n} Y_k$ form a martingale. Choose M such that

$$P\left\{ \sup_n |X_n| \leq M \right\} > 0;$$

this is possible since the series converges a.e.

Let $T = \inf\{n: |X_n| > M\}$; $T = \infty$ if there is no such n. If $T_n = T \wedge n$, then $\{X_{T_n}\}$ is a martingale, and just as in 6.8.4, $|X_{T_n}| \leq M + \sup_j |X_j - X_{j-1}| = M + Z$, where $E(Z^2) < \infty$ by hypothesis. It follows that the numbers $E(X_{T_n}^2)$ are uniformly bounded, so by 6.6.10, X_{T_n} coverages a.e. and in L^2. But by 6.8.5, $E(X_{T_n}^2) = \sum_{j=1}^{n} E[(X_{T_j} - X_{T_{j-1}})^2]$ (take $X_{T_0} = 0$). Since X_{T_n} is L^2-convergent, $E(X_{T_n}^2)$ approaches a finite limit; hence $\sum_{n=1}^{\infty} E[(X_{T_n} - X_{T_{n-1}})^2] < \infty$.

But $X_{T_n} - X_{T_{n-1}} = Y_n I_{\{T \geq n\}}$, so $\sum_{n=1}^{\infty} E(Y_n{}^2 I_{\{T \geq n\}}) < \infty$, and consequently

$$\sum_{n=1}^{\infty} E(Y_n{}^2 I_{\{T \geq n\}}|Y_1, \ldots, Y_{n-1}) < \infty \qquad \text{a.e.}$$

[To see this, note that if $Z_n \geq 0$ and $\sum_n E(Z_n) < \infty$, and $E\left(\sum_n Z_n\right) < \infty$, so $\sum_n Z_n < \infty$ a.e.; set $Z_n = E(Y_n{}^2 I_{\{T \geq n\}}|Y_1, \ldots, Y_{n-1})$.]

Now $I_{\{T \geq n\}} = I_{\{T \leq n-1\}^c}$, which is $\sigma(Y_1, \ldots, Y_{n-1})$-measurable; hence

$$\sum_{n=1}^{\infty} I_{\{T \geq n\}} E(Y_n{}^2|Y_1, \ldots, Y_{n-1}) = \sum_{n=1}^{\infty} I_{\{T \geq n\}} E(Y_n{}^2) < \infty \qquad \text{a.e.}$$

Pick an ω where the series converges and where $\sup_n |X_n(\omega)| \le M$ (see the beginning of the proof). The $T(\omega) = \infty \ge n$ for all n; it follows that $\sum_{n=1}^{\infty} E(Y_n^2) < \infty$. $\square$

We now have a partial solution to the general problems.

6.8.7   Theorem. Let $Y_1, Y_2, \ldots$ be independent random variable with 0 mean; assume $E[\sup_k Y_k^2] < \infty$. Then $\sum_{k=1}^{\infty} Y_k$ converges a.e. if and only if $\sum_{k=1}^{\infty} E(Y_k^2) < \infty$.

PROOF.   Apply 6.2.1 and 6.8.6. $\square$

We can now complete the solution to the random signs problem (see the discussion after 6.2.1). If $X_n = a_n Y_n$, where the Y_n are independent with $P\{Y_n = 1\} = P\{Y_n = -1\} = \frac{1}{2}$, and $\sum_n X_n$ converges a.e., then $a_n \to 0$, so the X_n are uniformly bounded. By 6.8.7, $\sum_n a_n^2 < \infty$.

Incidentally, there is a martingale proof of 6.2.1. If $X_n = \sum_{k=1}^{n} Y_k$, where the Y_k are independent with 0 mean and $\sum_{k=1}^{\infty} E(Y_k^2) < \infty$, then

$$E(|X_n|) \le [E(X_n^2)]^{1/2}$$

by the Cauchy–Schwarz inequality

$$= \left[\sum_{k=1}^{n} E(Y_k^2) \right]^{1/2}$$

since the Y_k are independent with 0 mean; hence orthogonal. Thus $\sup_n E(|X_n|) < \infty$; hence X_n coverages a.e.

The same argument also shows that if $\{X_n\}$ is a martingale and $\sum_n E(X_n - X_{n-1})^2 < \infty$ (so that the $X_n - X_{n-1}$ are orthogonal by 6.8.5, but not necessarily independent), then X_n coverages a.e.

We now drop the hypothesis of zero mean.

6.8.8   Theorem. Let Y_1, Y_2, be independent, uniformly bounded random variables. Then $\sum_{k=1}^{\infty} Y_k$ converges a.e. iff $\sum_{k=1}^{\infty} \operatorname{Var} Y_k < \infty$ and $\sum_{k=1}^{\infty} E(Y_k)$ converges.

PROOF.   "If": By 6.8.7, $\sum_{k=1}^{\infty} (Y_k - E(Y_k))$ converges a.e.; since $\sum_{k=1}^{\infty} E(Y_k)$ converges, we have $\sum_{k=1}^{\infty} Y_k$ convergent a.e.

"Only if": by *symmetrization.* Let $Y_1, Z_1, Y_2, Z_2, \ldots$ be independent, where for each j, Z_j has the same distribution as Y_j. Then $E(Y_j - Z_j) = E(Y_j) - E(Z_j) = 0$, $E[(Y_j - Z_j)^2] = \operatorname{Var}(Y_j - Z_j) = \operatorname{Var} Y_j + \operatorname{Var} Z_j = 2 \operatorname{Var} Y_j$. Since $\sum_j Y_j$ converges a.e., so does $\sum_j Z_j$. [To see this, note that $P\{(Y_1, \ldots, Y_n) \in B\} = P\{(Z_1, \ldots, Z_n) \in B\}$ for all n and all $B \in \mathscr{B}(\mathbb{R}^n)$; hence $P_Y = P_Z$.] It follows that $\sum_j (Y_j - Z_j)$ converges a.e., so by 6.8.7,

$\sum_j$ Var $Y_j < \infty$. Again by 6.8.7, $\sum_j(Y_j - E(Y_j))$ converges a.e.; hence $\sum_j E(Y_j)$ converges. $\square$

Finally, we obtain a general criterion for convergence of series of independent random variables.

6.8.9 Kolmogorov Three Series Theorem. Let $Y_1, Y_2, \dots$ be independent random variables. If $M > 0$, define

$$Y_j' = \begin{cases} Y_j & \text{if} \quad |Y_j| \le M, \\ 0 & \text{if} \quad |Y_j| > M. \end{cases}$$

(a) If $\sum_j Y_j$ converges a.e., then for any $M < o$, the three series $\sum_j P\{Y_j \ne Y_j'\}$, $\sum_j E(Y_j')$, $\sum_j$ Var Y_j' all converge.

(b) If for some $M > 0$, the three series converge, then $\sum_j Y_j$ converges a.e.

PROOF. (a) By hypothesis, $Y_j \to 0$ a.e., so eventually $Y_j = Y_j'$. Thus (a.e.) $Y_j \ne Y_j'$ for only finitely many j, that is, $P(\limsup_j\{Y_j \ne Y_j'\}) = 0$. By the second Borel–Cantelli lemma, $\sum_j P\{Y_j \ne Y_j'\} < \infty$. The other two series converge by 6.8.8.

(b) By 6.8.8, $\sum_j Y_j'$ converges a.e. Since $\sum_j P\{Y_j \ne Y_j'\} < \infty$, we have, almost surely, $Y_j = Y_j'$ eventually; hence $\sum Y_j$ converges a.e. $\square$

6.8.10 Branching Processes. As a final example we analyze in detail the branching process of 6.3.3(d). Recall that $X_0 = 1$, and if $X_n = k$, then $X_{n+1} = \sum_{i=1}^k Y_i$, where $Y_1, \dots, Y_k$ are independent and $P\{Y_i = p\} = P_l$, $l = 0, 1, \dots$. We assume that $m = E(Y_i) = \sum_{l=1}^\infty l p_l > 0$.

This excludes the degenerate case $p_0 = 1$ (in this case $X_n = 0$ for all $n \ge 1$). We also assume that $p_0 + p_1 < 1$. (If $p_0 + p_1 = 1$, then $X_n \le 1$ for all n. If $p_0 > 0$, then X_n is eventually 0 since $P\{X_n = 1 \text{ for all } n\} = \lim_{n\to\infty} p_1{}^n = 0$, and if $p_0 = 0$, then $X_n \equiv 1$.)

Case 1: $m < 1$. In this case, almost surely, X_n is 0 eventually; thus *the family name is extinguished with probability* 1.

For $E(X_{n+1}|X_n = k) = kE(Y_1) = km$; hence $E(X_{n+1}|X_n) = mX_n$. It follows that $E(X_{n+1}) = mE(X_n)$. It $m < 1$, then

$$E\left(\sum_{n=1}^\infty X_n\right) = \sum_{n=1}^\infty E(X_n) < \infty,$$

so $X_n \to 0$ a.e. But the X_n are integer-valued, and thus with probability 1, X_n is ultimately 0.

Case 2: $m > 1$. We show that with probability r, X_n is eventually 0, and with probability $1 - r$, $X_n \to \infty$, where r is the unique root in $[0, 1)$ of the equation $\sum_{j=0}^\infty p_j s^j = s$.

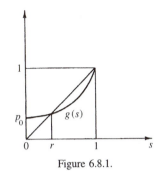

Figure 6.8.1.

Let $g(s) = \sum_{j=0}^{\infty} p_j s^j$, $0 \le s \le 1$. Consider $d[g(s) - s]/ds = g'(s) - 1$. We have $g'(0) - 1 = p_1 - 1 < 0$, $g'(1) - 1 = m - 1 > 0$, and since $p_0 + p_1 < 1$, $g'(s) - 1$ is strictly increasing. Since $g(s) - s = p_0$ when $s = 0$ and 0 when $s = 1$, $g(s) - s$ strictly decreases to a minimum occurring somewhere in $(0, 1)$, and then strictly increases to 0 at $s = 1$. If follows that $g(s) = s$ for exactly one $s \in [0, 1)$, say at $s = r$ (Fig. 6.8.1).

First assume $0 < r < 1$ (hence $p_0 > 0$). By 6.3.3(e), $\{r^{X_n}\}$ is a nonnegative martingale, and hence converges a.e. Since X_n is nonnegative integer-valued, this means that for almost every ω, $X_n(\omega)$ becomes constant (the constant depending on ω) or $X_n(\omega) \to \infty$.

Now $P\{X_n \text{ eventually constant}\} = \sum_{k=0}^{\infty} P\{X_n = k \text{ for sufficiently large } n\}$. If $k \ge 1$ and $P\{X_n = k \text{ eventually}\} > 0$, then $P\{X_{n=k} \text{ for all } n \ge N\} > 0$ for some N. But by the Markov property, this probability must be $P\{X_N = k\} \lim_{j \to \infty} q^j$, where $q = P\{X_{n+1} = k | X_n = k\}$. Now $q < 1$ since $P\{X_{n+1} = 0 | X_n = k\} = p_0^k > 0$. Thus $\lim_{j \to \infty} q^j = 0$, a contradiction.

Therefore $X_n \to X_\infty$ a.e., where $X_\infty = 0$ or ∞. Since $\{r^{X_n}, n = 0, 1, \ldots\}$ is bounded, the dominated convergence theorem gives $E(r^{X_n}) \to E(r^{X_\infty})$ $= 1 P\{X_\infty = 0\} + 0 P\{X_\infty = \infty\} = P\{X_\infty = 0\}$. But by the martingale property, $E(r^{X_n}) = E(r^{X_0}) = r$. Thus with probability r, X_n is eventually 0, and with probability $1 - r$, $X_n \to \infty$.

If $r = 0$, then $p_0 = 0$; hence $X_{n+1} \ge X_n \ge 1$, so that X_n increases to a limit X. But since the X_n are positive integer-valued,

$$P\{X < \infty\} = P\{X_n \quad \text{eventually constant}\}$$

$$= \sum_{k=1}^{\infty} P\{X_n = k \quad \text{eventually}\}$$

$$= 0 \quad \text{by the same argument as in the case} \quad 0 < r < 1.$$

(In the current situation, $q = P\{X_{n+1} = k | X_n = k\} = P\{Y_1 = \cdots = Y_k = 1\}$ $= p_1^k < 1$.) Thus $X_n \to \infty$ a.e., as desired.

Case 3: m = 1. Here, extinction occurs with probability 1. For by 6.3.3(d), $\{X_n\}$ is a nonnegative martingale, and hence converges a.e. to a finite limit. The analysis of case 2 shows that X_n cannot approach a nonzero constant on a set of positive probability, and hence $X_n \to 0$ a.e.; in other words, with probability 1, X_n is eventually 0.

Problems

1. Let $X_1, X_2, \ldots$ be arbitrary random variables, with $S_n = X_1 + \cdots + X_n$. What is the relation between the tail σ-field of the X_n and the tail σ-field of the S_n?

2. What happens in the branching process (see 6.8.10) if instead of $X_0 = 1$, X_0 is an arbitrary, positive integer-valued random variable?

3. (*Breiman's realistic gambling model*) Let X_n be a gambler's capital after n plays of a game of chance. Assume that the gambler has the option of betting or passing at each trial. If he passes at trial $n + 1$, then $X_{n+1} = X_n$, and if he bets, then we assume that $|X_{n+1} - X_n| \ge b > 0$; thus there is a minimum amount b that can be won or lost on a given trial. We do not spell out the gambler's strategy in detail; we simply assume that his strategy, together with the house rules, determine the distribution of $(X_0, X_1, \ldots)$. It is reasonable to assume that the game is unfavorable or at best fair; thus we take $\{X_n\}$ as a nonnegative supermartingale.

   Let T be the time of the last bet, that is, largest n such that $|X_{n+1} - X_n| \ge b$ ($T = \infty$ if there is no such n). Note that T is *not* a stopping time.

   (a) Show that T is a.e. finite.

   (b) Show that $E(X_T) \le E(X_0)$, so no system can increase the expected winning.

   (c) If the gambler's strategy is always to bet so long as his capital is at least b, show that (a.e.) X_n will eventually be less than b, in other words, *the persistent gambler goes broke with probability 1.*

   (d) If an unbiased coin is tossed independently over and over again, and we win a dollar for each head and lose a dollar for each tail, our accumulated capital must eventually reach $+1$ (see the end of Section 6.7). Suppose our strategy is simply to wait until we reach 1 and then quit, thus guaranteeing a profit. Why is this not realistic?

4. Let $Y_1, Y_2, \ldots$ be integrable random variables, and $\mathscr{F}_1, \mathscr{F}_2, \ldots$ an increasing sequence of sub σ-fields of $\mathscr{F}$. Assume that Y_k is $\mathscr{F}_k$-measurable for each k, and define

$$X_n = \sum_{k=1}^{n} [Y_k - E(Y_k | \mathscr{F}_{k-1})]$$

[take $\mathscr{F}_0 = \{\emptyset, \Omega\}$, so that $E(Y_1|\mathscr{F}_o) = E(Y_1)$]. Show that $\{X_n, \mathscr{F}_n\}$ is a martingale.

5. (*Lévy's extension of the Borel–Cantelli lemma*) Let $\{\mathscr{F}_n\}$ be an increasing sequence of σ-fields, and let $A_n \in \mathscr{F}_n, n = 1, 2, \ldots$. Define

$$q_j = P(A_j|\mathscr{F}_{j-1}), \qquad j = 1, 2, \ldots.$$

[take $\mathscr{F}_0 = \{\emptyset, \Omega\}$, so that $q_1 = P(A_1)$]. Show that (a.e.) infinitely many A_j occur iff $\sum_j q_j = \infty$, that is, $P[(\limsup_j A_j) \triangle \{\sum_j q_j = \infty\}] = 0$. Equivalently, $\sum_j I_{A_j}$ and $\sum_j q_j$ have essentially the same convergence set. [Apply 6.8.4 to $X_n = \sum_{j=1}^n (I_{A_j} - q_j)$.]

6.9 APPLICATIONS TO MARKOV CHAINS

In this section we apply martingale theory to the problem of classifying the states of a Markov chain. We must use a few basic properties of Markov chains: the reader who is unfamiliar with this subject may consult Ash (1970, Chapter 7). In particular, a state i is said to be *recurrent* iff starting at i there will be a return to i with probability 1; otherwise the state is *transient*. If C is a set of states such that every state in C can be reached (in a finite number of steps) from every other state, then all states in C are of the same type, recurrent or transient

We have the following criterion.

6.9.1 Theorem. Let $[p_{ij}]$ be the transition matrix of a Markov chain such that every state in the state space S is reachable from every other state (sometimes called an *irreducible chain*). Choose a fixed state, and label it 0 for convenience. The states are transient iff there is a nonconstant bounded $f\colon S \to R$ such that $\sum_{j \in S} p_{ij} f(j) = f(i)$ for all $i \neq 0$.

PROOF. Suppose such an f exists. By adding a constant to f we may assume that $f \geq 0$. Assume the initial state is $i \neq 0$, and let $\{X_n\}$ be the corresponding sequence of random variables. Let T be the time at which 0 is reached, and let $Y_n = X_{T \wedge n}, n = 0, 1 \ldots$; $\{Y_n\}$ can be realized as a Markov chain with the same initial distribution and transition matrix as $\{X_n\}$, except that 0 is now an absorbing state. In other words, the transition matrix for $\{Y_n\}$ is

$$\hat{p}_{ij} = p_{ij} \qquad \text{for all} \quad j \quad \text{if} \quad i \neq 0,$$

$$\hat{p}_{00} = 1.$$

Thus $\sum_{j\in S} \hat{p}_{ij} f(j) = f(i)$ for *all* $i \in S$. In matrix form, $\hat{\prod} f = f$; by induction $\hat{\prod}^n f = f$, that is,

$$\sum_{j\in S} \hat{p}_{ij}^{(n)} f(j) = f(i),$$

where $\hat{p}_{ij}^{(n)} = P\{Y_n = j | Y_0 = i\}$. But this says that $E[f(Y_n)|Y_0 = i] = f(i)$.

If the states of the original chain are recurrent, then 0 will be visited with probability 1; hence $Y_n \to 0$ a.e. By the dominated convergence theorem, $E[f(Y_n)|Y_0 = i] \to f(0)$. We conclude that $f(i) = f(0)$ for all i, contradicting the hypothesis that f is nonconstant.

Conversely, if the states are transient, we define $f: S \to R$ as follows. If $i \neq 0$, let $f(i) = f_{i0}$, the probability that, starting from i, 0 will eventually be reached; take $f(0) = 1$. Now in order ultimately to reach 0 from $i \neq 0$, we may either go directly to 0 at step 1, or go to a state $j \neq 0$ and then reach 0 at some time after the first step. It follows that

$$f(i) = \sum_{j\in S} p_{ij} f(j), \qquad i \neq 0.$$

(This may be formalized using the Markov property.)

Now f is clearly bounded, and $f_{i0} < 1$ for some $i \neq 0$, otherwise 0 would be a recurrent state. Thus f is nonconstant. $\square$

Martingale theory is used in deriving the following sufficient condition for recurrence.

6.9.2 Theorem. Let $[p_{ij}]$ be the transition matrix of an irreducible Markov chain whose state space S is the set of nonnegative integers. If there is a function $f: S \to R$ such that $f(i) \to \infty$ as $i \to \infty$, and $\sum_j p_{ij} f(j) \leq f(i)$ for all $i \neq 0$, then the chain is recurrent.

PROOF. As $f(i) \to \infty$ as $i \to \infty$, f is bounded below, so without loss of generality we may assume $f \geq 0$. Let the initial state be $i \neq 0$, and form the process $\{Y_n\}$ as in 6.9.1. The $\sum_{j\in S} \hat{p}_{ij} f(j) \leq f(i)$ for all i, which implies that $\{f(Y_n)\}$ is a nonnegative supermartingale, and hence converges a.e. to a finite limit. For

$$E[f(Y_n)|Y_0 = i_0, \dots, Y_{n-1} = i_{n-1}] = E[f(Y_n)|Y_{n-1} = i_{n-1}]$$

by the Markov property

$$= \sum_j \hat{p}_{i_{n-1}j} f(j)$$

$$\leq f(i_{n-1}).$$

Note also that

$$E[f(Y_n)|Y_0 = i] = \sum_{j \in S} \hat{p}_{ij}^{(n)} f(j) \le f(i) < \infty;$$

hence $\{f(Y_n)\}$ is integrable.

Assume the states transient. Then $f_{i0} < 1$ for some $i > 0$. Choose such an i as the initial state. Now $X_n \to \infty$ a.e. since a finite set of transient states cannot be visited infinitely often (see Ash, 1970, p. 223). This mean that with probability 1, $Y_n \to 0$ or ∞. But $P\{Y_n \to 0\} = f_{i0} < 1$, and hence $P\{Y_n \to \infty\} > 0$; this implies that $P\{f(Y_n) \to \infty\} > 0$, a contradiction.   □

The proof of 6.9.2 shows that if $[p_{ij}]$ is the transition matrix of a Markov chain $\{X_n\}$, f is a real-valued function on the state space, and

$$\sum_j p_{ij} f(j) \le f(i) \qquad \text{for all} \qquad i,$$

where the series is assumed to converge absolutely, then, with a fixed initial state i, the sequence $\{f(X_n)\}$ is a supermartingale. Similarly, replacement of "$\le$" by "$=$" in this equation yields a martingale, and replacement by "$\ge$" yields a submartingale.

We now apply 6.9.1 and 6.9.2 to a queueing process.

6.9.3   Example. Assume that customers are to be served at discrete times $t = 0, 1, \ldots$, and at most one customer can be served at a given time. Say there are X_t customers before the completion of service at time t, and in the interval $[t, t + 1)$, Y_t new customers arrive, where $P\{Y_t = k\} = p_k, k = 0, 1, \ldots$. The number of customers before completion of service at time $t + 1$ is

$$X_{t+1} = (X_t - 1)^+ + Y_t.$$

The queueing process may be represented as a Markov chain whose state space is the set of nonnegative integers and whose transition matrix is

$$
\Pi = \begin{array}{c} \\ 0 \\ 1 \\ 2 \\ 3 \\ \vdots \end{array}
\begin{array}{c} \begin{array}{cccccc} 0 & 1 & 2 & 3 & \cdots & \end{array} \\
\left[\begin{array}{cccccc}
p_0 & p_1 & p_2 & \cdots & & \\
p_0 & p_1 & p_2 & \cdots & & \\
0 & p_0 & p_1 & p_2 & \cdots & \\
0 & 0 & p_0 & p_1 & p_2 & \cdots \\
& & & & &
\end{array}\right].
\end{array}
$$

We assume that $p_0 > 0$ and $p_0 + p_1 < 1$, so that the chain is irreducible. We analyze the behavior of the chain using 6.9.1 and 6.9.2.

The equations $\sum_{j=0}^{\infty} p_{ij} f(j) = f(i), i > 0$, are

$$p_0 f(i-1) + p_1 f(i) + p_2 f(i+1) + \cdots + p_n f(i+n-1) + \cdots = f(i).$$
(1)

Let $m = E(Y_t) = \sum_{k=1}^{\infty} k p_k$; we show that the states are transient if $m > 1$, recurrent if $m \le 1$.

First assume $m > 1$; if $f(i) = r^i$, then (1) becomes

$$p_0 r^{i-1} + p_1 r^i + p_2 r^{i+1} + \cdots + p_n r^{i+n-1} + \cdots = r^i$$

or

$$\sum_{k=0}^{\infty} p_k r^k = r.$$

But this can be satisfied for some $r \in (0, 1)$ (see case 2 of 6.8.10). Thus $\{r^i\}$ is bounded and nonconstant, so by 6.9.1, the states are transient.

Now assume $m \le 1$, and let $f(i) = i$. Then if $i > 0$,

$$\sum_{j=0}^{\infty} p_{ij} f(j) = p_0(i-1) + p_1 i + p_2(i+1) + \cdots$$

$$= \sum_{k=i-1}^{\infty} k p_{k-i+1}$$

$$= \sum_{k=i-1}^{\infty} (k-i+1) p_{k-i+1} + i - 1$$

$$= \sum_{k=0}^{\infty} k p_k + i - 1 \le 1 + i - 1 = i = f(i).$$

By 6.9.2, the states are recurrent.

If i is a recurrent state and μ_i is the average length of time required to return to i when the initial state is i, then i is said to be *recurrent null* if $\mu_i = \infty$, *recurrent positive* if $\mu_i < \infty$. It can be shown that the states are recurrent null if $m = 1$, recurrent positive if $m < 1$ (see Karlin, 1966, pp. 74ff).

Problems

1. Consider the Markov chain in Section 6.3, Problem 2, and assume $q_0 < 1$. If $r > 1$ in (b), show that $X_n \to -\infty$ a.e.; hence the states are transient.

6.10 REFERENCES

Many classical results on strong laws of large numbers, martingales, and related topics may be found in Loève (1955) and Doob (1953). Another useful reference is Breiman (1968).

Martingale and Markov processes are important in the area of probabilistic potential theory. For an introduction to this subject, see Meyer (1966) and Kemeny, Snell, and Knapp (1966).

More recent works are by Shiryaev (1996) and Williams (1991).

7

THE CENTRAL LIMIT THEOREM

7.1 INTRODUCTION

If $X_1, X_2, \ldots$ are independent, identically distributed random variables with zero mean, and $S_n = X_1 + \cdots + X_n$, the strong law of large numbers states that S_n/n converges a.e. to 0. Thus given $\varepsilon > 0$, $|S_n/n|$ will be less than ε for large n; in other words, S_n will eventually be small in comparison with n. The strong law of large numbers gives no information about the distribution of S_n; the purpose of this chapter is to develop results (called versions of the *central limit theorem*) concerning the approximate distribution of S_n for large n. For example, if the X_n are iid with finite mean m and finite variance σ^2, then for large n, $(S_n - nm)/\sqrt{n}\,\sigma$ has, approximately, the normal distribution with mean 0 and variance 1.

There are two basic techniques that will be used. First is the theory of weak convergence. If $\mu, \mu_1, \mu_2, \ldots$ are finite measures on $\mathscr{B}(\mathbb{R})$, weak convergence of μ_n to μ means that $\int_{\mathbb{R}} f \, d\mu_n \to \int_{\mathbb{R}} f \, d\mu$ for every bounded continuous $f \colon \mathbb{R} \to \mathbb{R}$. If the corresponding (bounded) distribution functions are $F, f_1, f_2, \ldots$, the equivalent condition is that $F_n(a, b] \to F(a, b]$ at all continuity points of F. (See 2.8 for a discussion of weak convergence; in particular, recall that $+\infty$ and $-\infty$ are, by definition, continuity points of F.) We shall denote weak convergence by $\mu_n \xrightarrow{w} \mu$ or $F_n \xrightarrow{w} F$. Also, if B is a Borel subset of $\mathbb{R}$, the terms $F(B)$ and $\mu(B)$ will be synonomous.

Now assume that μ_n is the probability measure induced by a random variable X_n, $n = 0, 1, \ldots$ (with $\mu_0 = \mu$, $X_0 = X$). If $\mu_n \xrightarrow{w} \mu$, we say that the sequence $\{X_n\}$ *converges in distribution to* X, and write $X_n \xrightarrow{d} X$. Since $\int_{\mathbb{R}} f \, d\mu_n = E[f(X_n)]$, it follows that $X_n \xrightarrow{d} X$ iff $E[f(X_n)] \to E[f(X)]$ for all bounded continuous $f \colon \mathbb{R} \to \mathbb{R}$.

This in turn implies that if $X_n \xrightarrow{d} X$ and g is a continuous function from $\mathbb{R}$ to $\mathbb{R}$, then $g(X_n) \xrightarrow{d} g(X)$. In particular, if $X_n \xrightarrow{d} X$, then $X_n + c \xrightarrow{d} X + c$ for each real number c.

Notice that convergence in distribution is determined completely by the distribution functions, or equivalently by the induced probability measures, of the random variables. In particular, the random variables need not be defined on the same probability space. Note also that since the distribution function of a random variable always has the value 0 at $-\infty$ and the value 1 at $+\infty$, we have $X_n \xrightarrow{d} X$ iff $F_n(x) \to F(x)$ at all continuity points of F in $\mathbb{R}$.

Now by Theorem 2.8.1, $\mu_n \xrightarrow{w} \mu$ iff $\mu_n(A) \to \mu(A)$ for each Borel set A whose boundary ∂A has μ-measure 0. Thus $X_n \xrightarrow{d} X$ iff $P\{X_n \in A\} \to P\{X \in A\}$ for all Borel sets A such that $P\{X \in \partial A\} = 0$. This result justifies the terminology "convergence in distribution," for it says that if $X_n \xrightarrow{d} X$, then X_n and X have approximately the same distribution for large n. Of course it might seem more reasonable to require that $P\{X_n \in A\} \to P\{X \in A\}$ for *all* Borel sets A, but actually this is not so. For example, if X_n is uniformly distributed between 0 and $1/n$, that is, X_n has density $f_n(x) = n$, $0 \le x \le 1/n$, $f_n(x) = 0$ elsewhere, then for large n, X_n approximates a random variable X that is identically 0. But $P\{X_n = 0\} = 0$ for all n, and $P\{X = 0\} = 1$.

The second technique involves the use of characteristic functions, which we now define.

7.1.1 Definition. Let μ be a finite measure on $\mathscr{B}(\mathbb{R})$. The *characteristic function* of μ is the mapping from $\mathbb{R}$ to $\mathbb{C}$ given by

$$h(u) = \int_R e^{iux}\, d\mu(x), \qquad u \in \mathbb{R}.$$

Thus h is the Fourier transform of μ. If F is a distribution function corresponding to μ, we shall also write $h(u) = \int_{\mathbb{R}} e^{iux}\, dF(x)$, and call h the characteristic function of F (or of X if X is a random variable with distribution function F).

Characteristic functions are uniquely appropriate in the study of sums of independent random variables, because of the following result.

7.1.2 Theorem. Let $X_1, X_2, \ldots, X_n$ be independent random variables, and let $S_n = X_1 + \cdots + X_n$. Then the characteristic function of S_n is the product of the characteristic functions of the X_i.

Proof.

$$E(e^{iuS_n}) = E\left(\Pi_{j=1}^n e^{iuX_j}\right)$$

$$= \Pi_{j=1}^n E(e^{iuX_j}) \qquad \text{by independence.} \qquad \square$$

Theorem 7.1.2 allows us to compute the characteristic function of S_n, knowing only the distribution of the individual X_j's. In fact, once the characteristic function is known, the distribution function is determined.

7.1.3   Inversion Formula. If h is the characteristic function of the bounded distribution function F, and $F(a, b] = F(b) - F(a)$, then

$$F(a, b] = \lim_{c\to\infty} \frac{1}{2\pi} \int_{-c}^c \frac{e^{-iua} - e^{-iub}}{iu} h(u)\, du$$

for all points $a, b(a < b)$ at which F is continuous. If in addition, h is Lebesgue integrable on $(-\infty, \infty)$, then the function f given by

$$f(x) = \frac{1}{2\pi} \int_{-\infty}^\infty e^{-iux} h(u)\, du$$

is a density for F, that is, f is nonnegative and $F(x) = \int_{-\infty}^x f(t)\, dt$ for all x; furthermore, $F' = f$ everywhere. Thus in this case, f and h are "Fourier transform pairs":

$$h(u) = \int_{-\infty}^\infty e^{iux} f(x)\, dx,$$

$$f(x) = \frac{1}{2\pi} \int_{-\infty}^\infty e^{-iux} h(u)\, du.$$

If we are trying to compute the distribution of a sum S_n of independent random variables X_i, then Theorem 7.1.2 will be useful only if we can recover a distribution function from its characteristic function. In fact this is the case.

Proof.   Intuitively, if $h(u) = \int_{-\infty}^\infty e^{iux}\, dF(x)$, then $h(u) = \int_{-\infty}^\infty e^{iux} F'(x)\, dx$, so

$$F'(x) = \frac{1}{2\pi} \int_{-\infty}^\infty h(u) e^{-iux}\, du.$$

Thus

$$F(a, b] = \int_a^b F'(x)\, dx = \frac{1}{2\pi} \int_{-\infty}^\infty h(u) \left[\int_a^b e^{-iux}\, dx\right] du$$

and since

$$\int_a^b e^{-iux}\, dx = \frac{e^{-iua} - e^{-iub}}{iu},$$

this leads to the inversion formula. For a formal proof, let

$$I_c = \frac{1}{2\pi} \int_{-c}^c \frac{e^{-iua} - e^{-iub}}{iu} h(u)\, du, \qquad a < b.$$

Then

$$I_c = \frac{1}{2\pi} \int_{-c}^c \frac{e^{-iua} - e^{-iub}}{iu} \left[\int_{-\infty}^\infty e^{iux}\, dF(x) \right]\, du.$$

Now

$$\left| \frac{(e^{-iua} - e^{-iub})}{iu} e^{iux} \right| = \left| \frac{e^{-iua} - e^{-iub}}{iu} \right| = \left| \int_a^b e^{-iux}\, dx \right| \leq b - a$$

and

$$\int_{-c}^c \left[\int_{-\infty}^\infty (b - a)\, dF(x) \right]\, du = 2c(b - a)[F(\infty) - F(-\infty)] < \infty.$$

By Fubini's theorem, the order of integration may be interchanged to obtain

$$I_c = \frac{1}{2\pi} \int_{-\infty}^\infty \left[\int_{-c}^c \frac{(e^{-iua} - e^{-iub})}{iu} e^{iux}\, du \right]\, dF(x) = \int_{-\infty}^\infty J_c(x)\, dF(x),$$

where

$$J_c(x) = \frac{1}{2\pi} \int_{-c}^c \frac{\sin u(x - a) - \sin u(x - b)}{u}\, du.$$

[Note that

$$\int_{-c}^c \frac{\cos u(x - a) - \cos u(x - b)}{iu}\, du = 0$$

since the integrand is an odd function.] Let $v = u(x - a)$ and $w = u(x - b)$ to obtain

$$J_c(x) = \frac{1}{2\pi} \int_{-c(x-a)}^{c(x-a)} \frac{\sin v}{v}\, dv - \frac{1}{2\pi} \int_{-c(x-b)}^{c(x-b)} \frac{\sin w}{w}\, dw.$$

Now

$$\int_r^s \frac{\sin v}{v}\, dv \to \pi \qquad \text{as} \qquad s \to \infty \qquad \text{and} \qquad r \to -\infty,$$

and since the integral is continuous in r and s, it is bounded uniformly in r and s. Thus for some $M < \infty$, $|J_c(x)| \leq M$ for all c and x; furthermore, as $c \to \infty$, $J_c(x) \to J(x)$, where

$$
J(x) = \begin{cases} 0 & \text{if} & x < a \quad \text{or} \quad x > b, \\ 1 & \text{if} & a < x < b, \\ \tfrac{1}{2} & \text{if} & x = a \quad \text{or} \quad x = b. \end{cases}
$$

By the dominated convergence theorem,

$$
\lim_{c \to \infty} I_c = \int_{-\infty}^{\infty} J(x)\, dF(x)
$$

$$
= F(b^-) - F(a) + \tfrac{1}{2}[F(a) - F(a^-) + F(b) - F(b^-)]
$$

$$
= \frac{F(b) + F(b^-)}{2} - \frac{F(a) + F(a^-)}{2} = F(b) - F(a)
$$

if F is continuous at a and b. This proves the formula.

Now assume h integrable on $(-\infty, \infty)$. Let

$$
f(x) = \frac{1}{2\pi} \int_{-\infty}^{\infty} e^{-iux} h(u)\, du, \qquad -\infty < x < \infty.
$$

Since h is integrable, f is well-defined; furthermore, f is continuous by the dominated convergence theorem. Now by Fubini's theorem,

$$
\int_a^b f(x)\, dx = \frac{1}{2\pi} \int_{-\infty}^{\infty} h(u) \left[\int_a^b e^{-iux}\, dx \right] du
$$

$$
= \lim_{c \to \infty} \frac{1}{2\pi} \int_{-c}^{c} h(u) \left[\int_a^b e^{-iux}\, dx \right] du
$$

$$
= \lim_{c \to \infty} \frac{1}{2\pi} \int_{-c}^{c} \frac{e^{-iua} - e^{-iub}}{iu} h(u)\, du
$$

$$
= F(b) - F(a)
$$

by the inversion formula if a and b are continuity points of F. Thus $F(b) - F(a) = \int_a^b f(x)\, dx$ at continuity points of F.

But every point is a limit from above of continuity points since F is monotone and thus has only countably many discontinuities. Since the integral is a continuous function of its limits, it follows that $F(b) - F(a) = \int_a^b f(x)\, dx$ for all a and b.

Since f is continuous, it follows that F is differentiable everywhere and its derivative is f. Since F is increasing, f is everywhere nonnegative. Thus f is a density for F.   □

7.1.4   Corollary. Let P_1 and P_2 be probability measures (or more generally, finite measures) on $\mathscr{B}(\mathbb{R})$. If $\int_{\mathbb{R}} e^{iux}\, dP_1(x) = \int_{\mathbb{R}} e^{iux}\, dP_2(x)$ for all $u \in \mathbb{R}$, then $P_1 \equiv P_2$.

PROOF.   By the inversion formula 7.1.3, h determines F at all continuity points. But as in the proof of 7.1.3, every point is a limit from above of continuity points, and it follows that h determines F everywhere.   □

Various procedures involving Fourier or Laplace transforms may be used in actually computing the distribution of a random variable from its characteristic function (see Ash, 1970, Chapter 5).

Characteristic functions have the following basic properties.

7.1.5   Theorem. Let h be the characteristic function of the bounded distribution function F. Then

  (a) $|h(u)| \le h(0) = F(\infty) - F(-\infty)$ for all u;
  (b) h is continuous on $\mathbb{R}$;
  (c) $h(-u) = \overline{h(u)}$, the complex conjugate of $h(u)$;
  (d) $h(u)$ is real-valued for all u iff F is symmetric; that is, $\int_B dF(x) = \int_{-B} dF(x)$ for all Borel sets B, where $-B = \{-x:\ x \in B\}$.
  (e)   If $\int_{\mathbb{R}} |x|^r\, dF(x) < \infty$ for some positive integer r, then the rth derivative of h exists and is continuous on $\mathbb{R}$, and

$$h^{(r)}(u) = \int_{\mathbb{R}} (ix)^r e^{iux}\, dF(x).$$

PROOF.   Part (a) is clear since $|e^{iux}| = 1$, and (b) follows from the dominated convergence theorem (see Problem 1, 1.6). Part (c) follows from the fact that the conjugate of e^{iux} is e^{-iux}. To prove (d), assume h to be real-valued and, for now, let F be the distribution function of the random variable X. Now $E(e^{-iuX}) = h(-u) = \overline{h(u)} = h(u)$ as $h(u)$ is real, so that $-X$ has characteristic function h. By 7.1.4, X and $-X$ have the same distribution; hence $P\{X \in B\} = P\{X \in -B\}$, and F is therefore symmetric. In general, we may multiply F by a positive constant c such that $c(F(\infty) - F(-\infty)) = 1$. [If $F(\infty) - F(-\infty) = 0$, then $\int_B dF(x) = 0$ for all $B \in \mathscr{B}(\mathbb{R})$ and the result is trivial.] The above argument shows that $\int_B c\, dF(x) = \int_{-B} c\, dF(x)$, and hence F is symmetric.

Conversely, assume F symmetric, and let $g\colon \mathbb{R} \to \mathbb{R}$ be a bounded Borel measurable function. If g is odd [$g(-x) = -g(x)$ for all x], then

$\int_{\mathbb{R}} g(x)\,dF(x) = 0$; to prove this, approximate g by odd simple functions. In particular, $\int_{\mathbb{R}} \sin ux\,dF(x) = 0$, and h is real-valued.

Finally, we prove (e). Since $|(ix)^r e^{iux}| = |x|^r$ and $\int_{\mathbb{R}} |x|^r\,dF(x) < \infty$, we may differentiate $h(u) = \int_{\mathbb{R}} e^{iux}\,dF(x)$ r times under the integral sign (see 1.6, Problem 3 and 2.4, Problem 8), and the result follows. $\square$

Now suppose that h is a given function from $\mathbb{R}$ to $\mathbb{C}$, and we wish to determine whether or not h is the characteristic function of some bounded distribution function F. There are some practical approaches that often work. For example, if h fails to satisfy (a), (b), or (c) of 7.1.5, h cannot be a characteristic function. Also, suppose h is continuous and Lebesgue integrable on $\mathbb{R}$, and we compute

$$f(x) = \frac{1}{2\pi} \int_{\mathbb{R}} e^{-iux} h(u)\,du, \qquad x \in \mathbb{R}.$$

If f is Lebesgue integrable, then

$$h(u) = \int_{\mathbb{R}} e^{iux} f(x)\,dx, \qquad u \in \mathbb{R}.$$

(This is a standard Fourier transform result; see Rudin, 1966, p. 186.) Thus if f is everywhere nonnegative, then h is the characteristic function of a measure μ with density f, that is,

$$\mu(B) = \int_B f(x)\,dx, \qquad B \in \mathscr{B}(\mathbb{R}).$$

There is a general criterion for deciding whether or not h is a characteristic function. The result is of considerable importance in the development of second-order properties of stochastic processes. However, it is in general not useful when applied to explicit examples.

7.1.6 Bochner's Theorem.
If $h\colon \mathbb{R} \to \mathbb{C}$, then h is a characteristic function iff h is continuous at the origin and is *nonnegative definite*, in other words, for all $u_1, \ldots, u_n \in R$, $n = 1, 2, \ldots$, and all complex numbers $a_1, \ldots, a_n$,

$$\sum_{j,k=1}^{n} a_j h(u_j - u_k) \bar{a}_k$$

is real and nonnegative.

PARTIAL PROOF.   Assume h is a characteristic function; then h is continuous everywhere by 7.1.5(b). To prove nonnegative definiteness, write

$$0 \le \int_{-\infty}^{\infty} \left| \sum_{j=1}^{n} a_j e^{iu_j x} \right|^2 dF(x) = \sum_{j,k=1}^{n} \int_{-\infty}^{\infty} a_j e^{iu_j x} \bar{a}_k e^{-iu_k x} \, dF(x)$$

$$= \sum_{j,k=1}^{n} a_j h(u_j - u_k) \bar{a}_k.$$

The converse is considerably more difficult, and since the result will not be used in the text, the argument will be omitted. For a complete proof, see Loève (1955) or Ash and Gardner (1975). $\square$

We shall need the result that if X is normally distributed with mean m and variance σ^2, the characteristic function of X is

$$h(u) = \exp(ium) \exp(-\tfrac{1}{2}u^2\sigma^2).$$

For the computation, see Ash, 1970, p. 163.

Another basic result is the relation between convergence in distribution and convergence in probability.

7.1.7   Theorem. (a)   If X_n converges to X in probability, then X_n converges to X in distribution. (b)   A partial converse: If X_n converges in distribution to a constant c, then X_n converges in probability to c.

PROOF.   (a)   Let F_n be the distribution function of X_n, and F the distribution function of X. Then

$$F_n(x) = P\{X_n \le x\} = P\{X_n \le x, X > x + \varepsilon\} + P\{X_n \le x, X \le x + \varepsilon\}$$

$$\le P\{|X_n - X| \ge \varepsilon\} + P\{X \le x + \varepsilon\}$$

$$= P\{|X_n - X| \ge \varepsilon\} + F(x + \varepsilon), \qquad \text{and}$$

$$F(x - \varepsilon) = P\{X \le x - \varepsilon\} = P\{X \le x - \varepsilon, X_n > x\}$$

$$+ P\{X \le x - \varepsilon, X_n \le x\}$$

$$\le P\{|X_n - X| \ge \varepsilon\} + P\{X_n \le x\}$$

$$= P\{|X_n - X| \ge \varepsilon\} + F_n(x). \qquad \text{Thus}$$

$$F(x - \varepsilon) - P\{|X_n - X| \ge \varepsilon\} \le F_n(x) \le P\{|X_n - X| \ge \varepsilon\}$$

$$+ F(x + \varepsilon)$$

Since $X_n \xrightarrow{P} X$, we have $P\{|X_n - X| \ge \varepsilon\} \to 0$ as $n \to \infty$. If F is continuous at x then $F(x - \varepsilon)$ and $F(x + \varepsilon) \to F(x)$ as $\varepsilon \to 0$. Thus $F_n(x)$ is boxed between two quantities that can be made arbitrarily close to $F(x)$, so $F_n(x) \to F(x)$.

(b) $P\{|X_n - c| \geq \varepsilon\} = P\{X_n \geq c + \varepsilon\} + P\{X_n \leq c - \varepsilon\}$

$$= 1 - P\{X_n < c + \varepsilon\} + P\{X_n \leq c - \varepsilon\}$$

Now $P\{X_n \leq c + \frac{\varepsilon}{2}\} \leq P\{X_n < c + \varepsilon\}$ so

$$P\{|X_n - c| \geq \varepsilon\} \leq 1 - P\left\{X_n \leq c + \frac{\varepsilon}{2}\right\} + P\{X_n \leq c - \varepsilon\}$$

$$= 1 - F_n\left(c + \frac{\varepsilon}{2}\right) + F_n(c - \varepsilon)$$

But as long as $x \neq c$, $F_n(x)$ converges to the distribution function of the constant c, that is,

$$F_n(x) \to \begin{cases} 1 & \text{if } x > c \\ 0 & \text{if } x < c \end{cases}$$

Thus $F_n\left(c + \frac{\varepsilon}{2}\right) \to 1$ and $F_n(c - \varepsilon) \to 0$ as $n \to \infty$ and, therefore, $P\{|X_n - c| \geq \varepsilon\} \to 0$ as $n \to \infty$. $\square$

Problems

1. Give an example of a sequence of random variables X_n such that X_n converges in distribution to X, but X_n does not converge in probability to X.

2. The following application of Theorem 7.1.4 is useful in computations involving characteristic functions. Let f and g be nonnegative Borel measurable functions from $\mathbb{R}$ to $\mathbb{R}$, and assume that for some fixed real t, $\int_{-\infty}^{\infty} f(x)e^{-tx}\, dx < \infty$ and $\int_{-\infty}^{\infty} g(x)e^{-tx}\, dx < \infty$. If

$$\int_{-\infty}^{\infty} f(x)e^{-tx}e^{iux}\, dx = \int_{-\infty}^{\infty} g(x)e^{-tx}e^{iux}\, dx \qquad \text{for all} \qquad u \in \mathbb{R},$$

show that $f = g$ a.e. (Lebesgue measure).

3. If h_1 and h_2 are characteristic functions, show that $h_1 + h_2$ and Re h_1 are also characteristic functions. Is Im h_1 a characteristic function?

4. Let h be the characteristic function of the random variable X.
    (a) If $|h(u)| = 1$ for some $u \neq 0$, show that X has a lattice distribution, that is, with probability 1, X belongs to the set $\{a + nk: n \text{ an integer}\}$ for appropriate a and k $(=2\pi u^{-1})$. Conversely, if X has a lattice distribution, then $|h(u)| = 1$ for some $u \neq 0$.
    (b) If $|h(u)| = 1$ at two distinct points u and αu, where α is irrational, show that X is degenerate, that is, a.e. constant.

5. Let X be a random variable with $E(|X|^n) < \infty$ for some positive integer n. If h is the characteristic function of X, show that $h^{(k)}(0) = i^k E(X^k)$, $k = 0, \ldots, n$, and

(a) $h(u) = \sum_{k=0}^{n} \frac{E(X^k)}{k!}(iu)^k + o(u^n),$

(b) $\left| h(u) - \sum_{k=0}^{n-1} \frac{E(X^k)}{k!}(iu)^k \right| \leq \frac{|u|^n E(|X|^n)}{n!}.$

6.(a) If X is a random variable with $E(|X|^r) < \infty$ for all $r > 0$, show that the characteristic function of X is given by

$$h(u) = \sum_{n=0}^{\infty} \frac{E(X^n)}{n!}(iu)^n$$

within the interval of convergence of the series. This is the *moment-generating property* of characteristic functions.

(b) Give an example of a random variable X with $E(|X|^r) < \infty$ for all $r > 0$, such that the series

$$\sum_{n=0}^{\infty} \frac{E(X^n)}{n!}(iu)^n$$

converges only at $u = 0$.

7.(a) Let h be the characteristic function of the bounded distribution function F. Define

$$E_r h(u) = h(u+r), \qquad r \text{ real.}$$

Show that

$$\left(\frac{E_r - E_{-r}}{2r} \right)^2 h(0) = - \int_{-\infty}^{\infty} \left(\frac{\sin rx}{rx} \right)^2 x^2 \, dF(x).$$

$[(E_r - E_{-r})h(0) = h(r) - h(-r); (E_r - E_{-r})^2 h(0)$ means
$(E_r - E_{-r})(h(r) - h(-r)) = h(2r) - 2h(0) + h(-2r).]$

(b) If h'' exists and is finite at the origin, show that $\int_{-\infty}^{\infty} x^2 dF(x) < \infty$. (Use L'Hôpital's rule and Fatou's lemma.)

(c) If $h^{(2n)}(0)$ exists and is finite, show that $\int_{-\infty}^{\infty} x^{2n} dF(x) < \infty$ ($n = 1, 2, \ldots$).

[It is probably easier to use part (b) and an induction argument rather than to extend part (a).]

8. If X is a random variable, let $N(s) = E(e^{-sX})$, s complex; when $s = -iu$, we obtain the characteristic function of X. If N is analytic at the origin, show that $E(X^k)$ is finite for all $k > 0$, and

$$N(s) = \sum_{k=0}^{\infty} \frac{(-1)^k E(X^k)}{k!} s^k$$

within the circle of convergence of the series. In particular, $N^{(k)}(0) = (-1)^k E(X^k)$.

7.2 THE FUNDAMENTAL WEAK COMPACTNESS THEOREM

The basic connection between weak convergence and characteristic functions is essentially this. Let $\{F_n\}$ be a bounded sequence of distribution functions on $\mathbb{R}$ ("bounded" means that for some positive M, $F_n(\infty) - F_n(-\infty) \leq M$ for all n). Let $\{h_n\}$ be the corresponding sequence of characteristic functions. If F is a bounded distribution function with characteristic function h, then weak convergence of F_n to F is equivalent to pointwise convergence of h_n to h. In the course of developing this result, we must consider the following question. If $\{F_n\}$ is a bounded sequence of distribution functions, when will there exist a weakly convergent subsequence? Now any bounded sequence of real numbers has a convergent subsequence, so one might conjecture that any bounded sequence of distribution functions has a weakly convergent subsequence. In fact this is not true, but the following result comes close, in a sense.

7.2.1 Helly's Theorem. Let $F_1, F_2, \ldots$ be distribution functions on $\mathbb{R}$. Assume that $F_n(-\infty) = 0$ for all n, and $F_n(\infty) \leq M < \infty$ for all n. Then there is a distribution function F and a subsequence $\{F_{n_k}\}$ such that $F_{n_k}(x) \to F(x)$ for each $x \in \mathbb{R}$ at which F is continuous.

PROOF. Let $D = \{x_1, x_2, \ldots\}$ be a countable dense subset of $\mathbb{R}$. Since the sequence $\{F_n(x_1)\}$ is bounded, we can extract a subsequence $\{F_{1j}\}$ of $\{F_n\}$ with $F_{1j}(x_1)$ converging to a limit y_1 as $j \to \infty$. Since $\{F_{1j}(x_2)\}$ is bounded, there is a subsequence $\{F_{2j}\}$ of $\{F_{1j}\}$ such that $F_{2j}(x_2)$ approaches a limit y_2. Continuing inductively, we find subsequences $\{F_{mj}\}$ of $\{F_{m-1,j}\}$ with $F_{mj}(x_m) \to y_m$, $m = 1, 2, \ldots$ (of course all $|y_m|$ are bounded by M).

Define $F_D \colon D \to \mathbb{R}$ by $F_D(x_j) = y_j$, $j = 1, 2, \ldots$, and let $F_{n_k} = F_{kk}$, $k = 1, 2, \ldots$ (the "diagonal sequence"). Then $F_{n_k}(x) \to F_D(x), x \in D$. Since F_{n_k} is one of the original $F_n, x < y$ implies $F_{n_k}(x) \leq F_{n_k}(y)$; hence $F_D(x) \leq F_D(y)$. Define

$$F(x) = \inf\{F_D(y) \colon y \in D, y > x\}.$$

By definition, F is increasing. To prove that F is right-continuous, let $z_n \downarrow x$; then $F(z_n)$ approaches a limit $b \geq F(x)$. If $F(x) < b$, let $y_0 \in D$, $y_0 > x$, with

$F_D(y_0) < b$. For large n we have $x < z_n < y_0$, so $F(z_n) \le F_D(y_0) < b$. Thus $\lim F(z_n) < b$, a contradiction, and therefore $F(z_n) \to F(x)$.

Now we show that $F_{n_k}(x) \to F(x)$ at continuity points of F in $\mathbb{R}$. If $x < y \in D$, we have $\lim \sup_{k \to \infty} F_{n_k}(x) \le \lim \sup_{k \to \infty} F_{n_k}(y) = F_D(y)$; take the inf over $y \in D$, $y > x$ to obtain $\lim \sup_{k \to \infty} F_{n_k}(x) \le F(x)$.

If $x' < y < x$, $y \in D$, we have $F(x') \le F_D(y) = \lim_{k \to \infty} F_{n_k}(y) = \lim \inf_{k \to \infty} F_{n_k}(y) \le \lim \inf_{k \to \infty} F_{n_k}(x)$. Let $x' \to x$ to obtain $F(x^-) \le \lim \inf_{k \to \infty} F_{n_k}(x)$. Thus if $F(x) = F(x^-)$, we have $F_{n_k}(x) \to F(x)$. $\square$

It must be emphasized that Helly's Theorem does *not* say that $F_{n_k}(\infty) \to F(\infty)$. [Recall that $F(\infty)$ is $\lim_{x \to \infty} F(x)$; see the discussion after 1.4.2.] If every F_n is the distribution function of a random variable, so that $F_n(\infty) = 1$ for all n, it is possible for $F(\infty)$ to be strictly less than 1. (See the example that follows in 7.2.2.)

7.2.2   *Comments.* If instead of assuming that $F_n(-\infty) = 0$ and $F_n(\infty) \le M$ for all n, we assume that $F_n(\infty) - F_n(-\infty) \le M < \infty$ for all n, and 7.2.1 implies that there is a distribution function F and a subsequence $\{F_{n_k}\}$ with $F_{n_k}(a, b] \to F(a, b]$ for all $a, b \in \mathbb{R}$ at which F is continuous. [To see this, consider $G_n(x) = F_n(x) - F_n(-\infty)$.]

If $F_1, F_2, \dots$ are distribution functions on $\mathbb{R}^k$ [assumed monotone in each coordinate and 0 at $(-\infty, \dots, -\infty)$], and $F_n(\mathbb{R}^k) \le M < \infty$ for all n, there is a distribution function F and a subsequence $\{F_{n_k}\}$ converging to F at continuity points of F in $\mathbb{R}^k$. The proof is essentially the same as above.

There is no difficulty in constructing a bounded sequence of distribution functions with no weakly convergent subsequence. For example, let F_n be the distribution function of a random variable that is identically $n(F_n(x) = 1, x \ge n; F_n(x) = 0, x < n)$. If F_{n_k} converges weakly to F, then $F(a, b]$ must be 0 for all $a, b \in \mathbb{R}$, $a < b$; hence $F(\infty) - F(-\infty) = 0$. But $F_n(\infty) - F_n(-\infty) \equiv 1$, a contradiction.

A bounded sequence of distribution functions will always have a weakly convergent subsequence unless, as in the above example, too much mass escapes to infinity. Weak convergence requires $F_n(\infty) - F(\infty) \to F(\infty) - F(-\infty)$ (see 2.8.3 and 2.8.4). We now introduce a condition that guarantees that mass does not escape.

7.2.3   *Definition.* Let $\mathscr{M} = \{\mu_i, i \in I\}$ be a family of finite measures on the Borel sets of a metric space Ω. We say that $\mathscr{M}$ is *tight* iff for each $\varepsilon > 0$, there is a compact set $K \subset \Omega$ such that $\mu_i(\Omega - K) < \varepsilon$ for all i. (If $\Omega = \mathbb{R}^k$, the compact set can be replaced by an interval.) We say that $\mathscr{M}$ is *relatively compact* iff each sequence in $\mathscr{M}$ has a subsequence converging weakly to a finite measure on $\mathscr{B}(\Omega)$.

302 7 THE CENTRAL LIMIT THEOREM

If $\{F_i, i \in I\}$ is a family of distribution functions, tightness of $\{F_i\}$ means tightness of the associated family of measures, and similarly for relative compactness.

We may now prove the basic result.

7.2.4 *Prokhorov's Theorem.* Let $\mathscr{M} = \{F_i, i \in I\}$ be a family of distribution functions on $\mathbb{R}$, and assume that $F_i(\infty) - F_i(-\infty) \leq M < \infty$ for all i. Then the family is tight iff it is relatively compact.

PROOF. Assume $\mathscr{M}$ is tight, and let $F_1, F_2, \ldots$ be a sequence from $\mathscr{M}$. By 7.2.1 and 7.2.2, there is a distribution function F and a subsequence $\{F_{n_k}\}$ such that $F_{n_k}(a, b] \to F(a, b]$ for all $a, b \in \mathbb{R}$ at which F is continuous. Given $\varepsilon > 0$, let a and b be finite continuity points of F such that $F_n(\mathbb{R} - (a, b]) < \varepsilon$ for all n, and $F(\mathbb{R} - (a, b]) < \varepsilon$. If $x \in \mathbb{R}$ and x is a continuity point of F, then

$$F_n(\infty) - F_n(x) = F_n(\infty) - F_n(b) + F_n(b) - F_n(x).$$

But $F_{n_k}(b) - F_{n_k}(x) \to F(b) - F(x)$, $F_n(\infty) - F_n(b) < \varepsilon$ for all n, and $F(\infty) - F(b) < \varepsilon$. It follows that for sufficiently large k, $F_{n_k}(\infty) - F_{n_k}(x)$ differs from $F(\infty) - F(x)$ by less than 2ε. Therefore $F_{n_k}(\infty) - F_{n_k}(x) \to F(\infty) - F(x)$, and similarly $F_{n_k}(x) - F_{n_k}(-\infty) \to F(x) - F(-\infty)$. Thus $F_{n_k} \xrightarrow{w} F$, proving relative compactness.

Now if $\mathscr{M}$ is relatively compact but not tight, then for some $\varepsilon > 0$ we have, for each positive integer n, an $F_n \in \mathscr{M}$ with $F_n(\mathbb{R} - (-n, n)) \geq \varepsilon$. If $\{F_{n_k}\}$ is a subsequence converging weakly to F, then since $\mathbb{R} - (-n, n)$ is closed,

$$\limsup_k F_{n_k}(\mathbb{R} - (-n, n)) \leq F(\mathbb{R} - (-n, n)).$$

Thus $F(\mathbb{R} - (-n, n)) \geq \varepsilon$ for all n, and if we let $n \to \infty$, we obtain $0 \geq \varepsilon$, a contradiction. $\square$

Prokhorov's theorem yields the following result for sequences of random variables. If for each $n = 1, 2, \ldots, X_n$ is a random variable with distribution function F_n, and the sequence $\{F_n\}$ is tight, then there is a subsequence $\{F_{n_k}\}$ and a random variable X (possibly defined on a different probability space) with distribution function F, such that F_{n_k} converges weakly to F.

Prokhorov's theorem holds equally well for distribution functions on $\mathbb{R}^k$ (with $F_i(\mathbb{R}^k) \leq M < \infty$ for all i); the proof is essentially the same as above.

7.2.5 *Corollary.* Let $\{F_n\}$ be a bounded sequence of distribution functions on $\mathbb{R}$. If $\{F_n\}$ is tight and every weakly convergent subsequence of $\{F_n\}$ converges to the distribution function F, then $F_n \xrightarrow{w} F$.

PROOF.   If F_n does not converge weakly to F, then $\int_{\mathbb{R}} f(x)\,dF_n(x) \not\longrightarrow \int_{\mathbb{R}} f(x)\,dF(x)$ for some bounded continuous f. Hence there is an $\varepsilon > 0$ such that

$$\left| \int_{\mathbb{R}} f(x)\,dF_n(x) - \int_{\mathbb{R}} f(x)\,dF(x) \right| \geq \varepsilon$$

for infinitely many n, say for $n \in T$. By 7.2.4, $\{F_n, n \in T\}$ has a subsequence $\{F_{n_k}\}$ converging weakly to a distribution function G, and $G = F$ by hypothesis. But then $\int_{\mathbb{R}} f(x)\,dF_{n_k}(x) \to \int_{\mathbb{R}} f(x)\,dF(x)$, a contradiction. $\square$

7.2.6   Corollary. Let $\{F_n\}$ be a bounded sequence of distribution functions on R. If $\{F_n\}$ is tight, then $\{F_n\}$ converges weakly iff $\int_{\mathbb{R}} e^{iux}\,dF_n(x)$ approaches a finite limit as $n \to \infty$ for each $u \in \mathbb{R}$.

PROOF.   Assume $\int_{\mathbb{R}} e^{iux}\,dF_n(x)$ has a finite limit for all u. By 7.2.4, there is a subsequence $\{F_{n_k}\}$ converging weakly to a distribution function F. If F_n does not converge weakly to F, by 7.2.5, there is a subsequence $\{F_{m_k}\}$ converging weakly to a distribution function $G \neq F$. We know by hypothesis that $\int_{\mathbb{R}} e^{iux}\,dF_{n_k}(x)$ and $\int_{\mathbb{R}} e^{iux}\,dF_{m_k}(x)$ have the same limit as $k \to \infty$. Therefore $\int_{\mathbb{R}} e^{iux}\,dF(x) = \int_{\mathbb{R}} e^{iux}\,dG(x)$ for all $u \in \mathbb{R}$. By 7.1.4, $F = G$, a contradiction; thus $F_n \xrightarrow{w} F$. The converse follows from the definition of weak convergence. (In this proof, as in 7.2.5, distribution functions that differ by a constant have been identified.) $\square$

One more result is needed before we can relate weak convergence to convergence of characteristic functions.

7.2.7   Truncation Inequality. Let F be a bounded distribution function on $\mathbb{R}$, with characteristic function h. If $u > 0$, then for some constant $k > 0$.

$$\int_{|x| \geq 1/u} dF(x) \leq \frac{k}{u} \int_0^u [h(0) - \operatorname{Re} h(v)]\,dv.$$

PROOF.
$$\frac{1}{u} \int_0^u [h(0) - \operatorname{Re} h(v)]\,dv = \frac{1}{u} \int_0^u \int_{-\infty}^{\infty} (1 - \cos vx)\,dF(x)\,dv$$

$$= \int_{-\infty}^{\infty} \left[\frac{1}{u} \int_0^u (1 - \cos vx)\,dv \right] dF(x)$$

by Fubini's theorem

$$= \int_{-\infty}^{\infty} \left(1 - \frac{\sin ux}{ux} \right) dF(x)$$

$$\geq \inf_{|t| \geq 1} \left(1 - \frac{\sin t}{t} \right) \int_{|ux| \geq 1} dF(x)$$

$$= \frac{1}{k} \int_{|x| \geq 1/u} dF(x).   \square$$

In fact,

$$\inf_{|t|\geq 1} \left(1 - \frac{\sin t}{t}\right) = 1 - \sin 1 \geq \frac{1}{7},$$

so we may take $k = 7$.

7.2.8   Lévy's Theorem. Let $\{F_n\}$ be a bounded sequence of distribution functions on $\mathbb{R}$, and let $\{h_n\}$ be the corresponding sequence of characteristic functions. If $F_n \xrightarrow{w} F$, where F is a distribution function with characteristic function h, then $h_n(u) \to h(u)$ for all u. Conversely, if h_n converges pointwise to a complex-valued function h, where h is continuous at $u = 0$, then h is the characteristic function of some bounded distribution function F, and $F_n \xrightarrow{w} F$.

PROOF.   The first assertion follows from the definition of weak convergence, so assume $h_n(u) \to h(u)$ for all u, with h continuous at the origin. We claim that $\{F_n\}$ is tight. Using 7.2.7,

$$\int_{|x|\geq 1/u} dF_n(x) \leq \frac{k}{u} \int_0^u [h_n(0) - \operatorname{Re} h_n(v)] \, dv, \, u > 0$$

$$\to \frac{k}{u} \int_0^u [h(0) - \operatorname{Re} h(v)] \, dv \qquad \text{as} \qquad n \to \infty,$$

by the dominated convergence theorem.

Since h is continuous at 0,

$$\frac{k}{u} \int_0^u [h(0) - \operatorname{Re} h(v)] \, dv \to 0 \qquad \text{as} \qquad u \to 0;$$

hence, given $\varepsilon > 0$, we may choose u so small that

$$\int_{|x|\geq 1/u} dF_n(x) < \varepsilon \qquad \text{for all} \qquad n,$$

proving tightness. By 7.2.6, F_n converges weakly to a distribution function F; hence h_n converges pointwise to the characteristic function of F. But we know that $h_n \to h$, so that h is the characteristic function of F. $\square$

The following variation of Lévy's theorem is often useful.

7.2.9   Theorem. Let $\{F_n\}$ be a bounded sequence of distribution functions, and $\{h_n\}$ the corresponding sequence of characteristic functions. If F is a bounded distribution function with characteristic function h, then $F_n \xrightarrow{w} F$ iff $h_n(u) \to h(u)$ for all u, and in this case, h_n converges to h uniformly on bounded intervals.

PROOF.   If $F_n \xrightarrow{w} F$, then $h_n(u) \to h(u)$ for all u by definition of weak convergence. If $h_n(u) \to h(u)$ for all u, then by 7.2.8, F_n converges weakly to the distribution function whose characteristic function is h, namely, F.

Now let I be a bounded interval of R. Then

$$h_n(u + \delta) - h_n(u) = \int_{\mathbb{R}} (e^{i(u+\delta)x} - e^{iux})\, dF_n(x);$$

hence

$$|h_n(u + \delta) - h_n(u)| \le \int_I |e^{i\delta x} - 1|\, dF_n(x) + 2F_n(\mathbb{R} - I).$$

Since $F_n \xrightarrow{w} F$, $\{F_n\}$ is relatively compact, and therefore tight by 7.2.4. Thus if $\varepsilon > 0$ is given, we may choose I so that $2F_n(\mathbb{R} - I) < \varepsilon/2$ for all n. If $u \in \mathbb{R}$, and M is a bound on $\{F_n(\mathbb{R}), n = 1, 2, \ldots\}$, then

$$|h_n(u + \delta) - h_n(u)| \le M \sup_{x\in I} |e^{i\delta x} - 1| + \frac{\varepsilon}{2}$$

$$< \varepsilon \qquad \text{for small enough } \delta = \delta(\varepsilon).$$

Thus the h_n are equicontinuous on $\mathbb{R}$. But equicontinuity and pointwise convergence imply uniform convergence on compact sets (see Ash, 1993, 7.4, Problem 3); hence $h_n \to h$ uniformly on I. $\square$

Problems

1. If X has density $(1/\pi)(1 - \cos x)/x^2$, the characteristic function of X is $h(u) = 1 - |u|, |u| \le 1; h(u) = 0, |u| > 1$. (See Ash, 1970, p. 166, Problem 8.) It is possible to construct a different characteristic function f that agrees with h on $[-1, 1]$, as follows.
   Show that the Fourier series expansion on the interval $[-1, 1]$ of $f(u)$ $= 1 - |u|$ (extended periodically) is

$$f(u) = \frac{1}{2} + \sum_{n=-\infty}^{\infty} \frac{2}{\pi^2(2n + 1)^2} \exp(i(2n + 1)\pi u).$$

[Since f is continuous and of bounded variation, and $f(-1) = f(1)$, the Fourier series converges uniformly to f; see Titchmarsh (1939, p. 410).]

Thus if X is a discrete random variable with $P\{X = 0] = \frac{1}{2}$, $P\{X = (2n + 1)\pi\} = 2\pi^{-2}(2n + 1)^{-2}$, n an integer, then X has characteristic function f. Since f is periodic, we have $f \neq h$, as desired.

2. Using Problem 1, give examples to show that the following results are possible (the h's are characteristic functions of random variables).

   (a) $h_1 h_2 = h_1 h_3$ does not imply $h_2 = h_3$.
   (b) $h_n \to h$ on $[-1, 1]$ does not imply $h_n \to h$ everywhere.

3. Give an example of a sequence of characteristic functions converging pointwise to a function that is not a characteristic function.

4. (a) If $\{a_n\}$ is a sequence of complex numbers and $\exp(iua_n)$ converges to a (finite) limit $g(u)$ for almost all u in the open interval $I \subset \mathbb{R}$, show that $\{a_n\}$ converges.

   (b) If $X_n \equiv n$, so that $h_n(u) = e^{inu}$, $n = 1, 2, \ldots$, show that the sequence of characteristic functions h_n has no pointwise convergent subsequence. Thus the corresponding sequence of distribution functions has no weakly convergent subsequence, as was verified directly in 7.2.2.

5. Let $F_0, F_1, F_2, \ldots$ be distribution functions on $\mathbb{R}$, and assume $F_n(-\infty) = 0$, $n = 0, 1, \ldots$, $F_n(\infty) = 1$, $n = 1, 2, \ldots$. Give examples to show that the following situations are possible.

   (a) $F_n(x) \to F_0(x)$ for all $x \in \mathbb{R}$ at which F is continuous, but F_n does not converge weakly to F_0.
   (b) $F_n(b) - F_n(a) \to F_0(b) - F_0(a)$ for all $a, b \in \mathbb{R}$ at which F_0 is continuous, but $\lim_{n\to\infty} F_n(x)$ does not exist for any $x \in \mathbb{R}$. In particular, F_n does not converge weakly to F_0.

6. Let $F_0, F_1, F_2, \ldots$ be distribution functions on $\mathbb{R}$, and assume that $F_n(\infty) - F_n(-\infty) \leq M < \infty$ for all n, and $F_n(a, b] \to F_0(a, b]$ for all $a, b \in \mathbb{R}$ at which F_0 is continuous. Show that F_n converges weakly to F_0 iff $\{F_n\}$ is tight.

7. Let $F_0, F_1, F_2, \ldots$ be distribution functions on $\mathbb{R}$, and assume that $F_n(\infty) - F_n(-\infty) \leq M < \infty$ for all n. Show that F_n converges weakly to F_0 iff $F_n(a, b] \to F_0(a, b]$ for all $a, b \in \mathbb{R}$ at which F_0 is continuous, and $F_n(\infty) - F_n(-\infty) \to F_0(\infty) - F_0(-\infty)$.

8. Let $Y_1, Y_2, \ldots$ be independent random variables, and let $\mathscr{F}_n$ be the σ-field $\mathscr{F}(Y_1, \ldots, Y_n)$, $n = 1, 2, \ldots$. If h_k is the characteristic function of Y_k and

u is a fixed real number such that $h_k(u) \neq 0, k = 1, 2, \ldots$, define

$$X_n = \prod_{k=1}^{n} \frac{1}{h_k(u)} \exp(iuY_k).$$

(a) Show that $E(X_{n+1} | \mathscr{F}_n) = X_n$ a.e., and hence $\{X_n, \mathscr{F}_n\}$ is a complex-valued martingale. (In other words, $\{\text{Re } X_n, \mathscr{F}_n\}$ and $\{\text{Im } X_n, \mathscr{F}_n\}$ are martingales.)

(b) Assume that $\sum_{k=1}^{n} Y_k \xrightarrow{d} X$. Show that there is an open interval $I \subset \mathbb{R}$, with $0 \in I$, such that for each $u \in I$, $\exp[iu \sum_{k=1}^{n} Y_k(\omega)]$ converges for almost every ω.

(c) Under the hypothesis of part (b), show that for almost every ω, $\exp[iu \sum_{k=1}^{n} Y_k(\omega)]$ converges for almost every $u \in I$ (Lebesgue measure). Thus by Problem 4, $\sum_{k=1}^{\infty} Y_k(\omega)$ converges a.e. to a finite limit.

(d) Conclude that for a series of independent random variables, convergence in distribution, convergence in probability, and convergence almost everywhere are equivalent.

7.3 CONVERGENCE TO A NORMAL DISTRIBUTION

Let $X_1, X_2, \ldots$ be independent random variables, with each X_k having finite mean m_k and finite variance σ_k^2. Let $S_n = \sum_{k=1}^{n} X_k, n = 1, 2, \ldots$; then $E(S_n) = \sum_{k=1}^{n} m_k$, $\text{Var } S_n = c_n^2 = \sum_{k=1}^{n} \sigma_k^2$. We consider the normalized sum $T_n = c_n^{-1}(S_n - E(S_n))$, which has mean 0 and variance 1. [To avoid degeneracy, we assume that $c_n > 0$ for sufficiently large n. In fact the Lindeberg hypothesis (7.3.1) will force c_n to approach ∞ as $n \to \infty$.]

If X^* is a random variable having the normal distribution with mean 0 and variance 1, so that the distribution function of X^* is

$$F^*(x) = \frac{1}{\sqrt{2\pi}} \int_{-\infty}^{x} \exp\left[-\frac{1}{2}t^2\right] dt,$$

we ask for conditions under which T_n converges in distribution to X^*.

A long series of preliminary results were derived before a satisfactory solution was obtained, giving conditions that are sufficient and "almost" necessary for convergence to X^*. We consider sufficiency first.

7.3.1 Lindeberg's Theorem. Let $S_n = X_1 + \cdots + X_n, n = 1, 2, \ldots$, where the X_k are independent random variables with finite mean m_k and finite variance σ_k^2. Let $T_n = c_n^{-1}(S_n - E(S_n))$, where $c_n^2 = \operatorname{Var} S_n = \sum_{k=1}^n \sigma_k^2$, and let F_k be the distribution function of X_k. If for every $\varepsilon > 0$,

$$\frac{1}{c_n^2} \sum_{k=1}^n \int_{\{x:\ |x-m_k| \ge \varepsilon c_n\}} (x - m_k)^2 \, dF_k(x) \to 0 \qquad \text{as} \qquad n \to \infty,$$

then T_n converges in distribution to a random variable X^* that is normal with mean 0 and variance 1.

Before proving the theorem, we examine some of its implications. Lindeberg's theorem implies that $T_n \xrightarrow{d} X^*$ under any one of the following conditions.

1. *The uniformly bounded case.*   Assume $|X_k| \le M$ for all k, and $c_n \to \infty$. Then

$$\int_{\{x:\ |x-m_k| \ge \varepsilon c_n\}} (x - m_k)^2 \, dF_k(x) = E[(X_k - m_k)^2 I_{\{|X_k - m_k| \ge \varepsilon c_n\}}]$$

$$\le (2M)^2 P\{|X_k - m_k| \ge \varepsilon c_n\}$$

$$\le \frac{(2M)^2 \sigma_k^2}{\varepsilon^2 c_n^2}$$

by Chebyshev's inequality. Thus

$$\frac{1}{c_n^2} \sum_{k=1}^n \int_{\{x:\ |x-m_k| \ge \varepsilon c_n\}} (x - m_k)^2 \, dF_k(x) \le \frac{(2M)^2}{\varepsilon^2 c_n^2} \to 0.$$

2. *The identically distributed case.*   Assume that the X_k are iid, with finite mean m and finite variance $\sigma^2 > 0$. If F is the distribution function of the X_k, then

$$\frac{1}{c_n^2} \sum_{k=1}^n \int_{\{x:\ |x-m_k| \ge \varepsilon c_n\}} (x - m_k)^2 \, dF_k(x)$$

$$= \frac{1}{n\sigma^2} \sum_{k=1}^n \int_{\{x:\ |x-m| \ge \varepsilon \sigma \sqrt{n}\}} (x - m)^2 \, dF(x)$$

$$= \frac{1}{\sigma^2} \int_{\{x:\ |x-m| \ge \varepsilon \sigma \sqrt{n}\}} (x - m)^2 \, dF(x)$$

$$\to 0 \quad \text{since } \sigma^2 \text{ is finite and} \quad \{x:\ |x - m| \ge \varepsilon \sigma \sqrt{n}\} \downarrow \emptyset \quad \text{as} \quad n \to \infty.$$

3. *The Bernoulli case.*   Let S_n be the number of successes in n Bernoulli trials, with probability of success p on a given trial. We may write.

$$S_n = X_1 + \cdots + X_n,$$

where the X_k are independent and $P\{X_k = 1\} = p, P\{X_k = 0\} = q = 1 - p$. (We may take X_k as the indicator of a success on trial k.) Thus case 2 applies, with $m = E(X_k) = p, \sigma^2 = E(X_k^2) - [E(X_k)]^2 = p(1 - p), E(S_n) = nm = np, c_n^2 = n\sigma^2 = np(1 - p)$. Thus

$$T_n = \frac{S_n - np}{(npq)^{1/2}}$$

and $T_n \overset{d}{\longrightarrow} X^*$, that is, $P\{T_n \leq x\} \to F^*(x)$ for all x.

4. *Lyapunov's condition.*   Assume that

$$\frac{1}{c_n^{2+\delta}} \sum_{k=1}^{n} E[|X_k - m_k|^{2+\delta}] \to 0 \qquad \text{for some} \qquad \delta > 0.$$

Then

$$E[|X_k - m_k|^{2+\delta}] = \int_{-\infty}^{\infty} |x - m_k|^{2+\delta} \, dF_k(x)$$

$$\geq \int_{\{x: \, |x - m_k| \geq \varepsilon c_n\}} |x - m_k|^{\delta} |x - m_k|^2 \, dF_k(x)$$

$$\geq \varepsilon^{\delta} c_n^{\delta} \int_{\{x: \, |x - m_k| \geq \varepsilon c_n\}} (x - m_k)^2 \, dF_k(x).$$

Thus

$$\frac{1}{c_n^2} \sum_{k=1}^{n} \int_{\{x: \, |x - m_k| \geq \varepsilon c_n\}} (x - m_k)^2 \, dF_k(x) \leq \frac{1}{c_n^2} \sum_{k=1}^{n} \frac{E[|X_k - m_k|^{2+\delta}]}{\varepsilon^{\delta} c_n^{\delta}}$$

$$= \frac{\sum_{k=1}^{n} E[|X_k - m_k|^{2+\delta}]}{\varepsilon^{\delta} c_n^{2+\delta}}$$

$$\to 0.$$

PROOF OF THEOREM 7.3.1.   We may assume without loss of generality that all $m_k = 0$. For if we have proved the theorem under this restriction, let

$$X_k' = X_k - m_k \qquad \text{(so that } EX_k' \equiv 0\text{)}$$

$$S_n' = \sum_{k=1}^{n} X_k'$$

$$T_n' = \frac{S_n' - ES_n'}{c_n'} = \frac{S_n - ES_n}{c_n} = T_n.$$

Since

$$\int_{\{x: \, |x-m_k| \geq \varepsilon c_n\}} (x - m_k)^2 \, dF_k(x) = E[(X_k - m_k)^2 I_{\{|X_k - m_k| \geq \varepsilon c_n\}}]$$

$$= E[(X_k')^2 I_{\{|X_k'| \geq \varepsilon c_n'\}}]$$

$$= \int_{\{x: \, |x| \geq \varepsilon c_n'\}} x^2 \, dF_k'(x),$$

the Lindeberg hypothesis applies to the random variables X_k'; hence $T_n' \xrightarrow{d} X^*$. But then $T_n \xrightarrow{d} X^*$, as desired.

The following estimates will be needed. If y is any real number and z a complex number with $|z| \leq \frac{1}{2}$,

$$e^{iy} = 1 + iy + \frac{\theta y^2}{2}, \tag{1}$$

$$e^{iy} = 1 + iy - \frac{y^2}{2} + \frac{\theta_1 |y|^3}{6}, \tag{2}$$

where θ and θ_1 depend on y, and $|\theta| \leq 1$, $|\theta_1| \leq 1$;

$$\text{Log}(1 + z) = z + \theta' |z|^2, \tag{3}$$

where "Log" denotes the principal branch of the logarithm and $|\theta'| \leq 1$, θ' depending on z. (These formulas are exercises in calculus; for full details see Ash, 1970, p. 173.)

Throughout the proof, h_k will denote the characteristic function of X_k, and u will be a fixed real number. The characteristic function of $T_n = S_n/c_n$ is

$$h_{T_n}(u) = E(e^{iuT_n}) = E(e^{iuS_n/c_n}) = h_{S_n}\left(\frac{u}{c_n}\right) = \prod_{k=1}^{n} h_k\left(\frac{u}{c_n}\right). \tag{4}$$

By (1) and (2),

$$h_k(u) = \int_{-\infty}^{\infty} e^{iux} \, dF_k(x)$$

$$= \int_{|x| \geq \varepsilon c_n} \left(1 + iux + \frac{\theta u^2 x^2}{2}\right) dF_k(x)$$

$$+ \int_{|x| < \varepsilon c_n} \left(1 + iux - \frac{u^2 x^2}{2} + \frac{\theta_1 |u|^3 |x|^3}{6}\right) dF_k(x),$$

$|\theta|, |\theta_1| \leq 1, \theta, \theta_1$ depending on ux. Since

$$\int_{-\infty}^{\infty} iux \, dF_k(x) = iuEX_k = 0,$$

we may drop the terms involving iux. Thus

$$h_k\left(\frac{u}{c_n}\right) = 1 + \frac{u^2}{2c_n^2} \int_{|x| \geq \varepsilon c_n} \theta x^2 \, dF_k(x)$$

$$- \frac{u^2}{2c_n^2} \int_{|x| < \varepsilon c_n} x^2 \, dF_k(x) + \frac{|u|^3}{6c_n^3} \int_{|x| < \varepsilon c_n} \theta_1 |x|^3 \, dF_k(x). \quad (5)$$

Now

$$\left| \frac{1}{2} \int_{|x| \geq \varepsilon c_n} \theta x^2 \, dF_k(x) \right| \leq \frac{1}{2} \int_{|x| \geq \varepsilon c_n} x^2 \, dF_k(x);$$

hence

$$\frac{1}{2} \int_{|x| \geq \varepsilon c_n} \theta x^2 \, dF_k(x) = \theta_2 \int_{|x| \geq \varepsilon c_n} x^2 \, dF_k(x), \qquad \text{where} \qquad |\theta_2| \leq \frac{1}{2}.$$

Similarly,

$$\left| \frac{1}{6} \int_{|x| < \varepsilon c_n} \theta_1 |x|^3 \, dF_k(x) \right| \leq \frac{1}{6} \int_{|x| < \varepsilon c_n} |x|^3 \, dF_k(x)$$

$$\leq \frac{1}{6} \int_{|x| < \varepsilon c_n} \frac{\varepsilon c_n}{|x|} |x|^3 \, dF_k(x)$$

(note that $|x| < \varepsilon c_n$ implies $\varepsilon c_n / |x| \geq 1$). Hence

$$\frac{1}{6} \int_{|x| < \varepsilon c_n} \theta_1 |x|^3 \, dF_k(x) = \theta_3 \int_{|x| < \varepsilon c_n} \varepsilon c_n x^2 \, dF_k(x), \qquad |\theta_3| \leq \frac{1}{6}.$$

Thus if we set

$$\alpha_{nk} = \frac{1}{c_n^2} \int_{|x| \geq \varepsilon c_n} x^2 \, dF_k(x),$$

$$\beta_{nk} = \frac{1}{c_n^2} \int_{|x| < \varepsilon c_n} x^2 \, dF_k(x), \quad (\leq \varepsilon^2),$$

(5) becomes

$$h_k \left(\frac{u}{c_n} \right) = 1 + \gamma_{nk}, \tag{6}$$

where

$$\gamma_{nk} = \theta_2 u^2 \alpha_{nk} - \frac{u^2}{2} \beta_{nk} + |u|^3 \varepsilon \theta_3 \beta_{nk}. \tag{7}$$

Now $\sum_{k=1}^n \alpha_{nk} \to 0$ by hypothesis, and

$$\sum_{k=1}^n (\alpha_{nk} + \beta_{nk}) = \frac{1}{c_n^2} \sum_{k=1}^n \sigma_k^2 = 1.$$

Thus by (7) and the fact that $\beta_{nk} \leq \varepsilon^2$, we have

$$\max_k |\gamma_{nk}| \leq \frac{u^2 \varepsilon^2}{2} + |u|^3 \varepsilon \quad \text{and} \quad \sum_k |\gamma_{nk}| \leq \frac{u^2}{2} + |u|^3 \varepsilon$$

for sufficiently large n. By (3),

$$\frac{u^2}{2} + \sum_{k=1}^n \text{Log}(1 + \gamma_{nk}) = \frac{u^2}{2} + \sum_{k=1}^n (\gamma_{nk} + \theta' |\gamma_{nk}|^2), \qquad |\theta'| \leq 1.$$

Now

$$\sum_{k=1}^n |\gamma_{nk}|^2 \leq \left(\max_k |\gamma_{nk}| \right) \sum_{k=1}^n |\gamma_{nk}|, \tag{8}$$

and it follows from (8) that if $\delta > 0$ is given, and ε is chosen sufficiently small (depending on δ and u), then

$$\left| \frac{u^2}{2} + \sum_{k=1}^n \text{Log}(1 + \gamma_{nk}) \right|$$

will be less than δ for sufficiently large n. Since the exponential function is continuous, we obtain from (6) that

$$\exp\left(\frac{u^2}{2}\right) \prod_{k=1}^{n} h_k\left(\frac{u}{c_n}\right) \to 1,$$

in other words [see (4)], $h_{T_n}(u) \to \exp(-u^2/2)$, the characteristic function of X^*. By 7.2.9, $T_n \xrightarrow{d} X^*$. $\square$

There is another proof of the Lindeberg theorem that uses weak convergence directly, rather than via Lévy's theorem; see Billingsley (1968, p. 42).

We now show that the Lindeberg hypothesis is not necessary for convergence to X^*. If the Lindeberg condition holds, we claim that for all $\varepsilon > 0$,

$$P\left\{\frac{|X_k - m_k|}{c_n} \geq \varepsilon\right\} \to 0 \qquad \text{as} \qquad n \to \infty,$$

uniformly in k. This is referred to as *uniform asymptotic negligibility* (uan) of the random variables $(X_k - m_k)/c_n$. For

$$\frac{1}{c_n^2} \sum_{k=1}^{n} \int_{\{x:\, |x-m_k| \geq \varepsilon c_n\}} (x - m_k)^2 \, dF_k(x)$$

$$\geq \varepsilon^2 \sum_{k=1}^{n} P\{|X_k - m_k| \geq \varepsilon c_n\}$$

$$\geq \varepsilon^2 \max_{1 \leq k \leq n} P\{|X_k - m_k| \geq \varepsilon c_n\}.$$

Thus, intuitively, the contribution of each $(X_k - m_k)/c_n$ is small relative to the sum $(S_n - ES_n)/c_n$.

If we can construct an example of a sequence that is not uan but for which $T_n \xrightarrow{d} X^*$, we then have convergence to X^* without the Lindeberg condition; here is one possibility. Let $X_1, X_2, \ldots$ be independent, normally distributed random variables with zero mean. Let $\sigma_k^2 = 2^{k-2}$, $k \geq 2$; $\sigma_1^2 = 1$. Then $c_n^2 = \sum_{k=1}^{n} \sigma_k^2 = 2^{n-1}$; hence X_n/c_n is normal with mean 0 and variance $\frac{1}{2}$. Therefore

$$\max_{1 \leq k \leq n} P\left\{\left|\frac{X_k}{c_n}\right| \geq \varepsilon\right\} \geq P\left\{\left|\frac{X_n}{c_n}\right| \geq \varepsilon\right\},$$

which is a positive constant not depending on n. Thus the X_k/c_n are not uan, although $T_n \xrightarrow{d} X^*$. [In fact T_n is normal (0, 1) for all n.]

If we impose the uan requirement, the Lindeberg condition becomes necessary and sufficient for convergence to X^*.

7.3.2   Theorem. Let $X_1, X_2, \ldots$ be independent random variables, with each X_k having finite mean m_k and finite variance σ_k^2. Then the Lindeberg condition

$$\frac{1}{c_n^2} \sum_{k=1}^{n} \int_{\{x: \ |x-m_k| \geq \varepsilon c_n\}} (x - m_k)^2 \, dF_k(x) \to 0 \qquad \text{for all} \qquad \varepsilon > 0$$

holds if and only if $T_n \xrightarrow{d} X^*$ and the $(X_k - m_k)/c_n$ are uan.

PROOF.   The "only if" part follows from 7.3.1 and the above remarks. The "if" part, due to Feller (1950), is rather lengthy, and is proved in Appendix 3.   □

In the Lindeberg theorem, we have normalized the sum S_n in a special way, that is, we have considered a sequence of random variables $a_n^{-1}(S_n - b_n)$, where b_n is the mean and a_n the standard deviation of S_n. We might ask whether a different choice of constants a_n and b_n would produce different results, for example, convergence to a nonnormal random variable. Questions of this nature may be handled by the "theorem on convergence of types," which we now develop.

7.3.3   Definition.   Two random variables X and Y (or their distribution functions G and F) are said to be of the *same positive type* iff X and $a^{-1}(Y - b)$ have the same distribution for some real a and b, $a > 0$, that is, $G(x) = F(ax + b)$ for all x; X and Y are of the *same type* iff X and $a^{-1}(Y - b)$ have the same distribution for some real a, b with $a \neq 0$, not necessarily positive. (The notation $X \overset{d}{=} Y$ will indicate that X and Y have the same distribution.)

The notion of type is preserved under convergence in distribution, as the following basic theorem shows.

7.3.4   Convergence of Types Theorem.   (a)   Let $X_n \xrightarrow{d} X$, $Y_n \xrightarrow{d} Y$, where for each n, X_n and Y_n are of the same positive type, with $X_n \overset{d}{=} a_n^{-1}(Y_n - b_n)$, $a_n > 0$. Assume X and Y are *nondegenerate*, that is, not a.e. constant. Then there are real numbers a and b, with $a > 0$, such that $a_n \to a$, $b_n \to b$, and $X \overset{d}{=} a^{-1}(Y - b)$; thus X and Y are of the same positive type.

(b)   If $X_n \xrightarrow{d} X$, $Y_n \xrightarrow{d} Y$, where for each n, X_n and Y_n are of the same type, with $X_n \overset{d}{=} a_n^{-1}(Y_n - b_n)$, $a_n \neq 0$, and X and Y are nondegenerate, then there are real numbers a and b, with $a \neq 0$, such that $|a_n| \to |a|$ and $X \overset{d}{=} a^{-1}(Y - b)$; thus X and Y are of the same type.

The proof is an intricate exercise in real analysis, and is done in Appendix 4.

We can now consider the question raised earlier about the normalizing constants. If $(a_n')^{-1}(S_n - b_n') \overset{d}{\longrightarrow} Y$, and $a_n^{-1}(S_n - b_n) \overset{d}{\longrightarrow} X$, where Y is normal and X is nondegenerate, then X must be normal. For

$$a_n^{-1}(S_n - b_n) = \frac{a_n'}{a_n}[(a_n')^{-1}(S_n - b_n') - (a_n')^{-1}(b_n - b_n')];$$

hence by the convergence of types theorem, X and $a^{-1}(Y - b)$ have the same distribution for some a, b, with $a \neq 0$. But a brief computation then shows that X is normal. (To do this, look at characteristic functions, or use the technique of 4.9.4.) Thus if one set of constants produces a normal limit, all nondegenerate limits are normal.

The restriction that X be nondegenerate is necessary. For given any real number c, it is always possible to choose the constants a_n and b_n so that $a_n^{-1}(S_n - b_n) \overset{d}{\longrightarrow} c$ (see Problem 1).

There is a tricky aspect of Theorem 7.3.2 that is worth mentioning. Under the uan hypothesis, the Lindeberg condition is necessary and sufficient for $T_n \overset{d}{\longrightarrow} X^*$; thus if the $c_n^{-1}(X_k - m_k)$ are uan and the Lindeberg condition fails, $c_n^{-1}(S_n - E(S_n))$ cannot converge in distribution to a normal $(0, 1)$ random variable. However, it is possible for $c_n^{-1}(S_n - E(S_n))$ to converge in distribution to a random variable X that is not degenerate and not normal $(0, 1)$. In Problem 4, an example is given in which X is normal with a variance unequal to 1.

In fact there are examples of independent (but not identically distributed) random variables $X_1, X_2, \ldots$, each with finite mean and variance, such that for some constants a_n, b_n we have $a_n^{-1}(S_n - b_n) \overset{d}{\longrightarrow} X$, where X is not degenerate and not normal. The construction of such examples is quite elaborate, and we give the reference only: Gnedenko and Kolmogorov (1954, p. 152).

Problems

1. Let $X_1, X_2, \ldots$ be an arbitrary sequence of random variables. Given any real number c, show that constants a_n and $b_n (a_n > 0)$ can always be chosen so that $a_n^{-1}(X_1 + \cdots + X_n - b_n)$ converges in distribution to a random variable $X \equiv c$.

2. Give an example of sequences of random variables $\{X_n\}$ and $\{Y_n\}$ such that $X_n \overset{d}{=} a_n^{-1}(Y_n - b_n)$ for real numbers a_n and $b_n (a_n \neq 0)$, $X_n \overset{d}{\longrightarrow} X$, $Y_n \overset{d}{\longrightarrow} Y$ (so that X and Y are of the same type), but $|b_n|$ has no limit.

3. Let $\{X_{nk}, n = 1, 2, \ldots, k = 1, \ldots, n\}$ be a double sequence of random variables, and let h_{nk} be the characteristic function of X_{nk}. Show that the X_{nk} are uan; (in other words,

$$\max_{1 \leq k \leq n} P\{|X_{nk}| \geq \varepsilon\} \to 0$$

as $n \to \infty$ for every $\varepsilon > 0$) iff

$$\max_{1 \leq k \leq n} |h_{nk}(u) - 1| \to 0$$

as $n \to \infty$, and in this case, the convergence is uniform on any bounded interval. (Use 7.2.7 in the "if" part.)

4. Let $X_1, X_2, \ldots$ be independent random variables, defined as follows:

$X_1 = \pm 1$ with equal probability.
If $k > 1$, and c is a fixed real number greater than 1,

$$P\{X_k = 1\} = P\{X_k = -1\} = \frac{1}{2c},$$

$$P\{X_k = k\} = P\{X_k = -k\} = \frac{1}{2k^2}\left(1 - \frac{1}{c}\right),$$

$$P\{X_k = 0\} = 1 - \frac{1}{c} - \frac{1}{k^2}\left(1 - \frac{1}{c}\right).$$

Define

$$X_{nk}' = \begin{cases} X_k & \text{if} & |X_k| \leq \sqrt{n}, \\ 0 & \text{if} & |X_k| > \sqrt{n}. \end{cases}$$

Establish the following:

(a) The X_k/c_n satisfy the uan condition.
(b) The Lindeberg condition fails for the X_k, but holds for the X_{nk}'. Furthermore, if $S_n' = \sum_{k=1}^n X_{nk}'$, then $S_n'/c_n' \overset{d}{\longrightarrow}$ normal $(0, 1)$, where $(c_n')^2 = \text{Var } S_n' \sim n/c$.

(c) If $S_n = \sum_{k=1}^{n} X_k$, then $P\{S_n \neq S_n{}'\} \to 0$ as $n \to \infty$.

(d) $\sqrt{c}\, S_n/\sqrt{n} \xrightarrow{d}$ normal $(0, 1)$, but $S_n/\sqrt{n} \not\xrightarrow{d}$ normal $(0, 1)$.

7.4 STABLE DISTRIBUTIONS

If $X_1, X_2, \ldots$ are independent, identically distributed random variables, with finite mean m and finite variance σ^2, we know from the previous section that $(S_n - nm)/\sigma\sqrt{n} \xrightarrow{d} X^*$ normal $(0, 1)$; hence any limiting distribution of a sequence $a_n^{-1}(S_n - b_n)$ must be normal. If we drop the finite variance requirement, it is possible to obtain a nonnormal limit. For example, let the X_i have the Cauchy density $f(x) = \theta/\pi(x^2 + \theta^2)$, $x \in \mathbb{R}$, θ a fixed positive constant. The corresponding characteristic function is $h(u) = e^{-\theta|u|}$ (see Ash, 1970, p. 161, for the computation). Therefore S_n has characteristic function $[h(u)]^n = e^{-n\theta|u|}$; hence $n^{-1}S_n$ has characteristic function $[h(u/n)]^n = e^{-\theta|u|}$. Thus $n^{-1}S_n \xrightarrow{d} X$, where X has the Cauchy density with parameter θ. Since $E(|X|) = \infty$, this does not contradict the previous results.

The following investigation is suggested. Let $X_1, X_2, \ldots$ be iid random variables. If $a_n^{-1}(S_n - b_n) \xrightarrow{d} X$, what are the possible distributions of X? (We may assume that $a_n > 0$; for if negative a_n are allowed, we consider the two subsequences corresponding to $a_n > 0$ and $a_n < 0$.) In fact the possible limiting distributions may be completely characterized, as follows:

7.4.1 Definition. A random variable X (or its distribution function F, or its characteristic function h) is said to be *stable* iff, whenever $X_1, \ldots, X_n$ are iid random variables with distribution function F, then $S_n = X_1 + \cdots + X_n$ is of the same positive type as X; in other words, $X \overset{d}{=} a_n^{-1}(S_n - b_n)$, or equivalently $[h(u)]^n = \exp[ib_n u]h(a_n u)$, $u \in \mathbb{R}$, for appropriate $a_n > 0$ and b_n.

A sequence $\{a_n^{-1}(S_n - b_n)\}$, where $a_n > 0$ and $S_n = \sum_{k=1}^{n} X_k$, the X_k iid, is called a sequence of *normed sums*.

7.4.2 Theorem. The random variable X is stable iff there is a sequence of normed sums converging in distribution to X.

PROOF. If X is stable, let $X_1, \ldots, X_n$ be iid with $X_j \overset{d}{=} X$. Then $a_n^{-1}(S_n - b_n) \xrightarrow{d} X$ for appropriate $a_n > 0$ and b_n; in particular, $a_n^{-1}(S_n - b_n) \xrightarrow{d} X$.

Assume $X_1, X_2, \ldots$ iid, with $V_n = a_n^{-1}(S_n - b_n) \xrightarrow{d} X$. If X is degenerate, it is stable, so assume X nondegenerate. Fix the positive integer r, and define

$$S_n^{(1)} = X_1 + \cdots + X_n,$$
$$S_n^{(2)} = X_{n+1} + \cdots + X_{2n},$$
$$\vdots$$
$$S_n^{(r)} = X_{(r-1)n+1} + \cdots + X_{rn}.$$

Then let

$$W_n^{(r)} = \frac{S_n^{(1)} - b_n}{a_n} + \frac{S_n^{(2)} - b_n}{a_n} \cdots + \frac{S_n^{(r)} - b_n}{a_n} = Z_n^{(1)} + \cdots + Z_n^{(r)},$$

where $Z_n^{(1)}, \ldots, Z_n^{(r)}$ are independent. Now $Z_n^{(i)} \stackrel{d}{=} Z_n^{(1)}$ for all i; hence $Z_n^{(i)} \xrightarrow{d} X$ for each i. It follows from 7.2.9 that $W_n^{(r)} \xrightarrow{d} Z_1 + \cdots + Z_r$, where $Z_1, \ldots, Z_r$ are iid with $Z_i \stackrel{d}{=} X$. But we may also write

$$W_n^{(r)} = \frac{X_1 + \cdots + X_{rn} - rb_n}{a_n} = \frac{a_{rn}}{a_n} \left(\frac{X_1 + \cdots + X_{rn} - b_{rn}}{a_{rn}} \right) + \frac{b_{rn} - rb_n}{a_n}$$
$$= \alpha_n^{(r)} V_{rn} + \beta_n^{(r)},$$

where $\alpha_n^{(r)} = a_{rn}/a_n > 0$. To summarize:

$$V_{rn} = \frac{W_n^{(r)} - \beta_n^{(r)}}{\alpha_n^{(r)}}, \qquad V_{rn} \xrightarrow{d} X, \qquad W_n^{(r)} \xrightarrow{d} Z_1 + \cdots + Z_r.$$

By 7.3.4(a), $\alpha_n^{(r)}$ approaches a limit $\alpha_r > 0$, $\beta_n^{(r)}$ approaches a limit β_r, and $X \stackrel{d}{=} (\alpha_r)^{-1}(Z_1 + \cdots + Z_r - \beta_r)$. Thus X is stable. $\square$

In the above proof, we must have $\alpha_{rs} = \alpha_r \alpha_s$ for all positive integers r and s. For

$$\alpha_n^{(rs)} = \frac{a_{rsn}}{a_n} = \frac{a_{rn}}{a_n} \frac{a_{rsn}}{a_{rn}} = \alpha_n^{(r)} \alpha_{rn}^{(s)},$$

and we may let $n \to \infty$ to obtain $\alpha_{rs} = \alpha_r \alpha_s$.

7.4.3 Examples. It can be shown [see, for example, Breiman (1968, p. 204)] that X is stable iff its characteristic function h can be expressed as $h = e^g$, where g has one of the following two forms:

$$g(u) = iu\beta - d|u|^\alpha \left(1 + i\theta \frac{u}{|u|} \tan \frac{\pi}{2} \alpha \right),$$

$$0 < \alpha < 1 \quad \text{or} \quad 1 < \alpha \le 2, \quad \beta \in \mathbb{R}, \quad d \ge 0, \quad |\theta| \le 1, \quad (1)$$

$$g(u) = iu\beta - d|u| \left(1 + i\theta \frac{u}{|u|} \frac{2}{\pi} \ln|u| \right), \quad (2)$$

with β, d, θ as in (1). (Take $u/|u| = 0$ when $u = 0$.)

We shall prove only that a random variable with such a characteristic function must be stable. If $X_1, \ldots, X_n$ are iid, with each X_i having characteristic function $h = e^g$, with g of the form (1) or (2), let $\lambda = 1/\alpha$ in case (1), $\lambda = 1$ in case (2). Then in case (1),

$$g(n^\lambda u) = iun^\lambda \beta - dn|u|^\alpha \left[1 + i\theta \frac{u}{|u|} \tan \frac{\pi}{2} \alpha \right] = ng(u) - iu\beta(n - n^\lambda),$$

and in case (2),

$$g(n^\lambda u) = g(nu) = iun\beta - dn|u| \left[1 + i\theta \frac{u}{|u|} \frac{2}{\pi} (\ln n + \ln |u|) \right]$$

$$= ng(u) - iud\theta \frac{2}{\pi} n \ln n.$$

Thus in either case,

$$[h(u)]^n = \exp[ng(u)] = \exp[g(n^\lambda u)] \exp[ib_n u] = h(n^\lambda u) \exp[ib_n u],$$

where $b_n = \beta(n - n^\lambda)$ in case (1), and $b_n = d\theta(2/\pi)n \ln n$ in case (2). Therefore $S_n \stackrel{d}{=} a_n X + b_n$, with $a_n = n^\lambda = n^{1/\alpha}$; in particular, X is stable.

Now X has a symmetric distribution ($P\{X \in B\} = P\{-X \in B\}$ for all $B \in \mathcal{B}(\mathbb{R})$) iff h is real-valued [see 7.1.5(d)], so the general form of the symmetric stable characteristic function is

$$h(u) = \exp[-d|u|^\alpha], \quad d \ge 0, \quad 0 < \alpha \le 2. \quad (3)$$

When $d = 0$, we have $X \equiv 0$. [Similarly, if $d = 0$ in (1) or (2), then $X \equiv \beta$.] Thus assume $d > 0$. When $\alpha = 2$, X is normal $(0, 2d)$; when $\alpha = 1$, X has the Cauchy density with parameter d.

If X is stable (not necessarily symmetric) and $0 < \alpha \le 1$, then h is not differentiable at $u = 0$, so by 7.1.5(e), $E(|X|) = \infty$. In the symmetric case, $E(X)$ does not exist. For $X \overset{d}{=} -X$; hence $X^+ \overset{d}{=} (-X)^+ = X^-$, and therefore $E(X^+) = E(X^-)$, necessarily infinite. If $1 < \alpha < 2$, h can be differentiated once but not twice at $u = 0$, so that $E(X^2) = \infty$. This is to be expected, for if X has finite mean and variance, the fact that X can be obtained as a limit of a sequence of normed sums implies that X must be normal (see the opening paragraph of this section).

It can be shown (see Feller, 1966, p. 215) that if X is stable, X has a finite rth moment for all $r \in (0, \alpha)$.

Problem

1. The following problem shows that the functions $h(u) = \exp[-d|u|^\alpha]$, $d \ge 0$, $0 < \alpha \le 2$, are characteristic functions (see 7.4.3). Let $X_1, \ldots, X_n$ be independent random variables, each uniformly distributed between $-n$ and $+n$. Define

$$Y_n = k \sum_{i=1}^{n} \frac{\text{sgn } X_i}{|X_i|^r}, \qquad k > 0, \ r > \frac{1}{2}.$$

If the X_i are the positions of masses distributed at random on $[-n, n]$, then Y_n is the gravitational force at the origin, assuming an inverse rth power law.

(a) Show that the characteristic function of Y_n is

$$h_n(u) = \left(1 - \frac{1}{n}\left[\int_0^\infty [1 - \cos(kux^{-r})]\, dx - g(n)\right]\right)^n,$$

where $g(n) \to 0$ as $n \to \infty$.

(b) Show that $h_n(u) \to h(u) = \exp(-\int_0^\infty [1 - \cos(kux^{-r})]\, dx)$ as $n \to \infty$.

(c) Make the change of variable $y = |u|^{1/r} k^{1/r} x^{-1}$ to show that $h(u)$ is of the form $\exp[-d|u|^\alpha]$, $d > 0$, $0 < \alpha < 2$. (The case $\alpha = 2$ corresponds to a normal distribution, and $d = 0$ to a degenerate distribution, so these characteristic functions are automatically realizable.)

7.5 INFINITELY DIVISIBLE DISTRIBUTIONS

There are limit laws that do not fit into any of the categories we have considered so far. For example, let T_n be the number of successes in n Bernoulli trials, with probability p_n of success on a given trial. Then T_n has the binomial distribution:

$$P\{T_n = k\} = \binom{n}{k} p_n^k (1 - p_n)^{n-k}, \qquad k = 0, 1, \ldots, n.$$

If we let $n \to \infty$, $p_n \to 0$, with $np_n \to \lambda$, then

$$P\{T_n = k\} \to \frac{e^{-\lambda}\lambda^k}{k!}, \qquad k = 0, 1, \ldots$$

(see Ash, 1970, p. 95, for details). A discrete random variable X with $P\{X = k\} = e^{-\lambda}\lambda^k/k!, k = 0, 1, \ldots$, is said to have the *Poisson* distribution. In this case,

$$F_{T_n}(k) = P\{T_n \le k\} = \sum_{j=0}^{k} P\{T_n = j\} \to \sum_{j=0}^{k} P\{X = j\}$$

$$= P\{X \le k\} = F_X(k);$$

hence $T_n \xrightarrow{d} X$.

Now this can be regarded as a limit law for sums of independent random variables. We may represent T_n as $X_{n1} + X_{n2} + \cdots + X_{nn}$, where X_{ni}, the number of successes on trial i, or equivalently, the indicator of the event {successes on trial i}, is 1 with probability p_n and 0 with probability $1 - p_n$, and the X_{ni} are independent. The difference between this case and the previous ones is that we are no longer dealing with a *single sequence* of random variables; T_n is not simply $X_1 + \cdots + X_n$, where $X_1, X_2, \ldots$ are independent. Instead, for each n we have a different sequence $X_{n1}, \ldots, X_{nn}$.

We may construct a model that includes this case as well as all previous results, as follows. Consider a *triangular array*:

$$\begin{bmatrix} X_{11} & & & & \\ X_{21} & X_{22} & & & \\ X_{31} & X_{32} & X_{33} & & \\ \vdots & & & & \\ X_{n1} & X_{n2} & X_{n3} & \cdots & X_{nn} \\ \vdots & & & & \end{bmatrix}.$$

We assume that for each n, $X_{n1}, \ldots, X_{nn}$ are independent. (We say nothing as yet about any relation between rows.) We set $T_n = X_{n1} + \cdots + X_{nn}$; we want to investigate convergence in distribution of the sequence $\{T_n\}$.

Notice that if we are interested in sequences $\{a_n^{-1}(S_n - b_n)\}$, where S_n is the sum of independent random variables $X_1, \ldots, X_n$, we may construct an appropriate triangular array; take

$$X_{ni} = \frac{X_i}{a_n} - \frac{b_n}{na_n};$$

then

$$T_n = \sum_{i=1}^{n} X_{ni} = \frac{S_n - b_n}{a_n}.$$

Thus the triangular array scheme includes the previous models we have considered.

Note also that the Lindeberg theorem holds for triangular arrays. If the X_{nk} have finite mean m_{nk} and finite variance σ_{nk}^2,

$$c_n^2 = \text{Var}\left(\sum_{k=1}^{n} X_{nk}\right) = \sum_{k=1}^{n} \sigma_{nk}^2,$$

and for every $\varepsilon > 0$,

$$\frac{1}{c_n^2} \sum_{k=1}^{n} \int_{\{x:\ |x-m_{nk}| \geq \varepsilon c_n\}} (x - m_{nk})^2 \, dF_{nk}(x) \to 0 \qquad \text{as} \qquad n \to \infty,$$

then

$$c_n^{-1} \sum_{k=1}^{n} (X_{nk} - m_{nk}) \xrightarrow{d} X^* \qquad \text{normal } (0, 1).$$

The proof is the same as in 7.3.1, with the distribution function F_k replaced by F_{nk}.

A natural question is the characterization of the possible limiting distributions of a triangular array; this problem was solved for normed sums in Section 7.4. However, as it stands, the question is not sensible, even if we require that the triangular array come from a single sequence of random variables. Let X be an arbitrary random variable, and take $X_1 = X$, $X_n = 0$ for $n \geq 2$, $b_n \equiv 0$, $a_n \equiv 1$. Then $a_n^{-1}(S_n - b_n) \equiv X$, so any limit distribution is possible. Thus some restriction must be imposed.

One way to take care of this difficulty is to assume the hypothesis of uniform asymptotic negligibility, as we did in considering the converse of the Lindeberg theorem:

$$\max_{i \leq i \leq n} P\{|X_{ni}| \geq \varepsilon\} \to 0 \qquad \text{as} \qquad n \to \infty \qquad \text{for every} \qquad \varepsilon > 0.$$

However, we will sacrifice generality for simplicity, and assume that for each n, $X_{n1}, X_{n2}, \ldots, X_{nn}$ are identically distributed. We may then characterize the possible limiting distributions.

7.5.1 Definition. A random variable X (or its distribution function F, or its characteristic function h) is said to be *infinitely divisible* iff for each n, X has the same distribution as the sum of n independent, identically distributed random variables. In other words, for each n, we may write $h = (h_n)^n$, where h_n is the characteristic function of a random variable.

7.5.2   Theorem. The random variable X is infinitely divisible iff there is a triangular array, with $X_{n1}, \ldots, X_{nn}$ iid for each n, such that $T_n = \sum_{k=1}^n X_{nk} \overset{d}{\longrightarrow} X$.

PROOF. Let X be infinitely divisible. For each n, we may write $X \overset{d}{=} X_{n1} + \cdots + X_{nn}$, where the X_{ni} are iid. Then $T_n = \sum_{i=1}^n X_{ni} \equiv X$; hence $T_n \overset{d}{\longrightarrow} X$.

The converse is another application of Prokhorov's weak compactness theorem. Assume we have a triangular array with the X_{nk} iid for each n and $T_n \overset{d}{\longrightarrow} X$. Fix the positive integer r; then

$$T_{rn} = Z_n^{(1)} + \cdots + Z_n^{(r)},$$

where

$$Z_n^{(1)} = X_{rn,1} + \cdots + X_{rn,n},$$
$$Z_n^{(2)} = X_{rn,n+1} + \cdots + X_{rn,2n},$$
$$\vdots$$
$$Z_n^{(r)} = X_{rn,(r-1)n+1} + \cdots + X_{rn,rn}.$$

Since $T_{rn} \overset{d}{\longrightarrow} X$ as $n \to \infty$, it follows that $\{T_{rn}, n = 1, 2, \ldots\}$ is relatively compact. (This means that the associated sequence of distribution functions is relatively compact.) By 7.2.4, $\{T_{rn}, n = 1, 2, \ldots\}$ is tight. But

$$(P\{Z_n^{(1)} > z\})^r = P\{Z_n^{(1)} > z, \ldots, Z_n^{(r)} > z\}$$

$$\text{by independence of the } Z_n^{(i)}$$

$$\leq P\{T_{rn} > rz\}$$

and similarly,
$$(P\{Z_n^{(1)} < -z\})^r \leq P\{T_{rn} < -rz\}.$$

It follows that $\{Z_n^{(1)}, n = 1, 2, \ldots\}$ is tight, and hence relatively compact by 7.2.4. Thus we have a subsequence $\{Z_n^{(1)}, n = n_1, n_2, \ldots\}$ converging in distribution to a random variable Y. But the $Z_n^{(i)}$, $i = 1, \ldots, r$, are iid; hence $\{Z_n^{(i)}, n = n_1, n_2, \ldots\} \overset{d}{\longrightarrow} Y$. By 7.2.9, $T_{rn} \overset{d}{\longrightarrow} Y_1 + \cdots + Y_r$, where $Y_1, \ldots, Y_r$ are iid with $Y_i \overset{d}{=} Y$. But $T_{rn} \overset{d}{\longrightarrow} X$; hence $X \overset{d}{=} Y_1 + \cdots + Y_r$. $\square$

It can be shown (Gnedenko and Kolmogorov, 1954) that Theorem 7.5.1 still holds if the condition that for each n, the X_{ni} have the same distribution, is replaced by the uan condition.

7.5.3 Examples of Infinitely Divisible Random Variables. (a) Every *stable* random variable is infinitely divisible. This may be seen from the fact that every stable X is a limit in distribution of a sequence of normed sums, hence a limit of row sums of a triangular array in which the X_{ni}, $i = 1, 2, \ldots, n$, have the same distribution. Alternatively, if $X_1 + \cdots + X_n \overset{d}{=} a_n X + b_n$, then

$$X \overset{d}{=} \sum_{i=1}^{n} \left(\frac{X_i}{a_n} - \frac{b_n}{na_n} \right).$$

(b) A random variable of the *Poisson type* is infinitely divisible. Let Y have the Poisson distribution: $P\{Y = k\} = e^{-\lambda}\lambda^k/k!, k = 0, 1, \ldots$. The characteristic function of Y is

$$h(u) = e^{-\lambda} \sum_{k=0}^{\infty} \frac{(\lambda e^{iu})^k}{k!} = \exp[\lambda(e^{iu} - 1)]$$

and it follows that if $Y_1, \ldots, Y_n$ are independent, with Y_i Poisson with parameter λ_i, $i = 1, \ldots, n$, then $Y_1 + \cdots + Y_n$ is Poisson with parameter $\lambda_1 + \cdots + \lambda_n$. In particular, if Y is Poisson with parameter λ, then $Y \overset{d}{=} Y_1 + \cdots + Y_n$, where the Y_i are iid, each Poisson with parameter λ/n. Thus Y is infinitely divisible. Now if Y is infinitely divisible, so is $aY + b$ (a similar statement holds for stable random variables); hence a random variable of the Poisson type ($aY + b$, Y Poisson, $a \neq 0$) is infinitely divisible. The characteristic function of $aY + b$ is $\exp[ibu + \lambda(e^{iau} - 1)]$.

(c) A random variable with the *gamma distribution* is infinitely divisible. Let X have density

$$f(x) = \begin{cases} \dfrac{x^{\alpha-1}e^{-x/\beta}}{\Gamma(\alpha)\beta^\alpha} & x \geq 0, \\ 0, & x < 0, \end{cases}$$

where $\alpha, \beta > 0$. The characteristic function of X is

$$h(u) = (1 - i\beta u)^{-\alpha} = [(1 - i\beta u)^{-\alpha/n}]^n;$$

hence X is the sum of n independent gamma-distributed random variables with parameters α/n and β.

We now develop some general properties of infinitely divisible distributions.

7.5.4   Theorem. If h_1 and h_2 are infinitely divisible characteristic functions, so is $h_1 h_2$. If h is infinitely divisible, then $\bar{h}$, the complex conjugate of h, and $|h|^2$ are infinitely divisible.

PROOF.   If $h_i = (h_{in})^n$, $i = 1, 2$, then $h_1 h_2 = (h_{1n} h_{2n})^n$; since $h_{1n} h_{2n}$ is the characteristic function of the sum of two independent random variables with characteristic functions h_{1n} and h_{2n}, the first assertion is proved. If X has characteristic function h, then $-X$ has characteristic function $\bar{h}$ [see 7.1.5(c)]; thus if $h = (h_n)^n$, then $\bar{h} = (\bar{h}_n)^n$; hence $\bar{h}$ is infinitely divisible. Since $|h|^2 = h\bar{h}$, $|h|^2$ is also infinitely divisible. $\square$

If $\mathscr{C}$ is the entire class of characteristic functions of random variables, the proof of 7.5.4 shows that if $h_1, h_2 \in \mathscr{C}$; then $h_1 h_2 \in \mathscr{C}$.

Also, if $h \in \mathscr{C}$, then $\bar{h} \in \mathscr{C}$ and $|h|^2 \in \mathscr{C}$. Furthermore, 7.2.8 implies that if $h_n \in \mathscr{C}$, $n = 1, 2, \ldots$, and $h_n(u) \to h(u)$ for all u, where h is continuous at the origin, then $h \in \mathscr{C}$. A similar result holds for infinitely divisible characteristic functions.

7.5.5   Theorem. If h_n is an infinitely divisible characteristic function for each $n = 1, 2, \ldots$, and $h_n(u) \to h(u)$ for all u, where h is a characteristic function, then h is infinitely divisible.

PROOF.   Let Z_n be a random variable with characteristic function h_n, $n = 1, 2, \ldots$. If r is a fixed positive integer, then $Z_n \overset{d}{=} Z_n^{(1)} + \cdots + Z_n^{(r)}$, where the $Z_n^{(i)}$, $i = 1, \ldots, r$, are iid. If Z is a random variable with characteristic function h, then $Z_n \overset{d}{\longrightarrow} Z$ by 7.2.9, so that $\{Z_n\}$ is relatively compact, and hence tight by 7.2.4. Just as in the proof of 7.5.2, it follows that $\{Z_n^{(1)}\}$ is tight. By 7.2.4, we have a subsequence $\{Z_n^{(1)}, n = n_1, n_2, \ldots\}$ converging in distribution to a random variable Y; hence (again as in 7.5.2) $Z_n \overset{d}{\longrightarrow} Y_1 + \cdots + Y_r$, where $Y_1, \ldots, Y_r$ are iid with $Y_i \overset{d}{=} Y$. But $Z_n \overset{d}{\longrightarrow} Z$; hence $Z \overset{d}{=} Y_1 + \cdots + Y_r$. In other words, h is infinitely divisible. $\square$

Now if h is infinitely divisible, a uniqueness question arises; namely, can h be represented in two different ways as the nth power of a characteristic function? This is actually an exercise in complex variables, as follows. Let f and g be continuous complex-valued functions on the connected set S, with $f^n = g^n$; assume that $f(u) = g(u)$ for at least one $u \in S$. [In our case $S = \mathbb{R}$ and $f^n = g^n = h$; since f and g are characteristic functions of random variables, $f(0) = g(0) = 1$.] If f and g are never 0 on S, then $f \equiv g$. Note that $(f/g)^n = 1$, and therefore f/g is a continuous map of S into $\{\exp(i2\pi k/n), k = 0, 1, \ldots, n-1\}$. Since the image of S under f/g is connected, it must consist of a single point; thus f/g is a constant, necessarily 1 because f and g agree at one point. Thus the representation of h as the nth power of a characteristic function is unique, provided we can establish that an infinitely divisible characteristic function never vanishes.

7.5.6   Theorem. If h is an infinitely divisible characteristic function, then h is never 0.

PROOF.   If $h = (h_n)^n$, then $|h|^2 = |h_n|^{2n}$. Since $|h|^2$ is infinitely divisible by 7.5.4, we may as well assume that h and the h_n are real and nonnegative. Thus $h_n = h^{1/n} (= \exp[(1/n)\ln h])$, so if $h(u) > 0$, then $h_n(u) \to 1$, and if $h(u) = 0$, then $h_n(u) = 0$ for all n. But $h(0) = 1$; hence $h(u) > 0$ in some neighborhood of the origin. Thus h_n converges to a function g that is 1 in a neighborhood of the origin. By 7.2.8, g is a characteristic function, and hence continuous everywhere. But g takes on only the values 0 and 1, and hence $g \equiv 1$. Thus for any $u, h_n(u) \to 1$, so that $h_n(u) \neq 0$ for sufficiently large n. Therefore $h(u) = [h_n(u)]^n \neq 0$. $\square$

Example 7.5.3(b) is basic in the sense that random variables of the Poisson type can be used as building blocks for arbitrary infinitely divisible random variables.

7.5.7   Theorem. The random variable X is infinitely divisible iff there is a sequence of sums $\sum_{k=1}^{r(n)} X_{nk} \xrightarrow{d} X$, where for each n, the X_{nk} are independent (not necessarily identically distributed) and each X_{nk} is of the Poisson type.

PROOF.   The "if" part follows from 7.5.3(b), the first assertion of 7.5.4, and 7.5.5, so assume X infinitely divisible. Since h, the characteristic function of X, is continuous and never 0 (by 7.5.6), h has a continuous logarithm, to be denoted by $\log h$. If we specify that $\log h(0) = \log 1 = 0$, the logarithm is determined uniquely. (See Ash, 1971, p. 49ff., for a discussion of continuous logarithms.) If $h = (h_n)^n$, where h_n is a characteristic function, then

$$(h_n)^n = \left[\exp\left(\frac{1}{n}\log h\right)\right]^n ;$$

hence as in the discussion before 7.5.6,

$$h_n = \exp\left(\frac{1}{n}\log h\right),$$

so for any fixed $u \in \mathbb{R}$,

$$n(h_n(u) - 1) = n\left(\exp\left[\frac{1}{n}\log h(u)\right] - 1\right)$$

$$= n\left(\frac{1}{n}\log h(u) + o\left(\frac{1}{n}\right)\right) \qquad \text{since} \qquad e^z = 1 + z + o(z)$$

$$\to \log h(u) \qquad \text{as} \qquad n \to \infty.$$

Thus

$$\log h(u) = \lim_{n\to\infty} n(h_n(u) - 1) = \lim_{n\to\infty} n\int_{-\infty}^{\infty}(e^{iux} - 1)\,dF_n(x),$$

where F_n is the distribution function corresponding to h_n. It follows from the dominated convergence theorem that for each n we may select a positive number $m = m(n)$ such that $m \to \infty$ as $n \to \infty$ and

$$\left|\int_{-\infty}^{\infty}(e^{iux} - 1)\,dF_n(x) - \int_{-m}^{m}(e^{iux} - 1)\,dF_n(x)\right| \le \frac{1}{n^2} \qquad \text{for all} \qquad u,$$

and we may then choose a positive integer $r = r(n)$ such that

$$\left|\int_{-m}^{m}(e^{iux} - 1)\,dF_n(x) - \sum_{k=1}^{r}(e^{iux_k} - 1)[F_n(x_k) - F_n(x_{k-1})]\right| \le \frac{1}{n^2}$$

for all $u \in (-m, m)$, where $x_k = -m + 2mk/r$, $k = 0, 1, \ldots, r$.

It follows that we may obtain $h(u)$ as a pointwise limit of terms of the form

$$\prod_{k=1}^{r(n)}\exp[\lambda_{nk}(\exp(ia_{nk}u) - 1)],$$

where $\lambda_{nk} = n[F_n(x_k) - F_n(x_{k-1})]$ and $a_{nk} = x_k$. $\square$

We conclude this section by mentioning the *Lévy–Khintchine representation*: The characteristic function h is infinitely divisible iff

$$\log h(u) = iu\beta - \frac{u^2\sigma^2}{2} + \int_{\mathbb{R}}\left(e^{iux} - 1 - \frac{iux}{1 + x^2}\right)\frac{1 + x^2}{x^2}\,d\lambda(x),$$

where $\beta \in \mathbb{R}$, $\sigma^2 \ge 0$, and λ is a finite measure on $\mathscr{B}(\mathbb{R})$ such that $\lambda\{0\} = 0$.

The result is basic for a deeper study of the central limit theorem, in particular for deriving conditions for convergence to a particular infinitely divisible distribution, analogous to the results on normal convergence in 8.3. Full details are given by Gnedenko and Kolmogorov (1954). Proofs of the Lévy–Khintchine representation are also given by Chung (1968) and Tucker (1967).

Problems

1. The random variable X is said to have the *geometric distribution* iff $P\{X = k\} = q^{k-1}p, k = 1, 2, \ldots$, where $0 < p < 1, q = 1 - p$. Show that the associated characteristic function, given by

$$h(u) = pe^{iu} \sum_{k=1}^{\infty} (qe^{iu})^{k-1} = \frac{pe^{iu}}{1 - qe^{iu}},$$

   is infinitely divisible (use 7.5.7).

2. Let $g(s) = \sum_{n=1}^{\infty} n^{-s}$, Re $s > 1$, be the Riemann zeta function. The series converges uniformly for Re $s \geq 1 + \varepsilon$, any $\varepsilon > 0$; also,

$$g(s) = \prod_{k=1}^{\infty} \frac{1}{1 - p_k^{-s}},$$

   where p_n is the nth prime. (See Ash, 1971, Chapter 6 for details.) If c is a fixed real number greater than 1, show that $h(u) = g(c + iu)/g(c)$ is an infinitely divisible characteristic function (use 7.5.7).

3. Give an example of an infinitely divisible characteristic function that is not stable.

4. When characteristic functions are not easy to compute, the following technique is sometimes useful for actually finding the distribution of a sum of independent random variables.

   Let X and Y be independent random variables, and let $Z = X + Y$. If X, Y, and Z have distribution functions F_1, F_2, and F_3, show that F_3 is the *convolution* of F_1 and F_2 (notation: $F_3 = F_1 * F_2$), that is,

$$F_3(z) = \int_{-\infty}^{\infty} F_1(z - y) \, dF_2(y).$$

   F_3 is also the convolution of F_2 and F_1, that is,

$$F_3(z) = \int_{-\infty}^{\infty} F_2(z - x) \, dF_1(x).$$

If X [respectively, Y] has density f_1 [respectively, f_2], show that Z has density f_3, where

$$f_3(z) = \int_{-\infty}^{\infty} f_1(z-y)\,dF_2(y) = \int_{-\infty}^{\infty} f_2(z-x)\,dF_1(x).$$

[If both X and Y have densities, replace $dF_1(x)$ by $f_1(x)\,dx$ and $dF_2(y)$ by $f_2(y)\,dy$.]

Intuitively, the probability that X falls in $(x, x+dx)$ is $dF_1(x)$; given that $X = x$, we have $Z \leq z$ iff $Y \leq z - x$, and this happens with probability $F_2(z-x)$. Integrate over x to obtain the total probability that $Z \leq z$, namely, $F_3 = F_2 * F_1$. The other formulas have a similar interpretation.

Note also that convolution is associative, that is, $F_1 * (F_2 * F_3) = (F_1 * F_2) * F_3$. This is somewhat messy to prove directly, but a probabilistic interpretation makes it transparent. For if X_1, X_2, and X_3 are independent random variables with distribution functions F_1, F_2, and F_3, respectively, then $F_1 * (F_2 * F_3)$ is the distribution function of $X_1 + (X_2 + X_3)$, and $(F_1 * F_2) * F_3$ is the distribution function of $(X_1 + X_2) + X_3$.

Finally, we note that if F is the distribution function of a random variable, then F is infinitely divisible iff for each n there is a distribution function F_n (of a random variable) such that $F = F_n * F_n * \cdots * F_n$ (n times).

5. If $\lambda \geq 0$ and f is the characteristic function of a random variable, show that $\exp[\lambda(f-1)]$ is an infinitely divisible characteristic function.

7.6   UNIFORM CONVERGENCE IN THE CENTRAL LIMIT THEOREM

Let $X_1, X_2, \ldots$ be independent random variables with finite mean and variance, and suppose that $T_n = c_n^{-1}(S_n - E(S_n)) \xrightarrow{d} X^*$ normal $(0, 1)$. Very often the statement is made that "for large n, S_n is approximately normal with mean $a_n = E(S_n)$ and variance c_n^2," that is,

$$F_{S_n}(x) - \int_{-\infty}^{x} \frac{1}{\sqrt{2\pi}\,c_n} \exp\left[\frac{-(t-a_n)^2}{2c_n^2}\right] dt \to 0 \quad \text{as} \quad n \to \infty.$$

Let us try to prove this. If X is normal (a_n, c_n^2) and X^* is normal $(0, 1)$, then

$$|P\{S_n \leq x\} - P\{X \leq x\}| = \left|P\left\{T_n \leq \frac{x-a_n}{c_n}\right\} - P\left\{X^* \leq \frac{x-a_n}{c_n}\right\}\right|$$

$$= \left|F_{T_n}\left(\frac{x-a_n}{c_n}\right) - F^*\left(\frac{x-a_n}{c_n}\right)\right|$$

and this will approach 0 as $n \to \infty$ if $F_{T_n} \to F^*$ *uniformly* on $\mathbb{R}$. In fact this does happen; the proof rests on the following two results.

7.6.1   Theorem. Let $F, F_1, F_2, \ldots$ be bounded distribution functions on $\mathbb{R}$, S a dense subset of $\mathbb{R}$ containing all the discontinuity points of F.
   If

$$F_n(\infty) \to F(\infty),$$

$$F_n(-\infty) \to F(-\infty),$$

$$F_n(x) \to F(x) \qquad \text{for all} \qquad x \in S,$$

$$F_n(x^-) \to F(x^-) \qquad \text{for all} \qquad x \in S,$$

then $F_n \to F$ uniformly on $\mathbb{R}$.

PROOF.   Let $\varepsilon > 0$ be given. We wish to obtain a partition $-\infty = y_0 < y_1 < \cdots < y_m = \infty$ with

$$y_j \in S, \qquad\qquad 1 \leq j \leq m - 1,$$

$$F(y_{j+1}^-) < F(y_j) + \frac{\varepsilon}{2}, \qquad 0 \leq j \leq m - 1,$$

$$F(y_{j+1}) \geq F(y_j) + \frac{\varepsilon}{3}, \qquad 0 \leq j \leq m - 2$$

(take $y_m^- = \infty$).
   Set $y_0 = -\infty$ and define $z_1 = \sup\{x > y_0 \colon F(x) - F(y_0) \leq \varepsilon/3\}$. If $z_1 < \infty$, then $F(x) - F(y_0) \leq \varepsilon/3$ for $y_0 < x < z_1$; hence $F(z_1^-) \leq F(y_0) + (\varepsilon/3) < F(y_0) + (\varepsilon/2)$. Also, $F(z_1) \geq F(y_0) + (\varepsilon/3)$; for if not, $F(z_1') < F(y_0) + (\varepsilon/3)$ for some $z_1' > z_1$ (by right-continuity), contradicting the definition of z_1.
   Now if $F(z_1^-) < F(z_1)$, then $z_1 \in S$ by hypothesis, and we set $y_1 = z_1$. If $F(z_1^-) = F(z_1)$, then since S is dense and F is right-continuous, we can find $y_1 \in S$ such that $y_1 > z_1$ and $F(y_1) < F(y_0) + (\varepsilon/2)$. Thus in either case we obtain $y_1 \in S$ such that $F(y_1^-) < F(y_0) + (\varepsilon/2)$ and $F(y_1) \geq F(y_0) + (\varepsilon/3)$. Continue by defining $z_2 = \sup\{x > y_1 \colon F(x) - F(y_1) \leq \varepsilon/3\}$ and proceed as above. Since $F(y_{j+1}) \geq F(y_j) + (\varepsilon/3)$ and F is bounded, the process will terminate in a finite number of steps and produce the desired partition.
   Let $x \in \mathbb{R}$; say $y_j \leq x < y_{j+1}$. Then

$$F_n(x) - F(x) \leq F_n(y_{j+1}^-) - F(y_j)$$

$$< F(y_{j+1}^-) + \frac{\varepsilon}{2} - F(y_j)$$

$$\text{for large} \qquad n, \text{ since } y_{j+1} \in S \qquad \text{or} \qquad y_{j+1} = \infty$$

$$< \varepsilon.$$

Also,

$$F(x) - F_n(x) \le F(y_{j+1}^-) - F_n(y_j)$$

$$\le F(y_{j+1}^-) - F(y_j) + \frac{\varepsilon}{2}$$

$$\text{for large } n \text{ since } y_j \in S \quad \text{or} \quad y_j = -\infty$$

$$< \varepsilon.$$

The "large n" depends only on j and not on x, and it follows that $F_n \to F$ uniformly on $\mathbb{R}$. $\square$

7.6.2   Theorem. Let $F, F_1, F_2, \ldots$ be bounded distribution functions on $\mathbb{R}$. Assume that $F_n(-\infty) = 0$ for all n, $F(-\infty) = 0$, F is continuous everywhere, and F_n converges weakly to F. Then F_n converges to F uniformly on $\mathbb{R}$.

PROOF.   In 7.6.1, take $S = \mathbb{R}$. Since F_n converges weakly to F and F is continuous everywhere, $F_n(\infty) \to F(\infty)$ and $F_n(x) \to F(x)$ for all $x \in \mathbb{R}$. Since $F_n(-\infty) \to F(-\infty)$ by assumption, it remains to show that $F_n(x^-) \to F(x^-)$ for all x. If $y < x$,

$$F(y) = \lim_n F_n(y) \le \liminf_n F_n(x^-) \le \limsup_n F_n(x^-) \le \lim_n F_n(x) = F(x).$$

If $y \to x$, then $F(y) \to F(x)$ by continuity of F; hence $F_n(x^-) \to F(x)$ $= F(x^-)$. $\square$

Problem

1. (*Glivenko–Cantelli Theorem*) Let $X_1, X_2, \ldots$ be independent, identically distributed random variables with common distribution function F. Let $F_n(x, \omega), x \in R, \omega \in \Omega$, be the *empirical distribution function* of the X_i, based on n trials, that is,

$$F_n(x, \omega) = \frac{1}{n} \left[\begin{array}{l} \text{the number of terms among } X_1(\omega), \ldots, X_n(\omega) \\ \text{that are } \le x \end{array} \right]$$

$$= \frac{1}{n} \sum_{k=1}^n I_{\{X_k \le x\}}(\omega).$$

For example, if $n = 3, X_1(\omega) = 2, X_2(\omega) = \frac{1}{2}, X_3(\omega) = 7$, then

$$F_3(x, \omega) = \begin{cases} 0, & x < \frac{1}{2} \\ \frac{1}{3}, & \frac{1}{2} \le x < 2 \\ \frac{2}{3}, & 2 \le x < 7 \\ 1, & x \ge 7. \end{cases}$$

Intuitively, for large n, $F_n(x, \omega)$ should approximate $F(x)$. Show that there is a set A of probability 0 such that for $\omega \notin A$, $F_n(x, \omega) \to F(x)$ uniformly for $x \in \mathbb{R}$.

7.7 THE SKOROKHOD CONSTRUCTION AND OTHER CONVERGENCE THEOREMS

In this section we look at various results involving the interplay between convergence in distribution and other types of convergence. In particular, the Skorokhod construction frequently allows a very effective substitution of almost everywhere convergence for the weaker convergence in distribution. We begin with a result about convergence in distribution of sums, products, and quotients.

7.7.1 Slutsky's Theorem. If $X_n \xrightarrow{d} X$ and $Y_n \xrightarrow{d} c$ (hence $Y_n \xrightarrow{P} c$ by 7.1.7) then,

(a) $X_n + Y_n \xrightarrow{d} X + c$

(b) $X_n Y_n \xrightarrow{d} cX$

(c) $\dfrac{X_n}{Y_n} \xrightarrow{d} \dfrac{X}{c}$ if $c \neq 0.$

PROOF. (a) Let F_n be the distribution function of X_n, G_n the distribution function of $X_n + Y_n$, F the distribution function of X, and H the distribution function of $X + c$. If $X_n + Y_n \leq x$ and $|Y_n - c| < \varepsilon$, then $X_n \leq x - Y_n$ and $Y_n > c - \varepsilon$, so that $X_n \leq x - c + \varepsilon$. Similarly, if $X_n \leq x - c - \varepsilon$ and $|Y_n - c| < \varepsilon$ then $c > Y_n - \varepsilon$, so that $X_n \leq x - Y_n$. Therefore

$$G_n(x) \leq F_n(x - c + \varepsilon) + P\{|Y_n - c| \geq \varepsilon\}$$

and

$$F_n(x - c - \varepsilon) \leq G_n(x) + P\{|Y_n - c| \geq \varepsilon\}.$$

Let x be a continuity point of H, so that $x - c$ is a continuity point of F. Choose $\varepsilon > 0$ so that $x - c + \varepsilon$ and $x - c - \varepsilon$ are continuity points of F as well. Then

$$F(x - c - \varepsilon) \leq \liminf_{n \to \infty} G_n(x) \leq \limsup_{n \to \infty} G_n(x) \leq F(x - c + \varepsilon).$$

Since ε can be as small as we wish, it follows that $G_n(x) \to F(x - c) = H(x)$.

(b) We use the same notation as in (a), with $X_n + Y_n$ replaced by $X_n Y_n$ and $X + c$ replaced by cX. First assume that $c \neq 0$. Then we can take $c > 0$

without loss of generality, since $-Y_n \xrightarrow{d} -c$. The idea is essentially the same as in (a).

If $X_n Y_n \le x$ and $|Y_n - c| < \varepsilon$ then $X_n \le x/Y_n$ and $Y_n > c - \varepsilon$, so that $X_n \le \frac{x}{c-\varepsilon}$. Similarly, if $X_n \le \frac{x}{c+\varepsilon}$ and $|Y_n - c| < \varepsilon$ then $c + \varepsilon > Y_n$ and $X_n \le x/Y_n$. Therefore,

$$G_n(x) \le F_n\left(\frac{x}{c - \varepsilon}\right) + P\{|Y_n - c| \ge \varepsilon\}$$

and

$$F_n\left(\frac{x}{c + \varepsilon}\right) \le G_n(x) + P\{|Y_n - c| \ge \varepsilon\}.$$

Let x be a continuity point of H, so that x/c is a continuity point of F. Choose $\varepsilon > 0$ so that $x/c - \varepsilon$ and $x/c + \varepsilon$ are continuity points of F as well. Then

$$F\left(\frac{x}{c + \varepsilon}\right) \le \liminf_{n\to\infty} G_n(x) \le \limsup_{n\to\infty} G_n(x) \le F\left(\frac{x}{c - \varepsilon}\right),$$

and since ε can be taken as small as we wish, it follows that $G_n(x) \to F(x/c) = H(x)$. [We have assumed that $Y_n > 0$ in the above manipulations, and this causes no difficulty because $Y_n \ge \frac{1}{2}c > 0$ except on a set of small probability. A similar comment applies to the proof of (c) below.]

If $c = 0$ we must show that $X_n Y_n \xrightarrow{d} 0$. This may be accomplished by observing that $\{X \le M\} \uparrow \Omega$ as $M \to \infty$, and given $\varepsilon > 0$, we have $P\{|Y_n| < \varepsilon/M\} \to 1$ as $n \to \infty$. Thus for sufficiently large n, we have $|X_n Y_n| < \varepsilon$ except on a set of small probability.

(c)   As in (b) we can assume without loss of generality that $c > 0$. We use the same notation as in (a) with $X_n + Y_n$ replaced by X_n/Y_n and $X + c$ replaced by X/c. If $X_n/Y_n \le x$ and $|Y_n - c| < \varepsilon$ then $Y_n < c + \varepsilon$ and $X_n \le x(c + \varepsilon)$. Similarly, if $X_n \le x(c - \varepsilon)$ and $|Y_n - c| < \varepsilon$ then $c < Y_n + \varepsilon$ and $X_n \le xY_n$. Therefore,

$$G_n(x) \le F_n(x(c + \varepsilon)) + P\{|Y_n - c| \ge \varepsilon\}$$

and

$$F_n(x(c - \varepsilon)) \le G_n(x) + P\{|Y_n - c| \ge \varepsilon\}.$$

The argument is completed as in (b) with x/c replaced by cx, $x/(c - \varepsilon)$ by $x(c + \varepsilon)$ and $x/(c + \varepsilon)$ by $x(c - \varepsilon)$. $\square$

We now give the Skorokhod construction.

7.7.2 *Skorokhod's Theorem.* Assume that $X_n \xrightarrow{\;d\;} X$, and let F_n be the distribution function of X_n, with F the distribution function of X. Then there are random variables $Y_n\,(n = 1, 2, \ldots)$ and Y on $(0, 1)$ (with Borel sets and Lebesgue measure) such that Y_n has the same distribution as X_n, Y has the same distribution as X, and $Y_n(\omega) \to Y(\omega)$ for every $\omega \in (0, 1)$.

PROOF. The idea is that since F_n converges pointwise to F except possibly at discontinuity points of F, the inverse of F_n should converge to the inverse of F. Now whereas the distribution function F is increasing and right-continuous, for each $\omega \in (0, 1)$ there will be a minimum value of x, say $x = x_0$, such that $F(x) \geq \omega$, and we have $F(x) \geq \omega$ for all $x \geq x_0$, and $F(x) < \omega$ for all $x < x_0$. We define

$$Y(\omega) = x_0 = \min\{x\colon F(x) \geq \omega\}; \qquad \text{see Fig. 7.7.1.}$$

From the definition we have

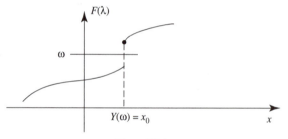

Figure 7.7.1.

(1) $Y(\omega) \leq x$ iff F has reached height ω at x or earlier iff $F(x) \geq \omega$. Similarly, we define $Y_n(\omega) = \min\{x\colon F_n(\omega) \geq \omega\}$; then $Y_n(\omega) \leq x$ iff $F_n(x) \geq \omega$.

If λ is Lebesgue measure then $\lambda\{\omega\colon Y_n(\omega) \leq x\} = \lambda\{\omega\colon \omega \leq F_n(x)\}$ $= F_n(x)$, so $Y_n \overset{d}{=} X_n$, and similarly $Y \overset{d}{=} X$. We will prove that

(2) $\liminf_{n\to\infty} Y_n(\omega) \geq Y(\omega)$.

Let $\omega \in (0, 1)$ and $\varepsilon > 0$. Choose a continuity point x of F such that $Y(\omega) - \varepsilon < x < Y(\omega)$; see Fig. 7.7.2. By (1), $Y(\omega) > x$ implies that $F(x) < \omega$, and since $F_n(x) \to F(x)$, we have $F_n(x) < \omega$ for all sufficiently large n. Again by (1), $Y_n(\omega) > x$ eventually. Thus $\liminf_{n\to\infty} Y_n(\omega) \geq x$, and since ε is arbitrary, the result follows.

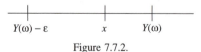

Figure 7.7.2.

The analogous result for $\limsup_{n\to\infty} Y_n(\omega)$ is more delicate because $\omega < F(x)$ is not equivalent to $Y(\omega) < x$. For example, see Fig. 7.7.1 with $x = x_0$.

(3) If Y is continuous at ω then $\limsup_{n\to\infty} Y_n(\omega) \le Y(\omega)$.

Let ω and ω' be in $(0, 1)$ with $\omega < \omega'$, and let $\varepsilon > 0$. Choose a continuity point y of F such that $Y(\omega') < y < Y(\omega') + \varepsilon$ (Fig. 7.7.3). Now $Y(\omega') < y$ implies that $F(Y(\omega')) \le F(y)$, and since $Y(\omega')$ is the first point at which F reaches height ω', we have $F(Y(\omega')) \ge \omega'$. Thus $\omega < \omega' \le F(Y(\omega')) \le F(y)$ (Fig. 7.7.4).

Figure 7.7.3.

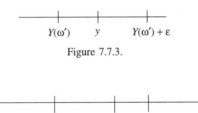

Figure 7.7.4.

But $F_n(y) \to F(y)$, so for all sufficiently large n, $F_n(y) \ge \omega$. By (1), $Y_n(\omega) \le y < Y(\omega') + \varepsilon$ (again see Fig. 7.7.3). Consequently,

$$\limsup_{n\to\infty} Y_n(\omega) \le Y(\omega') \to Y(\omega) \qquad \text{as} \qquad \omega' \to \omega,$$

and the result follows.

By definition, $Y(\omega)$ increases with ω and therefore has at most countably many discontinuities. Since a countable set has Lebesgue measure zero, we are free to change the definitions of Y_n and Y at points ω where Y is discontinuous. In particular, it is convenient to set $Y(\omega) = Y_n(\omega) = 0$ for all n. Skorokhod's theorem now follows from (2) and (3). $\square$

Part (c) of the next theorem gives a typical application of the Skorokhod construction [cf. 2.8.1(c)].

7.7.3 *Convergence of Transformed Sequences.* Let $X, X_1, X_2, \ldots$ be random variables on $(\Omega, \mathscr{F}, P)$ and let $g \colon \mathbb{R} \to \mathbb{R}$ be continuous a.e. $[P_X]$. Then

(a) $X_n \to X$ a.e. implies $g(X_n) \to g(X)$ a.e.

(b) $X_n \xrightarrow{P} X$ implies $g(X_n) \xrightarrow{P} g(X)$

(c) $X_n \xrightarrow{d} X$ implies $g(X_n) \xrightarrow{d} g(X)$.

PROOF. (a) Suppose that $X_n(\omega) \to X(\omega)$ for $\omega \in A$, where $P(A) = 1$, and g is continuous on B, where $P_X(B) = 1$. If $\omega \in A \cap X^{-1}(B)$ then $X_n(\omega) \to X(\omega)$ and g is continuous at $X(\omega)$, so $g(X_n(\omega)) \to g(X(\omega))$, Since $P(A \cap X^{-1}(B)) = 1$, the result follows.

(b) If $g(X_n)$ fails to converge in probability to $g(X)$, then there exist $\varepsilon > 0$ and $\delta > 0$, and a subsequence of $\{1, 2, \ldots\}$ such that on the *entire* subsequence,

$$P\{|g(X_n) - g(X)| \geq \varepsilon\} \geq \delta.$$

We can then extract a further subsequence on which $X_n \to X$ a.e., and consequently $g(X_n) \to g(X)$ a.e. by (a). On the second subsequence we must have $g(X_n) \xrightarrow{P} g(X)$, contradicting our choice of the first subsequence.

(c) By 7.7.2, there are random variables $Y_n \overset{d}{=} X_n$ and $Y \overset{d}{=} X$ with $Y_n \to Y$ a.e., and by (a), $g(Y_n) \to g(Y)$ a.e. By 7.1.7, $g(Y_n) \xrightarrow{d} g(Y)$. But $X_n \overset{d}{=} Y_n$ implies $g(X_n) \overset{d}{=} g(Y_n)$, and $X \overset{d}{=} Y$ implies $g(X) \overset{d}{=} g(Y)$. Therefore $g(X_n) \xrightarrow{d} g(X)$. $\square$

Problems

1. If $X_n \xrightarrow{d} X$, $a_n \to a$ and $b_n \to b$, show that $a_n X_n + b_n \xrightarrow{d} aX + b$.

2. If $X_n \xrightarrow{d} X$, show that $E(|X|) \leq \liminf_{n \to \infty} E(|X_n|)$.

3. If $X_n \xrightarrow{d} X$ and the X_n are uniformly integrable, show that X is integrable and $E(X_n) \to E(X)$.

7.8 THE k-DIMENSIONAL CENTRAL LIMIT THEOREM

Just as the sum of a large number of independent random variables is, under wide conditions, approximately normal, the sum of a large number of independent random vectors has, approximately, a Gaussian (also called multivariate normal) distribution. In this section we give a version of the central limit theorem for k-dimensional random vectors.

7.8.1 *Definitions and Comments.* First we recall some definitions and results from 1.4 and 2.8. On $\mathbb{R}^k$ we have a partial ordering: $x \leq y$ iff $x_i \leq y_i$ for every $i = 1, \ldots, k$, and $x < y$ iff $x_i < y_i$ for every $i = 1, \ldots, k$.

If $F \colon \mathbb{R}^k \to \mathbb{R}$, we say that F is *right-continuous* at x iff $F(x) = \lim_n F(u_n)$ for every sequence $u_1 \geq u_2 \geq \cdots \to x$. If F satisfies the property that $x \leq y$ implies $F(x) \leq F(y)$, we may assume in the definition of right-continuity that the u_n are all strictly greater than x.

If $F: \mathbb{R}^k \to \mathbb{R}$, we say that F is *increasing* iff $F(a, b] \geq 0$ whenever $a \leq b$. This is not equivalent to saying that $a \leq b$ implies $F(a) \leq F(b)$; neither condition implies the other in general.

An increasing, right-continuous function $F: \mathbb{R}^k \to \mathbb{R}$ is called a *distribution function* on $\mathbb{R}^k$. If F is a distribution function, there is a Lebesgue–Stieltjes measure μ on $(\mathbb{R}^k, \mathscr{B}(\mathbb{R}^k))$ such that $\mu(a, b] = F(a, b]$ for all $a, b \in R^k$. Conversely, if μ is a Lebesgue–Stieltjes measure on $(\mathbb{R}^k, \mathscr{B}(\mathbb{R}^k))$, there are several distribution functions F such that $\mu(a, b] = F(a, b]$ for all $a, b \in \mathbb{R}^k$. If μ is a *finite* measure, it is easiest to use the distribution function $G(x) = \mu(-\infty, x]$. Let us spell out in some detail the relation between μ and G.

7.8.2   Theorem. (a)   Let μ be a finite measure on $(\mathbb{R}^k, \mathscr{B}(\mathbb{R}^k))$, and let $G(x) = \mu(-\infty, x]$. Then G is a distribution function such that $a \leq b$ implies that $G(a) \leq G(b)$. Furthermore, for every $\varepsilon > 0$ there is an $A > 0$ such that if $x_j < -A$ for at least one coordinate x_j, then $G(x) \leq \varepsilon$.

(b)   Conversely, suppose that G is a distribution function on $\mathbb{R}^k$ with the property that for every $\varepsilon > 0$ there is an $A > 0$ such that if $x_j < -A$ for at least one j, then $G(x) \leq \varepsilon$. Then there is a unique Lebesgue–Stieltjes measure μ on $(\mathbb{R}^k, \mathscr{B}(\mathbb{R}^k))$ such that $G(x) = \mu(-\infty, x]$ for all x. [By (a), we have $G(a) \leq G(b)$ when $a \leq b$. The measure μ is finite iff $\sup_x G(x) < \infty$.]

PROOF.   (a)   By 1.4.8 and the discussion following it, G is a distribution function, and since μ is a (nonnegative) measure, $a \leq b$ implies $G(a) \leq G(b)$. Given $\varepsilon > 0$, choose $A > 0$ such that $\mu(\mathbb{R}^k - [-A, A]^k) \leq \varepsilon$. If $x_j < -A$ for at least one j, then $(-\infty, x] \subset \mathbb{R}^k - [-A, A]^k$, so that $G(x) \leq \varepsilon$.

(b)   By 1.4.9 there is a unique Lebesgue–Stieltjes measure μ on $(R^k, \mathscr{B}(\mathbb{R}^k))$ such that $\mu(a, b] = G(a, b]$ when $a \leq b$. If $a \to -\infty \in \overline{R}^k$, that is, each coordinate of a approaches $-\infty$, then $\mu(a, b] \to \mu(-\infty, b]$ by 1.2.7(a). But $G(a, b] \to G(b)$, since every term in 1.4.8(b) goes to 0 except for $G(b)$. Therefore $G(b) = \mu(-\infty, b]$. $\square$

In one dimension, G is continuous at x iff $\mu\{x\} = 0$. We can formulate this condition in such a way that it generalizes to the higher-dimensional case, as follows.

7.8.3   Theorem. Let $G(x) = \mu(-\infty, x]$, where μ is a finite measure on $(\mathbb{R}^k, \mathscr{B}(\mathbb{R}^k))$. Then G is continuous at x iff $\mu(\partial(-\infty, x]) = 0$ (where ∂ stands for boundary).

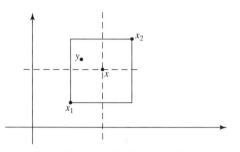

Figure 7.8.1. Proof of Theorem 7.8.3.

PROOF.   If G is continuous at x then $\mu(-\infty, x) = \lim\{G(y): y \uparrow x, y < x\}$ $= G(x)$, and it follows that $\mu(\partial(-\infty, x]) = 0$. Conversely, if $\mu(\partial(-\infty, x]) = 0$, then

$$\lim\{G(y): y \uparrow x, y < x\} = \lim\{G(z): z \downarrow x, z > x\} = G(x)$$

(by right-continuity of G).

Given $\varepsilon > 0$, there exist $x_1 < x$ and $x_2 > x$ such that $0 \le G(x) - G(x_1)$ $\le \varepsilon$ and $0 \le G(x_2) - G(x) \le \varepsilon$ (see Fig. 7.8.1). If $y \in [x_1, x_2]$, we have $G(x_1)$ $\le G(y) \le G(x_2)$, and therefore $G(y) \le G(x_2) \le G(x) + \varepsilon$ and $G(y) \ge G(x_1)$ $\ge G(x) - \varepsilon$. Thus $|G(y) - G(x)| \le \varepsilon$. □

We now consider the hyperplanes that are perpendicular to one of the coordinate axes, that is, sets of the form $\{x = (x_1, \ldots, x_k): x_i = a\}$, where $1 \le i \le k$ and $a \in \mathbb{R}$. As the measure μ is finite, only countably many of these hyperplanes can have strictly positive μ-measure (see 1.2, Problem 12). If H_μ is the set of these hyperplanes, we have the following result.

7.8.4   Theorem. If x does not belong to any of the hyperplanes in H_μ, then F is continuous at x.

PROOF.   This follows immediately from 7.8.3. □

Recall that a sequence of finite measures μ_n on $\mathbb{R}^k$ converges weakly to a finite measure μ iff $\int f d\mu_n \to \int f d\mu$ for every bounded continuous $f: \mathbb{R}^k \to \mathbb{R}$.

Theorem 2.8.1 gives several conditions equivalent to the definition of weak convergence, and as the theorem was proved for finite measures on the Borel sets of an arbitrary metric space, it remains valid in $\mathbb{R}^k$. We may also establish a k-dimensional analog of Theorem 2.8.4, as follows.

7.8.5   Theorem. Let $\mu, \mu_1, \mu_2, \ldots$ be finite measures on $(\mathbb{R}^k, \mathscr{B}(\mathbb{R}^k))$ with corresponding distribution functions $F, F_1, F_2, \ldots$, where we take $F_n(x)$

$= \mu_n(-\infty, x]$ and $F(x) = \mu(-\infty, x]$. The following conditions are equivalent:

(a) μ_n converges to μ weakly.

(b) $\mu_n(\mathbb{R}^k) \to \mu(\mathbb{R}^k)$ as $n \to \infty$ and $F_n(x) \to F(x)$ at each continuity point of F in $\mathbb{R}^k$.

(c) $\mu_n(\mathbb{R}^k) \to \mu(\mathbb{R}^k)$ as $n \to \infty$ and $F_n(x) \to F(x)$ at each point $x \in \mathbb{R}^k$ that does not lie in any of the hyperplanes of H_μ.

PROOF. (a) implies (b): Take $f \equiv 1$ to show that $\mu_n(\mathbb{R}^k) \to \mu(\mathbb{R}^k)$. If x is a continuity point of F, then $\mu(\partial(-\infty, x]) = 0$ by 7.8.3, so $F_n(x) \to F(x)$ by 2.8.1(e).

(b) implies (c): This follows from 7.8.4.

(c) implies (a): Given $\varepsilon > 0$ we can find a positive number T such that none of the vertices of $(-T, T]^k$ lie in a hyperplane of H_μ, and such that $\mu(\mathbb{R}^k - (-T, T]^k) < \varepsilon$. Then $\mu_n((-T, T]^k) \to \mu((-T, T]^k)$ and $\mu_n(\mathbb{R}^k) \to \mu(\mathbb{R}^k)$ by (c), and therefore $\mu_n(\mathbb{R}^k - (-T, T]^k) < 2\varepsilon$ for all sufficiently large n.

If f is a bounded continuous function on $\mathbb{R}^k$, it is uniformly continuous on $[-T, T]^k$, so using hyperplanes not in H_μ we can cut $(-T, T]^k$ into a finite number of rectangles $(u, v]$ such that if x and y belong to the same rectangle then $|f(x) - f(y)| < \varepsilon$. Thus we may approximate f on $(-T, T]^k$ by a function g such that g is constant on each $(u, v]$ and $|f - g| < \varepsilon$ on $(-T, T]^k$. Then, noting that $f = (f - g) + g$, we have

$$\left| \int_{\mathbb{R}^k} f \, d\mu_n - \int_{\mathbb{R}^k} f \, d\mu \right|$$

$$\leq \int_{\mathbb{R}^k - (-T, T]^k} (|f| \, d\mu_n + |f| \, d\mu)$$

$$+ \int_{(-T, T]^k} (|f - g| \, d\mu_n + |f - g| \, d\mu)$$

$$+ \left| \int_{(-T, T]^k} g \, d\mu_n - \int_{(-T, T]^k} g \, d\mu \right|.$$

The right side of this inequality is less than some constant multiple of ε for sufficiently large n, and the result follows. $\square$

We now begin the study of weak compactness on $\mathbb{R}^k$.

7.8.6 Helly's Theorem. Let $\mu_1, \mu_2, \ldots$ be finite measures on $\mathbb{R}^k$ with corresponding distribution functions $F_1, F_2, \ldots$. As above, we take $F_n(x) = \mu_n(-\infty, x]$. If $\mu_n(\mathbb{R}^k) \leq M < \infty$ for all n, there is a distribution function

F on $\mathbb{R}^k$ and a subsequence $\{F_{n_j}\}$ such that $F_{n_j}(x) \to F(x)$ at every continuity point x of F in $\mathbb{R}^k$.

PROOF. The procedure is essentially the same as in 7.2.1, but we must exercise some care in choosing the dense set D. Let M be a dense subset of $\mathbb{R}$, and take $D = M^k$. Then construct F_D and F exactly as in 7.2.1. We must show that F is increasing, that is, $F(a, b] \geq 0$ when $a \leq b$. To accomplish this we move the vertices of $(a, b]$ slightly to produce a new rectangle $(c, d]$ with $c, d \in D$ and $c \geq a, d \geq b$. All the vertices of $(c, d]$ are now in D, and if $|c - a|$ and $|d - b|$ are small enough, we have $|F(a, b] - F(c, d]| < \varepsilon$. But $F(c, d] = F_D(c, d] \geq 0$, and it follows that $F(a, b] \geq 0$. The proof that F is right continuous and that $F_{n_j}(x) \to F(x)$ when x is a continuity point of F in $\mathbb{R}^k$ is exactly the argument in 7.2.1. [As $F_n(x)$ is defined as $\mu_n(-\infty, x]$, we have $F_n(x) \leq F_n(y)$ if $x \leq y$; also, $F(x) \leq F(y)$ if $x \leq y$, by construction.] $\square$

In 7.2.3, we gave the definitions of tightness and relative compactness of a family of finite measures on the Borel sets of an arbitrary metric space, and this definition applies in $\mathbb{R}^k$. Thus, tightness means that given $\varepsilon > 0$, there is a rectangle whose complement has measure less than ε, uniformly throughout the family. Relative compactness means that every sequence from the family has a subsequence that converges weakly to a finite measure. Prokhorov's theorem also carries over.

7.8.7 Prokhorov's Theorem. Let $\mu_1, \mu_2, \ldots$ be finite measures on $(\mathbb{R}^k, \mathscr{B}(\mathbb{R}^k))$ with corresponding distribution functions $F_1, F_2, \ldots$ Again we assume that $F_n(x) = \mu_n(-\infty, x]$. Suppose that $\mu_n(\mathbb{R}^k) \leq M$ for all n. Then the sequence $\{\mu_n, n = 1, 2, \ldots\}$ is tight iff it is relatively compact.

PROOF. Relative compactness implies tightness just as in 7.2.4. To prove that tightness implies relative compactness, use Helly's theorem to get a subsequence $\{n_j\}$ and a distribution function F such that $F_{n_j}(x) \to F(x)$ at each continuity point x of F in $\mathbb{R}^k$. Tightness implies that given $\varepsilon > 0$, there exists $A > 0$ such that $\mu_n(\mathbb{R}^k - [-A, A]^k) \leq \varepsilon$ for all n. Therefore, if $x_j < -A$ for at least one j and x is a continuity point of F, we have $F_n(x) \leq \varepsilon$ for all n. Now the right-continuity of F implies that if $x_j < -A$ for at least one j, then $F(x) \leq \varepsilon$. By 7.8.2(b), there is a measure μ on $(\mathbb{R}^k, \mathscr{B}(\mathbb{R}^k))$ such that $F(x) = \mu(-\infty, x]$ for all x, and μ is finite because $\sup_x F(x) \leq M$. Tightness also implies that $\mu_{n_j}(\mathbb{R}^k) \to \mu(\mathbb{R}^k)$, and by 7.8.5, μ_{n_j} converges weakly to μ. This proves relative compactness. $\square$

7.8.8 Remarks. In Prokhorov's theorem, if the μ_n are probability measures, so is μ (since $\mu_{n_j}(\mathbb{R}^k) \to \mu(\mathbb{R}^k)$).

If the sequence of finite measures $\mu_1, \mu_2, \ldots$ is tight and there is a finite measure μ such that any weakly convergent subsequence of $\{\mu_n\}$ converges (weakly) to μ, then the entire sequence μ_n converges weakly to μ. For by Prokhorov's theorem there is at least one weakly convergent subsequence.

Characteristic functions can be defined in a natural way in $\mathbb{R}^k$.

7.8.9   Definition. Let μ be a finite measure on $\mathbb{R}^k$. The *characteristic function* of μ is the mapping from $\mathbb{R}^k$ to $\mathbb{C}$ given by

$$h(u) = \int_{\mathbb{R}^k} \exp(i <u, x>) \, d\mu(x), \qquad u \in \mathbb{R}^k,$$

where $<u, x> = \sum_{j=1}^{k} u_j x_j$.

If $\mu = P_X$, where X is a random vector $(X_1, \ldots, X_k)$, then $h(u) = E[\exp(i <u, X>)]$.

It follows that Theorem 7.1.2 can be extended to independent random vectors, in other words, the characteristic function of a sum of independent random vectors is the product of the individual characteristic functions.

As in one dimension, the characteristic function of a finite measure determines the measure uniquely, because of the following result.

7.8.10   Inversion Formula. If h is the characteristic function of a finite measure μ, and if $A = (a, b]$ is a bounded rectangle in $\mathbb{R}^k$ such that $\mu(\partial A) = 0$, then

$$\mu(A) = \lim_{c \to \infty} \frac{1}{(2\pi)^k} \int_{[-c,c]^k} \prod_{j=1}^{k} \left[\frac{\exp(-iu_j a_j) - \exp(-iu_j b_j)}{iu_j} \right] h(u) \, du.$$

PROOF.

$$\text{Let } I_c = \frac{1}{(2\pi)^k} \int_{[-c,c]^k} \prod_{j=1}^{k} \left[\frac{\exp(-iu_j a_j) - \exp(-iu_j b_j)}{iu_j} \right] h(u) \, du$$

$$= \frac{1}{(2\pi)^k} \int_{[-c,c]} \prod_{j=1}^{k} \left[\frac{\exp(-iu_j a_j) - \exp(-iu_j b_j)}{iu_j} \right]$$

$$\times \int_{\mathbb{R}^k} \exp \left[i \sum_{h=1}^{k} u_h x_h \right] d\mu(x)$$

$$= \int_{\mathbb{R}^k} \prod_{j=1}^{k} J_c(x_j, a_j, b_j) \, d\mu(x)$$

where $J_c(x_j, a_j, b_j) = \dfrac{1}{2\pi} \displaystyle\int_{-c}^{c} \dfrac{\sin u_j(x_j - a_j) - \sin u_j(x_j - b_j)}{u_j} \, du_j.$

As in 7.1.3 we have $\lim_{c \to \infty} J_c(x_j, a_j, b_j) = J(x_j, a_j, b_j)$, where J is 0 for $x_j \notin [a_j, b_j]$, 1 for $x_j \in (a_j, b_j)$ and $\frac{1}{2}$ for $x_j = a_j$ or b_j. Therefore,

$$\lim_{c \to \infty} I_c = \int_{\mathbb{R}^k} K(x, a, b) \, d\mu,$$

where K is 0 for $x \notin [a, b]$ and 1 for $x \in (a, b)$. On the boundary of $A = (a, b]$, K assumes values that are various powers of $\frac{1}{2}$. It follows that $\lim_{c \to \infty} I_c = \mu(A)$ if $\mu(\partial A) = 0$. $\square$

Here is the k-dimensional version of Levy's theorem 7.2.9.

7.8.11 Theorem. Let $\{\mu_n, n = 1, 2, \ldots\}$ be a sequence of measures on $\mathbb{R}^k$ such that $\mu_n(\mathbb{R}^k) \leq M < \infty$ for all n, and let h_n be the characteristic function of μ_n. Let μ be a finite measure on $\mathbb{R}^k$ with characteristic function h. Then $\mu_n \to \mu$ weakly iff $h_n(u) \to h(u)$ for every $u \in \mathbb{R}^k$.

PROOF. If $\mu_u \xrightarrow{w} \mu$ then $h_n(u) \to h(u)$ by 2.8.1, so assume $h_n(u) \to h(u)$ for all $u \in \mathbb{R}^k$. We consider vectors $u = te_j$, where t is real and the coordinate vector e_j has a 1 in position j and 0's elsewhere. Then

$$h_j^{(n)}(t) = h_n(te_j) = \int_{\mathbb{R}^k} e^{itx_j} \, d\mu_n(x)$$

is in fact the Fourier transform of the measure $m_j^{(n)}$ on $\mathbb{R}$ defined by

$$m_j^{(n)}(a, b] = \mu_n(P_j^{-1}(a, b]), \qquad a, b \in \mathbb{R},$$

where P_j is the projection $(x_1, \ldots, x_k) \to x_j$ on the jth coordinate axis. (Use 1.6.12 with $T = P_j$, $A = \mathbb{R}$ and $T^{-1}A = \mathbb{R}^k$.)

Since $h_j^{(n)}(t) \to h_j(t) = h(te_j)$ as $n \to \infty$, it follows from 7.2.8 that the $m_j^{(n)}, n \geq 1$, are tight measures on $\mathbb{R}$, so we can find a positive number r_j such that $m_j^{(n)}(R - [-r_j, r_j]) \leq \varepsilon$ for all n. Then $\mu_n(\mathbb{R}^k - \prod_{j=1}^{k}[-r_j, r_j]) \leq k\varepsilon$. Consequently, the $\mu_n, n \geq 1$, are tight measures on $\mathbb{R}^k$. The condition $h_n(u) \to h(u)$ assures that every weakly convergent subsequence of $\{\mu_n\}$ converges weakly to μ. By 7.8.8, $\mu_n \xrightarrow{w} \mu$. $\square$

The following result, which characterizes weak convergence of k-dimensional random vectors in terms of weak convergence on $\mathbb{R}$, is a key step in the proof of the multivariate central limit theorem.

7.8.12 Cramèr-Wold Device. Let $\{X_n = (X_{n1}, \ldots, X_{nk}), n \geq 1\}$ be a sequence of k-dimensional random vectors. Then the X_n converge weakly to the random vector $Y = (Y_1, \ldots, Y_k)$ if and only if

$$\sum_{j=1}^{k} u_j X_{nj} \xrightarrow{w} \sum_{j=1}^{k} u_j Y_j \qquad \text{for every} \qquad u = (u_1, \ldots, u_k) \text{ in } \mathbb{R}^k. \quad (1)$$

PROOF. Let h_n be the characteristic function of X_n, and h the characteristic function of Y. Then by 7.2.9, condition (1) is equivalent to

$$h_n(t(u_1, \ldots, u_k)) \to h(t(u_1, \ldots, u_k)) \qquad \text{for all} \qquad t \in \mathbb{R} \qquad \text{and} \qquad u \in \mathbb{R}^k. \quad (2)$$

But if (2) holds, then (take $t = 1$ and apply 7.8.11) $X_n \xrightarrow{w} Y$. Conversely, if $X_n \xrightarrow{w} Y$ then by 7.8.11, $h_n(v) \to h(v)$ for every $v \in \mathbb{R}^k$, in particular for $v = tu$. $\square$

If $X_1, \ldots, X_n$ are k-dimensional random vectors, with $X_r = (X_{r1}, \ldots, X_{rk})$, the *mean* of X_r is the vector $(m_{r1}, \ldots, m_{rk})$, where $m_{rj} = E(X_{rj})$. The *covariance* of X_r is the k by k matrix $\sum_r$ whose ij entry is $\text{Cov}(X_{ri}, X_{rj})$. If the X_j are iid, we may speak of the common mean $m = (m_1, \ldots, m_k)$ and covariance Σ. The following result uses some basic properties of the k-dimensional Gaussian distribution, which is discussed in Appendix 5.

7.8.13 A k-Dimensional Central Limit Theorem. Let $X_1, X_2, \ldots$ be iid k-dimensional random vectors with finite mean m and covariance Σ. If $S_n = \sum_{j=1}^{n} X_j$ then $\dfrac{S_n - nm}{\sqrt{n}}$ converges weakly to Y, where Y has a Gaussian distribution with mean 0 and covariance Σ.

PROOF. Let Y have a Gaussian distribution with mean 0 and covariance Σ. (There is such a random vector since Σ is symmetric and nonnegative definite.) By the Cramèr–Wold device, it is sufficient to show that for every u in $\mathbb{R}^k$,

$$T_n = \frac{1}{\sqrt{n}} \sum_{h=1}^{k} u_h \left(\sum_{j=1}^{n} X_{jh} - nm_h \right) \xrightarrow{w} \sum_{h=1}^{k} u_h Y_h.$$

But the random variables $Z_j = \sum_{h=1}^{k} u_h X_{jh}$ are iid with mean $\sum_{h=1}^{k} u_h m_h$ and variance $\sum_{h=1}^{k} \sum_{h'=1}^{k} u_h \sum_{hh'} u_{h'}$. [Note that $\text{Var } Z_j = \text{Cov}(Z_j, Z_j)$.]

Now,

$$T_n = \frac{\sum_{j=1}^{n} Z_j - nE(Z_1)}{\sqrt{n}},$$

and the one-dimensional central limit theorem says that T_n converges weakly to a normal distribution with mean 0 and variance $\sum_{h=1}^{k} \sum_{h'=1}^{k} u_h \sum_{hh'} u_{h'}$. But this is precisely the distribution of $\sum_{h=1}^{k} u_h Y_h$. $\square$

7.9   REFERENCES

An exhaustive treatment of the one-dimensional central limit theorem is given by Gnedenko and Kolmogorov (1954). Conditions are derived for convergence in distribution of the row sums of a triangular array to a given infinitely divisible distribution. If X is a given stable random variable with distribution function G and $X_1, X_2, \ldots$ are iid random variables with distribution function F, conditions on F are given that are necessary and sufficient for there to exist a sequence of normed sums (formed from the X_i) converging in distribution to X. (In this case, F is said to belong to the *domain of attraction* of G.) Also, results on rates of convergence are given; in other words, if $Y_n \xrightarrow{d} Y$, there are estimates as to how fast the distribution function of Y_n approaches that of Y.

Prokhorov's weak compactness theorem actually holds in a complete, separable metric space; see Billingsley (1968) for a proof. An n-dimensional extension of Lindeberg's theorem is given by Gikhman and Skorokhod (1969). Aspects of the central limit theorem for random variables with values in abstract spaces, for example, Hilbert spaces or locally compact groups, are discussed by Parthasarathy (1967).

For further applications to statistics, see Ferguson (1996) and Serfling (1980).

8

ERGODIC THEORY

8.1 INTRODUCTION

In Chapter 6 we proved the strong law of large numbers: if X_1, X_2, $\ldots, X_n, \ldots$ is a sequence of independent, identically distributed random variables with finite mean, then

$$S_n = \frac{X_1 + \cdots + X_n}{n} \to E(X_1) \qquad \text{a.e.}$$

In this chapter we will generalize this result and prove the basic "pointwise ergodic theorem."

The starting point for ergodic theory is the notion of a transformation that preserves the structure of the measure space, as defined below.

8.1.1 Definition. Let $(\Omega, \mathscr{F}, \mu)$ be a measurable space, and T a *measurable transformation* on $(\Omega, \mathscr{F}, \mu)$, that is, $T \colon (\Omega, \mathscr{F}) \to (\Omega, \mathscr{F})$.

The transformation T is said to be *measure-preserving* (we also say that T is μ-*preserving* or that T *preserves* μ) iff $\mu(T^{-1}A) = \mu(A)$ for all $A \in \mathscr{F}$.

(This implies that $\mu(T^{-k}A) = \mu(A)$ for all $A \in \mathscr{F}$ and all $k = 1, 2, \ldots,$ where $T^{-k}A = \{\omega \colon T^k\omega \in A\}$ and T^k is the composition of T with itself k times.)

The physical concept of a *flow* may be used to motivate the study of measure-preserving transformations. A flow may be regarded as a process in which a system of particles of a fluid (each point of the container corresponding to a particle) moves about under the action of an externally applied force. The force is assumed to be independent of time, so that, at least at discrete times $t = 0, 1, 2, \ldots,$ the flow can be described by a single (measurable) function T. If x is a point of the container, Tx is the position of the particle, originally at x, after one second has elapsed; thus $T^2x = T(Tx)$ is the position after two seconds, and so on. If A is a (Borel) subset of the container, then $T^{-1}(A)$ corresponds to the set of particles that will be in A after one application of T. If μ is Lebesgue measure (volume in this case) and the fluid is incompressible, it is reasonable to expect that $\mu(T^{-1}A) = \mu(A)$.

We consider some mathematical examples.

8.1.2   *Examples.*

1.   *Permutations.* Let Ω be a finite set $\{x_1, \ldots, x_n\}$, $n \geq 2$, with $\mathscr{F}$ consisting of all subsets of Ω. Let T be a cyclic permutation of Ω, say, $T(x_i) = x_{i+1}$, with indices reduced modulo n. Since $T^{-1}\{x_i\} = \{x_{i-1}\}$, T preserves μ iff $\mu\{x_i\}$ is constant for all i. Thus if μ is a probability measure P, then $P\{x_i\}$ must be $1/n$ for all i.

   More generally, if T is any permutation of Ω, T can be expressed as a product of disjoint cycles $C_1, \ldots, C_k$. Then T preserves μ iff within each cycle μ assigns equal weight to each point.

2.   *Translations.* Let $\Omega = \mathbb{R}$, $\mathscr{F} = \mathscr{B}(\mathbb{R})$, and let μ be Lebesgue measure. If $T(x) = x + c$, c constant, then T preserves μ because μ is translation-invariant.

3.   *Rotations of the circle.* Let Ω be the unit circle in the plane $\mathbb{R}^2$ (Ω can be identified with the interval $[0, 2\pi)$ under the correspondence $e^{i\theta} \to \theta$). Take $\mathscr{F}$ as the Borel sets, and $\mu = P = \lambda/2\pi$, where λ is arc length on the circle (or Lebesgue measure on $[0, 2\pi)$). Thus if A is a Borel subset of $[0, 2\pi)$, $\mu(A) = \int_A (2\pi)^{-1} d\theta$.

   Let α be fixed in $[0, 2\pi)$, and let T be rotation by α. Thus T is defined on Ω by $T(e^{i\theta}) = e^{i(\theta + \alpha)}$, or equivalently, on $[0, 2\pi)$ by $T(\theta) = \theta + \alpha$ (modulo 2π). As in Example 2, T preserves μ by the translation-invariance of Lebesgue measure.

4.   *One-sided shifts.* Let $\Omega = \mathbb{R}^\infty$, the collection of all sequences $s = (s_0, s_1, \ldots)$ of real numbers; take $\mathscr{F} = [\mathscr{B}(\mathbb{R})]^\infty$, and let μ be any probability measure P on $\mathscr{F}$. Define $T(s_0, s_1, s_2, \ldots) = (s_1, s_2, \ldots)$; T is called the *one-sided shift transformation.* Measures preserved by T are stationary in the sense of Definition 8.1.3 below.

5.   *Two-sided shifts.* Let Ω be the set of all doubly infinite sequences $s = (\ldots, s_{-1}, s_0, s_1, \ldots)$ of real numbers, $\mathscr{F}$ the σ-field generated by the measurable cylinders

$$\{s: (s_k, s_{k+1}, \ldots, s_{k+n-1}) \in B_n\},$$

$n = 1, 2, \ldots, k = 0, \pm 1, \pm 2, \ldots, B_n \in \mathscr{B}(\mathbb{R}^n)$. Let μ be any probability measure P on $\mathscr{F}$, and let T be the *two-sided shift* defined by

$$T(\ldots, s_{-1}; s_0, s_1, \ldots) = (\ldots, s_0; s_1, s_2, \ldots).$$

In other words, if the kth coordinate of s is s_k, the kth coordinate of $T(s)$ is s_{k+1}. As in Example 4, measures preserved by T are stationary in the sense of Definition 8.1.3.

In Example 4, the *coordinate variables* are defined as follows. If $\omega = (s_0, s_1, s_2, \ldots)$, then $X_k(\omega) = s_k$, $k = 0, 1, \ldots$. (A similar definition is made in Example 5.) If T is the one-sided shift (or, in Example 5, the two-sided shift), we have

$$X_k(T\omega) = X_{k+1}(\omega).$$

8.1.3 Definition. Let P be a probability on $\mathbb{R}^\infty$; P is *stationary* iff

$$P\{s\colon (s_0, s_1, \ldots, s_{n-1}) \in B_n\} = P\{s\colon (s_k, s_{k+1}, \ldots, s_{k+n-1}) \in B_n\}$$

for all $n, k = 1, 2, \ldots$, and all n-dimensional Borel sets B_n.
In the case of doubly infinite sequences, $k = 1, 2, \ldots$ is replaced by $k = \pm 1, \pm 2, \ldots$.

We show that T *preserves P iff P is stationary.*
First, note that

$$T^{-k}\{s\colon (s_0, \ldots, s_{n-1}) \in B_n\} = \{s\colon (s_k, \ldots, s_{k+n-1}) \in B_n\}.$$

If T preserves P, then

$$P(A) = P(T^{-1}A) = \ldots = P(T^{-k}A), \qquad A \in \mathscr{F},$$

and it follows that P is stationary.
Conversely, if P is stationary and $A = \{s\colon (s_0, \ldots, s_{n-1}) \in B_n\}$ is a measurable cylinder, then $T^{-1}(A) = \{s\colon (s_1, \ldots, s_n) \in B_n\}$. The class of sets $A \in \mathscr{F}$ such that $P(A) = P(T^{-1}A)$ is a monotone class containing the measurable cylinders, and hence coincides with $\mathscr{F}$. The result follows.

8.1.4 Definition. Let $(\Omega, \mathscr{F}, P)$ be a probability space and $X_0, X_1, \ldots,$ be a sequence of random variables. The sequence $X_0, X_1, \ldots$ is a *stationary sequence* iff for any $n = 0, 1, \ldots$ and any $k = 1, 2 \ldots$, $(X_0, \ldots, X_n)$ and $(X_k, \ldots, X_{n+k})$ have the same distribution.
In the case of doubly infinite sequences $(\ldots, X_{-1}, X_0, X_1, \ldots), k = 1, 2, \ldots$, is replaced by $k = \pm 1, \pm 2, \ldots$.

If $X_0, X_1, \ldots$ is a sequence of random variables defined on a probability space $(\Omega, \mathscr{F}, P)$ we can define on $(\mathbb{R}^\infty, [\mathscr{B}(\mathbb{R})]^\infty)$ a probability μ as follows: if B_n is a n-dimensional Borel set, we take

$$\mu_n\{s\colon (s_0, \ldots, s_{n-1}) \in B_n\} = P\{\omega\colon (X_0(\omega), \ldots, X_{n-1}(\omega)) \in B_n\}.$$

The probabilities μ_n are consistent in the sense of the Kolmogorov extension theorem 2.7.5, and define a probability μ on $(\mathbb{R}^\infty, [\mathscr{B}(\mathbb{R})]^\infty)$ characterized by

$$\mu\{s: (s_0, \ldots, s_{n-1}) \in B_n\} = P\{\omega: (X_0(\omega), \ldots, X_{n-1}(\omega)) \in B_n\}.$$

It is easy to verify that the sequence $X_0, X_1, \ldots$ is stationary iff the probability μ is a stationary probability on $\mathbb{R}^\infty$.

Note that the transformations of Examples 1, 2, 3, and 5 are *invertible* (measurable, one-to-one, onto, with T^{-1} measurable), while that of Example 4 is not invertible (it is not one-to-one).

We now consider a physical example to motivate the concept of an ergodic transformation. Suppose that rainfall data are collected at a very large number of observation points $a_0, a_1, \ldots$ at times $t = 0, 1, \ldots$. Assume that the statistical character of the observations at a_i is the same for all i. The observation at a_i is represented by a stationary sequence of random variables $X_{i0}, X_{i1}, \ldots$, with X_{in} the amount of rainfall at time n at a_i. Assume also that the a_i are "independent," in other words, the a_i correspond to a sequence of independent performances of a random experiment, where a performance means an observation of the entire sequence $(X_{i0}, X_{i1}, \ldots)$.

Suppose that the problem is to measure the average rainfall. Scientist A might take the following approach. He or she might take measurements at each observation point at a given time, say $t = 0$, and average the results. Scientist B might reason as follows. Since all observation points have the same statistical character, we can simply go to one observation point, take a large number of observations, say at $t = 0, 1, \ldots, n - 1$, and average the results. Scientist A is using what might be called a vertical measuring scheme, and Scientist B a horizontal scheme, as illustrated in the table; A's observations correspond to the first column, B's to the first row.

Observation point	Measurements			
	$t = 0$	1	2	$\cdots$
a_0	X_{00}	X_{01}	X_{02}	$\cdots$
a_1	X_{10}	X_{11}	X_{12}	$\cdots$
a_2	X_{20}	X_{21}	X_{23}	$\cdots$
$\vdots$				

Now A and B will not necessarily obtain the same result (not even "essentially" the same). For example, suppose that "nature" flips an unbiased coin at each observation point. If the coin comes up heads, the rain is one inch at each observation time; if the coin comes up tails, there is no rain at any time. Roughly half of A's observers will measure one inch of rainfall, and half will measure none. Thus A will arrive at an average rainfall of one-half inch. But B will either measure an average of one inch or no rain at all, and thus will not get the same answer.

Mathematically, B is computing a *time average*, namely (if we denote X_{0k} by X_k),

$$\frac{1}{n} \sum_{k=0}^{n-1} X_k(\omega) \qquad \text{for a } \textit{particular } \omega.$$

(Note also that if T is the one-sided shift, then $X_k(\omega) = X_0(T^k\omega)$.)

But A is observing an *ensemble average* at a *particular time*, namely, $(1/n) \sum_{i=0}^{n-1} Y_i$, where the $Y_i = X_{i0}$ are independent random variables, all having the same distribution as $Y_0 = X_0$. Thus A's result would approximate $E(X_0) = \int_\Omega X_0 \, dP$. For A and B to get the same answer, we must have

$$\frac{1}{n} \sum_{k=0}^{n-1} X_0(T^k\omega) \to \int_\Omega X_0 \, dP.$$

More generally, we might ask when it will be true that, for each integrable function f on $(\Omega, \mathscr{F}, P)$, we have

$$\frac{1}{n} \sum_{k=0}^{n-1} f(T^k\omega) \to \int_\Omega f \, dP \qquad (1)$$

at least for almost every ω. In particular, if f is an indicator I_A, the property to be verified is simply the convergence of the relative frequency of visits to A in the first n steps to the probability of A.

Now suppose that A is an "almost invariant" set. In other words, A and $T^{-1}A$ differ only by a set of measure 0. (In the case where nature flips an unbiased coin to determine rainfall, we may take $A = \{\omega: X_0(\omega) = 1\}$, so that $T^{-1}A = \{\omega: X_1(\omega) = 1\}$.) Then we have, almost everywhere,

$$\frac{1}{n} \sum_{k=0}^{n-1} I_A(T^k\omega) = I_A(\omega) \qquad \text{for all n.}$$

Thus the relative frequency of visits to A cannot converge to $P(A)$, except when $P(A) = 0$ or 1. Conversely, the pointwise ergodic theorem, to be proved in 8.3, implies that if every almost invariant set has probability 0 or 1, the convergence result in (1) holds. In the next section we shall prepare for the proof of this basic result.

One comment on terminology. In this chapter, we deal exclusively with real as opposed to complex-valued functions.

8.2 ERGODICITY AND MIXING

The following definitions are motivated by the analysis at the end of 8.1.

8.2.1 Definition. Let T be a measure-preserving transformation on $(\Omega, \mathscr{F}, \mu)$. A set $A \in \mathscr{F}$ is said to be *invariant* (under T) iff $A = T^{-1}A$, that is, $\omega \in A$ iff $T\omega \in A$; *almost invariant* iff A and $T^{-1}A$ differ by a set of measure 0, in other words, $\mu(A \triangle T^{-1}A) = 0$.

It is easily checked that the invariant sets form a σ-field, as do the almost invariant sets.

If $g\colon (\Omega, \mathscr{F}) \to (\mathbb{R}, \mathscr{B}(\mathbb{R}))$, g is said to be *invariant* iff $g(T\omega) = g(\omega)$ for all ω, *almost invariant* iff $g(T\omega) = g(\omega)$ for almost all ω. Note that a set is invariant (respectively almost invariant) iff its indicator is invariant (almost invariant).

The measure-preserving transformation T is said to be *ergodic* iff for every invariant set A, either $\mu(A) = 0$ or $\mu(\Omega - A) = 0$. In the case of a probability space, ergodicity means that each invariant set has probability 0 or 1.

Invariance may be replaced by almost invariance in the definition of ergodicity, as the following result shows.

8.2.2 Lemma. Let T be a measure-preserving transformation.

(a) If A is an almost invariant set, there is a (strictly) invariant set B such that $\mu(A \triangle B) = 0$.

(b) A measure-preserving transformation T is ergodic iff for each almost invariant set A, either $\mu(A) = 0$ or $\mu(\Omega - A) = 0$.

PROOF. Take $B = \limsup_n T^{-n}A$. Then $T^{-1}B = \limsup_n T^{-(n+1)}A = B$, hence B is invariant.

Now $A \triangle B \subset \bigcup_{k=0}^{\infty}(T^{-k}A \triangle T^{-(k+1)}A)$; for if $\omega \in A - B$ then $\omega \in T^{-n}A$ for only finitely many n, including $n = 0$. Thus $\omega \in T^{-k}A - T^{-(k+1)}A$ for some k. If $\omega \in B - A$, then $T^n\omega \in A$ for infinitely many n, but $\omega \notin A$. If $k+1$ is the smallest integer such that $T^{k+1}\omega \in A$, then $\omega \in T^{-(k+1)}A - T^{-k}A$.

Since $\mu(T^{-k}A \triangle T^{-(k+1)}A) = \mu(A \triangle T^{-1}A) = 0$, it follows that $\mu(A \triangle B) = 0$, proving (a).

To prove (b), let T be ergodic, and let A be almost invariant. If B is invariant and $\mu(A \triangle B) = 0$, then $\mu(A) = \mu(B)$ and $\mu(\Omega - A) = \mu(\Omega - B)$; hence $\mu(A) = 0$ or $\mu(\Omega - A) = 0$. The converse is clear since every invariant set is almost invariant. $\square$

We give another way of expressing ergodicity.

8.2.3 Lemma. Let T be a measure-preserving transformation on $(\Omega, \mathcal{F}, \mu)$. The following conditions are equivalent:
(a) T is ergodic.
(b) Every almost invariant function is a.e. constant.
(c) Every invariant function is a.e. constant.

PROOF. (a) *implies* (b): Let g be an almost invariant function. Then for each real λ, $A_\lambda = \{\omega: g(\omega) \le \lambda\}$ is an almost invariant set since $g(\omega) = g(T\omega)$ a.e. By (a) and 8.2.2, $\mu(A_\lambda) = 0$ or $\mu(A_\lambda^c) = 0$. Let $c = \sup\{\lambda: \mu(A_\lambda) = 0\}$; c is finite (ignoring the trivial case $\mu \equiv 0$) since $A_\lambda \uparrow \Omega$ as $\lambda \uparrow \infty$, and $A_\lambda^c \uparrow \Omega$ as $\lambda \downarrow -\infty$. Then

$$\mu\{\omega: g(\omega) < c\} = \mu\left[\bigcup_{n=1}^{\infty}\left\{\omega: g(\omega) \le c - \frac{1}{n}\right\}\right] = 0,$$

and similarly $\mu\{\omega: g(\omega) > c\} = 0$. Thus $g = c$ a.e.

(b) *implies* (c): Every invariant function is almost invariant.

(c) *implies* (a): If A is an invariant set, I_A is an invariant function, hence I_A is a.e. constant. If $I_A = 0$ a.e. then $\mu(A) = \int_\Omega I_A \, d\mu = 0$, and if $I_A = 1$ a.e. then $\mu(A^c) = \int_\Omega (1 - I_A) \, d\mu = 0$. □

Invariance and almost invariance can be defined in the same way for *extended* real-valued Borel measurable functions. Lemma 8.2.3 holds in this case also, with essentially the same proof.

The following characterization of almost invariance is often useful.

8.2.4 Lemma. Let T be a measure-preserving transformation. Assume μ is finite. A set $A \in \mathcal{F}$ is almost invariant iff either $\mu(T^{-1}A - A) = 0$ or $\mu(A - T^{-1}A) = 0$, that is, $T\omega \in A$ essentially implies $\omega \in A$, or $\omega \in A$ essentially implies $T\omega \in A$.

PROOF. We may write

$$\mu(A - T^{-1}A) = \mu(A) - \mu(A \cap T^{-1}A) = \mu(T^{-1}A) - \mu(A \cap T^{-1}A)$$
$$= \mu(T^{-1}A - A).$$

Thus $\mu(A \triangle T^{-1}A) = 2\mu(A - T^{-1}A) = 2\mu(T^{-1}A - A)$. □

Nonergodicity, that is, the existence of a nontrivial invariant set, indicates that T does not completely stir up the space. This concept of "stirring" may be developed as follows.

8.2.5 Definition. If T is a measure-preserving transformation on the *probability* space $(\Omega, \mathscr{F}, P)$, T is said to be *mixing* iff for all $A, B \in \mathscr{F}$,

$$\lim_{n \to \infty} P(A \cap T^{-n}B) = P(A)P(B).$$

The restriction to a probability measure is essential here. If, for example, T has the mixing property with respect to the measure μ, let $A = B = \Omega$ to obtain $\mu(\Omega) = [\mu(\Omega)]^2$. Thus if $\mu(\Omega)$ is finite it must be 1; if $\mu(\Omega) = \infty$, and A is a set in $\mathscr{F}$ with finite, strictly positive measure, take $B = \Omega$. Then $\mu(A \cap T^{-n}B) = \mu(A) < \infty$, but $\mu(A)\mu(B) = \infty$, a contradiction.

The mixing property has the following intuitive interpretation [this example is due to Halmos (1956)]. Regard the transformation T as defining a flow, as in the discussion after 8.1.1. Suppose, for example, that initially the container is filled with a liquid that is 90% gin, 10% vermouth, the "vermouth particles" occupying the set A, the "gin particles" the set $\Omega - A$. The externally applied force is due to a swizzle stick. The condition of the container is observed at times $t = 0, 1, 2, \ldots$. If B is any Borel subset of the container, let $P(B)$ be the volume of B divided by the volume of the container (so that $P(A) = 0.1$). It is reasonable to expect that if the mixing process is continued long enough, the percentage of vermouth in B should be approximately the same as the percentage in the entire container, namely, 10%. To translate this into mathematical terms note that if ω is a point of the container, and a particle is initially at ω, then $T^n\omega$ is the position of the particle n seconds later. Thus the set of vermouth particles that are in B at time $t = n$ is $\{\omega \in A : T^n\omega \in B\} = A \cap T^{-n}B$. The fraction of vermouth in B at time $t = n$ is $P(A \cap T^{-n}B)/P(B)$, and the mixing property is expressed by saying that $P(A \cap T^{-n}B)/P(B) \to P(A) = 0.1$.

Mixing is a stronger property than ergodicity, as we now prove.

8.2.6 Theorem. Let T be a mixing transformation on $(\Omega, \mathscr{F}, P)$. Then T is ergodic.

PROOF. Let B be an invariant set. If $A \in \mathscr{F}$, then since $B = T^{-n}B$, we have $P(A \cap B) = P(A \cap T^{-n}B)$ for all n. If we let $n \to \infty$ and invoke the mixing property, we obtain $P(A \cap B) = P(A)P(B)$. But since A is an arbirary set $\in \mathscr{F}$, we may take $A = B$, and thus $P(B) = [P(B)]^2$, hence $P(B) = 0$ or 1. $\square$

It is useful to observe that it is not necessary to verify the mixing condition for all the sets $A, B \in \mathscr{F}$, but only for A, B in a field $\mathscr{F}_0$ whose minimal σ-field is $\mathscr{F}$.

8.2.7 Theorem. Let T be a measure-preserving transformation on $(\Omega, \mathscr{F}, P)$. Let $\mathscr{F}_0$ be a field of subsets of Ω such that the σ-field generated by $\mathscr{F}_0$ is $\mathscr{F}$.

If the mixing condition holds for all $A, B \in \mathscr{F}_0$, it holds for all $A, B \in \mathscr{F}$ and hence T is mixing.

PROOF.   Let $A, B \in \mathscr{F}$, and find sets $A_k, B_k \in \mathscr{F}_0$ $(k = 1, 2, \ldots)$ such that $P(A \triangle A_k)$ and $P(B \triangle B_k) \to 0$ as $k \to \infty$ (see 1.3.11). Now

$$(A \cap T^{-n}B) \triangle (A_k \cap T^{-n}B_k) \subset (A \triangle A_k) \cup (T^{-n}(B \triangle B_k)),$$

so the probability of the set on the left is at most $P(A \triangle A_k) + P(B \triangle B_k)$, which approaches 0 as $k \to \infty$, uniformly in n. Thus $P(A_k \cap T^{-n}B_k) \to P(A \cap T^{-n}B)$ as $k \to \infty$, uniformly in n. By the hypothesis, $P(A_k \triangle T^{-n}B_k) \to P(A_k)P(B_k)$ as $n \to \infty$. Therefore by the standard double limit theorem,

$$\lim_{n \to \infty} P(A \cap T^{-n}B) = \lim_{n \to \infty} \lim_{k \to \infty} P(A_k \cap T^{-n}B_k)$$

$$= \lim_{k \to \infty} \lim_{n \to \infty} P(A_k \cap T^{-n}B_k) = P(A)P(B),$$

the desired result. $\square$

8.2.8   *Examples.*   We consider again the examples of 8.1.2.

1. *Permutations.* Let T be a permutation of $\Omega = \{x_1, \ldots, x_n\}$, $n \geq 2$, μ any measure on all subsets of Ω that assigns equal weight to each point within a given cycle of T. (Assume that $\mu\{x_i\} > 0$ for all i; when talking about the mixing property we also assume μ is a probability measure.)

    We claim that T is *ergodic iff* T *has only one cycle*; this follows because the only invariant sets are unions of cycles of T (and the empty set).

    But T *is never mixing.* Suppose that $\{x_1, \ldots, x_k\}$ is a cycle of T and $Tx_i = x_{i+1}$, with indices reduced modulo k. Assume $k \geq 2$; if $k = 1$, $\{x_1\}$ is a nontrivial invariant set. Let $A = B = \{x_i\}$. Then $A \cap T^{-n}B$ coincides with A if n is a multiple of k, and is the empty set otherwise. Thus $\lim_{n \to \infty} \mu(A \cap T^{-n}B)$ does not exist.

2. *Translations.* Let $T(x) = x + c$ on $\mathbb{R}$, with Borel sets and Lebesgue measure. Then T *is not ergodic*; $A = \bigcup_{n=-\infty}^{\infty} (nc, nc + c/2)$ is a nontrivial invariant set.

3. *Rotations of the circle.* Let T be rotation by α on the unit circle (or $T(\theta) = \theta + \alpha \pmod{2\pi}$ on $[0, 2\pi)$). We claim that T *is ergodic iff* $\alpha/2\pi$ *is irrational*, that is, iff $e^{i\alpha}$ is not a root of unity.

    Assume $\alpha/2\pi$ irrational, and let A be an invariant set in $\mathscr{F}$. Let a_n be the nth Fourier coefficient of the indicator function I_A, that is,

$$a_n = \frac{1}{2\pi} \int_0^{2\pi} I_A(\omega) e^{-in\omega} \, d\omega = \int_A e^{-in\omega} \, dP(\omega).$$

By 1.6.12,

$$a_n = \int_{T^{-1}A} e^{-in(\omega+\alpha)}\, dP(\omega) = e^{-in\alpha}\int_A e^{-in\omega}\, dP(\omega) = e^{-in\alpha} a_n$$

by invariance of A. Since $\alpha/2\pi$ is irrational, $e^{-in\alpha} \neq 1$ for $n \neq 0$, and it follows that $a_n = 0$ for $n \neq 0$. But $I_A \in L^2$, so that the Fourier series $\sum_{n=-\infty}^{\infty} a_n e^{in\omega}$ converges in L^2 to $I_A(\omega)$ (see 3.2, Problem 9). It follows that $I_A = a_0$ a.e., and therefore $P(A) = 0$ or 1. Thus T is ergodic.

Conversely, if $e^{i\alpha}$ is a root of unity, say, $e^{in\alpha} = 1$, then α is an integral multiple of $2\pi/n$. Let A be the union of the sectors $0 \leq \theta \leq \pi/n$, $2\pi/n \leq \theta \leq 3\pi/n$, $4\pi/n \leq \theta \leq 5\pi/n$, $\ldots, (2n-2)\pi/n \leq \theta \leq (2n-1)\pi/n$. Then A is invariant, but $P(A) = \frac{1}{2}$, so that T is not ergodic.

Now T is never mixing; to establish this we may assume that $\alpha/2\pi$ is irrational. Let $A = B = \{\theta\colon 0 \leq \theta \leq \pi\}$, corresponding to the upper semicircle. Given $\varepsilon > 0$, $e^{in\alpha}$ is within distance ε of $e^{i0} = 1$ for infinitely many n. (Extract a convergent subsequence $\{z_k\}$ from $\{e^{in\alpha}\}$, and select z_i and z_{i+j} such that $\mathrm{dist}(z_i, z_{i+j}) < \varepsilon$; i can be chosen larger than any preassigned positive integer. Then it is possible to form a chain that eventually goes entirely around the circle, with the distance between successive points less than ε.) It follows that A and $T^n A$ overlap except for a set of measure less than ε. Thus

$$P(A \cap T^{-n}B) = P(A \cap T^{-n}A) = P(T^n A \cap A)$$

$$\text{since } T \text{ is measure-preserving and invertible}$$

$$\geq P(A) - \varepsilon = \tfrac{1}{2} - \varepsilon > \tfrac{3}{8} \qquad \text{if} \qquad \varepsilon < \tfrac{1}{8}.$$

But $P(A)P(B) = [P(A)]^2 = \frac{1}{4}$, so the mixing property fails.

4. *One-sided and two-sided stationary processes.* Let T be the one-sided (or two-sided) shift transformation on the space of all infinite (or doubly infinite) sequences of real numbers. We consider only the case in which the coordinate random variables X_k are independent, that is, the measure P has the property that

$$P\{\omega\colon X_i(\omega) \in A_i, i = 1, 2, \ldots, n\} = \prod_{i=1}^{n} P\{\omega\colon X_i(\omega) \in A_i\}$$

for all real Borel sets $A_1 \ldots, A_n$ and all $n = 1, 2, \ldots$. In this case, T is *mixing* (hence ergodic). Let

$$A = \{\omega\colon (X_0(\omega), \ldots, X_{k-1}(\omega)) \in B_k\}$$

and

$$B = \{\omega: (X_0(\omega), \ldots, X_{r-1}(\omega)) \in B_r'\}$$

be measurable cylinders. For sufficiently large n we have $n > k - 1$, and therefore the indices defining the sets A and $T^{-n}B$ are distinct. Thus by independence,

$$P(A \cap T^{-n}B) = P\{\omega: (\omega_0, \ldots, \omega_{k-1}) \in B_k(\omega_n, \ldots, \omega_{n+r-1}) \in B_r'\}$$
$$= P(A)P(T^{-n}B) = P(A)P(B).$$

Thus the mixing condition holds for all measurable cylinders, and hence by 8.2.7, the mixing condition holds for all $A, B \in \mathscr{F}$, so that T is mixing.

Problems

1. If A is an almost invariant set, show that for almost every ω, $\omega \in A$ iff $T^n \omega \in A$ for all $n = 1, 2, \ldots$.

2. If μ is counting measure on the integers, show that $\omega \to \omega + 1$ is ergodic, but $\omega \to \omega + 2$ is not.

3. Let T be a measure-preserving transformation on $(\Omega, \mathscr{F}, \mu)$. Give examples to show that the following results are possible:

   (a) $A \in \mathscr{F}$ does not imply $T(A) \in \mathscr{F}$.
   (b) If $A \in \mathscr{F}$ and $T(A) \in \mathscr{F}$, $P(A)$ need not equal $P(TA)$.
   (c) If T is one-to-one onto, it need not be invertible (that is, T^{-1} need not be measurable).

4. (Jacobs, 1962) Let T be a measurable (but not necessarily measure-preserving) transformation on $(\Omega, \mathscr{F}, \mu)$. We say that T is *recurrent* iff for every $A \in \mathscr{F}$ and almost every $\omega \in A$, $T^n \omega \in A$ for some $n \geq 1$; T is *infinitely recurrent* iff for every $A \in \mathscr{F}$ and almost every $\omega \in A$, $T^n \omega \in A$ for infinitely many n. (In these definitions, the exceptional sets of measure 0 are allowed to depend on A.) A set $B \in \mathscr{F}$ is *wandering* iff $B, T^{-1}B, T^{-2}B, \ldots$ are disjoint; T is said to be *conservative* iff all wandering sets have measure 0. Finally, T is *incompressible* iff $A \subset T^{-1}A$ implies $\mu(T^{-1}A - A) = 0$. (Show that equivalently, $T^{-1}A \subset A$ implies $\mu(A - T^{-1}A) = 0$.)

   (a) Show that the following are equivalent:
       i. T is incompressible;
       ii. T is conservative;
       iii. T is recurrent;
       iv. T is infinitely recurrent.

   (One possible scheme is $(i) \Rightarrow (ii) \Rightarrow (iii) \Rightarrow (i)$, $(iv) \Rightarrow (iii)$, $(i) \Rightarrow (iv)$.)

(b) Show that $T(x) = x + 1$ on $\mathbb{R}$ (with Borel sets and Lebesgue measure) violates all four conditions of (a).

(c) (Poincaré) If T is measure-preserving and μ is finite, show that T is (infinitely) recurrent.

8.3 THE POINTWISE ERGODIC THEOREM

We will prove the pointwise ergodic theorem, which states that if T is a measure-preserving transformation on $(\Omega, \mathscr{F}, \mu)$ and $f \in L^1(\Omega, \mathscr{F}, \mu)$, then $n^{-1}(f(\omega) + f(T\omega) + \cdots + f(T^{n-1}\omega))$ converges to an integrable function $\hat{f}(\omega)$, for almost all ω.

It is possible to prove this result in a somewhat more general form (see the comments at the end of the chapter). The generalization is based on the fact that associated with a measure-preserving transformation is a positive contraction operator, as follows.

8.3.1 Theorem. If f is an extended real-valued Borel measurable function on $(\Omega, \mathscr{F}, \mu)$, and T is μ-preserving, let $\hat{T}f$ denote the function $f \circ T$. Then for every $p \in (0, \infty]$ we have $\|\hat{T}f\|_p = \|f\|_p$. Thus if we consider $\hat{T}$ as a linear operator on the Banach space $L^p(\Omega, \mathscr{F}, \mu)$, $1 \le p \le \infty$, then $\hat{T}$ is an *isometry* (a one-to-one, linear, norm-preserving map), in particular, $\hat{T}$ is a *contraction*, that is, $\|\hat{T}\| \le 1$ (in this case, $\|\hat{T}\| = 1$). Furthermore, $\hat{T}$ is *positive*, that is, if $f \ge 0$ a.e., then $\hat{T}f \ge 0$ a.e.

PROOF. By 1.6.12, $\int_\Omega |f(T\omega)|^p \, d\mu(\omega) = \int_\Omega |f(\omega)|^p \, d\mu(\omega)$, hence $\|\hat{T}f\|_p = \|f\|_p$, $0 < p < \infty$. If $p = \infty$, we have

$$\|\hat{T}f\|_\infty = \inf\{c \colon \mu\{|\hat{T}f| > c\} = 0\}$$
$$= \inf\{c \colon \mu\{\omega \colon |f(T\omega)| > c\} = 0\}$$
$$= \inf\{c \colon \mu T^{-1}\{\omega \colon |f(\omega)| > c\} = 0\}$$
$$= \inf\{c \colon \mu\{|f| > c\} = 0\} \qquad \text{since } T \text{ preserves } \mu$$
$$= \|f\|_\infty.$$

If $\mu(N) = 0$, and $f \ge 0$ on N^c, then $\mu(T^{-1}N) = 0$ and $\hat{T}f \ge 0$ on $(T^{-1}N)^c$, proving positivity. $\square$

The sequence of averages $f^{(n)}(\omega) = n^{-1}[f(\omega) + f(T\omega) + \cdots + f(T^{n-1}\omega)]$ can now be expressed in terms of $\hat{T}$ as

$$f^{(n)} = n^{-1}(f + \hat{T}f + \cdots + \hat{T}^{n-1}f)$$

where $\hat{T}^0 f = f$ and $\hat{T}^k$ is the composition of $\hat{T}$ with itself k times, $k \ge 1$.

In the three results to follow, T is a measure-preserving transformation on $(\Omega, \mathscr{F}, \mu)$, and $f: (\Omega, \mathscr{F}) \to (\overline{\mathbb{R}}, \mathscr{B}(\overline{\mathbb{R}}))$. Inspection of the proofs will show, however, that the results hold if $\hat{T}$ is replaced by an arbitrary positive contraction on L^1, not necessarily arising from a measure-preserving transformation.

8.3.2   Lemma. If $f \in L^1(\Omega, \mathscr{F}, \mu)$, let $f_0 = f$, and

$$f_n = max(f, f + \hat{T}f, \ldots, f + \hat{T}f + \cdots + \hat{T}^n f), \qquad n \geq 1.$$

Then

$$f_{n+1} \leq f + \hat{T} f_n^+, \qquad n = 0, 1, \ldots .$$

PROOF.   If $0 \leq m \leq n$, then $\sum_{k=0}^{m+1} \hat{T}^k f = f + \hat{T}\left(\sum_{k=0}^{m} \hat{T}^k f\right)$. Since $\sum_{k=0}^{m} \hat{T}^k f \leq f_n$, and $\hat{T}$ is positive, we have

$$\hat{T}\left(\sum_{k=0}^{m} \hat{T}^k f\right) \leq \hat{T} f_n \leq \hat{T} f_n^+.$$

Thus

$$\sum_{k=0}^{m+1} \hat{T}^k f \leq f + \hat{T} f_n^+, \qquad 0 \leq m \leq n.$$

Since $f \leq f + \hat{T} f_n^+$, we have $f_{n+1} \leq f + \hat{T} f_n^+$, as desired.   □

8.3.3   Lemma. Let f_n be defined as in 8.3.2, and assume $f \in L^1(\Omega, \mathscr{F}, \mu)$. If $A_n = \{f_n > 0\}$, then $\int_{A_n} f \, d\mu \geq 0$.

PROOF.   By 8.3.2,

$$\int_{A_n} f \, d\mu \geq \int_{A_n} f_{n+1} \, d\mu - \int_{A_n} (\hat{T} f_n^+) \, d\mu$$

$$\geq \int_{A_n} f_n \, d\mu - \int_{A_n} (\hat{T} f_n^+) \, d\mu$$

$$= \int_{\Omega} f_n^+ \, d\mu - \int_{A_n} (\hat{T} f_n^+) \, d\mu$$

$$\geq \int_{\Omega} f_n^+ \, d\mu - \int_{\Omega} (\hat{T} f_n^+) \, d\mu$$

$$= \|f_n^+\|_1 - \|\hat{T} f_n^+\|_1 \geq 0$$

since $\hat{T}$ is a contraction.   □

8.3.4   Maximal Ergodic Theorem. If $f \in L^1(\Omega, \mathscr{F}, \mu)$ and

$$A = \left\{ \omega: \ \sup_{n \geq 1} \sum_{k=0}^{n-1} (\hat{T}^k f)(\omega) > 0 \right\}$$

$$= \left\{ \omega: \ \sup_{n \geq 1} f^{(n)}(\omega) > 0 \right\}, \qquad \text{where} \qquad f^{(n)} = n^{-1} \sum_{k=0}^{n-1} \hat{T}^k f$$

then $\int_A f \, d\mu \geq 0$.

PROOF.   The sets A_n of 8.3.3 increase to A. $\square$

It will be convenient to isolate some of the technical difficulties in the proof of the pointwise ergodic theorem. Let $f \in L^1(\Omega, \mathscr{F}, \mu)$, and let $f^{(n)}(\omega) = n^{-1} \sum_{k=0}^{n-1} f(T^k \omega)$, $n = 1, 2, \ldots$, where T is μ-preserving. If $a < b$, define

$$C_{ab} = C_{ab}(f) = \left\{ \omega: \ \liminf_{n \to \infty} f^{(n)}(\omega) < a < b < \limsup_{n \to \infty} f^{(n)}(\omega) \right\},$$

$$N_b = \left\{ \omega: \ \sup_{n \geq 1} f^{(n)}(\omega) > b \right\};$$

note that C_{ab} is a subset of N_b.

Just as in 6.4.3 we can establish a.e. convergence of the sequence $f^{(n)}$ if we show that $\mu(C_{ab})$ is always 0. To do this, we may assume without loss of generality that $b > 0$. Note that

$$C_{ab}(f) = \{ \omega: \ \liminf - f^{(n)}(\omega) < -b < -a < \limsup - f^{(n)}(\omega) \}$$

$$= C_{-b, -a}(-f).$$

If $b \leq 0$, then $-a > 0$, and the argument below will show that $C_{-b, -a}(-f)$ has measure 0, and thus $\mu(C_{ab}) = 0$.

8.3.5   Lemma. The set C_{ab} has the following properties:

(a)   The set C_{ab} is almost invariant.
(b) $\mu(C_{ab}) < \infty$.
(c)   In fact $\mu(C_{ab}) = 0$.

PROOF.   (a)   We may write

$$f^{(n)}(T\omega) = \frac{1}{n}\sum_{k=0}^{n-1} f(T^{k+1}\omega)$$

$$= \frac{n+1}{n}\left[\frac{1}{n+1}\sum_{k=0}^{n} f(T^k\omega)\right] - \frac{f(\omega)}{n}$$

$$= \frac{n+1}{n} f^{(n+1)}(\omega) - \frac{f(\omega)}{n}.$$

Since $f \in L^1$, $f(\omega)$ is finite for almost all ω, hence

$$\liminf_{n\to\infty} f^{(n)}(\omega) = \liminf_{n\to\infty} f^{(n)}(T\omega),$$

and similarly for lim sup, except possibly on a set of measure 0. Thus, outside a set of measure 0, $\omega \in C_{ab}$ iff $T\omega \in C_{ab}$, and the result follows.

(b)   Let C be any set in $\mathscr{F}$ such that $C \subset C_{ab}$ and $\mu(C) < \infty$, and define $F_b = \{\omega: \sup_{n\geq 1}(f - bI_C)^{(n)}(\omega) > 0\}$. Note that if $\omega \in C_{ab}$, then $\omega \in N_b$, hence

$$\frac{1}{n}\sum_{k=0}^{n-1} f(T^k\omega) > b \qquad \text{for some } n$$

$$\geq \frac{1}{n}\sum_{k=0}^{n-1} bI_C(T^k\omega).$$

Thus $\omega \in F_b$; in particular, C is a subset of F_b. By the maximal ergodic theorem 8.3.4,

$$\int_{F_b}(f - bI_C)\,d\mu \geq 0.$$

Thus

$$\int_\Omega |f|\,d\mu \geq \int_{F_b} f\,d\mu \geq b\int_{F_b} I_C\,d\mu = b\mu(C \cap F_b) = b\mu(C)$$

so that

$$\mu(C) \leq b^{-1}\int_\Omega |f|\,d\mu < \infty.$$

Now note that C_{ab} is a subset of

$$\bigcup_{n=0}^{\infty}\{\omega: |f(T^n\omega)| > 0\};$$

By Theorem 1.6.12,

$$\int_\Omega |f(T^n\omega)|\,d\mu(\omega) = \int_\Omega |f(\omega)|\,d\mu(\omega) < \infty,$$

and it follows that $\{\omega\colon |f(T^n\omega)| > 0\}$ is a countable union of sets of finite measure, and therefore so is C_{ab} (see 2.2, Problem 2). Thus

$$\mu(C_{ab}) = \sup\{\mu(C)\colon C \in \mathscr{F}, C \subset C_{ab}, \mu(C) < \infty\}$$

$$\leq b^{-1}\int_\Omega |f|\,d\mu < \infty.$$

(c)   Since C_{ab} is almost invariant, T is a well-defined measure-preserving transformation on $(C_{ab}, \mathscr{F}_{ab}, \mu_{ab})$, where

$$\mathscr{F}_{ab} = \{A \in \mathscr{F}\colon A \subset C_{ab}\} = \{B \cap C_{ab}\colon B \in \mathscr{F}\},$$

and $\mu_{ab} = \mu$ restricted to C_{ab}. (Strictly speaking, if $D = C_{ab}\,\Delta\,T^{-1}C_{ab}$, then T is well defined on $C_{ab} - D$. Since $\mu(D) = 0$, this causes no difficulty. For example, we may redefine T as the identity on D, and then it will be well defined and measure-preserving on C_{ab}.)

The argument of part (b) may now be applied to T on C_{ab}. In particular, the equation $\int_{F_b}(f - bI_C)\,d\mu \geq 0$ now becomes

$$\int_{C_{ab}\cap F_b} (f - bI_C)\,d\mu \geq 0.$$

Since C_{ab} has finite measure, we may set $C = C_{ab}$, and since $C_{ab} \subset F_b$, we obtain

$$\int_{C_{ab}} (f - b)\,d\mu \geq 0.$$

Now let

$$F_{ab}{}' = \left\{\omega \in C_{ab}\colon \sup_{n\geq 1}(a - f)^{(n)}(\omega) > 0\right\}.$$

If $\omega \in C_{ab}$, then $f^{(n)}(\omega) < a$ for at least one n, hence $\omega \in F_{ab}'$; thus $C_{ab} = F_{ab}{}'$. Since C_{ab} has finite measure, constant functions are integrable on C_{ab}, and we may therefore apply the maximal ergodic theorem to obtain

$$\int_{C_{ab}} (a - f)\,d\mu \geq 0.$$

Thus we have, for $a < b$,

$$b\mu(C_{ab}) \leq \int_{C_{ab}} f \, d\mu \leq a\mu(C_{ab}),$$

a contradiction unless $\mu(C_{ab}) = 0$. $\square$

We may now prove the main result.

8.3.6 Pointwise Ergodic Theorem. Let T be a measure-preserving transformation on $(\Omega, \mathscr{F}, \mu)$, and let $f \in L^1(\Omega, \mathscr{F}, \mu)$. Then there is a function $\hat{f} \in L^1$ such that we have $n^{-1} \sum_{k=0}^{n-1} f(T^k\omega) \to \hat{f}(\omega)$ almost everywhere.

PROOF. Let $D = \{\omega: f^{(n)}(\omega)$ does not converge to a finite or infinite limit$\}$. Then $D = \bigcup \{C_{ab}(f): a < b, a, b$ rational$\}$. By 8.3.5, $\mu(D) = 0$, and hence $f^{(n)}(\omega)$ converges for almost all ω; we call the limit $\hat{f}(\omega)$. (Define $\hat{f} = 0$ on the exceptional set.) By Fatou's lemma,

$$\int_{\Omega} |\hat{f}| \, d\mu = \int_{\Omega} \lim_{n \to \infty} |f^{(n)}(\omega)| \, d\mu(\omega) \leq \liminf_{n \to \infty} \int_{\Omega} |f^{(n)}(\omega)| \, d\mu(\omega).$$

But

$$\int_{\Omega} |f^{(n)}| \, d\mu \leq \frac{1}{n} \sum_{k=0}^{n-1} \int_{\Omega} |f(T^k\omega)| \, d\mu(\omega)$$

$$= \frac{1}{n} \sum_{k=0}^{n-1} \int_{\Omega} |f| \, d\mu \qquad \text{by 1.6.12}$$

$$= \int_{\Omega} |f| \, d\mu < \infty,$$

and the theorem is proved. $\square$

We now look more closely at the convergence of the sequence $\{f^{(n)}\}$.

8.3.7 Theorem. If $\mu(\Omega) < \infty$ and $f \in L^p (1 \leq p < \infty)$, then $\hat{f} \in L^p$ and $f^{(n)} \xrightarrow{L^p} \hat{f}$.

PROOF. Since the finite-valued simple functions are dense in L^p, for each $\varepsilon > 0$ there is a bounded measurable function g such that $\|f - g\|_p < \varepsilon$. If $f_k(\omega) = f(T^k\omega)$, $g_k(\omega) = g(T^k\omega)$, and $|g| \le M$, then

$$|f^{(n)} - \hat{f}\|_p \le \left\|\frac{1}{n}\sum_{k=0}^{n-1}(f_k - g_k)\right\|_p + \left\|\left(\frac{1}{n}\sum_{k=0}^{n-1}g_k\right) - \hat{g}\right\|_p + \|\hat{g} - \hat{f}\|_p. \quad (1)$$

Since $|n^{-1}\sum_{k=0}^{n-1}g_k| \le M$ (hence $|\hat{g}| \le M$ a.e.), the second term on the right in (1) approaches zero as n approaches ∞, by the dominated convergence theorem. (The hypothesis that $\mu(\Omega) < \infty$ implies that the function constant at M is integrable.) By 1.6.12, $\|f_k - g_k\|_p = \|f - g\|_p < \varepsilon$, hence the first term is less than ε. Now

$$\int_\Omega |\hat{f} - \hat{g}|^p \, d\mu = \int_\Omega \lim_n \left|\frac{1}{n}\sum_{k=0}^{n-1}(f_k - g_k)\right|^p d\mu$$

$$\le \liminf_n \int_\Omega \left|\frac{1}{n}\sum_{k=0}^{n-1}(f_k - g_k)\right|^p d\mu$$

$$= \liminf_n \left\|\frac{1}{n}\sum_{k=0}^{n-1}(f_k - g_k)\right\|_p^p < \varepsilon^p.$$

Thus $\|f^{(n)} - \hat{f}\|_p < 2\varepsilon$ for large enough n, and the result follows. $\square$

If $p = 1$ in 8.3.7, the hypothesis that $\mu(\Omega) < \infty$ cannot be dropped (Problem 1). Also, the result fails for $p = \infty$, even if $\mu(\Omega) < \infty$ (Problem 2).

We can now identify the limit function $\hat{f}$. Theorem 8.3.9 indicates that although the pointwise ergodic theorem can be presented without reference to probability, some insight is lost in doing so.

8.3.8 Lemma. If $f \in L^1$, then $\hat{f}$ is almost invariant. Thus if $\mathscr{G}$ is the σ-field of almost invariant sets, then $\hat{f}: (\Omega, \mathscr{G}) \to (\mathbb{R}, \mathscr{B}(\mathbb{R}))$.

PROOF. If $f^{(n)}(\omega) \to \hat{f}(\omega)$ for $\omega \notin N$, where $\mu(N) = 0$, then $f^{(n)}(T\omega) \to \hat{f}(T\omega)$ for $\omega \notin T^{-1}N$, where $\mu(T^{-1}N) = 0$. But [see the proof of 8.3.5(a)]

$$f^{(n)}(T\omega) = \left(\frac{n+1}{n}\right)f^{(n+1)}(\omega) - \frac{f(\omega)}{n},$$

and $f(\omega)/n \to 0$ a.e. since $f \in L^1$. Thus $f^{(n)}(T\omega) \to \hat{f}(\omega)$ a.e. hence $\hat{f}(\omega) = \hat{f}(T\omega)$ a.e., proving $\hat{f}$ almost invariant.

If $B \in \mathscr{B}(\overline{\mathbb{R}})$ and $C = \{\omega: \hat{f}(\omega) \in B\}$, then if $\hat{f}(\omega) = \hat{f}(T\omega)$ we have $\omega \in C$ iff $T\omega \in C$. Thus C is almost invariant, and the proof is complete. $\Box$

8.3.9   Theorem. If $f \in L^1$, and A is an almost invariant set of finite measure, then $\int_A f \, d\mu = \int_A \hat{f} \, d\mu$. Thus in a probability space, $\hat{f} = E(f|\mathscr{G})$, where $\mathscr{G}$ is the σ-field of almost invariant sets.

PROOF.   Restrict T and μ to the almost invariant set A. Since $\mu(A) < \infty$, we have L^1 convergence (on A) by 8.3.7, hence $\int_A f^{(n)} \, d\mu \to \int_A \hat{f} \, d\mu$. But

$$\int_A f^{(n)} \, d\mu = \int_A f \, d\mu \qquad \text{by 1.6.12.} \quad \Box$$

In the ergodic case, $\hat{f}$ assumes a very special form.

8.3.10   Theorem. If T is ergodic and $f \in L^1$, then $\hat{f}$ is constant a.e. If $\mu(\Omega) = \infty$, the constant is $c = 0$; if $\mu(\Omega) < \infty$, we have

$$c = \frac{1}{\mu(\Omega)} \int_\Omega f \, d\mu.$$

Thus on a probability space, $\hat{f} = E(f)$ a.e.

PROOF.   By 8.3.8, $\hat{f}$ is almost invariant, so by 8.2.3(b), $\hat{f} = c$ a.e. If $\mu(\Omega) = \infty$, then c must be 0 because $\hat{f} \in L^1$ by 8.3.6. If $\mu(\Omega) < \infty$, then $c = [\mu(\Omega)]^{-1} \int_\Omega f \, d\mu$ by 8.3.9 (with $A = \Omega$). $\Box$

If T is ergodic and μ is finite, consider the case $f = I_A$. Then by 8.3.10, $\hat{f} = \mu(A)/\mu(\Omega)$ a.e., so that $n^{-1} \sum_{k=0}^{n-1} I_A(T^k\omega)$, the relative frequency of visits to A, converges a.e. to the relative mass of A (the probability of A if $\mu(\Omega) = 1$). Apparently we have a version of the strong law of large numbers, and in fact the pointwise ergodic theorem can be regarded as a generalization of this result. Let T be the one-sided shift transformation (see 8.1.2 Example 4 and 8.2.8 Example 4), with coordinate random variables X_k. If Z is an integrable random variable on $(\Omega, \mathscr{F}, P)$, then

$$Z^{(n)}(\omega) = n^{-1}[Z(\omega_0, \omega_1, \ldots) + Z(\omega_1, \omega_2, \ldots) + \cdots + Z(\omega_{n-1}, \omega_n, \ldots)],$$

in other words,

$$Z^{(n)} = n^{-1}[Z(X_0, X_1, \ldots) + Z(X_1, X_2, \ldots) + \cdots + Z(X_{n-1}, X_n, \ldots)].$$

By 8.3.9,

$$Z^{(n)} \xrightarrow{\text{a.e.}} E(Z \mid \mathscr{G})$$

where $\mathscr{G}$ is the σ-field of almost invariant sets; if T is ergodic, the limit is $E(Z)$ by 8.3.10. In particular, let $X_0, X_1, \ldots$ be iid random variables with finite expectation. If $Z(\omega) = \omega_0$, that is, $Z = X_0$, we obtain

$$n^{-1}(X_0 + \cdots + X_{n-1}) \to E(X_0) \qquad \text{a.e.,}$$

the iid case of the strong law of large numbers (see 6.2.5).

In the next section, it will be necessary to consider probability measures P_1 and P_2 that are each preserved by a fixed measurable transformation T on $(\Omega, \mathscr{F})$; P_i is said to be *ergodic* (relative to T) iff T is ergodic on $(\Omega, \mathscr{F}, P_i)$. If P_1 and P_2 are both ergodic, they must be identical or mutually singular.

8.3.11 Theorem. If P_1 and P_2 are ergodic probability measures relative to T, then either $P_1 \equiv P_2$ or $P_1 \perp P_2$.

PROOF. Suppose that $P_1(A) \neq P_2(A)$ for some $A \in \mathscr{F}$, and let

$$A_i = \{\omega \colon I_A^{(n)}(\omega) \to P_i(A)\}, \qquad i = 1, 2.$$

By 8.3.6 and 8.3.10, $P_1(A_1) = P_2(A_2) = 1$. But A_1 and A_2 are disjoint since $P_1(A) \neq P_2(A)$, hence $P_1 \perp P_2$. $\square$

Theorem 8.3.11 gives us a criterion for ergodicity (unfortunately impractical).

8.3.12 Theorem. The probability measure P is ergodic relative to T iff there is no probability measure P_1 preserved by T such that P_1 is absolutely continuous with respect to P but not identical to P.

PROOF. If $P_1 << P$, $P_1 \not\equiv P$ and P is ergodic, then so is P_1. If A is an invariant set, then $P(A) = 0$ or $P(A^c) = 0$ by ergodicity, hence $P_1(A) = 0$ or $P_1(A^c) = 0$ by absolute continuity. By 8.3.11, $P_1 \perp P$. But then P_1 is both absolutely continuous and singular with respect to P, hence P_1 is the zero measure, a contradiction.

Conversely, if P is not ergodic, let A be a T-invariant set with $0 < P(A) < 1$. Define $P_1(B) = P(B \mid A) = P(A \cap B)/P(A)$, $B \in \mathscr{F}$; then $P_1 << P$, and since $P_1(A) = 1 \neq P(A)$, $P_1 \not\equiv P$. Now

$$P_1(T^{-1}E) = \frac{P(T^{-1}E \cap A)}{P(A)}$$

$$= \frac{P(T^{-1}E \cap T^{-1}A)}{P(A)} \qquad \text{since } A \text{ is invariant}$$

$$= \frac{P(E \cap A)}{P(A)} \qquad \text{since } T \text{ preserves } P$$

$$= P_1(E).$$

Thus P_1 is preserved by T. □

Problems

1. Let $T(\omega) = \omega + 1$ on $\mathbb{R}$ (with Borel sets and Lebesgue measure). If f is the indicator of $(0, 1]$, show that $f^{(n)}$ does not converge in L^1 to $\hat{f}$.

2. Let $\Omega = \mathbb{R}^\infty$, $\mathscr{F} = [\mathscr{B}(\mathbb{R})]^\infty$, $X_n(\omega_0, \omega_1, \ldots) = \omega_n$, P the unique probability measure making the X_n independent with $P\{X_n = 0\} = P\{X_n = 1\} = \frac{1}{2}$. (In other words, consider an infinite sequence of Bernoulli trials with probability $\frac{1}{2}$ of success on a given trial.) If T is the one-sided shift and $f(\omega) = \omega_0$, show that $n^{-1} \sum_{k=0}^{n-1} f(T^k \omega)$ does not converge in L^∞ to $\hat{f}(\omega) = \frac{1}{2}$.

3. Let T be a measure-preserving transformation on $(\Omega, \mathscr{F}, \mu)$. If T is ergodic, we know that for every $f \in L^1$, $f^{(n)}$ converges a.e. to a constant. Conversely, if $\mu(\Omega) < \infty$ and if for every $f \in L^1$ there is a constant $c = c(f)$ such that $f^{(n)} \to c$ a.e., show that T is ergodic. Give a counterexample to this statement if $\mu(\Omega) = \infty$.

4. Let T be an ergodic measure-preserving transformation on $(\Omega, \mathscr{F}, \mu)$ with $\mu(\Omega) < \infty$. Let f be a real-valued Borel measurable function such that $\int_\Omega f \, d\mu$ exists. If $f \in L^1$, we know that $f^{(n)}$ converges a.e. to $[\mu(\Omega)]^{-1} \int_\Omega f \, d\mu$. Conversely, if $f^{(n)}$ converges a.e. to a finite limit, show that $f \in L^1$. (A special case of this result was considered in 6.2, Problem 1. Note also that the result fails when $\mu(\Omega) = \infty$; take $f \equiv c$.)

5. (*Mean Ergodic Theorem in a Hilbert Space*) Let U be a bounded linear operator on the Hilbert space H, and let U^* be the adjoint of U, defined by the requirement that $\langle Uf, g \rangle = \langle f, U^*g \rangle$ for all $f, g \in H$. ($f \to \langle Uf, g \rangle$ is a continuous linear functional on H, so by 3.3.4(a) there is a unique element $h \in H$ such that $\langle Uf, g \rangle = \langle f, h \rangle$ for every $f \in H$; h depends on g and we may write h as U^*g. It follows from the basic properties of the inner product that U^* is a bounded linear operator on H.)

    Establish the following results.

(a) The following conditions are equivalent (and define a unitary operator, that is, an invertible isometry).
    i. $UU^* = U^*U = I$, the identity operator on H.
    ii. U is one-to-one onto, and $\langle f, g \rangle = \langle Uf, Ug \rangle$ for all $f, g \in H$.
    iii. U is one-to-one onto, and $\|Uf\| = \|f\|$ for all $f \in H$.

(b) The following conditions are equivalent (and define an isometry).
    i. $U^*U = I$.
    ii. $\langle f, g \rangle = \langle Uf, Ug \rangle$ for all $f, g \in H$.
    iii. $\|Uf\| = \|f\|$ for all $f \in H$.

For the remainder of the problem, U is an isometry of H.

(c) If $f \in H$, then $Uf = f$ iff $U^*f = f$.

(d) Define $A_n = n^{-1}(I + U + U^2 + \cdots + U^{n-1})$; note that

$$\|A_n\| \leq n^{-1} \sum_{k=0}^{n-1} \|U\|^k = 1.$$

If $E = \{f \in H \colon \lim_{n \to \infty} A_n f \text{ exists (in } H)\}$, E is a closed subspace of H.

(e) Let M be the set of elements of H that are *invariant* under U, that is, $M = \{f \in H \colon Uf = f\}$; note that M is a closed subspace of H, by continuity of U. Let $N_0 = \{g - Ug \colon g \in H\}$. If we define $\hat{f} = \lim_{n \to \infty} A_n f$ (where the limit exists) then:

$$
\begin{array}{llll}
f \in M & \text{implies} & f \in E & \text{and} \quad \hat{f} = f; \\
f \in N_0 & \text{implies} & f \in E & \text{and} \quad \hat{f} = 0.
\end{array}
$$

(f) If $N = \overline{N_0}$, the closure of N_0, then H is the orthogonal direct sum of M and N (see the discussion after 3.2.11).

(g) (*Mean Ergodic Theorem*) Let U be an isometry of the Hilbert space H, and let P be the projection of H on the space M of all elements invariant under U. For every $f \in H$,

$$\frac{1}{n} \sum_{k=0}^{n-1} U^k f \to Pf.$$

If T is a measure-preserving transformation and $U = \hat{T}$, we obtain L^2 convergence of $f^{(n)}$ to $\hat{f}$.

(h) If, in addition, T is invertible and $S = T^{-1}$, then U is a unitary operator and $U^* = U^{-1} = \hat{S}$.

6. Let T be a measurable transformation on $(\Omega, \mathscr{F})$. Within the linear space of finite signed measures on $\mathscr{F}$, let K be the convex set of probability

measures preserved by T. Show that P is ergodic relative to T iff P is an extreme point of K, that is, P cannot be expressed as $\lambda_1 P_1 + \lambda_2 P_2$, with

$$\lambda_1, \lambda_2 > 0, \lambda_1 + \lambda_2 = 1, P_1, P_2 \in K, P_1 \not\equiv P_2.$$

7. This problem gives many conditions equivalent to ergodicity; in particular, some of the conditions involve convergence in probability rather than almost everywhere convergence.

Let T be a measure-preserving transformation on the probability space $(\Omega, \mathscr{F}, P)$, and let $\mathscr{F}_0$ be a field of sets whose minimal σ-field is $\mathscr{F}$. Show that the following conditions are equivalent:

(a) T is ergodic.
(b) $I_A^{(n)} \to P(A)$ a.e. for each $A \in \mathscr{F}$, where

$$I_A^{(n)}(\omega) = n^{-1} \sum_{k=0}^{n-1} I_A(T^k \omega).$$

(c) $I_A^{(n)} \to P(A)$ in probability for each $A \in \mathscr{F}$.
(d) $I_A^{(n)} \to P(A)$ in probability for each $A \in \mathscr{F}_0$.
(e) $I_A^{(n)} \to P(A)$ a.e. for each $A \in \mathscr{F}_0$.
(f) $n^{-1} \sum_{k=0}^{n-1} P(A \cap T^{-k}B) \to P(A)P(B)$ for all $A, B \in \mathscr{F}_0$.
(g) $n^{-1} \sum_{k=0}^{n-1} P(A \cap T^{-k}B) \to P(A)P(B)$ for all $A, B \in \mathscr{F}$.

If T is a one-sided shift transformation, we may take $\mathscr{F}_0$ to be the field of measurable cylinders. Furthermore, if the coordinate random variables take on only finitely many possible values, a measurable cylinder is a finite disjoint union of sets of the form $\{X_0 = i_0, \ldots, X_m = i_m\}$. Thus condition (d) is equivalent to the following statement:

For each $\alpha = (i_0, \ldots, i_m)$, $m = 0, 1, \ldots$, with the i_k belonging to the coordinate space, let N_α^n be the number of times that $i_0, \ldots, i_m$ occur in sequence in the first $n + m$ coordinates, that is,

$$N_\alpha^n(\omega) = \sum_{k=0}^{n-1} I_A(T^k \omega), \qquad \text{where} \qquad A = \{\omega\colon \omega_0 = i_0, \ldots, \omega_m = i_m\};$$

then $n^{-1} N_\alpha^n$ converges in probability to $p(\alpha) = P(A)$.

Note also that by the Kolmogorov extension theorem, given any one-sided shift with coordinate random variables X_n, there is a two-sided shift with coordinate random variables X_n' such that $(X_n', n \geq 0)$ and $(X_n, n \geq 0)$ have the same distribution. [For example, specify that (X'_{-8}, X'_{-3}, X'_6)

have the same distribution as (X_0, X_5, X_{14}).] Condition (d) shows that the one-sided shift is ergodic iff the corresponding two-sided shift is ergodic; by 8.2.7, the same is true if "ergodic" is replaced by "mixing."

8.4   APPLICATIONS TO MARKOV CHAINS

If T is a one-sided shift transformation and the coordinate random variables X_n are independent, we have seen that T is ergodic (we also say that the sequence $\{X_n\}$ is ergodic). In this section we exhibit a large class of examples in which T is ergodic, but the X_n are not independent.

First, we must consider some general properties of shift transformations. If $X_0, X_1, \ldots$ are the coordinate random variables of a one-sided shift, recall (see 6.2.6) that the *tail σ-field* of the X_n is defined by $\mathscr{F}_\infty = \bigcap_{n=0}^\infty \mathscr{F}_n$, where $\mathscr{F}_n = \mathscr{F}(X_n, X_{n+1}, \ldots)$, the smallest σ-field that makes X_i measurable for all $i \geq n$. We prove that the σ-field of almost invariant sets is essentially included in $\mathscr{F}_\infty$.

8.4.1   Theorem. If A is an almost invariant set, there is a set $B \in \mathscr{F}_\infty$ such that $P(A \triangle B) = 0$.

PROOF.   By 8.2.2(a), there is a strictly invariant set B with $P(A \triangle B) = 0$. Now if $C \in \mathscr{F}$ then $T^{-n}C \in \mathscr{F}_n$; for if $C = \{(X_0, X_1, \ldots) \in C'\}$, then $T^{-n}C = \{(X_n, X_{n+1}, \ldots) \in C'\}$. (Actually, $C = C'$ here since the X_n are coordinate random variables.) But $B = T^{-n}B$, so that $B \in \mathscr{F}_n$ for all n, hence $B \in \mathscr{F}_\infty$. $\square$

The proof of 8.4.1 shows that all strictly invariant sets belong to $\mathscr{F}_\infty$; however, an almost invariant set might not be in $\mathscr{F}_\infty$. For example, let $A = \{\omega: X_n(\omega) = X_0(\omega) \text{ for all } n\}$; then $A \subset T^{-1}A$, so that A is almost invariant by 8.2.4; but $A \notin \mathscr{F}_\infty$.

The following fact about conditional probabilities will be used.

8.4.2   Lemma. If T is a measure-preserving transformation on $(\Omega, \mathscr{F}, P)$, and $\mathscr{G}$ is a sub σ-field of $\mathscr{F}$, then for any $A \in \mathscr{F}$,

$$P(A \mid \mathscr{G})(T\omega) = P(T^{-1}A \mid T^{-1}\mathscr{G})(\omega) \qquad \text{a.e.}$$

In particular, if T is a shift transformation, then

$$P(A \mid X_n)(T\omega) = P(T^{-1}A \mid X_{n+1})(\omega) \qquad \text{a.e.}$$

PROOF.   If $T^{-1}B \in T^{-1}\mathscr{G}$, then

$$P(T^{-1}A \cap T^{-1}B) = \int_{T^{-1}B} P(T^{-1}A \mid T^{-1}\mathscr{G})\, dP.$$

But since T is measure-preserving,

$$P(T^{-1}A \cap T^{-1}B) = P(A \cap B) = \int_B P(A \mid \mathcal{G}) \, dP$$

$$= \int_{T^{-1}B} P(A \mid \mathcal{G})(T\omega) \, dP(\omega) \qquad \text{by } 1.6.12$$

Since $P(A \mid \mathcal{G})(T\omega)$ is $T^{-1}\mathcal{G}$-measurable, the result follows. $\square$

We may give an intuitive interpretation of 8.4.2. If T is a shift transformation and $\omega^* = (\omega_0^*, \omega_1^*, \ldots) \in \mathbb{R}^\infty$, then $P(T^{-1}A \mid X_{n+1})(\omega^*)$ is the probability that $\omega \in T^{-1}A$, given that the $(n+1)$th coordinate of ω is $X_{n+1}(\omega^*) = \omega_{n+1}^*$; $P(A \mid X_n)(T\omega^*)$ is the probability that $T\omega \in A$, given that the nth coordinate of $T\omega$ is $X_n(T\omega^*) = \omega_{n+1}^*$; these two expressions agree.

Our interest will be in sequences having the Markov property, defined as follows.

8.4.3 Definition. A sequence of random variables $\{X_n, n \geq 0\}$ is said to have the *Markov property* iff for each $B \in \mathcal{F}(X_{n+1}, X_{n+2}, \ldots)$, $n = 0, 1, \ldots$,

$$P(B \mid X_0, \ldots, X_n) = P(B \mid X_n) \qquad \text{a.e.}$$

If $\{X_n\}$ is a Markov chain (see 4.11, and Ash, 1970, Chapter 7), then

$$P\{X_{n+1} = i_{n+1}, \ldots, X_{n+k} = i_{n+k} \mid X_0 = i_0, \ldots, X_n = i_n\}$$
$$P\{X_{n+1} = i_{n+1}, \ldots, X_{n+k} = i_{n+k} \mid X_n = i_n\}$$

for all $i_0, \ldots, i_{n+k}$ in the state space. Thus the Markov property holds when B is a measurable cylinder, hence for all $B \in \mathcal{F}(X_{n+1}, X_{n+2}, \ldots)$ by the monotone class theorem.

In the Markov case, almost invariant sets assume a special form.

8.4.4 Theorem. Let the coordinate random variables X_n of a one-sided shift have the Markov property. If A is an almost invariant set, there is a set B of the form $\{X_0 \in C\}$, $C \in \mathcal{B}(\mathbb{R})$, such that $A = B$ a.e., that is, $P(A \triangle B) = 0$.

PROOF. By 8.4.1, there is a set B in the tail σ-field $\mathcal{F}_\infty$ such that $P(A \triangle B) = 0$. Since $B \in \mathcal{F}(X_{n+1}, X_{n+2}, \ldots)$ for all n, $P(B \mid X_0, \ldots, X_n) = P(B \mid X_n)$ a.e. Consequently, $P(A \mid X_0, \ldots, X_n) = P(A \mid X_n)$ a.e. Now

$$P(A \mid X_0, \ldots, X_n) \to P(A \mid X_0, X_1, \ldots) \qquad \text{a.e.}$$

(see 6.6.4); but

$$P(A \mid X_0, X_1, \ldots) = E(I_A \mid X_0, X_1, \ldots) = I_A \qquad \text{a.e.}$$

since $A \in \mathscr{F}(X_0, X_1, \ldots) = \mathscr{F}$. Therefore $P(A \mid X_n) \to I_A$ a.e.

Given $\varepsilon > 0$, let $H_n = \{\omega: |P(A \mid X_n)(\omega) - I_A(\omega)| \geq \varepsilon\}$. Then $P(H_n) \to 0$ since $P(A \mid X_n) \to I_A$ a.e., hence in probability. Now

$$T^{-1}(H_n) = \{\omega: |P(A \mid X_n)(T\omega) - I_A(T\omega)| \geq \varepsilon\}$$
$$= \{\omega: |P(A \mid X_{n+1})(\omega) - I_A(\omega)| \geq \varepsilon\} \qquad \text{a.e.}$$

by 8.4.2 and the almost invariance of A. Thus $T^{-1}(H_n) = H_{n+1}$ a.e. Since T preserves P, $P(H_n) = PT^{-1}(H_n) = P(H_{n+1})$ for all n. Since $P(H_n) \to 0$, we must have $P(H_n) \equiv 0$, so $P(A \mid X_n) = I_A$ a.e. for all n, in particular, $I_A = P(A \mid X_0)$ a.e.

Now $P(A \mid X_0)$ is $\mathscr{F}(X_0)$-measurable, hence can be expressed as $f(X_0)$ for some Borel measurable $f: \mathbb{R} \to \mathbb{R}$ [see 5.4.2(c)]. Thus (a.e.)

$$\omega \in A \quad \text{iff} \quad I_A(\omega) = 1$$
$$\text{iff} \quad f(X_0(\omega)) = 1$$
$$\text{iff} \quad X_0(\omega) \in C, \qquad \text{where} \qquad C = f^{-1}\{1\}. \quad \square$$

8.4.5   Corollary. Under the hypothesis of 8.4.4, a set $A \in \mathscr{F}$ is almost invariant iff A is of the form $\{X_n \in C$ for all $n\}$ for some $C \in \mathscr{B}(\mathbb{R})$.

PROOF. Let A be almost invariant. By 8.4.4, A is of the form $\{X_0 \in C\}$. But $A = T^{-1}A$ a.e., so $\{X_0 \in C\} = \{X_1 \in C\}$ a.e. Inductively, $A = \{X_n \in C$ for all $n\}$. Conversely, every set A of this form has $A \subset T^{-1}A$, hence is almost invariant by 8.2.4. $\square$

We therefore have the following criterion for ergodicity of a Markov sequence.

8.4.6   Theorem. If X_n has the Markov property, then $\{X_n\}$ is not ergodic iff there is a set $C \in \mathscr{B}(\mathbb{R})$ such that $0 < P\{X_n \in C$ for all $n\} < 1$.

PROOF. Apply 8.4.5 and 8.2.2(b). $\square$

We now apply these results to Markov chains. Let $\{X_n\}$ be a Markov chain; assume the initial distribution $\{v_i\}$ is a stationary distribution, so that $\{X_n\}$ is a stationary sequence and the machinery of ergodic theory is applicable. (See

Ash, 1970, pp. 236–240, for the appropriate background material on Markov chains.)

8.4.7   Theorem.
(a)   If there is exactly one positive recurrent class C, then $\{X_n\}$ is ergodic.
(b)   If there are at least two positive recurrent classes C_1 and C_2, and $\sum_{i \in C_1} v_i > 0$, $\sum_{i \in C_2} v_i > 0$, then $\{X_n\}$ is not ergodic.

PROOF.   (a)   Since the stationary distribution assigns probability 1 to C, we may as well assume that C is the entire space. Let D be a nonempty proper subset of C, and let $i \in D$, $j \in C - D$. By recurrence, if the initial state is i, then j must be visited; hence

$$P\{X_n \in D \text{ for all } n\} = \sum_i v_i P\{X_n \in D \text{ for all } n \mid X_0 = i\} = 0.$$

By 8.4.6, $\{X_n\}$ is ergodic.
   (b)   It is impossible to exit from a recurrent class, hence

$$P\{X_n \in C_1 \text{ for all } n\} = P\{X_0 \in C_1\} = \sum_{i \in C_1} v_i \in (0, 1).$$

By 8.4.6, $\{X_n\}$ is not ergodic. $\square$

   The case in which there are no positive recurrent classes is not discussed in 8.4.7 because in this case, there is no stationary distribution for the chain. Note that if there is exactly one positive recurrent class, the stationary distribution is unique.
   We now have many examples of ergodic sequences $\{X_n\}$ where the X_n need not be independent. For example, consider a finite Markov chain such that every state is reachable from every other state. The state space then forms a single equivalence class, necessarily recurrent positive. Thus if the initial distribution is the unique stationary distribution, the sequence $\{X_n\}$ is ergodic.
   Now suppose $\{X_n\}$ is an ergodic Markov chain, and assume that the entire space forms a positive recurrent class. Then by the pointwise ergodic theorem, if $f \in L^1(\Omega, \mathscr{F}, P)$, then

$$\frac{1}{n} \sum_{k=0}^{n-1} f(T^k \omega) \to E(f) \qquad \text{a.e.}$$

We have been assuming that the initial distribution $\{v_i\}$ is stationary, but this result holds regardless of the initial distribution. For

$$0 = P\left\{\frac{1}{n}\sum_{k=0}^{n-1} f(T^k\omega) \longrightarrow E(f)\right\}$$

$$= \sum_i v_i P\left\{\frac{1}{n}\sum_{k=0}^{n-1} f(T^k\omega) \longrightarrow E(f) \mid X_0 = i\right\}.$$

Since $v_i > 0$ for all i, we have $P\{n^{-1}\sum_{k=0}^{n-1} f(T^k\omega) \longrightarrow E(f) \mid X_0 = i\}$ $= 0$ for all i, and the result follows.

If a Markov chain has exactly one positive recurrent class, and therefore a unique stationary distribution $\{v_i\}$, then the mean recurrence time of state j, that is, the average number of steps required to return to j when the initial state is j, is the reciprocal of the probability v_j that the chain will be in state j at any particular time. This is intuitively reasonable; if, say, $v_j = \frac{1}{4}$, then in the long run, we are in state j on one out of four trials, so on the average it should take four steps to return to j.

We are going to prove a more general result of this type.

8.4.8 Theorem. Let T be a measure-preserving transformation on the probability space $(\Omega, \mathscr{F}, P)$. If $A \in \mathscr{F}$, let

$$A_k = \{\omega \in A:\ T^n\omega \notin A, n = 1, \ldots, k-1, T^k\omega \in A\};$$

thus A_k is the set of points $\omega \in A$ such that $T^n\omega$ returns to A for the first time at $n = k$.

Define the recurrence time of A by $r_A(\omega) = k$ if $\omega \in A_k$, $k = 1, 2, \ldots$; $r_A(\omega) = \infty$ if $\omega \in A - \bigcup_{k=1}^\infty A_k$ (define r_A arbitrarily on A^c). Then

$$\int_A r_A(\omega)\,dP(\omega) = P\left(\bigcup_{n=0}^\infty T^{-n}A\right).$$

Before giving the proof, let us go into more detail on the meaning of the theorem. If T is ergodic and $P(A) > 0$, let $E = \bigcup_{n=0}^\infty T^{-n}A$. Since $T^{-1}E \subset E$, E is almost invariant by 8.2.4, and since $P(E) \geq P(A) > 0$, $P(E)$ must be 1. Thus if $Q(B) = P(B \mid A) = P(B \cap A)/P(A)$, $B \in \mathscr{F}$, we have

$$\int_A r_A(\omega)\,dQ(\omega) = \frac{1}{P(A)}.$$

If T is a one-sided shift and $A = \{X_0 \in C\}$, then $\int_A r_A\,dQ$ is the average length of time required for X_n to return to the set C, given that the initial value X_0

belongs to C. Thus the mean recurrence time of C is the reciprocal of the probability that the process will be in C at any particular time.

PROOF. Let $C_k = T^{-k}A$, $B_k = C_k^c$, $k = 0, 1, \ldots$ (take $C_0 = A$). Then, with intersections written as products,

$$P(A_{k+1}) = P(C_0 B_1 \cdots B_k C_{k+1}), \qquad k \geq 1$$

$$= P(C_0 B_1 \cdots B_k) - P(C_0 B_1 \cdots B_{k+1})$$

$$= P(B_1 \cdots B_k) - P(B_0 B_1 \cdots B_k) - P(B_1 \cdots B_{k+1})$$

$$\quad + P(B_0 B_1 \cdots B_{k+1})$$

$$= P(B_0 \cdots B_{k-1}) - 2P(B_0 \cdots B_k) + P(B_0 \cdots B_{k+1})$$

$$\text{since } T \text{ preserves } P$$

$$= b_k - 2b_{k+1} + b_{k+2}$$

where $b_k = P(B_0 \cdots B_{k-1})$, $k \geq 1$. When $k = 0$ we have

$$P(A_1) = P(C_0 C_1)$$

$$= P(C_0) - P(C_0 B_1)$$

$$= 1 - P(B_0) - P(B_1) + P(B_0 B_1)$$

$$= 1 - 2P(B_0) + P(B_0 B_1),$$

hence we have $P(A_{k+1}) = b_k - 2b_{k+1} + b_{k+2}$ for all $k \geq 0$, if we take $b_0 = 1$. Now

$$\int_A r_A \, dP = \sum_{k=0}^{\infty} (k+1) P(A_{k+1})$$

$$= \lim_{n \to \infty} \left[\sum_{k=0}^{n-1} (k+1) b_k - 2 \sum_{k=1}^{n} k b_k + \sum_{k=2}^{n+1} (k-1) b_k \right]$$

$$= \lim_{n \to \infty} [1 - n(b_n - b_{n+1}) - b_n].$$

Now $b_n - b_{n+1} = P(B_0 \cdots B_{n-1} C_n) \geq 0$, and

$$b_n \to P\left(\bigcap_{k=0}^{\infty} B_k \right) = 1 - P\left(\bigcup_{k=0}^{\infty} C_k \right) = 1 - P\left(\bigcup_{n=0}^{\infty} T^{-n}A \right).$$

Since $1 - n(b_n - b_{n+1}) - b_n$ has a limit and b_n approaches a finite limit, $n(b_n - b_{n+1})$ must approach a finite limit. But the sets $B_0 \cdots B_{n-1} C_n$ are disjoint, so that $\sum_n (b_n - b_{n+1}) < \infty$; this implies that $n(b_n - b_{n+1}) \to 0$. $\square$

Problems

1. The one-sided shift transformation T is said to be a *Kolmogorov shift* (or to be *tail trivial*) iff the tail σ-field $\mathcal{F}_\infty$ consists only of sets with probability 0 or 1. It follows from 8.4.1 that a Kolmogorov shift is ergodic; show that, in fact, every Kolmogorov shift is mixing.

2. Let T be the shift transformation associated with a finite Markov chain where the initial distribution is stationary. Assume that $v_i > 0$ for all i. (If $v_i = 0$, then $P\{X_n = i \text{ for some } n\} \le \sum_n P\{X_n = i\} = 0$, so that i may as well be removed from the state space.)

    Show that T is mixing iff a steady-state distribution exists, that is, iff $p_{ij}^{(n)} \to q_j$ (independent of i) as $n \to \infty$, where $\sum_j q_j = 1$. (Necessarily, $q_j = v_j$, and the stationary distribution is unique; see Ash, 1970, pp. 236–237.)

3. Let $X_1, X_2, \ldots$ be iid random variables, and let $S_n = \sum_{k=1}^n X_k$, $n = 1, 2, \ldots$.

    (a) If $R_n(\omega)$ is the number of distinct points in the set $\{S_1(\omega), \ldots, S_n(\omega)\}$, and A is the event that S_n never returns to 0, that is, $A = \{S_1 \ne 0, S_2 \ne 0, \ldots\}$, show that $n^{-1}E(R_n) \to P(A)$. (*Hint:* Express R_n as $1 + \sum_{k=2}^n I_{B_k}$, where
    $$B_k = \{S_k \ne S_{k-1}, S_k \ne S_{k-2}, \ldots, S_k \ne S_1\}.)$$

    (b) For a fixed positive integer N, let $Z_k(\omega)$ be the number of distinct points in $\{S_{(k-1)N+1}(\omega), \ldots, S_{kN}(\omega)\}$, $k = 1, 2, \ldots$. Use the strong law of large numbers to show that
    $$\limsup_{n\to\infty} \frac{R_{nN}}{nN} \le \frac{E(Z_1)}{N} \quad \text{a.e.}$$

    (c) Show that $\limsup_{n\to\infty} n^{-1} R_n \le P(A)$ a.e.

    (d) Let V_k be the indicator of the set $\{S_{k+1} \ne S_k, S_{k+2} \ne S_k, \ldots\}$; thus $V_k = 1$ iff S_k is never revisited. Use the pointwise ergodic theorem to show that $n^{-1}(V_1 + \cdots + V_n) \to E(V_1)$ a.e.

    (e) Show that $\liminf_{n\to\infty} n^{-1} R_n \ge E(V_1)$ a.e., and conclude that $n^{-1} R_n \to P(A)$ a.e.

8.5   THE SHANNON–MCMILLAN THEOREM

We will apply the pointwise ergodic theorem to prove a basic result of information theory. Consider a shift transformation with coordinate random variables X_n taking values in a countable set. We define a rather unusual sequence of random variables $p(X_0, X_1, \ldots, X_{n-1})$ as follows. If $X_0(\omega) = i_0, \ldots, X_{n-1}(\omega) = i_{n-1}$, let

$$p(X_0(\omega), \ldots, X_{n-1}(\omega)) = P\{\omega': X_0(\omega') = i_0, \ldots, X_{n-1}(\omega') = i_{n-1}\}.$$

For example, if $P\{X_0 = 1\} = P\{X_0 = 2\} = \frac{1}{6}$, $P\{X_0 = 3\} = \frac{2}{3}$, then $p(X_0)$ has the value $\frac{1}{6}$ with probability $\frac{1}{3}$, namely, when $X_0 = 1$ or 2, and $p(X_0) = \frac{2}{3}$ with probability $\frac{2}{3}$, that is, when $X_0 = 3$.

Similarly, define the random variable $p(X_n \mid X_{n-1}, X_{n-2}, \ldots, X_{n-r})$ by specifying that if $X_n = i_n, X_{n-1} = i_{n-1}, \ldots, X_{n-r} = i_{n-r}$, then $p(X_n \mid X_{n-1}, \ldots, X_{n-r}) = P\{X_n = i_n \mid X_{n-1} = i_{n-1}, \ldots, X_{n-r} = i_{n-r})$ assuming $P(X_{n-1} = i_{n-1}, \ldots, X_{n-r} = i_{n-r}) > 0$.

The Shannon–McMillan theorem is an assertion about the convergence of $-n^{-1} \log p(X_0, \ldots, X_{n-1})$; in particular, if T is ergodic, we have convergence to a constant H, almost everywhere in L^1. Before turning to the proof, let us look at the intuitive interpretation of the convergence statement. For this, we require only that $-n^{-1} \log p(X_0, \ldots, X_{n-1})$ converge to H in probability. (It is traditional in information theory to use logs to the base 2, and we shall follow this practice here. Switching to natural logs involve only a multiplicative constant.)

Given $\varepsilon > 0$, $\delta > 0$, let

$$A_n = \left\{ \left| -\frac{1}{n} \log p(X_0, \ldots, X_{n-1}) - H \right| \leq \delta \right\};$$

thus if $\omega \in A_n$, then $p(X_0(\omega), \ldots, X_{n-1}(\omega))$ is between $2^{-n(H+\delta)}$ and $2^{-n(H-\delta)}$. If $-n^{-1} \log p(X_0, \ldots, X_{n-1}) \to H$ in probability, then $P(A_n) \geq 1 - \varepsilon$ for sufficiently large n.

Now let S_n be the set of all sequences of length n, with values in the coordinate space, coresponding to points in A_n, that is, S_n is the set of all sequences $(X_0(\omega), \ldots, X_{n-1}(\omega))$, $\omega \in A_n$. If $(i_0, \ldots, i_{n-1}) \in S_n$, then $2^{-n(H+\delta)} \leq P\{X_0 = i_0, \ldots, X_{n-1} = i_{n-1}\} \leq 2^{-n(H-\delta)}$; furthermore, $P\{(X_0, \ldots, X_{n-1}) \in S_n\} = P(A_n)$, so that $1 - \varepsilon \leq P\{(X_0, \ldots, X_{n-1}) \in S_n\} \leq 1$.

Thus each sequence in S_n has a probability between $2^{-n(H+\delta)}$ and $2^{-n(H-\delta)}$, and the total probability assigned to S_n is between $1 - \varepsilon$ and 1. Consequently, the maximum number of sequences in S_n is $1/2^{-n(H+\delta)} = 2^{n(H+\delta)}$, and the minimum number is $(1 - \varepsilon)/2^{-n(H-\delta)} = (1 - \varepsilon)2^{n(H-\delta)}$.

Thus, roughly, for large n there are approximatively 2^{nH} sequences of length n, each with probability approximatively 2^{-nH}; the remaining sequences are negligible, that is, have total probability at most ε. In information theory, this is referred to as the *asymptotic equipartition property*. If $\{X_n\}$ represents the output of an "information source" (such as a language) with r symbols, then of all the $r^n = 2^{n \log r}$ possible sequences of length n, only 2^{nH} can reasonably be expected to appear.

The number H will turn out to be the entropy of the sequence $\{X_n\}$, and we must discuss this concept before going any further.

8.5.1 Definition. If X is a discrete random variable (or random vector), define the *entropy* (also called the *uncertainty*) of X as

$$H(X) = -\sum_x p(x) \log p(x)$$

where $p(x) = P\{X = x\}$. If X and Y are discrete, define the *conditional entropy* of Y given $X = x$ as

$$H(Y \mid X = x) = -\sum_y p(y \mid x) \log p(y \mid x)$$

where $p(y \mid x) = P\{Y = y \mid X = x\}$; also define the *conditional entropy of Y given X* as a weighted average of the $H(Y \mid X = x)$, namely,

$$H(Y \mid X) = \sum_x p(x) H(Y \mid X = x)$$

$$= -\sum_{x,y} p(x, y) \log p(y \mid x).$$

The *joint entropy* of X and Y is defined by

$$H(X, Y) = -\sum_{x,y} p(x, y) \log p(x, y);$$

since $H(X, Y)$ is the entropy of the random vector (X, Y), nothing new is involved.

Note that entropy is always nonnegative; if the random variables are allowed to have a countably infinite set of values, an entropy of $+\infty$ may be obtained. Difficulties with events of probability zero are avoided by defining $0 \log 0 = 0$, $-\log 0 = +\infty$.

It is often convenient to express entropy in terms of the random variables $p(X)$ introduced at the beginning of the section; we have

$$H(X) = E[-\log p(X)], \qquad H(X \mid Y) = E[-\log p(Y \mid X)].$$

We now establish a few properties of entropy.

8.5.2   Theorem. Let X, Y, and Z be discrete random vectors with finite entropy.

(a) If $p_1, p_2, \ldots, q_1, q_2, \ldots$ are nonnegative numbers with $\sum_i p_i = \sum_i q_i = 1$, and either $-\sum_i p_i \log p_i < \infty$ or $-\sum_i p_i \log q_i < \infty$, then

$$-\sum_i p_i \log p_i \leq -\sum_i p_i \log q_i,$$

with equality iff $p_i = q_i$ for all i.

(b) $H(X, Y) \leq H(X) + H(Y)$, with equality iff X and Y are independent.

(c) $H(X, Y) = H(X) + H(Y \mid X) = H(Y) + H(X \mid Y)$.

(d) $H(Y \mid X) \leq H(Y)$, with equality iff X and Y are independent.

(e) If X takes on r possible values $x_1, \cdot, x_r$, then $H(X) \leq \log r$, with equality iff $p(x_i) = 1/r$, $i = 1, \ldots, r$.

(f) $H(Y, Z \mid X) \leq H(Y \mid X) + H(Z \mid X)$, with equality iff Y and Z are conditionally independent given X, that is,

$$p(y, z \mid x) = p(y \mid x) p(z \mid x) \qquad \text{for all} \qquad x, y, z.$$

(g) $H(Y, Z \mid X) = H(Y \mid X) + H(Z \mid X, Y)$.

(h) $H(Z \mid X, Y) \leq H(Z \mid X)$, with equality iff Y and Z are conditionally independent given X.

PROOF.   (a)   For convenience we switch to natural logs. Since $x - 1$ is the tangent to $\log x$ at $x = 1$, we have $\log x \leq x - 1$, with equality iff $x = 1$. Thus $\log(q_i/p_i) \leq (q_i/p_i) - 1$, with equality iff $p_i = q_i$; hence, even if p_i or $q_i = 0$,

$$p_i \log q_i = p_i \log(q_i/p_i) + p_i \log p_i \leq q_i - p_i + p_i \log p_i$$

with equality iff $p_i = q_i$. Sum over i to obtain the desired result. (Note that if $\sum_i p_i \log q_i = \sum_i p_i \log p_i$, then the sum is finite by hypothesis, so that $p_i = q_i$ for all i.)

(b)   By (a), $-\sum_{x,y} p(x, y) \log p(x, y) \leq -\sum_{x,y} p(x, y) \log p(x)p(y)$, with equality iff $p(x, y) = p(x)p(y)$ for all x, y.

(c)   This follows from the fact that $p(x, y) = p(x)p(y \mid x) = p(y)p(x \mid y)$.

(d)   This is immediate from (b) and (c).

(e)   Apply (a) with $p_i = p(x_i)$, and $q_i = 1/r$.

(f)   Since $H(Y \mid X) < \infty$ by (d), $H(Y \mid X = x) < \infty$ for each x (such that $p(x) > 0$), and similarly for $H(Z \mid X = x)$. The proof of (b) shows that

$$H(Y, Z \mid X = x) \leq H(Y \mid X = x) + H(Z \mid X = x),$$

with equality iff $p(y, z \mid x) = p(y \mid x)p(z \mid x)$ for all y, z. Multiply by $p(x)$ and sum over x to complete the proof.

(g)   This follows from $p(y, z \mid x) = p(y \mid x)p(z \mid x, y)$.

(h)   Apply (f) and (g). $\square$

We also need an entropy concept for stationary sequences.

8.5.3   Definitions and Comments. Let $X_0, X_1, \ldots$ be a stationary sequence of discrete random variables, and assume that $H(X_0) < \infty$. Define the *entropy* of the sequence as

$$H\{X_n\} = \lim_{n \to \infty} H(X_n \mid X_0, X_1, \ldots, X_{n-1}).$$

Now

$$H(X_{n+1} \mid X_0, \ldots, X_n) \leq H(X_{n+1} \mid X_1, \ldots, X_n) \qquad \text{by 8.5.2(h)}$$

$$= H(X_n \mid X_0, \ldots, X_{n-1}) \qquad \text{by stationarity;}$$

also by 8.5.2(h), $H(X_n \mid X_0, \ldots, X_{n-1}) \leq H(X_n) = H(X_0) < \infty$. Thus the limit defining $H\{X_n\}$ exists and is finite.

(To apply 8.5.2(h), it must be verified that $H(X_1, \ldots, X_n) < \infty$, but the proof of 8.5.2(b) shows that if $H(X_i) < \infty$, $i = 1, \ldots, n$, then

$$H(X_1, \ldots, X_n) \leq H(X_1) + \cdots + H(X_n),$$

with equality iff $X_1, \cdot, X_n$ are independent. In the present case, $H(X_i) = H(X_0) < \infty$ for all i.)

The entropy of $\{X_n\}$ may also be expressed as

$$H\{X_n\} = \lim_{n \to \infty} \frac{1}{n} H(X_0, \ldots, X_{n-1}).$$

To see this, observe that by induction using 8.5.2(c),

$$H(X_0, \ldots, X_{n-1}) = H(X_0) + H(X_1 \mid X_0) + H(X_2 \mid X_0, X_1)$$

$$+ \cdots + H(X_{n-1} \mid X_0, \ldots, X_{n-2}).$$

Thus $n^{-1}H(X_0, \ldots, X_{n-1})$ is the arithmetic average of a sequence converging to $H\{X_n\}$, and hence converges to $H\{X_n\}$.

We now begin the development of the Shannon–McMillan theorem. It will be convenient to consider a two-sided shift transformation T with coordinate random variables X_n, where the X_n take values in a countable set. It is assumed throughout that $H(X_0) < \infty$.

Martingale theory will be significant, as the following result suggests.

8.5.4 Theorem. Let $Y_0 = -\log p(X_0)$, $Y_k = -\log p(X_0 \mid X_{-1}, \ldots, X_{-k})$, $k \geq 1$. Then $\{Y_0, Y_1, \ldots\}$ is a nonnegative supermartingale, hence converges a.e. to an integrable limit function Y.

PROOF. Since $E(Y_k) = H(X_0 \mid X_{-1}, \ldots, X_{-k}) \leq H(X_0) < \infty$, the Y_k are integrable. Now since all random variables are discrete, we may write

$$E(Y_{n+1} \mid X_0 = x_0, \ldots, X_{-n} = x_{-n})$$

$$= -\sum_{x_{-(n+1)}} p(x_{-(n+1)} \mid x_0, \ldots, x_{-n}) \log p(x_0 \mid x_{-1}, \ldots, x_{-(n+1)})$$

$$= +\sum_{x_{-(n+1)}} \frac{p(x_0, \ldots, x_{-(n+1)})}{p(x_0, \ldots, x_{-n})} \log \frac{p(x_{-1}, \ldots, x_{-(n+1)})}{p(x_0, \ldots, x_{-(n+1)})}.$$

This is of the form $\sum_i \alpha_i \log x_i$, where $\alpha_i \geq 0$, $\sum_i \alpha_i = 1$, and hence is less than or equal to $\log\left(\sum_i \alpha_i x_i\right)$ by convexity. Thus

$$E(Y_{n+1} \mid X_0 = x_0, \ldots, X_{-n} = x_{-n})$$

$$\leq \log\left[\sum_{x_{-(n+1)}} \frac{p(x_{-1}, \ldots, x_{-(n+1)})}{p(x_0, \ldots, x_{-n})}\right]$$

$$= \log \frac{p(x_{-1}, \ldots, x_{-n})}{p(x_0, \ldots, x_{-n})}$$

$$= -\log p(x_0 \mid x_{-1}, \ldots, x_{-n}).$$

Therefore $E(Y_{n+1} \mid X_0, \ldots, X_{-n}) \leq Y_n$, and since Y_n is measurable relative to the σ-field $\mathscr{F}(X_0, \ldots, X_{-n})$, the result follows. $\square$

We will show that the random variables Y_n are uniformly integrable. It will be convenient to assume that the coordinate space is a subset of the positive integers; this amounts only to a relabeling, and can be done without changing the distribution of $(Y_0, Y_1, \ldots)$.

8.5.5   Lemma. If r is any fixed positive integer, the random variables $W_n = Y_n I_{\{X_0 \le r\}}$, $n = 0, 1, \ldots$, are uniformly integrable.

PROOF.   For any positive integer k,

$$\int_{\{W_n \ge k\}} W_n \, dP = \sum_{i=k}^{\infty} \int_{\{i \le W_n < i+1\}} W_n \, dP$$

$$\le \sum_{i=k}^{\infty} (i+1) P\{W_n \ge i\}.$$

But if $W_n \ge i$, then $Y_n \ge i$, and it follows that $p(X_0, \ldots, X_{-n}) \le 2^{-i} p(X_{-1}, \ldots, X_{-n})$. Thus

$$P\{W_n \ge i\} = \sum_{\{x_0, \ldots, x_{-n} : \ y_n \ge i, x_0 \le r\}} p(x_0, \ldots, x_{-n})$$

$$\le 2^{-i} \sum_{x_0 \le r} \left[\sum_{(x_{-1}, \ldots, x_{-n})} p(x_{-1}, \ldots, x_{-n}) \right]$$

$$\le r 2^{-i}.$$

Consequently,

$$\int_{\{W_n \ge k\}} W_n \, dP \le \sum_{i=k}^{\infty} r(i+1) 2^{-i} \to 0$$

as $k \to \infty$, uniformly in n. $\square$

8.5.6   Theorem. The random variables Y_n, $n = 0, 1, \ldots$, are uniformly integrable; thus in 8.5.4 we also have $Y_n \to Y$ in L^1.

PROOF.   For any positive integer k

$$\int_{\{Y_n \ge k\}} Y_n \, dP = \int_{\{Y_n \ge k, X_0 > r\}} Y_n \, dP + \int_{\{Y_n \ge k, X_0 \le r\}} Y_n \, dP$$

$$\le \int_{\{X_0 > r\}} Y_n \, dP + \int_{\{W_n \ge k\}} W_n \, dP.$$

By 8.5.5, the second integral on the right approaches 0 as $k \to \infty$, uniformly in n for a fixed r. Now

$$\int_{\{X_0 > r\}} Y_n \, dP = - \sum_{x_0 > r} \left[\sum_{x_{-1}, \dots, x_{-n}} p(x_0, \dots, x_{-n}) \right]$$
$$\times \log p(x_0 \mid x_{-1}, \dots x_{-n})$$
$$= - \sum_{x_{-1}, \dots, x_{-n}} p(x_{-1}, \dots, x_{-n})$$
$$\times \sum_{x_0 > r} p(x_0 \mid x_{-1}, \dots, x_{-n}) \log p(x_0 \mid x_{-1}, \dots, x_{-n}).$$

Write $\log p(x_0 \mid x_{-1}, \dots, x_{-n}) = \log p(x_0) + \log[p(x_0 \mid x_{-1}, \dots, x_{-n})/p(x_0)]$; the contribution due to $\log p(x_0)$ is

$$- \sum_{x_0 > r} \left[\sum_{x_{-1}, \dots, x_{-n}} p(x_0, \dots, x_{-n}) \log p(x_0) \right] = - \sum_{x_0 > r} p(x_0) \log p(x_0)$$

and the remaining contribution is

$$\sum_{x_{-1}, \dots, x_{-n}} p(x_{-1}, \dots, x_{-n}) \sum_{x_0 > r} p(x_0 \mid x_{-1}, \dots, x_{-n}) \log \frac{p(x_0)}{p(x_0 \mid x_{-1}, \dots, x_{-n})}.$$

An upper bound to this expression, obtained by switching to natural logs for convenience and using $\log x \le x - 1$, is

$$\sum_{x_0 > r} p(x_0) - \sum_{x_0 > r} p(x_0) = 0.$$

Therefore,

$$\int_{\{X_0 > r\}} Y_n \, dP \le - \sum_{x_0 > r} p(x_0) \log p(x_0) \to 0$$

as $r \to \infty$, uniformly in n, since $H(X_0) < \infty$.

If $\varepsilon > 0$ is given, we may choose a fixed r such that $\int_{\{X_0 > r\}} Y_n \, dP < \varepsilon/2$ for all n; then for sufficiently large k, $\int_{\{W_n \ge k\}} W_n \, dP < \varepsilon/2$ for all n, hence $\int_{\{Y_n \ge k\}} Y_n \, dP < \varepsilon$ for all n. $\square$

The other basic property of the Y_n that we need is that $\sup_n Y_n$ is integrable. We prove this after a preliminary lemma.

8.5.7   Lemma. If i is a positive integer, define $Y_n^{(i)} = -\log p(X_0 = i \mid X_{-1}, \ldots, X_{-n})$, that is, if $X_{-1} = i_{-1}, \ldots, X_{-n} = i_{-n}$, then

$$Y_n^{(i)} = -\log P\{X_0 = i \mid X_{-1} = i_{-1}, \ldots, X_{-n} = i_{-n}\}$$

(define $Y_0^{(i)} = -\log P\{X_0 = i\}$). If $\lambda \geq 0$, let

$$E_n(\lambda) = \left\{\max_{j<n} Y_j \leq \lambda < Y_n\right\}, \qquad E_n^{(i)}(\lambda) = \left\{\max_{j<n} Y_j^{(i)} \leq \lambda < Y_n^{(i)}\right\}.$$

If $A_i = \{X_0 = i\}$, then

$$P(E_n(\lambda) \cap A_i) = P(E_n^{(i)}(\lambda) \cap A_i) \leq 2^{-\lambda} P(E_n^{(i)}(\lambda)).$$

PROOF.   On A_i we have $Y_n = Y_n^{(i)}$, hence $E_n(\lambda) \cap A_i = E_n^{(i)}(\lambda) \cap A_i$. Now since $E_n^{(i)}(\lambda)$ belongs to the σ-field $\mathscr{F}(X_{-1}, \ldots, X_{-n})$,

$$P(A_i \cap E_n^{(i)}(\lambda)) = \int_{E_n^{(i)}(\lambda)} P(A_i \mid X_{-1}, \ldots, X_{-n}) \, dP$$

$$= \int_{E_n^{(i)}(\lambda)} p(X_0 = i \mid X_{-1}, \ldots, X_{-n}) \, dP$$

$$= \int_{E_n^{(i)}(\lambda)} 2^{-Y_n^{(i)}} \, dP \leq 2^{-\lambda} P(E_n^{(i)}(\lambda))$$

by definition of $E_n^{(i)}(\lambda)$. $\square$

8.5.8   Theorem. The random variables Y_n satisfy $E[\sup_k Y_k] < \infty$.

PROOF.   We may write

$$E\left[\sup_k Y_k\right] = \int_0^\infty P\left\{\sup_k Y_k > \lambda\right\} d\lambda \qquad \text{(see 6.2, Problem 2)}$$

$$\leq \sum_{r=0}^\infty P\left\{\sup_k Y_k > r\right\}$$

$$= \sum_{r=0}^\infty \sum_{n=0}^\infty P(E_n(r))$$

$$= \sum_{r=0}^\infty \sum_{n=0}^\infty \sum_i P(E_n^{(i)}(r) \cap A_i) \qquad \text{by 8.5.7.}$$

If the coordinate space is finite, there is no difficulty; by 8.5.7,

$$E\left[\sup_k Y_k\right] \leq \sum_{r,n=0}^{\infty} 2^{-r} \sum_i P(E_n^{(i)}(r)) < \infty$$

since the $E_n^{(i)}(r)$, $n = 0, 1, \ldots$, are disjoint and the sum over i is finite.

In the general case, assume without loss of generality that the numbers $p_i = P(A_i)$ decrease as i increases (if necessary, relabel the elements of the coordinate space). We then have $p_i \leq 1/i$ for all i, and if $p_i > 1/i$, then $p_1, \ldots, p_i$ are all greater than $1/i$, a contradiction. Let f be a function, to be specified later, from the nonnegative integers to the nonnegative reals, such that $\{r: f(r) < i\}$ is a finite set for each i. Then by 8.5.7,

$$\sum_{n=0}^{\infty}\left[\sum_{i\leq f(r)} P(E_n^{(i)}(r) \cap A_i)\right] \leq 2^{-r} \sum_{i\leq f(r)}\left[\sum_{n=0}^{\infty} P(E_n^{(i)})(r)\right] \leq 2^{-r} f(r)$$

by disjointness of the $E_n^{(i)}(r)$, $n = 0, 1, \ldots$.

Also by disjointness,

$$\sum_{n=0}^{\infty}\left[\sum_{i> f(r)} P(E_n^{(i)}(r) \cap A_i)\right] \leq \sum_{i> f(r)} P(A_i) = \sum_{i> f(r)} p_i.$$

Therefore,

$$E\left[\sup_k Y_k\right] \leq \sum_{r=0}^{\infty} 2^{-r} f(r) + \sum_{r=0}^{\infty}\left[\sum_{i> f(r)} p_i\right].$$

The second series is $\sum_{i=1}^{\infty} \sum_{f(r)<i} p_i = \sum_{i=1}^{\infty} f_0(i)p_i$, where $f_0(i)$ is the number of nonnegative integers r such that $f(r) < i$. If we set $f(r) = 2^r (r+1)^{-2}$, then the first series converges, and $f(r) < i$ iff $r < \log i + 2\log(r+1)$.

Choose a positive integer B such that $2x^{-1}\log(x+1) < \frac{1}{2}$ for all $x \geq B$. Then:

$$f(r) < i, \quad r \geq B \quad \text{implies} \quad r < \log i + r/2, \quad \text{that is,} \quad r < 2\log i,$$

and

$$f(r) < i, \quad r < B \quad \text{implies} \quad r < B.$$

Therefore, $f_0(i) \leq 2\log i + B$, $i = 1, 2, \ldots$, hence

$$\sum_{i=1}^{\infty} f_0(i) p_i \leq 2 \sum_{i=1}^{\infty} p_i \log i + B \sum_{i=1}^{\infty} p_i$$

$$\leq 2 \sum_{i=1}^{\infty} p_i \log \frac{1}{p_i} + B \qquad \text{since} \qquad p_i \leq \frac{1}{i}$$

$$= 2H(X_0) + B < \infty. \quad \square$$

We may now give the main result.

8.5.9   Shannon–McMillan Theorem. Let T be a two-sided shift transformation with discrete coordinate random variables X_n. Assume that $H(X_0)$ $< \infty$, and let H be the entropy of the sequence $\{X_n\}$.

If $Z_n = -n^{-1}\log p(X_0, \ldots, X_{n-1})$, there is an invariant random variable Z such that $E(Z) = H$ and $Z_n \to Z$, almost everywhere and in L^1. In particular, if T is ergodic, $Z = H$ a.e.

PROOF.   If the random variables Y_n are defined as in 8.5.4, we have

$$Z_n = -\frac{1}{n} \sum_{k=0}^{n-1} \log p(X_k \mid X_{k-1}, \ldots, X_0)$$

$$= \frac{1}{n} \sum_{k=0}^{n-1} Y_k(T^k)$$

$$= \frac{1}{n} \sum_{k=0}^{n-1} Y(T^k) + \frac{1}{n} \sum_{k=0}^{n-1} [Y_k(T^k) - Y(T^k)]. \qquad (1)$$

By 8.3.6, 8.3.7, and 8.3.8, the first series in (1) converges to an almost invariant function $\hat{Y}$, a.e. and in L^1. By 8.3.9, $E(\hat{Y}) = E(Y)$, and by 8.5.6, $E(Y) = \lim_{n\to\infty} E(Y_n) = H$. The second series converges to 0 in L^1 since $Y_k \to Y$ in L^1 by 8.5.6, and $\|Y_k(T^k) - Y(T^k)\|_1 = \|Y_k - Y\|_1$ by 1.6.12.

Thus if we take $Z = \hat{Y}$, all that remains is to show that $Z_n \to Z$ a.e., and this will follow if we show that the second series in (1) converges to 0 a.e. But for any positive integer N, the series is bounded in absolute value by

$$\frac{1}{n} \sum_{k=0}^{N-1} |Y_k(T^k) - Y(T^k)| + \frac{1}{n} \sum_{k=N}^{n-1} |Y_k(T^k) - Y(T^k)|. \qquad (2)$$

The first series approaches 0 a.e. as $n \to \infty$ since $Y_k(T^k) - Y(T^k)$ is integrable, hence finite a.e. If we define

$$G_N = \sup\{|Y_k - Y|: \ k \geq N\},$$

then $G_N \leq 2 \sup_k Y_k$, which is integrable by 8.5.8. Also, since $Y_k \to Y$ a.e., we have $G_N \to 0$ a.e. as $N \to \infty$, hence $E(G_N) \to 0$ by the dominated convergence theorem. But the second series in (2) (call it h_n) is bounded by $n^{-1} \sum_{k=N}^{n-1} G_N(T^k)$, which converges to $\hat{G}_N$ a.e. and in L^1, by 8.3.6 and 8.3.7.

Finally, given $\varepsilon > 0$, $\delta > 0$, we have $P\{\hat{G}_N \geq \varepsilon\} \leq \varepsilon^{-1} E(\hat{G}_N)$ by Chebyshev's inequality, and by 8.3.9, $E(\hat{G}_N) = E(G_N) \to 0$. If we choose N such that $\varepsilon^{-1} E(G_N) < \delta$, then $\hat{G}_N < \varepsilon$ on a set of probability greater than $1 - \delta$, hence $\limsup_{n \to \infty} h_n \leq \varepsilon$ on this set. Since ε and δ are arbitrary, it follows that $h_n \to 0$ a.e. $\square$

Since the Shannon–McMillan theorem involves only the random variables $X_n, n \geq 0$, it holds equally well for a one-sided shift. For an indication of how to prove this formally, see the discussion at the end of Problem 7 in 8.3.

Problems

1. (a)   In the Shannon–McMillan theorem, give an example in which the limit random variable Z is a.e. constant, but T is not ergodic.

   (b)   In the Shannon–McMillan theorem, give an example in which Z is not a.e. constant (of course, T cannot be ergodic).

2. If the coordinate space is finite and $1 \leq p < \infty$, show that the random variables Z_n^p are uniformly integrable; thus $Z_n \to Z$ in L^p.

3. [A short proof of a special case of the Shannon–McMillan theorem (Gallager, 1968).] Let $X_0, X_1, \ldots$ be a stationary sequence of discrete random variables, with $H(X_0) < \infty$. (In this problem, entropy will be expressed using natural logarithms for convenience.) Define an *mth-order approximation* to $p(X_0, \ldots, X_{n-1})$ by

$$q_m(X_0, \ldots, X_{n-1}) = p(X_0 \ , \ldots, X_{m-1}) p(X_m \mid X_0, \ldots, X_{m-1}) \cdots$$

$$p(X_{n-1} \mid X_{n-m-1}, \ldots, X_{n-2}), \quad m = 1, 2, \ldots; n > m.$$

Now note that $|\ln y| = \ln^+ y + \ln^- y = 2 \ln^+ y - \ln y \leq 2e^{-1} y - \ln y$, so that

$$E\left[\left| \ln \frac{q_m}{p} \right| \right] \leq \frac{2}{e} E\left(\frac{q_m}{p} \right) - E\left(\ln \frac{q_m}{p} \right)$$

where $q_m = q_m(X_0, \ldots, X_{n-1})$, $p = p(X_0, \ldots, X_{n-1})$.

(a)   Show that $E(q_m/p) \leq 1$ and

$$E\left(-\ln \frac{q_m}{p}\right) = H(X_0, \ldots, X_{m-1}) + (n-m)H(X_m \mid X_0, \ldots, X_{m-1})$$
$$- H(X_0, \ldots, X_{n-1}).$$

Thus

$$E\left[\frac{1}{n}\left|\ln \frac{q_m}{p}\right|\right] \leq \frac{2}{ne} + \frac{m\, H(X_0, \ldots, X_{m-1})}{n\quad m}$$
$$+ \left(1 - \frac{m}{n}\right) H(X_m \mid X_0, \ldots, X_{m-1}) - \frac{H(X_0, \ldots, X_{n-1})}{n}.$$

(b)   Assume $\{X_n\}$ ergodic, that is, the associated shift transformation is ergodic. Show that as $n \to \infty$, $-n^{-1}\ln q_m$ converges a.e. and in L^1 to $H(X_m \mid X_0, \ldots, X_{m-1})$.

(c)   Let H be the entropy of the ergodic sequence $\{X_n\}$. Given $\varepsilon > 0$, choose m such that

$$\left|\frac{H(X_0, \ldots, X_{m-1})}{m} - H\right| < \frac{\varepsilon}{2}$$

and

$$|H(X_m \mid X_0, \ldots, X_{m-1} - H| < \frac{\varepsilon}{2}.$$

Show that $E[|-n^{-1}\ln p(X_0, \ldots, X_{n-1}) - H|] < \varepsilon$ for sufficiently large n, thus establishing L^1 convergence in the Shannon–McMillan theorem under the hypothesis of ergodicity.

8.6   ENTROPY OF A TRANSFORMATION

We define in this section the notion of entropy of a measure-preserving transformation, and show that two isomorphic transformations have the same entropy. For Bernoulli shifts the converse is true: two Bernoulli shifts with the same entropy are isomorphic.

8.6.1   Definitions.

(a)   Let $(\Omega, \mathscr{F}, P)$ be a probability space and $\mathscr{A}$ a finite subfield of $\mathscr{F}$ (such an $\mathscr{A}$ is a σ-field) with atoms $A_1, \ldots, A_k$. The *entropy of* $\mathscr{A}$ is by definition

$$H(\mathscr{A}) = -\sum_{i=1}^{k} P(A_i) \log P(A_i).$$

(b) If $\mathscr{B}$ is another finite subfield of $\mathscr{F}$, with atoms $B_1, \ldots, B_h$, the *conditional entropy of $\mathscr{B}$ given $\mathscr{A}$* is

$$H(\mathscr{B} \mid \mathscr{A}) = - \sum_{i=1}^{k} \sum_{j=1}^{h} P(A_i \cap B_j) \log P(B_j \mid A_i).$$

There is in fact nothing new in those definitions. If the discrete random variables X and Y are defined by $X = i$ on A_i, $Y = j$ on B_j, we have $H(\mathscr{A}) = H(X)$ and $H(\mathscr{B} \mid \mathscr{A}) = H(Y \mid X)$.

Since we consider only *finite* subfields of $\mathscr{F}$, $H(\mathscr{A})$ and $H(\mathscr{B} \mid \mathscr{A})$ are always finite and the properties of entropy given in 8.5.2 can be rewritten in terms of finite subfields.

8.6.2 Theorem. Let $\mathscr{A}$, $\mathscr{B}$, $\mathscr{C}$ be finite subfields of $\mathscr{F}$. We have the following properties (where $\mathscr{A} \vee \mathscr{B}$ denotes the smallest (finite) field containing $\mathscr{A}$ and $\mathscr{B}$).

(a) $H(\mathscr{A} \vee \mathscr{B}) \le H(\mathscr{A}) + H(\mathscr{B})$ with equality iff $\mathscr{A}$ and $\mathscr{B}$ are independent.

(b) $H(\mathscr{A} \vee \mathscr{B}) = H(\mathscr{A}) + H(\mathscr{B} \mid \mathscr{A}) = H(\mathscr{B}) + H(\mathscr{A} \mid \mathscr{B})$.

(c) Consequently, $H(\mathscr{B}) \ge H(\mathscr{A})$ if $\mathscr{B} \supset \mathscr{A}$.

(d) $H(\mathscr{B} \mid \mathscr{A}) \le H(\mathscr{B})$ with equality iff $\mathscr{A}$ and $\mathscr{B}$ are independent.

(e) If $\mathscr{A}$ has k, atoms, $H(\mathscr{A}) \le \log k$, with equality iff each atom of $\mathscr{A}$ has probability $1/k$.

(f) $H(\mathscr{B} \vee \mathscr{C} \mid \mathscr{A}) \le H(\mathscr{B} \mid \mathscr{A}) + H(\mathscr{C} \mid \mathscr{A})$, with equality iff $\mathscr{B}$ and $\mathscr{C}$ are conditionally independent given $\mathscr{A}$.

(g) $H(\mathscr{B} \vee \mathscr{C} \mid \mathscr{A}) = H(\mathscr{B} \mid \mathscr{A}) + H(\mathscr{C} \mid \mathscr{A} \vee \mathscr{B}) = H(\mathscr{C} \mid \mathscr{A}) + H(\mathscr{B} \mid \mathscr{A} \vee \mathscr{C})$.

(h) Consequently, $H(\mathscr{C} \mid \mathscr{A}) \ge H(\mathscr{B} \mid \mathscr{A})$ if $\mathscr{C} \supset \mathscr{B}$.

(i) $H(\mathscr{C} \mid \mathscr{A} \vee \mathscr{B}) \le H(\mathscr{C} \mid \mathscr{A})$, with equality iff $\mathscr{B}$ and $\mathscr{C}$ are conditionally independent given $\mathscr{A}$.

(j) Consequently, $H(\mathscr{C} \mid \mathscr{B}) \le H(\mathscr{C} \mid \mathscr{A})$ if $\mathscr{A} \subset \mathscr{B}$.

Entropy is often said to be a measure of uncertainty. We give a few examples to justify this statement.

1. Consider two finite subfields $\mathscr{A}$ and $\mathscr{B}$ of $\mathscr{F}$ and assume that $\mathscr{B} \supset \mathscr{A}$. (Therefore each atom of $\mathscr{B}$ is a subset of an atom of $\mathscr{A}$.) If we know that the outcome ω is in the atom A_i of $\mathscr{A}$, we still do not know which atom of $\mathscr{B}$ contains ω; but knowing that ω is in the atom B_j of $\mathscr{B}$ is enough

to tell us which atom of $\mathscr{A}$ contains ω. This means that intuitively the uncertainty is greater for the subfield $\mathscr{B}$ than for the subfield $\mathscr{A}$. This agrees with the inequality $H(\mathscr{B}) \geq H(\mathscr{A})$.

2. If we roll a die, the uncertainty of the outcome is intuitively greatest when the die is unbiased, that is, when the 6 atoms corresponding to the 6 outcomes 1, 2, 3, 4, 5, 6 all have the same probability 1/6. This is exactly what 8.6.2(e) states.

3. The inequality 8.6.2(j) states that the more we know the less the uncertainty.

8.6.3 Remarks. Entropy is invariant under measure-preserving transformations: if T is a measure-preserving transformation on $(\Omega, \mathscr{F}, P)$, and if $\mathscr{A}$ and $\mathscr{B}$ are two finite subfields of $\mathscr{F}$, $T^{-n}\mathscr{A} = \{T^{-n}A \colon A \in \mathscr{A}\}$ and $T^{-n}\mathscr{B}$ are finite subfields of $\mathscr{F}$, $n = 1, 2, \ldots$. The measure-preserving property of T implies that $H(T^{-n}\mathscr{A}) = H(\mathscr{A})$ and $H(T^{-n}\mathscr{A} \mid T^{-n}\mathscr{B}) = H(\mathscr{A} \mid \mathscr{B})$, $n = 1, 2, \ldots$.

Because of the correspondence between measure-preserving transformations and stationary sequences, the entropy $H\{X_n\}$ of a stationary sequence can also be rewritten in terms of measure-preserving transformations.

8.6.4 Definition. If $A_1, \ldots, A_k$ are the atoms of $\mathscr{A}$, and X is the random variable defined by $X = i$ on A_i, the sequence $X_0 = X, X_1 = X \circ T, \ldots$, $X_n = X \circ T^n, \ldots$ is a stationary sequence and

$$H\left(T^{-n}\mathscr{A} \,\Bigg|\, \bigvee_{i=0}^{n-1} T^{-i}\mathscr{A}\right) = H(X_n \mid X_0, X_1, \ldots, X_{n-1}).$$

By 8.5.3, $\lim_{n \to \infty} H\left(T^{-n}\mathscr{A} \mid \bigvee_{i=0}^{n-1} T^{-i}\mathscr{A}\right) = \lim_{n \to \infty} H(X_n \mid X_0, X_1, \ldots,$ $X_{n-1}) = H\{X_n\}$ exists.

The *entropy of a measure-preserving transformation T with respect to a finite subfield* $\mathscr{A}$ is defined as

$$H(\mathscr{A}, T) = \lim_{n \to \infty} H\left(T^{-n}\mathscr{A} \,\Bigg|\, \bigvee_{i=0}^{n-1} T^{-i}\mathscr{A}\right).$$

The identity $H\{X_n\} = \lim_{n \to \infty} \dfrac{1}{n} H(X_0, \ldots, X_{n-1})$, proved in 8.5.3, can be written in terms of $H(\mathscr{A}, T)$.

8.6.5 Theorem.
(a) $H(\mathscr{A}, T) = \lim_{n \to \infty} \frac{1}{n} H\left(\bigvee_{i=0}^{n-1} T^{-i}\mathscr{A}\right)$.

(b) $H(\mathscr{A}, T) = \lim_{n \to \infty} H\left(\mathscr{A} \mid \bigvee_{i=1}^{n-1} T^{-i}\mathscr{A}\right)$.

PROOF. (a) is a consequence of 8.5.3.

(b) By 8.6.2(j) we have $H\left(\mathscr{A}\,|\,\bigvee_{i=1}^{n-1} T^{-i}\mathscr{A}\right) \geq H\left(\mathscr{A}\,|\,\bigvee_{i=1}^{n} T^{-i}\mathscr{A}\right) \geq 0$,

and therefore $\lim_{n\to\infty} H\left(\mathscr{A}\,|\,\bigvee_{i=0}^{n-1} T^{-i}\mathscr{A}\right)$ exists. Since

$$H\left(\bigvee_{i=0}^{n-1}\mathscr{A}\right) = H\left(\mathscr{A}\,\Big|\,\bigvee_{i=1}^{n-1} T^{-i}\mathscr{A}\right) + H\left(\bigvee_{i=1}^{n-1} T^{-i}\mathscr{A}\right) \qquad \text{by 8.6.2(b)}$$

$$= H\left(\mathscr{A}\,\Big|\,\bigvee_{i=1}^{n-1} T^{-i}\mathscr{A}\right) + H\left(\bigvee_{i=0}^{n-2} T^{-i}\mathscr{A}\right) \qquad \text{by 8.6.3}$$

$$= \left(\mathscr{A}\,\Big|\,\bigvee_{i=1}^{n-1} T^{-i}\mathscr{A}\right) + H\left(\mathscr{A}\,\Big|\,\bigvee_{i=1}^{n-2} T^{-i}\mathscr{A}\right) + \cdots$$

$$+ H(\mathscr{A}\,|\,T^{-1}\mathscr{A}) + H(T^{-1}\mathscr{A})$$

the argument in 8.5.3 shows that

$$\lim_{n\to\infty} H\left(\mathscr{A}\,\Big|\,\bigvee_{i=1}^{n-1} T^{-i}\mathscr{A}\right) = \lim_{n\to\infty} \frac{1}{n} H\left(\bigvee_{i=0}^{n-1} T^{-i}\mathscr{A}\right) = H(\mathscr{A},T). \qquad \square$$

8.6.6 Corollary. If T is an invertible (measurable, one-to-one, onto, with T^{-1} measurable) measure-preserving transformation, then

$$H(\mathscr{A},T) = \lim_{n\to\infty} H\left(\mathscr{A}\,\Big|\,\bigvee_{i=1}^{n} T^{i}\mathscr{A}\right) = \lim_{n\to\infty} H\left(T^{n}\mathscr{A}\,\Big|\,\bigvee_{i=1}^{n-1} T^{i}\mathscr{A}\right).$$

Notice that we can rewrite this corollary in terms of stationary sequences. Since T is invertible, we can define the two-sided stationary sequence $X_0 = X$, $X_n = X \circ T^n$, $n = \pm 1, \pm 2, \ldots$, where $X = i$ on the atom A_i of $\mathscr{A}$. The corollary then becomes $H\{X_n\} = \lim_{n\to\infty} H(X_0 \mid X_{-1}, \ldots, X_{-n})$ $= \lim_{n\to\infty} (X_{-n} \mid X_0, \ldots, X_{-(n-1)})$. This means that in the definition of $H\{X_n\}$ we can run the time backwards.

PROOF. By 8.6.3,

$$H\left(\mathscr{A}\,\Big|\,\bigvee_{i=1}^{n} T^{i}\mathscr{A}\right) = H\left(T^{-n}\mathscr{A}\,\Big|\,\bigvee_{i=-(n-1)}^{0} T^{i}\mathscr{A}\right)$$

$$= H\left(T^{-n}\mathscr{A}\,\Big|\,\bigvee_{i=0}^{n-1} T^{-i}\mathscr{A}\right),$$

and

$$H\left(T^n \mathscr{A} \,\middle|\, \bigvee_{i=1}^{n-1} T^i \mathscr{A}\right) = H\left(\mathscr{A} \,\middle|\, \bigvee_{i=-(n-1)}^{-1} T^i \mathscr{A}\right)$$

$$= H\left(\mathscr{A} \,\middle|\, \bigvee_{i=1}^{n-1} T^{-i} \mathscr{A}\right). \quad \square$$

8.6.7 Remark. If $\mathscr{A}$ and $\mathscr{B}$ are two finite subfields of $\mathscr{F}$, we have

(a) $H(\mathscr{A},T) \le H(\mathscr{B},T)$ if $\mathscr{A} \subset \mathscr{B}$.

(b) $H(\mathscr{A} \vee \mathscr{B}, T) \le H(\mathscr{A}, T) + H(\mathscr{B})$.

Here, (a) is an easy consequence of 8.6.2(c) and 8.6.5(a).
For (b), notice that

$$H\left(\bigvee_{i=0}^{n-1} T^{-i}(\mathscr{A} \vee \mathscr{B})\right) = H\left(\bigvee_{i=0}^{n-1} T^{-i}\mathscr{A} \vee \bigvee_{i=0}^{n-1} T^{-i}\mathscr{B}\right)$$

$$= H\left(\bigvee_{i=0}^{n-1} T^{-i}\mathscr{A}\right) + H\left(\bigvee_{i=0}^{n-1} T^{-i}\mathscr{B} \,\middle|\, \bigvee_{i=0}^{n-1} T^{-i}\mathscr{A}\right),$$

by 8.6.2(b).
By 8.6.2(f), (j), and 8.6.3 we have

$$H\left(\bigvee_{i=0}^{n-1} T^{-i}\mathscr{B} \,\middle|\, \bigvee_{i=0}^{n-1} T^{-i}\mathscr{A}\right)$$

$$\le \sum_{i=0}^{n-1} H\left(T^{-i}\mathscr{B} \,\middle|\, \bigvee_{i=0}^{n-1} T^{-i}\mathscr{A}\right)$$

$$\le \sum_{i=0}^{n-1} H(T^{-i}\mathscr{B} \mid T^{-i}\mathscr{A}) = nH(\mathscr{B} \mid \mathscr{A}).$$

We are now ready to define the entropy of a measure-preserving transformation in such a way that it is invariant under isomorphism. We first specify what we mean by isomorphic transformations.

8.6.8 Definition (Isomorphic Transformations). Let $(\Omega_1, \mathscr{F}_1, P_1, T_1)$ and $(\Omega_2, \mathscr{F}_2, P_2, T_2)$ be two probability spaces with measure-preserving transformations T_1 and T_2; $(\Omega_1, \mathscr{F}_1, P_1, T_1)$ and $(\Omega_2, \mathscr{F}_2, P_2, T_2)$ are isomorphic iff there exist $\Omega_1' \in \mathscr{F}_1$ and $\Omega_2' \in \mathscr{F}_2$, and a mapping $\Phi: \Omega_1' \to \Omega_2'$ such that:

(a) $P_1(\Omega_1') = P_2(\Omega_2') = 1$;

(b) the mapping Φ is one-to-one and onto; for any $B_1 \subset \Omega_1'$, $B_2 = \Phi(B_1)$ $\in \mathscr{F}_2$ iff $B_1 \in \mathscr{F}_1$;

(c) for any subset $B_1 \in \mathscr{F}_1$ of Ω_1', $P_2(\Phi(B_1)) = P_1(B_1)$;

(d) $T_1(\Omega_1') \subset \Omega_1'$ and $T_2(\Omega_2') \subset \Omega_2'$;

(e) $\Phi(T_1\omega) = T_2(\Phi\omega)$, for each ω in Ω_1'.

Properties (b) and (c) state that the probability spaces $(\Omega_1', \mathscr{F}_1', P_1)$ and $\Omega_2', \mathscr{F}_2' P_2)$ are isomorphic [where $\mathscr{F}_1'$ (resp. $\mathscr{F}_2'$) denotes the σ-field consisting of the subsets of Ω_1' (resp. Ω_2'), which are in $\mathscr{F}_1$ (resp. in $\mathscr{F}_2$)]. Property (d) is purely technical; it allows us to limit ourselves to what happens in Ω_1' and Ω_2'. Property (e) states that T_1 and T_2, considered as transformations on Ω_1' and Ω_2', are compatible with the isomorphism Φ. We could have defined the isomorphism of $(\Omega_1, \mathscr{F}_1, P_1, T_1)$ and $(\Omega_2, \mathscr{F}_2, P_2, T_2)$ by imposing that $\Omega_1' = \Omega_1$ and $\Omega_2' = \Omega_2$, but nothing is changed in the structure of a probability space if we add or remove a set of measure zero. It is therefore natural to add the property (a) to the definition.

8.6.9 Definition. Let T be a measure-preserving transformation on a probability space $(\Omega, \mathscr{F}, P)$; the *entropy of* T is

$$H(T) = \sup_{\mathscr{A}} H(\mathscr{A}, T) \qquad \text{where the sup is taken over}$$

$$\text{all finite subfields } \mathscr{A} \text{ of } \mathscr{F}.$$

8.6.10 Theorem. Let T_1 and T_2 be measure preserving transformations on $(\Omega_1, \mathscr{F}_1, P_1)$ and $(\Omega_2, \mathscr{F}_2, P_2)$, respectively. Suppose that $(\Omega_1, \mathscr{F}_1, P_1, T_1)$ and $(\Omega_2, \mathscr{F}_2, P_2, T_2)$ are isomorphic. Then $H(T_1) = H(T_2)$.

PROOF. The result is a consequence of the following remarks (we use the notations of 8.6.8).

(a) The entropies of T_1 in Ω_1 and Ω_1' are the same because $P(\Omega_1 - \Omega_1') = 0$. The same is true for the entropies of T_2 in Ω_2 and Ω_2'.

(b) For any finite subfield $\mathscr{A}_1$ of $\mathscr{F}_1'$, $\Phi\mathscr{A}_1$ is a finite subfield of $\mathscr{F}_2'$, and any finite subfield of $\mathscr{F}_2'$ is of this form.

(c) If $\mathscr{A}_1$ is a finite subfield of $\mathscr{F}_1'$

$$H(\mathscr{A}_1, T_1) = H(\Phi\mathscr{A}_1, T_2) \quad \square$$

To conclude this section we give two results, which in certain cases allow us to compute $H(T)$ easily.

8.6.11   Theorem. If T is a measure-preserving transformation on a probability space $(\Omega, \mathscr{F}, P)$, and if the finite subfield $\mathscr{A}$ of $\mathscr{F}$ is such that the subfields $T^{-n}\mathscr{A}$, $n = 0, 1, \ldots$ are independent, then $H(\mathscr{A}, T) = H(\mathscr{A})$. (If T is invertible, the independence of the $T^{-n}\mathscr{A}$ for $n = 0, 1, \ldots$ is equivalent to the independence of the $T^n\mathscr{A}$ for $n = 0, \pm 1, \pm 2, \ldots$ by the measure-preserving property of T.)

PROOF.   We have

$$H(\mathscr{A}, T) = \lim_{n \to \infty} \frac{1}{n} H \left(\bigvee_{i=0}^{n-1} T^{-i}\mathscr{A} \right) \qquad \text{by 8.6.5}$$

$$= \lim_{n \to \infty} \frac{1}{n} \left[H(\mathscr{A}) + H(T^{-1}\mathscr{A}) + \cdots + H(T^{-(n-1)}\mathscr{A}) \right]$$

$$\text{by 8.6.2(a)}$$

$$= \lim_{n \to \infty} \frac{1}{n} [H(\mathscr{A}) + H(\mathscr{A}) + \cdots + H(\mathscr{A})] \qquad \text{by 8.6.3}$$

$$= H(\mathscr{A}) \quad \square$$

We will now show that if $\bigcup_{n=-\infty}^{\infty} T^n\mathscr{A}$ generates $\mathscr{F}$, we have $H(T) = H(\mathscr{A}, T)$. We start by proving two lemmas.

8.6.12   Lemma. Let T be an invertible, measure-preserving transformation on the probability space $(\Omega, \mathscr{F}, P)$ and $\mathscr{A}$ be a finite subfield of $\mathscr{F}$. We have for every $n = 0, 1, \ldots,$

$$H \left(\bigvee_{i=-n}^{n} T^i\mathscr{A}, T \right) = H(\mathscr{A}, T).$$

PROOF.

$$H \left(\bigvee_{i=-n}^{n} T^i\mathscr{A}, T \right) = \lim_{k \to \infty} \frac{1}{k} H \left(\bigvee_{i=-n-k+1}^{n} T^i\mathscr{A} \right) \qquad \text{by 8.6.5(a)}$$

$$= \lim_{k \to \infty} \frac{1}{k} H \left(\bigvee_{i=0}^{2n+k-1} T^{-i}\mathscr{A} \right) \qquad \text{by 8.6.3}$$

$$= \lim_{N \to \infty} \frac{N}{N - 2n} \frac{1}{N} H \left(\bigvee_{i=0}^{N-1} T^{-i}\mathscr{A} \right)$$

$$= H(\mathscr{A}, T). \quad \square$$

8.6.13 Lemma. Let $\mathscr{A}$ and $\mathscr{B}$ be two finite subfields of $\mathscr{F}$. Assume that both $\mathscr{A}$ and $\mathscr{B}$ have k atoms, denoted by $A_1, \ldots, A_k$ and $B_1, \ldots, B_k$, respectively. Then for any $\varepsilon > 0$ there exists a $\delta > 0$ such that $\sum_{i=1}^k P(A_i \triangle B_i) \le \delta$ implies $|H(\mathscr{A}, T) - H(\mathscr{B}, T)| \le \varepsilon$.

PROOF. Let $C_0 = \bigcup_{i=1}^k (A_i \cap B_i)$ and $C_i = A_i - C_0$, $i = 1, \ldots, k$. We have $P(C_0) \ge 1 - \delta$. If $P(C_0) = 1$, then $P(A_i \triangle B_i) = 0$ for all i. In this case $H(\mathscr{A}, T) = H(\mathscr{B}, T)$. Therefore we can assume that $P(C_0) < 1$. If we apply 8.6.2(e) to the probability $P'(D) = P(D)/(1 - P(C_0))$ on $\Omega - C_0$, we see that

$$-\sum_{i=1}^k P'(C_i) \log P'(C_i) \le \log k.$$

Replacing $P'(C_i)$ by $P(C_i)/(1 - P(C_0))$ we get

$$-\sum_{i=1}^k P(C_i) \log P(C_i) \le (1 - P(C_0)) \log k - (1 - P(C_0)) \log(1 - P(C_0))$$
$$\le \delta \log k - \delta \log \delta$$

if $\delta < 1/e$. And, therefore, if $\mathscr{C}$ is the finite subfield generated by the C_i, $i = 0, \ldots, k$, we have

$$H(\mathscr{C}) \le -P(C_0) \log P(C_0) + \delta \log k - \delta \log \delta$$
$$\le -(1 - \delta) \log(1 - \delta) - \delta \log \delta + \delta \log k$$

if $\delta < 1/2$. We can choose δ in such a way that $H(\mathscr{C}) < \varepsilon$. We then have by 8.6.7(a)

$$H(\mathscr{A}, T) \le H(\mathscr{A} \vee \mathscr{B}, T) = H(\mathscr{B} \vee \mathscr{C}, T)$$

since $\mathscr{A} \vee \mathscr{B} = \mathscr{B} \vee \mathscr{C}$. By 8.6.7(b) we have $H(\mathscr{B} \vee \mathscr{C}, T) \le H(\mathscr{B}, T) + H(\mathscr{C})$. Therefore $H(\mathscr{A}, T) \le H(\mathscr{B}, T) + \varepsilon$, for sufficiently small δ. We can similarly prove that $H(\mathscr{B}, T) \le H(\mathscr{A}, T) + \varepsilon$, if δ is small enough. $\square$

8.6.14 Theorem Kolmogorov–Sinai. If T is an invertible, measure-preserving transformation on the probability space $(\Omega, \mathscr{F}, P)$, and if the finite subfield $\mathscr{A}$ is such that the σ-field generated by the field $\bigcup_{-\infty}^{\infty} T^n \mathscr{A}$ is equal to $\mathscr{F}$, then $H(T) = H(\mathscr{A}, T)$.

PROOF. Let $\mathscr{B}$ be a finite subfield of $\mathscr{F}$ with atoms $B_1, \ldots, B_k$. Given any $\delta > 0$ we can find $C_1 \ldots, C_k \in \bigcup_{-\infty}^{\infty} T^n \mathscr{A}$ such that $P(B_i \triangle C_i) \le \delta$,

$i = 1, \ldots, k$. Let N be such that all the C_i are in $\bigcup_{-N}^{N} T^n \mathscr{A}$. Let $C_1' = C_1$, $C_i' = C_i - \bigcup_{j=1}^{i-1} C_j$, and denote by $\mathscr{C}$ the subfield of $\mathscr{F}$ having the C_i' as atoms. We have $\sum_{i=1}^{k} |P(B_i) - P(C_i')| \leq k^2 \delta$. Given an $\varepsilon > 0$, 8.6.13 allows us to choose $\delta > 0$ such that $|H(\mathscr{B}, T) - H(\mathscr{C}, T)| \leq \varepsilon$. And we have

$$H(\mathscr{B}, T) \leq H(\mathscr{C}, T) + \varepsilon$$

$$\leq H \left(\bigvee_{i=-N}^{N} T^i \mathscr{A}, T \right) + \varepsilon$$

$$= H(\mathscr{A}, T) + \varepsilon \qquad \text{by 8.6.12.}$$

Since this is true for any finite subfield $\mathscr{B}$ of $\mathscr{F}$, we have $H(T) = H(\mathscr{A}, T)$. $\square$

8.6.15 Corollary. If T is an invertible, measure-preserving transformation on a probability space $(\Omega, \mathscr{F}, P)$, and $\mathscr{A}$ is a finite subfield of $\mathscr{F}$ such that the subfields $T^n \mathscr{A}$, $n = 0, \pm 1, \pm 2, \ldots$ are independent and the σ-field they generate is equal to $\mathscr{F}$, we have $H(T) = H(\mathscr{A})$.

8.7 BERNOULLI SHIFTS

We can now apply the results of 8.6 to compute the entropy of a Bernoulli shift.

8.7.1 Definition (Bernoulli Shift). Let S be the set $\{0, 1, \ldots, k-1\}$, $\mathscr{G}$ be the σ-field of all the subsets of S, and $\pi = (p_0, p_1, \ldots, p_{k-1})$ be the probability measure on $(S, \mathscr{G})$ such that $\pi(i) = p_i$, $i = 0, 1, \ldots, k-1$. We consider the probability space $(\Omega_\pi, \mathscr{F}_\pi, P_\pi)$ consisting of all doubly infinite sequences $\omega = (\ldots, \omega_{-1}, \omega_0, \omega_1, \ldots)$ of symbols $0, 1, \ldots, k-1$. The σ-field $\mathscr{F}_\pi$ is generated by the cylinders $C = \{\omega: \omega_i = a_i, -m \leq i \leq n\}$, where the a_i can be any elements of S. The probability of such a cylinder is $P_\pi(C) = \prod_{i=-m}^{n} \pi(a_i)$. (In other words, the coordinate variables $\alpha_n(\omega) = \omega_n$ generate $\mathscr{F}_\pi$ and are independent.)

The *Bernoulli shift with distribution* π is the transformation T_π defined by

$$\alpha_n(T_\pi \omega) = \alpha_{n+1}(\omega).$$

(The transformation T_π shifts the coordinates of ω to the left.) Here, T_π is measure-preserving and invertible because if C is a cylinder, $T^{-1}C$ and TC are also cylinders and have the same P_π-probability as C.

8.7.2 Theorem (Entropy of a Bernoulli Shift). The Bernoulli shift T_π with distribution $\pi = (p_0, \ldots, p_{k-1})$, has entropy $H(T_\pi) = -\sum_{i=0}^{k-1} p_i \log p_i$.

PROOF. Let $\mathscr{A}$ be the finite subfield generated by the projection α_0. Thus $\mathscr{A}$ is generated by the cylinders $\{\omega: \omega_0 = i\}$, $i = 0, \ldots, k-1$. The subfield $T_\pi^{-n}\mathscr{A}$ is generated by the projection α_n. Therefore the subfields $T_\pi^n\mathscr{A}$, $n = 0, \pm 1, \ldots$, are independent and generate $\mathscr{F}_\pi$, and according to 8.6.15

$$H(T_\pi) = H(\mathscr{A}) = -\sum_{i=0}^{k-1} p_i \log p_i.$$

Except for the trivial cases where $k = 1$, or the distribution π gives measure one to one of the symbols, the probability spaces $(\Omega_\pi, \mathscr{F}_\pi, P_\pi)$ are all isomorphic to the probability space $([0,1]^2, \mathscr{B}([0,1]^2), \mu)$ [μ is Lebesgue measure] and therefore isomorphic to each other. $\square$

We recall that two probability spaces $(\Omega_1, \mathscr{F}_1, P_1)$ and $(\Omega_2, \mathscr{F}_2, P_2)$ are isomorphic iff properties (a), (b), (c) of 8.6.0 are verified.

8.7.3 Theorem. For any integer $k > 1$, and any distribution π on $\{0, 1, \ldots, k-1\}$ such that $p_i \neq 1, i = 0, 1, \ldots, k-1$, the probability space $(\Omega_\pi, \mathscr{F}_\pi, P_\pi)$ is isomorphic to the probability space $([0,1]^2, \mathscr{B}([0,1]^2, \mu)$ where μ is Lebesgue measure.

If $k = 1$ or if one of the p_i is equal to 1, the probability space $(\Omega_\pi, \mathscr{F}_\pi, P_\pi)$ is essentially composed of one element and cannot be isomorphic to the unit square.

PROOF. If the distribution π gives the same probability $1/k$ to each of the k symbols, the proof is easy. For each point a of the unit square we write the expansion in base k of the x and y coordinates of a

$$x = .\omega_0\omega_1 \cdots, \qquad y = .\omega_{-1}\omega_{-2}\cdots,$$

and we define ϕa as

$$\phi a = \omega = (\ldots, \omega_{-2}, \omega_{-1}, \omega_0, \omega_1, \omega_2, \ldots).$$

If one of the coordinates of a is of the form n/k^m it has two expansions in base k (since $1/k^m = \sum_{i=m+1}^{\infty}(k-1)/k^i$), and ϕa is not well defined. We have to first remove such points from the unit square, but we are still left with a Borel set E of measure 1. It is easy to check then that $P_\pi(E) = 1$, and that ϕ is an isomorphism from E to ϕE.

If the p_i are not all equal, we have to modify the proof to make sure that $P_\pi(\phi A) = \mu(A)$ for all Borel subsets of E. We construct the image $(\ldots, \omega_{-1}, \omega_0, \omega_1, \ldots) = \phi(a)$ of a point a of $[0,1]^2$ in the following way. We cut the square $[0, 1]^2$ in k vertical open bands $A_0, A_1, \ldots, A_{k-1}$, with widths proportional to $p_0, p_1, \ldots, p_{k-1}$ so that

$$A_0 = \{(x, y): 0 < x < p_0\}, \qquad A_i = \left\{(x, y): \sum_{j=0}^{i-1} p_j < x < \sum_{j=0}^{i} p_j\right\}$$

(when the p_i are all equal to $1/k$, finding the first term of the expansion of x in base k requires dividing $[0, 1]^2$ in k equal vertical bands). If the point a is in A_i, we impose $\omega_0 = i$ for the 0th coordinate of ϕa. If the point a is on the vertical boundaries of one of the A_i we do not define ω_0.

We then divide the band A_i into k vertical open bands with widths proportional to $p_0, p_1, \ldots, p_{k-1}$; if the point a is in the jth band we impose $\omega_1 = j$ for the first coordinate of ϕa; again we do not try to define ω_1 when the point a falls on the vertical boundaries of one of those new bands. We iterate this procedure to define the values of ω_n for $n = 0, 1, \ldots$. To define the ω_{-n} for $n = 1, 2, \ldots$, we use the same procedure but we divide the square $[0, 1]^2$ into horizontal open bands.

Note that ϕa is not defined for the points a that fall on the boundary of one of the bands so defined. Since there are only a countable number of those bands, the function ϕ is defined on a set E of Lebesgue-measure 1. The function ϕ is not onto. The sequences that contain an infinite number of consecutive 0's or an infinite number of consecutive $k - 1$'s, are never images of elements of E. Let A_n be the set $\{\omega: \omega_i = 0 \quad \text{for all} \quad i \geq n\}$, and B_n be the set $\{\omega: \omega_i = k - 1 \quad \text{for all} \quad i \geq n\}$, and define the sets A_{-n} and B_{-n} similarly by replacing the condition $i \geq n$ by $i \leq -n$. The set of elements of Ω_π that are not images under ϕ of elements of E is $\bigcup_{n=-\infty}^{\infty} (A_n \cup B_n)$. Each A_n and each B_n is in $\mathscr{F}_\pi$ and has measure 0, therefore $\Omega' = \phi E$ is a measurable subset of P_π-measure one. Thus the restriction of ϕ to E is one-to-one.

By construction, the image of an open rectangle delimited by a vertical and a horizontal band constructed above is of the form $\Omega' \cap \{\omega: \omega_i = h_i, -m \leq i \leq n\}$, and has as P_π-measure the Lebesgue measure of the rectangle. Since the bands generate the Borel σ-field on the unit square, and the cylinders generate $\mathscr{F}_\pi$, the probability space $(\Omega_\pi, \mathscr{F}_\pi, P_\pi)$ is isomorphic to the unit square with Lebesgue measure. $\square$

Now that we know that essentially all the probability spaces $(S_\pi, \mathscr{F}_\pi, P_\pi)$ are isomorphic, we can start wondering whether some of the Bernoulli shifts are themselves isomorphic. If two Bernoulli shifts have different entropy, they cannot be isomorphic by 8.6.10. Thus the Bernoulli shift with distribution (1/3, 1/3, 1/3) is not isomorphic to the Bernoulli shift with distribution (1/4, 1/4, 1/4, 1/4). But what happens if two Bernoulli shifts have the same entropy? For example, the Bernoulli shifts with distributions (1/2, 1/8, 1/8, 1/8, 1/8) and (1/4, 1/4, 1/4, 1/4) both have entropy 2; are they isomorphic? The answer is given in the following theorem which we state without proof.

8.7.4   Theorem. If two Bernoulli shifts have the same entropy, they are isomorphic in the sense of 8.6.8.

8.8   REFERENCES

The interested reader can find elementary proofs of 8.7.4 (too long to include in this chapter) in Shields (1973) and in Cornfeld, Fomin, and Sinai (1982).

Some standard references on ergodic theory are Billingsley (1965), Halmos (1956), and Jacobs (1962). The pointwise ergodic theorem can be generalized in several ways. If T is a positive contraction operator on L^1, not necessarily arising from a measure-preserving transformation, one can investigate convergence of the sequence of averages $n^{-1}(f + Tf + \cdots + T^{n-1}f)$. In fact the arithmetic average can be replaced by a ratio of the form

$$(f + Tf + \cdots + T^{n-1}f)/(g + Tg + \cdots + T^{n-1}g),$$

where f and g belong to L^1. For results of this type (specifically, the Dunford–Schwartz ergodic theorem and the Chacon–Ornstein theorem), see Garsia (1970).

A discussion of the Shannon–McMillan theorem for a finite coordinate space may be found in Billingsley (1965). McMillan (1953) proved L^1 convergence and Breiman (1957) obtained a.e. convergence; Shannon's original paper (1948) considered convergence in probability for functions of a finite Markov chain. All these results were for finite coordinate spaces; the extension to the countable case is due to Chung (1961). For applications to information theory, see Ash (1965) and Gallager (1968).

The Shannon–McMillan theorem has been generalized to a more abstract setting. For a survey of this area and a unified approach to the various results, see Kieffer (1970).

The definition of the entropy of a transformation as an invariant by isomorphism is due to Kolmogorov (1958; 1959). The notion of Bernoulli shift can be generalized by allowing the σ-fields $\mathcal{A}$ to have a countable number of atoms. The entropy of such a shift can be infinite. The proof of the isomorphy of Bernoulli shifts with same entropy is due to Ornstein (1970a) for shifts with finite entropy, and for generalized Bernoulli shifts with finite or infinite entropy (Ornstein, 1970b).

The Ornstein isomorphism theorem for Bernoulli shifts has been applied by Gray (1975) to show the existence of a class of sliding-block noiseless source codes for a large class of ergodic sources.

9

BROWNIAN MOTION AND STOCHASTIC INTEGRALS

9.1 STOCHASTIC PROCESSES

A *stochastic process* on a probability space $(\Omega, \mathscr{F}, P)$ is a family of random variables $(X_t)_{t \in T}$, where the index set T can be any set. If $T = \mathbb{N}$, a stochastic process is simply a sequence of random variables X_n on the probability space $(\Omega, \mathscr{F}, P)$, and the indices $0, 1, \ldots$ may represent successive times. For instance, a gambler plays at times $0, 1, 2, \ldots$; X_0 is the initial fortune of the gambler and X_n is the fortune at time n. In this chapter we will consider only the case $T = \mathbb{R}^+ = [0, \infty)$. The index t can again be thought as denoting time and, for each fixed ω, the function $t \to X_t(\omega)$ is interpreted as the path of ω. We will say that the paths are continuous, or that the process is continuous (resp. right-continuous) if the functions $t \to X_t(\omega)$ are continuous (resp. right-continuous) for each ω.

Stochastic processes are often constructed as mathematical models. For instance, we may try to build a mathematical model for the number $N_t(\omega)$ of phone calls arriving at a switchboard in the interval of time $[0, t]$. Assume that the data support the following assumptions about the random variables N_t.

1. $N_0 = 0$ a.s.

2. If $0 \le t_1 < t_2 < \cdots < t_n$, the increments $N_{t_2} - N_{t_1}, \ldots, N_{t_n} - N_{t_{n-1}}$ are independent.

3. For $s < t$ the increment $N_t - N_s$ has a Poisson distribution with parameter $\lambda(t - s)$, where λ is the average number of calls in one unit of time.

One can construct the paths of the stochastic process $(N_t)_{t \in \mathbb{R}^+}$ (called the Poisson process) in the following way. Let $T_1 \le T_2 \le \cdots \le T_n \le \ldots$ be the successive arrival times of the calls, and $W_1 = T_1, W_2 = T_2 - T_1, \ldots, W_n = T_n - T_{n-1}, \ldots$ the waiting times between calls. Assume that the W_n are independent, identically distributed, each with exponential distribution with mean $1/\lambda$. The process $(N_t)_{t \ge 0}$ defined by $N_t = \sum_n I_{\{T_n \le t\}}$ satisfies conditions 1–3. It follows from the construction that, for almost every ω, the path $t \to$

$N_t(\omega)$ is a right-continuous, nondecreasing function that increases only by jumps of size 1. These are properties that we expect for the number of calls arriving in the interval $[0, t]$.

We can also use the Kolmogorov extension theorem to construct the Poisson process. Let $0 = t_1 < t_2 < \ldots < t_i < t_{i+1} < \cdots < t_n$; denote by S the set $\{t_1, \ldots, t_n\}$ and by $\mathscr{F}_S$ the σ-field of subsets of $\Omega = \mathbb{R}^{\mathbb{R}^+}$ generated by the coordinates $X_{t_1}, \ldots, X_{t_n}$. Let π_S be the probability on $(\Omega, \mathscr{F}_S)$ which makes $X_0 = 0$ a.s., and the $X_{t_k} - X_{t_{k-1}}$ independent and Poisson distributed with parameters $\lambda(t_k - t_{k-1})$, for $k = 2, \ldots, n$. If the π_S form a consistent system, then we can apply the Kolmogorov extension theorem 2.7.5 and construct a probability P on Ω for which the X_t satisfy conditions 1–3. The problem is that we now have to study the paths $t \to X_t$. Is it true that, outside of a set of P-probability 0, the functions $t \to X_t$ are right-continuous, nondecreasing and increase only by jumps of size 1? Unfortunately it is probably not true (see Problem 3); we have to modify each X_t on a set of probability 0 in such a way that the new paths satisfy all these intuitive properties. This will be the subject of Problem 7 in 9.2.

The above discussion leads to the notion of *version* of a process. Let $(X_t)_{t\geq 0}$ and $(Y_t)_{t\geq 0}$ be two stochastic processes on the same probability space $(\Omega, \mathscr{F}, P)$. The process $(Y_t)_{t\geq 0}$ is a *version* of the process $(X_t)_{t\geq 0}$ iff, for each $t \in \mathbb{R}^+$, we have $P(X_t = Y_t) = 1$. This does not imply the stronger condition $P(X_t = Y_t \forall t \in \mathbb{R}^+) = 1$ (the set $\{X_t = Y_t \forall t \in \mathbb{R}^+\}$ is generally not even measurable), but it does imply that, for any countable subset J of $\mathbb{R}^+$, $P(X_t = Y_t \forall t \in J) = 1$. In particular the processes $(X_t)_{t\geq 0}$ and $(Y_t)_{t\geq 0}$ have the same finite-dimensional distributions. The regularity of the paths of a process can always be destroyed by the wrong choice of version (see Problem 3).

In the next section we construct another stochastic process, Brownian motion $(B_t)_{t\in\mathbb{R}^+}$. The random variables B_t will satisfy conditions 1 and 2 above, but, for $s < t$, the distribution of $B_t - B_s$ will be normal with mean 0 and variance $t - s$. Since the process is constructed as a model for the movement of a particle, we would like its paths to be continuous. Unfortunately there is no intuitive construction of Brownian motion that gives us continuous paths; we will use the Kolmogorov extension theorem and then modify the process to obtain continuous paths.

We conclude this section with a few easy consequences of path regularity. Let $(X_t)_{t\geq 0}$ be a stochastic process; since X_t is a random variable, for each fixed t, the function $\omega \to X_t(\omega)$ is measurable from $(\Omega, \mathscr{F})$ to $(\mathbb{R}, \mathscr{B}(\mathbb{R}))$. The functions $t \to X_t(\omega)$ from $(\mathbb{R}^+, \mathscr{B}(\mathbb{R}^+))$ to $(\mathbb{R}, \mathscr{B}(\mathbb{R}))$ are not necessarily measurable. But it is the case if the paths are all regular enough, for example if, for each ω, the function $t \to X_t(\omega)$ is right-continuous.

The stochastic process $(X_t)_{t\geq 0}$ can be considered as a function X of the two variables t and ω. The process $(X_t)_{t\geq 0}$ is said to be *measurable* iff X is measurable as a function from $(\Omega \times \mathbb{R}^+, \mathscr{F} \times \mathscr{B}(\mathbb{R}^+))$ to $(\mathbb{R}, \mathscr{B}(\mathbb{R}))$; if the process $(X_t)_{t\geq 0}$ is measurable and $|X|$ is not too large, we can consider integrals of the form $E[\int_0^a X(t, \omega)\,dt]$ (Fubini's theorem). Again, right-continuity of the paths is sufficient to assure the measurability of X (see Problem 1).

Problems

1. Let $(X_t)_{t\geq 0}$ be a stochastic process. Show that, if the paths are right-continuous (or left-continuous), the process $(X_t)_{t\geq 0}$ is measurable. [Hint: if the paths of the process are right-continuous, consider the processes X_n defined by $X_n(t, \omega) = X((k+1)/n, \omega)$ if $k/n \leq t < (k+1)/n$, $k = 0, 1, \dots$.]

2. Let $(X_t)_{t\geq 0}$ and $(Y_t)_{t\geq 0}$ be two processes on $(\Omega, \mathscr{F}, P)$. Assume that $(X_t)_{t\geq 0}$ and $(Y_t)_{t\geq 0}$ are such that, for each $t \geq 0$, $P(X_t = Y_t) = 1$, and that they both have right-continuous paths. Show that the set $\{\omega\colon X_t(\omega) \neq Y_t(\omega)$ for at least one $t\}$ is measurable and has probability 0. (The same holds if both processes have left-continuous paths.)

3. Let $(X_t)_{t\geq 0}$ be a continuous process. Assume that the given probability space has a random variable Z defined on it such that $P(Z = a) = 0$ for every real number a. (If necessary enlarge the probability space.) Show that there exists a version $(Y_t)_{t\geq 0}$ of $(X_t)_{t\geq 0}$ such that the paths of $(Y_t)_{t\geq 0}$ are nowhere continuous. [Hint: let $f(t)$ be the function that is 1 on the rationals and 0 elsewhere. Consider the process $Y_t = X_t + f(Z+t)$.]

9.2 BROWNIAN MOTION

9.2.1 Definition. A *Brownian motion* is a stochastic process $(B_t)_{t\in\mathbb{R}^+}$ with the following properties.

1. The increments on disjoint intervals are independent: if $0 \leq t_1 < t_2 < \cdots < t_n$, the random variables $B_{t_2} - B_{t_1}, \dots, B_{t_n} - B_{t_{n-1}}$ are independent.

2. If $s < t$, the increment $B_t - B_s$ of the process on the interval $(s, t]$ is normally distributed with mean 0 and variance $t - s$.

3. The process starts a.s. at 0: $B_0 = 0$ with probability one.

4. The paths of the process B_t are all continuous.

9.2.2 Remarks. 1. The concept of Brownian motion goes back to the observation by the botanist Brown of the random movement of particles of pollen in water. Einstein and Smoluchovski showed that a good approximation for the projection of such a random movement on a line was given by the conditions in Definition 9.2.1, where the variance of $B_t - B_s$ is only assumed to be proportional to $t - s$. The coefficient of proportionality depends on the fluid. A rigorous proof of the existence of a continuous version was given by Wiener.

For this reason the Brownian motion is often called the Wiener process and denoted by W_t.

2. The probability space $(\Omega, \mathscr{F}, P)$ is not specified, and we are free to use whatever seems appropriate.

3. Conditions 1, 2, and 3 involve only the finite-dimensional distributions of the process. Condition 3 is mild. [If $(X_t)_{t\geq 0}$ is a process satisfying conditions 1, 2, 4, the process $Y_t = X_t - X_0$ is a Brownian motion.] The existence of a process satisfying 1, 2, 3 will be an easy consequence of the Kolmogorov extension theorem. It will be much harder to construct a process satisfying condition 4 as well.

4. Condition 4 is sometimes weakened to "outside of a set of probability 0 the paths of $(B_t)_{t\geq 0}$ are continuous." Probabilistically, it does not matter whether we work with the strong or the weak condition since we can always restrict ourselves to the set of probability 1 on which the paths are well behaved.

9.2.3 Theorem. There exists a probability P on the measurable space $(\mathbb{R}^{\mathbb{R}^+}, \mathscr{B}(\mathbb{R})^{\mathbb{R}^+})$ such that, for this probability, the coordinate functions X_t satisfy properties 1, 2, 3 of Definition 9.2.1.

PROOF. Let $0 = t_1 < t_2 < \cdots < t_i < t_{i+1} < \cdots < t_n$; denote by S the set $\{t_1, \ldots, t_n\}$ and by $\mathscr{F}_S$ the σ-field of subsets of $\Omega = \mathbb{R}^{\mathbb{R}^+}$ generated by the coordinates $X_{t_1}, \ldots, X_{t_n}$. Let π_S be the probability on $(\Omega, \mathscr{F}_S)$ which makes $X_0 = 0$ a.s. and the $X_{t_k} - X_{t_{k-1}}$ independent and normally distributed with mean 0 and variances $t_k - t_{k-1}$, for $k = 2, \ldots, n$. If the π_S form a consistent system we can apply the Kolmogorov extension theorem 2.7.5, and the existence of probability P is proven. To verify the consistency of the π_S, we have to show that, if S_i denotes the subset of S obtained by deleting the element t_i for one $i > 1$, the restriction of π_S to the σ-field $\mathscr{F}_{S_i}$ is π_{S_i}. We know that for π_S the random variables $X_{t_2} - X_{t_1}, \ldots, X_{t_i} - X_{t_{i-1}}$, $X_{t_{i+1}} - X_{t_i}, \ldots, X_{t_n} - X_{t_{n-1}}$ are independent and normally distributed with mean 0 and variances $t_2 - t_1, \ldots, t_i - t_{i-1}, t_{i+1} - t_i, \ldots, t_n - t_{n-1}$. We want to show that for π_S the random variables $X_{t_2} - X_{t_1}, \ldots, X_{t_{i+1}} - X_{t_{i-1}}, \ldots,$ $X_{t_n} - X_{t_{n-1}}$ are independent and normally distributed with mean 0 and variances $t_2 - t_1, \ldots, t_{i+1} - t_{i-1}, \ldots, t_n - t_{n-1}$. This is true since the random variable $X_{t_{i+1}} - X_{t_{i-1}}$ is the sum of the two independent and normally distributed random variables $X_{t_i} - X_{t_{i-1}}$ and $X_{t_{i+1}} - X_{t_i}$. $\square$

9.2.4 Remark. If P is the probability constructed in 9.2.3, then, by Chebyshev's inequality, $P(|X_t - X_s| > \varepsilon) \leq (t - s)/\varepsilon^2$. Therefore X_t converges in probability to X_s as $t \to s$. But convergence in probability is not enough; we want pointwise convergence. Assume the existence of a continuous version

$(Y_t)_{t\geq 0}$ of the process $(X_t)_{t\geq 0}$. Even if $(X_t)_{t\geq 0}$ has discontinuous paths, there will be traces of continuity left on the countable subsets of $\mathbb{R}^+$. If J is such a countable subset, then the process $(X_t)_{t\geq 0}$ will be almost surely continuous on J, since the set $\{X_t = Y_t \ \forall t \in J\}$ has probability one. Therefore our first step is to study the continuity of $(X_t)_{t\geq 0}$ on a countable dense subset of $\mathbb{R}^+$. We start with a lemma.

9.2.5 Lemma. Let $0 = t_0 < \cdots < t_n$. If the process $(X_t)_{t\geq 0}$ satisfies properties 1–3 of Definition 9.2.1, we have, for $a > 0$,

$$P\left(\max_{k=0,\ldots,n} X_{t_k} > a\right) \leq 2P(X_{t_n} > a)$$

and

$$P\left(\max_{k=0,\ldots,n} |X_{t_k}| > a\right) \leq 2P(|X_{t_n}| > a).$$

PROOF. Let T be defined on $\{\max_{k=0,\ldots,n} X_{t_k} > a\}$ as the first time t_i such that $X_{t_i} > a$. On the set $\{\max_{k=0,\ldots,n} X_{t_k} \leq a\}$, we take $T = \infty$. We have, using the fact that for $i = 0, \ldots, n - 1$, the $X_{t_n} - X_{t_i}$ have a symmetric distribution,

$$P\left(\max_{k=0,\ldots,n} X_{t_k} > a\right) = \sum_{i=0}^{n} P(T = t_i)$$

$$= 2\sum_{i=0}^{n-1} P(T = t_i)P(X_{t_n} - X_{t_i} > 0) + P(T = t_n).$$

Since the increment $X_{t_n} - X_{t_i}$ is independent of the σ-field $\sigma(X_{t_0}, \ldots, X_{t_i})$, which contains the set $\{T = t_i\}$, we obtain

$$P\left(\max_{k=0,\ldots,n} X_{t_k} > a\right) = 2\sum_{i=0}^{n-1} P(T = t_i, X_{t_n} - X_{t_i} > 0) + P(T = t_n)$$

$$\leq 2\sum_{i=0}^{n-1} P(T = t_i, X_{t_n} > a) + P(T = t_n, X_{t_n} > a)$$

$$\leq 2P(X_{t_n} > a).$$

For the second inequality it is sufficient to notice that the process $(-X_t)_{t\geq 0}$ also satisfies conditions 1–3 of 9.2.1, and that

$$P\left(\min_{k=0,\ldots,n} X_{t_k} < -a\right) = P\left(\max_{k=0,\ldots,n} -X_{t_k} > a\right) \leq 2P(-X_{t_n} > a)$$

$$= 2P(X_{t_n} < -a). \quad \square$$

9.2.6 Corollary. If the process $(X_t)_{t\geq0}$ satisfies conditions 1–3 of 9.2.1, and J is a countable subset of $[0, N]$, then, for $a > 0, P(\sup_{t\in J} |X_t| > a) \leq 2P(|X_N| > a)$.

9.2.7 Theorem. Let $(X_t)_{t\geq0}$ be the process constructed in 9.2.3. For every $N > 0$ the restriction of the process $(X_t)_{t\geq0}$ to the dyadic rationals $k/2^n$ is a.s. uniformly continuous on $[0, N]$.

PROOF. By rescaling the real line we can always assume that $N = 1$. Let S be the set of all dyadic rationals in $[0, 1]$. We have to show that

$$Y_n(\omega) = \sup_{t,s\in S, |t-s|\leq1/2^n} |X_t(\omega) - X_s(\omega)| \to 0 \qquad \text{a.s.}$$

and, since $Y_n \geq Y_{n+1}$, it is sufficient (by (2.5.3)) to show that, for all $a > 0, P(Y_n > a) \to 0$.

The random variables Y_n are hard to work with and the trick is to bound Y_n by $3\max_k Z_{n,k}$, where

$$Z_{n,k} = \sup_{t\in S\cap[k/2^n, (k+1)/2^n]} |X_t - X_{k/2^n}| \qquad k = 0, \ldots, 2^n - 1$$

and to use 9.2.6 to show that $P(\max_k Z_{n,k} > a) \to 0$ as $n \to \infty$. As the random variables $Z_{n,k}, k = 0, \ldots, 2^n - 1$ are identically distributed,

$$P(\max_k Z_{n,k} > a) = P\left(\bigcup_k \{Z_{n,k} > a\}\right) \leq \sum_{k=0}^{2^n-1} P(Z_{n,k} > a) = 2^n P(Z_{n,0} > a).$$

Now by 9.2.6,

$$P(Z_{n,0} > a) = P\left(\sup_{t\in S\cap[0,1/2^n]} |X_t| > a\right) \leq 2P(|X_{1/2^n}| > a),$$

so that

$$P\left(\max_k Z_{n,k} > a\right) \leq 2^{n+1} P(|X_{1/2^n}| > a).$$

This last quantity converges to 0 as $n \to \infty$ since $X_{1/2^n}$ has a normal distribution with mean 0 and variance $1/2^n$ (see 9.2.8 below). $\square$

9.2.8 Lemma. Let $a > 0, \beta > 0$ and assume that X_β is a normally distributed random variable with mean 0 and variance β. Then

$$\frac{1}{\beta}P(|X_\beta| > a) \to 0 \qquad \text{as } \beta \to 0.$$

PROOF.

$$\frac{1}{\beta}P(|X_\beta| > a) = \frac{2}{\beta\sqrt{2\pi\beta}} \int_a^\infty e^{-\frac{x^2}{2\beta}} dx = \frac{2}{\beta\sqrt{\pi}} \int_{a/\sqrt{2\beta}}^\infty e^{-y^2} dy$$

$$\leq \frac{2}{\beta\sqrt{\pi}} \int_{a/\sqrt{2\beta}}^\infty e^{-y} dy = \frac{2}{\beta\sqrt{\pi}} e^{-\frac{a}{\sqrt{2\beta}}},$$

if β is small enough to assure that $a/\sqrt{2\beta} > 1$. $\square$

We are now ready to construct a continuous version of the process $(X_t)_{t\geq 0}$.

9.2.9 Theorem. There exists a process satisfying the conditions of 9.2.1.

PROOF. With Y_n defined as in the proof of 9.2.7, let A be the set $\{Y_n \to 0\}$. The set A is measurable and has probability 1. On A we define, for all t, $B_t(\omega)$ as the limit of the random variables $X_s(\omega)$ as s approaches t along the dyadic rationals. On the complement of A we take $B_t(\omega) = 0$ for all t. By 9.2.7, the process $(B_t)_{t\geq 0}$ is continuous, and we just have to prove that $(B_t)_{t\geq 0}$ is a version of $(X_t)_{t\geq 0}$. Let $t \geq 0$, and (s_n) be a sequence of dyadic rationals converging to t. The random variables X_{s_n} converge a.s. to B_t and in probability to X_t (9.2.4), which is enough to assure that $B_t = X_t$ a.s. $\square$

9.2.10 Corollary. If $(B_t)_{t\geq 0}$ is a Brownian motion, then the functions $\sup_{t\leq t_0} B_t$ and $\sup_{t\leq t_0} |B_t|$ are measurable and satisfy, for $a > 0$, the following inequalities:

$$P(\sup_{t\leq t_0} B_t > a) \leq 2P(B_{t_0} > a) \qquad \text{and}$$

$$P(\sup_{t\leq t_0} |B_t| > a) \leq 2P(|B_{t_0}| > a).$$

PROOF. The continuity of the paths of $(B_t)_{t\geq 0}$ allows us to generalize 9.2.6 to $\mathbb{R}^+$. $\square$

Problems

1. Let $(X_t)_{t\geq 0}$ be a continuous process. Show that $(X_t)_{t\geq 0}$ is a Brownian motion iff, for any finite subset $\{t_1, \dots, t_n\}$ of $\mathbb{R}^+$, the random vector $(X_{t_1}, \dots, X_{t_n})$ is Gaussian with mean 0 and covariance $a_{i,j} = \min(t_i, t_j)$. (See Appendix 5 for the definition and properties of a Gaussian random vector.)

2. A stochastic process $(X_t)_{t\geq 0}$ is continuous in probability iff, for each $s \in \mathbb{R}^+$, $X_t \to X_s$ in probability as $t \to s$. Show that, if two processes

$(X_t)_{t \geq 0}$ and $(Y_t)_{t \geq 0}$ have the same finite-dimensional distributions, $(X_t)_{t \geq 0}$ is continuous in probability iff $(Y_t)_{t \geq 0}$ is.

3. Let $(B_t)_{t \geq 0}$ be a Brownian motion and $c > 0$. Show that the process

$$X_t = \frac{1}{c} B_{c^2 t}$$

is a Brownian motion.

4. Let $(B_t)_{t \geq 0}$ be a Brownian motion. Show that the process

$$X_t = \begin{cases} t B_{1/t}, & \text{if } t > 0, \\ 0, & \text{if } t = 0, \end{cases}$$

satisfies conditions 1–3 of Definition 9.2.1, and is continuous outside of a set of probability 0. The process $(X_t)_{t \geq 0}$ is therefore a Brownian motion and the properties of the Brownian motion at 0 and ∞ are related. [Hint: use Problem 1 to show that the process $(X_t)_{t \geq 0}$ satisfies conditions 1–3 of Definition 9.2.1, and then Theorem 9.2.7 to show the a.s. continuity at 0.]

5. Let $(B_t)_{t \geq 0}$ be a Brownian motion and $(X_t)_{t \geq 0}$ be the process defined as

$$X_t = B_t - t B_1, \qquad 0 \leq t \leq 1.$$

The process $(X_t)_{t \geq 0}$ is called a Brownian Bridge.
(a) Compute the means $E(X_t)$ and the covariances $E(X_s X_t)$. What is the distribution of X_t?
(b) Show that the process

$$Y_t = X_{1-t}, \qquad 0 \leq t \leq 1,$$

has the same finite-dimensional distributions as the process $(X_t)_{0 \leq t \leq 1}$.
(c) Let $U_1, \ldots, U_n$ be independent random variables with uniform distributions on $(0, 1)$ and let

$$F_n(t, \omega) = \frac{1}{n} \sum_{i=1}^{n} I_{(U_i(\omega) \leq t)}, \qquad 0 \leq t \leq 1.$$

$[F_n(t, \omega)$ is an empirical distribution for the uniform distribution on $(0, 1)$.] Show that the process

$$X_n(t, \omega) = \sqrt{n}(F_n(t, \omega) - t), \qquad 0 \leq t \leq 1,$$

has the same means and covariances as the Brownian Bridge.

    (d)   Show that, for each t, the random variables $X_n(t)$ converge in distribution to the random variable X_t.

6.   Let $(B_t)_{t \geq 0}$ be a Brownian motion. Show that $B_t/t \to 0$ a.s. as $t \to \infty$ as follows.

    (a)   Let $X_1, X_2, \ldots$ be identically distributed random variables. Show that if $E|X_1| < \infty$, then for all $a > 0$, $P(|X_n| > a\,n$ infinitely often$) = 0$.

    (b)   Show that $B_n/n \to 0$ a.s. as the integer $n \to \infty$.

    (c)   Let

$$X_n = \max_{n \leq t \leq n+1} |B_t - B_n|,$$

    show that $E|X_1| < \infty$.

    (d)   Conclude by noticing that for $n \leq t \leq n+1$

$$\left| \frac{B_t}{t} - \frac{B_n}{n} \right| \leq \frac{|B_n|}{n^2} + \frac{X_n}{n}.$$

    (e)   Show that we could have used Problem 4 to prove directly that $B_t/t \to 0$ a.s. as $t \to \infty$.

    [Hint: use the equality $E(X) = \int_0^\infty P(X > \lambda)\,d\lambda$ for nonnegative random variables X, the strong law of large numbers and 9.2.10.]

7.   The purpose of this exercise is to construct the Poisson process using the Kolmogorov extension theorem.

    (a)   Use the Kolmogorov Extension Theorem to construct a probability P on the measurable space $(\mathbb{R}^{\mathbb{R}^+}, \mathscr{B}(\mathbb{R})^{\mathbb{R}^+})$, such that, for this probability, the coordinate functions X_t satisfy properties i, ii, iii below:

        i. $X_0 = 0$ a.s.

        ii.   If $0 \leq t_1 < t_2 < \cdots < t_n$, the increments $X_{t_2} - X_{t_1}, \ldots, X_{t_n} - X_{t_{n-1}}$ are independent.

        iii.   For $s < t$, the increment $X_t - X_s$ has a Poisson distribution with parameter $\lambda(t - s)$.

    (b)   Let S be the set of dyadic rationals. Show that outside of a set of probability 0 we have

        i. $X_0 = 0$.

        ii.   The restrictions to S of the paths $t \to X_t(\omega)$ are nondecreasing, and take only nonnegative integer values.

    (c)   Modify the process $(X_t)_{t \geq 0}$ to obtain a right-continuous, integer valued process $(N_t)_{t \geq 0}$ satisfying the properties in question (a) and whose paths are nondecreasing.

    (d)   Show that, outside of a set of probability 0, the paths of $(N_t)_{t \geq 0}$ increase only by jumps of size 1.

    [Hint: Let $T > 0$ be fixed. To prove (d), examine the following limit: $\lim_n P(\sup_{j \leq 2^n T} (|N_{(j+1)/2^n} - N_{j/2^n}| \geq 2).]$

9.3 NOWHERE DIFFERENTIABILITY AND QUADRATIC VARIATION OF PATHS

The paths of the Brownian motion are continuous but they are nowhere differentiable.

9.3.1 Theorem. Let $(B_t)_{t\geq0}$ be a Brownian motion on a probability space $(\Omega, \mathscr{F}, P)$. Then the paths of $(B_t)_{t\geq0}$ are a.s. nowhere differentiable.

As stated above, the theorem is slightly misleading. The set $\{\omega\colon\ t \to B_t(\omega)$ is differentiable at least at one point$\}$ is not necessarily measurable, but we will show that it is included in a measurable set of probability 0.

PROOF.

We start by studying what differentiability at one point implies for a function. Let $f(t)$ be a real-valued function on $[0, \infty)$; assume that it has a derivative $f'(t_0)$ at a point t_0 and choose a real number $a > 0$ such that $|f'(t_0)| \leq a$. Then there exists a positive integer n_0 such that, for $n \geq n_0$,

$$|f(t) - f(t_0)| \leq 2a|t - t_0|, \qquad \text{if } |t - t_0| \leq 3/n. \tag{1}$$

For $n \geq n_0$ let k be the integer such that $(k - 1)/n \leq t_0 < k/n$. The points $(k - 1)/n, k/n, (k + 1)/n, (k + 2)/n$ are within distance $3/n$ of t_0; applying (1) and the triangle inequality we obtain, for example,

$$\left| f\left(\frac{k+2}{n}\right) - f\left(\frac{k+1}{n}\right) \right| \leq \left| f\left(\frac{k+2}{n}\right) - f(t_0) \right| + \left| f(t_0) - f\left(\frac{k+1}{n}\right) \right|$$

$$\leq 2a\left|\frac{k+2}{n} - t_0\right| + 2a\left|\frac{k+1}{n} - t_0\right| \leq \frac{10a}{n}.$$

Using the same method, we obtain similar inequalities for $|f(k + 1/n) - f(k/n)|$ and $|f(k/n) - f(k - 1/n)|$, which gives us

$$\max\left(\left| f\left(\frac{k+2}{n}\right) - f\left(\frac{k+1}{n}\right) \right|, \left| f\left(\frac{k+1}{n}\right) - f\left(\frac{k}{n}\right) \right|, \right.$$

$$\left. \left| f\left(\frac{k}{n}\right) - f\left(\frac{k-1}{n}\right) \right|\right) \leq \frac{10a}{n}.$$

Now we apply all this to the Brownian motion. Denote by X_k the random variable

$$X_k = \max\left(\left| B\left(\frac{k+2}{n}\right) - B\left(\frac{k+1}{n}\right) \right|, \left| B\left(\frac{k+1}{n}\right) - B\left(\frac{k}{n}\right) \right|, \right.$$

$$\left. \left| B\left(\frac{k}{n}\right) - B\left(\frac{k-1}{n}\right) \right|\right).$$

Let A be the set of ω's such that, somewhere on $[0, 1)$, the function $t \to B_t(\omega)$ has a derivative bounded in absolute value by a. According to the preceding discussion, any ω in A is, for n big enough, in A_n where A_n is the set of ω's such that for at least one k in $\{1, 2 \ldots, n\}$ we have $X_k(\omega) \leq (10a)/n$. If we show that $P(\liminf_n A_n) = 0$, we will have proved that, on $[0, 1)$, the paths have a.s. no derivative bounded by a in absolute value.

Since the increments of the Brownian motion on the intervals $(i/n, [(i+1)/n]$ are independent and identically distributed, we have

$$P(A_n) \leq \sum_{k=1}^n P\left(X_k \leq \frac{10a}{n}\right)$$

$$= nP\left(\max\left(\left|B\left(\frac{3}{n}\right) - B\left(\frac{2}{n}\right)\right|, \left|B\left(\frac{2}{n}\right) - B\left(\frac{1}{n}\right)\right|,\right.\right.$$

$$\left.\left|B\left(\frac{1}{n}\right)\right|\right) \leq \frac{10a}{n}\right)$$

$$= n\left[P\left(\left|B\left(\frac{1}{n}\right)\right| \leq \frac{10a}{n}\right)\right]^3$$

$$= n\left[\sqrt{\frac{n}{2\pi}} \int_{-10a/n}^{10a/n} e^{-nx^2/2}\, dx\right]^3$$

$$= n\left[\frac{1}{\sqrt{2\pi n}} \int_{-10a}^{10a} e^{-x^2/2n}\, dx\right]^3 \to 0.$$

We have been studying differentiability on the interval $[0, 1)$ with the bound a, but the same technique works on $[0, N)$ with bounds b as large as desired. The theorem is proved. $\square$

9.3.2   Remark. Theorem 9.3.1 can be generalized to upper, lower, right, and left derivatives: its proof can be modified to show that, outside of a set of probability 0, $\liminf_{t\to s, t<s}(B_t - B_s)/(t - s)$, $\limsup_{t\to s, t<s}(B_t - B_s)/(t - s)$, $\liminf_{t\to s, t>s}(B_t - B_s)/(t - s)$ and $\limsup_{t\to s, t>s}(B_t - B_s)/(t - s)$ are nowhere finite.

9.3.3   Corollary. Almost every path of the Brownian motion has infinite variation on every finite interval.

PROOF.   By 2.3.9, if a path has finite variation on an interval $[a, b]$, then it is almost everywhere differentiable on the interval. $\square$

If $[a, b]$ is an interval of $\mathbb{R}^+$, and $\mathscr{P} = \{t_0, \ldots, t_k\}$ is any partition of $[a, b]$, we have just shown that $\sup_{\mathscr{P}} \sum_i |B(t_{i+1}) - B(t_i)| = \infty$. What can we say about the *quadratic variation* $\lim_{\mathscr{P}} \sum_i |B(t_{i+1}) - B(t_i)|^2$ when the partitions $\mathscr{P}$ get finer?

9.3.4   *Theorem.* If $(B_t)_{t \geq 0}$ is a Brownian motion, $[a, b]$ is an interval of $\mathbb{R}^+$, and $\mathscr{P}_n = \{t_0^{(n)}, \ldots, t_k^{(n)}\}$ is a sequence of partitions of $[a, b]$, then

$$\sum_i \left| B(t_{i+1}^{(n)}) - B(t_i^{(n)}) \right|^2 \to (b - a) \quad \text{in } L^2 \quad \text{as} \quad \max_i \left| t_{i+1}^{(n)} - t_i^{(n)} \right| \to 0.$$

PROOF.

Let $S_n = \sum_i \left| B(t_{i+1}^{(n)}) - B(t_i^{(n)}) \right|^2$; we have

$$E(S_n - (b - a))^2 = E\left(\sum_i \left[(B(t_{i+1}^{(n)}) - B(t_i^{(n)}))^2 - (t_{i+1}^{(n)} - t_i^{(n)}) \right] \right)^2$$

$$= E \sum_{i,j} \left[(B(t_{i+1}^{(n)}) - B(t_i^{(n)}))^2 - (t_{i+1}^{(n)} - t_i^{(n)}) \right]$$

$$\left[(B(t_{j+1}^{(n)}) - B(t_j^{(n)}))^2 - (t_{j+1}^{(n)} - t_j^{(n)}) \right]$$

$$= E \sum_i \left[(B(t_{i+1}^{(n)}) - B(t_i^{(n)}))^2 - (t_{i+1}^{(n)} - t_i^{(n)}) \right]^2,$$

since the $\left(B(t_{i+1}^{(n)}) - B(t_i^{(n)}) \right)^2 - (t_{i+1}^{(n)} - t_i^{(n)})$ are independent random variables with mean 0. The random variables $B(t_{i+1}^{(n)}) - B(t_i^{(n)})$ are normally distributed with variance $t_{i+1}^{(n)} - t_i^{(n)}$; therefore, if Z denotes a normally distributed random variable with mean 0 and variance 1,

$$E(S_n - (b - a))^2 = E(Z^2 - 1)^2 \sum_i (t_{i+1}^{(n)} - t_i^{(n)})^2$$

$$\leq E(Z^2 - 1)^2 (b - a) \max_i \left| t_{i+1}^{(n)} - t_i^{(n)} \right| \to 0. \qquad \square$$

Problems

1. Show that, if the Brownian motion has a chord of slope greater than α on the interval $[0, 1]$, then the process $(X_t)_{t \geq 0}$ defined in Problem 3 of 9.2 has, on $[0, 1/c^2]$, a chord with slope greater than $c\alpha$. This shows intuitively why the Brownian motion is a.s. nowhere differentiable.

2. Let $\mathscr{P}_n$ and S_n be defined as in 9.3.4, and let $\|\mathscr{P}_n\| = \max_i \left| t_{i+1}^{(n)} - t_i^{(n)} \right|$. Show that, if for a sequence of partitions $\mathscr{P}_n$, $\sum_n \|\mathscr{P}_n\| < \infty$, then $S_n \to (b - a)$ a.s.

9.4   LAW OF THE ITERATED LOGARITHM

If $(B_t)_{t \geq 0}$ is a Brownian motion, we know that $B_t / t \to 0$ a.s. as $t \to \infty$ (Problem 6 of 9.2). In 9.4.4, we show that, more precisely, B_t stays asymptotically in between $-\sqrt{2t \log(\log t)}$ and $\sqrt{2t \log(\log t)}$ as $t \to \infty$. According to

Problem 4 of 9.2, the behaviors of B_t at 0 and ∞ are related, and it is not surprising that a similar asymptotic property is satisfied at 0 (see Theorem 9.4.3).

The following estimates will be useful in the proof of 9.4.3.

9.4.1 Lemma. Let X_t be a normally distributed random variable with mean 0 and variance t, and let a be strictly positive. Then

$$P(X_t > a) \le \frac{\sqrt{t}}{a\sqrt{2\pi}} e^{-a^2/2t}.$$

PROOF. Just notice that

$$P(X_t > a) = \frac{1}{\sqrt{2\pi t}} \int_a^\infty e^{-x^2/2t}\, dx$$

$$\le \frac{1}{a\sqrt{2\pi t}} \int_a^\infty x e^{-x^2/2t}\, dx$$

$$= \frac{\sqrt{t}}{a\sqrt{2\pi}} e^{-a^2/2t}. \quad \square$$

9.4.2 Lemma. The ratio of $\int_x^\infty e^{-s^2/2}\, ds$ and $(e^{-x^2/2})/x$ converges to 1 as x converges to ∞. (Thus the inequality of 9.4.1 is an asymptotic equality as $a \to \infty$.)

PROOF. Apply L'Hôpital's Rule. $\square$

9.4.3 Theorem. If B_t is a Brownian motion, then we have

$$\limsup_{t\to 0} \frac{B_t}{\sqrt{2t\log(\log 1/t)}} = 1 \qquad \text{a.s.}$$

and

$$\liminf_{t\to 0} \frac{B_t}{\sqrt{2t\log(\log 1/t)}} = -1 \qquad \text{a.s.}$$

PROOF. The second statement of the lemma is a consequence of the first because $-B_t$ is also a Brownian motion.

1. Let $u(t) = \sqrt{2t\log(\log 1/t)}$. We first show that

$$\limsup_{t\to 0} \frac{B_t}{\sqrt{2t\log(\log 1/t)}} \le 1 \qquad \text{a.s.}$$

This is equivalent to: for almost every ω, and any $\varepsilon > 0$, $B_t \le (1+\varepsilon)u(t)$ when t is near enough to 0.

Given $\varepsilon > 0$, choose $\alpha \in (0, 1)$ such that $\alpha(1+\varepsilon)^2 > 1$; consider the decreasing sequence $t_n = \alpha^n$, and let A_n be the set

$$A_n = \{\omega:\ B_t(\omega) > (1+\varepsilon)u(t) \qquad \text{for at least one } t \in (t_{n+1}, t_n]\}.$$

Using 9.2.10 and the fact that the function $u(t)$ is increasing for sufficiently small t, we obtain

$$P(A_n) \leq P(\sup_{t \leq t_n} B_t > (1 + \varepsilon)u(t_{n+1}))$$

$$\leq 2P(B_{t_n} > (1 + \varepsilon)u(t_{n+1})).$$

Therefore, by 9.4.1,

$$P(A_n) \leq \sqrt{\frac{2}{\pi}} \frac{\sqrt{t_n}}{(1 + \varepsilon)u(t_{n+1})} e^{-(1+\varepsilon)^2 u^2(t_{n+1})/2t_n}.$$

Substituting α^n for t_n we obtain the inequality

$$P(A_n) \leq \sqrt{\frac{2}{\pi}} \frac{1}{\sqrt{2(1 + \varepsilon)^2 \alpha \log\left[(n + 1)\log\frac{1}{\alpha}\right]}} \frac{1}{\left[(n + 1)\log\frac{1}{\alpha}\right]^{(1+\varepsilon)^2\alpha}}$$

$$\leq K \frac{1}{(n + 1)^\beta \sqrt{\log(n + 1)}},$$

where $\beta = (1 + \varepsilon)^2 \alpha > 1$. Therefore $\sum_n P(A_n) < \infty$ and, by the first Borel Cantelli Lemma 2.2.4, $P(\limsup_n A_n) = 0$. This assures that, for almost every ω, $B_t \leq (1 + \varepsilon)u(t)$ when t is near enough to 0.

2. We now show that $\limsup_{t \to 0} \left(B_t / \sqrt{2t \log(\log 1/t)}\right) \geq 1$ a.s. by showing that, for any $\delta > 0$, there exists a sequence $t_1 > t_2 > \ldots$ decreasing to 0 such that $P(B_{t_n} > (1 - \delta)u(t_n) \text{ i.o.}) = 1$. (Recall that the abbreviation i.o. stands for infinitely often.)

We now choose $\varepsilon > 0$ and $\alpha \in (0, 1)$ small enough to guarantee that $(1 - \varepsilon)^2 < (1 - \alpha)$, and $(1 - \varepsilon) - (1 + \varepsilon)\sqrt{\alpha} > 1 - \delta$. Let $t_n = \alpha^n$ and consider the independent random variables $X_n = B_{t_n} - B_{t_{n+1}}$. We want to estimate $\sum P(X_n > (1 - \varepsilon)u(t_n))$.

$$P(X_n > (1 - \varepsilon)u(t_n)) = P\left(\frac{X_n}{\sqrt{t_n - t_{n+1}}} > \frac{(1 - \varepsilon)u(t_n)}{\sqrt{t_n - t_{n+1}}}\right)$$

$$= P\left(\frac{X_n}{\sqrt{t_n - t_{n+1}}} > \frac{(1 - \varepsilon)\sqrt{2\log\left(n\log\frac{1}{\alpha}\right)}}{\sqrt{1 - \alpha}}\right).$$

Therefore, applying Lemma 9.4.2, we get, since $X_{t_n}/\sqrt{t_n - t_{n-1}}$ has a normal distribution with mean 0 and variance 1,

$$P(X_n > (1 - \varepsilon)u(t_n)) \sim \frac{K}{n^\gamma \sqrt{\log n}},$$

where $\gamma = (1 - \varepsilon)^2/(1 - \alpha) < 1$. This implies that $\sum_n P(X_n > (1 - \varepsilon)u(t_n)) = \infty$, and, by the second Borel Cantelli Lemma 6.1.5, we have

$$P(X_n > (1 - \varepsilon)u(t_n) \text{ i.o.}) = 1.$$

If we apply part 1 of the proof to the process $-B_t$, we see that

$$P(B_{t_{n+1}} \geq -(1 + \varepsilon)u(t_{n+1}) \text{ for } n \text{ large enough}) = 1,$$

and, therefore,

$$P\left(B_{t_n} \geq u(t_n)\left[(1 - \varepsilon) - (1 + \varepsilon)\frac{u(t_{n+1})}{u(t_n)}\right] \text{ i.o.}\right) = 1.$$

Since $u(t_{n+1})/u(t_n) \to \sqrt{\alpha}$ we obtain

$$P(B_{t_n} > (1 - \delta)u(t_n) \text{ i.o.}) = 1. \quad \square$$

We give now the result at ∞. We could prove the theorem directly, but it is easier to use Problem 4 of 9.2. If B_t is a Brownian motion, so is the process $Y_t = tB_{1/t}$, and the properties of the paths of the Brownian motion at 0 and ∞ are related.

9.4.4   Theorem. If B_t is a Brownian motion, then we have

$$\limsup_{t \to \infty} \frac{B_t}{\sqrt{2t \log(\log t)}} = 1 \qquad \text{a.s.}$$

and

$$\liminf_{t \to \infty} \frac{B_t}{\sqrt{2t \log(\log t)}} = -1 \qquad \text{a.s.}$$

PROOF. Apply Theorem 9.4.3 to the process $Y_u = uB_{1/u}$, and take $t = 1/u$. $\square$

Problem

1. Let $S_n = \sum_{k=1}^n Y_k$, where the Y_k are independent and each has a normal distribution with mean 0 and variance σ^2. Show that

$$\limsup_{n \to \infty} \frac{S_n}{\sqrt{2\sigma^2 n \log(\log n)}} = 1 \qquad \text{a.s.}$$

and

$$\liminf_{n \to \infty} \frac{S_n}{\sqrt{2\sigma^2 n \log(\log n)}} = -1 \qquad \text{a.s.}$$

9.5   THE MARKOV PROPERTY

The Markov property discussed in Chapter 4 can be expressed as follows: a process $(X_t)_{t \geq 0}$ satisfies the Markov property if the position of X_t, knowing what happened up to time s, $s < t$, depends on the value of X_s and not on the values of X_u, $u < s$. The Markov property of the Brownian motion is a corollary of the definition of the Brownian motion.

9.5.1   Theorem(Markov Property). Let $(B_t)_{t \geq 0}$ be a Brownian motion. Then if $s \geq 0$, the process $Y_t = B_{t+s} - B_s$ is a Brownian motion independent of the σ-field $\mathscr{B}_s = \sigma(B_u, u \leq s)$.

We would like to generalize this property by allowing s to be a stopping time. We give the definition of stopping times with respect to a general nondecreasing family of σ-fields since it will be needed in the problems, but, in this section, we are only interested in the family $(\mathscr{B}_t)_{t \geq 0}$.

9.5.2   Definition.   Let $(\Omega, \mathscr{F}, P)$ be a probability space and $(\mathscr{F}_t)_{t \geq 0}$ be a nondecreasing family of sub-σ-fields of $\mathscr{F}$. A nonnegative random variable T is a *stopping time* for the σ-fields $(\mathscr{F}_t)_{t \geq 0}$ iff, for each $t \geq 0$, $\{T \leq t\} \in \mathscr{F}_t$. (The random variable T is allowed to take the value $+\infty$.)

The σ-field $\mathscr{F}_T$ of events *prior* to T is

$$\mathscr{F}_T = \{A \in \mathscr{F} : A \cap \{T \leq t\} \in \mathscr{F}_t \ \forall t \geq 0\}.$$

It is also useful to consider the σ-field

$$\mathscr{F}_{T+} = \{A \in \mathscr{F} : A \cap \{T < t\} \in \mathscr{F}_t \ \forall t \geq 0\}.$$

These definitions are consistent with the definitions given in Chapter 6, since for the index set $\mathbb{N}$, the conditions $\{T = n\} \in \mathscr{F}_n$ for any n, and $\{T \leq n\} \in \mathscr{F}_n$ for any n, are equivalent. When the index set is $\mathbb{R}^+$, the two conditions are no longer equivalent, and the condition $\{T = t\} \in \mathscr{F}_t$ for any t is too weak to allow us to work with uncountable unions such as $\{T \leq t\} = \bigcup_{s \leq t} \{T = s\}$.

Constant times $T = t$ are stopping times. In that case, $\mathscr{F}_T = \mathscr{F}_t$ and $\mathscr{F}_{T+} = \mathscr{F}_{t+} = \bigcap_{s > t} \mathscr{F}_s$.

The following theorem gives some intuitive properties of stopping times and their associated σ-fields.

9.5.3   Theorem. Let $(\Omega, \mathscr{F}, P)$ be a probability space, $(\mathscr{F}_t)_{t \geq 0}$ a nondecreasing family of sub-σ-fields of $\mathscr{F}$, and S and T be two stopping times for the family $(\mathscr{F}_t)_{t \geq 0}$.

(a)   We have the inclusion $\mathscr{F}_T \subset \mathscr{F}_{T+}$.

(b) The random variable T is $\mathscr{F}_T$-measurable.
(c) If $S \leq T$, then $\mathscr{F}_S \subset \mathscr{F}_T$ and $\mathscr{F}_{S+} \subset \mathscr{F}_{T+}$.
(d) If $S < T$, then $\mathscr{F}_{S+} \subset \mathscr{F}_T$.
(e) If U is an $\mathscr{F}_T$-measurable random variable and $T \leq U$, then U is a stopping time.
(f) If S and T are two stopping times, then $S \vee T$ and $S \wedge T$ are stopping times and $\mathscr{F}_{S \wedge T} = \mathscr{F}_S \cap \mathscr{F}_T$.
(g) If S and T are two stopping times, then $S + T$ is also a stopping time.

PROOF.

(a) Let $A \in \mathscr{F}_T$. Then for any $t \geq 0$ the set $A \cap \{T < t\} = \bigcup_n (A \cap \{T \leq t - (1/n)\})$ is in $\mathscr{F}_t$.
(b) It is sufficient to show that, for any $t \geq 0$, the set $\{T \leq t\}$ is in $\mathscr{F}_T$. For any $s \geq 0$, we have

$$\{T \leq t\} \cap \{T \leq s\} = \{T \leq t \wedge s\} \in \mathscr{F}_s.$$

(c) Assume that $A \in \mathscr{F}_S$. Then $A \cap \{T \leq t\} = A \cap \{S \leq t\} \cap \{T \leq t\} \in \mathscr{F}_t$, and $A \in \mathscr{F}_T$. Similarly for the σ-fields $\mathscr{F}_{S+}$ and $\mathscr{F}_{T+}$.
(d) Use the equality $A \cap \{T \leq t\} = \bigcup_n (A \cap \{S < t - (1/n)\} \cap \{T \leq t\})$ to conclude.
(e) Use the equality $\{U \leq t\} = \{U \leq t\} \cap \{T \leq t\}$ to conclude.
(f) To show that $S \vee T$ is a stopping time, use the equality $\{S \vee T \leq t\} = \{S \leq t\} \cap \{T \leq t\}$. Similarly $\{S \wedge T \leq t\} = \{S \leq t\} \cup \{T \leq t\}$. Using (b) we see that $\mathscr{F}_{S \wedge T} \subset \mathscr{F}_S \cap \mathscr{F}_T$. Conversely, if $A \in \mathscr{F}_S \cap \mathscr{F}_T$, we have $A \cap \{S \wedge T \leq t\} = (A \cap \{S \leq t\}) \cup (A \cap \{T \leq t\}) \in \mathscr{F}_t$.
(g) Since S is $\mathscr{F}_S$-measurable, the sets $\{S \in A\} \cap \{S \leq t\}$ and $\{S \in A\} \cap \{S \leq t\} \cap \{T \leq t\}$ are in $\mathscr{F}_t$ for any set $A \in \mathscr{B}(\mathbb{R})$; which means that the restriction of S to the set $\{S \leq t\} \cap \{T \leq t\}$ is $\mathscr{F}_t$-measurable. A similar property holds for T. Therefore the restriction of $S + T$ to the set $\{S \leq t\} \cap \{T \leq t\}$ is $\mathscr{F}_t$-measurable, and the set $\{S + T \leq t\} = \{S + T \leq t\} \cap \{S \leq t\} \cap \{T \leq t\}$ is in $\mathscr{F}_t$. □

9.5.4 Remark. It is often assumed in the general theory of stochastic processes that the family of sub-σ-fields is right-continuous ($\mathscr{F}_t = \bigcap_{s > t} \mathscr{F}_s$ for any $t \geq 0$); in this case the property $\{T \leq t\} \in \mathscr{F}_t$ for any t is equivalent to $\{T < t\} \in \mathscr{F}_t$ for any t, and the σ-fields $\mathscr{F}_T$ and $\mathscr{F}_{T+}$ are the same. The family of natural σ-fields $(\mathscr{B}_t)_{t \geq 0}$ of the Brownian motion is not right-continuous, but we will see in Problem 2 that $\mathscr{B}_{t+}$ and $\mathscr{B}_t$ differ only by measurable sets of probability 0.

The following lemma is useful to verify that hitting times are stopping times.

9.5.5 Lemma. Let $g(t)$ be a continuous function on $[0, \infty)$ such that $g(0) = 0$. Let $a > 0$ and $T = \inf\{t > 0 \colon g(t) = a\} = \inf\{t > 0 \colon g(t) \geq a\}$, (we take $T = \infty$ if there is no such t). Then $T \leq t$ iff $\sup_{r \leq t, r \in \mathbb{Q}} g(r) \geq a$. (A similar result holds for $a < 0$.)

PROOF.   As the function g is continuous, $\sup_{r \leq t, r \in \mathbb{Q}} g(r) = \sup_{s \leq t} g(s)$ and $g(T) = a$. If $T \leq t$, then $\sup_{s \leq t} g(s) \geq g(T) \geq a$. Conversely if $\sup_{s \leq t} g(s) \geq a$ there exists a point t_0 in $[0, t]$ such that $g(t_0) = \sup_{s \leq t} g(s) \geq a$ (since g is continuous), and therefore $T \leq t$. □

9.5.6   Example. Let $(B_t)_{t \geq 0}$ be a Brownian motion and $a \neq 0$; the first hitting time of a is defined as $T = \inf\{t > 0 \colon B_t = a\}$. The Law of the Iterated Logarithm assures that the time T is almost surely finite, and the continuity of the Brownian motion assures that $B_T = a$ on $\{T < \infty\}$. If $a > 0$, then $\{T \leq t\} = \{\sup_{r \leq t, r \in \mathbb{Q}} B_r \geq a\} \in \mathscr{B}_t$, which shows that T is a stopping time for the family $(\mathscr{B}_t)_{t \geq 0}$. (A similar proof gives the result for $a < 0$.)

We want to generalize the Markov property to stopping times, but we first give a result on the measurability of B_T.

9.5.7   Theorem. Let $(B_t)_{t \geq 0}$ be a Brownian motion and T be a finite stopping time for the σ-fields $\mathscr{B}_t = \sigma(B_s, s \leq t)$, $t \geq 0$, then B_T is $\mathscr{B}_T$-measurable.

PROOF.   Define $X_n = B_{k/n}$ on $\{k/n \leq T < (k+1)/n\}$, $k = 0, 1, 2, \ldots$. Then X_n is $\mathscr{B}_T$-measurable, and $B_T = \lim_n X_n$. □

9.5.8   (Strong Markov Property). Let $(B_t)_{t \geq 0}$ be a Brownian motion and T be a finite stopping time for the family of σ-fields $(\mathscr{B}_t)_{t \geq 0}$. Then $X_t = B_{t+T} - B_T$ is a Brownian motion independent of the σ-field $\mathscr{B}_T$.

PROOF.   Consider the times T_n defined as follows:

$$T_n = \frac{k}{n} \quad \text{if} \quad \frac{k-1}{n} \leq T < \frac{k}{n}, k = 1, 2, \ldots .$$

The T_n are stopping times (9.5.3(e)) and, if $C \in \mathscr{B}_T$, the sets $C \cap \{T_n = k/n\}$ are in $\mathscr{F}_{k/n}$. Applying 9.5.1, we get, for $A \in \mathscr{B}(\mathbb{R})$,

$$P(\{B_{t+T_n} - B_{T_n} \in A\} \cap C) = \sum_{k=1}^{\infty} P\left(\{B_{t+k/n} - B_{k/n} \in A, T_n = \tfrac{k}{n}\} \cap C\right)$$

$$= \sum_{k=1}^{\infty} P\left(B_{t+k/n} - B_{k/n} \in A\right) P\left(\{T_n = \tfrac{k}{n}\} \cap C\right)$$

$$= P(B_t \in A)P(C).$$

We now restrict the sets A to be open intervals (a, b). When $n \to \infty$, the random variables $B_{t+T_n} - B_{T_n}$ converge a.s. to $B_{t+T} - B_T$; therefore $P(\{a < B_{t+T} - B_T < b\} \cap C) \leq \liminf P(\{a < B_{t+T_n} - B_{T_n} < b\} \cap C)$ $\leq \limsup P(\{a < B_{t+T_n} - B_{T_n} < b\} \cap C) \leq P(\{a \leq B_{t+T} - B_T \leq b\} \cap C)$. Letting n approach infinity, we get, for a and b such that $P(B_{t+T} - B_T = a)$ $= P(B_{t+T} - B_T = b) = 0$,

$$P(\{a < B_{t+T} - B_T < b\} \cap C) = P(a < B_t < b)P(C). \tag{1}$$

For any $\varepsilon > 0$, we have, since we have already verified the Strong Markov Property for the stopping times T_n,

$$P(B_{t+T} - B_T = a) \leq \liminf_n P(a - \varepsilon < B_{t+T_n} - B_{T_n} < a + \varepsilon)$$

$$= P(a - \varepsilon < B_t < a + \varepsilon).$$

Let $\varepsilon \to 0$ to obtain $P(B_{t+T} - B_T = a) = 0$, for any $a > 0$. Therefore equality (1) is satisfied for all a and $b \in \mathbb{R}$, and $B_{t+T} - B_T$ is a Brownian motion independent of $\mathscr{B}_T$. $\square$

9.5.9 Remark. The same proof shows that, more generally, if T is a finite stopping time for the family of σ-fields $\mathscr{B}_{t+}$, then $Y_t = B_{t+T} - B_T$ is a Brownian motion independent of the σ-field $\mathscr{B}_{T+}$: the stopping times T_n defined as above are still stopping times for the family $(\mathscr{B}_t)_{t \geq 0}$ and, as $T < T_n$, we have $\mathscr{B}_{T+} \subset \mathscr{B}_{T_n}$. Therefore the above proof is still valid if the set C is in $\mathscr{B}_{T+}$.

In particular, if T is a finite stopping time for the family $(\mathscr{B}_t)_{t \geq 0}$, then $Y_t = B_{t+T} - B_T$ is a Brownian motion independent of the σ-field $\mathscr{B}_{T+}$.

We now apply the strong Markov property to the problem of finding the distribution of the random variable $\sup_{s \leq t} B_s$. The idea is the following: let $a > 0$, and let T denote the first hitting time of a by the Brownian motion. Since the process $Y_u = B_{u+T} - B_T$ is a Brownian motion, we have, for any $u > 0$, $P(Y_u \geq 0) = P(Y_u \leq 0) = 1/2$. This property should still hold if we replace u by a time independent of the process $(Y_u)_{u \geq 0}$, for example $t - T$, and we should have $P(\sup_{s \leq t} B_s \geq a) = P(T \leq t) = 2P(T \leq t, Y_{t-T} \geq 0) = 2P(B_t \geq a)$. We now give a detailed proof.

9.5.10 Theorem. Let $(B_t)_{t \geq 0}$ be a Brownian motion and $a > 0$. Then

$$P(\sup_{s \leq t} B_s \geq a) = 2P(B_t \geq a).$$

PROOF. Let T be the first hitting time of a and define the stopping times T_n as in 9.5.8. Since the time T_n is independent of $B_{u+T_n} - B_{T_n}$, we have

$$P(B_t - B_{T_n} \geq 0, T_n < t) = \sum_k P\left(B_t - B_{k/n} \geq 0, T_n = \frac{k}{n} < t\right)$$

$$= \sum_{k<tn} P(B_t - B_{k/n} \geq 0)P\left(T_n = \frac{k}{n} < t\right)$$

$$= \tfrac{1}{2}P(T_n < t).$$

Since we let n approach ∞; as $P(B_t - B_T = 0) = P(B_t = a) = 0$ and $P(T = t) \leq P(B_t = a)$, an argument similar to the one given in 9.5.8 shows that

$$P(B_t \geq a) = P(B_t - B_T \geq 0, T \leq t) = \tfrac{1}{2}P(T \leq t) = \tfrac{1}{2}P(\sup_{s\leq t} B_s \geq a). \quad \square$$

Problems

1. Let $(B_t)_{t\geq 0}$ be a Brownian motion, $s < t$, $A \in \mathcal{B}(\mathbb{R})$ and $\mathcal{B}_s = \sigma(B_u, u \leq s)$. Show that

$$P(B_t \in A \mid \mathcal{B}_s) = P(B_t \in A \mid B_s) = p_{t-s}(B_s, A) \qquad \text{a.s.,}$$

    where $p_u(x, A) = \int_A (1/\sqrt{2\pi u}) \exp -(y - x)^2/2u \, dy$.

    (Hint: use the conditional density of $B_t - B_s$ given B_s.)
2. In this problem we will show that $\mathcal{B}_t$ and $\mathcal{B}_{t+}$ differ only by sets of probability 0.
    (a) If f is a measurable, bounded function from $\mathbb{R}$ into $\mathbb{R}$ define for $u > 0$,

$$p_u(x, f) = \int f(y) \frac{1}{\sqrt{2\pi u}} \exp -\frac{(y - x)^2}{2u} \, dy.$$

    Show that the function $(u, x) \to p_u(x, f)$ is continuous.
    (b) Use Problem 1 to show that if $s < t$ and f is a measurable, bounded function from $\mathbb{R}$ into $\mathbb{R}$, then

$$E[f(B_t) \mid \mathcal{B}_s] = p_{t-s}(B_s, f) \qquad \text{a.s.}$$

    (c) Show that if $s < t$ and f is a measurable, bounded function from $\mathbb{R}$ into $\mathbb{R}$, then

$$E[f(B_t) \mid \mathcal{B}_{s+}] = p_{t-s}(B_s, f) \qquad \text{a.s.}$$

(d) Show that if the functions $f_1, f_2, \ldots, f_m$ are measurable and bounded and if $s < t_1 < t_2 < \cdots < t_m$, then

$$E\left[\prod_{i=1}^{m} f_i(B_{t_i}) \mid \mathscr{B}_{s+}\right] = E\left[\prod_{i=1}^{m} f_i(B_{t_i}) \mid \mathscr{B}_s\right] \qquad \text{a.s.}$$

(e) Show that the above equality is still true when the t_i are not restricted to being larger than s.

(f) Now show that, if the random variable Z is bounded and $\sigma(B_u, u \geq 0)$-measurable, then

$$E[Z \mid \mathscr{B}_{s+}] = E[Z \mid \mathscr{B}_s].$$

This shows that a $\mathscr{B}_{s+}$-measurable random variable is a.s. equal to a $\mathscr{B}_s$-measurable random variable.

(g) Let $\mathscr{B}$ be the smallest *complete* σ-field generated by the random variables B_t, $t \geq 0$. Define $\mathscr{N} = \{A \in \mathscr{B} : P(A) = 0\}$ and let $\mathscr{B}_t'$ be the σ-field generated by $\mathscr{B}_t$ and $\mathscr{N}$. Show that the family of σ-fields $(\mathscr{B}_t')_{t \geq 0}$ is right-continuous.

(Hint: use 6.4.4 for (c) and (g).)

3. Let $(\Omega, \mathscr{F}, P)$ be a probability space and $(\mathscr{F}_t)_{t \geq 0}$ be a nondecreasing family of sub-σ-fields of $\mathscr{F}$. Let $\mathscr{G}_t = \mathscr{F}_{t+}$.
   (a) Show that a nonnegative random variable T is a stopping time for the family $(\mathscr{G}_t)_{t \geq 0}$ iff $\{T < t\} \in \mathscr{F}_t$, for any $t \geq 0$.
   (b) Show that, if T is a stopping time for the family $(\mathscr{F}_t)_{t \geq 0}$, we have $\mathscr{F}_{T+} = \mathscr{G}_T$.

4. Let $(\Omega, \mathscr{F}, P)$ be a probability space and $(\mathscr{F}_t)_{t \geq 0}$ be a nondecreasing family of sub-σ-fields of $\mathscr{F}$.
   (a) Show that, if T and S are two stopping times for the family $(\mathscr{F}_t)_{t \geq 0}$, the sets $\{S < T\}$, $\{S \leq T\}$, $\{S = T\}$, $\{S > T\}$, and $\{S \geq T\}$ are all in $\mathscr{F}_S \cap \mathscr{F}_T$.
   (b) Show that, if the set A is in $\mathscr{F}_S$, then the set $A \cap \{S \leq T\}$ is in $\mathscr{F}_T$.
   (c) Show that, if the set A is in $\mathscr{F}_{S+}$, then the set $A \cap \{S < T\}$ is in $\mathscr{F}_T$.

   (Hint: for (a) it suffices to prove that $(S < T)$ and $(S \leq T)$ are in $\mathscr{F}_T$. Use arguments similar to those of Theorem 9.5.3(f).)

   The above results generalize properties such as "if S is a stopping time, the sets $\{S < t\}$, $\{S = t\}$, $\{S \leq t\}$ are all in $\mathscr{F}_t$," and "if $A \in \mathscr{F}_S$, then $A \cap \{S \leq t\} \in \mathscr{F}_t$" to the case of a nonconstant stopping time T.

5. Let $(\Omega, \mathscr{F}, P)$ be a probability space, $(\mathscr{F}_t)_{t \geq 0}$ be a nondecreasing family of sub-σ-fields of $\mathscr{F}$, and (T_n) be a sequence of stopping times.
   (a) Show that $\sup_n T_n$ is a stopping time.

(b)   Show that, if the family $(\mathscr{F}_t)_{t\geq 0}$ is right-continuous, $\inf_n T_n$ is a
stopping time and $\mathscr{F}_T = \bigcap_n \mathscr{F}_{T_n}$. (Show first that $\mathscr{F}_{T+} = \mathscr{F}_T$.)

9.6 MARTINGALES

In Chapter 6, we defined the martingale and submartingale properties for
sequences $(X_n)_{n\geq 0}$ of random variables; in this section we generalize these notions to the case of an uncountable index set $\mathbb{R}^+$. Many results of Chapter 6 can
be generalized to right-continuous submartingales, but we will give only the results needed in Section 9.7. Two martingales will be essential in the definition
of stochastic integrals with respect to the Brownian motion, B_t and $B_t^2 - t$.

9.6.1 Definition.   Let $(\Omega,\ \mathscr{F},\ P)$ be a probability space, and $(\mathscr{F}_t)_{t\geq 0}$ be
a nondecreasing family of sub-σ-fields of $\mathscr{F}$. A family of random variables
$(X_t)_{t\geq 0}$ is a *martingale* with respect to $(\mathscr{F}_t)_{t\geq 0}$ iff

(a)   each X_t is $\mathscr{F}_t$-measurable and integrable, and
(b) $E[X_t \mid \mathscr{F}_s] = X_s$ a.s. if $0 \leq s < t$.

The notions of sub- and supermartingale can be similarly generalized.

If $\mathscr{F}_t = \sigma(X_s, s \leq t)$, we will say that $(X_t)_{t\geq 0}$ is a martingale without specifying the family of σ-fields.

9.6.2 Examples.   1. The Brownian motion $(B_t)_{t\geq 0}$ is a martingale: if $s < t$,
then, since the random variable $B_t - B_s$ is independent of $\mathscr{B}_s$ and has mean 0,

$$E[B_t - B_s \mid \mathscr{B}_s] = E[B_t - B_s] = 0.$$

2.   The process $Y_t = B_t^2 - t$ is also a martingale: we start by noticing that, if
$s < t$,

$$E[(B_t - B_s)^2 \mid \mathscr{B}_s] = E[(B_t - B_s)^2] = t - s,$$

and

$$E[(B_t - B_s)^2 \mid \mathscr{B}_s] = E[B_t^2 - 2B_t B_s + B_s^2 \mid \mathscr{B}_s] = E[B_t^2 - B_s^2 \mid \mathscr{B}_s],$$

so that

$$E[B_t^2 - B_s^2 \mid \mathscr{B}_s] = t - s.$$

To work with the uncountable index set $\mathbb{R}^+$, we need some regularity conditions on martingales. This is why our first step is to show that, under mild
conditions for the family of sub-σ-fields, we can find a right-continuous version of a martingale. The argument is similar to the proof of the existence of
a continuous version of the Brownian motion: we first study the martingale on
a countable dense subset S of $\mathbb{R}^+$, and we then construct a right-continuous
version by taking limits along S.

9.6.3 Theorem. Let $(\Omega, \mathscr{F}, P)$ be a probability space, and $(\mathscr{F}_t)_{t \geq 0}$ a nondecreasing *right-continuous* family of sub-σ-fields. We assume that the probability space is complete in the sense of 1.3.7, and that each $\mathscr{F}_t$ contains the set $\mathscr{N} = \{A \in \mathscr{F}\colon P(A) = 0\}$. Then any martingale $(X_t)_{t \geq 0}$ with respect to $(\mathscr{F}_t)_{t \geq 0}$ admits a right-continuous version.

PROOF.   (a) Let S be a countable dense subset of $\mathbb{R}^+$, a and b be two real numbers such that $a < b$, and let $S_n = S \cap [0, n]$. For any finite subset I of S_n, it follows from Theorem 6.4.2 that the number $U_{ab}(I)$ of upcrossing of (a, b) by the process $(X_t)_{t \in I}$ satisfies

$$E(U_{ab}(I)) \leq \frac{1}{b-a} E[(X_n - a)^+],$$

and therefore, for any $\lambda > 0$,

$$P(U_{ab}(I) \geq \lambda) \leq \frac{1}{\lambda(b-a)} E[(X_n - a)^+]. \tag{1}$$

Let I_k, $k = 1, 2, \ldots$ be an increasing sequence of finite subsets of S_n such that $\bigcup_k I_k = S_n$. The set $\{\lim_k U_{ab}(I_k) = \infty\}$ is measurable, has probability 0 and contains the set $A_{n,ab} = \{\omega\colon \exists t \in [0, n)$ such that $\liminf_{s \to t, s > t, s \in S} X_s < a < b < \limsup_{s \to t, s > t, s \in S} X_s\}$.   [If $P(\lim_k U_{ab}(I_k) = \infty) = \alpha > 0$, then, since the $U_{ab}(I_k)$ increase with k, for any $\lambda > 0$ there exists k such that $P(U_{ab}(I_k) \geq \lambda) > \alpha/2$. For λ large, this is contradictory to inequality (1).] Since $\mathscr{F}$ is complete for the probability P, $A_{n,ab}$ is measurable and $P(A_{n,ab}) = 0$.

(b) Let

$$A = \{\omega\colon \exists t \in [0, \infty) \quad \text{such that} \quad \liminf_{s \to t, s > t, s \in S} X_s < \limsup_{s \to t, s > t, s \in S} X_s\},$$

The set A is contained in the set $(\bigcup_{n,a,b} A_{n,ab})$ where the union is taken for all integer values of n, and all rationals $a < b$; since $(\bigcup_{n,a,b} A_{n,ab}), = 0$, A is measurable and has probability 0. Define the process

$$Y_t = \begin{cases} \lim_{s \to t, s > t, s \in S} X_s, & \text{on the complement of } A, \\ 0, & \text{on } A. \end{cases}$$

Since $\mathscr{F}_t = \mathscr{F}_{t+}$ contains all the sets in $\mathscr{N} = \{A \in \mathscr{F}\colon P(A) = 0\}$, the random variable Y_t is $\mathscr{F}_t$-measurable.

In the equality $X_s = E[X_n \mid \mathscr{F}_s]$, $s \leq n$, let s decrease to t; X_s converges a.s. to Y_t, and $E[X_n \mid \mathscr{F}_s]$ converges to $E[X_n \mid \mathscr{F}_{t+}] = E[X_n \mid \mathscr{F}_t] = X_t$ a.s. (6.4.4). Therefore $(Y_t)_{t \geq 0}$ is a right-continuous version of the martingale $(X_t)_{t \geq 0}$ and is itself a martingale. $\square$

We now give a couple of results necessary to prove 9.6.8, which will be essential in the construction of stochastic integrals.

9.6.4 Lemma. Let $(X_t)_{t \geq 0}$ be a right-continuous submartingale with respect to a family $(\mathscr{F}_t)_{t \geq 0}$ of sub-σ-fields. Then for any $\lambda > 0$ we have

$$\lambda P\{\sup_{s \leq t} X_s > \lambda\} \leq \int_{\{\sup_{s \leq t} X_s > \lambda\}} X_t \, dP.$$

PROOF.   We first give the proof when the index set is finite. We consider the submartingale on the index set $t_1 < t_2 < \cdots < t_k$. Let T be the stopping time defined by

$$T = \begin{cases} \min\{t_i : X_{t_i} > \lambda\}, & \text{on the set } \{\max_i X_{t_i} > \lambda\} \\ t_k, & \text{on the set } \{\max_i X_{t_i} \leq \lambda\}. \end{cases}$$

We have

$$\lambda P(\max_i X_{t_i} > \lambda) \leq \sum_{i=1}^{k-1} E[X_{t_i} I_{\{T=t_i\}}] + E[X_{t_k} I_{\{T=t_k, \max_i X_{t_i} > \lambda\}}]$$

$$\leq \sum_{i=1}^{k-1} E[X_{t_k} I_{\{T=t_i\}}] + E[X_{t_k} I_{\{T=t_k, \max_i X_{t_i} > \lambda\}}]$$

$$= \int_{\{\max_i X_{t_i} > \lambda\}} X_{t_k} \, dP.$$

Consider an increasing sequence of finite set I_k such that $\bigcup_k I_k = \mathbb{Q}_t$, where $\mathbb{Q}_t = \{\text{all rationals smaller than } t\} \cup \{t\}$. Since the random variable X_t is integrable, and the sets $\{\max_{t_i \in I_k} X_{t_i} > \lambda\}$ increase to the set $\{\sup_{s \in \mathbb{Q}_t} X_s > \lambda\}$, we obtain

$$\lambda P\{\sup_{s \in \mathbb{Q}_t} X_s > \lambda\} \leq \int_{\{\sup_{s \in \mathbb{Q}_t} X_s > \lambda\}} X_t \, dP. \tag{1}$$

Since the submartingale $(X_t)_{t \geq 0}$ is right-continuous, we have

$$\sup_{s \leq t} X_s = \sup_{s \in \mathbb{Q}_t} X_s,$$

and

$$\lambda P\{\sup_{s \leq t} X_s > \lambda\} \leq \int_{\{\sup_{s \leq t} X_s > \lambda\}} X_t \, dP. \quad \square$$

9.6.5 Lemma. If $(X_t)_{t\geq 0}$ is a right-continuous martingale with respect to $(\mathscr{F}_t)_{t\geq 0}$, and each X_t is in L^2, then we have

$$\| \sup_{s\leq t} |X_s| \|_2 \leq 2\|X_t\|_2.$$

PROOF. Let $Y = \sup_{s\leq t} |X_s|$. The process $|X_t|$ is a submartingale, and therefore, by Lemma 9.6.4, we have

$$\lambda P(Y > \lambda) \leq \int_{\{Y>\lambda\}} |X_t|\, dP.$$

Since we do not know yet that $\|Y\|_2 < \infty$, we have to work with the random variables $Y_n = Y \wedge n$. The set $\{Y_n > \lambda\}$ is empty if $\lambda \geq n$, and coincides with $\{Y > \lambda\}$ if $\lambda < n$. Therefore Y_n satisfies

$$\lambda P(Y_n > \lambda) \leq \int_{\{Y_n>\lambda\}} |X_t|\, dP. \tag{1}$$

Let F_n be the distribution function of Y_n. Since $s^2 = \int_0^s 2\lambda\, d\lambda$, it follows that

$$E[Y_n^2] = \int_0^\infty s^2\, dF_n(s) = \int_0^\infty \int_0^s 2\lambda\, d\lambda\, dF_n(s)$$

$$= \int_0^\infty 2\lambda P(Y_n > \lambda)\, d\lambda \qquad \text{by Fubini's Theorem}$$

$$\leq 2 \int_0^\infty \int_\Omega |X_t| I_{\{Y_n>\lambda\}}\, dP\, d\lambda \qquad \text{by (1)}$$

$$= 2 \int_\Omega |X_t| \int_0^\infty I_{\{Y_n>\lambda\}}\, d\lambda\, dP$$

$$= 2E[|X_t|Y_n] \leq 2\|X_t\|_2\|Y_n\|_2,$$

and

$$\|Y_n\|_2 \leq 2\|X_t\|_2.$$

We finish by letting n approach ∞. $\square$

We now assume that the hypotheses of 9.6.3 are satisfied: $(\Omega, \mathscr{F}, P)$ is a probability space, and $(\mathscr{F}_t)_{t\geq 0}$ a nondecreasing *right-continuous* family of sub-σ-fields. The probability space is complete in the sense of 1.3.7, and each $\mathscr{F}_t$ contains the set $\mathscr{N} = \{A \in \mathscr{F}: P(A) = 0\}$.

9.6.6 Definition. We denote by $\mathcal{M}_a$ the vector space of all processes X $= (X_t)_{0 \le t \le a}$ such that $(X_t)_{0 \le t \le a}$ is a $(\mathcal{F}_t)_{0 \le t \le a}$ right-continuous martingale and X_a is in L^2.

If two elements X and Y of $\mathcal{M}_a$ are versions of each other, the set $\{\omega \colon \exists t \in [0, a]$ such that $X_t(\omega) \ne Y_t(\omega)\}$ is measurable and has probability 0 (see 9.1). In this case, we will not distinguish between the processes X and Y.
We consider on $\mathcal{M}_a$ the inner product $\langle X, Y \rangle = E(X_a Y_a)$.

9.6.7 Theorem.
(a) If $\langle X, X \rangle = 0$, the martingale X is indistinguishable from the process identically equal to 0. Therefore $\|X\|_{\mathcal{M}_a} = \langle X, X \rangle$ defines a norm on $\mathcal{M}_a$.
(b) If a sequence X_n, $n = 1, 2 \ldots$, converges to X in $\mathcal{M}_a$, then $\sup_{t \le a} |X_{n,t} - X_t| \to 0$ in L^2. (In particular, for any $0 \le t \le a$, $X_{n,t} \to X_t$ in L^2.)

PROOF. This follows from Lemma 9.6.5. □

9.6.8 Theorem. The space $\mathcal{M}_a$ is a Hilbert space for the inner product $\langle X, Y \rangle = E[X_a Y_a]$, and the subspace of continuous martingales is closed in $\mathcal{M}_a$.

PROOF. Let X_n be a Cauchy sequence in $\mathcal{M}_a$. The sequence of random variables $X_{n,a}$ is a Cauchy sequence in L^2; let Z_a be its L^2-limit, and $Z_t = E(Z_a \mid \mathcal{F}_t), 0 \le t \le a$. The martingale $(Z_t)_{0 \le t \le a}$ admits a right-continuous version $Y = (Y_t)_{0 \le t \le a}$ (9.6.3), and the X_n converge to Y in $\mathcal{M}_a$.
Assume now that the martingales X_n are continuous, and let X be their right-continuous limit in $\mathcal{M}_a$. We can assume that $\sum_n E[(X_{n,a} - X_a)^2] < \infty$ (if necessary use a subsequence). Applying successively 9.6.5 and Chebyshev's inequality, we have

$$\sum_n E[\sup_{t \le a} |X_{n,t} - X_t|^2] < \infty$$

and

$$\sum_n P(\sup_{t \le a} |X_{n,t} - X_t| \ge \varepsilon) < \infty,$$

for any $\varepsilon > 0$. By the first Borel-Cantelli lemma we get

$$P(\limsup_n \{\sup_{t \le a} |X_{n,t} - X_t| \ge \varepsilon\}) = 0,$$

and, taking a sequence $\varepsilon_k \to 0$, we see that

$$P(\{\sup_{t \le a} |X_{n,t} - X_t| \longmapsto 0\}) \le \sum_k P(\limsup_n \{\sup_{t \le a} |X_{n,t} - X_t| \ge \varepsilon_k\}) = 0.$$

For almost all ω, the right-continuous path $t \to X_t(\omega)$, $t \in [0, a]$, is a uniform limit of the continuous paths $t \to X_{n,t}(\omega)$ and the martingale X is indistinguishable on $[0, a]$ from a continuous martingale. $\square$

Problems

1. Let $(X_t)_{t \ge 0}$ be a process with independent increments such that $X_t \in L^1$ for all $t \ge 0$. Show that $Y_t = X_t - E(X_t)$ is a martingale.

2. (a) Let $(B_t)_{t \ge 0}$ be a Brownian motion and $\mathscr{B}_t = \sigma(B_s, s \le t)$. Show that, for any $u \in \mathbb{R}$, the process

$$Y_t^{(u)} = \exp iuB_t + \tfrac{1}{2}u^2 t$$

is a martingale for the family $(\mathscr{B}_t)_{t \ge 0}$.

   (b) Conversely, let $(X_t)_{t \ge 0}$ be a continuous stochastic process such that $X_0 = 0$ and for any $u \in \mathbb{R}$ the process

$$Y_t^{(u)} = \exp iuX_t + \tfrac{1}{2}u^2 t$$

is a martingale with respect to the family $(\mathscr{F}_t = \sigma(X_s, s \le t))_{t \ge 0}$. Show that $(X_t)_{t \ge 0}$ is a Brownian motion.

3. Let $(\Omega, \mathscr{F}, P)$ be a probability space, $(\mathscr{F}_t)_{t \ge 0}$ a nondecreasing right-continuous family of sub-σ-fields and $(X_t)_{t \ge 0}$ a right-continuous martingale.

   (a) Show that, if T is a stopping time bounded by t_0, then X_T is $\mathscr{F}_T$-measurable and $X_T = E[X_{t_0} \mid \mathscr{F}_T]$ a.s.

   (b) Show that, if T is a bounded stopping time, then the process $Y_t = X_{t \wedge T}$, $t \ge 0$, is a martingale.

   (Hint: use the stopping times $T_n = k/n$ if $(k-1)/n \le T < k/n, k = 1, 2, \ldots$, Corollary 6.4.4 and Problem 5 of 9.5.)

4. Let $(\Omega, \mathscr{F}, P)$ be a probability space, $(\mathscr{F}_t)_{t \ge 0}$ a nondecreasing right-continuous family of sub-σ-fields and $(X_t)_{t \ge 0}$ a right-continuous martingale. We assume that the random variables X_t, $t \ge 0$, are uniformly integrable.

   (a) Show that $X_\infty = \lim_{t \to \infty} X_t$ exists a.s. and that $(X_t)_{0 \le t \le \infty}$ is a martingale.

   (b) Show that, for any stopping time (finite or not), X_T is $\mathscr{F}_T$-measurable, integrable and $X_T = E[X_\infty \mid \mathscr{F}_T]$ a.s.

   (Hint: the inequalities and methods of 6.4 extend easily to right-continuous martingales.)

9.7   Itô Integrals

One possible mathematical model for the price of shares in the stock market is the Brownian motion: the price of one share of a particular stock is assumed to equal $a + \lambda B_t(\omega)$ at time t. At times $0 = t_0 < t_1 < \cdots < t_{n-1} < t_n = t$, an investor decides the amount $f(t_i, \omega)$ of shares of the stock to have during time (t_i, t_{i+1}). Then the gain in the time interval $(0, t)$ is

$$\lambda \sum_{i=0}^{n-1} f(t_i, \omega)(B_{t_{i+1}}(\omega) - B_{t_i}(\omega)).$$

At each time t_i, the investor knows only what has happened up to time t_i. Therefore each random variable $f(t_i, .)$ should be $\mathscr{F}_{t_i}$-measurable. What happens when the time intervals between consecutive decisions become smaller? Does the above sum have a limit? That is, can we define integrals of the form $\int f(t, \omega) \, dB_t(\omega)$? If the paths of the Brownian motion had a.s. finite variations on finite intervals, we could use 2.3.3 and 1.4.4; for almost all ω, the function $t \to B_t(\omega)$ would be a difference of two continuous, nondecreasing functions and $\int f(t, \omega) \, dB_t(\omega)$ could be defined, for each ω, as a Lebesgue-Stieltjes integral. Unfortunately, this is not the case and we have to use martingale theory to get around the difficulty. The symbol $\int_0^t f(s, \omega) \, dB_s$, $t \le a$ will no longer be defined path by path, but as a limit in $\mathscr{M}_a$.

Let $(B_t)_{t \ge 0}$ be a Brownian motion on a probability space $(\Omega, \mathscr{F}, P)$, and let $\mathscr{B}$ denote the smallest complete σ-field generated by all the random variables B_s, $s \ge 0$. We will work with the σ-fields $\mathscr{B}_t'$, where $\mathscr{B}_t'$ is the σ-field generated by $\mathscr{B}_t$ and $\mathscr{N} = \{A \in \mathscr{B}: P(A) = 0\}$. The family $(\mathscr{B}_t')_{t \ge 0}$ is right-continuous (9.5 Problem 2); therefore the probability space $(\Omega, \mathscr{B}, P)$ and the family $(\mathscr{B}_t')_{t \ge 0}$ satisfy the hypotheses of Theorems 9.6.3 and 9.6.8.

Which processes $f(t, \omega)$ do we want to integrate? If we want $\int f(t, \omega) \, dB_t$ to be a random variable, the process f should be measurable in both variables. We will also assume, as in the example of the investor, that, for each t, the random variable $\omega \to f(t, \omega)$ does not depend on the future. Therefore we consider only processes $f(t, \omega)$ satisfying the following conditions:

(a)   the process $f(t, \omega)$ is measurable, and
(b)   the process $f(t, \omega)$ is *non-anticipating* (or *adapted*) — for each t the random variable $\omega \to f(t, \omega)$ is $\mathscr{B}_t'$-measurable.

For simple processes $f(t, \omega)$ we know what the integral should be.

(a)   If $f(t, \omega) = I_{(u,v]}(t)$, then $\int_0^a f(t, \omega) \, dB_t = B_{v \wedge a} - B_{u \wedge a}$.
(b)   If $f(t, \omega) = I_{(u,v]}(t)\phi(\omega)$, then $\int_0^a f(t, \omega) \, dB_t = \phi(\omega)(B_{v \wedge a} - B_{u \wedge a})$.

For the moment we choose $a > 0$ and we restrict ourselves to processes that vanish outside of $[0, a] \times \Omega$.

9.7.1 Definition.

(a) We denote by $\mathscr{S}_a$ the set of processes $f(t, \omega) = I_{[0]}(t)\phi_0(\omega) + \sum_{i=1}^{n-1} I_{(t_i, t_{i+1}]}(t)\phi_i(\omega)$ such that

    (i) $0 = t_1 < t_2, \ldots, t_n = a$ is a finite partition of $[0, a]$,
    (ii) ϕ_0 is $\mathscr{B}_0'$-measurable,
    (iii) for $i \geq 1$, ϕ_i is $\mathscr{B}_{t_i}'$-measurable, and
    (iv) each ϕ_i is in L^2.

For such a process we define, for $0 \leq t \leq a$,

$$Y_t = \int_0^t f(s, \omega)\, dB_s(\omega) = \sum_{i=1}^{n-1} \phi_i(\omega)(B_{t_{i+1} \wedge t}(\omega) - B_{t_i \wedge t}(\omega)).$$

(b) We denote by $\mathscr{M}_a$ the set of measurable, nonanticipating processes $g(t, \omega)$ which vanish outside the set $[0, a] \times \Omega$, and such that $E[\int_0^a g^2(s, \omega)\, ds] < \infty$.

For processes in $\mathscr{S}_a$ we have the following straightforward lemma.

9.7.2 Lemma. If f and g are two processes in $\mathscr{S}_a$, then the processes $Y_t = \int_0^t f(s)\, dB_s$ and $Z_t = \int_0^t g(s)\, dB_s$, $0 \leq t \leq a$, satisfy the following properties.

(a) $(Y_t)_{0 \leq t \leq a}$ and $(Z_t)_{0 \leq t \leq a}$ are continuous martingales for the family of σ-fields $(\mathscr{B}_t')_{0 \leq t \leq a}$.

(b) The martingales Y and Z are in $\mathscr{M}_a$ and, for $t \leq a$, we have $E[Y_t Z_t] = E[\int_0^t f(s)g(s)\, ds]$.

(c) In particular $E[Y_t^2] = E[\int_0^t f^2(s)\, ds]$, $t \leq a$, and, in $\mathscr{M}_a$,

$$\langle Y, Z \rangle = E\left[\int_0^a f(s)g(s)\, ds\right].$$

(d) If $f \in \mathscr{S}_a$ and $0 \leq t \leq a$ we have $\int_0^t f(s)\, dB_s = \int_0^a f(s)I_{[0,t]}(s)\, dB_s$.

PROOF. If the ϕ_i's are in L^2 and are $\mathscr{B}_{t_i}'$ measurable, then we have

$$E[\phi_i(B_{t_{i+1}} - B_{t_i}) \mid \mathscr{B}_{t_i}'] = \phi_i E[(B_{t_{i+1}} - B_{t_i}) \mid \mathscr{B}_{t_i}'] = 0,$$

$$E[\phi_i^2(B_{t_{i+1}} - B_{t_i})^2] = E[\phi_i^2 E[(B_{t_{i+1}} - B_{t_i})^2 \mid \mathscr{B}_{t_i}']] = E[\phi_i^2(t_{i+1} - t_i)]$$

and for $i < j$,

$$E[\phi_i(B_{t_{i+1}} - B_{t_i})\phi_j(B_{t_{j+1}} - B_{t_j})]$$
$$= E[\phi_i(B_{t_{i+1}} - B_{t_i})\phi_j E[(B_{t_{j+1}} - B_{t_j}) \mid \mathscr{B}_{t_j}']] = 0. \qquad \square$$

9.7.3   Theorem.

(a)   The mapping $f \to (Y_t = \int_0^t f(s)\,dB_s)_{0 \le t \le a}$ is an isometry from $\mathscr{S}_a$ to $\mathscr{M}_a$, where on $\mathscr{S}_a$ we use the norm of the Hilbert space $L^2([0, a] \times \Omega, \mathscr{B}([0, a]) \times \mathscr{F}, ds \times dP)$; the mapping can therefore be extended to the closure of $\mathscr{S}_a$.

(b)   The closure of $\mathscr{S}_a$ in $L^2([0, a] \times \Omega, \mathscr{B}([0, a]) \times \mathscr{F}, ds \times dP)$ contains $\mathscr{A}_a$.

(c)   If $f \in \mathscr{A}_a$ the process $Y_t = \int_0^t f(s)\,dB_s$, thus defined as a limit in $\mathscr{M}_a$, is a continuous martingale, and, for each $t \le a$, $E[Y_t^2] = E[\int_0^t f^2(s)\,ds]$.

(d)   If f and g are two processes in $\mathscr{A}_a$, the processes $Y_t = \int_0^t f(s)\,dB_s$ and $Z_t = \int_0^t g(s)\,dB_s$, $0 \le t \le a$, satisfy $E[Y_t Z_t] = E[\int_0^t f(s)g(s)\,ds]$.

(e)   If $f \in \mathscr{A}_a$ and $t \le a$ we have $\int_0^t f(s)\,dB_s = \int_0^a f(s)I_{[0,t]}(s)\,dB_s$ a.s.

PROOF.   (a) is a consequence of 9.7.2, and (c) a consequence of 9.6.8. Properties (d) and (e) extend from $\mathscr{S}_a$ to its closure, so we just have to prove (b). We will follow the proof in Doob (1953).

1.   Let $f(t)$ be a bounded Lebesgue-measurable function of t, which vanishes outside a finite interval. Then, by 2.4.14, for every $\varepsilon > 0$ there exists a continuous, bounded function f_ε which vanishes outside a finite interval and such that

$$\int_{-\infty}^{\infty} |f(t) - f_\varepsilon(t)|^2 \, dt \le \varepsilon^2.$$

[By 2.4.14 there exists a continuous function $\tilde{f}_\varepsilon$ such that $\int_{-\infty}^{\infty} |f(t) - \tilde{f}_\varepsilon(t)|^2 \, dt \le \varepsilon^2$. Assume that the function f vanishes outside a finite interval $[a, b]$. Let k be the continuous function which is 1 in $[a, b]$, 0 outside of $[a-1, b+1]$ and linear elsewhere. The function $f_\varepsilon = k\tilde{f}_\varepsilon$ vanishes outside a finite interval and $\int_{-\infty}^{\infty} |f(t) - f_\varepsilon(t)|^2 \, dt \le \int_{-\infty}^{\infty} |f(t) - \tilde{f}_\varepsilon(t)|^2 \, dt \le \varepsilon^2$.]

Using Minkowski's inequality and the dominated convergence theorem, we obtain

$$\limsup_{h \to 0} \left[\int_{-\infty}^{\infty} |f(t + h) - f(t)|^2 \, dt \right]^{1/2}$$

$$\le \limsup_{h \to 0} \left[\int_{-\infty}^{\infty} |f_\varepsilon(t + h) - f_\varepsilon(t)|^2 \, dt \right]^{1/2} + 2\varepsilon = 2\varepsilon$$

so that

$$\lim_{h \to 0} \int_{-\infty}^{\infty} |f(t + h) - f(t)|^2 \, dt = 0.$$

2. Assume now that $g(t, \omega)$ is a bounded, measurable, nonanticipating process that vanishes for $t \notin [0, a]$; we want to show that g is in the closure of $\mathscr{S}_a$.

We partition $\mathbb{R}$ into intervals $(j/2^n, (j+1)/2^n]$ and consider the process

$$g_{n,s}(t, \omega) = g(\alpha_n(t - s) + s, \omega) \quad \text{where } \alpha_n(u) = j/2^n \text{ on } (j/2^n, (j+1)/2^n].$$

For fixed n and s the process $g_{n,s}(t, \omega)$ is equal to $g(s + j/2^n, \omega)$ on the interval $(s + j/2^n, s + (j+1)/2^n]$. Therefore the restriction of $g_{n,s}(t, \omega)$ to $[0, a] \times \Omega$ belongs to $\mathscr{S}_a$. We now show that there exists an s and a sequence n_j such that

$$\lim_j E\left[\int_0^a |g_{n_j,s}(t, \omega) - g(t, \omega)|^2 \, dt\right]$$
$$\leq \lim_j E\left[\int_{-\infty}^\infty |g_{n_j,s}(t, \omega) - g(t, \omega)|^2 \, dt\right] \to 0.$$

This will show that the process g is in the closure of $\mathscr{S}_a$.

Part 1 assures that for each ω

$$\lim_{h \to 0} \int_{-\infty}^\infty |g(s + h, \omega) - g(s, \omega)|^2 \, ds = 0.$$

Since Lebesgue measure is translation-invariant we have, for each fixed t and ω,

$$\lim_{h \to 0} \int_{-\infty}^\infty |g(s + t + h, \omega) - g(s + t, \omega)|^2 \, ds = 0.$$

We can replace h by $\alpha_n(t) - t$, which approaches 0 as $n \to \infty$ and obtain

$$\lim_{n \to \infty} \int_{-\infty}^\infty |g(\alpha_n(t) + s, \omega) - g(t + s, \omega)|^2 \, ds = 0.$$

Since all the processes considered are bounded and vanish outside a fixed finite interval, we have

$$\lim_{n \to \infty} \int_{-\infty}^\infty E\left[\int_{-\infty}^\infty |g(\alpha_n(t) + s, \omega) - g(t + s, \omega)|^2 \, dt\right] ds$$
$$= \lim_{n \to \infty} E\left[\int_{-\infty}^\infty \int_{-\infty}^\infty |g(\alpha_n(t) + s, \omega) - g(t + s, \omega)|^2 \, ds \, dt\right] = 0.$$

Therefore there exists a subsequence n_j such that

$$\lim_{j \to \infty} E\left[\int_{-\infty}^\infty |g(\alpha_{n_j}(t) + s, \omega) - g(t + s, \omega)|^2 \, dt\right] = 0 \qquad \text{a.s. in } s.$$

We have now shown that there exists a subsequence n_j and at least one s such that

$$\lim_{j\to\infty} E\left[\int_{-\infty}^{\infty} |g(\alpha_{n_j}(t-s)+s,\omega)-g(t,\omega)|^2\,dt\right]$$

$$= \lim_{j\to\infty} E\left[\int_{-\infty}^{\infty} |g(\alpha_{n_j}(t)+s,\omega)-g(t+s,\omega)|^2\,dt\right] = 0,$$

and g is in the closure of $\mathscr{S}_a$.

3. If the process g on $[0,a]$ is measurable, nonanticipating and satisfies

$$E\left[\int_0^a g^2(t)\,dt\right] < \infty,$$

we define the processes

$$g_n = \begin{cases} g, & \text{if } |g| \le n, \\ 0, & \text{otherwise.} \end{cases}$$

According to part 2, each g_n is in the closure of $\mathscr{S}_a$. The dominated convergence theorem assures that

$$E\left[\int_0^a |g(t)-g_n(t)|^2\,dt\right] \to 0,$$

and g is in the closure of $\mathscr{S}_a$. $\square$

The stochastic integrals $Y_t = \int_0^t g(s)\,dB_s$ that we have thus defined are limits in L^2, so the random variables Y_t are only determined a.s. Theorem 9.7.3 assures the existence of continuous versions of $(Y_t)_{0\le t\le a}$, and, when we talk about stochastic integrals, we always assume that we are working with a continuous version; it does not matter which, since two continuous versions are indistinguishable.

We now generalize the definition of stochastic integrals to the interval $[0,\infty)$.

9.7.4 Theorem. Let $\mathscr{A}$ be the set of measurable, nonanticipating processes $g(t,\omega)$ on $[0,\infty) \times \Omega$ such that $E\left[\int_0^a g(t)^2\,dt\right] < \infty$, for any $a \ge 0$.

(a) There exists a continuous martingale $(Y_t)_{t\ge0}$ such that $Y_t = \int_0^t g(s)\,dB_s$ a.s. for any $t \ge 0$.

(b) If f and g are in $\mathscr{A}$, then $\int_0^t (\alpha f + \beta g)(s)\,dB_s = \alpha \int_0^t f(s)\,dB_s + \beta \int_0^t g(s)\,dB_s$ a.s. (α and β are constants).

(c)   If $f \in \mathcal{A}$, then $\int_0^t f(s)\,dB_s = \int_0^a f(s)I_{[0,t]}(s)\,dB_s$ a.s. for $t \leq a$.

(d)   If f and g are in $\mathcal{A}$, then for any t,

$$E\left[\left(\int_0^t f(s)\,dB_s\right)\left(\int_0^t g(s)\,dB_s\right)\right] = E\left[\int_0^t f(s)g(s)\,ds\right].$$

PROOF.   Property (a): For $n \geq 0$ and $0 \leq t \leq n$, we define $Y_{(n,t)} = \int_0^t f(s)I_{[0,n]}(s)\,dB_s$ using the results in 9.7.3. This is possible since $fI[0,n]$ is in $\mathcal{A}_n$. Property (e) of Theorem 9.7.3 assures that, on $[0,n]$, the processes $Y_{(n+1,t)}$ and $Y_{(n,t)}$ are indistinguishable, and we can define the stochastic integrals $Y_t = \int_0^t f(s)\,dB_s$ for any $t \geq 0$ by taking $Y_t = Y_{(n,t)}$ on $(n-1, n]$. The process Y_t is then a martingale that is continuous except possibly for ω in the set $\bigcup_n\{Y_{(n,n-1)} \neq Y_{(n-1,n-1)}\}$, which has probability 0. We can therefore change the values of the Y_t's on this set (for example take $Y_t = 0$) to get a continuous version of the stochastic integrals.

The other properties are true for $g \in \mathcal{S}_a$; they extend to $g \in \mathcal{A}_a$, and $g \in \mathcal{A}$. □

Problems

1.   Let T be a stopping time for the family of σ-fields $(\mathcal{B}'_t)_{t \geq 0}$, and denote by $[0, T]$ the stochastic interval $\{(t, \omega): 0 \leq t \leq T(\omega)\}$.
   (a)   Show that the left-continuous process $I_{[0,T]}$ is measurable and non-anticipating. Therefore if $f \in \mathcal{A}$ we can consider the two processes $Y_t = \int_0^t f\,dB_s$, and $Z_t = \int_0^t fI_{[0,T]}\,dB_s$.
   (b)   Show that, if $a \geq 0$ and $f \in \mathcal{S}_a$, then the processes $(Y_{t \wedge T})_{0 \leq t \leq a}$ and $(Z_t)_{0 \leq t \leq a}$ are indistinguishable. (Show it first for a stopping time that takes only a finite number of values.)
   (c)   Extend this property to the processes $f \in \mathcal{A}_a$ and then to the processes $f \in \mathcal{A}$.

2.   Let $(\Omega, \mathcal{F}, P)$ be a probability space, and $(\mathcal{F}_t)_{t \geq 0}$ a nondecreasing, right-continuous family of sub-σ-fields of $\mathcal{F}$. A process $f(s, \omega)$ defined on $\mathbb{R}^+ \times \Omega$ is *progressively measurable* iff, for any $t \geq 0$, the restriction of $f(s, \omega)$ to $[0, t] \times \Omega$ is $\mathcal{B}([0, t]) \times \mathcal{F}_t$ measurable.
   (a)   Show that a progressively measurable process is measurable and non-anticipating.
   (b)   Show that a right-continuous and nonanticipating process is progressively measurable.

3.   Let $f(t, \omega)$ be a progressively measurable process such that, for each $t \geq 0$, $\int_0^t f^2(s, \omega)\,ds < \infty$ a.s. We want to extend the notion of stochastic integral to the process f.

(a) For each integer $n > 0$, let $T_n(\omega) = \inf\{t: \int_0^t f^2(s, \omega)\, ds \geq n\}$.
    $[T_n = \infty$ if the set $\{t: \int_0^t f^2(s, \omega)\, ds \geq n\}$ is empty.] Show that the
    T_n are stopping times for the family of σ-fields $(\mathscr{B}_t')_{t\geq 0}$, and that
    $\lim_n T_n = \infty$ a.s.

(b) We denote by $[0, T_n]$ the stochastic interval $\{(t, \omega): 0 \leq t \leq T_n(\omega)\}$.
    Show that $fI_{[0,T_n]}$ is in $\mathscr{A}$.

(c) Let $Y_{(n,t)} = \int_0^t fI_{[0,T_n]}\, dB_s$ as defined in Theorem 9.7.4. Show that
    the processes $Y_{(n,t\wedge T_n)}$ and $Y_{(n+1,t\wedge T_n)}$ are indistinguishable.

(d) Show that there exists a continuous process $(Y_t)_{t\geq 0}$ such that, for
    each n, the processes $(Y_{t\wedge T_n})_{t\geq 0}$ and $(Y_{(n,t\wedge T_n)})_{t\geq 0}$ are indistinguish-
    able.

    We now define $\int_0^t f(s)\, dB_s$ by $\int_0^t f(s)\, dB_s = Y_t$. [The process
    $(Y_t)_{t\geq 0}$ is not necessarily a martingale, but it is a local martingale
    in the sense that there exists a nondecreasing sequence of stopping
    times $T_1 \leq T_2 \leq \cdots \leq T_n \leq \ldots$ such that $\lim_n T_n = \infty$ and each
    $(Y_{t\wedge T_n})_{t\geq 0}$ is a martingale.]

(e) Let T be a stopping time such that $fI_{[0,T]} \in \mathscr{A}$. Let $Z_t = \int_0^t$
    $fI_{[0,T]}\, dB_s$, as defined in 9.7.4. Use Problem 1 to show that the
    processes $(Z_t)_{t\geq 0}$ and $(Y_{t\wedge T})_{t\geq 0}$ are indistinguishable.

(f) Let S_n, $n = 0, 1, \ldots$, be a nondecreasing sequence of stopping times
    such that $\lim_n S_n = \infty$ and each process $fI_{[0,S_n]}$ is in $\mathscr{A}$. Just as
    in (d), we could use the stopping times S_n to define the integrals
    $\int_0^t f\, dB_s$. Show that the integrals do not depend on the sequence of
    stopping times used in the construction.

4. Generalize Problem 3 to measurable, nonanticipating processes such that,
   for each $t \geq 0$, $\int_0^t f^2(s, \omega)\, ds < \infty$ a.s. [Use an argument similar to the
   proof of 9.7.3 to show that the integral $\int_0^t f^2(s, \omega)\, ds$ is $\mathscr{F}_t$-measurable.]

9.8 ITÔ'S DIFFERENTIATION FORMULA

Assume that a function $f(t)$, defined on $\mathbb{R}$, can be expanded in a Taylor
series. We know that $f(t) = \int_0^t f'(s)\, ds$. We can heuristically justify this statement; if $0 = t_1 < t_2 < \cdots < t_n = t$ is a partition of $[0, t]$, then

$$f(t) - f(0) = \sum_i [f(t_{i+1}) - f(t_i)]$$

$$= \sum_i \left[f'(t_i)(t_{i+1} - t_i) + \sum_{n=2}^{\infty} \frac{f^{(n)}(t_i)}{n!}(t_{i+1} - t_i)^n \right].$$

The first term $\sum_i f'(t_i)(t_{i+1} - t_i)$ converges to $\int_0^t f'(s)\, ds$, and, for $n \geq 2$,
each term $\sum_i (f^{(n)}(t_i)/n!)(t_{i+1} - t_i)^n$ should go to 0, since $\sum_i |t_{i+1} - t_i|^n$
$\leq \max_i |t_{i+1} - t_i|^{n-1} \sum_i |t_{i+1} - t_i| \to 0$ when the partition of $[0, t]$ gets finer.

Therefore, if everything goes well, the sum over n of those terms also goes to 0. This heuristic reasoning would still apply if we considered a continuous function $G(s)$ having finite variation on any finite interval:

$$f(G(t)) - f(G(0)) = \sum_i [f(G(t_{i+1})) - f(G(t_i))]$$

$$= \sum_i \left[f'(G(t_i))(G(t_{i+1}) - G(t_i)) + \sum_{n=2}^{\infty} \frac{f^{(n)}(G(t_i))}{n!} (G(t_{i+1}) - G(t_i))^n \right].$$

The first term converges to the Lebesgue-Stieltjes integral $\int_0^t f'(G(s)) \, dG(s)$, and, for $n \geq 2$, $\sum_i f^{(n)}(G(t_i))(G(t_{i+1}) - G(t_i))^n$ should go to 0, since $\sum_i |G(t_{i+1}) - G(t_i)|^n \leq \max_i |G(t_{i+1}) - G(t_i)|^{n-1} \sum_i |G(t_{i+1}) - G(t_i)|$, and $\sum_i |G(t_{i+1}) - G(t_i)|$ remains bounded when the partition becomes finer (G is continuous and has finite variation on $[0, t]$).

What happens when we replace $G(s)$ by the paths B_s of the Brownian motion? The first term $\sum_i f'(B_{t_i})(B(t_{i+1}) - B(t_i))$ converges to $\int_0^t f'(B(s)) \, dB(s)$. The term $\sum_i f''(B_{t_i})/2(B(t_{i+1}) - B(t_i))^2$ no longer converges to 0, since the variation of the paths is a.s. infinite. The sum $\sum_i (B(t_{i+1}) - B(t_i))^2$ converges in L^2 to t, so $\sum_i f''(B_{t_i})/2(B(t_{i+1}) - B(t_i))^2$ probably converges to $\frac{1}{2} \int_0^t f''(B_s) \, ds$. For $n \geq 3$, $\sum_i f^{(n)}(B_{t_i})(B(t_{i+1}) - B(t_i))^n$ should go to 0, since $\sum_i |B(t_{i+1}) - B(t_i)|^n \leq \max_i |B(t_{i+1}) - B(t_i)|^{n-2} \sum_i (B(t_{i+1}) - B(t_i))^2$. Now that we have intuitively derived Itô's formula, we are ready for the rigorous proof.

9.8.1 (Itô's Differentiation Formula). Let f be a continuous function on $\mathbb{R}$. We assume that the first and second derivatives f' and f'' exist and are continuous. Then, for any $t \geq 0$,

$$f(B_t) - f(B_0) = \int_0^t f'(B_s) \, dB_s + \frac{1}{2} \int_0^t f''(B_s) \, ds \qquad \text{a.s.}$$

[Since the processes $Y_t = f(B_t) - f(B_0)$ and $X_t = \int_0^t f'(B_s) \, dB_s + \frac{1}{2} \int_0^t f''(B_s) \, ds$ are continuous, they will be indistinguishable.]

PROOF.   1.   The first step is to make sure that the integrals used in the statement of Theorem 9.8.1 exist. The process $f'(B_s(\omega))$ is nonanticipating and continuous; therefore it is progressively measurable (Problem 2 of 9.7). For fixed ω, the function $f'(B_s(\omega))$ is continuous in s, therefore, on an interval $[0, t]$, it is bounded by a constant $C(\omega, t)$, and $\int_0^t f'(B_s(\omega))^2 \, ds$ is finite. Consequently, we can, by Problem 3 of 9.7, define a continuous version of the stochastic integrals $\int_0^t f'(B_s)\,dB_s$. A similar (in fact easier) argument shows that the Lebesgue-Stieljes integrals $\int_0^t f''(B_s)\,ds$ exist.

2. Let $t \geq 0$ and $t_0 = 0 < t_1 < \cdots < t_n = t$ be a partition $\mathcal{P}$ of $[0, t]$. We denote by $\|\mathcal{P}\|$ the quantity $\max_i |t_{i+1} - t_i|$. Since the second derivative f'' is continuous, there exists a value $\alpha_i(\omega)$ in between $B_{t_i}(\omega)$ and $B_{t_{i+1}}(\omega)$ such that

$$f(B_{t_{i+1}}) - f(B_{t_i}) = f'(B_{t_i})(B_{t_{i+1}} - B_{t_i}) + \tfrac{1}{2}f''(\alpha_i)(B_{t_{i+1}} - B_{t_i})^2.$$

There exists $S_i(\omega) \in [t_i, t_{i+1}]$ such that $B_{S_i}(\omega) = \alpha_i(\omega)$ [this is due to the continuity of the function $s \to B_s(\omega)$; the random variable S_i is not necessarily a stopping time], and we have

$$f(B_t) - f(B_0) = \sum_i [f(B_{t_{i+1}}) - f(B_{t_i})]$$

$$= \sum_i f'(B_{t_i})(B_{t_{i+1}} - B_{t_i}) + \sum_i \tfrac{1}{2}f''(B_{S_i})(B_{t_{i+1}} - B_{t_i})^2.$$

We study separately the limit of each sum as the partition gets finer.

3. We want to show that $Z_t = \sum_i f'(B_{t_i})(B_{t_{i+1}} - B_{t_i})$ converges to $U_t = \int_0^t f'(B_s)\,dB_s$. The process $\sum_i f'(B_{t_i})I_{(t_i,t_{i+1}]}(s)$ converges a.s. to the process $f'(B_s)I_{(0,t]}(s)$, but, since $\sup_{\omega,s\leq t}|f'(B_s)|$ is not necessarily finite, we are not sure that

$$E\left[\int_0^t \left(\sum_i f'(B_{t_i})I_{(t_i,t_{i+1}]}(s) - f'(B_s)I_{(0,t]}(s)\right)^2 ds\right] \to 0.$$

Consider the times T_m, $m \geq 1$, defined as follows:

$$T_m = \begin{cases} \inf\{s:\ s > 0, |B_s| \geq m\}, & \text{if} \quad |B_s| \geq m \quad \text{for some} \quad s > 0 \\ \infty, & \text{if} \quad \{s:\ s > 0, |B_s| \geq m\} = \emptyset. \end{cases}$$

The set $\{T_m \leq t\} = \{\sup_{r\in\mathbb{Q},r\leq t}|B_r| \geq m\}$ (9.5.5) is in $\mathcal{B}_t'$, and the T_m form a nondecreasing sequence of stopping times that converges to ∞. For each m, the function $f'(x)$ is bounded on $[-m, m]$, therefore

$$E\left[\int_0^t I_{[0,T_m]}\left(\sum_i f'(B_{t_i})I_{(t_i,t_{i+1}]}(s) - f'(B_s)I_{(0,t]}(s)\right)^2 ds\right] \to 0,$$

when $\|\mathcal{P}\| \to 0$; therefore $Z_{t\wedge T_m} \to U_{t\wedge T_m}$ in L^2 and in probability [we used Problem 3(e) of 9.7]. The terms $Z_t - Z_{t\wedge T_m}$ and $U_t - U_{t\wedge T_m}$ are zero except on the set $\{T_m < t\}$. By first choosing m such that $P(T_m < t) < \varepsilon$, and then $\|\mathscr{P}\|$ small enough to have $P(|Z_{t \wedge T_m} - U_{t \wedge T_m}| > \varepsilon) < \varepsilon$, we obtain

$$P(|Z_t - U_t| > \varepsilon) \le P(|Z_{t \wedge T_m} - U_{t \wedge T_m}| > \varepsilon) + P(T_m < t) \le 2\varepsilon.$$

Therefore Z_t converges in probability to U_t.

4. We now study the term $\sum_i f''(B_{S_i})(B_{t_{i+1}} - B_{t_i})^2$. We want to show that if the sequence of partition is well chosen, it converges to $\int_0^t f''(B_s)\,ds$ in probability.

For partitions $\mathscr{P}$ of $[0, t]$, let us consider the random variables

$$X(\mathscr{P}) = \sum_i (f''(B_{S_i}) - f''(B_{t_i}))(B_{t_{i+1}} - B_{t_i})^2,$$

$$Y(\mathscr{P}) = \sum_i f''(B_{t_i})[(B_{t_{i+1}} - B_{t_i})^2 - (t_{i+1} - t_i)].$$

Since $\sum_i f''(B_{t_i})(t_{i+1} - t_i)$ converges for each ω to $\int_0^t f''(B_s)\,ds$, it is enough to show that $\sum_i f''(B_{S_i})(B_{t_{i+1}} - B_{t_i})^2 - \sum_i f''(B_{t_i})(t_{i+1} - t_i) = X(\mathscr{P}) + Y(\mathscr{P})$ converges to 0 in probability.

We first study the term $X(\mathscr{P})$.

$$|X(\mathscr{P})| \le \max_i |f''(B_{S_i}) - f''(B_{t_i})| \sum_i (B_{t_{i+1}} - B_{t_i})^2.$$

According to 9.3.4, we can choose a sequence of partitions $\mathscr{P}_n$, $n \ge 1$, of $[0, t]$ such that $\|\mathscr{P}_n\| \to 0$, and $\sum_i (B_{t_{i+1}} - B_{t_i})^2$ converges a.s. to t (L^2 convergence implies the existence of a subsequence which converges a.s.). For each fixed ω, the continuous function $s \to f''(B_s)$ is uniformly continuous on $[0, t]$, and $\max_i |f''(B_{S_i}) - f''(B_{t_i})| \to 0$ when $\|\mathscr{P}_n\|$ approaches 0. Therefore $X(\mathscr{P}_n) \to 0$ a.s. and in probability.

We now study the term $Y(\mathscr{P})$. Let T_m, $m \ge 1$, be the stopping times defined in part 3 of the proof. For each m, we have

$$|\sup_s f''(B_{s \wedge T_m})| \le \sup_{-m \le x \le m} |f''(x)| = K_m.$$

We consider the random variables

$$Y(\mathscr{P}, T_m) = \sum_i f''(B_{t_i \wedge T_m})[(B_{t_{i+1} \wedge T_m} - B_{t_i \wedge T_m})^2 - (t_{i+1} \wedge T_m - t_i \wedge T_m)],$$

and denote by V_i the random variable

$$V_i = f''(B_{t_i \wedge T_m})[(B_{t_{i+1} \wedge T_m} - B_{t_i \wedge T_m})^2 - (t_{i+1} \wedge T_m - t_i \wedge T_m)].$$

We have

$$E[Y(\mathcal{P}, T_m)^2] = E\left[\left(\sum_i V_i\right)^2\right] = E\left[\sum_i V_i^2\right] + 2\sum_{i<j} E[V_i V_j].$$

Since the processes $B_{s \wedge T_m}$ and $U_s = B_{s \wedge T_m}^2 - s \wedge T_m$, $0 \le s \le t$, are martingales for the family of σ-fields $\mathcal{B}'_{s \wedge T_m}$ (apply Problem 3 of 9.6 to the martingales B_s, $B_s^2 - s$ and the bounded stopping time $T_m \wedge t$), we have

$$E[(B_{t_{j+1} \wedge T_m} - B_{t_j \wedge T_m})^2 \mid \mathcal{B}'_{t_j \wedge T_m}]$$

$$= E[B_{t_{j+1} \wedge T_m}^2 - 2B_{t_{j+1} \wedge T_m} B_{t_j \wedge T_m} + B_{t_j \wedge T_m}^2 \mid \mathcal{B}'_{t_j \wedge T_m}]$$

$$= E[B_{t_{j+1} \wedge T_m}^2 - B_{t_j \wedge T_m}^2 \mid \mathcal{B}'_{t_j \wedge T_m}]$$

$$= E[t_{j+1} \wedge T_m - t_j \wedge T_m \mid \mathcal{B}'_{t_j \wedge T_m}].$$

Therefore, for $i < j$,

$$E[V_i V_j] = E[V_i E[V_j \mid \mathcal{B}'_{t_j \wedge T_m}]] = 0,$$

and

$$E[Y(\mathcal{P}, T_m)^2] = E\left[\sum_i V_i^2\right]$$

$$\le K_m^2 E \sum_i [(B_{t_{i+1} \wedge T_m} - B_{t_i \wedge T_m})^2 - (t_{i+1} \wedge T_m - t_i \wedge T_m)]^2$$

$$\le K_m^2 E \sum_i [(B_{t_{i+1}} - B_{t_i})^2 - (t_{i+1} - t_i)]^2 \to 0$$

(see the proof of 9.3.4). Hence $Y(\mathcal{P}, T_m)$ converges to 0 in L^2 and in probability when $\|\mathcal{P}\| \to 0$. Since $\{Y(\mathcal{P}) \ne Y(\mathcal{P}, T_m)\} \subset \{T_m < t\}$, we can show, as in Part 3, that $Y(\mathcal{P}) \to 0$ in probability when $\|\mathcal{P}\| \to 0$. $\square$

9.8.2 Corollary. In particular we have

$$B_t^2 = \int_0^t 2B_s \, dB_s + t.$$

(We already knew that $B_t^2 - t$ is a martingale, Itô's Formula gives us the exact form of the martingale.)

Problem

1. Let $\alpha(t, \omega)$ and $\beta(t, \omega)$ be two continuous, nonanticipating processes.

   (a) Show that the process $X_t = \int_0^t \alpha(s, \omega) \, dB_s + \int_0^t \beta(s, \omega) \, ds$ is continuous and nonanticipating.

   (b) Show that if the function $f(x)$, $x \in \mathbb{R}$, has a continuous second derivative, then we have, for each t,

   $$f(X_t) = f(X_0) + \int_0^t f'(X_s)\alpha(s) \, dB_s + \int_0^t f'(X_s)\beta(s) \, ds$$

   $$+ \frac{1}{2} \int_0^t f''(X_s)\alpha^2(s) \, ds \qquad \text{a.s.}$$

(Hint: use an argument similar to the proof of 9.8.1.)

9.9 REFERENCES

We have proved Itô's formula in its simplest form; generalizations and applications to stochastic differential equations can be found in Wong (1973), McKean (1969), Itô and McKean (1965) and the chapter on diffusion in Gikhman and Skorokhod (1969). A general treatment of martingales and stochastic integrals with respect to martingales is given in Dellacherie and Meyer (1980).

APPENDICES

Appendix 1   THE SYMMETRIC RANDOM WALK IN $\mathbb{R}^K$: A PRECISE ANALYSIS FOR $K = 1$ AND $K = 2$, AN INFORMAL APPROACH FOR $K \geq 3$

Consider a Markov chain with transition probabilities p_{ij}, and let $p_{ij}^{(n)}$ be the probability, starting from state i, that the process will be in state j at time n, that is, after n transitions. Let $f_{ii}^{(n)}$ be the probability, starting from i, that the first return to i will occur at time n. We then have

$$p_{ii}^{(n)} = \sum_{k=1}^{n} f_{ii}^{(k)} p_{ii}^{(n-k)}.$$

A1.1   Theorem.

(a) $\displaystyle \sum_{n=1}^{N} p_{ii}^{(n)} \leq \sum_{k=1}^{N} f_{ii}^{(k)} \sum_{r=0}^{N} p_{ii}^{(r)}$

(b) State i is recurrent if and only if $\displaystyle \sum_{n} p_{ii}^{(n)} = \infty$.

PROOF.

(a) $\displaystyle \sum_{n=1}^{N} p_{ii}^{(n)} = \sum_{n=1}^{N} \sum_{k=1}^{n} f_{ii}^{(k)} p_{ii}^{(n-k)} = \sum_{k=1}^{N} f_{ii}^{(k)} \sum_{n=k}^{N} p_{ii}^{(n-k)}$

$$\leq \sum_{k=1}^{N} f_{ii}^{(k)} \sum_{r=0}^{N} p_{ii}^{(r)}$$

(b) If $\sum_{n} p_{ii}^{(n)} < \infty$ then by the Borel–Cantelli Lemma, i is visited only finitely many times and therefore must be transient. If the series diverges, let

f_{ii} be the probability, starting from i, that there will ever be a return to i. Then

$$f_{ii} = \sum_{k=1}^{\infty} f_{ii}^{(k)} \geq \sum_{k=1}^{N} f_{ii}^{(k)} \geq \frac{\sum_{n=1}^{N} p_{ii}^{(n)}}{\sum_{r=0}^{N} p_{ii}^{(r)}}$$

by part (a), and the fraction approaches 1 as $N \to \infty$. Therefore i is recurrent. $\square$

The *symmetric random walk in* $\mathbb{R}^k$ is a Markov chain whose state space consists of k-tuples of integers. At each transition, exactly one coordinate changes, by ± 1 with equal probability. The probability of moving from $(a_1, \ldots, a_i, \ldots, a_k)$ to $(a_1, \ldots, a_i + 1, \ldots, a_k)$, as well as the probability of moving from $(a_1, \ldots, a_i, \ldots, a_k)$ to $(a_1, \ldots, a_i - 1, \ldots, a_k)$, is $\frac{1}{2}k$ for $i = 1, \ldots, k$. Thus in one dimension, we have a particle that moves right or left with equal probability. In two dimensions, the particle moves right, left, up or down, with probability $\frac{1}{4}$ in each case. In three dimensions, a particle at (x, y, z) can move to $(x + 1, y, z)$, $(x - 1, y, z)$, $(x, y + 1, z)$, $(x, y - 1, z)$, $(x, y, z + 1)$, or $(x, y, z - 1)$, and each outcome has probability $\frac{1}{6}$.

A1.2   Theorem.

(a)   When $k = 1$ we have $p_{ii}^{(2n)} = \binom{2n}{n} \left(\frac{1}{2}\right)^{2n}$, and the states are recurrent.

(b)   When $k = 2$ we have $p_{ii}^{(2n)} = \left[\left(\frac{1}{2}\right)^{2n} \binom{2n}{n}\right]^2$, and again the states are recurrent.

(c)   When $k \geq 3$, the states are transient.

PROOF.   (a)   Starting from i, the particle will be at i after $2n$ steps iff it has taken exactly n steps to the right and n steps to the left. Since the n positions where a step to the right occurs can be chosen in $\binom{2n}{n}$ ways, the formula for $p_{ii}^{(2n)}$ follows. Now $\binom{2n}{n} = \frac{(2n)!}{n!n!}$, and by Stirling's formula,

$$(2n)! \sim (2n)^{2n} e^{-2n} \sqrt{2\pi 2n} \text{ and } n! \sim n^n e^{-n} \sqrt{2\pi n}. \text{ Thus}$$

$$\binom{2n}{n} \sim \frac{2^{2n}}{\sqrt{n\pi}} \text{ and therefore } p_{ii}^{(2n)} \sim \frac{1}{\sqrt{n\pi}}. \text{ As } \sum \frac{1}{\sqrt{n}} = \infty,$$

the states are recurrent.

(b)   Starting from i, the particle will be at i after $2n$ steps iff for some t between 0 and n, it has taken exactly t steps right, t steps left, $n - t$ steps up and $n - t$ steps down. Therefore,

$$p_{ii}^{(2n)} = \sum_{t=0}^{n} \frac{(2n)!}{t!t!(n-t)!(n-t)!} \left(\frac{1}{4}\right)^{2n}.$$

Multiply and divide by $n!n!$ to obtain

$$p_{ii}^{(2n)} = \left(\frac{1}{4}\right)^{2n} \binom{2n}{n} \sum_{t=0}^{n} \binom{n}{t} \binom{n}{n-t} \quad \left[\text{recall that } \binom{n}{t} = \binom{n}{n-t}\right].$$

Now in selecting n objects out of $2n$ we can choose t from the first n and $n - t$ from the second n, so the summation is simply $\binom{2n}{n}$. This establishes the desired formula for $p_{ii}^{(2n)}$. By part (a), $p_{ii}^{(2n)} \sim 1/n\pi$, and again the states are recurrent.

We now argue intuitively to justify that for all $k \geq 3$, $p_{ii}^{(2kn)} = 0(n^{-k/2})$, that is, for some constant C we have $p_{ii}^{(2kn)} \leq C/n^{k/2}$ for all sufficiently large n. In a sequence of $2kn$ steps, roughly $2n$ will involve a change in the first coordinate, and in the remaining $2(k-1)n$ steps, a coordinate other than the first will change. Thus we can think of decomposing the k-dimensional walk into two subwalks of dimensions 1 and $k - 1$, and both subwalks must be back in their initial state at the end. We can then complete the heuristic argument as follows to show that the states are transient for all $k \geq 3$. [A formal proof by induction can be constructed by defining $b_k^{(2n)}$ as the probability that a symmetric k-dimensional random walk is back at its original state after $2n$ steps, and then showing that

$$b_k^{(2n)} = \sum_{s=0}^{n} \binom{2n}{2s} \left(\frac{1}{k}\right)^{2s} b_1^{(2s)} \left(1 - \frac{1}{k}\right)^{2n-2s} b_{k-1}^{(2n-2s)}.$$

(There must be $2s$ steps in which the first coordinate changes and is back at its original value at time $2n$.)]

As we have seen in the proof of Theorem A1.2, $p_{ii}^{(2n)} \sim 1/\sqrt{n\pi}$ for the 1-dimensional walk, and for the $(k-1)$-dimensional walk we have (by induction hypothesis):

$$p_{ii}^{(2(k-1)n)} = 0([(k-1)n]^{-(k-1)/2}) = 0(n^{-(k-1)/2}).$$

Thus for the k-dimensional walk,

$$p_{ii}^{(2kn)} = 0(n^{-1/2}n^{-(k-1)/2}) = 0(n^{-k/2}).$$

Since $\sum n^{-k/2} < \infty$ for $k \geq 3$, the states are transient. $\quad\square$

Appendix 2 SEMICONTINUOUS FUNCTIONS

If $f_1, f_2, \ldots$ are continuous maps from the metric space Ω to the extended reals $\overline{\mathbb{R}}$, and $f_n(x)$ increases to a limit $f(x)$ for each x, f need not be continuous; however, f is lower semicontinuous. Functions of this type play an important role in many aspects of analysis and probability.

A2.1 Definition. Let Ω be a metric space. The function $f: \Omega \to \overline{\mathbb{R}}$ is said to be *lower semicontinuous* (LSC) on Ω iff $\{x \in \Omega: f(x) > a\}$ is open in Ω for each $a \in \overline{\mathbb{R}}$, *upper semicontinuous* (USC) on Ω iff $\{x \in \Omega: f(x) < a\}$ is open in Ω for each $a \in \overline{\mathbb{R}}$. Thus f is LSC iff $-f$ is USC. Note that f is continuous iff it is both LSC and USC.

We have the following criterion for semicontinuity.

A2.2 Theorem. The function f is LSC on Ω iff, for each sequence $\{x_n\}$ converging to a point $x \in \Omega$, we have $\liminf_n f(x_n) \geq f(x)$, where $\liminf_n f(x_n)$ means $\sup_n \inf_{k \geq n} f(x_k)$. Hence f is USC iff $\limsup_n f(x_n) \leq f(x)$ when $x_n \to x$.

PROOF. Let f be LSC. If $x_n \to x$ and $b < f(x)$, then $x \in f^{-1}(b, \infty]$, an open subset of Ω, hence eventually $x_n \in f^{-1}(b, \infty]$, that is $f(x_n) > b$ eventually. Thus $\liminf_n f(x_n) \geq f(x)$. Conversely if $x_n \to x$ implies

$$\liminf_n f(x_n) \geq f(x),$$

we show that $V = \{x: f(x) \geq a\}$ is open. Let $x_n \to x$, where $f(x) > a$. Then $\liminf_n f(x_n) > a$, hence $f(x_n) > a$ eventually, that is, $x_n \in V$ eventually. Thus V is open. $\square$

We now prove a few properties of semicontinuous functions.

A2.3 Theorem. Let f be LSC on the compact metric space Ω. Then f attains its infimum. (Hence if f is USC on the compact metric space Ω, f attains its supremum.)

PROOF. If $b = \inf f$, there is a sequence of points $x_n \in \Omega$ with $f(x_n) \to b$. By compactness, we have a subsequence x_{nk} converging to some $x \in \Omega$. Since f is LSC, $\liminf_k f(x_{nk}) \geq f(x)$. But $f(x_{nk}) \to b$, so that $f(x) \leq b$; consequently $f(x) = b$. $\square$

A2.4 Theorem. If f_i is LSC on Ω for each $i \in I$, then $\sup_i f_i$ is LSC; if I is finite, then $\min_i f_i$ is LSC. (Hence if f_i is USC for each i, then $\inf_i f_i$ is USC, and if I is finite, then $\max_i f_i$ is USC.)

PROOF. Let $f = \sup_i f_i$; then $\{x: f(x) > a\} = \bigcup_{i \in I}\{x: f_i(x) > a\}$; hence $\{x: f(x) > a\}$ is open. If $g = \min(f_1, f_2, \ldots, f_n)$, then

$$\{x: g(x) > a\} = \bigcap_{i=1}^{n}\{x: f_i(x) > a\}$$

is open. □

A2.5 Theorem. Let $f: \Omega \to \overline{R}$, Ω any metric space, f arbitrary. Define

$$\underline{f}(x) = \liminf_{y \to x} f(y), \qquad x \in \Omega;$$

that is,

$$\underline{f}(x) = \sup_V \inf_{y \in V} f(y),$$

where V ranges over all open balls with center at x and radius $1/n$, $n = 1, 2, \ldots$. Then $\underline{f}$ is LSC on Ω and $\underline{f} \leq f$; furthermore, if g is LSC on Ω and $g \leq f$, then $g \leq \underline{f}$. Thus $\underline{f}$, called the *lower envelope* of f, is the sup of all LSC functions that are less than or equal to f (there is always at least one such function, namely the function constant at $-\infty$).

Similarly, if $\overline{f}(x) = \limsup_{y \to x} f(y) = \inf_V \sup_{y \in V} f(y)$, then $\overline{f}$, the *upper envelope* of f, is USC and $\overline{f} \geq f$; in fact $\overline{f}$ is the inf of all USC functions that are greater than or equal to f.

PROOF. It suffices to consider $\underline{f}$. Let $\{x_n\}$ be a sequence in Ω with $x_n \to x$ and $\liminf_n \underline{f}(x_n) < b < \underline{f}(x)$. If V is a neighborhood of x, we can choose n such that $x_n \in V$ and $\underline{f}(x_n) < b$. Since V is also a neighborhood of x_n, we have

$$b > \underline{f}(x_n) \geq \inf_{y \in V} f(y),$$

so

$$\underline{f}(x) = \sup_V \inf_{y \in V} f(y) \leq b < \underline{f}(x),$$

a contradiction. By A2.2, $\underline{f}$ is LSC, and $\underline{f} \leq f$ by definition of $\underline{f}$. Finally if g is LSC, $g \leq f$, then $\underline{f}(x) = \liminf_{y \to x} f(y) \geq \liminf_{y \to x} g(y) \geq g(x)$ since g is LSC. [If $\sup_V \inf_{y \in V} g(y) < b < g(x)$, then for each V pick $x_V \in V$ with $g(x_V) < b$. If we do this for $V = V_n = B(x, 1/n)$, $n = 1, 2, \ldots$, the $x_n = x_{V_n}$ form a sequence converging to x, while $\liminf_V g(x_V) \leq b < g(x)$, contradicting A2.2.] □

A2.6   Theorem. Let Ω be a metric space, f a LSC function on Ω. There is a sequence of continuous functions $f_n: \Omega \to \overline{R}$ such that $f_n \uparrow f$. (Thus if f is USC, there is a sequence of continuous functions $f_n \downarrow f$.) If $|f| \le M < \infty$, the f_n may be chosen so that $|f_n| \le M$ for all n.

PROOF.   (Following Hausdorff, 1962). First assume $f \ge 0$ and finite-valued. If d is the metric on Ω, define $g(x) = \inf\{f(z) + t\,d(x, z): z \in \Omega\}$, where $t > 0$ is fixed; then $0 \le g \le f$ since $g(x) \le f(x) + t\,d(x, x) = f(x)$.

If $x, y \in \Omega$, then $f(z) + t\,d(x, z) \le f(z) + t\,d(y, z) + t\,d(x, y)$. Take the inf over z to obtain $g(x) \le g(y) + t\,d(x, y)$. By symmetry,

$$|g(x) - g(y)| \le td(x, y),$$

hence g is continuous on Ω.

Now set $t = n$; in other words let $f_n(x) = \inf\{f(z) + nd(x, z): z \in \Omega\}$. Then $0 \le f_n \uparrow h \le f$. But given $\varepsilon > 0$, for each n we can choose $z_n \in \Omega$ such that

$$f_n(x) + \varepsilon > f(z_n) + nd(x, z_n) \ge nd(x, z_n).$$

But $f_n(x) + \varepsilon \le f(x) + \varepsilon$, and it follows that $d(x, z_n) \to 0$. Since f is LSC, $\liminf_{n\to\infty} f(z_n) \ge f(x)$; thus $f(z_n) > f(x) - \varepsilon$ eventually. But now

$$f_n(x) > f(z_n) - \varepsilon + nd(x, z_n) \ge f(z_n) - \varepsilon$$
$$> f(x) - 2\varepsilon \qquad \text{for large enough } n.$$

It follows that $0 \le f_n \uparrow f$. If $|f| \le M < \infty$, then $f + M$ is LSC and nonnegative; if $0 \le g_n \uparrow f + M$, then $f_n = g_n - M \uparrow f$ and $|f_n| \le M$.

In general, let $h(x) = \frac{1}{2}\pi + \arctan x$, $x \in \mathbb{R}$; then h is an order-preserving homeomorphism of $\overline{\mathbb{R}}$ and $[0, \pi]$. If f is LSC, then $h \circ f$ is finite-valued, LSC, and nonnegative, so that we can find continuous functions g_n such that $g_n \uparrow h \circ f$. Let $f_n = h^{-1} \circ g_n$; then $f_n \uparrow f$. $\square$

Appendix 3   COMPLETION OF THE PROOF OF THEOREM 7.3.2

The "if" part of the theorem remains to be proved. We may assume without loss of generality that $EX_k \equiv 0$. The uan condition is equivalent to the statement that $h_k(u/c_n) \to 1$ as $n \to \infty$ uniformly in $k = 1, \ldots, n$, that is,

$$\max_{1 \le k \le n} \left| 1 - h_k\left(\frac{u}{c_n}\right) \right| \to 0 \qquad \text{as} \qquad n \to \infty.$$

Furthermore, in this case

$$\max_{1 \leq k \leq n} \left| 1 - h_k \left(\frac{u}{c_n} \right) \right| \to 0$$

uniformly for u in any bounded interval (see Problem 3, Section 7.3).

(a)   We show first that

$$\frac{u^2}{2} + \sum_{k=1}^{n} \left[h_k \left(\frac{u}{c_n} \right) - 1 \right] \to 0 \qquad \text{as} \qquad n \to \infty.$$

Let h_Y denote the characteristic function of the random variable Y. By the normal convergence,

$$h_{T_n}(u) = \prod_{k=1}^{n} h_k \left(\frac{u}{c_n} \right) = \prod_{k=1}^{n} \left[1 + \left(h_k \left(\frac{u}{c_n} \right) - 1 \right) \right] \to \exp \left[\frac{-u^2}{2} \right]$$

and the convergence is uniform for u in any bounded interval (see 7.2.9). If "Log" denotes the principal branch of the logarithm, then

$$\sum_{k=1}^{n} \text{Log} \left[1 + \left(h_k \left(\frac{u}{c_n} \right) - 1 \right) \right]$$

is the (necessarily unique) continuous logarithm of $h_{T_n}(u)$ having the value 0 at $u = 0$. Thus

$$\sum_{k=1}^{n} \text{Log} \left[1 + \left(h_k \left(\frac{u}{c_n} \right) - 1 \right) \right] \to \frac{-u^2}{2}.$$

Therefore,

$$\frac{u^2}{2} + \sum_{k=1}^{n} \left[h_k \left(\frac{u}{c_n} \right) - 1 \right] + \sum_{k=1}^{n} \theta_k \left| h_k \left(\frac{u}{c_n} \right) - 1 \right|^2 \to 0,$$

where each $|\theta_k|$ is at most 1. But if $a_n + b_n \to 0$, then $\limsup_{n \to \infty} |a_n| = \limsup_{n \to \infty} |b_n|$ (eventually $|a_n| < |b_n| + \varepsilon$, and thus $\limsup_{n \to \infty} |a_n| \leq \limsup_{n \to \infty} |b_n|$); hence

$$\limsup_n \left| \frac{u^2}{2} + \sum_{k=1}^{n} \left[h_k \left(\frac{u}{c_n} \right) - 1 \right] \right|$$

$$= \limsup_n \left| \sum_{k=1}^{n} \theta_k \left| h_k \left(\frac{u}{c_n} \right) - 1 \right|^2 \right|$$

$$\leq \limsup_n \max_{1 \leq k \leq n} \left| h_k \left(\frac{u}{c_n} \right) - 1 \right| \sum_{k=1}^{n} \left| h_k \left(\frac{u}{c_n} \right) - 1 \right|.$$

Now

$$\max_{1 \leq k \leq n} \left| h_k \left(\frac{u}{c_n} \right) - 1 \right| \to 0$$

and

$$\sum_{k=1}^{n} \left| h_k \left(\frac{u}{c_n} \right) - 1 \right| = \sum_{k=1}^{n} \left| \int_{-\infty}^{\infty} \left[\exp \left(\frac{iux}{c_n} \right) - 1 \right] dF_k(x) \right|$$

$$= \sum_{k=1}^{n} \left| \int_{-\infty}^{\infty} \theta \left(\frac{u^2 x^2}{2c_n^2} \right) dF_k(x) \right|$$

$$\leq \frac{u^2}{2c_n^2} \sum_{k=1}^{n} \sigma_k^2 = \frac{u^2}{2} < \infty.$$

This proves (a).

(b)   For any $\varepsilon > 0$,

$$\limsup_{n \to \infty} \left[1 - \frac{1}{c_n^2} \sum_{k=1}^{n} \int_{|x| < \varepsilon c_n} x^2 \, dF_k(x) \right] \leq \frac{4}{\varepsilon^2 u^2}.$$

For by (a),

$$\frac{u^2}{2} - \sum_{k=1}^{n} \mathrm{Re} \left[1 - h_k \left(\frac{u}{c_n} \right) \right] \to 0,$$

that is,

$$\frac{u^2}{2} - \sum_{k=1}^{n} \int_{-\infty}^{\infty} \left(1 - \cos \frac{ux}{c_n} \right) dF_k(x) \to 0,$$

or equivalently,

$$\frac{u^2}{2} - \sum_{k=1}^{n} \int_{|x| < \varepsilon c_n} \left(1 - \cos \frac{ux}{c_n} \right) dF_k(x)$$

$$- \sum_{k=1}^{n} \int_{|x| \geq \varepsilon c_n} \left(1 - \cos \frac{ux}{c_n} \right) dF_k(x) \to 0.$$

Again, noting that $a_n + b_n \to 0$ implies $\lim \sup_{n\to\infty} |a_n| = \lim \sup_{n\to\infty} |b_n|$, we have

$$\lim_{n\to\infty} \sup \left| \frac{u^2}{2} - \sum_{k=1}^{n} \int_{|x|<\varepsilon c_n} \left(1 - \cos \frac{ux}{c_n}\right) dF_k(x) \right|$$

$$= \lim_{n\to\infty} \sup \left| \sum_{k=1}^{n} \int_{|x|\geq\varepsilon c_n} \left(1 - \cos \frac{ux}{c_n}\right) dF_k(x) \right|. \qquad (1)$$

But

$$\sum_{k=1}^{n} \int_{|x|<\varepsilon c_n} \left(1 - \cos \frac{ux}{c_n}\right) dF_k(x) \leq \sum_{k=1}^{n} \int_{|x|<\varepsilon c_n} \frac{u^2 x^2}{2c_n^2} dF_k(x)$$

$$\leq \frac{u^2}{2c_n^2} \sum_{k=1}^{n} \sigma_k^2 = \frac{u^2}{2}$$

and thus the absolute values around the "lim sup" terms in (1) may be dropped. [This also shows that the quantity inside the brackets in (b) is nonnegative.] Consequently,

$$\lim_{n\to\infty} \sup \left(\frac{u^2}{2} \left[1 - \frac{1}{c_n^2} \sum_{k=1}^{n} \int_{|x|<\varepsilon c_n} x^2 dF_k(x) \right] \right)$$

$$\leq \lim_{n\to\infty} \sup \left[\frac{u^2}{2} - \sum_{k=1}^{n} \int_{|x|<\varepsilon c_n} \left(1 - \cos \frac{ux}{c_n}\right) dF_k(x) \right]$$

$$= \lim_{n\to\infty} \sup \sum_{k=1}^{n} \int_{|x|\geq\varepsilon c_n} \left(1 - \cos \frac{ux}{c_n}\right) dF_k(x) \qquad \text{by (1)}.$$

But

$$\sum_{k=1}^{n} \int_{|x|\geq\varepsilon c_n} \left(1 - \cos \frac{ux}{c_n}\right) dF_k(x) \leq 2 \sum_{k=1}^{n} \int_{|x|\geq\varepsilon c_n} dF_k(x)$$

$$= 2 \sum_{k=1}^{n} P\{|X_k| \geq \varepsilon c_n\}$$

$$\leq 2 \sum_{k=1}^{n} \frac{\sigma_k^2}{\varepsilon^2 c_n^2} \qquad \text{by Chebyshev's inequality}$$

$$= \frac{2}{\varepsilon^2}, \qquad \text{proving (b)}.$$

The proof of the theorem is completed by letting $u \to \infty$ in (b). $\qquad \square$

Appendix 4   PROOF OF THE CONVERGENCE OF TYPES THEOREM **7.3.4**

(a)   Let G_n, G, F_n, F be the distribution functions of X_n, X, Y_n, Y, respectively.

We may select convergent subsequences $\{a_{n_k}\}$, $\{b_{n_k}\}$ such that

$$a_{n_k} \to a, \qquad b_{n_k} \to b, \qquad \text{where} \qquad 0 \le a \le \infty, \qquad -\infty \le b \le \infty.$$

We first show that $a < \infty$.

Suppose that $a = \infty$. Let $E = \{x \in R: \limsup_{k \to \infty}(a_{n_k}x + b_{n_k}) < \infty\}$ and let $c = \sup E$. (Take $c = -\infty$ if $E = 0$.)

(i)   If $x \in \mathbb{R}$, $x < c$, then $a_{n_k}x + b_{n_k} \to -\infty$.

PROOF.   Let $x < u < c$, with $u \in E$ (u exists since $c = \sup E$). Then $u \in E$, $x < u$ implies $x \in E$ by definition of E.

Now

$$a_{n_k}x + b_{n_k} = a_{n_k}(x - u) + a_{n_k}u + b_{n_k}; \limsup_{k \to \infty}(a_{n_k}u + b_{n_k}) < \infty$$

since $u \in E$, and $a_{n_k}(x - u) \to -\infty$ since $a_{n_k} \to a = \infty$ and $x - u < 0$. This proves (i).

(ii)   If $x \in \mathbb{R}$, $x < c$, then $G(x) = 0$.

PROOF.   Let $\varepsilon > 0$ be arbitrary. Choose $z \in \mathbb{R}$ such that $F(z) < \varepsilon$ and z is a continuity point of F. Then $F_{n_k}(z) \to F(z)$, so eventually $F_{n_k}(z) < \varepsilon$. By (i), eventually $a_{n_k}x + b_{n_k} < z$, so

$$G_{n_k}(x) = F_{n_k}(a_{n_k}x + b_{n_k}) \le F_{n_k}(z) < \varepsilon$$

for large enough k. Thus $G_{n_k}(x) \to 0$, $x < c$. Let $x < x' < c$, x' a continuity point of G. Then $G(x') = \lim_{k \to \infty} G_{n_k}(x') = 0$, and since $G(x) \le G(x')$, we have $G(x) = 0$, proving (ii).

(iii)   If $x \in \mathbb{R}$, $x > c$, then $G(x) = 1$.

PROOF.   By definition of E, we find a subsequence $\{r_j\}$ such that $a_{r_j}x + b_{r_j} \to \infty$. Choose a continuity point w of F such that $F(w) > 1 - \varepsilon$. Eventually, $a_{r_j}x + b_{r_j} > w$, so $G_{r_j}(x) = F_{r_j}(a_{r_j}x + b_{r_j}) \ge F_{r_j}(w) > 1 - \varepsilon$. Thus $G_{r_j}(x) \to 1$ for $x > c$. If $c < y < x$, y a continuity point of G, we have $G(x) \ge G(y) = \lim_{j \to \infty} G_{r_j}(y) = 1$, proving (iii).

It follows from (ii) and (iii) that G is degenerate, a contradiction.

Next we show that b is finite.

If $b_{n_k} \to \infty$, then $a_{n_k} x + b_{n_k} \to \infty$ for every $x \in \mathbb{R}$, so the argument of (iii) may be repeated to show $G(x) \equiv 1$, a contradiction. If $b_{n_k} \to -\infty$, then $a_{n_k} x + b_{n_k} \to -\infty$ for each $x \in \mathbb{R}$, so the argument of (ii) shows $G(x) \equiv 0$, a contradiction.

Now we show that $a > 0$.

Let x be a continuity point of G, and let $\varepsilon_1, \varepsilon_2 > 0$ be such that $ax + b + \varepsilon_1$ and $ax + b - \varepsilon_2$ are continuity points of F. Then $a_{n_k} x + b_{n_k} \to ax + b$, so eventually $ax + b - \varepsilon_2 \leq a_{n_k} x + b_{n_k} \leq ax + b + \varepsilon_1$; hence

$$F_{n_k}(ax + b - \varepsilon_2) \leq F_{n_k}(a_{n_k} x + b_{n_k}) = G_{n_k}(x) \leq F_{n_k}(ax + b + \varepsilon_1).$$

Let $k \to \infty$ to obtain $F(ax + b - \varepsilon_2) \leq G(x) \leq F(ax + b + \varepsilon_1)$. Since ε_1 and ε_2 may be chosen arbitrarily small, $F(ax + b)^- \leq G(x) \leq F(ax + b)$ for all continuity points x of G.

If $a = 0$, then $F(b^-) \leq G(x) \leq F(b)$ for all continuity points, and hence for all $x \in \mathbb{R}$. But then $F(b) = 1$, $F(b^-) = 0$ because $G(\infty) = 1$, $G(-\infty) = 0$. Thus F is degenerate at b, a contradiction.

Finally, if G is continuous at x and F is continuous at $ax + b$, we have just seen that $F(ax + b)^- \leq G(x) \leq F(ax + b)$, so $G(x) = F(ax + b)$. Since there are only countably many real numbers y such that G is discontinuous at y or F is discontinuous at $ay + b$, it follows that $G(x) = F(ax + b)$ for all $x \in \mathbb{R}$.

Now if we have other convergent subsequences $a_{mi} \to a'$, $b_{mi} \to b'$, the above argument shows that $0 < a' < \infty$, $-\infty < b' < \infty$, and $G(x) = F(ax + b) = F(a'x + b')$ for all $x \in \mathbb{R}$, so that $a^{-1}(Y - b) \overset{d}{=} (a')^{-1}(Y - b')$. Random variables with the same distribution have the same characteristic function, therefore

$$\exp\left(\frac{-iub}{a}\right) h_Y\left(\frac{u}{a}\right) = \exp\left(\frac{-iub'}{a'}\right) h_Y\left(\frac{u}{a'}\right) \qquad \text{for all} \qquad u.$$

Say $a < a'$, and set $k = a/a'$. Let $v = u/a$ to obtain

$$|h_Y(v)| = \left| h_Y\left(\frac{av}{a'}\right) \right| = |h_Y(kv)|$$

for all v. Thus

$$|h_Y(v)| = |h_Y(kv)| = |h_Y(k^2 v)| = \cdots = |h_Y(k^n v)| \to |h_Y(0)| = 1.$$

It follows that Y is degenerate, a contradiction (see Problem 4, Section 7.1). Thus $a = a'$, so $e^{-iub} = e^{-iub'}$ for all sufficiently small u; hence $b = b'$. Therefore, $a_n \to a$, $b_n \to b$, proving 7.3.4(a).

Comment. If $a^{-1}(Y-b)\stackrel{d}{=}(a')^{-1}(Y-b')$, where a and a' are nonzero but not necessarily positive, then if $|a|<|a'|$ and we set $k=a/a'$, we obtain a contradiction as above. We can conclude that $|a|=|a'|$ but not that $a=a'$.

(b) If $a_n>0$, X_n and Y_n are of the same positive type, and if $a_n<0$, $X_n\stackrel{d}{=}-a_n^{-1}(-Y_n+b_n)$; hence X_n and $-Y_n$ are of the same positive type.

Let $S_1=\{n:a_n>0\}$, $S_2=\{n:a_n<0\}$. If S_1 is infinite, part (a) shows that X and $a^{-1}(Y-b)$ have the same distribution for some real a, b, $a>0$, and

$$\lim_{\substack{n\to\infty\\n\in S_1}}a_n=a,\qquad \lim_{\substack{n\to\infty\\n\in S_1}}b_n=b.$$

Now suppose that S_2 is infinite. Then $Y_n\stackrel{d}{\longrightarrow}Y$ implies $-Y_n\stackrel{d}{\longrightarrow}-Y$ (use 7.2.9), and it follows from part (a) that for some real a', b', with $a'<0$, we have $X\stackrel{d}{=}-(a')^{-1}(-Y+b')$, and

$$\lim_{\substack{n\to\infty\\n\in S_2}}a_n=a',\qquad \lim_{\substack{n\to\infty\\n\in S_2}}b_n=b'.$$

Now there are three possibilities:

Case 1. S_1 and S_2 are both infinite. Then since $a^{-1}(Y-b)\stackrel{d}{=}(a')^{-1}(Y-b')\stackrel{d}{=}X$, we have $|a|=|a'|$ [see the comment after the proof of (a) here]. Thus $|a_n|\to|a|$ and the result follows.

Case 2. S_1 is infinite, S_2 finite. Then $a_n\to a$, $b_n\to b$, and $X\stackrel{d}{=}a^{-1}(Y-b)$, proving the result.

Case 3. S_1 is finite, S_2 infinite. Then $a_n\to a'$, $b_n\to b'$, and $X\stackrel{d}{=}(a')^{-1}(Y-b')$, and the result follows. □

Appendix 5 THE MULTIVARIATE NORMAL DISTRIBUTION

In this appendix, u will denote a column vector with components $u_1,\dots,u_n$, x a column vector with components $x_1,\dots,x_n$, and X a random (column) vector with components $X_1,\dots,X_n$. The superscript t will indicate the transpose of a matrix. To avoid awkward special cases, we agree that normal with mean μ and variance 0 will mean degenerate at μ.

A5.1 Definition. The random variables $X_1,\dots,X_n$ are said to be *jointly Gaussian* (or the random vector $X=(X_1,\dots,X_n)$ is said to be *Gaussian*) if the characteristic function of X is

$$h(u_1,\dots,u_n)=\exp[iu^tb]\exp[-\tfrac{1}{2}u^tKu]\qquad(1)$$

where the b_i are arbitrary real numbers and K is a symmetric, nonnegative definite matrix (with real coefficients).

A much more concrete interpretation is possible.

A5.2   ***Theorem.*** The random vector X is Gaussian if and only if X can be expressed as $AY + b$ where the Y_i are independent normal random variables with 0 mean.

PROOF.   If $X = AY + b$, then $E[\exp(iu'X)] = \exp[iu'b]E[\exp(iu'AY)]$. But

$$E[\exp(iv'Y)] = E\left[\prod_{k=1}^{n} \exp(iv_k Y_k)\right] = \prod_{k=1}^{n} E[\exp(iv_k Y_k)]$$

$$= \exp\left[-\frac{1}{2}\sum_{k=1}^{n}\lambda_k v_k^2\right] = \exp\left[-\frac{1}{2}v'DV\right]$$

where D is a diagonal matrix whose entries are $\lambda_k = \mathrm{Var}\, Y_k$, $k = 1, \ldots, n$. Set $V = A'u$; the characteristic function of X is then given by

$$h(u_1, \ldots, u_n) = \exp[iu'b - \tfrac{1}{2}u'KU]$$

where $K = ADA'$. Since the diagonal matrix D is symmetric, so is K, and K is nonnegative definite as well since $u'Ku = v'Dv = \sum_{k=1}^{n}\lambda_k v_k^2 \geq 0$.

Conversely, assume that X is Gaussian, with a characteristic function given by (1) above. Let A be an orthogonal matrix such that $A'KA = D$, a diagonal matrix whose entries are the eigenvalues λ_k of K. Let

$$Y = A'(X - b).$$

Then

$$E[\exp(iu'Y)] = \exp[-iu'A'b]\exp[iu'A'X]$$

$$= \exp[-\tfrac{1}{2}V'Kv] \qquad \text{where} \qquad V = Au.$$

Thus

$$E[\exp(iu'Y)] = \exp\left[-\frac{1}{2}u'A'KAu\right] = \exp\left[-\frac{1}{2}u'Du\right]$$

$$= \exp\left[-\frac{1}{2}\sum_{k=1}^{n}\lambda_k u_k^2\right].$$

The form of the characteristic function of Y shows that $Y_1, \ldots, Y_n$ are independent, and Y_k is normal with mean 0 and variance λ_k. Since A is orthogonal, we have $A^t = A^{-1}$, so that $X = AY + b$. □

A5.3 Corollary. For any column vector b and symmetric, nonnegative definite matrix K, there is always a Gaussian vector X whose characteristic function is given by A5.1 with the prescribed b and K.

PROOF. Let A be an orthogonal matrix with $A^t K A = D$, the diagonal matrix of eigenvalues of K; then $K = ADA^t$. Let $X = AY + b$, where the Y_k are independent and normal $(0, \lambda_k)$. The first part of the proof of Theorem A5.2 shows that the characteristic function of X has the desired form. □

We may give a probabilistic interpretation of the vector b and the matrix K.

A5.4 Theorem. If X is Gaussian with characteristic function given by A5.1, then $E(X) = b$, in other words, b_j is the expectation of X_j, $j = 1, \ldots, n$. Furthermore, K is the covariance matrix of the X_j, that is, $K_{jk} = Cov(X_j, X_k)$ for all j, k.

PROOF. Let $X = AY + b$ as in Theorem A5.2. Since the Y_j have finite second moments (in fact finite moments of all orders), so do the X_j. By linearity of the expectation we have $E(X) = b$. If A is any matrix, we denote by $E(A)$ the matrix whose jk entry is $E(a_{jk})$. Then the covariance matrix of the X_j can be written as

$$E[(X - b)(X - b)^t] = E[AYY^t A^t] = AE[YY^t]A^t = ADA^t,$$

where D is a diagonal matrix with entries $\lambda_k = \text{Var } Y_k$. But by the first part of the proof of the Theorem, $ADA^t = K$. □

We now show that if the covariance matrix K is nonsingular, then X has a density.

A5.5 Theorem. Let X be Gaussian with mean vector b and covariance matrix K. If K is nonsingular, then the $X_j - b_j$ are linearly independent, that is,

$$\text{if} \quad \sum_{j=1}^{n} c_j(X_j - b_j) = 0 \text{ a.e.,} \quad \text{then} \quad c_j = 0 \quad \text{for all} \quad j.$$

Furthermore, X has density f, where

$$f(X) = (2\pi)^{-n/2}(\det K)^{-1/2} \exp[-\tfrac{1}{2}(x - b)^t K^{-1}(x - b)].$$

PROOF. Let A be an orthogonal matrix such that $A^t K A = D$, and let $X = AY + b$ as in Theorem A5.2. If K is nonsingular, then every eigenvalue λ_k is strictly positive, so Y has density g, where

$$g(y) = (2\pi)^{-n/2}(\lambda_1 \ldots \lambda_n)^{-1/2} \exp\left[-\frac{1}{2}\sum_{k=1}^{n}\frac{y_k^2}{\lambda_k}\right]$$

$$= (2\pi)^{-n/2}(\det K)^{-1/2} \exp\left[-\frac{1}{2}y^t D^{-1} y\right].$$

Now the Jacobian of the transformation $X = Ay + b$ is $\det A$, which is ± 1 because A is orthogonal. Since $y = A^t(x - b)$, X has density f, where

$$f(x) = (2\pi)^{-n/2}(\det K)^{-1/2} \exp[-\tfrac{1}{2}(x - b)^t AD^{-1}A^t(x - b)].$$

But $A^t K A = D$, and it follows that $D^{-1} = A^t K^{-1} A$, and therefore $AD^{-1}A^t = K^{-1}$.

Now if $c^t(X - b) = \sum_{j=1}^{n} c_j(X_j - b_j) = 0$ a.e., then

$$0 = E[|c^t(X - b)|^2] = E[c^t(X - b)(X - b)^t c]$$

$$= c^t E[(X - b)(X - b)^t]c = c^t K C.$$

But K is nonsingular, and therefore positive definite, so c_j must be 0 for all j. $\square$

If K is singular, the last part of the proof of Theorem A5.5 shows that the $X_j - b_j$ are linearly dependent. For $c^t K c = c^t A D A^t c = \sum_{k=1}^{n} \lambda_k a_k^2$ where $a = A^t c$. Since at least one λ_k must be 0, we can choose a nonzero a, and hence a nonzero c, such that $c^t(X - b) = 0$ a.e. If, say, $X_1 - b_1, \ldots, X_r - b_r$ form a maximal linearly independent subset of $\{X_1 - b_1, \ldots, X_n - b_n\}$, then $(X_1, \ldots, X_r)$ has a density of the form given in Theorem A5.5, with K replaced by the submatrix determined by the first r rows and the first r columns of K. The remaining random variables $X_j - b_j$, $j = r + 1, \ldots, n$ can be expressed (on a set of probability 1) as linear combinations of the $X_j - b_j$, $1 \leq j \leq r$.

The result that nonsingularity of K is equivalent to linear independence of the $X_j - b_j$ holds for arbitrary random variables with finite second moments, as the above analysis shows.

A5.6   Theorem. (a)   If X is Gaussian and $Z = CX$ where C is any n by n matrix, then Z is Gaussian.

(b) If $X_1, \ldots, X_n$ are jointly Gaussian, then so are $X_1, \ldots, X_r$ for any $r \leq n$.

(c) If $X_1, \ldots, X_n$ are jointly Gaussian, then for any constants $c_1, \ldots, c_n$, $\sum_{j=1}^{n} c_j X_j$ is a normally distributed random variable.

PROOF. (a) Let $X = AY + b$ where the Y_j are independent normal random variables with 0 mean. Then $Z = CAY + Cb$, which is Gaussian by Theorem A5.2.

(b) In part (a) take $C = [I\, 0]$ where I is an r by r identity matrix.

(c) In part (a) take $C = [c_1\, c_2 \ldots c_n]$. □

A5.7 Example. Let X be normal $(0, 1)$ and define Y as follows. Let Z take on the values 1 and 0 with equal probability, with X and Z independent. If $Z = 1$, set $Y = X$, and if $Z = 0$, take $Y = -X$. Then with probability $\frac{1}{2}$ we have $X + Y = 0$, so that $X + Y$ is certainly not Gaussian. By Theorem A5.6(c), X and Y are not jointly Gaussian. However, X is Gaussian by assumption, and Y is also Gaussian because $-X$ is normal $(0, 1)$ and therefore,

$$P\{Y \leq y\} = \tfrac{1}{2}P\{X \leq y\} + \tfrac{1}{2}P\{-X \leq y\} = P\{X \leq y\}.$$

Thus the converse assertion fails in both A5.6(b) and A5.6(c).

A5.8 Theorem. If $X_1, \ldots, X_n$ are jointly Gaussian and the X_j are *uncorrelated*, that is, the covariance of X_j and X_k is 0 for every $j \neq k$, then $X_1, \ldots, X_n$ are independent.

PROOF. The covariance matrix is diagonal with entries $\lambda_j = \operatorname{Var} X_j$. We may assume with loss of generality that all λ_j are strictly positive, because if some $\lambda_j = 0$ then X_j is constant a.e. and can be deleted. Then K is nonsingular and K^{-1} is diagonal with entries $1/\lambda_j$. By Theorem A5.5, the joint density of $X_1, \ldots, X_n$ is

$$f(x_1, \ldots, x_n) = (2\pi)^{-n/2}(\lambda_1 \ldots \lambda_n)^{-1/2} \exp\left[-\frac{1}{2}\sum_{j=1}^{n} \frac{(x_j - b_j)^2}{\lambda_j} \right].$$

The form of the density shows that the X_j are independent, with X_j normal with mean b_j and variance λ_j. □

Bibliography

Apostol, T.M. (1957). *Mathematical Analysis.* Reading, MA: Addison–Wesley.

Ash, C. (1993). *The Probability Tutoring Book.* Piscataway, NJ: IEEE Press.

Ash, R.B. (1993). *Real Variables with Basic Metric Topology.* Piscataway, NJ: IEEE Press.

Ash, R.B. (1971). *Complex Variables.* New York: Academic Press.

Ash, R.B. (1970). *Basic Probability Theory.* New York: Wiley.

Ash, R.B. (1965). *Information Theory.* New York: Wiley; republished (1990). New York: Dover.

Ash, R.B. and Gardner, M.F. (1975). *Topics in Stochastic Processes.* New York: Academic Press.

Bachman, G. and Narici, L. (1966). *Functional Analysis.* New York: Academic Press.

Billingsley, P. (1968). *Convergence of Probability Measures.* New York: Wiley.

Billingsley, P. (1965). *Ergodic Theory and Information.* New York: Wiley.

Breiman, L. (1968). *Probability.* Reading, MA: Addison–Wesley.

Breiman, L. (1957). The individual ergodic theorem of information theory. *Ann. Math. Statist.,* **28** (3): 809–811; correction, **31** (3): 809–810.

Brooks, J.K. (1971). The Lebesgue decomposition theorem for measures. American Mathematical Monthly, **78**: 660–661.

Chung, K.L. (1968). *A Course in Probability Theory.* New York: Harcourt.

Chung, K.L. (1961). A note on the ergodic theorem of information theory. *Ann. Math. Statist.,* **32**: 612–614.

Conway, J.B. (1990). *A Course in Functional Analysis.* 2nd edition. New York: Springer Verlag.

Cornfeld, I.P., Fomin, S.V., and Sinai, Y.G. (1982). *Ergodic Theory.* New York: Springer Verlag.

Dellacherie, C. and Meyer, P.A. (1980). *Probabilités et Potential, Théorie des Martingales.* Paris: Hermann.

Doob, J.L. (1994). *Measure Theory.* New York: Springer Verlag.

Doob, J.L. (1953). *Stochastic Processes.* New York: Wiley.

Doob, S.L. (1940). Regularity properties of certain families of chance variables. *Trans. Amer. Math. Soc.* **47**: 455–486.

Dunford, N. and Schwartz, J.T. (1958). *Linear Operators.* New York: Wiley (Interscience); part 1; part 2 (1963); part 3 (1970).

Feller, W. (1966). *Introduction to Probability Theory,* vol. II. New York: Wiley.

Feller, W. (1950). *Introduction to Probability Theory,* vol. I. New York: Wiley.

Ferguson, T.S. (1996). *A Course in Large Sample Theory.* New York: Chapman and Hall.

Folland, G.B. (1984). *Real Analysis.* New York: Wiley (Interscience).

Gallager, R.G. (1968). *Information Theory and Reliable Communication.* New York: Wiley.

Garsia, A.M. (1970). *Topics in Almost Everywhere Convergence.* Chicago: Markham.

Gikhman, I.I. and Skorokhod, A.V. (1969). *Introduction to the Theory of Random Processes.* Philadelphia: Saunders.

Gnedenko, B.V. and Kolmogorov, A.N. (1954). *Limit Distributions for Sums of Independent Random Variables.* Reading, MA: Addison–Wesley.

Gray, R.M. (1975). Sliding-block source coding. *IEEE Trans. Inform. Theory,* **21**: 357–368.

Halmos, P.R. (1956). *Lectures on Ergodic Theory.* Tokyo: Mathematical Society of Japan.

Halmos, P.R. (1951). *Introduction to Hilbert Space.* New York: Chelsea.

Halmos, P.R. (1950). *Measure Theory.* Princeton, NJ: Van Nostrand-Reinhold.

Hausdorff, F. (1962). *Set Theory.* New York: Chelsea.

Itô, K. and McKean, H.P., Jr. (1965). *Diffusion Processes and Their Sample Paths*. New York: Academic Press.

Jacobs, K. (1962). *Lecture Notes on Ergodic Theory*. Aarhus, Denmark: Univ. of Aarhus.

Karlin, S. (1966). *A First Course in Stochastic Processes*. New York: Academic Press.

Kelley, J.L. and Namioka, L. (1963). *Linear Topological Spaces*. Princeton, NJ: Van Nostrand-Reinhold.

Kemeny, J.G., Snell, J.L., and Knapp, A.W. (1966). *Denumerable Markov Chains*. Princeton, NJ: Van Nostrand-Reinhold.

Kieffer, J. (1970). *A Generalization of the Shannon-McMillan Theorem and Its Application to Information Theory*. Thesis, University of Illinois.

Kolmogorov, A.N. (1959). Entropy per unit time as a metric invariant of automorphisms. *Dokl. Akad. Nauk, SSSR*, **124**: 754–755.

Kolmogorov, A.N. (1958). A new invariant for transitive dynamical systems. *Dokl. Akad. Nauk, SSSR*, **119**: 861–864.

Lévy, P. (1937). *Théorie de l'addition des variables aléatoires*. Paris: Gauthier-Villars.

Liusternik, L. and Sobolev, V. (1961). *Elements of Functional Analysis*. New York: Ungar.

Loeve, M. (1955). *Probability Theory*. 2nd edition (1960); 3rd edition (1963). New York: Van Nostrand-Reinhold.

McKean, H.P., Jr. (1969). *Stochastic Integrals*. New York: Academic Press.

McMillan, B. (1953). The basic theorem of information theory. *Ann. Math. Statist.*, **24** (2): 196–219.

Meyer, (1966). *Probability and Potentials*. Waltham, MA: Blaisdell.

Neveu, J. (1965). *Mathematical Foundations of the Calculus of Probability*. San Francisco: Holden-Day.

Ornstein, D.S. (1974). *Ergodic Theory, Randomness, and Dynamical Systems*. New Haven: Yale University Press.

Ornstein, D.S. (1970a). Bernoulli shifts with the same entropy are isomorphic. *Adv. in Math.*, **4**: 337–352.

Ornstein, D.S. (1970b). Two Bernoulli shifts with the same entropy are isomorphic. *Adv. in Math.*, **5**: 339–348.

Parthasarathy, K. (1967). *Probability Measures on Metric Spaces*. New York: Academic Press.

Parzen, E. (1960). *Modern Probability Theory*. New York: Wiley.

Rao, B.V. (1969). *Bull. Amer. Math. Soc.*, **75**: 614.

Ross, S. (1993). *A First Course in Probability*. 5th edition, New York: Academic Press.

Royden, H.L. (1963). *Real Analysis*. New York: Macmillan. 2nd edition (1968).

Rudin, W. (1966). *Real and Complex Analysis*. New York: McGraw-Hill.

Schaefer, H. (1966). *Topological Vector Spaces*. New York: Macmillan.

Serfling, R.J. (1980). *Approximation Theorems of Mathematical Statistics*. New York: Wiley.

Shannon, C.E. (1948). A mathematical theory of communication. *Bell Syst. Tech. J.*, **27**: 379–423; 623–656. Reprinted in Shannon, C.E. and W. Weaver (1949). *The Mathematical Theory of Communication*. Urbana, IL: University of Illinois Press.

Shields, P. (1973). *The Theory of Bernoulli Shifts*. Chicago: The University Press of Chicago.

Shiryaev, A.N. (1996). *Probability*. New York: Springer Verlag.

Taylor, A.E. (1958). *Introduction to Functional Analysis*. New York: Wiley.

Titchmarsh, E.C. (1939). *The Theory of Functions*. London: Oxford Univ. Press.

Tucker, H.G. (1967). *A Graduate Course in Probability*. New York: Academic Press.

Williams, D. (1991). *Probability and Martingales*. Cambridge: Cambridge Univ. Press.

Wojtaszczyk, P. (1991). *Banach Spaces for Analysts*. Cambridge: Cambridge Univ. Press.

Wong, E. (1971). *Stochastic Processes in Information and Dynamical Systems*. New York: McGraw-Hill.

Yosida, K. (1968). *Functional Analysis*. New York: Springer Verlag.

SOLUTIONS TO PROBLEMS

CHAPTER 1

Section 1.1

2. We have $\limsup_n A_n = (-1, 1]$, $\liminf_n A_n = \{0\}$.

3. Using $\limsup_n A_n = \{\omega\colon \omega \in A_n \text{ for infinitely many } n\}$, $\liminf_n A_n = \{\omega\colon \omega \in A_n \text{ for all but finitely many } n\}$, we obtain

$$\liminf_n A_n = \{(x, y)\colon x^2 + y^2 < 1\},$$

$$\limsup_n A_n = \{(x, y)\colon x^2 + y^2 \le 1\} - \{(0, 1), (0, -1)\}.$$

4. If $x = \limsup_{n \to \infty} x_n$, then $\limsup_n A_n$ is either $(-\infty, x)$ or $(-\infty, x]$. If $y \in A_n$ for infinitely many n, then $x_n > y$ for infinitely many n; hence $x \ge y$. Thus $\limsup_n A_n \subset (-\infty, x]$. But if $y < x$, then $x_n > y$ for infinitely many n, so $y \in \limsup_n A_n$. Thus $(-\infty, x) \subset \limsup_n A_n$, and the result follows. The same result is valid for $\liminf$; the above analysis applies, with "eventually" replacing "for infinitely many n."

Section 1.2

4. (a) If $-\infty \le a < b < c < \infty$, then $\mu(a, c] = \mu(a, b] + \mu(b, c]$, and $\mu(a, \infty) = \mu(a, b] + \mu(b, \infty)$; finite additivity follows quickly. If $A_n = (-\infty, n]$, then $A_n \uparrow \mathbb{R}$, but $\mu(A_n) = n \nrightarrow (\mathbb{R}) = 0$. Thus μ is not continuous from below, hence not countably additive.

(b) Finiteness of μ follows from the definition; since $\mu(-\infty, n] \to \infty$, μ is unbounded.

5. We have $\mu(\bigcup_{i=1}^{\infty} A_i) \ge \mu(\bigcup_{i=1}^{n} A_i) = \sum_{i=1}^{n} \mu(A_i)$ for all n; let $n \to \infty$ to obtain the desired result.

8. The minimal σ-field $\mathscr{F}$ (which is also the minimal field) consists of the collection $\mathscr{G}$ of all (finite) unions of sets of the form $B_1 \cap B_2 \cap \cdots \cap B_n$, where B_i is either A_i or A_i^c. Any σ-field containing $A_1, \ldots, A_n$ must contain

all sets in $\mathscr{G}$; hence $\mathscr{G} \subset \mathscr{F}$. But $\mathscr{G}$ is a σ-field; hence $\mathscr{F} \subset \mathscr{G}$. Since there are 2^n disjoint sets of the form $B_1 \cap \cdots \cap B_n$, and each such set may or may not be included in a typical set in $\mathscr{F}$, $\mathscr{F}$ has at most 2^{2^n} members. The upper bound is attained if all sets $B_1 \cap \cdots \cap B_n$ are nonempty. When $n = 2$, the sets are $\emptyset$, Ω, $A_1 \cap A_2^c$, $A_1 \cap A_2^c$, $A_1^c \cap A_2$, $A_1^c \cap A_2^c$, along with all sets that can be generated from these by taking unions 2 and 3 at a time.

9. (a) As in Problem 8, any field over $\mathscr{C}$ must contain all sets in $\mathscr{G}$; hence $\mathscr{G} \subset \mathscr{F}$. But $\mathscr{G}$ is a field; hence $\mathscr{F} \subset \mathscr{G}$. For if $A_i = \bigcap_{j=1}^r B_{ij}$, then $(\bigcup_{i=1}^n A_i)^c = \bigcap_{i=1}^n \bigcup_{j=1}^r B_{ij}^c$, which belongs to $\mathscr{G}$ because of the distributive law $A \cap (B \cup C) = (A \cap B) \cup (A \cap C)$.

    (b) Note that the complement of a finite intersection $\bigcap_{j=1}^r B_{ij}$ belongs to $\mathscr{D}$; for example, if $B_1, B_2 \in \mathscr{C}$, then

$$(B_1 \cap B_2^c)^c = B_1^c \cup B_2$$
$$= (B_1^c \cap B_2) \cup (B_1^c \cap B_2^c) \cup (B_2 \cap B_1) \cup (B_2 \cap B_1^c) \in \mathscr{D}.$$

Now $\mathscr{D}$ is closed under finite intersection by the distributive law, and it follows from this and the above remark that $\mathscr{D}$ is closed under complementation and is therefore a field. Just as in the proof that $\mathscr{F} = \mathscr{G}$, we find that $\mathscr{F} = \mathscr{D}$.

    (c) This is immediate from (a) and (b).

11. (a) Let $A_n \in \mathscr{S}$, $n = 1, 2, \ldots$. Then A_n belongs to some $\mathscr{C}_{a_n}$, and we may assume $\alpha_1 \leq \alpha_2 \leq \cdots$, so $\mathscr{C}_{a_1} \subset \mathscr{C}_{a_2} \cdots$. Let $\alpha = \sup_n \alpha_n < \beta_1$. Then all $\mathscr{C}_{a_n} \subset \mathscr{C}_\alpha$, hence all $A_n \in \mathscr{C}_\alpha$. Thus $\bigcup_n A_n \in \mathscr{C}_{\alpha+1} \subset \mathscr{S}$, so $\bigcup_n A_n \in \mathscr{S}$. If $A \in \mathscr{S}$, then A belongs to some $\mathscr{C}_\alpha$; hence $A^c \in \mathscr{C}_{\alpha+1} \subset \mathscr{S}$.

    (b) We have card $\mathscr{C}_\alpha \leq c$ for all α. This is true for $\alpha = 0$, by hypothesis. If it is true for all $\beta < \alpha$, then $\bigcup_{\beta < \alpha} \mathscr{C}_\beta$ has cardinality at most (card α)$c = c$. Now if $\mathscr{D}$ has cardinality c, then $\mathscr{D}'$ has cardinality at most $c^{\aleph_0} = (2^{\aleph_0})^{\aleph_0} = 2^{\aleph_0} = c$. Thus card $\mathscr{C}_\alpha \leq c$. It follows that $\bigcup_{\alpha < \beta_1} \mathscr{C}_\alpha$ has cardinality at most c.

Section 1.3

3. (a) Since $\lambda(\emptyset) = 0$ we have $\Omega \in \mathscr{M}$, and $\mathscr{M}$ is clearly closed under complementation. If $E, F \in \mathscr{M}$ and $A \subset \Omega$, then

$$\lambda[A \cap (E \cup F)] = \lambda[A \cap (E \cup F) \cap E]$$
$$+ \lambda[A \cap (E \cup F) \cap E^c] \qquad \text{since} \qquad E \in \mathscr{M}$$
$$= \lambda(A \cap E) + \lambda(A \cap F \cap E^c).$$

Thus

$$\lambda[A \cap (E \cup F)] + \lambda[A \cap (E \cup F)^c]$$
$$= \lambda(A \cap E) + \lambda(A \cap E^c \cap F) + \lambda(A \cap E^c \cap F^c)$$
$$= \lambda(A \cap E) + \lambda(A \cap E^c) \qquad \text{since} \qquad F \in \mathcal{M}$$
$$= \lambda(A) \qquad \text{since} \qquad E \in \mathcal{M}.$$

This proves that $\mathcal{M}$ is a field. Also, if E and F are disjoint we have

$$\lambda[A \cap (E \cup F)] = \lambda[A \cap (E \cup F) \cap E] + \lambda[A \cap (E \cup F) \cap E^c]$$
$$= \lambda(A \cap E) + \lambda(A \cap F \cap E^c)$$
$$= \lambda(A \cap E) + \lambda(A \cap F) \qquad \text{since} \qquad E \cap F = \emptyset.$$

Now if the E_n are disjoint sets in $\mathcal{M}$ and $F_n = \bigcup_{i=1}^{n} E_i \uparrow E$, then

$$\lambda(A) = \lambda(A \cap F_n) + \lambda(A \cap F_n^c)$$
$$\text{since } F_n \text{ belongs to the field } \mathcal{M}$$
$$\geq \lambda(A \cap F_n) + \lambda(A \cap E^c)$$
$$\text{since } E^c \subset F_n^c \text{ and } \lambda \text{ is monotone}$$
$$= \sum_{i=1}^{n} \lambda(A \cap E_i) + \lambda(A \cap E^c)$$

by what we have proved above.

Since n is arbitrary,

$$\lambda(A) \geq \sum_{n=1}^{\infty} \lambda(A \cap E_n) + \lambda(A \cap E^c)$$
$$\geq \lambda(A \cap E) + \lambda(A \cap E^c) \qquad \text{by countable subadditivity of } \lambda.$$

Thus $E \in \mathcal{M}$, proving that $\mathcal{M}$ is a σ-field.

Now $\lambda(A \cap E) + \lambda(A \cap E^c) \geq \lambda(A)$ by subadditivity, hence

$$\lambda(A) = \sum_{n=1}^{\infty} \lambda(A \cap E_n) + \lambda(A \cap E^c).$$

Replace A by $A \cap E$ to obtain $\lambda(A \cap E) = \sum_{n=1}^{\infty} \lambda(A \cap E_n)$, as desired.

(b) All properties are immediate except for countable subadditivity. If $A = \bigcup_{n=1}^{\infty} A_n$, we must show that $\mu^*(A) \leq \sum_{n=1}^{\infty} \mu^*(A_n)$, and we may assume that $\mu^*(A_n) < \infty$ for all n. Given $\varepsilon > 0$, we may choose sets $E_{nk} \in \mathscr{F}_0$ with $A_n \subset \bigcup_k E_{nk}$ and $\sum_k \mu(E_{nk}) \leq \mu^*(A_n) + \varepsilon 2^{-n}$. Then $A \subset \bigcup_{n,k} E_{nk}$ and $\sum_{n,k} \mu(E_{nk}) \leq \sum_n \mu^*(A_n) + \varepsilon$. Thus $\mu^*(A) \leq \sum_n \mu^*(A_n) + \varepsilon$, ε arbitrary.

Now if $A \in \mathscr{F}_0$, then $\mu^*(A) \leq \mu(A)$ by definition of μ^*, and if $A \subset \bigcup_n E_n$, $E_n \in \mathscr{F}_0$, then $\mu(A) \leq \sum_n \mu(E_n)$ by 1.2.5 and 1.3.1. Take the infimum over all such coverings of A to obtain $\mu(A) \leq \mu^*(A)$; hence $\mu^* = \mu$ on $\mathscr{F}_0$.

(c) If $F \in \mathscr{F}_0$, $A \subset \Omega$, we must show that $\mu^*(A) \geq \mu^*(A \cap F) + \mu^*(A \cap F^c)$; we may assume $\mu^*(A) < \infty$. Given $\varepsilon > 0$, there are sets $E_n \in \mathscr{F}_0$ with $A \subset \bigcup_n E_n$ and $\sum_{n=1}^{\infty} \mu(E_n) \leq \mu^*(A) + \varepsilon$. Now

$$\mu^*(A \cap F) \leq \mu^* \left(\bigcup_n (E_n \cap F) \right) \qquad \text{by monotonicity}$$

$$\leq \sum_n \mu(E_n \cap F)$$

since μ^* is countably subadditive and $\mu^* = \mu$ on $\mathscr{F}_0$.

Similarly,
$$\mu^*(A \cap F^c) \leq \sum_n \mu(E_n \cap F^c).$$

Thus
$$\mu^*(A \cap F) + \mu^*(A \cap F^c) \leq \sum_n \mu(E_n) \leq \mu^*(A) + \varepsilon,$$

and the result follows.

(d) If $A = B \cup N$, where $B \in \sigma(\mathscr{F}_0)$, $N \subset M \in \sigma(\mathscr{F}_0)$, $\mu^*(M) = 0$, then $B \in \mathscr{M}$ [note $\mathscr{F}_0 \subset \mathscr{M}$ and $\mathscr{M}$ is a σ-field, so $\sigma(\mathscr{F}_0) \subset \mathscr{M}$]. Also, any set C with $\mu^*(C) = 0$ belongs to $\mathscr{M}$ by definition of μ^*-measurability; hence $A \in \mathscr{M}$. Therefore the completion of $\sigma(\mathscr{F}_0)$ is included in $\mathscr{M}$.

Now assume μ σ-finite on $\mathscr{F}_0$, and let $A \in \mathscr{M}$. If Ω is the disjoint union of sets $A_n \in \mathscr{F}_0$ with $\mu(A_n) < \infty$, then by definition of μ^*, there is a set $B_n \in \sigma(\mathscr{F}_0)$ such that $A \cap A_n \subset B_n$ and $\mu^*(B_n - (A \cap A_n)) = 0$. [Note that if $A \notin \mathscr{M}$ we obtain only $\mu^*(B_n) = \mu^*(A \cap A_n)$; however, if $A \in \mathscr{M}$ (so that $A \cap A_n$ also belongs to $\mathscr{M}$), we have

$$\mu^*(B_n - (A \cap A_n)) = \mu^*(B_n) - \mu^*(A \cap A_n) = 0.]$$

If $B = \bigcup_n B_n$, then $B \in \sigma(\mathscr{F}_0), A \subset B$, and $\mu^*(B - A) = 0$. This argument applied to A^c yields a set $C \in \sigma(\mathscr{F}_0)$ with $C \subset A$ and $\mu^*(A - C) = 0$. Therefore, $A = C \cup (A - C)$ with $C \in \sigma(\mathscr{F}_0), A - C \subset B - C \in \sigma(\mathscr{F}_0)$, and $\mu^*(B - C) = \mu^*(B - A) + \mu^*(A - C) = 0$. Thus A belongs to the completion of $\sigma(\mathscr{F}_0)$ relative to μ^*.

Section 1.4

1. Using the formulas of 1.4.5, the following results are obtained:

   (a) 3;   (b) 8.5;   (c) 5;

   (d) 7.25;   (e) $\mu\left(\frac{1}{2}, \infty\right) + \mu\left(-\infty, -\frac{1}{2}\right) = 7.25$.

4. Let $C_k = \{x \in \mathbb{R}^n: -k < x_i \le k, i = 1, \ldots, n\}$. Then μ is finite on C_k; hence the Borel subsets B of C_k such that $\mu(a + B) = \mu(B)$ form a monotone class including the field of finite disjoint unions of right-semiclosed intervals in C_k; hence all Borel subsets of C_k belong to the class (see 1.2.2). If $B \in \mathscr{B}(\mathbb{R}^n)$, then $B \cap C_k \uparrow B$; hence $a + (B \cap C_k) \uparrow a + B$, and it follows that $\mu(a + B) = \mu(B)$.

   Now if $B \in \overline{\mathscr{B}}(\mathbb{R}^n)$, then $B = A \cup C, A \in \mathscr{B}(\mathbb{R}^n), C \subset D \in \mathscr{B}(\mathbb{R}^n)$, with $\mu(D) = 0$. Thus $a + B = (a + A) \cup (a + C)$, and, by Problem 3, $a + A \in \mathscr{B}(\mathbb{R}^n), a + C \subset a + D \in \mathscr{B}(\mathbb{R}^n)$. By what we have proved above, $\mu(a + D) = \mu(D) = 0$; hence $a + B \in \overline{\mathscr{B}}(\mathbb{R}^n)$ and $\mu(a + B) = \mu(B)$.

5. Let A be the unit cube $\{x \in \mathbb{R}^n: 0 < x_i \le 1, i = 1, \ldots, n\}$, and let $c = \mu(A)$. For any positive integer, r, we may divide each edge of A into r equal parts, so that A is decomposed into r^n subcubes $A_1, \ldots, A_{r^n}$, each with volume r^{-n}. By translation-invariance, $\mu(A_i)$ is the same for all i, so if λ is Lebesgue measure, we have

$$\mu(A_i) = r^{-n}\mu(A) = r^{-n}c = r^{-n}c\lambda(A) = c\lambda(A_i), \qquad i = 1, \ldots, r^n.$$

   Now any subinterval I of the unit cube can be expressed as the limit of an increasing sequence of sets B_k, where each B_k is a finite disjoint union of subcubes of the above type. Thus $\mu = c\lambda$ on subintervals of the unit cube, and hence on all Borel subsets of the unit cube by the Carathéodory extension theorem. Since $\mathbb{R}^n$ is a countable disjoint union of cubes, it follows that $\mu = c\lambda$ on $\mathscr{B}(\mathbb{R}^n)$.

6. (a) If $r + x_1 = s + x_2, x_1, x_2 \in A$, then x_1 is equivalent to x_2, so that $x_1 = x_2$ since A was constructed by taking one member from each distinct B_x. Thus $r = s$, a contradiction.

   If $x \in \mathbb{R}$, then $x \in B_x$; if y is the member of B_x that belongs to A, then $x - y$ is a rational number r, hence $x \in r + A$.

(b)   If $0 \le r \le 1$, then $r + A \subset [0, 2]$; thus

$$\sum \{\mu(r + A): \ 0 \le r \le 1, \quad r \text{ rational}\}$$

$$= \mu \left(\bigcup \{r + A: \ 0 \le r \le 1, \quad r \text{ rational}\} \right) \quad \text{by (a)}$$

$$\le \mu[0, 2] < \infty.$$

But $\mu(r + A) = \mu(A)$ by Problem 4; hence $\mu(r + A)$ must be 0 for all r. Since $\mathbb{R}$ is a countable union of sets $r + A$ by (a), $\mu(\mathbb{R}) = 0$, a contradiction.

8.   Let $F(x, y) = 1$ if $x + y \ge 0$; $F(x, y) = 0$ if $x + y < 0$. If $a_1 = 0$, $b_1 = 1, a_2 = -1, b_2 = 0$, then

$$\Delta_{b_1 a_1} \Delta_{b_2 a_2} F(x, y) = F(b_1, b_2) - F(a_1, b_2)$$

$$- F(b_1, a_2) + F(a_1, a_2)$$

$$= 1 - 1 - 1 + 0 = -1;$$

hence F is not a distribution function. Other examples: $F(x, y) = \max(x, y)$, $F(x, y) = [x + y]$, the largest integer less than or equal to $x + y$.

Section 1.5

2.   If $B \in \mathscr{B}(\mathbb{R})$,

$$\{\omega: \ h(\omega) \in B\} = \{\omega \in A: \ h(\omega) \in B\} \cup \{\omega \in A^c: \ h(\omega) \in B\}$$

$$= [A \cap f^{-1}(B)] \cup [A^c \cap g^{-1}(B)],$$

which belongs to $\mathscr{F}$ since f and g are Borel measurable.

5.   (a) $\{x: f \text{ is discontinuous at } x\} = \bigcup_{n=1}^{\infty} D_n$, where $D_n = \{x \in \mathbb{R}^k:$ for all $\delta > 0$, there exist $x_1, x_2 \in \mathbb{R}^k$ such that $|x_1 - x| < \delta$ and $|x_2 - x| < \delta$, but $|f(x_1) - f(x_2)| \ge 1/n\}$. We show that the D_n are closed. Let $\{x_\alpha\}$ be a sequence of points in D_n with $x_\alpha \to x$. If $\delta > 0$ and $N = \{y: |y - x| < \delta\}$, then $x_\alpha \in N$ for large α, and since $x_\alpha \in D_n$, there are points x_{a_1} and $x_{a_2} \in N$ such that $|f(x_{a_1}) - f(x_{a_2})| \ge 1/n$. Thus $|x_{a_1} - x| < \delta$, $|x_{a_2} - x| < \delta$, but $|f(x_{a_1}) - f(x_{a_2})| \ge 1/n$, so that $x \in D_n$.

The result is true for a function from an arbitrary topological space S to a metric space (T, d). Take $D_n = \{x \in S:$ for every neighborhood N of x, there exist $x_1, x_2 \in N$ such that $d(f(x_1), f(x_2)) \ge 1/n\}$. (The above proof goes through with "sequence" replaced by "net.")

The result is false if no assumptions are made about the topology of the range space. For example, let $\Omega = \{1, 2, 3\}$, with open sets

$\emptyset, \Omega$, and $\{1\}$. Define $f: \Omega \to \Omega$ by $f(1) = f(3) = 1$, $f(2) = 2$.
Then the set of discontinuities is $\{3\}$, which is not an F_σ.

(b)   This follows from part (a) because the irrationals I cannot be expressed as a countable union of closed sets C_n. If this were possible,
then each C_n would have empty interior since every nonempty open
set contains rational points. But then I is of category 1 in $\mathbb{R}$, and
since $Q = \mathbb{R} - I$ is of category 1 in $\mathbb{R}$, it follows that $\mathbb{R}$ is of category
1 in itself, contradicting the Baire category theorem.

6.   By Problem 11 in 1.4, there are c Borel subsets of $\mathbb{R}^n$; hence there are only
c simple functions on $\mathbb{R}^n$. Since a Borel measurable function is the limit
of a sequence of simple functions, there are $c^{\aleph_0} = c$ Borel measurable
functions from $\mathbb{R}^n$ to $\mathbb{R}$. By 1.5.8, there are only c Borel measurable
functions from $\mathbb{R}^n$ to $\mathbb{R}^k$.

7.   (a)   Since the P_n are measures, $\sum_k P_n(A_k) = P_n(\Omega) = 1$, and it follows
quickly that the a_{nk} satisfy the hypotheses of Steinhaus' lemma. If
$\{x_n\}$ is the sequence given by the lemma, let $S = \{k: x_k = 1\}$ and
let B be the union of the sets $A_k, k \in S$. Then

$$t_n = \frac{1}{1-\alpha} \sum_{k \in S} [P_n(A_k) - P(A_k)] = \frac{1}{1-\alpha}\left[P_n(B) - \sum_{k \in S} P(A_k)\right],$$

and it follows that t_n converges, a contradiction. Thus P is a probabil-
ity measure. If $B_k \in \mathscr{F}, B_k \downarrow \emptyset$, then given $\varepsilon > 0$, we have $P(A_k) < \varepsilon$
for large k, say, for $k \geq k_0$. Thus $P_n(A_{k_0}) < \varepsilon$ for large n, say, for
$n \geq n_0$. Since the A_k decrease, we have $\sup_{n \geq n_0} P_n(A_k) \leq \varepsilon$ for $k \geq
k_0$, and since $A_k \downarrow \emptyset$, there is a k_1 such that for $n = 1, 2, \ldots, n_0 -
1, P_n(A_k) < \varepsilon$ for $k \geq k_1$. Thus $\sup_n P_n(A_k) \leq \varepsilon, k \geq \max(k_0, k_1)$.

(b)   Without loss of generality, assume $P_n(\Omega) \leq 1$ for all n. Add a point
(call it ∞) to the space and set $P_n\{\infty\} = 1 - P_n(\Omega) \to 1 - P(\Omega) =
P\{\infty\}$. The P_n are now probability measures, and the result follows
from part (a).

Section 1.6

2. $\int_\Omega \sum_{n=1}^\infty |f_n|\, d\mu = \sum_{n=1}^\infty \int_\Omega |f_n|\, d\mu < \infty$; hence $\sum_{n=1}^\infty |f_n|$ is integrable
and therefore finite a.e. Thus $\sum_{n=1}^\infty f_n$ converges a.e. to a finite-valued
function g.

Let $g_n = \sum_{k=1}^n f_k$. Then $|g_n| \leq \sum_{k=1}^\infty |f_k|$, an integrable function. By
the dominated convergence theorem, $\int_\Omega g_n\, d\mu \to \int_\Omega g\, d\mu$, that is,

$$\sum_{n=1}^\infty \int_\Omega f_n\, d\mu = \int_\Omega \sum_{n=1}^\infty f_n\, d\mu.$$

3. Let $x_0 \in (c, d)$, and let $x_n \to x_0, x_n \neq x_0$. Then

$$\frac{1}{(x_n - x_0)} \left[\int_a^b f(x_n, y)\, dy - \int_a^b f(x_0, y)\, dy \right]$$
$$= \int_a^b \left[\frac{f(x_n, y) - f(x_0, y)}{x_n - x_0} \right] dy.$$

By the mean value theorem,

$$\frac{f(x_n, y) - f(x_0, y)}{x_n - x_0} = f_1(\lambda_n, y)$$

for some $\lambda_n = \lambda_n(y)$ between x_n and x_0. By hypothesis, $|f_1(\lambda_n, y)| \leq h(y)$, where h is integrable, and the result now follows from the dominated convergence theorem (since $[f(x_n, y) - f(x_0, y)]/[x_n - x_0] \to f_1(x_0, y)$, $f_1(x, \cdot)$ is Borel measurable for each x).

8. Let μ be Lebesgue measure. If f is an indicator I_B, $B \in \mathcal{B}(\mathbb{R})$, the result to be proved states that $\mu(B) = \mu(a + B)$, which holds by translation-invariance of μ (Problem 4 in 1.4). The passage to nonnegative simple functions, nonnegative measurable functions, and arbitrary measurable functions is done as in 1.6.12.

Section 1.7

2. (a) If f is Riemann–Stieltjes integrable, $\alpha = f = \beta$ a.e. $[\mu]$ as in 1.7.1(a). Thus the set of discontinuities of f is a subset of a set of μ-measure 0, together with the endpoints of the subintervals of the P_k. Take a different sequence of partitions having the original endpoints as interior points to conclude that f is continuous a.e. $[\mu]$. Conversely, if f is continuous a.e. $[\mu]$, then $\alpha = f = \beta$ a.e. $[\mu]$. [The result that f is continuous at x implies $\alpha(x) = f(x) = \beta(x)$ is true even if x is an endpoint.] As in 1.7.1(a), f is Riemann–Stieltjes integrable.
   (b) This is done exactly as in 1.7.1(b).

3. (a) By definition of the improper Riemann integral, f must be Riemann integrable (hence continuous a.e.) on each bounded interval, and the result follows. For the counterexample to the converse, take $f(x) = 1$, $n \leq x < n + 1$, n an even integer; $f(x) = -1$, $n \leq x < n + 1$, n an odd integer. Then the limit of $r_{ab}(f)$ does not exist. (Alternatively, take $f(x)$ identically 1; then $r_{ab}(f) \to +\infty$ as $a \to -\infty$, $b \to \infty$.)
   (b) Define

$$f_n(x) = \begin{cases} f(x) & \text{if } -n \leq x \leq n, \\ 0 & \text{elsewhere.} \end{cases}$$

Then $f_n \uparrow f$; hence f is measurable relative to the completed σ-field; also $\int_\Omega f_n \, d\mu \uparrow \int_\Omega f \, d\mu$ by the monotone convergence theorem. But $\int_\Omega f_n \, d\mu = r_{-n,n}(f)$ by 1.7.1(b), and $r_{-n,n}(f) \to r(f)$ by hypothesis; the result follows.

For the counterexample, take

$$f(x) = \begin{cases} \dfrac{(-1)^n}{n+1}, & n \le x < n+1, \qquad n = 0, 1, \ldots, \\ 0, & x < 0. \end{cases}$$

We have $r(f) = 1 - \frac{1}{2} + \frac{1}{3} - \cdots$, but

$$\int_{\mathbb{R}} |f| \, d\mu = 1 + \frac{1}{2} + \frac{1}{3} + \cdots = \infty,$$

so that f is not Lebesgue integrable on $\mathbb{R}$.

CHAPTER 2

Section 2.1

2. Let $D = \{\omega: f(\omega) < 0\}$; then $\lambda(A \cap D) \le 0$, $\lambda(A \cap D^c) \ge 0$ for all $A \in \mathscr{F}$. By 2.1.3(d),

$$\lambda^+(A) = \lambda(A \cap D^c) = \int_A f^+ \, d\mu$$

since $f^+ = f$ on D^c and $f^+ = 0$ on D. Similarly,

$$\lambda^-(A) = -\lambda(A \cap D) = \int_A f^- \, d\mu$$

since $f^- = -f$ on D, and $f^- = 0$ on D^c. The result follows.

4. If $E_1, \ldots, E_n$ are disjoint sets in $\mathscr{F}$, with all $E_i \subset A$,

$$\sum_{i=1}^n |\lambda(E_i)| = \sum_{i=1}^n |\lambda^+(E_i) - \lambda^-(E_i)| \le \sum_{i=1}^n [\lambda^+(E_i) + \lambda^-(E_i)]$$

$$= |\lambda| \left(\bigcup_{i=1}^n E_i \right) \le |\lambda|(A).$$

Thus the sup of the terms $\sum_{i=1}^n |\lambda(E_i)|$ is at most $|\lambda|(A)$. But

$$|\lambda|(A) = \lambda^+(A) + \lambda^-(A)$$
$$= \lambda(A \cap D^c) - \lambda(A \cap D)$$

$$= |\lambda(A \cap D^c)| + |\lambda(A \cap D)|$$

$$\text{since } \lambda(A \cap D^c) \geq 0 \quad \text{and} \quad \lambda(A \cap D) \leq 0$$

$$= |\lambda(E_1)| + |\lambda(E_2)|$$

and the result follows.

Section 2.2

2. Let $A_n = \{\omega: |g(\omega)| \geq 1/n\}$, $n = 1, 2, \ldots$, so that $A = \bigcup_{n=1}^{\infty} A_n$. Now

$$\infty > \int_{A_n} |g| \, d\mu \geq \frac{1}{n} \mu(A_n);$$

hence $\mu(A_n) < \infty$. For the example, let μ be Lebesgue measure on $\mathscr{B}(\mathbb{R})$, and let $g(x)$ be any strictly positive integrable function, such as $g(x) = e^{-|x|}$. In this case, $A = \mathbb{R}$, so that $\mu(A) = \infty$.

4. If f is an indicator I_A, the result is true by hypothesis. If f is a nonnegative simple function $\sum_{j=1}^{n} x_j I_{A_j}$, the A_j disjoint sets in $\mathscr{F}$, then

$$\int_{\Omega} f \, d\lambda = \sum_{j=1}^{n} x_j \lambda(A_j) = \sum_{j=1}^{n} x_j \int_{A_j} g \, d\mu = \sum_{j=1}^{n} x_j \int_{\Omega} I_{A_j} g \, d\mu$$

$$= \int_{\Omega} fg \, d\mu \qquad \text{by the additivity theorem.}$$

If f is a nonnegative Borel measurable function, let $f_1, f_2, \ldots$ be nonnegative simple functions increasing to f. By what we have just proved, $\int_{\Omega} f_n d\lambda = \int_{\Omega} f_n g \, d\mu$; hence $\int_{\Omega} f \, d\lambda = \int_{\Omega} fg \, d\mu$ by the monotone convergence theorem. Finally, if f is an arbitrary Borel measurable function, write $f = f^+ - f^-$. By what we have just proved,

$$\int_{\Omega} f^+ \, d\lambda = \int_{\Omega} f^+ g \, d\mu, \qquad \int_{\Omega} f^- \, d\lambda = \int_{\Omega} f^- g \, d\mu$$

and the result follows from the additivity theorem.

6. (a) In the definition of $|\lambda|$, we may assume without loss of generality that the E_i partition A. If A is the disjoint union of sets $A_1, A_2, \ldots, \in \mathscr{F}$, then

$$\sum_{j=1}^{n} |\lambda(E_j)| = \sum_{j=1}^{n} \left| \sum_{i=1}^{\infty} \lambda(E_j \cap A_i) \right| \leq \sum_{j=1}^{n} \sum_{i=1}^{\infty} |\lambda(E_j \cap A_i)|$$

$$= \sum_{i=1}^{\infty} \sum_{j=1}^{n} |\lambda(E_j \cap A_i)| \leq \sum_{i=1}^{\infty} |\lambda|(A_i).$$

Thus $|\lambda|(A) \leq \sum_{i=1}^{\infty} |\lambda|(A_i)$. Now to show the reverse inequality, we may assume $|\lambda|(A) < \infty$; hence $|\lambda|(A_i) \leq |\lambda|(A) < \infty$. For each i, there is a partition $\{E_{i1}, \ldots, E_{in_i}\}$ of A_i such that

$$\sum_{j=1}^{n_i} |\lambda(E_{ij})| > |\lambda|(A_i) - \frac{\varepsilon}{2^i}, \qquad \varepsilon > 0 \qquad \text{preassigned.}$$

Then for any n,

$$|\lambda|(A) \geq \sum_{i=1}^{n} \sum_{j=1}^{n_i} |\lambda(E_{ij})| \geq \sum_{i=1}^{n} |\lambda|(A_i) - \varepsilon.$$

Since n and ε are arbitrary, the result follows.

(b) If $E_1, \ldots, E_n$ are disjoint measurable subsets of A,

$$\sum_{i=1}^{n} |(\lambda_1 + \lambda_2)(E_i)| \leq \sum_{i=1}^{n} |\lambda_1(E_i)| + \sum_{i=1}^{n} |\lambda_2(E_i)|$$

$$\leq |\lambda_1|(A) + |\lambda_2|(A),$$

proving $|\lambda_1 + \lambda_2| \leq |\lambda_1| + |\lambda_2|$; $|a\lambda| = |a| \, |\lambda|$ is immediate from the definition of total variation.

(c) If $\mu(A_i) = 0$ and $|\lambda_i|(A_i^c) = 0$, $i = 1, 2$, then $\mu(A_1 \cup A_2) = 0$ and by (b), $|\lambda_1 + \lambda_2|(A_1^c \cap A_2^c) \leq |\lambda_1|(A_1^c) + |\lambda_2|(A_2^c) = 0$.

(d) This has been established when λ is real (see 2.2.5), so assume λ complex, say, $\lambda = \lambda_1 + i\lambda_2$. If $\mu(A) = 0$, then $\lambda_1(A) = \lambda_2(A) = 0$; hence $\lambda \ll \mu$ implies $\lambda_1 \ll \mu$ and $\lambda_2 \ll \mu$. By 2.2.5(b), $|\lambda_1| \ll \mu$, $|\lambda_2| \ll \mu$; hence by (b), $|\lambda| \ll \mu$. The converse is clear since $|\lambda(A)| \leq |\lambda|(A)$.

(e) The proof is the same as in 2.2.5(c).

(f) See 2.2.5(d).

(g) The "if" part is done as in 2.2.5(e); for the "only if" part, let $\mu(A_n) \to 0$. Since $|\lambda| \ll \mu$ by (d), $|\lambda|(A_n) \to 0$ by 2.2.5(e); hence $\lambda(A_n) \to 0$.

Section 2.3

2. We have

$$\int_a^b f'(x) \, dx = \int_a^b \lim_{h \to 0} \left[\frac{f(x+h) - f(x)}{h} \right] dx$$

where we may assume $h > 0$

$$\leq \liminf_{h \to 0} \int_a^b \left[\frac{f(x+h) - f(x)}{h} \right] dx$$

$$= \liminf_{h \to 0} \frac{1}{h} \left[\int_b^{b+h} f(x)\,dx - \int_a^{a+h} f(x)\,dx \right]$$

[define $f(x) = f(b), \quad x > b; \quad f(x) = f(a), \quad x < a$]

$$\leq \liminf_{h \to 0} \frac{1}{h}[hf(b+h) - hf(a)] \qquad \text{since } f \text{ is increasing}$$

$$= f(b) - f(a).$$

Alternatively, let μ be the Lebesgue–Stieltjes measure corresponding to f (adjusted so as to be right continuous), and let m be Lebesgue measure. Write $\mu = \mu_1 + \mu_2$, where $\mu_1 << m$ and $\mu_2 \perp m$. By 2.3.8 and 2.3.9,

$$\int_a^b f'(x)\,dx = \int_a^b f'\,dm$$

$$= \int_a^b d\mu\,dm$$

$$= \int_a^b \frac{d\mu_1}{dm}\,dm = \mu_1(a,b] \leq \mu(a,b] = f(b) - f(a).$$

6.   (a)   Since A is linear and m is translation-invariant, so is λ; hence by Problem 5 in 1.4, $\lambda = c(A)m$ for some constant $c(A)$. Now if A_1 and A_2 are linear transformations on $\mathbb{R}^k$, then

$$m(A_1 A_2 E) = c(A_1)m(A_2 E) = c(A_1)c(A_2)m(E)$$

and

$$m(A_1 A_2 E) = c(A_1 A_2)m(E);$$

hence

$$c(A_1 A_2) = c(A_1)c(A_2).$$

Since $\det(A_1 A_2) = \det A_1 \det A_2$, it suffices to assume that A corresponds to an elementary row operation. Now if Q is the unit cube $\{x: 0 < x_i \leq 1, i = 1, \ldots, k\}$, then $m(Q) = 1$; hence $c(A) = m(A(Q))$. If $e_1, \ldots, e_k$ is the standard basis for $\mathbb{R}^k$, then A falls into one of the following three categories:

(1) $Ae_i = e_j, \ Ae_j = e_i, \ Ae_k = e_k, \ k \neq i$ or j. Then $c(A) = 1 = |\det A|$.

(2) $Ae_i = ke_i, \ Ae_j = e_j, \ j \neq i$. Then $c(A) = |k| = |\det A|$.

(3) $Ae_i = e_i + e_j, Ae_k = e_k, k \neq i$. Then $\det A = 1$ and $c(A)$ is 1 also, by the following argument. We may assume $i = 1$, $j = 2$. Then

$$A(Q) = \left\{ A \sum_{i=1}^{k} a_i e_i \colon 0 < a_i \leq 1, \quad i = 1, \ldots, k \right\}$$

$$= \left\{ y = \sum_{i=1}^{k} b_i e_i \colon 0 < b_i \leq 1, \quad i \neq 2, \quad b_1 < b_2 \leq b_1 + 1 \right\}.$$

If $B_1 = \{y \in A(Q) \colon b_2 \leq 1\}$, $B_2 = \{y \in A(Q) \colon b_2 > 1\}$, and $B_2 - e_2 = \{y - e_2 \colon y \in B_2\}$, then $B_1 \cap (B_2 - e_2) = \emptyset$; for if $y \in B_1$, then $b_1 < b_2$ and if $y + e_2 \in B_2 \subset A(Q)$, then $b_2 + 1 \leq b_1 + 1$. Therefore,

$$c(A) = m(A(Q)) = m(B_1) + m(B_2) = m(B_1) + m(B_2 - e_2)$$
$$= m(B_1 \cup (B_2 - e_2)) = m(Q) = 1.$$

(b)   Fix $x \in V$ and define $S(y) = T(x + y) - T(x)$, $y \in \{z - x \colon z \in V\}$. Then if C is a cube containing 0, we have

$$T(x + C) = \{T(x + y) \colon y \in C\} = T(x) + \{S(y) \colon y \in C\}$$
$$= T(x) + S(C).$$

Therefore, $m(T(x + C)) = m(S(C))$, so that differentiability of T at x is equivalent to differentiability of S at 0, and the derivatives, if they exist, are equal. Also, the Jacobian matrix of S at 0 coincides with the Jacobian matrix of T at x, and the result follows.

(c)   Let $A = A(0)$, and define $S(x) = A^{-1}(T(x))$, $x \in V$; the Jacobian matrix of S at 0 is the identity matrix, and $S(0) = A^{-1}(0) = 0$. If we show that the measure given by $m(S(E))$, $E \in \mathscr{B}(V)$, is differentiable at 0 with derivative 1, then

$$\left| \frac{m(S(C))}{m(C)} - 1 \right| < \frac{\varepsilon}{|\det A|}$$

for sufficiently small cubes C containing 0. Thus by (a), $m(T(C)) = m(AS(C)) = |\det A| m(S(C))$; hence

$$\left| \frac{m(T(C))}{m(C)} - |\det A| \right| = \left| \frac{m(S(C))}{m(C)} - 1 \right| |\det A| < \varepsilon,$$

so that μ is differentiable at 0 with derivative $|\det A|$.

(d) (i) If $x \in \overline{C}$, then $|x| \leq \sqrt{k}\,\beta < \delta$; hence $|T(x) - x| \leq \alpha\beta = \frac{1}{2}(\beta_2 - \beta)$. Therefore $T(x) \in C_2$.

(ii) If $x \in \partial C$, then $|T(x) - x| \leq \alpha\beta$ as above, and $\alpha\beta = \frac{1}{2}(\beta - \beta_1)$. Therefore $T(x) \notin C_1$.

(iii) We have $|T(x) - x| \leq \alpha\beta$, and

$$\alpha\beta = \frac{\alpha\beta_1}{1 - 2\alpha} \leq \frac{\frac{1}{4}}{1 - \frac{1}{2}}\beta_1 = \frac{1}{2}\beta_1,$$

and the result follows.

(iv) If $y \in C_1 - T(C)$ but $y \in T(\overline{C})$, then $y \in T(\partial C)$, contradicting (ii).

Now C_1 is the disjoint union of the sets $C_1 \cap T(C)$ and $C_1 - T(C)$; the first set is open since T is an open map, and the second set is open by (iv). By (iii), $C_1 \cap T(C) \neq \emptyset$, so by connectedness of C_1, we have $C_1 = C_1 \cap T(C)$, that is, $C_1 \subset T(C)$. Also, $T(C) \subset C_2$ by (i). Therefore $m(C_1) \leq m(T(C)) \leq m(C_2)$, that is,

$$(1 - 2\alpha)^k \beta^k \leq m(T(C)) \leq (1 + 2\alpha)^k \beta^k.$$

Thus $1 - \varepsilon < (1 - 2\alpha)^k < m(T(C))/m(C) \leq (1 + 2\alpha)^k < 1 + \varepsilon$ as desired.

(e) Assume $m(E) = 0$ and $\lambda(E) > 0$. By 1.4.11, and the fact that

$$E = \bigcup_{n,j=1}^{\infty} E \cap \left\{ \sup_{C_r, r<1/j} \frac{\lambda(C_r)}{m(C_r)} < n \right\},$$

we can find a compact set K and positive integers n and j such that $m(K) = 0$, $\lambda(K) > 0$, and $\lambda(C) < nm(C)$ for all open cubes C containing a point of K and having diameter less than $1/j$. If $\varepsilon > 0$, choose an open set $D \supset K$ such that $m(D) < \varepsilon$.

Now partition $\mathbb{R}^k$ into disjoint (partially closed) cubes B of diameter less than $1/j$ and small enough so that if $B \cap K \neq \emptyset$, then $B \subset D$. If the cubes that meet K are $B_1, \ldots, B_t$, we may find open cubes $C_i \supset B_i, 1 \leq i \leq t$, such that $m(C_i) < 2m(B_i)$ and diam $C_i < 1/j$. Then

$$\lambda(K) \leq \sum_{i=1}^{t} \lambda(B_i) \leq \sum_{i=1}^{t} \lambda(C_i) < n \sum_{i=1}^{t} m(C_i) < 2n \sum_{i=1}^{t} m(B_i)$$
$$\leq 2nm(D) < 2n\varepsilon.$$

Since ε is arbitrary, $\lambda(K) = 0$, a contradiction.

(f) If f is an indicator I_B, let $E = T^{-1}(B)$, so that $B = T(E)$. Then

$$\int_W f(y)\,dy = m(T(E)) \quad \text{and} \quad \int_V f(T(x))|J(x)|\,dx = \int_E |J(x)|\,dx,$$

and the result follows from part (e). The usual passage to simple functions, nonnegative measurable functions and arbitrary measurable functions completes the proof.

Section 2.4

1. (a) If $f = \{a_1, \ldots, a_n, 0, 0, \ldots\}$, then $\int_\Omega f\,d\mu = \sum_{k=1}^n a_k$ by definition of the integral of a simple function. If $f = \{a_n, n = 1, 2, \ldots\}$, with all $a_n \geq 0$, $\int_\Omega f\,d\mu = \sum_{n=1}^\infty a_n$ by the result for simple functions and the monotone convergence theorem. If the a_n are real numbers, then $\int_\Omega f\,d\mu = \int_\Omega f^+\,d\mu - \int_\Omega f^-\,d\mu = \sum_{n=1}^\infty a_n^+ - \sum_{n=1}^\infty a_n^-$ if this is not of the form $+\infty - \infty$. Finally, if the a_n are complex,

$$\int_\Omega f\,d\mu = \sum_{n=1}^\infty \operatorname{Re} a_n + i \sum_{n=1}^\infty \operatorname{Im} a_n$$

provided $\operatorname{Re} f$ and $\operatorname{Im} f$ are integrable; since $|\operatorname{Re} a_n|, |\operatorname{Im} a_n| \leq |a_n| \leq |\operatorname{Re} a_n| + |\operatorname{Im} a_n|$, this is equivalent to $\sum_{n=1}^\infty |a_n| < \infty$.

(b) If $f(\alpha) = 0$ except for $\alpha \in F$, F finite, then $\int_\Omega f\,d\mu = \sum_\alpha f(\alpha)$ by definition of the integral of a simple function. If $f \geq 0$, then $\int_\Omega f\,d\mu \geq \int_F f\,d\mu$ for all finite F; hence $\int_\Omega f\,d\mu \geq \sum_\alpha f(\alpha)$. If $f(\alpha) > 0$ for uncountably many α, then $\sum_\alpha f(\alpha) = \infty$; hence $\int_\Omega f\,d\mu = \infty$ also. If $f(\alpha) > 0$ for only countably many α, then $\int_\Omega f\,d\mu = \sum_\alpha f(\alpha)$ by the monotone convergence theorem. The remainder of the proof is as in part (a).

8. Apply Hölder's inequality with $g = 1$, f replaced by $|f|^r$, $p = s/r$, $1/q = 1 - r/s$, to obtain

$$\int_\Omega |f|^r\,d\mu \leq \left(\int_\Omega |f|^{rp}\,d\mu\right)^{1/p} \left(\int_\Omega 1\,d\mu\right)^{1/q}.$$

Therefore $\|f\|_r \leq \|f\|_s [\mu(\Omega)]^{1/rq}$, as desired.

9. We have $\int_\Omega |f|^p\,d\mu \leq \|f\|_\infty^p \mu(\Omega)$, so $\limsup_{p\to\infty} \|f\|_p \leq \|f\|_\infty$. Now let $\varepsilon > 0$, $A = \{\omega: |f(\omega)| \geq \|f\|_\infty - \varepsilon\}$ (assuming $\|f\|_\infty < \infty$). Then

$$\int_\Omega |f|^p\,d\mu \geq \int_A |f|^p\,d\mu \geq (\|f\|_\infty - \varepsilon)^p \mu(A).$$

Since $\mu(A) > 0$ by definition of $\|f\|_\infty$, we have $\liminf_{p\to\infty} \|f\|_p$ $\geq \|f\|_\infty$. If $\|f\|_\infty = \infty$, let $A = \{\omega: |f(\omega)| \geq M\}$ and show that

$$\liminf_{p\to\infty} \|f\|_p \geq M;$$

since M can be arbitrarily large, the result follows.

If $\mu(\Omega) = \infty$, it is still true that $\liminf_{p\to\infty} \|f\|_p \geq \|f\|_\infty$; if $\mu(A) = \infty$ in the above argument, then $\|f\|_p = \infty$ for all $p < \infty$. However, if μ is Lebesgue measure on $\mathscr{B}(\mathbb{R})$ and $f(x) = 1$ for $n \leq x \leq n + (1/n)$, $n = 1, 2, \ldots$, and $f(x) = 0$ elsewhere, then $\|f\|_p = \infty$ for $p < \infty$, but $\|f\|_\infty = 1$.

11. (a) We have $|\int_E (f - z)\,d\mu| \leq \int_E |f - z|\,d\mu$. But $\omega \in E$ implies $f(\omega) \in D$; hence $|f(\omega) - z| \leq r$; and thus $\int_E |f - z|\,d\mu \leq r\mu(E)$. If $\mu(E) > 0$, then

$$\left| \frac{1}{\mu(E)} \int_E f\,d\mu - z \right| = \left| \frac{1}{\mu(E)} \int_E (f - z)\,d\mu \right| \leq r;$$

hence $[1/\mu(E)] \int_E f\,d\mu \in D \subset S^c$, a contradiction. Therefore $\mu(E) = 0$, that is, $\mu\{\omega: f(\omega) \in D\} = 0$. Since $\{\omega: f(\omega) \notin S\}$ is a countable union of sets $f^{-1}(D)$, the result follows.

(b) Let $\mu = |\lambda|$; if $E_1, \ldots, E_n$ are disjoint measurable subsets of A_r,

$$\sum_{j=1}^n |\lambda(E_j)| = \sum_{j=1}^n \left| \int_{E_j} h\,d\mu \right| \leq \sum_{j=1}^n \int_{E_j} |h|\,d\mu$$

$$\leq r \sum_{j=1}^n \mu(E_j) \leq r\mu(A_r).$$

Thus $\mu(A_r) \leq r\mu(A_r)$, and since $0 < r < 1$, we have $\mu(A_r) = 0$. If $A = \{\omega: |h(\omega)| < 1\} = \bigcup\{A_r: 0 < r < 1, r \text{ rational}\}$, then $\mu(A) = 0$, so that $|h| \geq 1$ a.e.

Now if $\mu(E) > 0$, then $[1/\mu(E)] \int_E h\,d\mu = \lambda(E)/\mu(E) \in S$, where $S = \{z \in \mathbb{C}: |z| \leq 1\}$. By (a), $h(\omega) \in S$ for almost every ω, so $|h| \leq 1$ a.e. $[|\lambda|]$.

(c) If $E \in \mathscr{F}$, $\int_\Omega I_E h\,d|\lambda| = \int_E h\,d|\lambda| = \lambda(E)$ by (b); also, $\int_\Omega I_E g\,d\mu = \int_E g\,d\mu = \lambda(E)$ by definition of λ. It follows immediately that $\int_\Omega f h\,d|\lambda| = \int_\Omega f g\,d\mu$ when f is a complex-valued simple function. If f is a bounded, complex-valued Borel measurable function, by 1.5.5(b) there are simple functions $f_n \to f$ with $|f_n| \leq |f|$. By the dominated convergence theorem, $\int_\Omega f h\,d|\lambda| = \int_\Omega f g\,d\mu$. If $f = \bar{h} I_E$, we obtain $|\lambda|(E) = \int_E \bar{h} g\,d\mu$.

(d)   In (c), $|\lambda|(E) \geq 0$ for all E; hence $\bar{h}g \geq 0$ a.e. $[\mu]$ by 1.6.11. But if $g(\omega) = |g(\omega)|e^{i\theta(\omega)}$ and $h(\omega) = e^{i\varphi(\omega)}$, then $e^{i(\theta-\varphi)} = 1$ a.e. on $\{g \neq 0\}$, so that $\bar{h}g = |g|$ a.e., as desired.

12.   If $I_{(a,b)}$ can be approximated in L^∞ by continuous functions, let $0 < \varepsilon < \frac{1}{2}$ and let f be a continuous function such that

$$\|I_{(a,b)} - f\|_\infty \leq \varepsilon;$$

hence $|I_{(a,b)} - f| \leq \varepsilon$ a.e. For every $\delta > 0$, there are points $x \in (a, a + \delta)$ and $y \in (a - \delta, a)$ such that $|1 - f(x)| \leq \varepsilon$ and $|f(y)| \leq \varepsilon$. Consequently, $\limsup_{x \to a^+} f(x) \geq 1 - \varepsilon$ and $\liminf_{x \to a^-} f(x) \leq \varepsilon$, contradicting continuity of f.

Section 2.5

4.   Let $B_{jk\delta} = \{|f_j - f_k| \geq \delta\}$, $B_\delta = \bigcap_{n=1}^{\infty} \bigcup_{j,k=n}^{\infty} B_{jk\delta}$. Then

$$\bigcup_{j,k=n}^{\infty} B_{jk\delta} \downarrow B_\delta,$$

and the proof proceeds just as in 2.5.4.

5.   Let $\{f_{n_k}\}$ be a subsequence converging a.e., necessarily to f by Problem 1. By 1.6.9, f is μ-integrable. Now if $\int_\Omega f_n \, d\mu \not\longrightarrow \int_\Omega f \, d\mu$, then for some $\varepsilon > 0$, we have $|\int_\Omega f_n \, d\mu - \int_\Omega f \, d\mu| \geq \varepsilon$ for n in some subsequence $\{m_k\}$. But we may then extract a subsequence $\{f_{r_j}\}$ of $\{f_{m_k}\}$ converging to f a.e., so that $\int_\Omega f_{r_j} \, d\mu \to \int_\Omega f \, d\mu$ by 1.6.9, a contradiction.

Section 2.6

4.   By Fubini's theorem,

$$\mu(C) = \iint_\Omega I_C \, d\mu = \int_{\Omega_1} \left[\int_{\Omega_2} I_C \, d\mu_2\right] d\mu_1 = \int_{\Omega_1} \mu_2(C(\omega_1)) \, d\mu_1(\omega_1).$$

Similarly, $\mu(C) = \int_{\Omega_2} \mu_1(C(\omega_2)) \, d\mu_2(\omega_2)$. The result follows since $f \geq 0$, $\int_\Omega f = 0$ implies $f = 0$ a.e.

7.   (a)   Let

$$A_{nk} = \left\{x \in \Omega_1 : \frac{k-1}{n} \leq f(x) < \frac{k}{n}\right\},$$

$$B_{nk} = \left\{y \in \Omega_2 : \frac{k-1}{n} \leq y < \frac{k}{n}\right\}$$

$(n = 1, 2, \ldots, k = 1, 2, \ldots, n$; when $k = n$, include the right end-point as well). Then

$$G = \bigcap_{n=1}^{\infty} \bigcup_{k=1}^{n} (A_{nk} \times B_{nk}) \in \mathscr{F}.$$

If f is only defined on a subset of Ω_1, replace Ω_1 by the domain of f in the definition of A_{nk}.

(b) Assume $B \subset C_1$. Each $x \in \Omega_1$ is countable and $(x, y) \in B$ implies $y \leq x$; hence there are only countably many points $y_{x1}, y_{x2}, \ldots \in \Omega_2$ such that $(x, y_{xn}) \in B$. (If there are only finitely many points $y_{x1}, \ldots, y_{xn}$, take $y_{xk} = y_{xn}$ for $k \geq n$.)

Thus $B = \bigcup_{n=1}^{\infty} G_n$, where G_n is the graph of the function f_n defined by $f_n(x) = y_{xn}$. By part (a), $B \in \mathscr{F}$. (Note that $y_{xn} \in \Omega_2$, which may be identified with $[0, 1]$; thus (a) applies.) If $B \subset C_2$, each $y \in \Omega_2$ is countable, and $(x, y) \in B$ implies $x < y$, so there are only countably many points $x_{yn} \in \Omega_1$ such that $(x_{yn}, y) \in B$. The result follows as above.

(c) If $F \subset \Omega$, then $F = (F \cap C_1) \cup (F \cap C_2)$ and the result follows from part (b).

9. Assuming the continuum hypothesis, we may replace $[0, 1]$ by the first uncountable ordinal β_1. Thus we may take $\Omega_1 = \Omega_2 = \beta_1$, $\mathscr{F}_1 = \mathscr{F}_2$ = the image of the Borel sets under the correspondence of $[0, 1]$ with β_1, and $\mu_1 = \mu_2 =$ Lebesgue measure. Let $f = I_C$, where $C = \{(x, y) \in \Omega_1 \times \Omega_2: y \leq x\}$ and the ordering "$\leq$" is taken in β_1, not $[0, 1]$. For each x, $\{y: f(x, y) = 1\}$ is countable, and for each y, $\{x: f(x, y) = 0\}$ is countable; it follows that f is measurable in each coordinate separately. Now $\int_0^1 f(x, y)\, dy = 0$ for all x; hence $\int_0^1 \left[\int_0^1 f(x, y)\, dy\right] dx = 0$. But $\int_0^1 [1 - f(x, y)]\, dx = 0$ for all y; hence $\int_0^1 \left[\int_0^1 f(x, y)\, dx\right] dy = 1$. It follows that f is not jointly measurable, for if so, the iterated integrals would be equal by Fubini's theorem.

Section 2.7

1. Let $\mathscr{G}$ be the smallest σ-field containing the measurable rectangles. Then $\mathscr{C} \subset \prod_{j=1}^{\infty} \mathscr{F}_j$ since a measurable rectangle is a measurable cylinder. But the class of sets $A \subset \prod_{j=1}^{n} \Omega_j$ such that $\{\omega \in \Omega: (\omega_1, \ldots, \omega_n) \in A\} \in \mathscr{G}$ is a σ-field that contains the measurable rectangles of $\prod_{j=1}^{n} \Omega_j$, and hence contains all sets in $\prod_{j=1}^{n} \mathscr{F}_j$. Thus all measurable cylinders belong to $\mathscr{G}$, so $\prod_{j=1}^{\infty} \mathscr{F}_j \subset \mathscr{G}$.

5.

$$\{x \in \mathbb{R}^\infty : f(x) = n\} = \left\{ x : \sum_{i=1}^{k} x_i < 1, \quad k = 1, 2, \ldots, n-1, \sum_{i=1}^{n} x_i \geq 1 \right\}$$

$$\text{if} \qquad n = 1, 2, \ldots,$$

$$\{x \in \mathbb{R}^\infty : f(x) = \infty\} = \left\{ x : \sum_{i=1}^{n} x_i < 1, \quad n = 1, 2, \ldots \right\}.$$

In each case we have a finite or countable intersection of measurable cylinders.

CHAPTER 3

Section 3.2

6. (a) Let $z \in K$, a compact subset of U. If $r < d_0$, a standard application of the Cauchy integral formula yields

$$f(z) = \frac{1}{2\pi} \int_0^{2\pi} f(z + re^{i\theta}) \, d\theta.$$

Thus if $0 < d < d_0$,

$$f(z) \frac{d^2}{2} = \int_0^d r f(z) \, dr = \frac{1}{2\pi} \int_0^d r \, dr \int_0^{2\pi} f(z + re^{i\theta}) \, d\theta,$$

or

$$f(z) = \frac{1}{\pi d^2} \int_{\theta=0}^{2\pi} \int_{r=0}^{d} f(z + re^{i\theta}) r \, dr \, d\theta$$

$$= \frac{1}{\pi d^2} \iint_D f(x + iy) \, dx \, dy$$

where D is the disk with center at z and radius r. An application of the Cauchy–Schwarz inequality to the functions 1 and f shows that

$$|f(z)| \leq \frac{1}{\pi d^2} (\pi d^2)^{1/2} \|f\|,$$

as desired.

(b) If $f_1, f_2, \ldots$ is a Cauchy sequence in $H(U)$, part (a) shows that f_n converges uniformly on compact subsets to a function f analytic on U. But $H(U) \subset L^2(\Omega, \mathcal{F}, \mu)$ where $\Omega = U$, $\mathcal{F}$ is the class of Borel sets, and μ is Lebesgue measure; hence f_n converges in L^2 to a function $g \in H(U)$. Since a subsequence of $\{f_n\}$ converges to g a.e., we have $f = g$ a.e. Therefore $f \in H(U)$ and $f_n \to f$ in L^2, that is, in the $H(U)$ norm.

7. (a) If $0 \leq r < 1$,

$$\frac{1}{2\pi} \int_0^{2\pi} |f(re^{i\theta})|^2 \, d\theta = \frac{1}{2\pi} \int_0^{2\pi} \sum_n a_n r^n e^{in\theta} \sum_m \bar{a}_m r^m e^{-im\theta} \, d\theta$$

$$= \frac{1}{2\pi} \sum_{n,m} a_n \bar{a}_m r^{n+m} \int_0^{2\pi} e^{i(n-m)\theta} \, d\theta$$

since the Taylor series converges uniformly on compact subsets of D

$$= \sum_{n=0}^{\infty} |a_n|^2 r^{2n},$$

which increases to $\sum_{n=0}^{\infty} |a_n|^2$ as r increases to 1.

(b) $$\iint_D |f(x+iy)|^2 \, dx \, dy = \int_0^1 r \, dr \int_0^{2\pi} |f(re^{i\theta})|^2 \, d\theta$$

$$\leq 2\pi N^2(f) \int_0^1 r \, dr = \pi N^2(f).$$

(c) $$\iint_D |f_n(x+iy)|^2 \, dx \, dy = \int_0^{2\pi} \int_0^1 r^{2n} r \, dr \, d\theta = \frac{2\pi}{(2n+2)} \to 0,$$

but

$$\frac{1}{2\pi} \int_0^{2\pi} |f_n(re^{i\theta})|^2 \, d\theta = \frac{1}{2\pi} \int_0^{2\pi} r^{2n} \, d\theta = r^{2n};$$

hence $N(f_n) = 1$ for all n.

(d) If $\{f_n\}$ is a Cauchy sequence in H^2, part (b) shows that $\{f_n\}$ is Cauchy in $H(D)$. By Problem 6, f_n converges uniformly on compact subsets and in $H(D)$ to a function f analytic on D. Now if

$0 < r_0 < 1$, $\varepsilon > 0$, the Cauchy property in H^2 gives

$$\frac{1}{2\pi} \int_0^{2\pi} |f_n(re^{i\theta}) - f_m(re^{i\theta})|^2 \, d\theta \le \varepsilon$$

for $r \le r_0$ and n, m exceeding some integer $N(\varepsilon)$. Let $m \to \infty$; since $f_m \to f$ uniformly for $|z| \le r_0$,

$$\frac{1}{2\pi} \int_0^{2\pi} |f_n(re^{i\theta}) - f(re^{i\theta})|^2 \, d\theta \le \varepsilon, \qquad r \le r_0, \qquad n \ge N(\varepsilon).$$

Since r_0 may be chosen arbitrarily close to 1, $N^2(f_n - f) \le \varepsilon$ for $n \ge N(\varepsilon)$, proving completeness.

(e) Since e_n corresponds to $(0, \ldots, 0, 1, 0, \ldots)$, with the 1 in position n, in the isometric isomorphism between H^2 and a subspace of l^2, the e_n are orthonormal. Now if $f \in H^2$, with Taylor coefficients a_n, $n = 0, 1, \ldots$, then $\langle f, e_n \rangle = a_n$, again by the isometric isomorphism. Thus

$$N^2(f) = \sum_{n=0}^{\infty} |a_n|^2 = \sum_{n=0}^{\infty} |\langle f, e_n \rangle|^2$$

and the result follows from 3.2.13(f).

(f) $\langle e_n, e_m \rangle = \displaystyle\iint_D e_n(x + iy)\bar{e}_m(x + iy) \, dx \, dy$

$\qquad = \displaystyle\iint_D e_n(re^{i\theta})\bar{e}_m(re^{i\theta}) r \, dr \, d\theta$

$\qquad = [(2n + 2)(2m + 2)]^{1/2} \left(\dfrac{1}{2\pi}\right) \displaystyle\int_0^{2\pi} \int_0^1 r^{n+m} e^{i(n-m)\theta} r \, dr \, d\theta$

$\qquad = \begin{cases} 0, & n \ne m, \\ 1, & n = m. \end{cases}$

Thus the e_n are orthonormal. Now if $f \in H(D)$ with

$$f(z) = \sum_{n=0}^{\infty} a_n z^n,$$

then

$$\int_0^{2\pi} \int_0^{r_0} f(re^{i\theta}) \bar{e}_m (re^{i\theta}) r \, dr \, d\theta$$

$$= \sum_{n=0}^{\infty} a_n \int_0^{2\pi} \int_0^{r_0} r^n e^{in\theta} \left(\frac{2m+2}{2\pi} \right)^{1/2} r^m e^{-im\theta} r \, dr \, d\theta$$

since the Taylor series converges uniformly on compact subsets of D

$$= a_m \left(\frac{2m+2}{2\pi} \right)^{1/2} \frac{r_0^{2m+2}}{2m+2} 2\pi = a_m \left(\frac{2\pi}{2m+2} \right)^{1/2} r_0^{2m+2}.$$

Now $f\bar{e}_m$ is integrable on D (by the Cauchy–Schwarz inequality), so we may let $r_0 \to 1$ and invoke the dominated convergence theorem to obtain

$$\langle f, e_m \rangle = a_m \left(\frac{2\pi}{2m+2} \right)^{1/2}.$$

But the same argument with e_m replaced by f shows that

$$\|f\|_{H(D)}^2 = \lim_{r_0 \to 1} \sum_{n,m=0}^{\infty} a_n \bar{a}_m \int_0^{2\pi} \int_0^{r_0} r^n e^{in\theta} r^m e^{-im\theta} r \, dr \, d\theta$$

$$= \sum_{n=0}^{\infty} \frac{|a_n|^2 2\pi}{2n+2}.$$

The result now follows from 3.2.13(f).

9. (a) Let g be a continuous complex-valued function on $[0, 2\pi]$ with $g(0) = g(2\pi)$. Then $g(t) = h(e^{it})$, where $h(z)$ is continuous on $\{z \in \mathbb{C}: |z| = 1\}$. By the Stone–Weierstrass theorem, h can be uniformly approximated by functions of the form $\sum_{k=-n}^{n} c_k z^k$. For the algebra generated by z^n, $n = 0, \pm 1, \pm 2, \ldots$, separates points, contains the constant functions, and contains the complex conjugate of each of its members since $\bar{z} = 1/z$ for $|z| = 1$.

Thus $g(t)$ can be uniformly approximated (hence approximated in L^2) by trigonometric polynomials $\sum_{k=-n}^{n} c_k e^{ikt}$. Since any continuous function on $[0, 2\pi]$ can be approximated in L^2 by a continuous function with $g(0) = g(2\pi)$, and the continuous functions are dense in L^2, the trigonometric polynomials are dense in L^2.

(b) By 3.2.6, $\int_0^{2\pi} |f(t) - \sum_{k=-n}^{n} c_k e^{ikt}|^2 \, dt$ is minimized when $c_k = a_k$ $= (1/2\pi) \int_0^{2\pi} f(t) e^{-ikt} \, dt$. Furthermore, *some* sequence of trigono- metric polynomials converges to f in L^2 since the trigonometric polynomials are dense. The result follows.

(c) This follows from part (a) and 3.2.13(c), or, equally well, from part (b) and 3.2.13(d).

10. (a) Let $\{e_n\}$ be an infinite orthonormal subset of H. Take $M = \{x_1, x_2, \ldots\}$, where $x_n = (1 + 1/n)e_n$, $n = 1, 2, \ldots$. To show that M is closed, we compute, for $n \neq m$,

$$\|x_n - x_m\|^2 = \left\|\left(1 + \frac{1}{n}\right)e_n - \left(1 + \frac{1}{m}\right)e_m\right\|^2$$

$$= \left(1 + \frac{1}{n}\right)^2 + \left(1 + \frac{1}{m}\right)^2 \geq 2.$$

Thus if $y_n \in M$, $y_n \to y$, then $y_n = y$ eventually, so $y \in M$. Since $\|x_n\|^2 = 1 + (1/n)$, M has no element of minimum norm.

(b) Let M be a nonempty closed subset of the finite-dimensional space H. If $x \in H$ and $N = M \cap \{y: \|x - y\| \leq n\}$, then $N \neq \emptyset$ for some n. Since $y \to \|x - y\|$, $y \in N$, is continuous and N is compact, $\inf\{\|x - y\|: y \in N\} = \|x - y_0\|$ for some $y_0 \in N \subset M$. But the inf over N is the same as the inf over M; for if $y \in M$, $y \notin N$, then $\|x - y_0\| \leq n < \|x - y\|$. Note that y_0 need not be unique; for ex- ample, let $H = \mathbb{R}$, $M = \{-1, 1\}$, $x = 0$.

Section 3.3

3. Since $\int_a^b |K(s, t)| \, dt$ is continuous in s, it assumes a maximum at some point $u \in [a, b]$. If $K(u, t) = r(t)e^{i\theta(t)}$, $r(t) \geq 0$, let $z(t) = e^{-i\theta(t)}$. Let x_1, $x_2, \ldots$ be a sequence in $C[a, b]$ such that $\int_a^b |x_n(t) - z(t)| \, dt \to 0$; we may assume that $|x_n(t)| \leq 1$ for all n and t (see 2.4.14). Since K is bounded,

$$\left|\int_a^b K(s, t)z(t) \, dt\right| = \lim_{n \to \infty} \left|\int_a^b K(s, t)x_n(t) \, dt\right| = \lim_{n \to \infty} |(Ax_n)(s)| \leq \|A\|.$$

Set $s = u$ to obtain

$$\int_a^b |K(u, t)| \, dt \leq \|A\|,$$

as desired.

7. (a) If $x \in L$, then

$$\|x\|_1 = \left\| \sum_{i=1}^{n} x_i e_i \right\|_1 \leq \sum_{i=1}^{n} |x_i| \|e_i\|_1 \leq \left(\max_i \|e_i\|_1 \right) \sum_{i=1}^{n} |x_i|.$$

But $\sum_{i=1}^{n} |x_i| \leq \sqrt{n} \|x\|_2$, so we may take $k = \sqrt{n} \max_i \|e_i\|_1$.

(b) Let $S = \{x \colon \|x\|_2 = 1\}$. Since $(L, \| \ \|_2)$ is isometrically isomorphic to $\mathbb{C}^n$, S is compact in the norm $\| \ \|_2$. Now the map $x \to \|x\|_1$ is a continuous real-valued function on $(L, \| \ \|_1)$, and by part (a), the topology induced by $\| \ \|_1$ is weaker than the topology induced by $\| \ \|_2$. Thus the map is continuous on $(L, \| \ \|_2)$; hence it attains a minimum on S, necessarily positive since $x \in S$ implies $x \neq 0$.

(c) If $x \in L$, $x \neq 0$, let $y = x/\|x\|_2$; then $\|y\|_1 \geq m\|y\|_2$ by (b); hence $\|x\|_1 \geq m\|x\|_2$. By (a) and Problem 6(b), $\| \ \|_1$ and $\| \ \|_2$ induce the same topology.

(d) By the above results, the map $T \colon \sum_{i=1}^{n} x_i e_i \to (x_1, \ldots, x_n)$ is a one-to-one onto, linear, bicontinuous map of L and $\mathbb{C}^n$ [note that $\|\sum_{i=1}^{n} x_i e_i\|_2$ is the Euclidean norm of $(x_1, \ldots, x_n)$ in $\mathbb{C}^n$]. If $y_j \in L$, $y_j \to y \in M$, then $y_j - y_k \to 0$ as $j, k \to \infty$; hence $T(y_j - y_k) = Ty_j - Ty_k \to 0$. Thus $\{Ty_j\}$ is a Cauchy sequence in $\mathbb{C}^n$. If $Ty_j \to z \in \mathbb{C}^n$, then $y_j \to T^{-1}z \in L$.

9. For (a) implies (b), see Problem 7; if (c) holds, then $\{x \colon \|x\| \leq \varepsilon\}$ is compact for small enough $\varepsilon > 0$; hence every closed ball is compact (note that the map $x \to kx$ is a homeomorphism). But any closed bounded set is a subset of a closed ball, and hence is compact.

To prove that (f) implies (a), choose $x_1 \in L$ such that $\|x_1\| = 1$. Suppose we have chosen $x_1, \ldots, x_k \in L$ such that $\|x_i\| = 1$ and $\|x_i - x_j\| \geq \frac{1}{2}$ for $i, j = 1, \ldots, k$, $i \neq j$. If L is not finite-dimensional, then $S\{x_1, \ldots, x_k\}$ is a proper subspace of L, necessarily closed, by Problem 7(d). By Problem 8, we can find $x_{k+1} \in L$ with $\|x_{k+1}\| = 1$ and $\|x_i - x_{k+1}\| \geq \frac{1}{2}$, $i = 1, \ldots, k$. The sequence $x_1, x_2, \ldots$ satisfies $\|x_n\| = 1$ for all n, but $\|x_n - x_m\| \geq \frac{1}{2}$ for $n \neq m$; hence the unit sphere cannot possibly be covered by a finite number of balls of radius less than $\frac{1}{4}$.

11. (a) Define $\lambda(A) = f(I_A)$, $A \in \mathscr{F}$. If $A_1, A_2, \ldots$ are disjoint sets in $\mathscr{F}$ whose union is A, then $\lambda(A) = \sum_{i=1}^{\infty} f(I_{A_i})$ since f is continuous and $\sum_{i=1}^{n} I_{A_i} \xrightarrow{L^p} I_A$. [Note that

$$\int_{\Omega} \left| \sum_{i=1}^{n} I_{A_i} - I_A \right|^p d\mu = \sum_{i=n+1}^{\infty} \mu(A_i) \to 0$$

by finiteness of μ.] Thus λ is a complex measure on $\mathscr{F}$. If $\mu(A) = 0$, then $I_A = 0$ a.e. $[\mu]$, so we may write $I_A \xrightarrow{L^p} 0$ and use the continuity of f to obtain $\lambda(A) = 0$. By the Radon–Nikodym theorem, we have $\lambda(A) = \int_A y \, d\mu$ for some μ-integrable y. Thus $f(x) = \int_\Omega xy \, d\mu$ when x is an indicator; hence when x is a simple function. Since f is continuous, y is μ-integrable, and the finite-valued simple functions are dense in L^p, the result holds when x is a bounded Borel measurable function.

Now let $y_1, y_2, \ldots$ be nonnegative, finite-valued, simple functions increasing to $|y|$. Then

$$\|y_n\|_q^q = \int_\Omega y_n^q \, d\mu \leq \int_\Omega y_n^{q-1} |y| \, d\mu = \int_\Omega y_n^{q-1} e^{-i \arg y} y \, d\mu$$

$$= f(y_n^{q-1} e^{-i \arg y}) \quad \text{since} \quad y_n^{q-1} e^{-i \arg y} \text{ is bounded}$$

$$\leq \|f\| \, \|y_n^{q-1}\|_p = \|f\| \, \|y_n\|_q^{q/p} \quad \text{since} \quad (q-1)p = q.$$

Thus $\|y_n\|_q \leq \|f\|$; hence by the monotone convergence theorem, $\|y\|_q \leq \|f\|$; in particular, $y \in L^q$. But now Hölder's inequality and the fact that finite-valued simple functions are dense in L^p yield $f(x) = \int_\Omega xy \, d\mu$ for all $x \in L^p$. Hölder's inequality also gives $\|f\| \leq \|y\|_q$; hence $\|f\| = \|y\|_q$. If y_1 is another such function, then $g(x) = \int_\Omega x(y - y_1) \, d\mu = 0$ for all $x \in L^p$. By the above argument, $\|y - y_1\|_q = 0$, so $y = y_1$ a.e. $[\mu]$.

(b)   (i)   If, say, $y_A - y_B > 0$ on the set $F \subset A \cap B$, let $x = I_F$; then $xI_A = xI_B$; hence $\int_\Omega x(y_A - y_B) \, d\mu = 0$, that is,

$$\int_F (y_A - y_B) \, d\mu = 0.$$

But then $\mu(F) = 0$.

(ii)   Since $y_{A_n \cup A_m} = y_{A_n}$ a.e. on A_n we have

$$\|y_{A_n}\|_q^q \leq \|y_{A_n \cup A_m}\|_q^q = \|y_{A_n}\|_q^q + \int_{A_m - A_n} |y_{A_m}|^q \, d\mu.$$

Since $\|y_{A_n}\|_q^q$ approaches k^q as $n \to \infty$, so does $\|y_{A_n \cup A_m}\|_q^q$, and it follows that $\int_{A_m - A_n} |y_{A_m}|^q \, d\mu \to 0$ as $n \to \infty$. By symmetry, we may interchange m and n to obtain

$$\int_\Omega |y_{A_m} - y_{A_n}|^q \, d\mu = \int_{A_m - A_n} |y_{A_m}|^q \, d\mu$$

$$+ \int_{A_n - A_m} |y_{A_n}|^q \, d\mu \to 0$$

as $n, m \to \infty$. Thus y_{A_n} converges in L^q to a limit y, and since $\|y_{A_n}\|_q \leq \|f\|$ for all n, $\|y\|_q \leq \|f\|$. If $\{B_n\}$ is another sequence of sets with $\|y_{B_n}\|_q \to k$, the above argument with A_m replaced by B_n shows that $\|y_{A_n} - y_{B_n}\|_q \to 0$; hence $y_{B_n} \to y$ also.

(iii) Let $A \in \mathcal{F}$, $\mu(A) < \infty$. In (ii) we may take all $A_n \supset A$, so that $y_{A_n} = y_A$ a.e. on A; hence $y = y_A$ a.e. on A. Thus if $x = I_A$, then $f(x) = f(xI_A) = \int_\Omega x y_A \, d\mu = \int_\Omega x y \, d\mu$. It follows that $f(x) = \int_\Omega x y \, d\mu$ if x is simple. [If $\mu(B) = \infty$, then x must be 0 on B since $x \in L^p$.] Since $y \in L^q$, the continuity of f and Hölder's inequality extend this result to all $x \in L^p$.

(c) The argument of (a) yields a μ-integrable y such that $f(x) = \int_\Omega x y \, d\mu$ for all bounded Borel measurable x. Let $B = \{\omega: |y(\omega)| \geq k\}$; then

$$k\mu(B) \leq \int_B |y| \, d\mu = \int_\Omega I_B e^{-i \, arg \, y} y \, d\mu$$

$$= f(I_B e^{-i \, arg \, y}) \leq \|f\| \, \|I_B\|_1 = \|f\|\mu(B).$$

Thus if $k > \|f\|$, we have $\mu(B) = 0$, proving that $y \in L^\infty$ and $\|y\|_\infty \leq \|f\|$. As in (a), we obtain $f(x) = \int_\Omega x y \, d\mu$ for all $x \in L^1$, $\|f\| = \|y\|_\infty$, and y is essentially unique.

(d) Part (i) of (b) holds, with the same proof. Now if Ω is the union of disjoint sets A_n, with $\mu(A_n) < \infty$, define y on Ω by taking $y = y_{A_n}$ on A_n. Since $\|y_{A_n}\|_\infty \leq \|f\|$ for all n, we have $y \in L^\infty$ and $\|y\|_\infty \leq \|f\|$. If $x \in L^1$, then $\sum_{i=1}^n x I_{A_i} \xrightarrow{L^1} x$; hence

$$f(x) = \sum_{n=1}^\infty f(xI_{A_n}) = \sum_{n=1}^\infty \int_\Omega x y_{A_n} \, d\mu$$

$$= \sum_{n=1}^\infty \int_{A_n} x y \, d\mu = \int_\Omega x y \, d\mu.$$

Since $\|f\| \leq \|y\|_\infty$ by Hölder's inequality (with $p = 1$, $q = \infty$), the result follows.

Section 3.4

3. Let $\{y_n\}$ be a Cauchy sequence in M, and let x_0 be any element of L with norm 1. By 3.4.5 (c), there is an $f \in L^*$ with $\|f\| = 1$ and $f(x_0) = \|x_0\| = 1$; we define $A_n \in [L, M]$ by $A_n x = f(x)y_n$. Then $\|(A_n - A_m)x\| = |f(x)| \, \|y_n - y_m\| \leq \|y_n - y_m\| \, \|x\|$, so that $\|A_n - A_m\| \leq \|y_n - y_m\|$

$\rightarrow 0$. By hypothesis, the A_n converge uniformly to some $A \in [L, M]$; there-fore, $\|y_n - Ax_0\| = \|A_n x_0 - Ax_0\| \leq \|A_n - A\| \rightarrow 0$.

7. Let L be the set of all bounded scalar-valued functions on Ω, with sup norm, and let M be the subspace of L consisting of simple functions

$$x = \sum_j x_j I_{A_j},$$

where the A_j are disjoint sets in $\mathscr{F}_0$. Define g on M by

$$g(x) = \sum_j x_j \mu_0(A_j).$$

Now $|g(x)| \leq \max_j |x_j| \sum_j \mu_0(A_j) \leq \max_j |x_j| \mu_0(\Omega) = \mu_0(\Omega) \|x\|$; hence $\|g\| \leq \mu_0(\Omega) < \infty$. By the Hahn–Banach theorem, g has an extension to a continuous linear functional f on L, with $\|f\| = \|g\|$. Define $\mu(A) = f(I_A), A \subset \Omega$. Since f is linear, μ is finitely additive, and since f is an extension of g, μ is an extension of μ_0. Now if $\mu(A) < 0$, then

$$f(I_{A^c}) = \mu(A^c) = \mu(\Omega) - \mu(A) > \mu(\Omega).$$

But $\|I_{A^c}\| = 1$, so that $\|f\| > \mu(\Omega)$, a contradiction.

8. Since $A_n x \rightarrow Ax$ for each x, $\sup_n \|A_n x\| < \infty$. By the uniform bounded-ness principle, $\sup_n \|A_n\| = M < \infty$; hence

$$\|Ax\| \lim_{n \to \infty} \|A_n x\| = \liminf_{n \to \infty} \|A_n x\|$$

$$\leq \|x\| \liminf_{n \to \infty} \|A_n\| \leq M \|x\|.$$

12. Let L be the set of all complex-valued functions x on $[0, 1]$ with a continu-ous derivative x', M the set of all continuous complex-valued functions on $[0, 1]$, with the sup norm on L and M. If $Ax = x', x \in L$, then A is a linear map of L onto M, and A is closed. If $x_n \rightarrow x$ and $x_n' \rightarrow y$, then since convergence relative to the sup norm is uniform convergence, $\int_0^t x_n'(s) ds \rightarrow \int_0^t y(s) ds = z(t)$. Thus $x_n(t) - x_n(0) \rightarrow z(t)$; hence $x(t) - x(0) = z(t)$. Therefore $x' = z' = y$. But A is unbounded, for if $x_n(t) = \sin nt$, then $\|x_n\| = 1$, $\|Ax_n\| \rightarrow \infty$.

13. By the open mapping theorem, A is open; hence $A\{x \in L: \|x\| < 1\}$ is a neighborhood of 0 in M; say, $\{y \in M: \|y\| < \delta\} \subset A\{x \in L: \|x\| < 1\}$. If $y \in M, y \neq 0$, then $\delta y/2\|y\|$ has norm less than δ, hence equals Az for some $z \in L$, $\|z\| < 1$. If $x = (2\|y\|/\delta)z$, then $Ax = y$ and $\|x\| < 2\|y\|/\delta$. Thus we may take $k = 2/\delta$.

CHAPTER 4

Section 4.8

4. To prove (a), let $A, B \in \mathscr{C}_{i_1}, B \subset A$. If $C_{i_k} \in \mathscr{C}_{i_k}, k = 2, \ldots, n$, then (using a product notation for intersection)

$$P[(A - B)C_{i_2} \cdots C_{i_n}] = P(AC_{i_2} \cdots C_{i_n}) - P(BC_{i_2} \cdots C_{i_n})$$

$$= [P(A) - P(B)] \prod_{k=2}^{n} P(C_{i_k})$$

$$= P(A - B)P(C_{i_2}) \cdots P(C_{i_n}).$$

Thus $A - B$ can be added to C_{i_1} while preserving independence. Since i_1 is arbitrary, (a) follows. The proofs of (b) and (c) are quite similar, and (d) follows from (a), (b), and (c).

Now let $\mathscr{C}_1 = \{A, B\}$, $\mathscr{C}_2 = \{C\}$, where A and C are independent, and B and C are independent. Since $P[(A \cap B) \cap C]$ need not equal $P(A \cap B)P(C)$ [see 4.3.3(b)], $A \cap B$ cannot be added to $\mathscr{C}_1$.

Finally, we show that the $\sigma(\mathscr{C}_i)$ are independent iff each $\mathscr{C}_i$ is closed under finite intersection. Fix i, and consider the collection of classes $\mathscr{A}_i$ such that $\mathscr{A}_i$ is closed under finite intersection, $\mathscr{C}_i \subset \mathscr{A}_i$, and $\mathscr{A}_i$ and the $\mathscr{C}_j, j \neq i$, are independent. Partially order the $\mathscr{A}_i$ by inclusion. Each chain has an upper bound (the union of the chain) so there is a maximal class $\mathscr{D}_i$. Since $\mathscr{D}_i$ is closed under finite intersection, it is closed under *arbitrary* differences by (a) $(A - B = A - (A \cap B))$; hence by (a), (b), and (c), $\mathscr{D}_i$ is a σ-field. Thus $\sigma(\mathscr{C}_i) \subset \mathscr{D}_i$, and consequently $\mathscr{C}_i$ can be replaced by $\sigma(\mathscr{C}_i)$ while preserving independence. Since i is arbitrary, the result follows.

Section 4.9

4. $P\{Y \in B\} = P\{X \in g^{-1}(B)\} = \int_{g^{-1}(B)} f_1(x) \, dx$. Under the transformation $x = h(y), y = g(x)$, this becomes $\int_B f_1(h(y)) |J_h(y)| \, dy$ (see Problem 6, Section 2.3). The result follows.

6. Let $A_j = \{X_j > 2 + T_1\}, T_1 = \min_i X_i$. Then $X_0 = \sum_{j=1}^{n} I_{A_j}$, where I_{A_j} is the indicator of A_j. Therefore $E(X_0) = \sum_{j=1}^{n} E(I_{A_j}) = \sum_{j=1}^{n} P(A_j)$, where $E(X_0) = \int_\Omega X_0 \, dP$. By symmetry,

$$E(X_0) = nP\{X_2 > 2 + T_1\}$$

$$= n \sum_{\substack{j=1 \\ j \neq 2}}^{n} P\{T_1 = X_j, X_2 > 2 + X_j\}$$

$$= n(n-1)P\{T_1 = X_1, X_2 > 2 + X_1\}$$

$$= n(n-1)P\{X_2 > 2 + X_1, X_3 > X_1, \dots, X_n > X_1\}$$

$$= n(n-1) \int_{-\infty}^{\infty} f(x_1)\, dx_1 \int_{2+x_1}^{\infty} f(x_2)\, dx_2 \int_{x_1}^{\infty} f(x_3)\, dx_3$$

$$\cdots \int_{x_1}^{\infty} f(x_n)\, dx_n$$

$$= n(n-1) \int_{-\infty}^{\infty} f(x_1)[1 - F(x_1)]^{n-2}[1 - F(2 + x_1)]\, dx_1.$$

7. Define F^{-1} and X as suggested. Then if $0 < y < 1$, $F^{-1}(y) \le x$ iff $y \le F(x)$. For if $F^{-1}(y) > x$, we can find $x_0 > x$ such that $F(x_0) < y$, and thus $F(x) \le F(x_0) < y$; if $F(x) < y$, by right continuity we can find $x_0 > x$ such that $F(x_0) < y$; therefore $F^{-1}(y) \ge x_0 > x$. Now

$$P\{\omega\colon X(\omega) \le x\} = P\{\omega\colon F^{-1}(\omega) \le x\} = P\{\omega\colon \omega \le F(x)\} = F(x).$$

[This also shows that X is measurable, and that

$$X(\omega) = \min\{x\colon F(x) \ge \omega\}, \qquad 0 < \omega < 1.]$$

Section 4.10

2. Separate F into discrete and absolutely continuous parts: $F = F_1 + F_2$, where

$$F_1(x) = \begin{cases} 0, & x < 2, \\ \frac{1}{3}, & x \ge 2, \end{cases}$$

$$F_2(x) = \int_{-\infty}^{x} f_2(t)\, dt$$

where

$$f_2(x) = \begin{cases} 0, & x < 2 \quad \text{or} \quad x > 3, \\ \frac{2}{3}, & 2 \le x \le 3. \end{cases}$$

Thus

$$E(X^2) = \int_{-\infty}^{\infty} x^2\, dF(x) = \int_{-\infty}^{\infty} x^2\, dF_1(x) + \int_{-\infty}^{\infty} x^2\, dF_2(x)$$

$$= 2^2 \left(\frac{1}{3}\right) + \int_{-\infty}^{\infty} x^2 f_2(x)\, dx$$

$$= \frac{4}{3} + \frac{2}{3} \int_{2}^{3} x^2\, dx = \frac{50}{9}.$$

Section 4.11

1.   (a)   Assume $Y_i \colon (\Omega, \mathscr{F}) \to (\Omega_i, \mathscr{F}_i)$, $i = 1, 2, \ldots$. By hypothesis,

$$P\{(Y_1, \ldots, Y_n) \in A, (Y_{n+1}, Y_{n+2}, \ldots) \in B\}$$
$$= P\{(Y_1, \ldots, Y_n) \in A\} P\{(Y_{n+1}, Y_{n+2}, \ldots) \in B\}$$

if A is a measurable rectangle in $\prod_{i=1}^{n} \mathscr{F}_i$ and B is a measurable rectangle in $\prod_{i=n+1}^{\infty} \mathscr{F}_i$, the formula is still valid if A and B are finite disjoint unions of measurable rectangles. Two applications of the monotone class theorem establish this result for all $A \in \prod_{i=1}^{n} \mathscr{F}_i$ and $B \in \prod_{i=n+1}^{\infty} \mathscr{F}_i$.

   (b)   Let $\mathscr{C}$ be the class of sets $B \in \mathscr{S}^{\infty}$ such that

$$P\{(Y_1, Y_2, \ldots) \in B\} = P\{(Y_n, Y_{n+1}, \ldots) \in B\}.$$

Since the Y_i are independent and P_{Y_i} is the same for all i, all measurable rectangles belong to $\mathscr{C}$; hence $\mathscr{C}$ contains all finite disjoint unions of measurable rectangles. But $\mathscr{C}$ is a monotone class; hence $\mathscr{C} = \mathscr{S}^{\infty}$, and the result follows.

CHAPTER 5

Section 5.3

1.   (a)   The conditional density of X given Y is $h(x|y) = f(x, y)/f_2(y)$, where $f_2(y) = \int_{-\infty}^{\infty} f(x, y)\,dx$. Thus

$$E(g(X)|Y = y) = \int_{-\infty}^{\infty} g(x)h(x|y)\,dx,$$

assuming $E[g(X)]$ exists [cf. 5.3.5(c), Eq. (5)].

   (b) $E(Y|A) = E(YI_A)/P(A)$ [see 5.3.5(b)]. Now

$$P(A) = \int_{x \in B} \int_{y=-\infty}^{\infty} f(x, y)\,dx\,dy$$

and

$$E(YI_A) = \int_{-\infty}^{\infty} \int_{-\infty}^{\infty} yI_A(x, y)f(x, y)\,dx\,dy$$
$$= \int_{x \in B} \int_{y=-\infty}^{\infty} yf(x, y)\,dx\,dy.$$

(c) $E(X|A) = E(XI_A)/P(A)$, where

$$P(A) = \iint\limits_{x+y\in B} f(x, y)\, dx\, dy$$

and

$$E(XI_A) = \iint\limits_{x+y\in B} x f(x, y)\, dx\, dy.$$

3.   By 5.3.1, $P\{X \in A, Y \in B\} = \sum_{x\in A} P\{X = x\} \int_B h(y|x)\, dy$. Thus (take $A = \mathbb{R}$) Y has density $f(y) = \sum_x P\{X = x\} h(y|x)$. Now define

$$P\{X \in A|Y = y\} = \sum_{x\in A} P\{X = x|Y = y\},$$

where $P\{X = x|Y = y\}$ is as specified in the problem. Then

$$\int_B P\{X \in A|Y = y\}\, dP_Y(y) = \int_B P\{X \in A|Y = y\} f(y)\, dy$$

$$= \int_B \sum_{x\in A} P\{X = x\} h(y|x)\, dy$$

$$= \sum_{x\in A} P\{X = x\} \int_B h(y|x)\, dy$$

$$= P\{X \in A, Y \in B\}.$$

The result follows from 5.3.1.

Section 5.4

1.   (a)   We show that $\{(X(\omega), Z(\omega)): \omega \in \Omega\}$ is a function; f may then be defined arbitrarily off $X(\Omega)$. If we do not have a function, then there are points $\omega_1, \omega_2 \in \Omega$ with $X(\omega_1) = X(\omega_2)$ but $Z(\omega_1) \neq Z(\omega_2)$. Let $C_1, C_2 \in \mathscr{F}''$, $C_1 \cap C_2 = \emptyset$, with $Z(\omega_1) \in C_1$, $Z(\omega_2) \in C_2$. Now $Z^{-1}(C_1) \cap Z^{-1}(C_2) = \emptyset$, and since $C_1, C_2 \in \mathscr{F}''$, we have $Z^{-1}(C_j) = X^{-1}(B_j)$ for some $B_j \in \mathscr{F}'$, $j = 1, 2$. Now $Z(\omega_1) \in C_1$; hence $\omega_1 \in Z^{-1}(C_1) = X^{-1}(B_1)$; therefore $X(\omega_1) \in B_1$. But $X(\omega_1) \notin B_2$, for if so, $Z(\omega_1) \in C_2$ as well as C_1. Similarly, $X(\omega_2) \in B_2$, $X(\omega_2) \notin B_1$. But $X(\omega_1) = X(\omega_2)$, and this contradicts the fact that $(B_1 - B_2) \cap (B_2 - B_1)$ is always empty.

(b)   Let $\Omega_0 = X(\Omega)$. If $C \in \mathscr{F}''$, then

$$Z^{-1}(C) = X^{-1}(f^{-1}(C)) = X^{-1}(f^{-1}(C) \cap \Omega_0).$$

But
$$Z^{-1}(C) = X^{-1}(B) = X^{-1}(B \cap \Omega_0) \qquad \text{for some} \qquad B \in \mathscr{F}'.$$

Since X maps onto Ω_0, $X[X^{-1}(A)] = A$ for any $A \subset \Omega_0$; hence $f^{-1}(C)$
$\cap \Omega_0 = B \cap \Omega_0 \in \mathscr{F}'$. But if $f(\Omega' - \Omega_0) = \{a\}$, then

$$f^{-1}(C) \cap (\Omega' - \Omega_0) = \begin{cases} \varnothing & \text{if } a \notin C, \\ \Omega' - \Omega_0 & \text{if } a \in C. \end{cases}$$

Therefore $f^{-1}(C) \cap (\Omega' - \Omega_0) \in \mathscr{F}'$; hence $f^{-1}(C) \in \mathscr{F}'$.

Section 5.5

2. $$\int_{\{X \in A, Z \in B\}} Y \, dP = E[Y(I_A \circ X)(I_B \circ Z)]$$

$$= E[Y(I_A \circ X)]E[I_B \circ Z] \qquad \text{by independence}$$

$$= E[E(Y(I_A \circ X)|X)]E[I_B \circ Z]$$

$$= E[(I_A \circ X)E(Y|X)]E[I_B \circ Z]$$

$$= E[(I_A \circ X)E(Y|X)E(I_B \circ Z)] \qquad \text{by independence}$$

$$= \int_{\{X \in A, Z \in B\}} E(Y|X) \, dP.$$

Thus
$$\int_{\{(X,Z) \in C\}} Y \, dP = \int_{\{(X,Z) \in C\}} E(Y|X) \, dP \tag{1}$$

for C a measurable rectangle $A \times B$, $A \in \mathscr{F}'$, $B \in \mathscr{F}''$ (where X: $(\Omega, \mathscr{F})$ $\to (\Omega', \mathscr{F}')$, Z: $(\Omega, \mathscr{F}) \to (\Omega'', \mathscr{F}'')$). By the monotone class theorem, (1) holds for all $C \in \mathscr{F}' \times \mathscr{F}''$. [Integrability of Y is used in showing that if (1) holds for $C_1, C_2, \ldots$ and $C_n \downarrow C$, then (1) holds for C.]

Section 5.6

1. (a) Let $g(x, y) = I_{A_0}(x)I_{B_0}(y)$ be the indicator of a measurable rectangle $A_0 \times B_0 \in \mathscr{F}' \times \mathscr{F}''$. Then

$$\int_{\{X \in A\}} g(X, Y) \, dP = P\{X \in A \cap A_0, Y \in B_0\}$$

$$= \int_{A \cap A_0} P\{Y \in B_0 | X = x\} \, dP_X(x)$$

$$= \int_A I_{A_0}(x)P_x(B_0) \, dP_X(x).$$

Thus

$$E[g(X, Y)|X = x] = I_{A_0}(x)P_x(B_0)$$

$$= \int_{\Omega''} I_{A_0}(x)I_{B_0}(y) \, dP_x(y)$$

$$= \int_{\Omega''} g(x, y) \, dP_x(y) \qquad (\text{a.e. } [P_X]).$$

Thus the result holds for g of this type. We proceed to indicators of arbitrary sets in $\mathscr{F}' \times \mathscr{F}''$ using the monotone class theorem, and then to arbitrary g in the usual way.

(b) By (a),

$$P\{(X, Y) \in C|X = x\} = E[I_C(X, Y)|X = x]$$

$$= \int_{\Omega''} I_C(x, y) \, dP_x(y)$$

$$= P_x(C(x)) \qquad (\text{a.e. } [P_X]).$$

(c) $P\{X \in \Omega', (X, Y) \in C\} = \int_{\Omega'} P\{(X, Y) \in C|X = x\} \, dP_X(x)$ by definition of conditional probability. The result follows from (b).

3. (a) Since $\mu(E) > 0$ and $\delta < 1$, there is an open set $V \supset E$ such that $\mu(V) \leq \delta^{-1}\mu(E)$; V is a disjoint union of open intervals I_n. Then

$$\delta \sum_n \mu(I_n) = \delta\mu(V) \leq \mu(E) = \mu(E \cap V) = \sum_n \mu(E \cap I_n). \quad (1)$$

Therefore $\delta\mu(I_n) \leq \mu(E \cap I_n)$ for some n. [Note that

$$\sum_n \mu(I_n) = \mu(V) \leq \delta^{-1}\mu(E) < \infty,$$

so it is not possible to have both sums infinite in (1).]

(b) By (a) there is an open interval I such that $\mu(E \cap I) \geq \frac{3}{4}\mu(I)$. We show that $\left(-\frac{1}{2}\mu(I), \frac{1}{2}\mu(I)\right) \subset D(E)$. Let $|x| < \frac{1}{2}\mu(I)$. If $E \cap I$ and $(E \cap I) + x$ are disjoint, the measure of their union is $2\mu(E \cap I) \geq \frac{3}{2}\mu(I)$. But $(E \cap I) \cup [(E \cap I) + x] \subset I \cup (I + x)$, an open interval of length less than $\mu(I) + \frac{1}{2}\mu(I) = \frac{3}{2}\mu(I)$, a contradiction. Thus there is an element $y \in (E \cap I) \cap [(E \cap I) + x]$. But then $y \in E$ and $y = z + x$ for some $z \in E$; hence $x \in D(E)$, as desired.

CHAPTER 5 **489**

(c) Since the circle is compact, there is a subsequence converging to a point v on the circle. Given any positive integer N, choose z_n such that $n \geq N$ and $|z_n - v| < \varepsilon/2$; then pick $z_{n+k} (k > 0)$ such that $|z_{n+k} - v| < \varepsilon/2$. Then $0 < |z_n - z_{n+k}| < \varepsilon$. (Note that $z_n \neq z_{n+k}$ since $\alpha/2\pi$ is irrational.) Thus $z_n, z_{n+k}, z_{n+2k}, \ldots$ form a chain that eventually goes entirely around the circle, with the distance between successive points less than ε. Thus, given N, we can find $z_r, r \geq N$, such that $|z_r - z| < \varepsilon$. The result follows.

(d) Since $C = \{1 + x \colon x \in B\}$, it suffices to consider B. But B is dense iff the set of numbers $n\zeta$, n an integer, reduced modulo 2, is dense in $[0, 2)$. Equivalently (consider $\theta \to e^{i\pi\theta}, 0 \leq \theta < 2$), $\{e^{in\alpha} \colon n \text{ an integer}\}$ is dense in the circle if α/π is irrational. But in this case $\alpha/2\pi$ is also irrational, and the result follows from (c).

(e) Let $F \in \mathscr{B}(\mathbb{R})$, $F \subset E_0$. We claim that $D(F) \cap A \subset \{0\}$. For if $x, y \in F$ and $x - y \in A$, then $x \sim y$; but $x, y \in E_0$; hence $x = y$ by definition of E_0. Now assume $\mu(F) > 0$. $D(F)$ includes a neighborhood of 0 by (b), so that $(0, a) \subset D(F)$ for some $a > 0$. Since A is dense by (c), we have $(0, a) \cap A \neq \emptyset$, contradicting $D(F) \cap A \subset \{0\}$. Thus $\mu(F)$ must be 0, so that if E_0 is Lebesgue measurable, then $\mu(E_0) = 0$.

Now if $x \in \mathbb{R}$, then x is equivalent to some $y \in E_0$; hence $x - y \in A$. Therefore $\mathbb{R} = \bigcup\{E_0 + a \colon a \in A\}$. But if $y + a_1 = z + a_2$, where $y, z \in E_0$, $a_1, a_2 \in A$, then $y - z = a_2 - a_1 \in A$. (Note that A is a group under addition.) Thus $y \sim z$; but since $y, z \in E_0$, $y = z$ and therefore $a_1 = a_2$. Thus the sets $E_0 + a, a \in A$, are disjoint.

Finally, assume E_0 Lebesgue measurable. Then $\mu(E_0 + a) = \mu(E_0)$ by translation-invariance of Lebesgue measure. Since A is countable, the preceding paragraph implies that $\mu(\mathbb{R}) = 0$, a contradiction.

(f) If $x \in \mathbb{R}$, then $x = y + a$ for some $y \in E_0, a \in A$ [see the argument of (e)]. Since $A = B \cup C$, it follows that $\mathbb{R} = M \cup M'$. Let F be a Borel subset of M. We claim that $D(F) \cap C \subset \{0\}$. Let $x, y \in F$ with $x - y \in C \subset A$. Then $x \sim y$, and $x = z_1 + b_1$, $y = z_2 + b_2$, where $z_1, z_2 \in E_0$, $b_1, b_2 \in B$. It follows that $z_1 - z_2 = x - y + b_2 - b_1 \in A$; hence $z_1 = z_2$. But then $x - y = b_1 - b_2 \in B \cap C$, so $x = y$. Since C is dense by (d), the same argument as in (e) shows that $\mu(F) = 0$. Finally, since $M' = \{x + 1 \colon x \in M\}$, any Borel subset of M' has Lebesgue measure 0, by translation-invariance.

(g) The first statement follows from (f). If $E \cap M \subset G \subset E$, then $E - G \subset E - M \subset E \cap M'$, so the second statement follows from (f) also.

4. (a) If $(B_1 \cap H) \cup (B_2 \cap H^c) = (B_1' \cap H) \cup (B_2' \cap H^c)$, then $B_1 \cap H = B_1' \cap H$, $B_2 \cap H^c = B_2' \cap H^c$. If, say, $\mu(B_1 - B_1') > 0$, then $B_1 - B_1'$ is not a subset of H^c since H^c has inner Lebesgue measure 0,

so there is an $x \in (B_1 - B_1') \cap H$, contradicting $B_1 \cap H = B_1' \cap H$. Thus $\mu(B_1 - B_1') = 0$; a symmetrical argument shows that $\mu(B_1' - B_1) = 0$.

(b) If $B \in \mathcal{G}$,

$$P(H \cap B) = \frac{1}{2}\mu(B) = \frac{1}{2}P(B) = \int_B \frac{1}{2}\, dP.$$

Thus $P(H|\mathcal{G}) = \frac{1}{2}$ a.e. But $P(H|\mathcal{G}) = Q(\cdot, H)$ a.e.; hence $Q(\omega, H) = \frac{1}{2}$ a.e. Similarly $Q(\omega, H^c) = \frac{1}{2}$ a.e.

(c) If $B, B_1 \in \mathcal{G}$, $P(B \cap B_1) = \int_{B_1} I_B\, dP$, so $P(B|\mathcal{G}) = I_B$ a.e.

(d) By (b) and (c), there is a set $N \in \mathcal{F}$ with $P(N) = 0$ such that for $\omega \notin N$, $Q(\omega, H) = Q(\omega, H^c) = \frac{1}{2}$ and $Q(\omega, B) = I_B(\omega)$ for all intervals B with rational endpoints. For any such ω, $Q(\omega, \{\omega\}) = 1$. [Consider a sequence of rational intervals decreasing to $\{\omega\}$ and use the fact that $Q(\omega, \cdot)$ is a probability measure.] But if $\omega \in H$, then $Q(\omega, \{\omega\}) \leq Q(\omega, H) = \frac{1}{2}$, and if $\omega \notin H$, then $Q(\omega, \{\omega\}) \leq Q(\omega, H^c) = \frac{1}{2}$, a contradiction.

CHAPTER 6

Section 6.1

1. (a) $\dfrac{\sum_{j,k=1}^n P(A_j \cap A_k)}{\left[\sum_{k=1}^n P(A_k)\right]^2}$

$$= \frac{\sum_{j \neq k} P(A_j)P(A_k) + \sum_{k=1}^n P(A_k)}{\sum_{j,k=1}^n P(A_j)P(A_k)}$$

$$= \frac{\sum_{j,k=1}^n P(A_j)P(A_k) + \sum_{k=1}^n P(A_k) - \sum_{k=1}^n (P(A_k))^2}{\sum_{j,k=1}^n P(A_j)P(A_k)},$$

Now

$$0 \leq \frac{\sum_{k=1}^n P(A_k)(1 - P(A_k))}{\left[\sum_{k=1}^n P(A_k)\right]^2} \leq \frac{\sum_{k=1}^n P(A_k)}{\left[\sum_{k=1}^n P(A_k)\right]^2} \to 0 \quad \text{as} \quad n \to \infty;$$

hence the lim inf condition is satisfied.

(b) $P\left\{\left|\displaystyle\sum_{k=1}^n I_k - \sum_{k=1}^n P(A_k)\right| > \varepsilon \sum_{k=1}^n P(A_k)\right\} \leq \dfrac{\mathrm{Var}\left(\sum_{k=1}^n I_k\right)}{\varepsilon^2 \left[\sum_{k=1}^n P(A_k)\right]^2}.$

Now

$$\text{Var}\left(\sum_{k=1}^{n} I_k\right) = E\left[\left(\sum_{k=1}^{n} I_k\right)^2\right] - \left[\sum_{k=1}^{n} E(I_k)\right]^2$$

$$= \sum_{j,k=1}^{n} P(A_j \cap A_k) - \left[\sum_{k=1}^{n} P(A_k)\right]^2.$$

The "lim inf" hypothesis implies the desired result.

(c) Let

$$d_n = P\left\{\sum_{k=1}^{n} I_k < \frac{1}{2}\sum_{k=1}^{n} P(A_k)\right\}$$

$$\leq P\left\{\left|\sum_{k=1}^{n} I_k - \sum_{k=1}^{n} P(A_k)\right| > \frac{1}{2}\sum_{k=1}^{n} P(A_k)\right\}.$$

Then $\liminf_{n\to\infty} d_n = 0$ by (b). Thus we can find integers $n_1 < n_2 < \cdots$ such that $\sum_{j=1}^{\infty} d_{n_j} < \infty$. By the Borel–Cantelli lemma, with probability 1 we have

$$\sum_{k=1}^{n_j} I_k < \frac{1}{2}\sum_{k=1}^{n_j} P(A_k)$$

for only finitely many j, that is,

$$\sum_{k=1}^{n_j} I_k \geq \frac{1}{2}\sum_{k=1}^{n_j} P(A_k)$$

for large enough j.

(d) By (c), $\sum_{k=1}^{\infty} I_k$ diverges a.e.; hence with probability 1, infinitely many A_n occur; thus $P(\limsup_n A_n) = 1$.

Section 6.2

1. Let $S_n = \sum_{k=1}^{n} X_k$. If S_n/n converges a.e. to a finite limit, then

$$\frac{X_n}{n} = \frac{S_n - S_{n-1}}{n} \to 0 \qquad \text{a.e.}$$

By the second Borel–Cantelli lemma, $\sum_{n=1}^{\infty} P\{|X_n| \geq n\} < \infty$. Since all X_n have the same distribution, $\sum_{n=1}^{\infty} P\{|X_1| \geq n\} < \infty$, so by 6.2.4, $E(X_1)$ is finite. But then $S_n/n \to E(X_1)$ a.e. by 6.2.5.

4. Let $X = (X_1, X_2, \ldots)$. Then $g(X) = g(X(T))$ whenever T permutes finitely many coordinates. Thus $\{g(X) \leq a\}$ is symmetric, and hence by the Hewitt–Savage zero-one law has probability 0 or 1. If $c = \inf\{a: P\{g(X) \leq a\} = 1\}$, then $g(X) = c$ a.e.

7. (a) Let x be an r-adic rational with r-adic expansion. $i_1 i_2 \cdots i_n 0\, 0 \cdots$. Then $P\{x < X \leq x + r^{-n}\} = P\{X_1 = i_1, \ldots, X_n = i_n\} = r^{-n}$. (Note that $P\{X = x\} = \prod_{k=1}^{\infty} a_k$, with $a_k = r^{-n}$ for all k; thus $P\{X = x\} = 0$.) It follows that if λ is Lebesgue measure, $P\{X \in I\} = \lambda(I)$ for every r-adic interval $I \subset [0, 1]$, and hence for every interval $I \subset [0, 1]$ by continuity. Thus $P\{X \leq y\} = y, 0 \leq y \leq 1$, as desired.

(b) Fix r and i, and let $A_{ri} = \{x \in [0, 1]:$ the relative frequency of i in the first n digits of the r-adic expansion of x converges to $1/r\}$. Then, if Y_k is the indicator of $\{X_k = i\}$,

$$P\{X \in A_{ri}\} = P\left\{\frac{1}{n}(Y_1 + \cdots + Y_n) \to \frac{1}{r}\right\} = 1$$

by the strong law of large numbers. If

$$A = \bigcap_{r=2}^{\infty} \bigcap_{i=0}^{r-1} A_{ri},$$

it follows that $P_X(A) = P\{X \in A\} = 1$. But P_X is Lebesgue measure by part (a), and the result follows.

(c) We may write

$$\int_0^1 R_n(x)\, dx = \int_0^1 R_n(x)\, dP_X(x)$$
$$= E[R_n(X)] = E[2X_n - 1] = 0,$$

and similarly,

$$\int_0^1 R_n(x)R_m(x)\, dx = E[R_n(X)R_m(X)].$$

Since $R_n(X)$ and $R_m(X)$ are independent for $n \neq m$, and $R_n^2(X) \equiv 1$, the result follows.

Section 6.3

3. Each A_{nj} is a countable union of sets $A_{n+1,k}$; hence

$$\int_{A_{nj}} X_{n+1} \, dP = \sum_k \int_{A_{n+1,k}} X_{n+1} \, dP$$

$$= \sum_k \lambda(A_{n+1,k}) \qquad \text{by definition of } X_{n+1}$$

$$= \lambda(A_{nj}) \qquad \text{since } \lambda \text{ is countably additive}$$

$$= \int_{A_{nj}} X_n \, dP \qquad \text{by definition of } X_n.$$

5. $E(Y_n) = \int_{\{p_n(x)>0\}} \dfrac{q_n(x)}{p_n(x)} p_n(x) \, dx \le \int_{\mathbb{R}^n} q_n(x) \, dx = 1.$

Let $A = \{(X_1, \dots, X_n) \in B\}$, $B \in \mathscr{B}(\mathbb{R}^n)$. Then $\int_A Y_{n+1} \, dP = E(Y_{n+1} I_A)$; hence

$$\int_A Y_{n+1} \, dP = \int_{\{x \in B, \, p_{n+1}(x')>0\}} \frac{q_{n+1}(x')}{p_{n+1}(x')} p_{n+1}(x') \, dx', \qquad (1)$$

where $x' = (x_1, \dots, x_{n+1}) \in \mathbb{R}^{n+1}$, $x = (x_1, \dots, x_n)$, the first n coordinates of x'. Now $p_n(x) = \int_{-\infty}^{\infty} p_{n+1}(x') \, dx_{n+1}$; thus if $x \in B$ and $p_n(x) = 0$, then $p_{n+1}(x') = 0$ except for x_{n+1} in a set of Lebesgue measure 0, so the integration of $q_{n+1}(x')$ with respect to x_{n+1} in (1) will be 0. In other words,

$$\int_{\{x \in B, \, p_n(x)=0, \, p_{n+1}(x')>0\}} q_{n+1}(x') \, dx' = 0.$$

Therefore the right side of (1) becomes

$$\int_{\{x \in B, \, p_n(x)>0, \, p_{n+1}(x')>0\}} q_{n+1}(x') \, dx'$$

$$\le \int_{\{x \in B, \, p_n(x)>0\}} q_{n+1}(x') \, dx'$$

$$= \int_{\{x \in B, \, p_n(x)>0\}} q_n(x) \, dx$$

$$= \int_{\{x \in B, \, p_n(x)>0\}} \frac{q_n(x)}{p_n(x)} p_n(x) \, dx$$

$$= E(Y_n I_A) = \int_A Y_n \, dP,$$

proving the supermartingale property.

6.   (a) $A_n = \sum_{i=0}^{n-1} X_i - \sum_{i=1}^{n} E(X_i|\mathscr{F}_{i-1})$

$$Y_n = \sum_{i=0}^{n} X_i - \sum_{i=1}^{n} E(X_i|\mathscr{F}_{i-1}),$$

so $Y_n - A_n = X_n$.

(b) $E(Y_n|\mathscr{F}_{n-1}) = \sum_{i=0}^{n-1} X_i + E(X_n|\mathscr{F}_{n-1}) - \sum_{i=1}^{n} E[E(X_i|\mathscr{F}_{i-1})|\mathscr{F}_{n-1}]$

$$= Y_{n-1} + E(X_n|\mathscr{F}_{n-1}) - E(X_n|\mathscr{F}_{n-1})$$

$$= Y_{n-1}.$$

(c) $A_{n+1} - A_n = X_n - E(X_{n+1}|\mathscr{F}_n) \geq 0$ a.e.

Section 6.4

1.   (a) $E(X_{n+1}|X_n = 0) = p_{n+1}(a_{n+1} - a_{n+1}) + (1 - 2p_{n+1})0 = 0,$

$E(X_{n+1}|X_n = a_n) = a_n,$ $E(X_{n+1}|X_n = -a_n) = -a_n,$

proving the martingale property. Since for all ω, either $X_n(\omega) = 0$ for all n or for some j, $X_n(\omega) = a_j$ for $n \geq j$, X_n converges everywhere.

(b) $E(|X_2|) = 2p_2a_2,$ $E(|X_3|) = 2p_2a_2 + (1 - 2p_2)2p_3a_3,$

$E(|X_4|) = 2p_2a_2 + (1 - 2p_2)2p_3a_3 + (1 - 2p_2)(1 - 2p_3)2p_4a_4,$

and so on. Thus

$$\lim_{k \to \infty} E(|X_k|) \geq \left[\prod_{k=1}^{\infty} (1 - 2p_k) \right] 2 \sum_{k=2}^{\infty} p_k a_k.$$

The infinite product is greater than 0 since $\sum p_k < \infty$; hence $E(|X_k|) \to \infty$.

4.   By definition of the problem, $E(X_{n+1}|X_1, \ldots, X_n) = E(X_{n+1}|X_n)$. If before the nth drawing there are r balls in the urn, c of them white,

$$E\left(X_{n+1}|X_n = \frac{c}{r}\right) = \left(\frac{c}{r}\right)\left(\frac{c+1}{r+1}\right) + \left(1 - \frac{c}{r}\right)\frac{c}{r+1} = \frac{c}{r}.$$

Thus $E(X_{n+1}|X_n) = X_n$. Since $|X_n| \leq 1$ for all n, $X_n \to X_\infty$ a.e. By the dominated convergence theorem, $E(X_\infty) = \lim_n E(X_n) = E(X_1)$.

Section 6.5

1. Since $|f_n|^p \leq 2^{p-1}(|f_n - f|^p + |f|^p)$, $p \geq 1$, and $|f_n|^p \leq |f_n - f|^p + |f|^p$, $p \leq 1$, by 6.5.3 it suffices to show that the $|f_n - f|^p$ are uniformly integrable. Now

$$\int_A |f_n - f|^p \, d\mu \to 0 \qquad \text{as} \qquad \mu(A) \to 0$$

for any fixed n, and

$$\int_A |f_n - f|^p \, d\mu \leq \int_\Omega |f_n - f|^p \, d\mu \to 0 \qquad \text{as} \qquad n \to \infty$$

by the L^p-convergence. It follows that the integrals of $|f_n - f|^p$ are uniformly continuous and uniformly bounded; the result follows from 6.5.3.

Section 6.6

2. (a) If the X_n are uniformly integrable, $E(X_n) \to E(X_\infty)$ by 6.5.5. Conversely assume $E(X_n) \to E(X_\infty)$. If $\vee$ stands for max and $\wedge$ for min, we have $|X_n - X_\infty| = (X_n \vee X_\infty) - (X_n \wedge X_\infty)$ and $X_n + X_\infty = (X_n \vee X_\infty) + (X_n \wedge X_\infty)$. By hypothesis, $E(X_n + X_\infty) \to 2E(X_\infty)$, and by the dominated convergence theorem, $E(X_n \wedge X_\infty) \to E(X_\infty)$. Hence $E(X_n \vee X_\infty) \to E(X_\infty)$, so $E(|X_n - X_\infty|) \to E(X_\infty) - E(X_\infty) = 0$. Thus $X_n \to X_\infty$ in L^1, and it follows that the X_n are uniformly integrable. (See Problem 1.6.5; in general, L^p convergence of $\{f_n\}$ implies uniform integrability of $\{|f_n|^p\}$.)

   (b) If $A \in \mathscr{F}_n$, $n \leq m$, then $\int_A X_n \geq \int_A X_m$. Let $m \to \infty$; by Fatou's lemma,

$$\int_A X_\infty = \int_A \lim_m X_m \leq \liminf_m \int_A X_m \leq \int_A X_n.$$

   (c) By Fatou's lemma, $E(X_\infty) = E(\lim_n X_n) \leq \liminf_n E(X_n) = 0$.

Section 6.7

1. $|X_T| = X_T^+ + X_T^- = 2X_T^+ - X_T$; hence $E(|X_T|) \leq 2E(X_T^+) - E(X_1)$ by 6.7.3. But $\{X_1^+, \ldots, X_n^+\}$ is a submartingale by 6.3.6(a); hence $E(X_T^+) \leq E(X_n^+)$ by 6.7.3, as desired.

3. (a) Define T as indicated. By 6.7.3, $\{X_T, X_n\}$ is a submartingale; hence

$$E(X_n) \geq E(X_T) = \int_{\{\max X_i \geq \lambda\}} X_T \, dP + \int_{\{\max X_i < \lambda\}} X_T \, dP$$

$$\geq \lambda P\{\max X_i \geq \lambda\} + \int_{\{\max X_i < \lambda\}} X_n \, dP$$

and the result follows.

(b) By 6.7.3, $\{X_1, X_T\}$ is a supermartingale; hence

$$E(X_1) \geq E(X_T) = \int_{\{\max X_i \geq \lambda\}} X_T \, dP + \int_{\{\max X_i < \lambda\}} X_T \, dP$$

$$\geq \lambda P\{\max X_i \geq \lambda\} + \int_{\{\max X_i < \lambda\}} X_n \, dP.$$

Since $-X_n \leq (-X_n)^+ = X_n^-$, the result follows.

(c) Since

$$\left\{ \max_{1 \leq i \leq n} X_i \geq \lambda + \frac{1}{k} \right\} \uparrow \left\{ \sup_n X_n > \lambda \right\} \qquad \text{as} \qquad n, k \to \infty,$$

the result follows from (a) and (b). Note also that the same inequalities hold with $\{\sup_n X_n > \lambda\}$ replaced by $\{\sup_n X_n \geq \lambda\}$; this follows because

$$\left\{ \sup_n X_n > \lambda - \frac{1}{k} \right\} \downarrow \left\{ \sup_n X_n \geq \lambda \right\}.$$

5. (a) Since $X_n - nm = \sum_{k=1}^n (Y_k - m)$, and $E(Y_k - m) = 0$, $\{X_n - nm\}$ is a martingale. By 6.7.3, $E(X_1 - m) = E(X_{T_n} - T_n m)$; hence $E(X_{T_n}) = mE(T_n)$. Since $Y_j \geq 0$, $X_{T_n} \uparrow X_T$ as $n \to \infty$, so by the monotone convergence theorem, $E(X_T) = mE(T)$.

(b) We write $X_n = \sum_{j=1}^n Y_j^+ - \sum_{j=1}^n Y_j^- = X_n' - X_n''$. By (a), $E(X_T')$ $= E(Y_1^+)E(T)$, $E(X_T'') = E(Y_1^-)E(T)$. Since $E(T)$ is finite, so are $E(X_T')$ and $E(X_T'')$; hence $E(|X_T|) < \infty$ and

$$E(X_T) = E(X_T') - E(X_T'') = [E(Y_1^+) - E(Y_1^-)]E(T) = mE(T).$$

(c) To prove (a), observe that if all $Y_j \geq 0$, then

$$E(X_T) = \sum_{n=1}^\infty E(X_n I_{\{T=n\}})$$

$$= \sum_{n=1}^\infty E(X_n) P\{T = n\} \qquad \text{by independence}$$

$$= m \sum_{n=1}^{\infty} n P\{T = n\}$$

$$= m E(T).$$

Part (b) is proved just as above.

Section 6.8

3. (a) $\{X_n\}$ converges a.e. to a finite limit; hence

$$P\{|X_{n+1} - X_n| \geq b \quad \text{for infinitely many} \quad n\} = 0,$$

so, a.e., $X_{n+1} = X_n$ eventually.

(b) Since $X_n \geq 0$, $X_T I_{\{T \leq n\}} \uparrow X_T$; hence by the monotone convergence theorem,

$$E(X_T) = \lim_{n \to \infty} \int_{\{T \leq n\}} X_T = \lim_{n} \int_{\{T \leq n\}} X_n$$

since on $\{T = k\}$, $k \leq n$, we have $X_T = X_k = X_{k+1} = \cdots = X_n$. But

$$\lim_{n} \int_{\{T \leq n\}} X_n \leq \lim\sup_{n} \int_{\Omega} X_n \leq E(X_0)$$

by the supermartingale property.

(c) T is the time at which the betting stops. In this case, T is also the time of going broke. By (a), T is a.e. finite, and the result follows.

(d) Realistically, there is a limit on what we can lose. In practice, what we are doing is starting with a capital of $x > 0$, and stopping when we reach $x + 1$, provided we have not been wiped out (reduced to zero) beforehand. The probability of reaching $x + 1$ before 0 is $x/(x+1) < 1$, and

$$E(X_T) = \frac{x}{x+1}(x+1) + \frac{1}{x+1}(0) = x = E(X_0)$$

(See Ash, 1970, 6.2 for details.)

5. (i) If $\sum_{j=1}^{\infty} I_{A_j} = \infty$ and $\sum_{j=1}^{\infty} q_j < \infty$, then $X_n \to \infty$.

(ii) If $\sum_{j=1}^{\infty} I_{A_j} < \infty$ and $\sum_{j=1}^{\infty} q_j = \infty$, then $X_n \to -\infty$.

But $\{X_n\}$ is a martingale by Problem 4; hence by 6.8.4, X_n converges a.e. to a finite limit on $\{\sup X_n < \infty$ or $\inf X_n > -\infty\}$. In case (i), $\inf X_n > -\infty$ and in case (ii), $\sup X_n < \infty$, so we have a contradiction unless the sets

$$\left\{ \sum I_{A_j} = \infty, \quad \sum q_j < \infty \right\} \quad \text{and} \quad \left\{ \sum I_{A_j} < \infty, \quad \sum q_j = \infty \right\}$$

have probability 0. (Note that $|I_{A_j} - q_j| \leq 2$ so 6.8.4 actually applies.)

CHAPTER 7

Section 7.1

4. (a) If $|h(u)| = 1$, then $h(u) = e^{iua}$ for some a; hence $e^{iua} = \int_{\mathbb{R}} e^{iux} \, dF(x)$, or $1 = \int_{\mathbb{R}} e^{iu(x-a)} \, dF(x)$. Take real parts to obtain $\int_{\mathbb{R}} [1 - \cos u(x - a)] dF(x) = 0$. Since the integrand is nonnegative, we have $\cos u(x - a) = 1$ a.e. $[P_X]$. But $\cos u(x - a) = 1$ iff $x = a + 2\pi n u^{-1}$, n an integer, so X has a lattice distribution. The converse is proved by reversing the argument.

   (b) By part (a),

$$P\{X = a + 2\pi n u^{-1} \qquad \text{for some integer } n\}$$
$$= P\{X = b + 2\pi m(\alpha u)^{-1} \qquad \text{for some integer } m\} = 1$$

for appropriate real numbers a and b. If X is nondegenerate, the lattices $\{a + 2\pi n u^{-1}: n \text{ an integer}\}$ and $\{b + 2\pi m(\alpha u)^{-1}: m \text{ an integer}\}$ must have at least two points in common, and this implies that $2\pi u^{-1}$ and $2\pi(\alpha u)^{-1}$ are rationally related. Thus α is a rational number, a contradiction.

5. (a) By 7.1.5(e), h has n continuous derivatives on $\mathbb{R}$ and $h^{(k)}(0) = i^k E(X^k)$, $k = 0, 1, \ldots, n$. Now if $h: I \to C$, where I is an interval of R containing 0, and h has n continuous derivatives on I, then for $u \in I$,

$$h(u) = \sum_{k=0}^{n-1} \frac{h^{(k)}(0)}{k!} u^k + u^n \int_0^1 h^{(n)}(ut) \frac{(1-t)^{n-1}}{(n-1)!} \, dt.$$

(This is an exercise in calculus; see Ash, 1970, p. 172 for details.) Add and subtract

$$\frac{h^{(n)}(0)u^n}{n!} = u^n \int_0^1 h^{(n)}(0) \frac{(1-t)^{n-1}}{(n-1)!} \, dt$$

from the above equation to obtain

$$h(u) = \sum_{k=0}^{n} \frac{h^{(k)}(0)}{k!} u^k + R_n(u),$$

where

$$R_n(u) = u^n \int_0^1 [h^{(n)}(ut) - h^{(n)}(0)] \frac{(1-t)^{n-1}}{(n-1)!} \, dt.$$

Since $R_n(u)/u^n \to 0$ as $u \to 0$ by the dominated convergence theorem, the result follows.

(b) By 7.1.5(e), h has n continuous derivatives, so as in part (a),

$$h(u) - \sum_{k=0}^{n-1} \frac{h^{(k)}(0)}{k!} u^k = u^n \int_0^1 h^{(n)}(ut) \frac{(1-t)^{n-1}}{(n-1)!} \, dt.$$

Now $h^{(k)}(u) = \int_{\mathbb{R}} (ix)^k e^{iux} \, dF(x)$ by 7.1.5(e); hence $|h^{(n)}(ut)| \leq E(|X|^n)$; the result follows.

7. (a) $\left(\dfrac{E_r - E_{-r}}{2r} \right)^2 h(0) = \dfrac{1}{4r^2} (h(2r) - 2h(0) + h(-2r))$

$$= \int_{-\infty}^{\infty} \left(\frac{e^{irx} - e^{-irx}}{2r} \right)^2 dF(x)$$

$$= -\int_{-\infty}^{\infty} \left(\frac{\sin rx}{rx} \right)^2 x^2 \, dF(x).$$

(b) By L'Hôspital's rule,

$$\lim_{r \to 0} \frac{1}{4r^2} [h(2r) - 2h(0) + h(-2r)]$$

$$= \lim_{r \to 0} \frac{2h'(2r) - 2h'(-2r)}{8r}$$

$$= \lim_{r \to 0} \frac{h'(2r) - h'(0) + h'(0) - h'(-2r)}{2(2r)} = h''(0)$$

Thus

$$-h''(0) = \lim_{r \to 0} \int_{-\infty}^{\infty} \left(\frac{\sin rx}{rx} \right)^2 x^2 \, dF(x)$$

$$\geq \int_{-\infty}^{\infty} \lim_{r \to 0} \left(\frac{\sin rx}{rx} \right)^2 x^2 \, dF(x) \qquad \text{by Fatou's lemma}$$

$$= \int_{-\infty}^{\infty} x^2 \, dF(x).$$

(c) Assume the result holds up to the integer n, and assume $h^{(2n+2)}(0)$ exists and is finite. Then $\int_{-\infty}^{\infty} x^{2n} \, dF(x) < \infty$, so by 7.1.5(e),

$$h^{(2n)}(u) = \int_{-\infty}^{\infty} (ix)^{2n} e^{iux} \, dF(x)$$

or

$$(-1)^n h^{(2n)}(u) = \int_{-\infty}^{\infty} e^{iux} \, dG(x),$$

where

$$G(x) = \int_{-\infty}^{x} t^{2n} \, dF(t).$$

Since G is a bounded distribution function with characteristic function $(-1)^n h^{(2n)}$, part (b) shows that $\int_{-\infty}^{\infty} x^2 \, dG(x) < \infty$, that is, $\int_{-\infty}^{\infty} x^{2n+2} dF(x) < \infty$. The result follows by induction.

Section 7.2

4. (a) Choose $a, b \in I$ such that $L = \int_a^b g(u) \, du \neq 0$. (If this is not possible, the integral of g is 0 on all subintervals of I; hence on all Borel subsets of I, and therefore $g = 0$ a.e., contradicting $|\exp(iua_n)| = 1$.) We may assume that $\exp(iua_n)$ converges when $u = a$ and $u = b$ (since $\int_a^b g(u) \, du$ is continuous in a and b). Now

$$ia_n = ia_n \frac{\int_a^b \exp(iua_n) \, du}{\int_a^b \exp(iua_n) \, du} = \frac{\exp(iba_n) - \exp(iaa_n)}{\int_a^b \exp(iua_n) \, du} \to \frac{g(b) - g(a)}{L},$$

   (b) This is immediate from (a).

5. (a) Let $F_n(x) = 1, x \geq n$; $F_n(x) = 0, x < n$ (corresponding to a random variable $X_n \equiv n$). Let $F_0(x) = 0$ for all x. Then $F_n(x) \to F_0(x)$ for all $x \in \mathbb{R} \cup \{-\infty\}$, but $F_n(\infty) \not\longrightarrow F_0(\infty)$, so F_n does not converge weakly to F_0.

   (b) If n is even, let $F_n(x) = 1, x \geq n$; $F_n(x) = 0, x < n$ (corresponding to $X_n \equiv n$). If n is odd, let $F_n(x) = 1, x \geq -n$; $F_n(x) = 0, x < -n$ ($X_n \equiv -n$). Let $F_0(x) \equiv 0$. Then $F_n(a, b] \to F_0(a, b]$ for all $a, b \in \mathbb{R}$, but $F_n(-\infty, \infty] \not\longrightarrow F_0(-\infty, \infty]$. Furthermore, $\lim_{n \to \infty} F_n(x)$ does not exist for any $x \in \mathbb{R}$.

6. If F_n converges weakly to F_0, then $\{F_n\}$ is relatively compact, and hence tight by 7.2.4. Thus assume $\{F_n\}$ tight. Given $\varepsilon > 0$, let a and b be finite continuity points of F_0 such that $F_n(\mathbb{R} - (a, b]) < \varepsilon$ for all n. Then

$$\limsup_{n \to \infty} F_n(\mathbb{R}) \leq \varepsilon + \limsup_{n \to \infty} F_n(a, b]$$

$$= \varepsilon + F_0(a, b]$$

$$\leq \varepsilon + F_0(\mathbb{R}).$$

But $F_n(\mathbb{R}) \geq F_n(a, b]$; hence

$$\liminf_{n\to\infty} F_n(\mathbb{R}) \geq F_0(a, b].$$

Since ε is arbitrary and b may be taken arbitrarily large and a arbitrarily small, we have $F_n(\mathbb{R}) \to F_0(\mathbb{R})$. A similar argument shows that $F_n(-\infty, b] \to F_0(-\infty, b]$ and $F_n(a, \infty] \to F_0(a, \infty]$ if a and b are finite continuity points of F_0. Therefore F_n converges weakly to F_0.

8. (a) $E(X_{n+1}|\mathscr{F}_n) = E(X_n h_{n+1}^{-1}(u) \exp(iuY_{n+1})|\mathscr{F}_n)$

$$= X_n h_{n+1}^{-1}(u) E[\exp(iuY_{n+1})]$$

since X_n is $\mathscr{F}_n$-measurable and the Y_k are independent

$$= X_n.$$

(b) By hypothesis, $\prod_{k=1}^{n} h_k \to h_X$ uniformly on bounded intervals. Thus if I is a bounded open interval containing 0 on which $|h_X| \geq \delta > 0$, then $\prod_{k=1}^{n} h_k$ is bounded away from 0 on I. Thus for any fixed $u \in I$, $\{X_n, \mathscr{F}_n\}$ is a bounded martingale, and hence converges a.e. But

$$X_n(\omega) = \left[\prod_{k=1}^{n} h_k(u)\right]^{-1} \exp\left[iu \sum_{k=1}^{n} Y_k(\omega)\right],$$

and the result follows.

(c) Let C be the set of pairs (u, ω), $u \in I$, $\omega \in \Omega$, such that

$$\exp\left[iu \sum_{k=1}^{n} Y_k(\omega)\right]$$

fails to converge. By (b), $\{\omega: (u, \omega) \in C\}$ has probability 0 for each $u \in I$, so by Problem 4, Section 2.6, $\{u: (u, \omega) \in C\}$ has Lebesgue measure 0 for almost every ω.

(d) Convergence a.e. implies convergence in probability since a probability measure is finite, and convergence in probability implies convergence in distribution by 7.1.7. By parts (b) and (c), convergence in distribution implies convergence a.e.

Section 7.3

3. Let the X_{nk} be uan. Then

$$\max_{1\le k\le n} |h_{nk}(u) - 1| = \max_{1\le k\le n} \left| \int_{-\infty}^{\infty} (e^{iux} - 1)\, dF_{nk}(x) \right|$$

$$\le \max_{1\le k\le n} \int_{|x|<\varepsilon} |e^{iux} - 1|\, dF_{nk}(x)$$

$$+ \max_{1\le k\le n} \int_{|x|\ge\varepsilon} |e^{iux} - 1|\, dF_{nk}(x).$$

Now $|e^{iux} - 1| \le |ux|$; hence

$$\max_{1\le k\le n} |h_{nk}(u) - 1| \le \max_{1\le k\le n} \int_{|x|<\varepsilon} |ux|\, dF_{nk}(x)$$

$$+ \max_{1\le k\le n} \int_{|x|\ge\varepsilon} 2\, dF_{nk}(x)$$

$$\le |u|\varepsilon + 2 \max_{1\le k\le n} P\{|X_{nk}| \ge \varepsilon\}.$$

The second term approaches 0 as $n \to \infty$ for any $\varepsilon > 0$ by the uan hypothesis, and thus

$$\max_{1\le k\le n} |h_{nk}(u) - 1| \to 0 \qquad \text{as} \qquad n \to \infty$$

uniformly for u in a bounded interval.
    Conversely, assume

$$\max_{1\le k\le n} |h_{nk}(u) - 1| \to 0.$$

By the truncation inequality 7.2.7,

$$\max_{1\le k\le n} P\{|X_{nk}| \ge \varepsilon\} = \max_{1\le k\le n} \int_{|x|\ge\varepsilon} dF_{nk}(x)$$

$$\le \max_{1\le k\le n} 7\varepsilon \int_0^{1/\varepsilon} [1 - \operatorname{Re} h_{nk}(v)]\, dv$$

$$\le 7\varepsilon \int_0^{1/\varepsilon} \max_{1\le k\le n} |1 - h_{nk}(v)|\, dv$$

since $|1 - \operatorname{Re} z| = |\operatorname{Re}(1 - z)| \le |1 - z|$.

The integral approaches 0 as $n \to \infty$ by the hypothesis and the dominated convergence theorem, and the result follows.

4. (a) Var $X_k = 1$ for all n, so $c_n = \sqrt{n}$. Now, for a given $\varepsilon > 0$, if n is large enough so that $\varepsilon\sqrt{n} > 1$, then

$$P\left\{\left|\frac{X_k}{c_n}\right| \geq \varepsilon\right\} = P\{|X_k| \geq \varepsilon\sqrt{n}\}$$

$$= \begin{cases} 0 & \text{if} \quad k < \varepsilon\sqrt{n}, \\ \dfrac{1}{k^2}\left(1 - \dfrac{1}{c}\right) & \text{if} \quad k \geq \varepsilon\sqrt{n}. \end{cases}$$

Thus

$$\max_{1 \leq k \leq n} P\left\{\left|\frac{X_k}{c_n}\right| \geq \varepsilon\right\} \leq \frac{1}{\varepsilon^2 n}\left(1 - \frac{1}{c}\right) \to 0.$$

(b) $\dfrac{1}{c_n^2}\displaystyle\sum_{k=1}^{n}\int_{|x|\geq\varepsilon c_n} x^2\, dF_k(x) = \dfrac{1}{n}\displaystyle\sum_{k=1}^{n} E\left[X_k^2 I_{\{|X_k|\geq\varepsilon\sqrt{n}\}}\right].$

Again let n be large enough so that $\varepsilon\sqrt{n} > 1$. We obtain

$$\frac{1}{n}\sum_{k=\varepsilon\sqrt{n}}^{n} k^2 P\{|X_k| = k\} \sim \frac{1}{n}(n - \varepsilon\sqrt{n})\left(1 - \frac{1}{c}\right) \to 1 - \frac{1}{c} > 0.$$

Thus the Lindeberg condition fails for the X_k. Now

$$\text{Var } X_{nk}' = E[X_k^2 I_{\{|X_k|\leq\sqrt{n}\}}]$$

$$= E(X_k^2) = 1 \qquad \text{if} \quad k \leq \sqrt{n}$$

$$= P\{|X_k| = 1\} = \frac{1}{c} \qquad \text{if} \quad k > \sqrt{n}.$$

Thus

$$(c_n')^2 = \sum_{k=1}^{n} \text{Var } X_{nk}' = [\sqrt{n}] + \frac{1}{c}(n - [\sqrt{n}]) \sim \frac{n}{c}.$$

The Lindeberg sum for the X_{nk}' is

$$\frac{1}{(c_n')^2} \sum_{k=1}^{n} E[(X_{nk}')^2 I_{\{|X_{nk}'| \geq \varepsilon c_n'\}}]$$

$$\sim \frac{c}{n} \sum_{k=\varepsilon c_n'}^{\sqrt{n}} k^2 P\{|X_k| = k\}$$

$$\sim \frac{c}{n} \left(\sqrt{n} - \varepsilon\sqrt{\frac{n}{c}}\right)\left(1 - \frac{1}{c}\right) \to 0.$$

By 7.3.1, $S_n'/c_n' \xrightarrow{d}$ normal $(0, 1)$.

(c) $P\{S_n \neq S_n'\} \leq \sum_{k=1}^{n} P\{X_k \neq X_{nk}'\} \leq \sum_{k=1}^{n} P\{|X_k| > \sqrt{n}\}$

$$\leq \sum_{k=\sqrt{n}}^{n} P\{|X_k| = k\}$$

$$\to 0 \qquad \text{as} \qquad n \to \infty$$

since

$$P\{|X_k| = k\} = \frac{1}{k^2}\left(1 - \frac{1}{c}\right) \qquad \text{and} \qquad \sum_{k=1}^{\infty} \frac{1}{k^2} < \infty.$$

(d) If $Y_n \xrightarrow{d}$ and $a_n \to 1$, then $a_n Y_n \xrightarrow{d} Y$; for if h_n is the characteristic function of Y_n and h is the characteristic function of Y, we
have $h_n \to h$ uniformly on bounded intervals; hence $h_n(a_n u)$
$\to h(u)$. Now $\sqrt{c} S_n'/\sqrt{n} = a_n(S_n'/c_n')$, where $a_n \to 1$, so that
$(\sqrt{c}/\sqrt{n})S_n' \xrightarrow{d}$ normal $(0, 1)$ by (b). Also, if $Y_n \xrightarrow{d} Y$ and
$P\{Y_n \neq Y_n'\} \to 0$, then $Y_n' \xrightarrow{d} Y$ because

$$P\{Y_n \leq y\} = P\{Y_n \leq y, Y_n' \leq y\} + P\{Y_n \leq y, Y_n' > y\}$$
$$\leq P\{Y_n' \leq y\} + P\{Y_n \neq Y_n'\}.$$

Thus by (c), $\sqrt{c} S_n/\sqrt{n} \xrightarrow{d}$ normal $(0, 1)$. But by (a) and (b),
$S_n/c_n = S_n/\sqrt{n} \not\xrightarrow{d}$ normal $(0, 1)$.

Section 7.4

1. (a) $E[\exp(iuk(\operatorname{sgn} X_1)|X_1|^{-r})]$

$$= \int_{-n}^{n} \frac{1}{2n} \exp(iuk(\operatorname{sgn} x)|x|^{-r}) \, dx$$

$$= \frac{1}{n} \int_{0}^{n} \cos(ukx^{-r}) \, dx$$

$$= 1 - \frac{1}{n} \int_{0}^{n} [1 - \cos(ukx^{-r})] \, dx$$

$$= 1 - \frac{1}{n} \left[\int_{0}^{\infty} [1 - \cos(ukx^{-r})] \, dx - g(n) \right],$$

where $g(n) = \int_{n}^{\infty} [1 - \cos(ukx^{-r})] \, dx \to 0$ as $n \to \infty$ since

$$1 - \cos(ukx^{-r}) = 2 \sin^2 \tfrac{1}{2} ukx^{-r} \sim cx^{-2r}, \qquad 2r > 1.$$

The result follows from Theorem 7.1.2.

(b) Let $I = \int_{0}^{\infty} [1 - \cos(kux^{-r})] \, dx$; then [see Eq. (3) of the proof of 7.3.1],

$$n \ln \left(1 - \frac{1}{n}[I - g(n)] \right) = g(n) - I + \frac{\theta}{n}|I - g(n)|^2 \to -I;$$

hence $h_n(u) \to e^{-I}$.

(c) We have

$$dy = -|u|^{1/r} k^{1/r} x^{-2} \, dx = -|u|^{-1/r} k^{-1/r} y^2 \, dx;$$

hence

$$h(u) = \exp \left(-|u|^{1/r} k^{1/r} \int_{0}^{\infty} (1 - \cos y^r) y^{-2} \, dy \right).$$

This is of the form $\exp[-d|u|^{\alpha}]$, $d > 0$, $0 < \alpha < 2$.

Section 7.5

1. A logarithm of $h(u)$ is given by

$$iu + \operatorname{Log}(1 - q) - \operatorname{Log}(1 - qe^{iu}) = iu + \sum_{k=1}^{\infty} \frac{q^k}{k}(e^{iuk} - 1).$$

Thus

$$h(u) = \lim_{n \to \infty} \exp \left(\sum_{k=1}^{n} \left[\frac{iu}{n} + \frac{q^k}{k} (e^{iuk} - 1) \right] \right).$$

But

$$\exp \left[\frac{iu}{n} + \frac{q^k}{k} (e^{iuk} - 1) \right]$$

is of the Poisson type [see 7.5.3(b)] with $\lambda = q^k/k$, $a = k$, $b = 1/n$, and the result follows from 7.5.7.

4. $F_3(z) = P_{XY}\{(x, y) \in \mathbb{R}^2 : x + y \leq z\}$

$$= \iint_{x+y \leq z} dP_{XY}(x, y)$$

$$= \int_{-\infty}^{\infty} \int_{-\infty}^{\infty} I_{\{x+y \leq z\}} \, dP_X(x) \, dP_Y(y)$$

$$= \int_{-\infty}^{\infty} P_X\{x : x \leq z - y\} \, dP_Y(y)$$

$$= \int_{-\infty}^{\infty} F_1(z - y) \, dF_2(y)$$

$$= \int_{-\infty}^{\infty} F_2(z - x) \, dF_1(x) \qquad \text{by a symmetrical argument.}$$

In the case where X has a density,

$$F_3(z) = \int_{-\infty}^{\infty} F_1(z - y) \, dF_2(y) = \int_{-\infty}^{\infty} \left(\int_{-\infty}^{z-y} f_1(x) \, dx \right) dF_2(y).$$

Let $x = u - y$ to obtain

$$F_3(z) = \int_{-\infty}^{\infty} \left(\int_{-\infty}^{z} f_1(u - y) \, du \right) dF_2(y)$$

$$= \int_{-\infty}^{z} \left(\int_{-\infty}^{\infty} f_1(u - y) \, dF_2(y) \right) du,$$

and the result follows.

Section 7.6

1. By the strong law of large numbers, $F_n(x, \omega)$ converges a.e. to

$$E[I_{\{X_1 \leq x\}}] = F(x), \quad \text{and} \quad F_n(x^-, \omega) = \frac{1}{n} \sum_{k=1}^{n} I_{\{X_k < x\}}(\omega) \to F(x^-) \quad \text{a.e.}$$

(This holds for each fixed $x \in \mathbb{R}$, and the exceptional set of measure 0 depends on x.)

Assume that $F_n(x, \omega) \to F(x)$ and $F_n(x^-, \omega) \to F(x^-)$ for $\omega \notin A_x$, where $P(A_x) = 0$. Let S be a countable dense subset of $\mathbb{R}$ containing all discontinuity points of F. If $A = \bigcup \{A_x : x \in S\}$, then $P(A) = 0$ since S is countable. If $\omega \notin A$, then $F_n(x, \omega) \to F(x)$ and $F_n(x^-, \omega) \to F(x^-)$ for all $x \in S$. Furthermore, $F_n(\infty, \omega) \equiv 1$, $F(\infty) = 1$, $F_n(-\infty, \omega) \equiv 0$, $F(-\infty) = 0$. By 7.6.1, $F_n(x, \omega) \to F(x)$ uniformly for $x \in \mathbb{R}$.

CHAPTER 8

Section 8.2

4. If $T^{-1}A \subset A$, then $A^c \subset T^{-1}A^c$, and conversely; also, $A - T^{-1}A = T^{-1}A^c - A^c$. This shows that the two definitions of incompressibility are equivalent.

(a) (i) *implies* (ii): Let A be wandering, and let $B = \bigcup_{n=0}^{\infty} T^{-n}A$. Then $T^{-1}B = \bigcup_{n=1}^{\infty} T^{-n}A \subset B$, so by (1), $\mu(B - T^{-1}B) = 0$. But $B - T^{-1}B = A$ since the $T^{-n}A$ are disjoint; hence $\mu(A) = 0$.

(ii) *implies* (iii): Let $A^{(r)} = A \cap \bigcup_{n=1}^{\infty} T^{-n}A = \{\omega \in A : T^n\omega \in A \text{ for some } n \geq 1\}$. If $C = A - A^{(r)}$, then $T^{-n}C = \{\omega : T^n\omega \in A, \text{ but } T^k\omega \notin A, k > n\}$. Thus the $T^{-n}C$ are disjoint, so that C is wandering. By (2), $\mu(C) = 0$, and therefore T is recurrent.

(iii) *implies* (i): Let $T^{-1}A \subset A$. Then (by induction) $T^{-n}A \subset T^{-1}A$, $n \geq 1$. Thus $T^{-1}A = \bigcup_{n=1}^{\infty} T^{-n}A$. Now

$$A - T^{-1}A = A - \bigcup_{n=1}^{\infty} T^{-n}A = A - A^{(r)},$$

which has measure 0 by (3). Thus $\mu(A - T^{-1}A) = 0$, proving (1).

(iv) *implies* (iii): Obvious.

(i) *implies* (iv): Let

$$A^{(i)} = A \cap \limsup_{n \geq 1} T^{-n}A = \{\omega \in A : T^n\omega \in A$$

for infinitely many $n \geq 1\}$.

If $A \in \mathscr{F}$, let $B = \bigcup_{n=0}^{\infty} T^{-n}A$. Then $T^{-1}B \subset B$, hence by (1), $\mu(B - T^{-1}B) = 0$. Similarly, $T^{-(k+1)}B \subset T^{-k}B$, hence

$$\mu(T^{-k}B - T^{-(k+1)}B) = 0.$$

But

$$T^{-k}B - T^{-(k+1)}B$$

$$= \bigcup_{n=k}^{\infty} T^{-n}A - \bigcup_{n=k+1}^{\infty} T^{-n}A$$

$$= \{\omega: T^n\omega \text{ enters } A \text{ for the last time at } n = k\}.$$

Since

$$A - A^{(i)} = A \cap \bigcup_{k=0}^{\infty} \{\omega: T^n\omega \in A \text{ for the last time at } n = k\},$$

we have

$$\mu(A - A^{(i)}) \le \sum_{k=0}^{\infty} \mu(T^{-k}B - T^{-(k+1)}B) = 0.$$

(b) Any interval of length less than 1 is a nontrivial wandering set.

(c) Let A be a wandering set; since

$$\sum_{n=0}^{\infty} \mu(A) = \sum_{n=0}^{\infty} \mu(T^{-n}A) = \mu\left(\bigcup_{n=0}^{\infty} T^{-n}A\right) \le \mu(\Omega) < \infty,$$

$\mu(A)$ must be 0. Thus T is conservative, hence infinitely recurrent.

Section 8.3

5. (a) (i) *implies* (ii): By hypothesis, U has a left and a right inverse, hence U is one-to-one onto; also, $\langle f, g \rangle = \langle f, U^*Ug \rangle = \langle Uf, Ug \rangle$.

(ii) *implies* (iii): Take $g = f$.

(iii) *implies* (ii): Use the polarization identity (3.2.17).

(ii) *implies* (i): Write $\langle U^*Uf, g \rangle = \langle Uf, U^{**}g \rangle = \langle Uf, Ug \rangle = \langle f, g \rangle$. Since f and g are arbitrary, $U^*U = I$. But then $(UU^*)U = U$; so if U is onto, then $UU^* = I$.

(b) This is done exactly as in (a).

(c) If $Uf = f$, then $U^*Uf = U^*f$, hence by (b), $f = U^*f$. Conversely, if $U^*f = f$, then

$$\|Uf - f\|^2 = \|Uf\|^2 - \langle f, Uf \rangle - \langle Uf, f \rangle + \|f\|^2$$

$$= 2\|f\|^2 - \langle U^*f, f \rangle - \langle f, U^*f \rangle$$

$$= 2\|f\|^2 - 2\langle f, f \rangle \qquad \text{by hypothesis}$$

$$= 0.$$

(d) If $f_k \in E$, $f_k \to f$, then

$$\|A_m f - A_n f\| \le \|A_m f - A_m f_k\| + \|A_m f_k - A_n f_k\| + \|A_n f_k - A_n f\|$$
$$\le 2\|f - f_k\| + \|A_m f_k - A_n f_k\|,$$

and it follows that $\{A_n f\}$ is a Cauchy sequence, so that $f \in E$, proving E closed. It is immediate that E is a subspace.

(e) If $f \in M$, then $A_n f \equiv f$, so $\hat{f} = f$. If $f = g - Ug \in N_0$, then $A_n f = n^{-1}(g - U^n g)$, hence $\|A_n f\| \le 2n^{-1}\|g\| \to 0$.

(f) If $h \in H$, then

$$\begin{aligned}
h \perp N \quad &\text{iff} \quad h \perp N_0 \\
&\text{iff} \quad \langle h, g - Ug \rangle = 0 \quad \text{for all} \quad g \in H \\
&\text{iff} \quad \langle h - U^* h, g \rangle = 0 \quad \text{for all} \quad g \in H \\
&\text{iff} \quad U^* h = h \\
&\text{iff} \quad Uh = h \quad \text{by (c)} \\
&\text{iff} \quad h \in M.
\end{aligned}$$

Thus $M = \overset{\perp}{N}$, and the result follows.

(g) Since $E = H$, $A_n f$ converges to a limit $\hat{f}$. Write $f = f_1 + f_2$, where $f_1 \in M$, $f_2 \in N$. Now $A_n f_1 \to f_1$ by (e), and also $A_n f_2 \to 0$. Choose $g \in N_0$ such that $\|f_2 - g\| < \varepsilon$; then

$$\|A_n f_2\| \le \|A_n(f_2 - g)\| + \|A_n g\| \le \|f_2 - g\| + \|A_n g\| < \varepsilon + \|A_n g\|.$$

By (e), $A_n g \to 0$, so $A_n f_2 \to 0$. Therefore $A_n f \to f_1 = Pf$.

(h) By definition of $\hat{S}$ and $\hat{T}$, we have $\hat{S}\hat{T} = \hat{T}\hat{S} = I$. By 8.3.1, $\hat{S}$ and $\hat{T}$ are isometries, and they are invertible, they must be unitary operators. By (a), if $U = \hat{T}$, then $U^* = \hat{S}$.

6. If P is not an extreme point, so that a representation of the given form is possible, then P_1 is preserved by T, $P_1 \ll P$, and $P_1 \not\equiv P$ (if $P_1 \equiv P$, then $(1 - \lambda_1)P_1 = \lambda_2 P_2$, so that P_1 and P_2, being probability measures, must be identical). By 8.3.12, P is not ergodic.

Conversely, assume that P is not ergodic. If A is an invariant set with $0 < P(A) < 1$, then for each $B \in \mathscr{F}$,

$$P(B) = P(A)P(B \mid A) + P(A^c)P(B \mid A^c)$$
$$= \lambda_1 P_1(B) + \lambda_2 P_2(B).$$

By the end of the proof of 8.3.12, P_1 and P_2 are preserved by T, hence $P_1, P_2 \in K$. If $P_1 \equiv P_2$, then $P \equiv P_1$; but $P_1(A) = P(A \mid A) = 1 \ne P(A)$. Therefore $P_1 \not\equiv P_2$, so that P is not an extreme point.

7. (a) *implies* (b): Apply the pointwise ergodic theorem.

(b) *implies* (c), (c) *implies* (d): Obvious.

(d) *implies* (e): By the pointwise ergodic theorem, $I_A^{(n)}$ converges to $\hat{I}_A$ almost everywhere, hence in probability. By (d), $\hat{I}_A = P(A)$ a.e.

(e) *implies* (f): Since $I_B^{(n)} \to \hat{I}_B$ a.e., we may multiply by I_A and integrate to obtain, by the dominated convergence theorem,

$$\frac{1}{n} \sum_{k=0}^{n-1} P(A \cap T^{-k}B) \to E(I_A\hat{I}_B).$$

But if $\mathscr{G}$ is the σ-field of almost invariant sets,

$$E(I_A\hat{I}_B) = E[E(I_A\hat{I}_B)|\mathscr{G}]$$
$$= E[\hat{I}_B E(I_A|\mathscr{G})] \qquad \text{by 8.3.8}$$
$$= E(\hat{I}_A\hat{I}_B) \qquad \text{by 8.3.9.}$$

Under the hypothesis (e), $\hat{I}_A = P(A)$ a.e., $\hat{I}_B = P(B)$ a.e., proving (f).

(f) *implies* (g): If $A, B \in \mathscr{F}$ and $\varepsilon > 0$, choose $A_0, B_0 \in \mathscr{F}_0$ such that $P(A \triangle A_0)$, $P(B \triangle B_0)$, and $P[(A \cap T^{-k}B) \triangle (A_0 \cap T^{-k}B_0)]$ are less than ε for all k (see the proof of 8.2.7). Then

$$\left| \frac{1}{n} \sum_{k=0}^{n-1} P(A \cap T^{-k}B) - P(A)P(B) \right|$$

$$\leq \left| \frac{1}{n} \sum_{k=0}^{n-1} [P(A \cap T^{-k}B) - P(A_0 \cap T^{-k}B_0)] \right|$$

$$+ \left| \frac{1}{n} \sum_{k=0}^{n-1} [P(A_0 \cap T^{-k}B_0) - P(A_0)P(B_0)] \right|$$

$$+ |P(A_0)P(B_0) - P(A)P(B)|.$$

The first term is less than ε for all n, and the second is less than ε for large n by (f). Since the third term is less than 2ε, the result follows.

(g) *implies* (a): Let A be an invariant set, and set $B = A$; then $n^{-1}\sum_{k=0}^{n-1} P(A \cap T^{-k}B) = P(A)$. By (g), $P(A) = [P(A)]^2$, hence $P(A) = 0$ or 1; thus T is ergodic.

Section 8.4

3. (a) We have $E(R_n) = 1 + \sum_{k=2}^{n} P(B_k)$, and

$$P(B_k) = P\{X_k \neq 0, X_k + X_{k-1} \neq 0, \ldots, X_k + \cdots + X_2 \neq 0\}$$
$$= P\{S_1 \neq 0, S_2 \neq 0, \ldots, S_{k-1} \neq 0\}$$

since the X_i are iid.

   Thus $n^{-1}E(R_n)$ is the arithmetic average of a sequence converging to $P(A)$, hence $n^{-1}E(R_n) \to P(A)$.

(b) The Z_k are iid, with $|Z_k| \leq N$. Since $R_{nN} \leq Z_1 + \cdots + Z_n$, the strong law of large numbers yields the desired result.

(c) For any positive integer n, $(k-1)N + 1 \leq n \leq kN$ for some k, hence $|R_n - R_{kN}| \leq N$; therefore

$$\frac{R_n}{n} \leq \frac{R_{kN}}{kN}\frac{kN}{n} + \frac{N}{n}.$$

   Since $kN/n \to 1$ as $n \to \infty$, $\limsup_{n\to\infty} n^{-1}R_n \leq N^{-1}E(Z_1)$ by (b). Now $Z_1 = R_N$ and N is arbitrary; thus the result follows from (a).

(d) Since $V_k = 1$ if $X_{k+1} \neq 0, X_{k+1} + X_{k+2} \neq 0, \ldots$, and $V_k = 0$ otherwise, V_k can be expressed as $g(X_k, X_{k+1}, \ldots)$ where g: $R^\infty \to R^\infty$, measurable relative to $[\mathscr{B}(R)]^\infty$. The pointwise ergodic theorem therefore applies.

(e) The sum $\sum_{i=1}^{n} V_i$ is the number of states visited in the first n steps that are never revisited. If $i < j$, and S_i and S_j are never revisited, then $S_i \neq S_j$; thus $\sum_{i=1}^{n} V_i \leq R_n$. By (d), $\liminf_{n\to\infty} n^{-1}R_n \geq E(V_1)$ a.e. But since the X_i are iid,

$$E(V_1) = P\{X_1 \neq 0, X_1 + X_2 \neq 0, \ldots\} = P(A).$$

Section 8.5

1. Let $\{X_n\}$ and $\{X_n'\}$ be discrete ergodic sequences with entropies H and H'. Flip a coin; if the result is heads, let $X_n'' = X_n$ for all n, and if tails, let $X_n'' = X_n'$ for all n. If p is the probability of heads, the limit random variable is H with probability p, and H' with probability $1 - p$, so that the entropy of $\{X_n''\}$ is $pH + (1-p)H'$. In part (a), choose $H = H'$, and in part (b), choose $H \neq H'$. There is no problem in realizing these choices; for example, if the X_n are independent and take on r values with equal probability, then $H = \log r$.

Index

ISBN 0-12-065202-1